# High-Temperature Ordered Intermetallic Alloys

MATERIALS RESEARCH SOCIETY SYMPOSIA PROCEEDINGS

ISSN 0272 - 9172

Volume 1—Laser and Electron-Beam Solid Interactions and Materials Processing, J. F. Gibbons, L. D. Hess, T. W. Sigmon, 1981

Volume 2—Defects in Semiconductors, J. Narayan, T. Y. Tan, 1981

Volume 3—Nuclear and Electron Resonance Spectroscopies Applied to Materials Science, E. N. Kaufmann, G. K. Shenoy, 1981

Volume 4—Laser and Electron-Beam Interactions with Solids, B. R. Appleton, G. K. Celler, 1982

Volume 5—Grain Boundaries in Semiconductors, H. J. Leamy, G. E. Pike, C. H. Seager, 1982

Volume 6—Scientific Basis for Nuclear Waste Management, S. V. Topp, 1982

Volume 7—Metastable Materials Formation by Ion Implantation, S. T. Picraux, W. J. Choyke, 1982

Volume 8—Rapidly Solidified Amorphous and Crystalline Alloys, B. H. Kear, B. C. Giessen, M. Cohen, 1982

Volume 9—Materials Processing in the Reduced Gravity Environment of Space, G. E. Rindone, 1982

Volume 10—Thin Films and Interfaces, P. S. Ho, K.-N. Tu, 1982

Volume 11—Scientific Basis for Nuclear Waste Management V, W. Lutze, 1982

Volume 12—In Situ Composites IV, F. D. Lemkey, H. E. Cline, M. McLean, 1982

Volume 13—Laser Solid Interactions and Transient Thermal Processing of Materials, J. Narayan, W. L. Brown, R. A. Lemons, 1983

Volume 14—Defects in Semiconductors II, S. Mahajan, J. W. Corbett, 1983

Volume 15—Scientific Basis for Nuclear Waste Management VI, D. G. Brookins, 1983

Volume 16—Nuclear Radiation Detector Materials, E. E. Haller, H. W. Kraner, W. A. Higinbotham, 1983

Volume 17—Laser Diagnostics and Photochemical Processing for Semiconductor Devices, R. M. Osgood, S. R. J. Brueck, H. R. Schlossberg, 1983

Volume 18—Interfaces and Contacts, R. Ludeke, K. Rose, 1983

Volume 19—Alloy Phase Diagrams, L. H. Bennett, T. B. Massalski, B. C. Giessen, 1983

Volume 20—Intercalated Graphite, M. S. Dresselhaus, G. Dresselhaus, J. E. Fischer, M. J. Moran, 1983

Volume 21—Phase Transformations in Solids, T. Tsakalakos, 1984

Volume 22—High Pressure in Science and Technology, C. Homan, R. K. MacCrone, E. Whalley, 1984

Volume 23—Energy Beam-Solid Interactions and Transient Thermal Processing, J. C. C. Fan, N. M. Johnson, 1984

Volume 24—Defect Properties and Processing of High-Technology Nonmetallic Materials, J. H. Crawford, Jr., Y. Chen, W. A. Sibley, 1984

## MATERIALS RESEARCH SOCIETY SYMPOSIA PROCEEDINGS

Volume 25—Thin Films and Interfaces II, J. E. E. Baglin, D. R. Campbell, W. K. Chu, 1984

Volume 26—Scientific Basis for Nuclear Waste Management VII, G. L. McVay, 1984

Volume 27—Ion Implantation and Ion Beam Processing of Materials, G. K. Hubler, O. W. Holland, C. R. Clayton, C. W. White, 1984

Volume 28—Rapidly Solidified Metastable Materials, B. H. Kear, B. C. Giessen, 1984

Volume 29—Laser-Controlled Chemical Processing of Surfaces, A. W. Johnson, D. J. Ehrlich, H. R. Schlossberg, 1984

Volume 30—Plasma Processing and Synthesis of Materials, J. Szekely, D. Apelian, 1984

Volume 31—Electron Microscopy of Materials, W. Krakow, D. Smith, L. W. Hobbs, 1984

Volume 32—Better Ceramics Through Chemistry, C. J. Brinker, D. E. Clark, D. R. Ulrich, 1984

Volume 33—Comparison of Thin Film Transistor and SOI Technologies, H. W. Lam, M. J. Thompson, 1984

Volume 34—Physical Metallurgy of Cast Iron, H. Fredriksson, M. Hillerts, 1985

Volume 35—Energy Beam-Solid Interactions and Transient Thermal Processing/1984, D. K. Biegelsen, G. Rozgonyi, C. Shank, 1985

Volume 36—Impurity Diffusion and Gettering in Silicon, R. B. Fair, C. W. Pearce, J. Washburn, 1985

Volume 37—Layered Structures, Epitaxy and Interfaces, J. M. Gibson, L. R. Dawson, 1985

Volume 38—Plasma Synthesis and Etching of Electronic Materials, R. P. H. Chang, B. Abeles, 1985

Volume 39—High-Temperature Ordered Intermetallic Alloys, C. C. Koch, C. T. Liu, N. S. Stoloff, 1985

Volume 40—Electronic Packaging Materials Science, E. A. Giess, K.-N. Tu, D. R. Uhlmann, 1985

Volume 41— Advanced Photon and Particle Techniques for the Characterization of Defects in Solids, J. B. Roberto, R. W. Carpenter, M. C. Wittels, 1985

Volume 42—Very High Strength Cement-Based Materials, J. F. Young, 1985

Volume 43—Coal Combustion and Conversion Wastes: Characterization, Utilization, and Disposal, G. J. McCarthy, R. J. Lauf, 1985

Volume 44—Scientific Basis for Nuclear Waste Management VIII, C. M. Jantzen, J. A. Stone, R. C. Ewing, 1985

MATERIALS RESEARCH SOCIETY SYMPOSIA PROCEEDINGS VOLUME 39

# High-Temperature Ordered Intermetallic Alloys

Symposium held November 26-28, 1984, Boston, Massachusetts, U. S. A.

EDITORS:

C. C. Koch
North Carolina State University, Raleigh, North Carolina, U. S. A.

C. T. Liu
Oak Ridge National Laboratory, Oak Ridge, Tennessee, U. S. A.

N. S. Stoloff
Rensselaer Polytechnic Institute, Troy, New York, U. S. A.

MRS MATERIALS RESEARCH SOCIETY
Pittsburgh, Pennsylvania

Symposium Sponsors:
Department of Energy, Division of Energy, Conversion and Utilization Technologies
Office of Naval Research



Published by:

Materials Research Society
9800 McKnight Road, Suite 327
Pittsburgh, Pennsylvania 15237
telephone (412) 367-3003

Library of Congress Cataloging in Publication Data

Main entry under title:

High-temperature ordered intermetallic alloys.

(Materials Research Society symposia proceedings ; v. 39)

Includes bibliographies and indexes.

1. Heat resistant alloys——Congresses. 2. Order-disorder in alloys——Congresses. 3. Intermetallic compounds——Congresses. I. Koch, C. C. II. Liu, C. T. (Chain Tsuan), 1937- III. Stoloff , N. S. IV. Materials Research Society. V. Series.

TA485.H56 1985 620.1'617 85-11488
ISBN 0-931837-04-9

Manufactured in the United States of America

Manufactured by Publishers Choice Book Mfg. Co.
Mars, Pennsylvania 16046

# Contents



*Invited Speaker

## PART IV: MECHANICAL BEHAVIOR



*Invited Speaker

*Invited Speaker

## PART VI: PHYSICAL AND METALLURGICAL PROPERTIES



*Invited Speaker

# Preface

For several decades the special properties of ordered intermetallic alloys have intrigued alloy designers. In particular, the aluminides offered the potential for high temperature structural applications due to good strength combined with oxidation resistance. Attainment of this potential has been hampered by the brittle behavior and low fracture toughness of these materials. However, in recent years significant improvement in the ductility and toughness of, at first, titanium aluminides, then iron and nickel aluminides, has been achieved by metallurgical techniques such as processing control, grain refinement, and microalloying. These successes have stimulated new interest in ordered intermetallics - in particular the aluminides.

The Materials Research Society symposium on High-Temperature Ordered Intermetallic Alloys provided a forum to discuss the alloy chemistry, physical metallurgy, and mechanical behavior of ordered intermetallic alloys which have potential to be used as new structural materials. The three-day symposium, held November 26-28, 1984 at the Marriott Hotel/Copley Place, Boston, Massachusetts, contained invited review papers and contributed research-in-progress presentations in the areas of theory of alloy ordering, microstructural features and phase transformation, mechanical behavior, alloy design, and physical properties. This was the first international conference on this subject in over a decade and the first such conference held in the United States in fifteen years.

All the papers in this symposium proceedings were reviewed. The symposium session chairmen organized the review of papers in their respective sessions. This review process slowed down the publishing time but improved the overall quality of the papers included in the book. Special thanks are given to the session chairmen who performed this duty; D. E. Polk, D. P. Pope, and J. O. Stiegler. Thanks are also extended to the individual reviewers, most of whom were selected from the contributors to this symposium.

Our appreciation is given to the agencies which sponsored this symposium and to the individuals in those agencies who worked to make this support possible. These were J. A. Carpenter, Jr., J. J. Eberhardt, and A. C. Schaffhauser, Energy Conversion and Utilization Technologies (ECUT), Department of Energy, and D. E. Polk, Office of Naval Research.

C. C. Koch
C. T. Liu
N. S. Stoloff

April, 1985

PART I

# Keynote Lecture

# ORDERED ALLOYS FOR HIGH TEMPERATURE APPLICATIONS

NORMAN S. STOLOFF, Materials Engineering Department, Rensselaer Polytechnic Institute, Troy, NY 12181

## ABSTRACT

This paper will provide an overview of the current status of the development of ordered alloys for high temperature structural applications. The physical and mechanical properties of ordered alloys will be reviewed with particular emphasis on aluminides and $(FeNiCo)_3V$ alloys. Alloy theory, slip systems, yielding, strain hardening, fatigue, wear resistance and processing are among the subjects to be covered. Current research programs in the U.S. and abroad will be outlined, and the kinds of scientific and application-oriented research needed in the future will be discussed.

## INTRODUCTION

This symposium is the first international meeting on the structure and properties of ordered alloys since the Conference on Order-Disorder Transformation in Alloys held in Tubingen, Germany on 3-6 September, 1973.[1] Previously, another important meeting on the subject had been held in 1969 at Bolton Landing.[2] Since the early 1970's, the pace of work on ordered alloys and intermetallic compounds slackened, as a result of lack of progress in improving either ductility or creep resistance of these otherwise very intriguing alloys.

Interest in utilizing ordered alloys for structural applications was reawakened in this country when researchers at Wright-Patterson Air Force Base[3] (together with contractors sponsored by the Air Force).discovered that ductility and strength improvements could be achieved in TiAl- and $Ti_3Al$-base alloys using a combination of powder metallurgy and alloying techniques. Later work on the titanium aluminides utilized ingot metallurgy. The development of rapid solidification methods (Rapid Solidification Rate and melt spinning, in particular) led to renewed interest in the iron and nickel aluminides. Finally, several exciting discoveries were reported that permitted substantial ductility to be achieved in cast and wrought ordered alloys previously known to be extremely brittle. The first of these accomplishments involved replacing cobalt in $Co_3V$ with nickel[4] and then iron[5], leading to a series of face-centered cubic $L1_2$-type superlattices with extensive ductility at ambient temperatures. Shortly after the first reports of this achievement at Oak Ridge National Laboratory (ORNL), it was reported in Japan that polycrystalline $Ni_3Al$ could be made ductile by adding small quantities of boron[6]. Later this work was confirmed at ORNL[7] and General Electric Co.[8] and the critical composition range over which boron was beneficial was identified. These discoveries, together with the national search for replacements for strategic metals such as cobalt and chromium and the need to develop energy-efficient propulsion systems, have in the past year or two stimulated much additional work - largely in the area of improving low temperature ductility and increasing high temperature strength - in this country, see Table I, Japan, and Western Europe. Soviet publications on ordered alloys have been voluminous, but there is little visible evidence to date of Soviet progress in developing such alloys for structural applications; rather, Soviet research has emphasized alloy theory, yielding and strain-hardening phenomena, and the use of various imaging techniques to aid in the study of crystal structure and crystal defects.

The characteristics that render ordered alloys so interesting as bases

TABLE I

Compilation of Research Efforts on Ordered Alloys in the United States[14]

| Laboratory | Source of Funding | System Studied | Principal Investigators |
|---|---|---|---|
| NASA-Lewis | NASA | FeAl | D. Whittenberger |
| AFWAL | AF | $Fe_3Al$,TiAl,$Ti_3Al$ | H. Lipsitt |
| Systems Res. Lab | AF | $Fe_3Al$,FeAl,$Fe_3Si$ | M.G. Mendiratta |
| Gen. Elec. R&D Center | AF | $Ni_3Al$+B | S.C. Huang |
| Pratt & Whitney | | | |
| E. Hartford | - | $Ni_3Al$ | D. Duhl |
| W. Palm Beach | AF | $Fe_3Al$+B,TiAl | E. Slaughter |
| E. Hartford | AF | TiAl | M. Blackburn |
| Oak Ridge National Lab | DOE | $(FeNi)_3V$,$Ni_3Al$+B | C.T. Liu |
| | ONR | $Fe_3Al$ | C.T. Liu |
| Olin Corp. | Internal | $Fe_3Al$ | --- |
| Marko Materials | DARPA/AMMRC | FeAl+B | R.V. Ray |
| Univ. of Pennsylvania | NSF | $Ni_3Al$ | D.P. Pope |
| | DOE-ORNL | $(FeNi)_3V$ | D.P. Pope |
| Dartmouth Univ. | DOE-ORNL | $Ni_3Al$,$Fe_3Al$ | E. Schulson |
| | NASA | NiAl | E. Schulson |
| Vanderbilt Univ. | DOE-ORNL | $(FeNi)_3V$,$Ni_3Al$ | J.J. Wert |
| Case Western Univ. | AF | $Fe_3Al$+B | K. Vedula |
| | NASA | FeAl,NiAl | K. Vedula |
| Stanford Univ. | NASA | CoAl,NiAl | W. Nix |
| Texas A&M Univ. | NASA | NiAl,CoAl,FeAl | A. Wolfenden |
| Rensselaer Polytechnic Institute | DOE-ORNL | $(FeNi)_3V$,$Ni_3Al$+B | N.S. Stoloff |
| | ONR | $Fe_3Al$,$Ni_3Al$ | N.S. Stoloff |
| | NSF | FeCo-V,$Ni_3Fe$ | N.S. Stoloff |
| Univ. of Michigan | NSF | NiAl | R. Gibala |
| Northwestern Univ. | NSF | Au-Ni | J.B. Cohen |
| North Carolina State | DOE-ORNL | $Ni_3Al$-oxides | C. Koch |
| North Carolina State | NASA | $Ni_3Al$+B, Fe-(Ni,Mn)-Al-C | C. Koch |
| Lehigh University | DOE-ORNL | $(FeNi)_3V$ | Y.T. Chou |
| Notre Dame | | $Ni_2Cr$ | L.P. Lehman |
| New Mexico State | | $Ni_3Ge$ | H-r. Pak |

for alloy development will be discussed below as will their current status and applications. Several recent reviews should be consulted for additional background information on both single phase and multiphase ordered alloys[9-14].

## CHARACTERISTICS OF ORDERED ALLOYS

The formation of long-range order in alloy systems frequently produces a significant effect on mechanical properties, including elastic constants, yield and tensile strengths, strain-hardening rates, ductility, and resistance to cyclic or static (creep) deformation[15-18].

In some ordered systems, there is a change in crystal structure at the order-disorder transformation, leading to property changes dependent chiefly on stresses generated by the transformation. More widely studied and significant, however, are the properties of alloys in which: (1) there is no lattice change with ordering but only a rearrangement of atoms on lattice sites, (2) atoms rearrange from a body centered or face centered cubic to simple cubic, or (3) no order-disorder transition occurs below the melting point of the alloy. Table II identifies the most common superlattice types and presents information concerning dislocation types[18].

# TABLE II

## Dislocation Morphologies in Some Ordered Alloys

TABLE 1 Dislocation Morphologies in Some Ordered Alloys

| Superlattice Type (Strukturbericht Designation) | Chemical Designation | Unit Cell Dimensions | Alloy Types | Superlattice Dislocation Type | Burgers Vector of Each Dislocation | Antiphase Boundary Type[a] |
|---|---|---|---|---|---|---|
| B2 | CsCl | $a_o$ | NiAl, AgMg<br>AuZn | 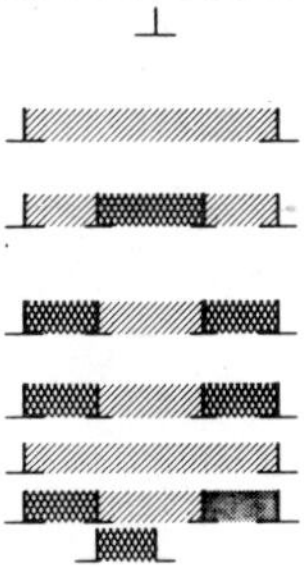 ⊥ | $a_o\langle 100\rangle$ | None |
| | | | CuZn, FeCo<br>FeAl, FeRh<br>NiAl, AgMg<br>AuZn | | $1/2a_o\langle 111\rangle$ | NN |
| $DO_3$ | $Fe_3Al$ | $a_o$ | $Fe_3Al$, $Fe_3Si$<br>$Fe_3B$ | | $1/4a_o\langle 111\rangle$ | NN<br>NNN |
| $L1_2$ | $Cu_3Au$ | $a_o$ | $Cu_3Au$, $Ni_3Mn$<br>$Ni_3Al$, $Ni_3Fe$<br>$Cu_3Pd$, $Ni_3Ti$<br>$Ag_3Mg$, $Ni_3Ta$<br>$Ni_3Si$, $Cu_3Pt$<br>$Ni_3Ga$ | | $1/6a_o\langle 112\rangle$ | NN<br>NN+SF |
| $DO_{19}$ | $Mg_3Cd$ | $a_o$<br>$c_o$ | $Mg_3Cd$ | | $1/6a_o\langle 10\bar{1}0\rangle$<br>$1/2a_o\langle 2\bar{1}\bar{1}0\rangle$ | NN<br>SF<br>NNN |
| $L1_o$ | CuAu | $a_o$<br>$c_o$ | CuAu, CoPt<br>FePt | | $1/6a_o\langle 112\rangle$ | NN<br>NN+SF<br>SF |

(Source: Marcinkowski 1974)

[a]NN, nearest neighbor; NNN, next-nearest neighbor; SF, stacking fault.

## A. Superlattice Dislocations and Antiphase Boundaries

Since an ordinary (unit) dislocation moving in a superlattice cannot recreate the crystal structure in its wake, disorder in the form of an antiphase boundary (APB) will result from the motion of such dislocations. The additional energy of the APB can be eliminated, however, by motion of dislocations in groups such that no net change in order occurs behind the dislocations. These groups, which consist of two or more dislocations connected by a strip of APB or other planar fault, are known as superlattice dislocations. Within the superlattice dislocation, each unit dislocation may further dissociate into its constituent partial dislocations (as listed for various superlattices in Table II). The appearance of slip bands intersecting a polished surface is quite different in alloys in the ordered and disordered states. Many ordered alloys reveal diffuse slip on only one or a few slip systems and cross slip is restricted, thereby leading to brittleness, whereas coarse slip steps are observed in the disordered condition. It is the motion of superlattice dislocations and their interactions with each other and with obstacles such as grain boundaries, precipitate particles, or grown-in antiphase boundaries that control the mechanical behavior of most ordered alloys. Notable exceptions to this behavior are $Ni_4Mo$[19], which rarely exhibits superlattice dislocations at room temperature and hypo-stoichiometric $Fe_3Al$, which has been suggested to deform by unit dislocations[20].

Antiphase boundaries, which are produced during heat treatment, separate domains that may be perfectly ordered within themselves but are out of step with one another. During the early stages of an isothermal ordering treatment, antiphase domains (APDs) generally are very small and may not be in contact with each other. However, with continued annealing, the APDs will impinge on each other and may grow to a size limited by the specimen size (in single crystals) or the grain size (in polycrystals). In some cases, (e.g., $Ni_3Mn$ and

$Ni_3Fe$), however, the kinetics of domain growth are very sluggish, and domain sizes are limited to about 500$\mu_m$[21]. In some alloys, for example in $Cu_3Au$[22], domains can give rise to considerable strengthening. Also, the nature of the domain structure can give important information about fault energies, as discussed below.

The antiphase boundary energy in $L1_2$ alloys has been calculated to be anisotropic, with the lowest energy on {001} planes, taking into account first nearest neighbor (NN) interactions only[23]. Accordingly, there is a driving force for an APB to lie on cube planes in these alloys, leading to important consequences with regard to yielding and strain hardening behavior. When this occurs, the electron microscope shows a maze pattern (as in $Cu_3Au$ and $Cu_3Pt$) rather than the "swirl" pattern that is associated with superlattices possessing isotropic domain networks ($Ni_3Fe$, $Ni_3Mn$, $Fe_3Al$). The morphology of an APB in $Cu_3Au$ can be changed from a maze to a swirl pattern by adding 5 at% Ni[24]; this is perhaps due to a reduction in the electron/atom ratio, hence, in the relative phase stability of $L1_2$ versus long-period superlattices in which periodic {001} APBs are characteristic. Recently, swirl pattern APDs have been noted in rapidly quenched $Ni_3Al$-X alloys in which $Ni_3Al$-type compounds are formed[25] as well as in binary $Ni_3Al$[26]; however, after short annealing treatments, these swirl patterns change to maze patterns and then quickly grow to large sizes.

The slip vector in B2-type superlattices varies from one alloy to another (Table II) depending on factors such as APB energy, atom ratios, atomic size ratios, and elastic energies of dislocations[18]. As in bcc metals, <111> are usually the slip vectors, and slip occurs on planes of types {110}, {112}, and {123}; however, several B2 superlattices display <100> slip. In FeAl single crystals there is a transition from <111> to <100> as temperature increases[27]. In $L1_2$ superlattices, on the other hand, the slip plane may change with temperature, as in $Ni_3Al$, which deforms on {111} planes at low temperatures and on {100} planes at high temperatures[28]. Cross slip from {111} to {100} is believed responsible for an anomalous increase in strain hardening rate of $Cu_3Au$ with increasing temperature[29,30]. The slip vector, however, is <110> at all test temperatures.

## Departures from Stoichiometry

Departures from perfect order cause strengthening through means not directly connected with the APB. Deviations in stoichiometry from the ideal AB or $A_3B$ composition often lead to strengthening at low test temperatures, and, for this reason, slowly cooled alloys of the Fe-Co, Ni-Al, Ni-Ga, Mg-Cd, or Cu-Au systems reveal minima in room-temperature hardness at stoichiometric compositions. Hardness data obtained at 77°K and room temperature for compositions near stoichiometric $Ni_3Ga$ and $Ni_3Al$ are shown in Figure 1[31]. Note the more rapid hardening with excess of the minority element. In systems with a very high ordering energy, the hardening at nonstoichiometric compositions might arise from lattice defects that are required to maintain the number of electrons per unit cell at a constant value required for phase stability[32]. In such cases, the vacancies should enhance plastic flow at high temperature, when the defects are mobile. Such behavior was reported for $Ni_3Al$ by Guard and Westbrook[32], who showed decreased flow stresses at 800°C in alloys containing an excess of either Al or Ni. However, Noguchi and co-workers[31] have demonstrated significant hardening at elevated temperatures in aluminum-rich $Ni_3Al$ (Figure 2) and in gallium-rich $Ni_3Ga$. Solute-induced hardening was attributed to changes in probability of cross slip from {111} to {100} plane with composition. Moreover, Aoki and Izumi[33] showed that off-stoichiometric $Ni_3Al$ does not contain excess vacancies. It is clear that the nature and effects of lattice defects as a function of stoichiometry are an important research subject that remains to be addressed adequately.

Changes in stoichiometry influence not only the yield strength but also

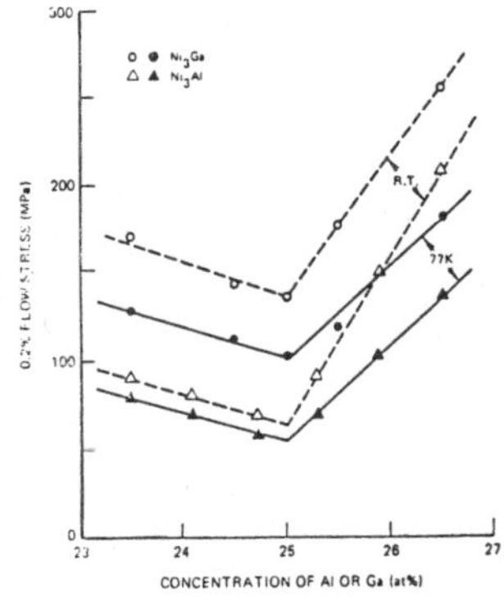


Fig. 1 Effects of composition and temperature on strength of $Ni_3Ga$ and $Ni_3Al$[31].

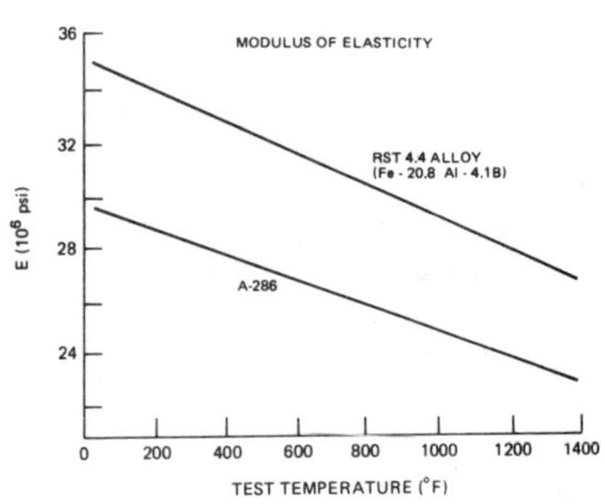


Fig. 2 Elastic moduli of Fe-20.8w%Al-4.1%B and A-286[36].

the dislocation substructure and fracture behavior. For example, unit dislocations are suggested to be responsible for plastic deformation in hypo-stoichiometric $Fe_3Al$ whereas paired dislocations are found for Al>26%[20]. Further, hypo-stoichiometric $Ni_3Al$ and $Fe_3Al$ are more ductile than hyper-stoichiometric compositions.

## PHYSICAL AND MECHANICAL PROPERTIES

### A. Elastic Constants

Usually, the change in elastic constants with order is small (in the range of 3 to 10 percent) and, therefore, changes in other mechanical properties arising from ordering are seldom controlled by effects on elastic constants(16). Rather, the principal effects of ordering on mechanical properties arise from changes in dislocation configurations due to the added constraints that ordering places on nucleation and motion of dislocations.

Some representative Young's modulli, melting ($T_m$) and critical ($T_c$) temperatures, and densities are shown in Table III for a number of ordered alloys, including the aluminides[14]. Modulus data for several of the B2 aluminides[34] and TiAl[35] and $Ti_3Al$[35] suggest that high stiffness/density ratios are characteristic of the aluminides. For example, the specific modulus of a boride-strengthened $Fe_3Al$ alloy recently has been reported to be 50 percent higher than that of A-286 (a widely used Fe-base alloy containing $Ni_3Ti$ precipitates) between 25°C and 760°C (Figure 2)[36].

### B. Yielding

Ordering exerts a significant influence on initial plastic deformation as well as on strain hardening of many single crystals. Initial yielding generally occurs at a lower stress in the fully ordered condition, but then more rapid strain hardening causes the ultimate strength to exceed that of disordered material. The interaction of unit dislocations with short-range-ordered regions produced by typical "disordering" treatments (quenching from above $T_c$) gives rise to a higher initial flow stress than for fully ordered material in which

TABLE III

Properties of Intermetallic Compounds

| Alloy | Structure | Young's Modulus ($10^6$ psi) | $T_m$(°C) | $T_c$(°C) | Density (g/cc) |
|---|---|---|---|---|---|
| TiAl | $L1_o$ | 25.5 | 1460 | 1460 | 3.91* |
| $Ti_3Al$ | $DO_{19}$ | 21.0 | 1600 | 1100 | 4.2** |
| NiAl | $B_2$ | 42.7[a] | 1640 | 1640 | 5.86* |
| $Ni_3Al$ | $L1_2$ | 25.9[e] | 1390 | 1390 | 7.50* |
| FeAl | B2 | 37.8[b] | 1250-1400 | 1250-1400 | 5.56* |
| $Fe_3Al$ | $DO_3$ | 20.4[c] | 1540 | 540 | 6.72* |
| CoAl | B2 | 42.7[b] | 1648 | 1648 | 6.14* |
| $Zr_3Al$ | $L1_2$ | 19.6[f] | 1400 | 975 | 5.76* |
| $Fe_3Si$ | $DO_3$ | 39.4[d] | 1270 | 1270 | 7.25* |
| $Co_3V$ | hex | --- | 1400 | 1070 | 7.92** |
| $(Fe_{22}Co_{78})_3V$ | $L1_2$ | --- | 1400 | 950 | 7.80** |
| $(Fe_{60}Ni_{40})_3(V_{96}Ti_4)$ | $L1_2$ | --- | 1400 | 680 | 7.60** |

*Calculated from lattice parameter data.
**Estimated.
[a]Schafrik 1977
[b]Wolfenden 1983
[c]Morgand et al. 1968
[d]H. A. Lipsitt unpublished
[e]Stoloff and Davies 1965
[f]Turner et al. 1978

superlattice dislocations interact with large APD. A similar behavior is noted in many, but by no means all, superlattices. For example, ordering increases the yield strength of $Ni_3Mn$ because of sluggish domain growth in that alloy[16].

Strength (and ductility) data for representative aluminides appear in Table IV. Note that strengths generally are not very high at room temperature. However, one of the striking features of the plastic deformation of most ordered alloys, particularly a large number of those with the $L1_2$ structure, is an anomalous sharp rise in flow stress with increasing temperature, as is shown in Fig. 3 for several $Ni_3Al$ alloys[31]. The flow stress peak occurs in single crystals as well as in polycrystals and its position with respect to temperature in $Ni_3Al$ is a function of crystal orientation[37] and alloy content[38]. In the case of $(Fe,Co)_3V$ alloys, however, there is no orientation dependence of CRSS at temperatures near $T_c$[39]. Alloying with zirconium and hafnium is especially effective in raising high temperature strength of $Ni_3Al$; the density- compensated yield strength of these alloys is higher at 850°C than those of commercial superalloys, as shown in Figure 4[26]. However, an asymmetry between flow stresses measured in tension and in compression has been reported [40], and this should be taken into account when comparing different sets of yield stress data. In a few cases (e.g., $Fe_3Ge$), there is no anomalous behavior with increasing temperature[41].

Increases in flow stress with temperature also have been noted in long range ordered alloys of other cubic or hexagonal crystal structures. There is undoubtedly more than one mechanism responsible for this behavior since the peak strength is sometimes associated with the order-disorder temperature, $T_c$, or with the temperature of a transition from one ordered structure to another (e.g., $Fe_3Al$); in other cases, it bears no apparent relation to a transformation (e.g., $Ni_3Al$, β CuZn, and $Ni_3Ge$). A continuous range of behavior between anomalous and regular systems can be produced by alloying. The addition of iron to $Ni_3Ge$ causes the flow stress peak to disappear at an iron content of about 28 at%. Suzuki and co-workers[41] have suggested that three factors control such behavior: APB and stacking fault energy on {111} and APB energy on {100} planes. A high anisotropy of energy between {100} and

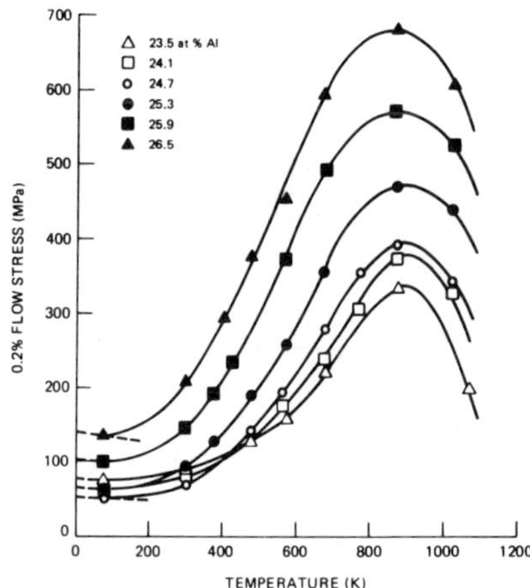


Fig. 3 Effect of composition on temperature dependence of yielding in $Ni_3Al$ alloys[31].

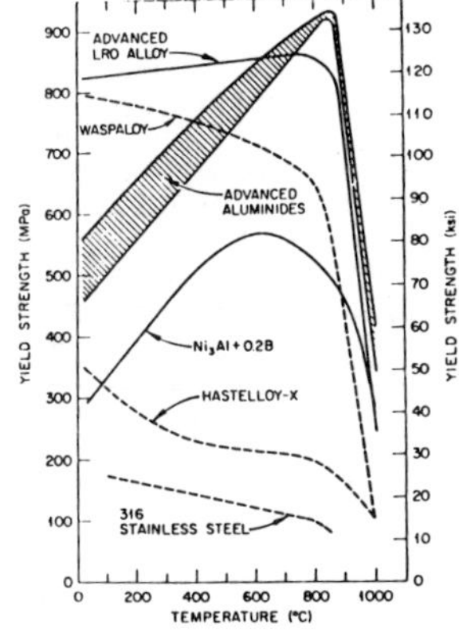


Fig. 4 Strength of advanced $Ni_3Al$ and LRO alloys compared to conventional alloys[26].

{111} planes is believed responsible for the strength anomaly in $L1_2$ alloys[11]. This anisotropy leads to cross slip of superpartials onto {100} planes to minimize APB energy (Figure 5), thereby creating obstacles to further plastic flow[42]. The obstacle density increases with increasing liklihood of cross slip onto {100}, whether through raising temperature or testing crystals favorably created for cross slip.

In the case of alloys such as FeCo-2% V, where the flow stress peak occurs at the order-disorder transition temperature, and $Fe_3Al$, where the flow stress peak occurs just under the temperature for a transition between two ordered structures, the anomaly has been attributed to a transition from plastic deformation by superlattice dislocations at temperature below that of the peak to deformation by unit dislocations at temperatures above that of the peak[43]. However, this mechanism cannot apply to $Fe_3Al$ in which the aluminum content is less than 25 at%, since only unit dislocations have been observed in the $DO_3$ condition[20]. Further, Inouye[44] has reported that two phase $\alpha + DO_3$ or $\alpha$ + B2 microstructures may be responsible for the flow stress peaks in $Fe_3Al$ annealed near $T_c$. Fujita et al[45], have suggested that changes in the anisotropy between {010} and {111} APB energies produced by alloying can alter the mechanism of yield in $Cu_3Au$. Other mechanisms for the flow stress peak are discussed in the literature[46,46], but it is clear that additional work on identifying dislocation configurations and phase relationships in the temperature range of the observed flow stress peaks is needed.

## C. Strain Hardening

Long-range-ordered alloys usually exhibit high strain-hardening rates compared to their disordered counterparts. For $L1_2$ superlattices, the strain-hardening rate can double with order at temperatures near 298°K, whereas lesser increments in rate are noted in cubic B2 or $DO_3$ systems (e.g., FeCo-2%V and $Fe_3Al$); for hcp $(DO_{19})Mg_3Cd$, there is no change in strain-hardening rate with order at 298°K[16]. The role of superlattice dislocations in work hardening is controversial. For example, one model is based on the formation of "super-jogs" by the intersection of superlattice dislocations and the difficulty of dragging these super-jogs during continuing plastic deformation[48].

Since these super-jogs usually are not seen in thin foils of deformed alloys, an alternative mechanism based on the cross slip of superpartials on {100} planes (as first suggested by Kear and Wilsdorf)[42] (Figure 5) also has been proposed to account for rapid hardening[16]. Experimental data obtained to date, including the observation that the rate of work hardening in ordered $Cu_3Au$ increases with temperature, generally favor the latter model; this point has been described in more detail in a recent paper by Pope and Ezz[11].

High strain hardening rates induced by long range order may permit attainment of very high strengths through cold working operations or thermal-mechanical treatments, as was shown in Figure 4 for $Ni_3Al$+B[26]. Wear-resistance also should be enhanced by rapid strain hardening, permitting possible replacement of cobalt-base alloys for such applications.

## D. Fracture

### 1. General Features

Polycrystalline ordered alloys often are brittle when tested in tension, although considerable plasticity may be displayed by single crystals or by polycrystals tested in compression. In the Fe-Co system, brittleness and ordering fall in the same composition range, 35 to 65 percent cobalt. The addition of a few percent of vanadium or chromium increases ductility of alloys tested in the quenched condition, perhaps as a result of the slowing of ordering kinetics (rapid quenching of binary FeCo always leaves some residual order). Recently, small additions of molybdenum, tungsten, tantalum, niobium, carbon, and nickel also have been shown to improve workability (ductility) of equiatomic Fe-Co alloys[50], as will be disucssed in a later section.

A direct relationship between ductility and the degree of long-range order has been demonstrated in FeCo-2%V[16][37]. Varying composition in binary alloys or adding ternary solutes influences the ductility of other alloys with the B2 structure[51]. Solute effects may occur, in some cases, by interaction of solute elements with interstitial impurities or by affecting slip character. For example, nickel promotes wavy slip in β-CuZn and reduced brittleness, whereas manganese suppresses wavy slip and raises the ductile-to-brittle transition temperature[52]. In the Fe-Al system, a sharply reduced ductility is observed as the aluminum content approaches 25 at% (Figure 6)[53]. Alloys with 25 to 50 at% Al have usually been reported to be completely brittle at room temperature when processed by conventional ingot techniques. Fracture was generally reported to be intergranular in $Fe_3Al$, but recent work has shown that the ductility of alloys with aluminum contents in the range 23 to 35 at% can be as much as 5 to 7 percent at room temperature[58,59], see Table IV, and fracture is generally transgranular; nevertheless, all of the iron aluminides display extreme notch sensitivity[54]. In addition, Fe-20 at% Al-5%Si exhibits a small amount of plasticity at room temperature; failure was by transgranular cleavage, with a transition to intergranular cavity fracture at higher test temperatures. $Ti_3Al$ is brittle below about 600°C due to the absence of c+a slip at low temperatures[56], TiAl also suffers from brittleness due to pinning of partial dislocations at low temperatures[57].

Only in the case of hcp $Mg_3Cd$ has long-range order been reported to increase ductility[60]. In this alloy, slip occurs on more slip systems in the ordered than in the disordered state[55], whereas in all other alloys studied to date, ordering either reduces the number of available slip systems or restricts cross slip of screw dislocations from one slip plane to another. For this reason, the brittleness of many polycrystalline non-cubic ordered alloys has been attributed to a lack of sufficient slip systems to provide compatibility of deformation at grain boundaries under arbitrary states of

TABLE IV

Aluminide Mechanical Properties

| Alloy at % | Condition | g.s. (μm) | 0.2%$\sigma_y$ (MPa) | UTS (MPa) | $\varepsilon_F$ (%) | Fracture Mode | Ref. |
|---|---|---|---|---|---|---|---|
| 25°C | | | | | | | |
| Fe-31Al | Cast extruded | --- | --- | --- | 5.6 | TG | 59 |
| Fe-35Al | RSR | --- | --- | --- | 7 | Cleavage | 58 |
| Fe-50Al | RSR | --- | --- | --- | 0 | IG | 75 |
| Ni-24Al+0.02B | Wrought[a] | 30 | 300 | 1300 | 54 | Dimples | 63 |
| $Ni_3Al$+1.5Hf+B | Wrought[a] | 30 | 460 | 1460 | 47 | Dimples | 26 |
| Ni-25Al | --- | --- | 90 | --- | --- | --- | 110 |
| Ti-54.2Al | PM | 25 | -- | 488 | 0.1 | Cleavage | 57 |
| Ni-50Al | PM | 100 | 220 | --- | 0 | Cleavage | 71 |
| $Ni_3Al$+1at%B | Melt spun | --- | 690 | --- | --- | TG | 8 |
| Ni-25Al | Poly. | --- | 230 | --- | --- | IG | 8 |
| Ni-50Al | extruded | 50 | --- | --- | 4-5 | --- | 111 |
| Ni-25Al | poly. | 10 | --- | --- | >15 | --- | 112 |
| 500°C | | | | | | | |
| Fe-35 | RSR | --- | --- | --- | --- | Dimples | 58 |
| Fe-50 | RSR | --- | --- | --- | "high" | --- | 75 |
| Ni-24Al+B | Wrought[a] | 30 | 540 | 1100 | 42 | Dimples | 63 |
| $Ni_3Al$+1.5Hf+B | Wrought[a] | 30 | 700 | 1270 | 36 | Dimples | 26 |
| Fe-40Al | Wrought[a] | 100 | --- | 500 | 30(RA) | --- | 113 |
| Fe-30.9Al-15.1B | RSR/PM | --- | 813 | 992 | 5.4 | --- | 36 |
| Fe-32Al-2Mo-2V-15B | RSR/PM | --- | 947 | 1118 | 2.5 | --- | 36 |
| Fe-28Al-12.1B | RSR/PM | 2 | 813 | 967 | 7.6 | --- | 36 |

[a]Recrystallized

strain[37]. In the case of AgMg and the aluminides, the segregation of impurities to grain boundaries has been suggested to be the predominant factor[61]. However, recent work on high-purity NiAl and $Ni_3Al$ has shown that impurity segregation in these alloys is minimal[71], yet they remain brittle.

2. Improved Ductility by Microalloying

Considerable effort is being devoted to improving the ductility of several aluminides (FeAl, $Fe_3Al$, NiAl, $Ni_3Al$) through such diverse techniques as grain refinement through thermal-mechanical treatment, microalloying with boron, and various rapid solidification techniques. A particularly promising development is the demonstration of high ductility of $Ni_3Al$ doped with small quantities of boron[6-8], see Fig. 7. Boron segregates preferentially to grain boundaries[26], even in melt spun material[64]. Unfortunately, the beneficial effect of boron seems to be lost when Al≥25 at% [7-13]. The ductility of polycrystalline $Ni_3Al$+B decreases substantially with increasing temperature;[63, 64] consequently, cold working with alternate anneals was considered preferable to hot reduction of $Ni_3Al$+B ingots[62]. Boron changes the fracture mode of FeAl from intergranular to cleavage, but little improvement in ductility accompanies this change in fracture mode[65]. Little success has been obtained through microalloying with respect to the ductility of $Fe_3Al$ and NiAl, but this may be a consequence of missing a relatively narrow range of compositions in which ductility can be improved.

The ability to improve ductility of $Ni_3Al$ by microalloying has led to a breakthrough in alloy development schemes for this system. Advanced alloys based on solid solution additions of hafnium and iron have demonstrated very high strengths, see Fig. 4, relative to commercial alloys, coupled with about 10% lower density. Further improvements are achieved by cold work[26].

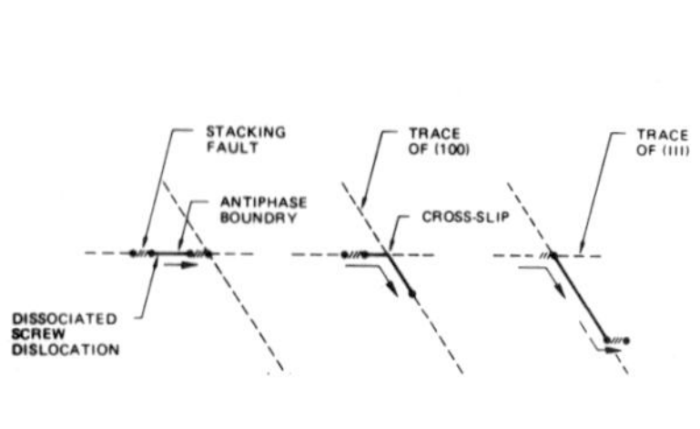


Fig. 5 Cross slip of leading dissociated screw dislocation from (111) to (001) to lower APB energy[42].

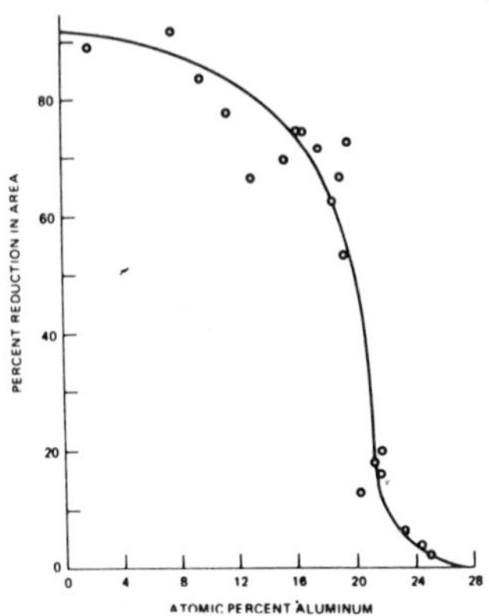


Fig. 6 Ductility of Fe-Al alloys at 477°K[53].

There exists now an opportunity to dispersion strengthen $Ni_3Al$ or to utilize $Ni_3Al$ as the matrix of mechanically incorporated composites. Koch[66] has already begun a program on mechanical alloying of $Ni_3Al$+B, and other programs will undoubtedly soon follow.

3. Macroalloying

Liu and co-workers[4,5,67], have utilized a combination of macro and micro alloying principles to achieve ductility in ordered alloys based on tetragonal $Ni_3V$ and hexagonal $Co_3V$ without resorting to rapid solidification. The unmodified alloys are very brittle in the ordered state. Work on ternary $(Co,Ni)_3V$ alloys showed that considerable ductility could be achieved by control of the ordered structure through alloying to control the electron/atom ($^e/a$) ratio, as first suggested by Beck[68]. In particular, replacing cobalt in $Co_3V$ with both iron and nickel produced an $L1_2$ superlattice. Several alloys of varying cobalt, iron and nickel content have been shown to be ductile at room temperature, see Fig. 8[26], and to exhibit tensile and fatigue strengths superior to those of conventional single-phase alloys (e.g., austenitic stainless steels, Hastelloy X) at temperatures of 600 or 800°C, Fig. 4[10,26]. Ductility problems at elevated temperatures have been alleviated through addition of small amounts of reactive solutes such as titanium and rare earths to the alloys. The effect of these is to suppress intergranular cracking in both monotonic and cyclic experiments.

4. Processing

Schulson and Barker[69] recently reported that an elongation of more than 40 percent can be achieved in Ni-49 at% Al at 400°C when the grain size is reduced to less than 20μm, Fig. 9, while the ductility at 25°C by this treatment was 3 to 5 percent. (A summary of ductility data for aluminides at 25°C and 500°C is presented in Table IV).

Schulson[70] has suggested, utilizing dislocation pileup concepts, that a critical grain diameter, $d_c$, exists in polycrystals, above which brittle fracture occurs. For an unnotched specimen loaded in tension

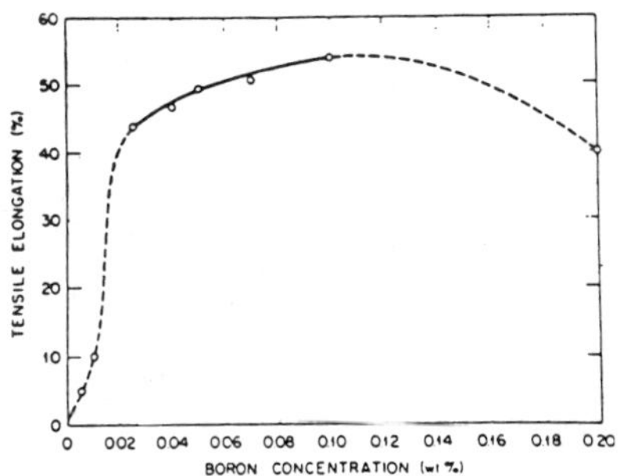


Fig. 7 Effect of boron on ductility of $Ni_3Al$[63].

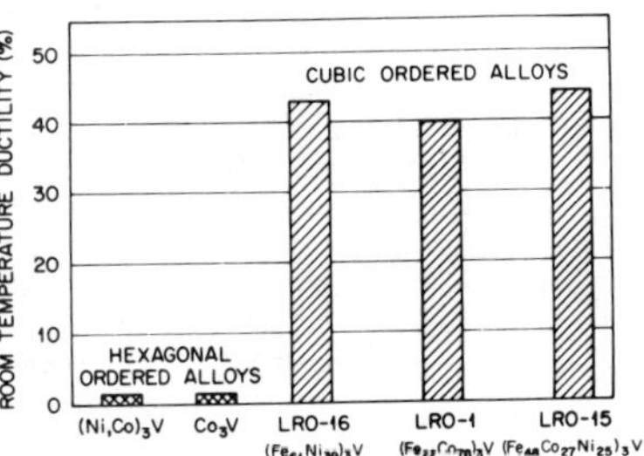


Fig. 8 Ductility of LRO alloys[26].

$$d_c = Y\left(\frac{K_{I_c} - ky}{\sigma_i}\right)^2 \tag{1}$$

where Y is a geometrial factor of the order of unity, $K_{I_c}$ is the fracture toughness, ky is the Hall-Petch slope, and $\sigma_i$ is the Hall-Petch intercept. For NiAl, $d_c$ is estimated to be about 50μm; polycrystals of this grain size do exhibit slight ductility at 25°C.

There have been several reports in the literature of a beneficial effect of partial recrystallization on ductility of B2-type alloys. For example, fully ordered FeCo-2%V exhibits about 5 percent elongation or reduction in area at grain sizes in the range 12 to 100μm. When partially recrystallized, however, the ductility of ordered FeCo-2%V increases to about 20 percent[51]. Recently, Schulson[71] has reported a small increase in ductility by partially recrystallizing NiAl. Another means of improving ductility through control of microstructure is to utilize rapid solidification techniques such as melt spinning.

It has been shown by Inoue and co-workers[25,72] that melt-spun $L1_2$ compounds of type Ni-Al-X (X=Cr, Mn, Fe, Co or Si) are both strong and ductile, whereas their conventionally solidified counterparts are extremely brittle. The most effective strengtheners were chromium and silicon. Silicon also has been reported to be an effective strengthener in conventionally prepared $Ni_3Al$[31]. However, the high strength of the melt-spun alloys decreases significantly after subsequent annealing (e.g., after annealing at 1000°C for 1 hr), see Fig. 10, and the strength loss is accompanied by a drastic drop in ductility and a change from transgranular to intergranular cracking[25]. The change was attributed to an increased degree of long-range order and segregation of solutes to grain boundaries. Koch and co-workers[73] also have been able to produce ductile $Ni_3Al$ containing small additions of boron using an arc hammer apparatus, but the quench rates could not exceed a critical value if ductility was to be retained. No ductility was achieved without boron. Taub and co-workers[8], on the other hand, have reported that boron has a beneficial effect on the ductility of melt-spun $Ni_3Al$, especially for Al$\leq$25a%. It is clear that small differences in composition often play a significant role in determining ductility.

RSR processing via the P/M route also has been employed to produce intermetallic compounds with improved ductility. For example, off-stoichiometric FeAl (35%Al)[58] with a room temperature ductility of 7 percent has been produced by the RSR process. $Fe_3Al$ containing sufficient quantities of boron to produce borides also has been prepared by RSR[76] and FeAl has been prepared by melt spinning followed by pulverization and powder processing[36], but relatively little properties data have been reported in the open literature. The microstructure of $Ni_3Al$ and $Ni_3Al$-B-Ti prepared by RSR also has been studied[77]. A summary of representative rapid solidification processing efforts reported to date appears in Table V.

The benefits of rapid solidification on the ductility of aluminides at low temperatures have been suggested to be the following: elimination of macrosegregation and grain boundary segregation (both RSR and melt spinning); production of very fine grained materials (∿3μm grain size by melt spinning, 1-2μm by RSR); and reduction in the degree of order as an interim step during processing (melt spinning).

However, the investigations reported above also have demonstrated certain disadvantages with respect to high-temperature mechanical properties: loss of strength and ductility of melt-spun $Ni_3Al$-type alloys after annealing[70]; elevated temperature intergranular fracture in RSR-processed Fe-Al alloys[75,76]; inadequate stress rupture strength of melt-spun and PM-processed Fe-Al-B alloys at temperatures above 537°C (1000°C)[36].

A further disadvantage of the melt spinning process is the restriction of product to thin sections (e.g., 0.1 mm thickness), and the sensitivity of the ribbons to surface imperfections produced upon solidification. However, these disadvantages can be eliminated if the melt-spun foils are pulverized and then processed by PM techniques, as has been done with Fe-Al alloys containing borides[36]. This material is brittle at room temperature and, when inclusions are present, exhibits poor low cycle fatigue

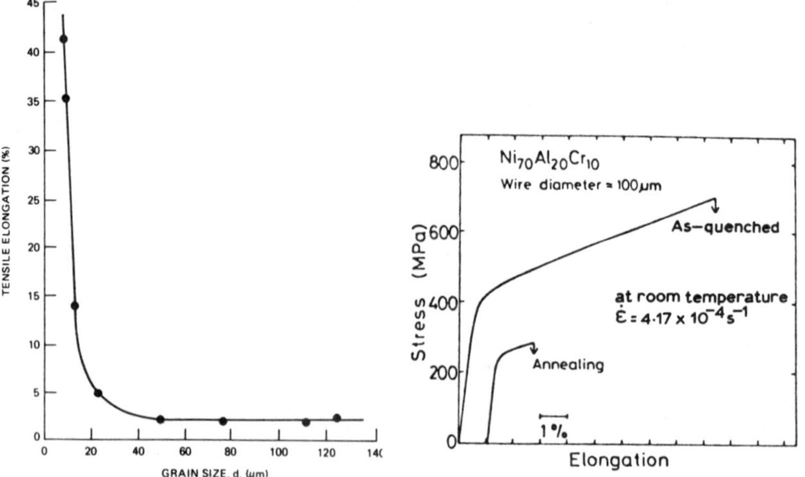


Fig. 9 Grain size dependence of ductility in NiAl at 673°K, $\dot{\varepsilon}=10^{-4}$ $s^{-1}$[69].

Fig. 10 Stress-strain behavior of melt-spun Ni-Al-Cr[25].

TABLE V

Summary of results of experiments on rapid-solidification-rate (RSR) processing of aluminides

| Alloy | Method | Remarks | Ref. |
|---|---|---|---|
| Ni–Al–X | Melt spin | Ductile at 25°C, unstable at 1000°C | 25 |
| $Ni_3Al$–B | Arc hammer | Ductile | 63 |
| FeAl borides | Melt spin and P/M | Brittle to 540°C | 36 |
| Fe–35at.-%Al | RSR | Cracking is transgranular at 25°C, intergranular at 600°C | 58 |
| Fe–50at.-%Al | RSR | Brittle to 400°C; intergranular cracking | 75 |
| $Ni_3Al$–B | Melt spin | Ductility varies with B and Al contents | 73 |
| $Fe_3Al$–B | RSR | Borides lead to strengthening | 76 |

resistance at 600°C[114].

5. Fabricability

The growing number of ordered alloys which display appreciable ductility has led to increasing attention to fabricability of such alloys. Aluminides in bulk may be prepared by powder processing, by arc melting and casting, followed by either high temperature extrusion or by alternate cold rolling and recrystallization treatments[26]. Drop casting has proven to be a useful means of refining grain size and reducing segregation during solidification. In addition, melt spinning of ribbons has been successfully demonstrated. The addition of alloying elements to $Ni_3Al$+B has been shown to have an appreciable effect on ductility. Small amounts of Fe (<2 at %) cause severe cracking during cold rolling, while alloys containing large quantities of Fe (6-15%) were readily rolled to sheet[62]. Small amounts of Hf, Zr and Mn also improve cold formability of $Ni_3Al$, perhaps due to removal of sulfur from grain boundaries by sulfide formation. Additions of 0.1-0.2%C also result in fabricable alloys.

The influence of small additions of ternary elements on workability of equiatomic FeCo alloys has been examined by Kawahara[50]. Carbon, vanadium, chromium and molybdenum are among those solutes that improved hot forgeability, while aluminum, beryllium, boron and copper are typical of those that are ineffective. The effective solutes combine with cobalt to form a $Co_3X$ compound that can be produced only by diffusion. It was postulated that ordering of the lattice in the vicinity of precipitates was prevented by the increased localized depletion of cobalt. In turn, these local zones of disorder were postulated to increase ductility.

## E. Creep

Diffusion-controlled phenomena such as recovery, recrystallization, creep, and oxidation are suppressed by long-range order. The improvement in steady-state creep resistance with increasing order is manifested by an increase in the activation enthalpy, $\Delta H$[16,79,80]. The steady-state creep rate of alloys often can be expressed as:

$$\dot{\varepsilon} = A\left(\frac{\sigma}{G}\right)^n e^{-\Delta H/RT} \tag{2}$$

where A is a constant, T is absolute temperature, R is the gas constant, $\sigma$ is the applied stress, G is the shear modulus, and n is a constant of the order of 4-6. Tracer experiments to determine activation energies for diffusion in ordered systems have provided values in close agreement with activation energies for creep[16]. Since self-diffusion depends on vacancy interchange with atoms, the activation enthalpy is the sum of the enthalpies of formation and migration for vacancies. Both the enthalpies of formation and migration increase with the formation of long-range order, resulting in higher activation enthalpies for diffusion and for creep, and, therefore, lower creep rates. In the case of LRO $(Fe,Co,Ni)_3V$-type alloys, however, formation of long range order causes only a small increase in $\Delta H$, even though creep rate is decreased by two orders of magnitude[10].

Creep resistance of unalloyed NiAl[81,82] and $Ni_3Al$[38] tends to be poor relative to that of commercial high temperature alloys, see Fig. 11[74] However, alloying of boron-modified $Ni_3Al$ with hafnium and zirconium provides creep resistance comparable to or better than that of Waspaloy, a nickel-base alloy widely used in gas turbines[26]. Advanced LRO alloys, in which aluminum replaces vanadium, also display good creep resistance, although not quite as high as that of $Ni_3Al$+B (Table VI)[26]. These results, which were obtained at a very early stage of development of both classes of alloys, suggest the desirability of further intensive research on these systems.

Limited stress rupture data for Fe-20.8w%Al-4.1%B (30.9 at% Al-15.2%B)

TABLE VI

Comparison of Creep Properties of Advanced LRO Alloys and Nickel Aluminides with Commercial Alloys

| Alloys | Steady-State Creep Rate ($10^{-6}$h) | Rupture Life or Test Time (h) |
|---|---|---|
| **At 20,000 psi, 760°C** | | |
| Boron-doped $Ni_3Al$ | 34 | 500 |
| Advanced $Ni_3Al$ aluminides | 3 | 600[a] |
| Advanced LRO alloys $(Fe, Ni)_3(V,Al)$ | 7 | 500[a] |
| Type 316 stainless steel | 8,500 | 65 |
| Hastelloy X | 1,300 | 200 |
| **At 40,000 psi, 760°C** | | |
| Advanced $Ni_3Al$ aluminides | 20 | 600[a] |
| Advanced LRO alloys $(Fe,Ni)_3(V,Al)$ | 30 | 600[a] |
| Waspaloy | | 1000 |
| **At 94,000 psi, 760°C** | | |
| MAR-M-246 | 133 | 1000 |

[a]The test was stopped (without rupture) at the time indicated.
SOURCE: private communication with C. T. Liu, Oak Ridge National Laboratory, 1983.

show that at 538°C (1000°F) the 100 hour specific creep strength is comparable to that of A-286, an Fe-Ni superalloy[36]. However, with increasing temperature the stress rupture strength of the very fine-grained rapidly solidified Fe-Al-B alloys drops at a faster rate than for the conventionally processed, coarser-grained A-286. Control of grain size is expected to be a problem in providing balanced creep and tensile properties of <u>all</u> rapidly solidified alloys.

Experiments on the slow strain rate behavior of Fe-39.8at%Al (B2 structure) produced by powder metallurgy indicate that extrusion temperature can affect high temperature deformation mechanisms. Grain size refinement to about 10 μm produced effective strengthening to at least 0.75 of Tm, the melting temperature[85].

## F. Fatigue

The supression of cross slip or reduction in number of available slip systems with long-range order that occurs in most alloys suggests a diminished probability of crack nucleation under cyclic loading. In the few systems for which room-temperature fatigue data have been published (e.g., FeCo-2%V, $Ni_3Mn$, and several LRO alloys, ordering does, indeed, lead to an increase in high-cycle (stress-controlled) fatigue life)[78,86]. A comparison of HCF life of several ordered alloys with a number of iron and nickel base superalloys appears in Fig. 12. It is noteworthy that, although the crack path is predominantly TG in both $Ni_3Al$+B and $Fe_3Al$, crack initiation occurs at grain boundaries in $Fe_3Al$, and the overload region is entirely IG in $Ni_3Al$, see Figs. 13a) and b), respectively. The most detailed account of fatigue resistance in an ordered alloy is that of Williams and Smith[87], who reported on the high-cycle behavior of β-brass at 25°C. The marked elastic anisotropy in this alloy produces severe stress concentrations at grain boundaries, leading to intercrystalline crack initiation. The fatigue limit of this alloy is reduced when the sample is tested in a 3 percent aqueous solution of NaCl.

Crack propagation data for ordered alloys are not yet generally available in the literature although recent work has shown that the crack growth resistance of a $(FeNi)_3V$ alloy, LRO-60 is superior to that of conventional alloys at both 25°C, and 600°C, see Fig. 14[78]. Ordering has a significant

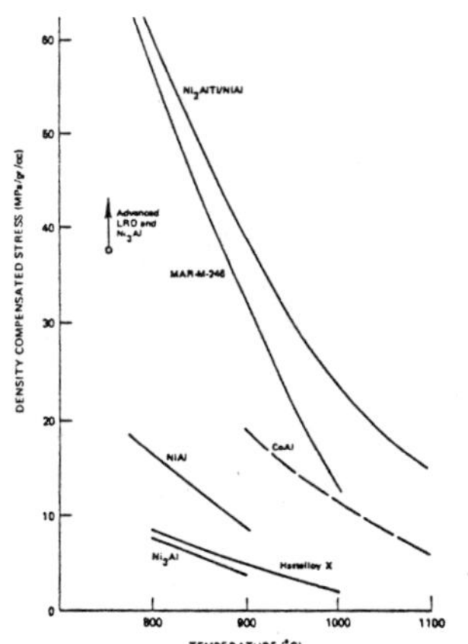


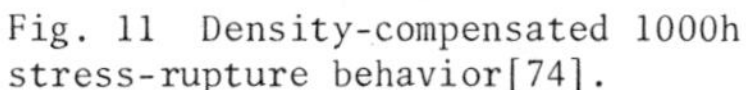
Fig. 11 Density-compensated 1000h stress-rupture behavior[74].

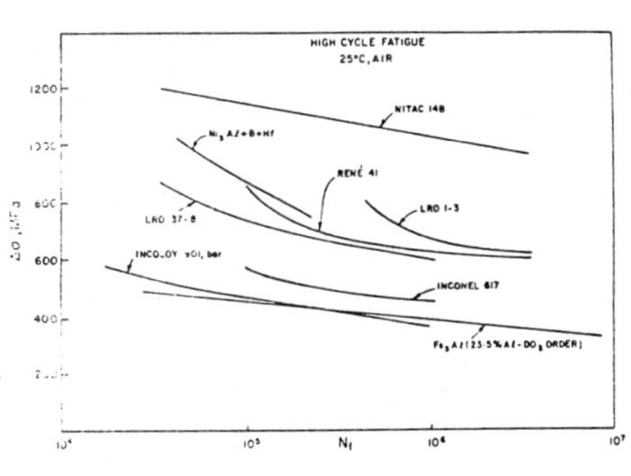


Fig. 12 High cycle fatigue of ordered alloys and superalloys.

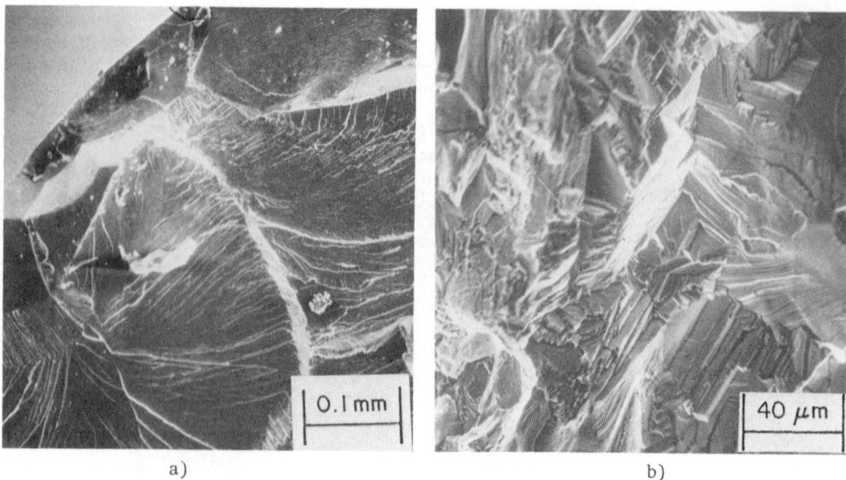


Fig. 13 Fracture surfaces in high cycle fatigue a) $Fe_3Al$, showing IG crack origin, TG cleavage b) $Ni_3Al$+B+Hf, showing TG cleavage

effect in slowing fatigue crack growth at low $\Delta K$ levels in LRO-60, see Fig. 15. This behavior may be explained by a suggestion of McClintock[88], in which the crack growth increment per cycle, $da/dN = \delta$, the crack opening displacement.

$$da/dN = \delta = \frac{K^2}{E\sigma_{ys}} \quad (3)$$

where E = Young's modulus and $\sigma_{ys}$ is the yield stress.
At threshold, $\Delta K = \Delta K_{Th}$, and $\delta$ is assumed to be of the order of the Burger's vector, b; thus:

$$\Delta K_{Th} = \sqrt{E\sigma_{ys} b} \quad (4)$$

Upon ordering, superlattice dislocations are formed in $L1_2$ alloys such as LRO-60, leading to an effective doubling of the Burger's vector, and a significant increase in predicted threshold, since E and $\sigma_{ys}$ are little changed by LRO. The data in Fig. 15 are in accord with Eq. 4.

Data on high-temperature fatigue resistance are few, with work reported for only four systems: TiAl[89], $Ti_3Al$[89,90], $(CoNi)_3V$[91], and $Cu_3Au$[92]. In each case there is a tendency towards increased intergranular crack propagation as temperature increases. Compatibility stresses at grain boundaries due to a lack of five independent slip systems was the cause of intergranular cracking in $Ti_3Al$[90]. In the cobalt-base LRO alloys this tendency can be reduced by doping with Ti[91].

There are few data available on low-cycle fatigue (LCF) of ordered alloys. LCF resistance of $Cu_3Au$ is little affected by long-range order although fully ordered crystals cyclically harden much more rapidly than disordered crystals[84]. A cyclic strain-hardening exponent, n'=0.36, has been reported, but this value is higher than that observed for most materials. Ordering has little effect on the fracture mode of $Cu_3Au$. Limited LCF data at 650°C for Fe-20.8w%Al-4.1%B have recently been reported[36]. Comparisons with several other engineering alloys, including an $(Fe,Ni)_3V$ alloy (LRO-49) are shown in Fig. 16. Pak and Hsiung[93] have studied the LCF properties of $Ni_3Ge$ at room temperature. Neither dislocation cells nor ladder structure was observed, in contrast to the behavior of Cu single crystals cycled to saturation[105]. It has been suggested previously that persistent slip band formation may be supressed by long range order, thereby resulting in delayed crack initiation[94].

## ENVIRONMENTAL EFFECTS

### A. Cracking Phenomena

The previously described changes in slip character caused by long-range order, most notably reduced cross slip in alloys with low APB energies, suggest that there should be a significant effect of LRO on resistance to environmental cracking, since increased susceptibility to stress corrosion cracking of some austenitic steels deforming by planar glide has been noted[95] (although this association has been disputed)[96]. Nevertheless, few studies of the influence of LRO on environmental cracking have been reported. In the case of Hastelloy B, attempts to link LRO with increased susceptibility to hydrogen embrittlement were inconclusive[97] although LRO had been suggested to be the cause of embrittlement[98]. Increased embrittlement of FeCo-2%V and an $(Fe,Ni)_3V$ alloy (LRO-42) in the presence of hydrogen when the alloys were ordered has been demonstrated by means of tensile, delayed failure, and fatigue tests[99]. A possible explanation for enhanced susceptibility to hydrogen embrittlement by ordering is through the transport of hydrogen over long distances by superlattice dislocations confined to the original slip planes. However, no marked change in slip character with ordering has been noted by ordinary metallographic observations in these alloys. Recently, intergranular fracture accompanied by a severe loss of ductility also has been noted in $Ni_3Al$+B tested in tension in the presence of hydrogen at room temperature, see Fig. 17[100]. The embrittlement was shown to be reversible after an outgassing treatment at 200°C. Further, embrittling effects of hydrogen could be eliminated by a 24 hr pre-oxidation treatment which prevented subsequent ingress of hydrogen into the material. The role of boron in this phenomenon is being intensively investigated, in view of previous reports that boron supresses hydrogen

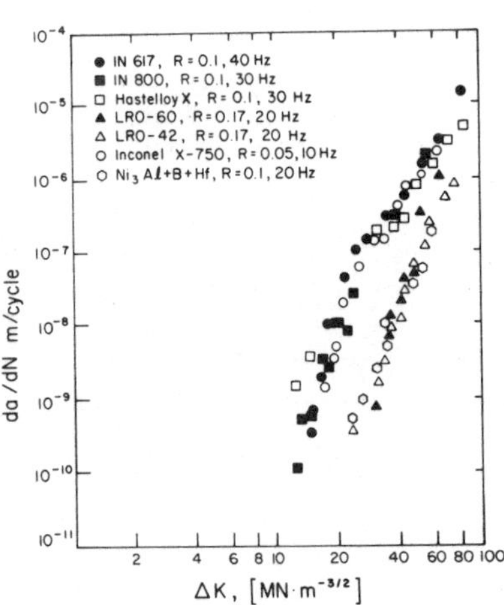

Fig. 14 Crack growth of several alloys at 298°K[78].

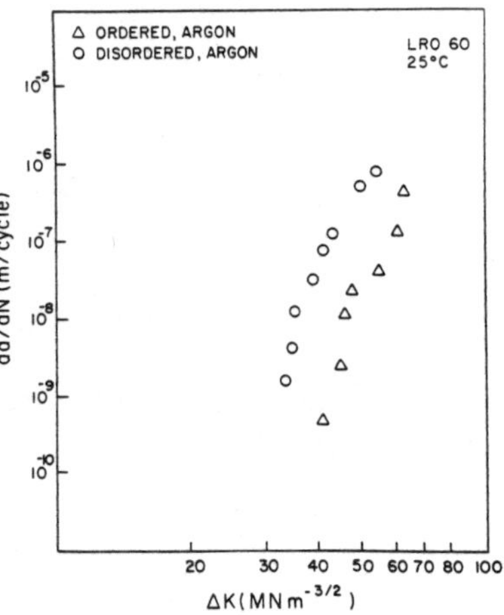

Fig. 15 Ordering and crack growth in LRO-60[78].

embrittlement in unalloyed nickel[101]. However, it appears that models of ductility improvement in $Ni_3Al$ due to boron, based on the Messmer-Briant model[102] are untenable.

B. Aqueous Corrosion

Corrosion studies of ordered alloys have been largely confined to Fe-Si-Al alloys and to Ni-Mo alloys. Silicon has a pronounced beneficial effect on corrosion resistance of iron in 10% $H_2SO_4$ at 80°C[103]. The corrosion resistance of stoichiometric $Fe_3Si$ has been attributed to an adherent protective film of $SiO_2$[104]. The critical current density of the 14.5% w/Si alloy is two orders of magnitude lower than that of more dilute Fe-Si alloys; passitivity initiates at a lower potential and is displayed over a much wider range of potential. Aluminum further reduces the passivation potential of Fe-Si-Al alloys: an Fe-8 (Al+Si) alloy passivates, while Fe-8%Si does not.

Ni-20%Mo alloys containing Fe, Cr and C have been available commercially since the early 1930's. These alloys provide excellent resistance to acid chloride environments, provided that oxidizing species (dissolved oxygen and ferric ions) are monitored at low levels[13]. The corrosion rates for Hastelloy B2 as well as Ni-24%Mo are much lower than for Ni and dilute Ni-Mo alloys in aerated and deaerated $H_2SO_4$ and HCl. Polarization curves for Ni-20%Mo in deaerated IN $H_2SO_4$ and IN HCl indicate corrosion potentials of 0.1 and 0.15V, which lie between values for pure Ni and Mo. There is no tendency for passivation with either solution.

Systematic corrosion experiments on other ordered alloys systems have rarely been reported in the literature. There are indications, however, that aqueous corrosion resistance of iron aluminides may be high at ordinary temperatures.

C. Oxidation

The superior oxidation resistance of aluminides based upon Ni, Co and Fe has been extensively exploited through coatings on gas turbine hardware. Although brittle, aluminides such as NiAl readily form an adherent layer of $Al_2O_3$ which is protective under oxidizing conditions to temperatures in

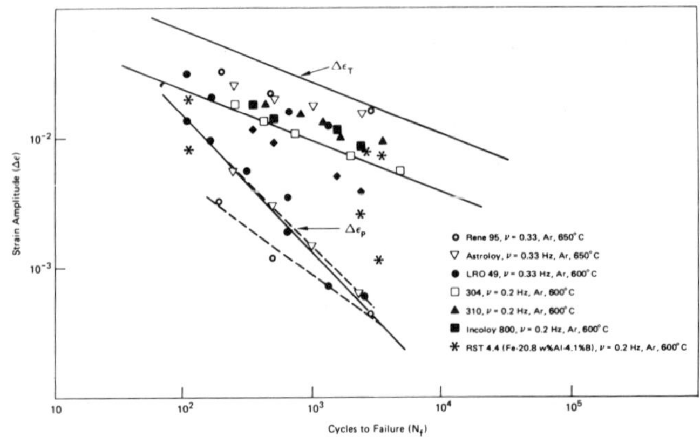


Fig. 16 Low cycle fatigue of several alloys.

excess of 1000°C. In the case of the iron-aluminum system, Kanthal (TM) alloys have been utilized for heating elements for many decades.

Interest in utilizing aluminides and other ordered alloys in structural applications has led to several investigations of oxidation and corrosion behavior. Limited work to date has shown that hafnium-modified $Ni_3Al$+B displays good resistance to oxidation at 1000°C[26]. see Fig. 18. It appears that, although slightly more weight gain occurs in the hafnium-containing alloys, oxide scale adhesion is improved. Considerable weight loss and severe flaking of oxide occurred in the iron-containing alloy. The oxidation resistance at 1000°C of hafnium-modified $Ni_3Al$ is superior to that of Hastelloy X and 316 stainless steel. This superiority is attributed to the formation of a compact, adherent film of $Al_2O_3$ on the aluminides.

Limited experiments also have been carried out on oxidation and corrosion resistance of LRO $(Fe,Co,Ni)_3V$ alloys in air, superheated steam, oxidizing-sulfidizing environments and in liquid lithium. The lack of Al or Cr in these alloys was the cause of spalling subsequent to extensive oxidation in air at temperatures above 600°C. However, the same alloys exhibited excellent resistance to superheated steam corrosion at 540°C[106]. LRO alloys cntaining up to 40%Ni showed little weight loss and no change in microstructure after exposure in static lithium at 500°C and 600°C for several thousand hours in containers of the same alloys. The weight loss of these alloys compared very favorably to that of Incoloy 800H under the same conditions. However, when dissimilar metals such as stainless or ferritic steels encapsulated the lithium, severe weight loss occurred, perhaps due to mass transport of nickel between dissimilar metals[106].

## WEAR RESISTANCE

The pronounced effects of ordering on the mechanical and physical properties of alloys suggest that wear resistance also might be influenced by long-range order. It has been shown, for example, that the coefficient of adhesion is changed by long-range order[107]. For Cu-Au and Pt-Co alloys, the coefficients are lower for the ordered conditions. These observations have been made by Buckley[108] for the frictional behavior of Cu-Au alloys

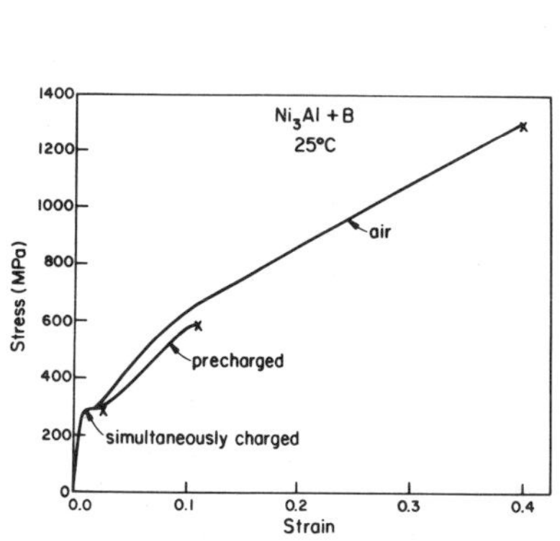


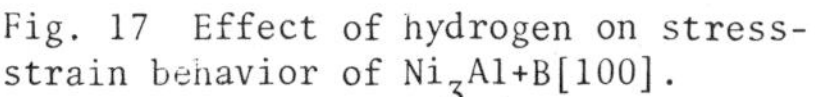

Fig. 17 Effect of hydrogen on stress-strain behavior of $Ni_3Al$+B[100].

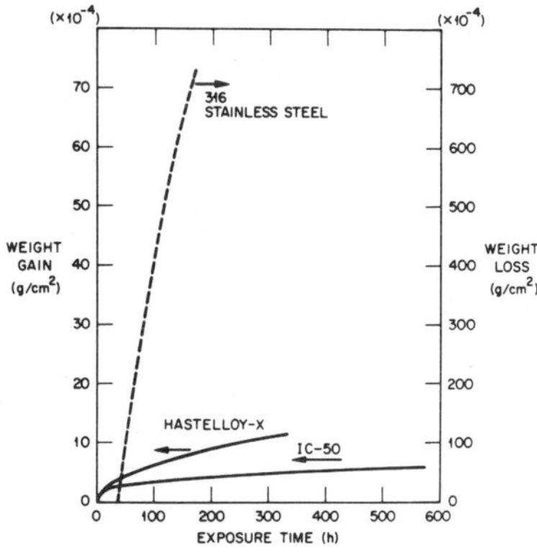


Fig. 18 Oxidation of $Ni_3Al$ (IC50) compared to other alloys[105].

in vacuum. Although little work has been done on the effects of long range order on erosion behavior, work on $Cu_3Au$[109] has shown that the finer slip in the ordered material enhances the erosion resistance.

## CURRENT RESEARCH EFFORTS AND SUGGESTIONS FOR THE FUTURE

This introductory paper has reviewed the unique properties that make ordered alloys extremely attractive for structural use. Among them are a high specific modulus, especially at elevated temperatures, high strength at elevated temperatures, high strain-hardening rates, and low self-diffusion rates with resulting low creep rates and high recrystallization temperatures. A major problem with most ordered alloys has been a tendency for low ductility; however, recent work performed in the United States and Japan has shown that there are a number of reasons for the brittleness of ordered alloys, and the reasons can be quite varied depending on the alloy system. Furthermore, it has been clearly demonstrated that once the reasons for the brittleness of a given alloy have been identified, the ductility can, in many cases, be significatnly improved, and/or the fracture path may be changed. These improvements now make it possible to consider ordered alloys for a much greater range of structural uses. Nevertheless, many problems remain, as outlined below. Much of our current progress has been measured in terms of increases in tensile strength and ductility on unnotched bars tested under laboratory conditions on small heats. Little is known about the inherent notch sensitivity and crack propagation resistance of the aluminides. Preliminary work on crack growth in $Ni_3Al$ and LRO alloys is promising due to the extremely slow growth rates recorded at low $\Delta K$. However, $Fe_3Al$, a much more brittle alloy, behaves similarly to conventional alloys, except for a lower $K_{I_c}$. Data on toughness, LCF, creep and thermal fatigue remain to be reported. There is as yet no satisfactory explanation for the ductilitizing effect of boron in $Ni_3Al$, especially in the light of recent work that demonstrates extreme susceptibility of this alloy to intergranular fracture in the presence of hydrogen. The absence of similar ductilizing effects in $Ni_3Al$ with Al≥25% and other intermetallics such as NiAl, $Fe_3Al$ and FeAl remains to be explained, although it has been noted that boron segregations to grain boundaries decreases as Al≥25%[115]. Other basic scientific questions remain: the phase diagram of the Fe-Al system is still controversial. We are just beginning to understand the importance of phase relationship in the Fe-Al system as they pertain to yielding phenomena. There is as yet no common basis for understanding the anomalous rise in strength of many intermetallics, with the possible exception of $Ni_3Al$. Little systematic work has been done on the influence of ternary elements on APB energies, dislocation arrays and yielding and strain hardening phenomena. Studies of the role of point defects in intermetallic deformation and fracture behavior are needed, especially in view of the possible linkage of defects with the stoichiometric anomalies noted in $Ni_3Al$+B. Further, although oxidation resistance of the aluminides appears to be good, little is known about their resistance of hot corrosion, as under gas turbine operating conditions. In view of the many uncertainities cited above, much remains to be done before intermetallics can finally realize their great potential.

## ACKNOWLEDGMENTS

The author is grateful to his present graduate students, A.K. Kuruvilla, S.J. Choe, G. Fuchs and J. Beckman for their contributions to our understanding of the behavior of intermetallic compounds. Further, the author wishes to acknowledge the financial support of the Dept. of Energy under the Energy Conversion and Utilization Technologies (E-Cut) Program, the Office of Naval Research for support under Contract N00014-84-K-0276 and the National Science Foundation for support under Grant No. DMR-8409593.

## REFERENCES


1. H. Warlimont, Ed., Order Disorder Transformations in Alloys, (Springer-Verlag, Berlin, 1974.

2. B.H. Kear, et al, Eds., Ordered Alloys - Physical Metallurgy and Structural Applications, Claitor's Publ. Div., Baton Rouge, LA, 1970.

3. H. Lipsitt, Aviation Week, 81, Jan. 26, 1976.

4. C.T. Liu, Met. Trans., 4, 1743 (1973).

5. C.T. Liu, J. Nucl. Mat. 85-86, 907 (1979).

6. K. Aoki and O. Izumi, Nippon Kinzoku Gakkaishi, 43, 1190 (1979).

7. C.T. Liu, C.L. White, C.C. Koch and E.H. Lee, Preparation of Ductile Nickel Aluminides for High Temperature Use, in Proceedings of the Conference on High Temperature Materials Chemistry, Princeton, New Jersey: Electrochemical Society, in press (1984).

8. A.I. Taub, S.C. Huang, Met. Trans. A, 15A, 399 (1984).

9. N.S. Stoloff, Int. Met. Rev. 29, 123 (1984).

10. C.T. Liu, Int. Met. Rev. 29, 169, (1984).

11. D.P. Pope and S.S. Ezz, Int. Met. Rev. 29, 136 (1984).

12. E.M. Schulson, Int. Met. Rev. 29, 195 (1984).

13. C.R. Brooks, J.E. Spruiell and E.E. Stansbury, Int. Met. Rev., 29, 210 (1984).

14. "Structural Uses for Ductile Ordered Alloys", NMAB-419, National Academy Press, Washington, D.C. (1984).

15. A. Lawley in Intermetallic Compounds, Ed. by J.H. Westbrook, New York: (John Wiley and Sons, New York, 1967), pp. 464-490.

16. N.S. Stoloff and R.G. Davies, The Mechanical Properties of Ordered Alloys, Prog. in Mat. Sci., 13, 1 (1966).

17. N.S. Stoloff, in Strengthening Methods in Crystals, Ed. A. Kelly and R.B. Nicholson, (Elsevier, New York, 1971) pp. 193-259.

18. M.J. Marcinkowski, in Treatise on Materials Science and Technology, Ed. H. Herman, (Academic Press, New York, 1974) pp. 333-

19. L.A. Nesbit and D.E. Laughlin, Acta Met., 29, 989 (1980).

20. P. Morgand, P. Mouturat and G. Sainfort, Acta Met. 16, 867 (1968).

21. Y. Calvayrac and M. Fayard, Phys. Stat. Sol. 17, 407 (1973).

22. G.W. Ardley, Acta Met. 3, 525 (1955).

23. P.A. Flinn, Trans. TMS-AIME, 218, 145 (1960).

24. M. Yodogawa, D.M. Wee, Y. Oya and T. Suzuki, Scripta Met, 14, 849 (1980).

25. A. Inoue, H. Tomioku and T. Masumoto, Met. Trans. A, 14A, 1367 (1983).

26. C.T. Liu, Oak Ridge National Lab, unpublished

27. M.G. Mendiratta, H.M. Kim and H.A. Lipsitt, Met. Trans. A, 15A, 394 (1984).

28. P.H. Thornton, R.G. Davies and T.L. Johnston, Met. Trans., 1, 207 (1970).

29. D.E. Mikkola, J.B. Cohen, Acta Met, 14, 105 (1966).

30. R.G. Davies and N.S. Stoloff, Phil. Mag., 12, 297 (1965).

31. O. Noguchi, Y. Oya and T. Suzuki, Met. Trans. A, 12A, 1647 (1981).

32. R.W. Guard and J.H. Westbrook, Trans. Met. Soc. AIME, 215, 807 (1959).

33. K. Aoki and O. Izumi, Phys. Stat. Sol. (a) 32, 657 (1975).

34. A. Wolfenden and M. Marmouche, J. Metals 35 (12), 90 (1983).

35. R.E. Schrafik, Met. Trans. A 8A, 1003 (1977).

36. R. Ray, V. Panchanathan and S. Isserow, J. Metals 35(6), 30 (1983).

37. E. Kuramoto and D.P. Pope, Acta Met. 26, 207 (1978).

38. R.G. Davies and T.L. Johnson in Ordered Alloys-Structural Application and Physical Metallurgy, B.H. Kear, et al, Eds., Claitor's Publ. Div., Baton Rouge, LA, pp. 447- (1970).

39. Y. Umakoski, J.R. Boland, D.P. Pope and V. Vitek, Mat. Sci. Eng., 64, 27, (1984).

40. S.S. Ezz, D.P. Pope and V. Paidar, Acta Met., 30, 921, (1984).

41. T. Suzuki, Y. Oya and D.M. Wee, Acta Met., 28, 301 (1980).

42. B.H. Kear and H.G.F. Wilsdorf, Trans. TMS-AIME 224, 382 (1962).

43. N.S. Stoloff and R.G. Davies, Acta Met, 12, 473, (1964).

44. H. Inoue, Oak Ridge National Lab, this meeting.

45. A. Fujita, Y. Mishima and T. Suzuki, J. Mat. Sci., 18, 1881 (1983).

46. S. Hanada, S. Watanabe, T. Sato and I. Izumi, Scripta Met, 15, 1345, (1981).

47. G.J.L. Van der Wegen, P.M. Bronsveld and J. Th. M. De Hosson, Acta Met., 30, 1537 (1982).

48. A.E. Vidoz and L.M. Brown, Phil. Mag. 7, 1167 (1962).

49. S.K. Ehlers and M.G. Mendiratta, J. Metals, 34(12), 80 (1982).

50. K. Kawahara, J. Mat. Sci. 18, 1709 (1983).

51. N.S. Stoloff and I.L. Dillamore in Ordered Alloys, Structural Application and Physical Metallurgy, by B.H. Kear, et al, Eds., Claitor's Publ. Div., pp. 525-544 Baton Rouge, LA (1970).

52. M.M. Shea and N.S. Stoloff, Met. Trans. 5, 755 (1974).

53. M.J. Marcinkowski, M.E. Taylor and F.X. Kayser, J. Mat. Sci., 10, 406 (1975).

54. G. Fuchs and N.S. Stoloff, Rensselaer Polytechnic Inst., unpublished (1984).

55. M.J. Blackburn, Trans. TMS-AIME, 239, 600 (1967).

56. R. Schafrik, H.A. Lipsitt and D. Schectman, AFWAL, Wright Patterson AFB, Ohio, unpublished.

57. H.A. Lipsitt, D. Shechtman and R.E. Schafrik, Met. Trans. A 6A, 1991 (1975).

58. D.K. Chaterjee and M.G. Mendiratta, J. Metals 33(12), 5 (1981).

59. S.K. Ehlers and M.G. Mendiratta, J. Metals 33(12), 5 (1981).

60. N.S. Stoloff and R.G. Davies, Trans. ASM 57, 247 (1964).

61. J.H. Westbrook and D.L. Wood, J. Inst. Met. 91, 174 (1963).

62. C.T. Liu, et al, ORNL 6067, Oak Ridge National Laboratory (1983).

63. C.T. Liu and C.C. Koch, Development of Ductile Polycrystalline $Ni_3Al$ for High Temperature Applications in Proceedings of the Conference on Trends in Critical Materials Requirements for Steels of the Future, Vanderbilt Univ., Nashville, TN (1982).

64. A.I. Taub, S.C. Huang and K.M. Chang, unpublished.

65. K. Vedula, abstract TMS-AIME Fall Meeting, Detroit, MI, J. of Met., (1984).

66. C.C. Koch, North Carolina State Univ., unpublished (1984).

67. C.T. Liu and H. Inoue, Met. Trans. A 10A, 1515 (1979).

68. P.A. Beck, Adv. X-ray Anal. 12, 1 (1969).

69. E.M. Schulson, and D.R. Barker, Scripta Met., 17, 519 (1983).

70. E.M. Schulson, Res. Mecanica Letters 1, 111, (1983).

71. E.M. Schulson, COSAM Program Overview, p. 175. NASA TN 830006, National Aeronautics and Space Administration, Washington, D.C. (1982).

72. A. Inoue, H. Tomioku and T. Masumoto, Met. Trans A 14A, 1367 (1983).

73. C.C. Koch, J.A. Horton, C.T. Liu, O.B. Cavin and J.O Scarbrough, in Rapid Solidification Processing Principles and Technologies III, R. Mehrabian, Ed., National Bureau of Standards (1983) pp. 264-269.

74. "Structural Uses for Ductile Ordered Alloys", NMAB-419, National Academy Press, Washington, D.C., p. 74 (1984).

75. D.K. Chaterjee and M.G. Mendiratta, J. Met 33(12), 6 (1981).

76. E.R. Slaughter and S.K. Das in Proc. of Second International Conference on Rapid Solidification Processing, Claitor's Publ. Div., Baton Rouge, LA, pp. 354-363 (1980).

77. I. Baker, F. Ishishita and E.M. Schulson, Proc. MRS Symp. on Rapid Solidification Processing, Nov. 14-17, Boston, MA, in press (1983).

78. A.K. Kuruvilla and N.S. Stoloff, Met. Trans. A, in press (1984).

79. A. Lawley, J.A. Coll and R.W. Cahn, Trans. Met. Soc. AIME 218, 166 (1960).

80. R.R. Vandervoort, A.K. Mukherjee and J.E. Dorn, Trans. ASM 59, 930 (1966).

81. P.R. Strutt and R.A. Dodd, in Ordered Alloys: Structural Applications and Physical Metallurgy, B.H. Kear, et al, Eds., Claitor's Publ. Div., Baton Rouge, LA, pp. 475-503 (1969).

82. G. Leverant and D. Duhl, quoted in ref. 81.

83. T.L. Johnston, R.G. Davies and N.S. Stoloff, Phil. Mag. 12, 305 (1965).

84. K.H. Chien and E.A. Starke, Acta Met., 23, 1173 (1975).

85. J.D. Whittenberger, Mat. Sci. and Eng. 57, 77 (1983).

86. R.C. Boettner, N.S. Stoloff and R.G. Davies, Trans. TMS-AIME, 236, 131 (1966).

87. H.D. Williams and G.C. Smith, Phil. Mag. 13, 835 (1966).

88. F. McClintock quoted in E. Tschegg and S. Stanzl, Acta Met. 33, 33 (1981).

89. S.M.L. Sastry and H.A. Lipsitt, Met. Trans. A, 8A, 299 (1977).

90. S.M.L. Sastry and H.A. Lipsitt, Acta Met. 25, 1279 (1977).

91. S. Ashok, K. Kain, J. Tartaglia and N.S. Stoloff, Met. Trans. A, 14A, 1997 (1983).

92. A. Gittins, Met. Sci. J. 2, 114 (1968).

93. H.-r.Pak, New Mexico State Univ., unpublished.

94. N.S. Stoloff, unpublished proposal to National Science Foundation, (1983).

95. S. Barnatt, Corrosion 18, 322 (1962).

96. M.N. Saxena and R.A. Dodd, in Environment-Sensitive Mechanical Behavior, Ed. A.R.C. Westwood and N.S. Stoloff, (Gordon and Breach, New York 1966), pp. 455-479.

97. B.J. Berkowitz and C. Miller, Met. Trans. A, 11A, 1877 (1980).

98. A.I. Asphahani, in Proceedings of the Second International Congress on Hydrogen in Metals, No. 4, Sec. C, Paper 2, (Pergamon Press, New York, 1977).

99. A.K. Kuruvilla, S. Ashok and N.S. Stoloff, in Proceedings of the Third International Congress on Hydrogen in Metals, No. 4 (Pergamon Press, New York, 1980) pp. 629-633.

100. A.K. Kuruvilla and N.S. Stoloff, Scripta Met, in press (1984).

101. Y. Ogino and T. Yamasaki, Scripta Met, 15, 821 (1981).

102. R.P. Messmer and C.L. Briant, Acta Met, 30, 457 (1982).

103. S. Wakefield, F.H. Beck and G.W. Powell, Corrosion, 40, 190 (1984).

104. M.G. Fontana and N.D. Greene, Corrosion Engineering (McGraw Hill Book Co., New York, 1967) p. 161.

105. H. Mughrabi in ASTM STP 675 (complete reference to be supplied).

106. J.C. Griess, C.T. Liu and J.H. Devan, Oak Ridge National Lab, unpublished (1984) (See also ref. 10).

107. J.A. Bailey and M.E. Sikorski, Wear, 14, 181 (1969).

108. D.H. Buckley, Influence of Order-Disorder Transformation on Friction Characteristics of Copper-Gold Alloys in Vacuum, NASA TN D-2985, Washington, D.C., NASA, Washington, D.C. (1965).

R.N. Wright and D.E. Mikkola, Mat. Sci. Eng. 26, 263 (1976).

110. D.M. Wee, O. Noguchi, Y. Oya and T. Suzuki, Trans. Japan Inst. Met. 21, 237 (1980).

111. A.G. Rozner and R.J. Wasilewski, J. Inst. Met. 94, 69 (1966).

112. K. Aoki and O. Izumi, Trans. Japan Inst. Met. 19, 203 (1978).

113. G. Sainfort, Fragilite et Effects de L'Irradiation. Presses Universitaires de France, 187 (1967).

114. A.K. Kuruvilla and N.S. Stoloff, unpublished (1984).

115. C.T. Liu and C.W. White, Oak Ridge National Lab, unpublished (1984).

PART II

# Theory of Order and Phase Stability

# THEORIES OF ALLOY ORDERING - OVERVIEW AND FORECAST

PHILIP C. CLAPP
Department of Metallurgy and Institute of Materials Science, University of Connecticut, Storrs, CT 06268

## ABSTRACT

The task of any alloy ordering theory simply put is to predict from a small amount of experimental data the ordering phase transitions, their temperature, composition and crystal structure in some selected alloy system. Ideally, the experimental information needed would be not more than the atomic numbers of the alloy components involved. This paper attempts to assess the progress that has been made towards that objective, the various techniques that have been used to provide solutions and the probable avenues of most rapid progress in the next ten years. The major areas that will be reviewed are Ising model approaches, many body interaction approaches, computer simulation techniques, and first principle calculations. It will be argued that we are on the verge of some quite exciting breakthroughs in the field of alloy phase prediction.

## INTRODUCTION

Almost any theory of alloy ordering starts from the Ising model [1] which was developed in 1925 to explain ferro-magnetism. The statistical mechanics for describing ordering in alloys is very similar to that of ferro-magnetic ordering and the equation for the case of a multicomponent alloy can be written as follows:

$$E\{\sigma\} = \frac{1}{2} \sum_{\alpha,\beta=1}^{M} \sum_{i,j=1}^{N} V_{ij}^{\alpha\beta} \sigma_i^{\alpha} \sigma_j^{\beta} \qquad (1)$$

$\sigma_i^{\alpha}$ refers to the occupation of the ith lattice site (out of N possibilities) by the $\alpha$th chemical species (out of M possibilities). If the $\alpha$ species is on site i, then $\sigma_i^{\alpha} = 1$, otherwise $\sigma_i^{\alpha} = 0$. $V_{ij}^{\alpha\beta}$ is the energy of interaction between an $\alpha$ species atom on site i and a $\beta$ species atom on site j. The factor 1/2 is to correct for the double counting of pairs in the summation.

The usual assumption is that the $V_{ij}$'s are temperature independent and that the lattice is a regular rigid structure. If the $V_{ij}$'s are known or determinable from experiment then the energy of any configuration of the alloy can be calculated from the above equation. If the alloy configurations can be enumerated, then a statistical mechanics program can be carried out to evaluate the free energy of the system. This can be done by calculating the partition function:

$$Z = \sum_{\{\sigma\}} e^{-\beta E\{\sigma\}} \; ; \; \beta=1/kT \qquad (2)$$

$$F = -kT\ln Z \qquad (3)$$

Thus allowing the calculation of the free energy of the system as a function of temperature, T. This in turn, would allow the prediction of the phase diagram of the alloy system and also a number of other thermodynamic properties.

It might be immediately asked whether it is realistic to hope that these calculations could be done. The answer would have to be "yes", but with very severe limitations. Successful examples are the calculation by Ising [1] in 1925 for a one-dimensional chain of atoms with nearest neighbor interactions. Onsager [2] succeeded in calculating most of the statistical properties of a two-dimensional square lattice with nearest neighbor interactions in 1944, which led, in part, to the Nobel prize. Wilson [3] has developed a method called "Renormalization Group Theory" to deal with three-dimensional lattices for the purposes of calculating statistical properties very close to the ordering temperature in cases of higher order transformations. This effort was also worth a Nobel prize.

Despite these very impressive exact calculations, one can still ask how these calculations can be connected to real alloy systems. In the first place one would certainly want to include more than nearest neighbor interactions in the calculations. Secondly, elastic distortions from the regular lattice must be allowed which would imply that the interactions between neighbors need to depend on relative distance. A third consideration is that many body interactions be allowed such as three-body, four-body and higher order cases. Only when all of these factors have been included can we believe that we are realistically representing some actual alloy system and are able to make accurate predictions of phase diagrams. This review paper attempts to provide a perspective of the progress that has been made in this general program.

Overview

This overall program is perhaps easiest to understand by breaking it into its component parts. One major area of effort has been the determination of lowest energy ordered structures for a given lattice and a given set of interaction parameters. One can think of this as the task of determining the phase diagram at the absolute zero of temperature. A second part of the program is the determination of the effects of finite temperature on the phases. This requires that one be able to evaluate the entropy and free energy at different temperatures. A third task has been to go beyond the Ising model. This has involved incorporating elastic strain into the calculations, including many body interactions, and efforts to solve the energy of an alloy configuration from "first principles". We will describe progress in each of these areas in the order listed.

A) Lowest Energy Ordered Structures

Clapp and Moss [4-7] developed a method for determining the ground state of FCC or BCC lattices at AB or $A_3B$ compositions with first, second and third neighbor interactions by using the Fourier transform of the interaction parameters. This can be written as follows:

$$V(k) = \sum_{r_{ij}} V(r_{ij}) e^{ik \cdot r_{ij}} \quad (4)$$

They found that the minima of V(k) coincide with the diffraction peaks of the miminum energy ordered structure, if such a structure can be constructed. In some cases, compositional conflicts or coordination conflicts prevent a structure from being realized. From this type of consideration they generated structure maps which would predict the lowest energy ordered structure within any given range of interaction parameter.

Richards, Allen and Cahn [8,9] generalized this approach to include alloys of any binary composition on a FCC or BCC lattice. Kanamori, Kudo and Katsura [10-12] carried this work even further to include interactions up to fourth nearest neighbors and also considered HCP lattices up to second nearest neighbor interactions. In regard to other types of lattices, other than those mentioned above, it must be said that not very much has been done to evaluate their ground state configurations. Since most alloy ordering systems fall into the lattice types that have been examined, perhaps this is not a great lack at the present.

B) Finite Temperature Phase Stability

Cowley [13] developed a statistically approximate expression relating the short range order parameters of a disordered alloy with the pair-wise interaction energies. Following upon this work Clapp and Moss [4-5] obtained an approximate relationship between the Fourier transform of the interaction energies and the Fourier transform of the short range order (SRO) parameters which could be directly related to the short range order diffuse intensity obtained by x-ray or neutron scattering. This equation made it possible to directly relate the experimentally obtained data to the interaction parameters in any alloy system in a relatively simple manner, and has been carried out for quite a number of alloy systems to date. This equation is given as follows:

$$\alpha(k) = \frac{C}{1+2C_A C_B \beta V(k)} \tag{5}$$

where $C_A$, $C_B$ are the atom fractions of A, B atoms in the alloy, C is a normalization constant, $\beta$ is the temperature factor $(kT)^{-1}$, V(k) is the interaction Fourier transform and $\alpha(k)$ is the SRO parameter transform.

A similar equation was developed independently by Krivoglaz [14] differing only in the value of the constant in the numerator. The Clapp-Moss equation can also be used to predict the ordered structure and the temperature at which the ordered structure will occur once the values of V(k) have been determined from the short range order diffuse intensity at some chosen temperature. This is done by determining at what temperature the denominator of the expression vanishes, indicating that the short range intensity at some point in reciprocal space will now become infinite thus signaling the appearance of a Bragg peak. The positions of these Bragg peaks will occur wherever V(k) has its minima and this in turn indicates the structure of the ordered phase. This prediction of the ordering temperature is equivalent in accuracy to mean field theory but, of course, provides the additional information of what ordered structure could be expected to emerge in the alloy. It may also happen that the minima of V(k) will show that no ordered phase will appear but rather segregation will occur. For example, an analysis of the copper-nickel system led to such a prediction [7]. Wilkins [15] showed that the statistical error involved in the Clapp-Moss formula could be reduced if the fit to the data was done in terms of the inter-atomic interaction ratios $V_n/V_1$.

The cluster variation method (CVM) has gained a much increased use over the last decade or so in calculating entropies and free energies of alloy phases. The method was first developed by Kikuchi [16] and has subsequently been improved by Kikuchi, deFontaine and Sanchez [17] as well as other workers. The input in this method is a specification of the lattice type, the lattice composition and the values of the inter-action energies as a function of distance in the lattice. The output, after a considerable amount of calculation on a high speed computer, is the free energy as a function of temperature of different structures and of the disordered phase. It then becomes possible to make detailed predictions of the phase diagram in this manner. However, the reverse process of going from the experimental data to the value of the input parameters is more difficult. One way of dealing with this difficulty is to use equation (5) to obtain the approximate values of the inter-action energies from the short range order data and then to insert those values in a CVM calculation to predict the whole phase diagram as a function of temperature. Another way just recently developed and reviewed by deFontaine in this volume (see Eqs. 31-33) has been dubbed the "inverse CVM". It uses the CVM approximation for the free energy written in terms of all the multiplet correlations contained in the assumed cluster (pair, triplet, etc.). Inserting the experimental data for the pair correlations in this free energy expression and then maximizing it with respect to the unknown multiplet correlations allows a determination of both the effective pair energies and multiplet correlations through a set of simultaneous non-linear equations.

An extremely powerful mathematical method for dealing with the statistical mechanics of ordering processes where the transformation is not first order in character has become known as "renormalization group theory". This approach has been developed by Wilson, Kadanoff and Fisher [3,18,19] primarily. It is a method that has created a great deal of excitement because of its considerable accuracy in predicting most thermodynamic details of interest near the transition temperature but unfortunately can not tell us very much about the behavior well away from the transition temperature nor about cases where the ordering occurs with a latent heat. In reality most of the alloy ordering processes that are of practical importance fall outside the range of applicability of this theory. It is to be hoped however, that renormalization group theory can be extended in the near future to deal with cases which involve first order transformations. This is currently an area of very intense effort.

## C) Beyond The Ising Model

Since most alloys are made up of a mixture of atoms of different sizes the elastic distortions that will occur in the alloy provide an important energy term which must be included in any realistic assessment of the phase diagram. Cook and deFontaine [20] provided an interesting foundation for including elastic strain terms in the development of a microscopic elasticity theory. This theory included terms which coupled displacements of atoms from the regular lattice sites and near-neighbor site occupations, as well as the traditional Ising model interactions. As a result their theory, depending on the size and magnitude of the different interaction parameters, could describe an ordering transformation, spinodal decomposition or continuous ordering (which is a combination of segregation and ordering occurring simultaneously). However, since there are three sets of interaction parameters instead of the one set in the Ising model approximations the parameters of this theory are far more difficult to extract from the experimental data. Recently there has been considerable progress in this direction and an alloy of composition Au-40Ni has been analyzed in complete detail by Wu and Cohen [21,22] who have been able to determine all of the parameters of the Cook-deFontaine theory for this particular system.

Another direction that has been taken to deal with the question of elastic strain effects is an approach involving empirical intermetallic potentials. This is work primarily carried out by Machlin [23-27] who used an expression of the following form for the interaction between atoms of different types in an alloy as a function of their interatomic separation:

$$V_{ij}^{\alpha\beta}(r_{ij}) = \frac{-A^{\alpha\beta}}{(r_{ij})^4} + \frac{B^{\alpha\beta}}{(r_{ij})^8} \qquad (6)$$

where $A^{\alpha\beta}$, $B^{\alpha\beta}$ are constants dependent on the alloy species $\alpha,\beta$ and $r_{ij}$ is the interatomic separation. Machlin determined the adjustable parameters of this interaction from lattice parameters and cohesive energy data that was available for the pure elements and a selected set of intermetallic compounds. He was then able to make extensive predictions of the lattice parameter, the cohesive energy and the relative stability of various intermetallic cubic phases: FCC, BCC, $L1_2$, $L1_0$, A15, C15, etc. and also some non-cubic phases. As impressive as this accomplishment is, it is difficult to know what degree of accuracy applies in any particular case, or how to relate it to the more fundamental calculations which are now becoming possible. As such, it is a very useful semiquantitative approach which provides considerable intuitive information for the alloy designer, but must be used within relatively sensible error ranges. The interactions that Machlin determines by his semi-empirical approach could then be used in one of the thermodynamic theories to make predictions of the phase diagram as a function of temperature, if this was desired.

Some effort has been made to include many body interactions in the calculation of phase diagrams. The energy as a function of alloy configuration would now look as follows:

$$E\{\sigma\} = \frac{1}{2} \sum_{\alpha,\beta=1}^{M} \sum_{i,j=1}^{N} V_{ij}^{\alpha\beta} \sigma_i^{\alpha} \sigma_j^{\beta} + \frac{1}{3!} \sum_{\alpha,\beta,\gamma=1}^{M} \sum_{i,j,k=1}^{N} V_{ijk}^{\alpha\beta\gamma} \sigma_i^{\alpha} \sigma_j^{\beta} \sigma_k^{\gamma} \cdots \qquad (7)$$

These higher body interactions are indicated from fundamental calculations of alloys (such as pseudo-potential methods) and also the fits to phonon dispersion curves indicate the presence of many body interactions. With this motivation in mind deFontaine and Kikuchi[(28)] used the assumption of nearest neighbor interactions but included pair-wise, three-body and four-body forces for the particular case of the copper-gold alloy system. They used the CVM to calculate the entire phase diagram with just two adjustable parameters which were chosen to make the ordering temperature at the $Cu_3Au$ and at the CuAu compositions match the experimental data. No other adjustable parameters were involved and the prediction for the phase diagram compares very favorably with the actual phase diagram.

Perhaps the most exciting development in phase diagram prediction that has come on the scene in the last few years are the so-called first principles calculations. These involve calculating the total lattice energy, including the energy of all of the electrons. The input for this technique is simply the atomic number of the species involved and the type of lattice that the atoms are arranged on. The actual method of calculating the total energy involves the use of what has become known as the density functional approximation and is combined with older methods of approximately solving the many-body Schrödinger equation such as the spherical APW method, the KKR-CPA method or the LCAO method. The output from these calculations provides the electronic density of states, the total lattice energy as a function of lattice parameter, bulk modulus, phonon dispersion curves, equilibrium lattice parameter, elastic constants, shape of the Fermi surface, etc. all at 0°K. One interesting example of this type of calculation is for pure silicon carried out by Yin and Cohen [29]. Their results show the energy of each of a number of possible silicon phases as a function of atomic volume. This calculation was carried out by combining the density functional approximation with pseudo-potential calculations. As can be seen from Fig. 1 of their paper the lowest energy structure is in fact the phase which is found in nature. Furthermore, if the application of pressure at absolute zero is considered, then by drawing a tie line tangentially between the diamond structure and the beta-tin structure a prediction can be made for the density at which there will be a transition to a two-phase mixture of those two phases. The authors show in their paper that this estimate is fairly close to the observed structural high-pressure transition. It is also impressive that the correct high-pressure phase is predicted and that its density is fairly accurately determined as well. This calculation also demonstrates the relative stability of the other principal types of phases which is potentially very useful information for those who are trying to create metastable phases by various processing methods.

Another impressive example is the calculation by Connolly and Williams [30] of the Cu-Ag alloy system. They performed their calculation using the density functional approximation combined with the ASW method and found the total lattice energy of each of a set of ordered alloys on a face centered cubic lattice for either the $L1_2$ structure or $L1_0$ structure as well as for the pure elements. The results showed that a two-phase mixture of the pure elements will always have a lower energy than any of the ordered structures. Thus their calculation predicted that the Ag-Cu system should be a segregating one over its entire composition range which agrees remarkably well with the actual phase diagram. They have also analyzed their data in terms of an effective nearest neighbor pair interaction, a nearest neighbor three-body interaction and a nearest neighbor four body interaction and found that the three-body and four-body interactions were much weaker than the two-body interactions. There are however, additional parameters which incorporate the effect of elastic strain and are given by the $V_0$ and $V_1$ terms in the following equation:

$$E(r) = V_0(r)+V_1(r)\langle\sigma_0\rangle+V_2(r)\langle\sigma_0\sigma_1\rangle+V_3(r)\langle\sigma_0\sigma_1\sigma_2\rangle+V_4(r)\langle\sigma_0\sigma_1\sigma_2\sigma_3\rangle \quad (8)$$

The $V_0$ and $V_1$ are very substantial in magnitude as is their dependence on the lattice parameter. Since the value of $V_2$ is positive, this factor taken alone would lead one to predict that the system would tend to order which would be the case if the system were confined to have a constant lattice parameter in the vicinity of the CuAg minimum. However, once the energies involved with the strain of the two different sized atoms is allowed to take its effect, then the system is able to lower its energy by phase separation into two phases of different lattice parameters producing a very different result from that predicted by any rigid lattice theory.

To bring this review full circle, it is interesting to note a first principles calculation by Wadsworth, Gyorffy and Stocks [31] who used the KKR-CPA method to predict the short range scattering that should be produced by the Fermi surface in copper-palladium alloys. They obtained a very satisfactory fit between the predicted shape of the short range order scattering and the systematic changes observed in the experimental data as the composition of the alloys was varied. Furthermore they were able to analyze their calculations in the form of an equation relating $\alpha(k)$ to a two-body response function which can be regarded as a generalization of the two body interaction $V(k)$ of the Clapp-Moss formula.

Thus it seems that a fundamentally new interpretation arising out of first principles calculations can be made for an old equation.

As a general forecast of work to come in the area of predicting alloy phase diagrams, it would seem that the first principles calculations will be of enormous value when guided by semi-empirical intuition. It may be expected that the impact of this approach on predictions will be felt very visibly over the next decade now that high speed computers are capable of carrying out the calculations in a reasonable amount of time. However, that still leaves the entropy problem to be dealt with. The best approaches to this problem will probably be computer statistics calculations by means of either Monte Carlo, Molecular Dynamics or CVM. The real value in these efforts to predict alloy phase diagrams will undoubtedly be in those areas where it is very difficult to carry out the experiment directly. For instance, these predictions should be of considerable benefit in forecasting metastable phases that might be produced by rapid solidification techniques, predicting phase stability at very high temperatures and/or pressures, or predictions for multi-component alloy phase diagrams where the number of components is substantially greater than two or three.

As a simple example it would now appear possible to carry out a complete first principles calculation for the case of metallic hydrogen to determine the precise pressure at which it would be stable, and also to be able to do a stability analysis to determine the temperature at which it could remain stable. This would be of considerable interest for superconducting technology because this form of hydrogen is thought to have a very high superconducting transition temperature, and would potentially be of great practical value.

## References

An excellent general review of many of the topics of this paper is given in: D. de Fontaine, Solid State Physics 34, 73 (1979).

1) E. Ising, Z. f. Physik 31, 253(1925).
2) L. Onsager, Phys. Rev. 65, 117 (1944).
3) K. G. Wilson, Phys. Rev. B4, 3174, 3184 (1971).
4) P. C. Clapp, Physics Letters 13, 305 (1964).
5) P. C. Clapp & S. C. Moss, Phys. Rev. 142, 418 (1966).
6) P. C. Clapp & S. C. Moss, Phys. Rev. 171, 754 (1968).
7) S. C. Moss & P. C. Clapp, Phys. Rev. 171, 764 (1968).

8) M. J. Richards & J. W. Cahn, Acta Met. 19, 1263 (1971).
9) S. M. Allen & J. W. Cahn, Scripta Met. 7, 1261 (1973); Acta Met. 20, 423 (1972).
10) J. Kanamori, Prog. Theor. Phys. 35, 66 (1966).
11) T. Kudo & S. Katsura, Prog. Theor. Phys. 56, 435 (1976).
12) J. Kanamori & Y. Kakehashi, J. Phys. (Paris) 38, C7-274 (1977).
13) J. M. Cowley, Phys. Rev. 77, 669 (1950); 120, 1648 (1960).
14) M. A. Krivoglaz & A. A. Smirnov, "The Theory of Order-Disorder in Alloys", MacDonald, London (1964).
15) S. W. Wilkins, Phys. Rev. B2, 3935 (1970).
16) R. Kikuchi, Phys. Rev. 81, 988 (1951).
17) J. M. Sanchez & D. de Fontaine, Phys. Rev. B17, 2926 (1978).
18) K. G. Wilson & M. E. Fisher, Phys., Rev. Letts. 28, 240 (1972).
19) L. P. Kadanoff, Physics 2, 263 (1966).
20) H. E. Cook & D. de Fontaine, Acta Met. 17, 915 (1969); 18, 189 (1970).
21) T. B. Wu, J. B. Cohen & W. Yelon, Acta Met 30, 2065 (1982).
22) T. B. Wu & J. B. Cohen, Acta Met 31, 1929 (1983).
23) E. S. Machlin, Acta Met. 22, 95 (1974).
24) E. S. Machlin, Acta Met. 22, 109 (1974).
25) E. S. Machlin, Acta Met. 22, 367 (1974).
26) E. S. Machlin, Acta Met. 22, 1433 (1974).
27) E. S. Machlin, Acta Met. 24, 543 (1976).
28) D. de Fontaine & R. Kikuchi in "Applications of Phase Diagrams in Metallurgy & Ceramics" N.B.S. Special Publication 496 (G. C. Carter, ed.) p. 999 (1978).
29) M. T. Yin & M. L. Cohen, Phys. Rev. Letts. 45, 1004 (1980).
30) J. W. D. Connolly & A. R. Williams, Phys. Rev. B 27, 5169 (1983).
31) J. Wadsworth, B. L. Gyorffy & G. M. Stocks - to be published.

# ON THE FEASIBILITY OF AB INITIO CALCULATIONS OF ORDERING ALLOY PHASE DIAGRAMS

D. de Fontaine

University of California
Department of Materials Science and Mineral Engineering
Berkeley, CA 94720

ABSTRACT

Many stable or metastable intermetallic phases useful to the alloy designer have crystal structures which are ordered superstructures of a parent disordered phase. A highly reliable statistical mechanical method (CVM) has now been developed for calculating such superstructure phase equilibria derived from say, the fcc parent lattice. To obtain phase diagrams, one needs certain physical parameters, such as effective pair interaction ratios. It is possible, in principle, to extract these parameters from band structure calculations in the coherent potential approximation (CPA), particularly from recently developed cluster-CPA techniques. If sufficient accuracy can be achieved, truly first-principles phase diagram calculations may soon become feasible.

## 1. INTRODUCTION

As explained elsewhere in these conference proceedings, particular ordered structures are found to produce excellent mechanical properties in certain classes of alloys. The alloy designer would of course like to predict what sorts of alloying elements and heat treatments will promote which structures. If we restrict attention to superstructures of a given parent lattice, say the $L1_2$ structure in fcc, the problem appears to be amenable to theoretical solution. In other words, it may be possible, in favorable cases, to predict what ordered phases will be stable (or at least metastable) under what conditions, by performing purely first-principles calculations, i.e., ideally, from a knowledge of the atomic numbers of the elements considered, the temperature and the average composition.

The purpose of this paper is to summarize recent results which presently lead one to believe that such first-principles calculations may soon become quite feasible. Both band structure and configurational entropy aspects will be briefly covered. Possible experimental techniques which are available for verification of the theoretical predictions will be mentioned.

The present treatment will be essentially qualitative. For more detailed accounts, the reader is referred to the cited references. Background can be found, for example, in the author's early review paper [1].

## 2. STATE OF ORDER

In general, crystalline alloy phases are **disordered systems.** If we disregard, for simplicity, small atomic displacements from lattice sites, the disorder is essentially **compositional.** Perfect stoichiometric compounds (complete order) and completely disordered solid solutions (complete disorder) are extreme special cases of the general one under consideration here: that of **partial order.**

How does one describe such states? Perfect order is completely described by specifying the crystallographic unit cell. Total disorder is completely described by the average concentration since, by definition, the occupation probability of each lattice point is the same, namely the average concentration c ($\equiv x_B$ of B, say, in an A-B alloy), independently of the environment, since all correlations vanish. Hence, one may state that the **unit cell** is the appropriate descriptive unit for complete order, and the **lattice point** (occupied by an "average atom") is the appropriate one for complete disorder.

For partial order (or disorder), the appropriate descriptive unit is the **cluster** of lattice points. If the cluster is the whole crystal, we obtain an exact but impractical description of the state of order. Hence, only small clusters are retained, the smallest one, beyond the point, being the pair ($1^{st}$, $2^{nd}$, $3^{rd}$...neighbors). One can also envisage triplets of lattice point, quadruplets (tetrahedra,...), etc. States of order can then be described by a so-called "cluster algebra" [2] which makes use of **multiplet correlation functions** $\xi$ defined as follows:

$$\xi_{m,s} = \langle \sigma_1 \sigma_2 \ldots \sigma_m \rangle \qquad (1)$$

for a cluster of m lattice points, the subscript denoting the site in the cluster. In Eq.(1), the $\sigma$ are occupation variables taking value +1 or −1 depending on the occupation of the site, A or B. The brackets indicate an average of the σ product over the whole crystal, or better, over an ensemble of systems, and the index s labels the type of m-point cluster.

A cluster (of n points, of type r, say) may be populated by A and B atoms in a certain way, i.e., it may have a certain **configuration** (J), with resulting frequency of occurrence, or probability, or cluster concentration x given by [2,3]

$$x_{n,r}(J) = \frac{1}{2^n} [1 + \sum_{m,s} \nu_{n,r;m,s}(J)\ \xi_{m,s}] \tag{2}$$

where the (rectangular) matrix ν has elements given by sums of products of plus ones and minus ones. The nature of the ν-matrix is entirely fixed by the nature of the crystal lattice and by the largest clusters (and their sub-clusters) used. The multiplet correlations ξ form a set of linearly independent variables which completely describe the state of order at the level of approximation desired. These variables are thus the fundamental ones not only for the problem at hand, but also for calculating the internal energy, determining the ordered ground states and for calculating the configurational free energy of the system at arbitrary temperature, as will be seen in Sects. 3, 4 and 5.

## 3. INTERNAL ENERGY

It was stated above that a complete and exact description of the state of order could be given by specifying the statistical weights X(J) of all possible configurations J of the whole crystal, J denoting a given distribution of +1 and −1 over the $N=N_A+N_B$ points of the crystal, $N_A$ and $N_B$ being the fixed number of A and B atoms in the crystal. Let configuration J have energy E(J). For fixed concentration, the expectation value of the energy is then given by a canonical ensemble average over all $M = N!/(N_A!N_B!)$ configurations $J_o$ having concentration $c=N_B/N$ :

$$\langle E\rangle = \sum_{J_o} X(J_o)\ E(J_o), \tag{3}$$

This expression is exact; it can be written in terms of independent variables ξ by means of Eq. (2) applied to X, the "cluster concentration" of the whole crystal [3,4]. Upon substitution of Eq. (2), written for the whole crystal, into (3), one finds, after subtracting the energy of the completely disordered state:

$$\langle E\rangle = E_o + \sum_{m,s} \varepsilon_{m,s}\ \ \delta\xi_{m,s} \tag{4}$$

in which

$$E_o = \frac{1}{M} \sum_{J_o} E(J_o) \tag{5}$$

is the energy of the disordered state of composition c, and where $\delta\xi_{m,s} = \xi_{m,s} - \xi^\circ_{m,s}$ is the difference of correlation functions between actual (ξ) and disordered (ξ°) states, the **effective interaction parameters** being defined by

$$\varepsilon_{m,s} = \frac{1}{2^N} \sum_{J_o} \nu_{m,s}(J_o)\ E(J_o)\ . \tag{6}$$

The physical meaning of these interaction parameters can be illustrated for the case of pairs (m=2). We have, for the particular pair joining points p and p':

$$\varepsilon_{pp'} = \frac{1}{2^2} \sum_{\sigma_p=\pm1} \sum_{\sigma_{p'}=\pm1} \sigma_p\sigma_{p'}\frac{1}{2^{N'}} \sum_{J'} E_{\sigma_p\sigma_{p'}}(J') = \frac{1}{4}(V_{++}-V_{+-}-V_{-+}+V_{--}) \quad (7)$$

with

$$V_{ij} = \frac{1}{2^{N'}} \sum_{J'} E_{ij}(J'), \quad (8)$$

with N' = N-2. Because of translational symmetry, the effective interaction does not depend on the location of the pair, only on its type s, so we have:

$$E_{2,s} = -\frac{1}{2} V(s) \quad (9)$$

with

$$V = V_{AB} - \frac{V_{AA} + V_{BB}}{2} \quad (10)$$

in the familiar notation (see for example Ref. [1]), where, for instance, + occupation represents an A atom, and - a B atom. It is apparent from the definition (8) that the quantity $V_{ij}$ represents the total energy of an ij pair embedded in an average, random medium. The difference V thus represents a net or **exchange interaction** which is typically a very small fraction of the $V_{ij}$ energies themselves. Hence, Eq. (10) cannot be used to compute the effective ordering energies themselves, since the V (or ε) are expressed as very small differences of almost identical very large numbers. Various schemes which have been suggested for getting around this difficulty will be reviewed, below.

Note that nowhere have we tried to express the cohesive energy as a sum of pair energies: according to Eq. (4), the total energy is written as sum of $E_o$, the cohesive energy of a random solid solution, plus a much smaller term, the ordering energy:

$$E = \sum_r N_r \varepsilon_r \xi_r \quad (11)$$

where $N_r$ is now (N/2) times the number of neighbors in coordination shell r about a lattice point, the summation being limited, in practice, to a few coordination shells only. It is then clear, from Eqs. (7) and (8), that the effective interactions ε (or V) must depend on the average medium through the set of configurations $J_o$, hence they must be concentration dependent. In Sect. 7, we shall examine how such interaction parameters might be obtained from electronic band structure calculations. For the next two sections, however, we shall regard the pair interaction parameters as given constants. All of the physics of phase diagram prediction in the present context is contained in the sets of ε (or V) parameters.

## 4. ORDERED GROUND STATES

Before introducing the temperature dependence of alloy phase stability, it is necessary to determine what ordered superstructures are to be expected at absolute zero, in other words, the question "what ordered ground states will be found for a given set of V parameters and for various compositions" must be answered. The correct superstructures are those which will minimize the ordering energy (11). Since, by Eq. (11), or (12), E is linear form, one must solve a linear programming problem, as the $\xi$ variables are subject to certain constraints. First, the values of the correlations must lie between +1 and -1, then, since cluster concentrations must be non-negative, we have, by Eq. (2), the additional constraints:

$$1 + \sum_{m,s} \nu_{n,r;m,s}(J)\, \xi_{m,s} \geq 0 \qquad (13)$$

for all configuraitons J of the largest clusters (n,r) used. It turns out [6] that inequalities (13) define a "configuration polytope" in multidimensional $\xi$-space, the ratios of which, in principle, have $\xi$-coordinates which uniquely define the correct ordered ground states. Unfortunately, inequalities (13) may yield vertices which produce "non-constructible" structures, so that a "tighter" set of inequalities must be derived, which is not always feasible to do.

Early methods and bibliography are reviewed in Ref. [1] and [6]. All methods are based on linear programming techniques, although the inequalities and techniques of minimization may differ. Inequalities are derived through "cluster" considerations, as in (13). Completely solved problems include the ground states of fcc and bcc lattices with first and second neighbor pair interactions [7-9]. The case of the fcc lattice with $V_1$, $V_2$, $V_3$ and $V_4$ interactions has been partially solved [7], and so has the case of the fcc lattice with $V_1$, $V_2$, tetrahedron and octahedron multiatom interactions [6]. Recently [10], a very elegant computer code was developed and applied to the bcc lattice with $V_1$, $V_2$, $V_3$, and $V_5$ pair interactions.

Many experimentally observed ordered fcc superstructures, such as the $L1_0$, $L1_2$, $L1_1$, $DO_{22}$, $Pt_2Mo$, etc...have been predicted by the ground state analysis limited to $V_1$ and $V_2$. Other structures, such as the $D1_a$, require, in addition, 4th neighbor pair interactions for stability, as shown by Kanamori [7]. The complete list of fcc ground states and the ranges of the ratio $\alpha = V_2/V_1$ over which the various structures are stable have been given in many publications, such as [1], [3] and [6]-[9], with perspective drawings of the unit cells of the less symmetric structures illustrated in Ref. [11].

## 5. FREE ENERGY

It was seen that, at absolute zero, the state of order of, say, a binary solid solution is given by a vector $\xi$ at the origin of multidimensional $\xi$-space and end point on the surface of the configurational polytope: at vertices in case of stoichiometric superstructures, at edges and faces, away from stoichiometry. At higher temperatures, the configurational entropy contribution favors some disorder so that the vector $\xi$ no longer touches the surface of the polytope, but lies somewhere inside it. To determine equilibrium configurations, i.e. equilibrium values of the multiplet correlations, it is necessary to minimize the ordering free energy

$$F = E - TS \qquad (14)$$

where E, in general, is given by Eq. (11).

It was first shown by Kikuchi [12] that the configurational entropy S could be expressed by an increasingly accurate hierarchy of cluster approximations, known as the Cluster Variation Method (CVM). As the name implies, the CVM is a variational technique, meaning that the free energy is written as a functional of cluster concentrations x, which functional must then be minimized with respect to the unknown x. The CVM entropy is found to be given approximately by a sum of partial "cluster" entropies $s_\ell$:

$$S = N \sum \gamma_\ell s_\ell \tag{15}$$

where the collective index $\ell$ has been used for (m,s), the sum in Eq. (15) running over all clusters and sub-clusters retained in the state-of-order description. The coefficients $\gamma$ are positive or negative integers which may be obtained by geometrical considerations [12] or by recursive formulas [13,14]. The partial entropies are given by:

$$s_\ell = - k_B \sum_J x_\ell(J) \ln x_\ell(J), \tag{16}$$

$k_B$ being Boltzmann's constant.

Since by Eq. (2), the cluster concentrations are linear functions of the multiplet correlations $\xi$, the free energy (14) can finally be written as a functional of the independent $\xi$ variables

$$F = F(\xi_1 \ldots \xi_\ell \ldots \xi_L) \tag{17}$$

or, more explicitly, in the pair-energy approximation, for the free energy per lattice point,

$$f \equiv \frac{F}{N} = \sum_r \omega_r \varepsilon_r \xi_r + k_B T \sum_\ell \gamma_\ell \sum_J x_\ell(J) \ln x_\ell(J) \tag{18}$$

in which $\omega_r$ is half the r-shell coordination number. Equation (18) is written for the disordered state, in which all lattice points are equivalent. In ordered states, a distinction must be made between various sublattices, on each of which points are equivalent. Thus, for ordered phases, the free energy expression (18) must be amended, as explained elsewhere [15], though no new principles are introduced.

Equilibrium is determined by minimizing (17) with respect to the (independent) variables $\xi$. Usually, this is carried out by Newton-Raphson iteration [2,15]. Actually, one minimizes the "grand potential" [16]:

$$w = f - \mu\xi \tag{19}$$

where, for any ordered phase, $\xi$ is a sum of point correlation variables over the sublattices, and $\mu$ is an appropriate difference of chemical potentials. For given $\mu$ and temperature T, the minimization then returns equilibrium values of the $\xi$ correlations and hence, by Eq. (2), all desired cluster concentrations, including the average concentrations $x_A$ and $x_B$.

## 6. PHASE DIAGRAMS

When an analytic form of the free energy, as Eq. (18), is known, much of the thermodynamics properties of the system can be derived, such as configurational energy, entropy, specific heat, etc. The vibrational contribution to these quantities is not available from the present model; however, it is reasonable to assume that, although strictly thermal effects can be as large or larger than the configurational ones, their differences, from one phase to another, at given temperature, will be small - recall that the various ordered phases considered here (intermetallics) are superlattices of a given parent lattice.

A critical test for a free energy model is to see how good a (temperature-composition) phase diagram can be derived from it, and much effort has been expended over the last few years in calculating phase diagrams from CVM free energies.

The first such calculation was that of Van Baal [17] who considered first-neighbor pair interactions ($V_1$) on an fcc lattice, the tetrahedron cluster (T) approximation of the CVM being used in the entropy expression. Next, tetrahedron interactions were included in the internal energy expression (11), and a very good likeness of the Cu-Au phase diagram was produced [18]. Calcluated and experimentaly determined diagrams are also reproduced in Ref. [1], and compared to that obtained by the older Bragg-Williams (BW) model.

Next, second neighbor pair interactions were introduced so that both tetrahedron and octahedron (TO) clusters had to be used in the CVM entropy. The ferromagnetic (clustering) transition temperature at zero field (50/50 composition) calculated by the TO-CVM was found to differ by only 2% from that obtained by the best available high-temperature expansion [2]. By contrast, the BW approximation gives a result which is off by 22%.

The TO-CVM has been used to calculate a number of phase diagrams featuring the fcc ordered ground states proven to be stable for various values of the ratio $\alpha = V_2/V_1$. Hence, $\alpha$ is the only physical parameter which enters the calculations. In all, seven diagrams were computed for values $\alpha$ = 0.0, 0.25, 0.35, 0.45, 0.55, -0.2, -1.0. These prototype ordering phase diagrams have recently been collected in an overview article [11], details of the calculations being given in Refs. [15], [19] and [20]. Since no real alloy system is expected to be well modeled by ordering energies depending on constant $V_1$ and $V_2$ pair interactions, the accuracy of the TO-CVM prototype diagrams could only be tested against available Monte-Carlo (MC) simulations of ordering in the fcc Ising model with first and second neighbor interactions. In general, MC and CVM results agree closely, particularly at first-order transitions [11,20]. The CVM tends to overestimate second-order transition temperatures, particularly away from stoichiometry. It should be noted, however, that the Monte Carlo results for $\alpha$ = 0.0 are still somewhat controversial [21,22].

Only one prototype phase diagram is shown here, that for $\alpha$ = 0.35 (Fig. 1, see Ref. [19] for a detailed description). Ordered superstructures labeled $A_5B$, $A_3B$, $A_2B$ and $A_2B_2$ are, respectively, monoclinic, tetragonal, orthorhombic and tetragonal superstructures of fcc [11]. The dashed line is the stability limit of the disordered phase (or ordering spinodal [1,23]), at which (in the present case) a $\langle 1\,{}^1\!/_2\,0\rangle$ ordering wave just becomes unstable [1, 3, 23, 24]. Second-order phase transitions occur where phase boundaries (full lines) and instabilities (dotted lines) coincide.

Recently, Sanchez and collaborators have extended the CVM calculations to include ordering energies obtained by Lennard-Jones potentials. This procedure effectively results in concentration-dependent ordering energy parameters; the lattice parameter also varies with concentration. This method has been successfully applied to mapping the experimentally determined $Ni_3Al$ phase boundaries [25], and to producing "incoherent" prototype phase diagrams exhibiting fcc and bcc superstructures and miscibility gaps. The interesting problem of completing "ionic" and magnetic ordering has also been investigated [27]. A new formulation [14] of the CVM and various applications have also been reviewed recently by Sanchez and co-workers [28].

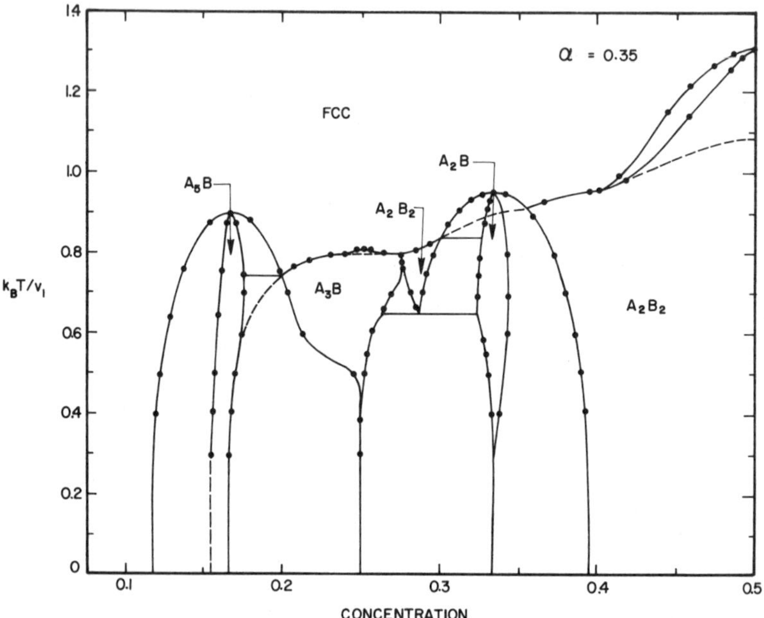


XBL 846-2311

Fig. 1 Prototype ordering phase diagram calculated by the TO aproximation of the CVM with $\alpha = V_2/V_1 = 0.35$ [15]. Dashed line is <1 $^1/_2$ 0> instability.

## 7. ELECTRONIC BAND STRUCTURE CALCULATIONS

From the foregoing, it is apparent that the statistical mechanical machinery is in place for calculating reliable configurational free energies, and hence realistic phase diagrams. To make the calculations truly predictive, it would be necessary to derive the $V_1$, $V_2$, ... interactions from first principles. It was seen that the "cluster" concept was essential for describing states of partial order, for determining ground states superstructures, and for calculating the configurational entropy. It is therefore expected that clusters may play an important role in fundamental electronic energy calculations as well.

### 7.1. Basic Formulation

It is a fairly straightforward matter to formulate the problem of calculating the cohesive energy of a metal or alloy: one writes down the total Hamiltonian in the form [29]:

$$H_t = K_e + K_n + V_{en} + V_{ee} + V_{nn} \tag{20}$$

when $K_e$ and $K_n$ are the kinetic energies of the electrons (e) and nuclei (n), respectively, $V_{en}$ is the electron-nuclei interaction, $V_{ee}$ is the electron-electron interaction, and $V_{nn}$ is the nuclei-nuclei interaction. The $K_n$ term, which can be handled semi-classically, contributes to the vibrational entropy and will not concern us here. Likewise, the Coulomb nuclear repulsion term (practically, an ionic repulsion term) can be treated classically and added on after the quantum mechanical calculations have been performed. The terms $K_e$ must be written in differential operator form and combined with $V_{en}$ and $V_{ee}$ to produce a strictly "electronic" Hamiltonian H, to be used in the Schroedinger equation:

$$H\Psi = E\Psi \tag{21}$$

where $\Psi$ is the wave function and E the associated energy eigenstate.

Without drastic simplifications, Eq. (21) cannot be solved, the $V_{ee}$ electron-electron interaction representing a major difficulty. Several schemes are now available for reducing the problem to an equivalent one-electron problem, the most popular one being the local density functional method (see for example Ref. [29]). Even so, the **absolute** magnitude of certain important physical parameters are predicted incorrectly unless more elaborate corrections are carried out [30]. It is believed, however, that these corrections will not be required in the calculation of **relative** quantities such as ordering energies, which we are concerned with here.

Solving Eq. (21) in an effective one-electron approximation with periodic potential (for pure or fully ordered crystals) still presents a formidable problem. Various approximate methods are available, the most useful one for alloy calculations being the tight binding (TB) and Korringa-Kohn-Rostoker (KKR) methods. These and other techniques are described in Ref. [29], and also by Pettifor, in a particularly readable account [36]. Basically, the tight binding is used as an approximation valid for transition metals if it is assumed that nearly all of the binding energy is provided by electrons in the d band. The KKR is of more general applicability (s, p, d electrons) and is based on multiple scattering theory. If f electrons participate in the bonding (heavy nuclei), relativistic effects must be taken into account, so that the appropriate differential equation to solve is no longer the Schroedinger, but the Dirac equation, thereby further complicating the problem.

Actually, for cohesive energy calculations, the wave function itself is not required; what is needed is the eigenvalue spectrum of the operator H, i.e., the **density of states** (DOS), the spectrum being virtually continuous. The total electronic energy is then obtained by filling up the DOS with electrons according to the Fermi-Dirac distribution. Techniques for carrying out this computational program are now under control. Impressive applications of such calculations to elemental solids or stoichiometric compounds are, for example, those of Moruzzi, Williams and Janak [31] or those of Cohen and collaborators [32] who use the pseudopotential method. Accurate absolute zero values have been obtained for the lattice parameter, elastic moduli and cohesive energy. Usually, the crystal structure is assumed, but in some cases [33], it has been possible to perform the calculations for various crystal structures to see which one gave the lowest energy; often, though not always, the correct structure was predicted. High accuracy is required, of course, as energy differences between competing structures are usually very small. In favorable cases, then, a true **first principles calculation** was performed: physical properties were derived from a knowledge of the atomic number Z alone.

## 7.2. Single-Site CPA

For pure crystals and stoichiometric compounds, all of the crystallographic information is contained in the unit cell, and translation symmetry reduces the computational problem to manageable proportions. In the case of alloys, the disordered nature of the system precludes the use of such simplifications. Since it is out of the question to perform the required quantum mechnical calculations on all possible configurations of a (large) representative region of the disordered crystal, it is imperative to attempt to restore translational symmetry by suitable averaging techniques.

As was mentioned in connection with the state of order description (Sect. 2), if complete disorder is assumed, it is sufficient to regard the lattice sites as being occupied by identical "average atoms". In this "point" approximation, the question is: how does one define an average atom? In the regular solution or Bragg-Williams free energy models, each lattice site is assumed to be occupied by an average atom of concentration $x_B$ (or $x_A$), given by the average concentration on the given lattice or sublattice. In electron theory a lattice site must be occupied by an "average atom" having sensible physical properties, in other words, one must associate with each site an appropriate potential $V_{av}(\mathbf{r})$, function of position $\mathbf{r}$, which can properly scatter electrons. Following the BW idea, the simplest procedure consists in setting:

$$V_{av}(r) = x_A V_A(r) + x_B V_B(r), \qquad (22)$$

where $V_A$ and $V_B$ are the A and B atomic potential, respectively. Such is the so-called **virtual crystal approximation** which has been shown to yield unreliable results, particularly for transition metal alloys [29].

Electronic structure calculation on fully disordered systems improved considerably with the advent of the coherent potential approximation (CPA). For a review, see Ref. [34], for example. The density of states can be obtained directly from the Green's function, the operator that is inverse to the Hamiltonian of Eq. (21). Actually, an average Green's function <G> is required, i.e., averaged over all configurations. The function <G> is then written as the inverse of an effective Hamiltonian $H_{eff}$ which is expressed in terms of an effective potential w(r), which is the same at every site. In such a way, an effective medium is created which possesses translational symmetry. The problem now consists in calculating the scattering of an electron off a real atomic potential, $V_A$ or $V_B$, imbedded in the CPA medium. The effective potential w is chosen in such a way that the replacement of the average potential by a real one at a given site causes no change in the scattering. This requirement leads to the basic CPA equations which must be solved self-consistently.

### 7.3. Cluster CPA

The CPA has been found to give satisfactory results in those cases for which comparison could be made with exact calculations [29, 34]; it is thus regarded as the best available single-site approximation, valid for complete compositional disorder. As such, this approximation cannot begin to do justice to the complexities of order-disorder phenomena any more than the regular solution model of statistical thermodynamics is able to handle order-disorder reactions.

It is possible to treat long-range order through the so-called "inhomogeneous CPA" which is to the "point" (homogeneous) CPA what the Bragg-Williams model is to the regular solution: the trick is to subdivide the lattice of the disordered state into distinct sub-lattices, and perform "point" CPA's on each. The inhomogeneous CPA is thus still a single-site approximation, so that short-range order cannot be treated in this framework.

As anticipated, the richness of order-disorder phenomena can only be produced from a theory which can handle states of partial order, i.e., from theories which are based on cluster methods. Ideally, one would like to do for electronic theory what the CVM has done for the statistical aspects of ionic ordering. Despite repeated efforts, a fully self-consistent Cluster-CVM remains, as of this writing, an elusive goal. A fundamental difficulty is that quantum mechanical operators must be dealt with rather than the simple scalars of the ordinary CVM. Nevertheless, some progress has been made, as will be now briefly described.

Gonis and Freeman [37] list three types of cluster-CPA models: (I) cluster theories in which the average medium reflects the cluster configurations in a self-consistent way, (II) cluster theories in which the cluster is embedded in a medium determined in a non self-consistent way, and (III) cluster theories in which the cluster is embedded in a medium that is determined in some self-consistent way which involves clusters smaller than the embedded one. If the medium cluster is a single-site, the average medium is then that of the single-site CPA.

The latter method has been used to calculate the DOS of a one-dimensional "alloy" exhibiting various degrees of short-range order (SRO), described by either positive (clustering) or negative (ordering) nearest neighbor pair correlations. Clusters of various sizes (one to nine points) were embedded in both single-site and pair-site CPA media. Calculations were performed for a single-band tight-binding Hamiltonian and compared to exact results obtained by direct summation of energy eigenvalues for systems of various configurations having chosen SRO values. It was shown [37] that, particularly with the larger clusters, the embedded cluster method reproduced the considerable complexities of the exact DOS to a remarkable extent.

Gonis, Stocks, Butler and Winter [38] used method III in an fcc binary alloy choosing, as embedded cluster, a central site and its 12 nearest neighbors (cuboctahedron), the effective medium being that of the self-consistent single-site CPA. A muffin-tin Hamiltonian was used and the calculations were performed in the KKR framework, the complications of electron-electron interactions having been neglected. The authors believe, as was surmised above, the $V_{ee}$ corrections to be relatively unimportant when ordering energies are required. To express the average internal energy given above by Eq. (2) in terms of the embedded cluster formalism, let us focus on a particular (central) fcc lattice point ($p_o$) of the crystal and its coordination shell of first nearest neighbors. The expectation value of the energy can then be written as:

$$\langle E\rangle = \sum_J X(J)E(J) = \sum_j x(j) \sum_{J'} X(j,J')\, E(j,J') \qquad (23)$$

where x(j) is the probability of finding, in the ensemble of systems, a 13-point cluster centered at $p_o$ having configuration j, X(j,J') being the probability of observing configuration J' outside the central cluster, given that the latter has configuration j. If it is assumed that correlations do not extend much beyond the first coordination shell, we may take this conditional probability to be approximately independent of j, so that we may write:

$$\langle E\rangle = N \sum [x_{A,z}E_{A,z} + x_{B,z}E_{B,z}] \qquad (24)$$

where

$$E_{i,z} = \frac{1}{N} \sum_{J'} X(J')\, E(i,z;J') \qquad (25)$$

and where the index j has been replaced by the double index i (= A or B, for the central atom) and z (denoting the first-shell configuration). In Eq. (24), $x_{i,z}$ has the same meaning as x(j) in Eq. (23).

The energy defined by Eq. (25) is the energy per site of a crystal having atom i at the point considered when surrounded by a particular first-shell configuration z. That energy can be calculated from the corresponding local DOS $n_{i,z}$ as follows [39].

$$E_{i,z} = \int_{-\infty}^{E_F} E\, n_{i,z}(E)\, dE \qquad (26)$$

where $E_F$ is the Fermi energy. In the single-site approximation, Eq. (24) would give simply:

$$\langle E\rangle = x_A E_{A,z} + x_B E_{B,z} = E_o, \qquad (27)$$

where $E_o$ is the random-solution energy defined by Eq. (5). Note that, despite a superficial resemblence, Eqs. (27) and (22) mean very different things. An expression similar to the one in square brackets in Eq. (24) has been used by Gonis, Butler and Stocks [39] to evaluate approximate ordering energies in a 50/50 AgPd alloy. To obtain correct energies, in partially disordered states, one would have to know, at given temperature and average concentrations, all probabilities $x_{i,z}$, of which there are 288 distinct ones [40], and calculate all corresponding energies $E_{i,z}$. This is a formidable undertaking which, not surprisingly, has not been carried out.

Gonis et al. [38, 39] did obtain, however, some extremely interesting and promising results for the local DOS $n_{i,z}$ for selected first-shell configurations z in AgPd. The various computed $n_{i,z}$ clearly exhibited the expected results for partially ordered systems. In particular, the pure-Ag or pure Pd cluster DOS resembled those of the pure elements, while that of, say, Pd surrounded by an all-Ag shell exhibited a single peak characteristic of a single impurity embedded in a pure medium, though somewhat broadened by the disorder [38].

## 7.4. Generalized Perturbation Method

These are encouraging results; unfortunately it is not clear how to extract directly from the cluster CPA the various $V_n$ effective interaction parameters required by statistical thermodynamic theories, such as the CVM. Hence, the generalized perturbation method (GPM) of Gautier and Ducastelle [42-47] may turn out to be more convenient. In this approach, the energy of a given alloy configuration is expressed by a "cluster expansion" of the reference medium E [5]:

$$
\begin{aligned}
E(\{\gamma\}) = E_o &+ \frac{1}{2}\sum_{i\neq j} V(p_i p_j)\gamma(p_i)\gamma(p_j) \\
&+ \frac{1}{3}\sum_{\substack{i\neq j\\ j\neq k\\ k\neq i}} V(p_i p_j p_k)\gamma(p_i)\gamma(p_j)\gamma(p_k) \qquad (28)\\
&+ \frac{1}{4}\sum_{\substack{i\neq j, j\neq k\\ k\neq \ell, \ell\neq i}} V(p_i p_j p_k p_\ell)\gamma(p_i)\gamma(p_j)\gamma(p_k)\gamma(p_\ell) + \ldots
\end{aligned}
$$

In this expansion, the occupation variables are defined [1] as the deviation from the average concentration ($c \equiv x_B$) of the occupation (1 or 0) of site $p_i$, and the V are effective cluster interactions. By taking an ensemble average of (28), one recovers an expression similar to Eq. (4), with the ensemble averaged $\gamma$'s related linearly to the correlations $\xi$ [Eq. (1)] and pair interactions $V_{ij} \equiv \langle V(p_i p_j)\rangle$ having the same meaning as the quantities defined by Eqs. (7) - (10). The multisite interactions differ somewhat from the effective cluster interactions of Eqs. (4) and (6): the restrictions on the summations in Eq. (28) indicate that the multisite interactions involve paths along clusters' of points, over which steps may be retraced. The ensemble average of $E_o$ is just $E_o$ itself, the energy of the reference medium, calculated as that of the single-site CPA medium; it is identifed with the corresponding term in Eq. (4), the energy of the completely disordered state. Both in Eqs. (4) and (28), the reference energy $E_o$ is seen to depend on concentration, as do the interactions V themselves arising, as they do, from the expansion of the reference state. It is indeed because the energy of the reference medium is in some sense "close" to that of the actual configuration considered that expansion (28) converges rapidly. Calculations based on a tight-binding Hamiltonian have shown [46] that pair interactions are dominant. In particular, for fcc transition metal substitutional alloys, it was found that, in general, $|V_1| \gg (|V_2|, |V_3|, |V_4|) \gg (|V_5|, |V_6|)$ (the index indicating first, second...neighbor separation) [46], [48], but with $V_4$ (in magnitude) of the order of $V_2$ but larger than $V_3$. For bcc, one has, in general, $|V_1| > |V_2| \gg |V_3| \cong |V_5| > |V_4|$. Ducastelle and co-workers [48, 49, 50] have also shown by performing moment expansions of the density of states, that the $V_n$ pair interactions must necessarily change sign a number of times as a function of d-band filling. This implies that, for certain ranges of concentration in certain transition metal alloys, e.g., $V_1$ must become very small so that $V_2$, say, will become the dominant interaction.

Studies by the French school have also shown that multisite interactions are generally negligible in magnitude compared to pair interactions. In certain cases, however, cluster interactions involving self-retraced path may become important. Most importantly, calculations have shown that, although $V_1$ is generally expected to be larger than $V_2$ for substitutional alloys, the reverse can be true for off-stoichiometric interstitial compounds [48, 49]. This explains why such structures as "$A_5B$" [Fig. 1] or "CuPt" (see, for example, Ref. [11]), which require $V_1$, in magnitude, comparable to $V_2$, are far more common in interstitial sublattices than in substitutional fcc lattices.

In the TB approximation, it is in fact possible to express the pair interactions in transition metal alloys analytically as [44]

$$V_n = W\, V_n(c,\, N_d,\, \delta_d,\, \delta_{nd}) \tag{29}$$

where c is the average concentration, $N_d$ is the number of electrons in the d band, and where $\delta_d$ and $\delta_{nd}$ are the so-called diagonal and off-diagonal disorder. The average d band width W appears in Eq. (29) merely as a scale factor. In this approximation, Eq. (29) thus enables one to predict transition metal alloy ground state superstructures, since these are known as a function of $V_1 \ldots V_4$ from the work of Kanamori and co-workers [9]. In this way, Gautier and collaborators [44, 47] have produced a remarkable structural map showing predicted regions of stability of various common fcc superstructures as a function of the parameters $N_d$ and $\delta_d$, for various concentrations c. These extremely important results are also summarized by Pettifor [36]. Agreement with experimental data is, in most cases, satisfactory.

The possiblity offered by Eq. (29) of computing the $V_n$ pair interactions directly from band structures is an exciting one, leading to ordered ground states and even phase diagram predictions. It would be desirable, however, to check the pair interaction values experimentally, or the band structure itself. Ways of doing this are described briefly below.

## 8. EXPERIMENTAL VERIFICATION

Ideally, one would like to obtain direct experimental evidence of the band structure. This is a very difficult undertaking, especially in the case of alloys; but two techniques are becoming available for carrying out such a program: angle-resolved photoemission and positron annihilation spectroscopy.

An application of the former method has been used by Jordan et al. [51] to ordered and disordered $Cu_3Au$. DOS spectra were obtained at various temperatures and compared to those calculated **ab initio** by fully relativistic (ordered state) and semi-relativistic (disordered state) KKR CPA formalisms. The disappearence, at higher temperatures, of small ordered state peaks were correlated in both theoretical and experimental DOS [51]. The technique is thus promising, but suffers from the serious drawback that only electron states near the surface of the specimen are sampled which, because of alloy surface segregation problems, may not be representative of the bulk.

Positron annihilation spectroscopy does not have the latter disadvantage and appears to be quite well suited for the study of the Fermi surface of metals and, particularly, of alloys (for reviews, see Refs. [52] and [53]). However, studies of electronic structure in general will require careful differentiation of effects due to point defects, particularly vacancies.

Values of the effective interaction parameters $V_n$ themselves may be determined by comparing experimental and theoretical values of short-range order intensity (ISRO) fluctuations in disordered solid solutions, particularly just above a transition temperature. The art of measuring ISRO entails measurements of (x-ray or Neutron) diffuse diffracted intensity in substantial regions of reciprocal space, followed by fairly complex corrections for displacement effects. For a review of the current methods, see Ref. [54], for example.

Theoretical expressions for ISRO can be obtained by applying the fluctuation-dissipation theorem to whatever analytical free energy model one wishes to consider. At any point $\mathbf{k}$ in the first Brillouin zone, $I_{SRO}(\mathbf{k})$ is proportional to the expectation value of the Fourier transform $X_1$ of the point correlation function $\xi_1$ [55]:

$$I_{SRO}(\mathbf{k}) \sim \langle|X_1(\mathbf{k})|^2\rangle = \frac{k_B T}{N} (F^{-1})_{11} \tag{30}$$

where $(F^{-1})_{11}$ is the first element of the inverse of the Fourier transform of the matrix of second derivatives of the free energy with respect to correlation variables. Equation (30) is exact, and would yield exact results if an exact free energy were available. For the one-dimensional Ising model with nearest neighbor interactions, the condition of conservation of integrated intensity is rigorously obeyed [55] if the CVM free energy is used. For higher dimensional systems, the CVM integrated intensity no longer has the required property, but does significantly better than that derived by BW models, such as the Krivoglaz and Clapp-Moss formulas (for a review of BW derivations, see for example Ref. [1], and references cited therein).

Recently, CVM and BW calculated ISRO were compared for the fcc Ising model with first ($V_1$) and second ($V_2$) neighbor (constant) pair interactions. Contours of constant intensity in a (001) section of reciprocal space are shown in Fig. 2, calculated according to the Sanchez formula (CVM, left frame) and the Krivoglaz-Clapp-Moss (KCM) forumla (BW, right frame), for the case of $\alpha = V_2/V_1 = 0.35$, at temperature $T = T_0/0.95$, where $T_0$ is the instability temperature, indicated by the dashed line at the chosen concentration of $c = 0.5$ in the phase diagram of Fig. 1. It is seen that the CVM diffuse intensity peaks at $\langle 1\ {}^1/_2\ 0\rangle$ are much sharper than those of the BW calculation. The differences may not appear that dramatic, but the sharpness of the CVM peaks insure a more nearly constant integrated intensity at all temperatures except very close to the instability $T_0$. Furthermore, the ISRO calculations (Fig. 2) are completely consistent with the calculated phase diagram (Fig. 1) which, at least for $\alpha = 0.25$, is known to agree closely with that obtained by Monte Carlo simulations (see Ref. [11]).

Up to now, the KCM has been used exclusively for extracting pair interactions from diffuse intensity measurements: the procedure consists in choosing the correct set of $V_n$ in an attempt to reproduce theoretically the experimentally determined intensity contours (see Ref. [56], for example). A simpler method would be to use the author's fcc and bcc "eigenvalue tables" [23] to fit major and minor axes of closed intensity contours about

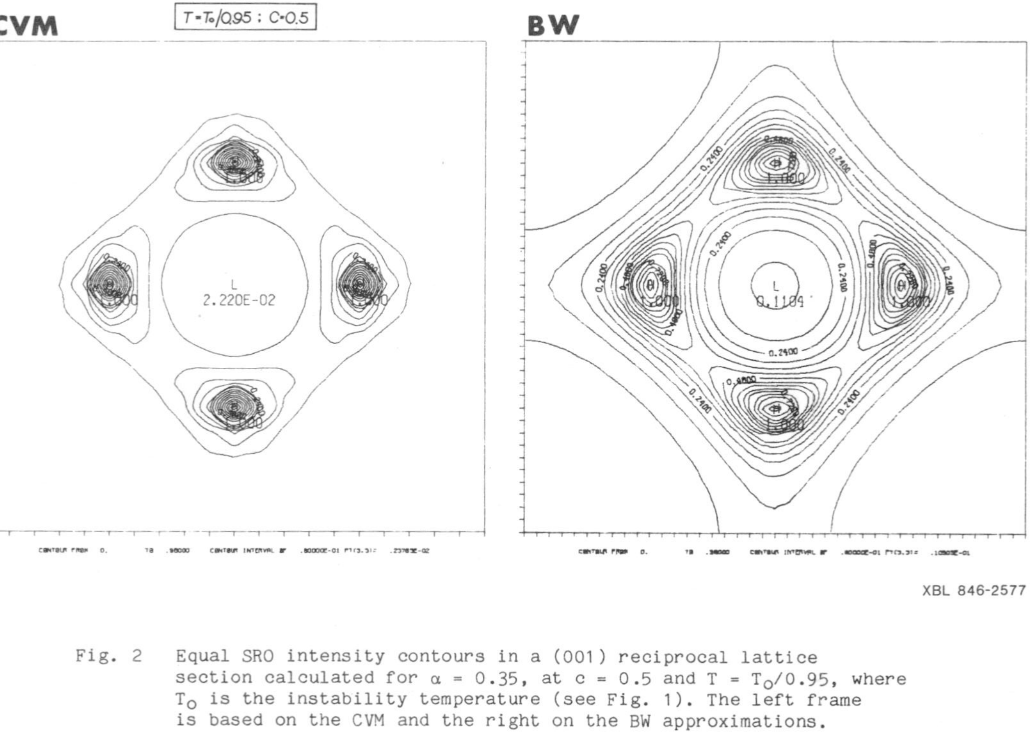


Fig. 2 Equal SRO intensity contours in a (001) reciprocal lattice section calculated for $\alpha = 0.35$, at $c = 0.5$ and $T = T_0/0.95$, where $T_0$ is the instability temperature (see Fig. 1). The left frame is based on the CVM and the right on the BW approximations.

diffuse intensity peaks. The resulting $V_n$ values are only as good as the BW model itself, however, which is not good at all for fcc-based ordering systems.

A very interesting comparison of back-calculated $V_n$ values obtained by CVM and BW methods has just been performed. The technique used has been called the "inverse CVM" [57], and was first suggested to the author by E. Van Royen [58]. It is, in a sense, an extension of Clapp's Probability Variation method (PVM) [40], itself an offshoot of Kikuchi's CVM (see, for example pg. 247 of Ref. [1]). As in the PVM, one must assume that the internal (ordering) energy can be expressed by means of pair interactions alone, as in Eq. (12), the V's and ε's being related simply by Eq. (9). At equilibrium, the corresponding CVM free energy, Eq. (18) must be minimized, with pair correlations $\xi_r$ being known from experiment, but with unknown interaction parameters $\varepsilon_r$. Equation (18) can be written equivalently, by means of Eq. (2), as

$$f = \sum_r \omega_r \varepsilon_r \xi_r - T\, s(\xi_1 \ldots \xi_n, \zeta_1 \ldots \zeta_L) \qquad (31)$$

where s designates the configurational entropy per lattice site, expressed as a function of the n (known) pair correlations $\xi_n$ and L (unknown) multiplet correlations $\xi_\ell$. The equilibrium conditions are [2]

$$\frac{\partial f}{\partial \xi_r} = \omega_r \varepsilon_r - T \frac{\partial s}{\partial \xi_r} = 0 \qquad (32a)$$

$$\frac{\partial f}{\partial \zeta_\ell} = - T \frac{\partial s}{\partial \zeta_\ell} = 0 \quad , \qquad (32b)$$

since the correlation variables are independent. Equations (32b) represent a maximum entropy condition, just as in the PVM, which return the equilibrium multiplet correlations. These values are then inserted into (32a) along with given $\xi_r$, yielding s*, say, from which the required interactions (ε or V) can be obtained directly:

$$\varepsilon_r = \frac{T}{\omega_r} \frac{\partial s^*}{\partial \xi_r} \quad (r = 1, \ldots n) \qquad (33)$$

In a recent study [57], [59], pair correlations in an fcc lattice were determined "experimentally" by Monte Carlo simulation with up to fourth-neighbor pair correlations. The set $V_1 \ldots V_4$ was then back-calculated by the inverse CVM in the "quadruple tetrahedron" approximation. The CVM interactions were found to agree much more closely with the MC input values than did those calculated by the BW model.

In another recent development, Gyorffy and Stocks [60] have performed first-principles KKR-CPA calculations of ISRO in one-dimensional disordered systems and have also confirmed the existence of flat portions in the (3-Dim) Fermi surface of CuPd alloys for certain ranges of average concentrations. This latter calculation predicts, in the disordered state, the wavelength on long-period modulations as a function of concentration in excellent agreement with experimental observations.

It is clear that diffraction methods, coupled with CVM ISRO analysis in partially ordered crystals can, in principle, yield reliable values of $V_n$ interactions, which may then be compared to theoretical values obtained, for instance, by the Generalized Perturbation Method of Gautier, Ducastelle and co-workers. It is also expected that synchrotron radiation will play an increasingly important role (a) by providing high-intensity, high-resolution ISRO data and (b) through angle-resolved photoemission spectroscopy, by providing insight into the electronic density of states itself.

## 9. DISCUSSION: TOWARDS A GENERAL ALLOY THEORY

It should be clear from the foregoing that, if the unit cell is the fundamental structural unit for elemental crystals and stoichiometric compounds, and the lattice point that for completely disordered solid solutions, the **cluster** of points is the fundamental unit for partially ordered crystals. Cluster methods are indispensable to characterize the state of order (Sect. 2), to predict ground states of order (Sect. 4), to approximate the configurational entropy (Sect. 5). In combination, these cluster methods culminate in the calculation of "prototype" ordering phase diagrams (Sect. 6) constructed with given, fixed sets of pair interactions, $V_n$.

The configurational energy also can be expressed in terms of cluster expansions (Sect. 3). If one could calculate reliably cluster interactions, pairs primarily, from electronic band structure theories, one would obtain, from first principles, the free energy, configurational entropy, specific heat, SRO, LRO, diffuse intensity (Sect. 8), phase diagrams, etc.. for those alloy systems which can be described fairly well by Ising models. Presently, the most critical problem of course remains that of calculating the DOS of alloys with sufficient accuracy.

It is not surprising that cluster methods play an important role in electronic energy computations. In fact, a parallel can be established between statistical models and band structure methods (Sect. 7):

| | | | |
|---|---|---|---|
| Level 1: | Regular Solution Model | ←→ | Single-Site CPA |
| Level 2: | Bragg-Williams Model | ←→ | Inhomogeneous CPA |
| Level 3: | Cluster Variation Method | ←→ | Cluster CPA, or Generalized Perturbation Method. |

A complete **Alloy Theory** must operate at Level 3; it should integrate the CVM and the CPA, ideally in a self consistent loop, as was done initially by Kittler and Falicov [61,62] on simple models. The CVM is well under control, although, in practice, calculations are limited to small clusters, and numerical convergence of the algorithms often poses serious problems. On the electronic theory side, the CPA appears to be very promising as a means of calculating the all-important $V_n$ effective pair interactions which constitute the major input for ground state calculations and the CVM free energy. For phase diagram computations, the average energy $E_o$ must also be known as a function of concentration and perhaps atomic volume. Thus far, the GPM has been formulated in the tight-binding approximation, limiting its application strictly to transition metal alloys for which only d-electron states need be considered. Furthermore, the alloy elements must be non-magnetic, unless one wished to perform CVM calculations involving both magnetic and ionic ordering. Finally, T-B calculations are model ones, hence not truly of the "first-principles" category.

By contrast, the KKR-CPA used in conjunction with a muffin-tin Hamiltonian can be regarded as a truly first-principles method, not limited to transition metal alloys. Codes have even been developed for the heavier elements which require a relativistic approach [63]. Unfortunately, even in the single-site CPA version, calculations are very laborious and require considerable supercomputer time. Cluster CPA codes are even more elaborate and, despite recent successes, do not yield directly the effective $V_n$ interactions. What would be required is the implementation of the generalized perturbation method in the KKR-CPA framework, which is not a trivial undertaking [64].

Still, if, as a first approximation, the effect of atomic displacements can be neglected (dynamic and static), the exciting possibility exists today of predicting thermodynamic properties and phase diagrams, for certain classes of binary alloys (AB), strictly from first principles, i.e., from a knowledge of the atomic numbers $Z_A$ and $Z_B$ only. Best candidate systems include those phase equilibria involving only superstructures of a given parent lattice. This limitation may not be as restrictive as it may appear: even in alloy systems featuring stable "incoherent" intermetallic compounds, the first product of solid state precipitation reactions generally consist of "coherent" superstructure phases, for which little experimental information is available, but which often determine final and are thus of crucial importance for the alloy designer.

In conclusion, a General Alloy Theory is an idea whose time has come.

## ACKNOWLEDGEMENTS

The author has benefited greatly from many useful conversations with, among others, Dr. W. H. Butler, F. Ducastelle, L. M. Falicov, J. S. Faulkner, A. Finel, J.-P. Gaspard, F. Gautier, B. L. Gyorffy, A. Gonis, D. Gratias, T. Mohri, J. M. Sanchez, R. W. Siegel, G. M. Stocks and P. Turchi. Early work on prototype phase diagrams was supported by the U.S. Army Research Office (Durham). Portions of the present work were supported by the Director, Office of Energy Research, Office of Basic Energy Sciences, Materials Sciences Division of the U.S. Department of Energy under Contract No. DE-AC03-76SF00098, LBL No. 18773.

## REFERENCES

1. D. de Fontaine, Solid State Phys., H. Ehrenreich, F. Seitz and D. Turnbull, Eds., 34, pp. 73-294, Academic Press (1979).

2. J. M. Sanchez and D. de Fontaine, Phys. Rev. B 17, 2926 (1978).

3. D. de Fontaine, Modulated Structure Materials, NATO ASI Series E, No. 83, T. Tsakalakos, Ed., pp. 43-80, Martinus Nijhoff, Publ. (1984).

4. J. M. Sanchez, private communication.

5. A. Bieber and F. Gautier, J. of the Phys. Soc. of Jap., 53, 2061, 1984.

6. J. M. Sanchez and D. de Fontaine, Structure and Bonding in Crystals II, M. O'Keeffe and A. Navrotsky, Eds., pp. 117-132, Academic Press, 1981.

7. J. Kanamori, Progr. Theor. Phys. 35, 66 (1966).

8. S. M. Allen and J. W. Cahn, Acta Metall. 20, 423 (1972): Scripta Metall. 7, 1261 (1973).

9. J. Kanamori and Y. Kakehashi, J. Phys. (Paris) 38, C7-274 (1977).

10. A. Finel and F. Ducastelle, Phase Tranformations in Solids, MRC Symposium Proceedings, 21, pp. 293-298, North-Holland (1984).

11. T. Mohri, J. M. Sanchez and D. de Fontaine, Acta Metall., in press.

12. R. Kikuchi, Phys. Rev. 81, 988 (1951).

13. J. A. Barker, Proc. Rog. Soc. A, 216, 45 (1953).

14. J. M. Sanchez, F. Ducastelle and D. Gratias, preprint.

15. J. M. Sanchez and D. de Fontaine, Phys. Rev. B 21, 216 (1980).

16. R. Kikuchi, J. Chem. Phys. 60, 1071 (1974).

17. C. M. Van Baal, Physica (Utrecht) 64, 571 (1973).

18. D. de Fontaine and R. Kikuchi, Applications of Phase Diagrams in Metallurgy and Ceramics, Ed. by G. C. Carter, National Bureau of Standards SP-496 (1978), pp. 967-998.

19. J. M. Sanchez and D. de Fontaine, Phys. Rev. B 25, 1759 (1982).

20. J. M. Sanchez, D. de Fontaine and W. Teitler, Phys. Rev. B 26, 1465 (1982).

21. K. Binder, J. L. Lebowitz, M. K. Phani and M. H. Kalos, Acta Metall. 29, 1655 (1981).

22. G. Inden, private communication.

23. D. de Fontaine, Acta Metall. 23, 553 (1975).

24. T. Mohri, J. M. Sanchez and D. de Fontaine, preprint.

25. J. M. Sanchez, J. R. Barefoot, R. N. Jarrett and J. K. Tien, Acta Metall. 32, 1519 (1984).

26. C. Sigli and J. M. Sanchez, CALPHAD journal, in press.

27. J. M. Sanchez and C. H. Lin, Phys. Rev. B 30 (1984).

28. J. M. Sanchez, R. N. Jarrett, C. Sigli and J. K. Tien in High-Termperature Alloys: Theory and Design, J. O. Stiegler, Ed., Oak Ridge National Lab. (in press).

29. J. S. Faulkner, Progr. Mat. Science, J. W. Christian, P. Haasen and T. B. Massalski, Eds., pp. 1-187, Pergamon Press (1982).

30. S. Louie, private communication.

31. V. L. Moruzzi, A. R. Williams and J. F. Janak, Phys. Rev. B 15, 2854 (1977).

32. M. L. Cohen, Physica Scripta, T1, 5 (1982).

33. A. R. Williams, discussion in Theory of Alloy Phase Formation, L. H. Bennett, Ed. p. 63, Met. Soc. AIME (1980).

34. H. Ehrenreich and L. M. Schwartz, Solid State Phys., H. Ehrenreich, F. Seitz and D. Turnbull, Eds., SSP 31, pp. 149-280 (1976).

35. F. Gautier, F. Ducastelle and J. Giner, Phil. Mag. 31, 1373 (1975).

36. D. G. Pettifor in Physical Metallurgy, R. W. Cahn and P. Haasen, Eds., pp. 74-152, Elsevier (1983).

37. A. Gonis and A. J. Freeman, preprint.

38. A. Gonis, G. M. Stocks, W. H. Butler and H. Winter, Phys. Rev. B 29, 555 (1984).

39. A. Gonis, W. H. Butler and G. M. Stocks, Phys. Rev. Letters 50, 1482 (1983).

40. P. C. Clapp, Phys. Rev. B 4, 255 (1971).

41. F. Ducastelle, J. Phys. C 8, 3297 (1975).

42. F. Ducastelle and F. Gautier, J. Phys. F 6, 2039 (1976).

43. G. Treglia and F. Ducastelle, J. Phys. F 10, 2137 (1980).

44. A. Bieber and F. Gautier, Solid State Comm. 38, 1219 (1981).

45. A. Bieber, F. Ducastelle, F. Gautier, G. Treglia and P. Turchi, Solid State Comm. 45, 585 (1983).

46. A. Bieber and F. Gautier, Physica 107, B 71 (1981).

47. F. Gautier, High-Temperature Alloys: Theory and Design, J. D. Stiegler, Ed., Oak Ridge National Lab. (in press).

48. F. Ducastelle, Phase Transformations in Solids, MRS Symposium Proceedings Vol. 21, pp. 375-380.

49. J. P. Landesman, P. Turchi, F. Ducastelle and G. Treglia, Phase Transformations in Solids, T. Tsakalakos, Ed., Mat. Res. Symposium Proc. Vol 21, pp. 363-368, Elsevier (1984).

50. A. Bieber, F. Gautier, G. Treglia and F. Ducastelle, Solid State Comm. 39, 149 (1981).

51. R. G. Jordan, G. S. Sohal, B. L. Gyorffy, P. Durham, W. M. Temmerman and P. Weinberger, preprint.

52. R. W. Siegel in Ann. Rev. Mater. Sci. 10, 393-425 (1980).

53. S. Berko in Positron Solid State Physics, Int. School of Physics "Enrico Fermi", W. Brandt and A. Dupasquier, Eds., North Holland (1983).

54. P. Georgopoulos and J. B. Cohen, in Modulated Structure Materials, NATO ASI Series E., No 83, T. Tsakalakos, Ed., pp. 265-284).

55. J. M. Sanchez, Physica 111A, 200 (1982).

56. P. Georgopoulos and J. B. Cohen, J. Phys. (Paris) 38, C7 MI (1977).

57. J. P. Gaspard and D. Gratias, private communication.

58. E. Van Royen and J. de Hosson, unpublished work at the University of Groningen, The Netherlands.

59. P. Cenedese and J. P. Gaspard, Phase Tranformations in Solids, MRC Symposium Proceedings Vol. 21, pp. 51-55, North-Holland (1984).

60. B. L. Gyorffy and G. M. Stocks, Phys. Rev. Letters 50, 374 (1983).

61. R. C. Kittler and L. M. Falicov, J. Phys. C 9, 4259 (1976).

62. L. M. Falicov and R. C. Kittler, Theory of Alloy Phase Formation, L. H. Bennett, Ed., pp. 303-325.

63. P. Weinberger, private communication.

64. W. H. Butler, private communication.

INTERATOMIC INTERACTIONS IN TRANSITION AND NOBLE METAL ALLOYS AND THE STABILITY OF THE ORDERED STATE

DEMENTYEV V.M., KOZLOV E.V. AND KOLODEZNAYA S.F.*
* Institute of Civil Engineers, Solyanaya Sq., 2, 634003 Tomsk, USSR

ABSTRACT

The nature and the role of the pair interatomic interaction of the noble and transition metal ordering alloys are considered. The determining role in the electron susceptibility in the behaviour of the interatomic interaction is shown. The important role of the Fermi surface local geometry in the definition of superstructure periods is established. Illustrating calculations are made. It is shown that in the noble and transition metal alloys with complicated Fermi surface the oscillations of the pair interatomic interaction are of essentially nonfriedel character.

INTRODUCTION

The pair ordering energy

$$V(R) = V_{AA}(\vec{R}) + V_{BB}(\vec{R}) - 2V_{AB}(\vec{R}) \qquad (1)$$

and its Fourier transform:

$$V(\vec{k}) = \frac{1}{N}\sum_{R} V(\vec{R})e^{i\vec{k}\vec{R}} \qquad (2)$$

are the main characteristics of ordering binary alloys. The modern quantum theory of an atomic ordering must make it possible to calculate these values from the first principles. At present the pseudopotential method is such a theory. But the majority of ordering alloys are the noble and transition metal alloys. So the pseudopotential method must be generalized for this broad class of alloys, i.e. it must take into account such electron structure effects as the s-d- hybridization, the resonance character of d-states, the overlapping the external d-states and others. On the other hand, it turned out that to understand the nature of ordering energy, physics of phase order-disorder transition it is necessary to exceed the limits of the tradition in a pseudopotential method Fermi surface approximation, i.e., the theory must take into account the real Fermi surface topology and its local geometry characteristics.

The main results of the authors on elaborating such a theory on the basis of the generalization of Moriarty's pseudo Creens function method [ 1 ] are summed up in the review [2 ]. In the present paper the role of the pair interaction and the Fermi surface form in the stabilization of the ordering state as well as the choice of the type of the forming superstructures are considered.

ALLOY ENERGY AND ENERGY-WAVE-NUMBER CHARACTERISTICS

The total energy of the alloy can be expressed in the following

$$E = E_0 + E_{bs} + E_{ol} + E_{es} \qquad (3)$$

where $E_0$ is the part independent on the structure of the electronic energy, $E_{bs}$ -band-structure energy, $E_{over}$ - overlapping energy, $E_{es}$ - electrostatic energy (see detailed expressions for partial contributions in [2].

The basis of the application of the pseudopotential method in the atomic ordering theory is the possibility to obtain the obvious dependence of the total alloy energy on the long - and short - range order parameters and simulteneously the possibility of the transition to an effective interatomic interaction.

All physical characteristics of transition and noble metal alloys are expressed by energy - wave - number characteristics of the pseudopotential method )( in a second order of the perturbation theory)

$$\bar{F}(\vec{q}) = c_A^2 F_{AA}(\vec{q}) + c_B^2 F_{BB}(\vec{q}) + 2c_A c_B F_{AB}(\vec{q}) \qquad (4)$$

$$F(\vec{q}) = F_{AA}(\vec{q}) + F_{BB}(\vec{q}) - 2F_{AB}(\vec{q}) \qquad (5)$$

$$F_{\mu\nu}(\vec{q}) = \frac{4\pi Z_\mu^* Z_\nu^* e^2}{\Omega_0 q^2} e^{-q^2/4\xi} + \omega_\mu(\vec{q})\,\omega_\nu(\vec{q})\,\chi(\vec{q})\,\varepsilon(\vec{q}) +$$

$$+ \omega_\mu(\vec{q})\, g_\nu(\vec{q})\, \pi_\nu(\vec{q}) + \omega_\nu(\vec{q})\, g_\mu(\vec{q})\, \pi_\mu(\vec{q}) +$$

$$+ g_\mu(\vec{q})\, g_\nu(\vec{q})\, \pi_{\mu\nu}(\vec{q}) \qquad (6)$$

$c_\mu, g_\mu(\vec{q}), Z_\mu^*$ are the concentration, hybridization potential and effective valence of $\mu$ ion ( $\mu$ = A, B )

$$\chi(\vec{q}) = \frac{1}{N}\sum_{\vec{k}} \frac{\theta(E_F - k^2)}{k^2 - |\vec{k}+\vec{q}|^2} \qquad (7)$$

is susceptibility of collectivized states

$$\pi_\mu(\vec{q}) = \frac{1}{N}\sum_{\vec{k}} \frac{(k^2 - E_d^{(\mu)})\,\vartheta(k^2 - E_F)}{(k^2 - |\vec{k}+\vec{q}|^2)\left[(k^2 - E_d^{(\mu)})^2 + (\omega_d^{(\mu)})^2\right]} \qquad (8)$$

$$\Pi_{\mu\nu}(\vec{q}) = \frac{1}{N}\sum_{\vec{k}} \frac{(k^2 - E_d^{(\mu)})(k^2 - E_d^{(\nu)})\vartheta(k^2 - E_F)}{(k^2 - |\vec{k}+\vec{q}|^2)[(k^2 - E_d^{(\mu)})^2 + (\omega_d^{(\mu)})^2][(k^2 - E_d^{(\nu)})^2 + (\omega_d^{(\nu)})^2]} \tag{9}$$

- susceptibilities of S - d - hybridized states

$$\mathcal{E}(\vec{q}) = 1 - [1 - G(\vec{q})]\chi(\vec{q})(8\pi e^2/\Omega_0 q^2) \tag{10}$$

- dielectric function; $E_d^{(\mu)}$ and $\omega_d^{(\mu)}$ are the position and the half width of the d - resonances; $\theta(x)$ - Havyside's function, $E_F$ - Fermi energy:

$$\vartheta(k^2 - E_F) = \theta(E_F - k^2) - \frac{1}{\pi}\left(arctg\,\frac{E_F - E_d^{(\mu)}}{\omega_d^{(\mu)}} + \frac{\pi}{2}\right)$$

The configurational energy for the $LI_2$ superstructure is expressed in the following:

$$E_K = \sum_{\vec{q}=\vec{K}_n'} \bar{F}(\vec{q}) + \frac{1}{16}\eta^2 \sum_{\vec{q}=\vec{K}_n''} F(\vec{q}) + \frac{1}{N}\sum_{\vec{q}\neq 0}\Big\{(c_A c_B - \frac{3}{16}\eta^2) + \sum_{\rho\neq 0}\mathcal{E}(\vec{\rho})\cos(\vec{q}\cdot\vec{\rho})\Big\}F(\vec{q}) \tag{11}$$

where $K_n'$ and $K_n''$ are the reciprocal lattice vectors, $\eta$ is the long - range order parameter, $\mathcal{E}(\vec{\rho})$ is the pair correlation parameter. The first term in (11) is the energy of the virtual crystal. The main contribution in the mixing energy is the term $(c_A c_B - \frac{3}{16}\eta^2)$ which vanishes in stoxiometry under the total order. Terms, $\sim\eta^2$ are responsible for the specific ordering energy. The short - range order is stabilized by the contribution depending on $\mathcal{E}(\vec{\rho})$ Expressions like (11) are the basis in the analysis of the energy stability and the separate contributions in the ordering energy for the concrete superstructures. The direct calculation [3], made for the $Cu_3Au$ alloy, showed a good efficiency of the expression (11) ; the calculated heat of the transition coincides with the experimental one.

## SINGULARITIES IN ELECTRON SUSCEPTIBILITY AND FOURIER TRANSFORM OF ORDERING ENERGY

It is known that the type of the superstructure formed are ear defined by the position of the absolute minimum of the Fourier transform of the ordering energy V(k) [4]. In the pseudopotential method V(k) has the form [5]:

$$V(\vec{k}) = 2\sum_{K_n} F(\vec{k} + \vec{K}_n) - \frac{2\Omega_0}{\pi^2}\int F(\vec{q})\,d\vec{q} \tag{12}$$

The first term determines the dependence of ordering energy on vector k and correspondingly the type of the superstructure formed. The next term can determine the sign of V(k) and so the inclination of the alloy to the ordering

In the Fermi sphere approximation for the $\chi(\vec{q})$ and $\mathcal{J}_{\mu}(\vec{q})$ the minimum V(k) can be formed because of the difference of the formfactors of components $\Delta w(\vec{q})$. And therefore the reason for the ordering of the simple metals alloys is in the difference of the atomic properties of components. But the reason of the ordering may consist in the special properties of the susceptibilities $\chi(\vec{q})$ and $\mathcal{J}_{\mu}(\vec{q})$ connected with the Fermi surface geometry. The Fermi surface approximation for the susceptibility $\chi(\vec{q})$ (7) was used. In this case $\chi(\vec{q})$ has the well-known Kohn singularity at $q=2\,k_{F0}$ ( $k_{F0}$ is the Fermi surface radius) (Fig. 1):

$$\chi(q) \sim S(q-2k_{F_0})\ln|q-2k_{F_0}| \qquad (13)$$

This smooth singularity (only the derivative shows a logarithmic infinity at $q=2\,k_{F0}$ is greatly intensified with the deviation Fermi surface form from the spherical. These strong singularities occur in $\chi(\vec{q})$, when vector q connects two sections on the Fermi surface whose normals are antiparallel [7,8]. The most singular contribution in $\chi(\vec{q})$ occurs for the flat section with the S area, perpendicular to $q_z$ with the principle radii of curvature being $R_1$, $R_2$ (at a distance of $k_0$ from the beginning of coordinates) (Fig. 1.)

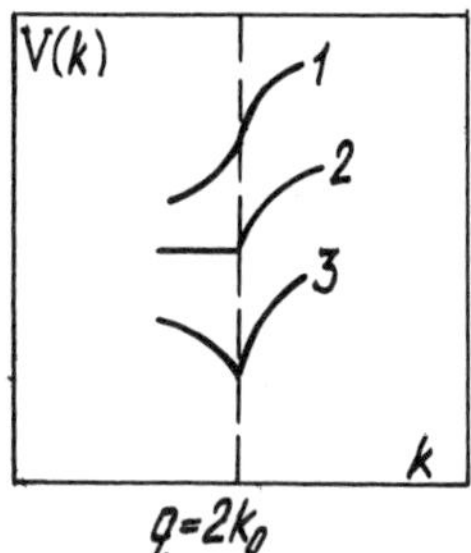

Fig. 1. $V(k)$ behaviour near the singularity points

$$\chi(q_z) \sim S\ln|q_z-2k_0|/\sqrt{R_1R_2} \qquad (14)$$

It can be seen from (12) and (14) that only the presence of Fermi surface sections with the small curvature leads to the formation of the minimum of V(k). In the noble and transition metal alloys the presence of d - resonance states leads to the appearence of new singularities differing from Kohn's singularities. Thus, for the noble metal alloy with the flat section

$$\Pi_\mu (q_z) \sim S(q_z - 2\sqrt{E_d^{(\mu)}})^{-1} \tag{15}$$

This singularity erodes in the transition metal alloy because of nonzero width of d - resonances

$$\Pi_\mu (q_z) \sim S\left[(q_z - 4E_d^{(\mu)})^2 + 16(\omega_d^{(\mu)})^2\right]^{-1} \tag{16}$$

(there is a minimum for

$$q_z = 2\sqrt{E_d^{(\mu)} + \omega_d^{(\mu)}} \quad \text{(Fig. 2.)}$$

All these singularities lead to the new minimums of V(k) and thus determine the ordering state structure.

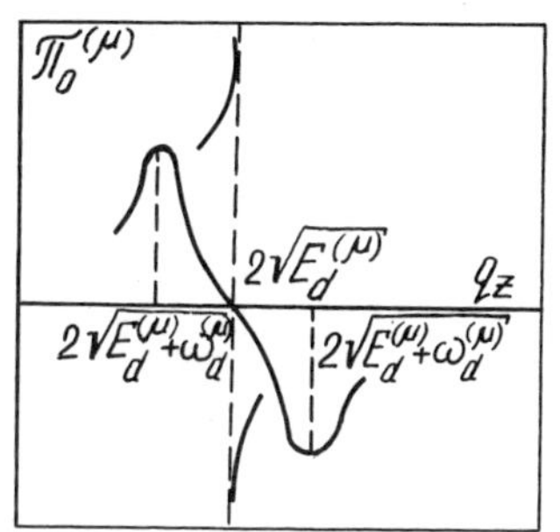


Fig. 2. The erosion of singularities at $q = 2\sqrt{E_d^{(\mu)}}$ in the transition metal alloys

At present it is clear that the modern quantum - statistical atomic ordering theory must take into account the many - body interatomic interactions. The construction of such a theory is possible in the pseudopotential method, but with the account of more higher order of the perturbation theory than the second one [ 9, 10 ]. These higher orders make the definite contribution in the pair interaction as well. The contribution of many - body interactions intensifies the noncentral character of the interatomic forces. On the other hand, the account, e.g., of the third order of disturbance theory leads to the appearance of the new minimums of caused by the singularities in the tripole $\Lambda^{(3)}(\vec{q}_1, \vec{q}_2, \vec{q}_3)$ [11]. They can be another reason for the formation of the ordering state and are significant in manyvalence metal alloys, when the first vectors of the reciprocal lattice are $|\vec{K}_n| < 2\,k_{Fo}$.

## FERMI SURFACE CALCULATION AND THE SUPERSTRUCTURE FORECASTING

The calculation made for the ordering alloys of the Cu-Au system, show that the whole number of characteristics can be calculated within the approximation of Fermi surface in quite good agreement with the experiment. Those characteristics are the heat and the energy of transformation [3] the diffuse scattering intensity, which are due to the short - range order [12] and the interatomic potentials [13]

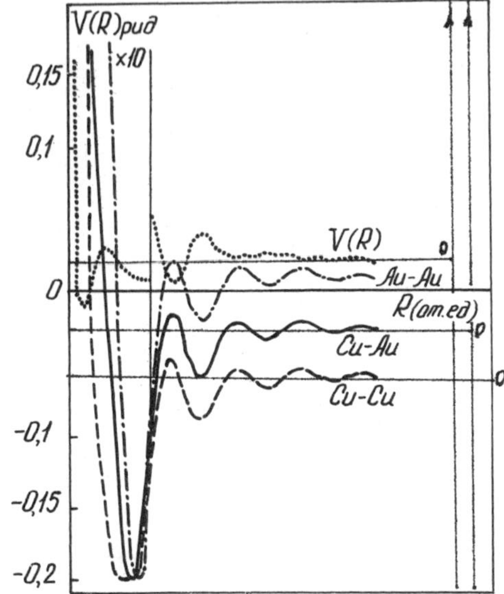


Fig. 3. The potentials of the interatomic interaction $V_{Cu-Cu}(R), V_{Cu-Au}(R), V_{Au-Au}(R)$, $V(R)$ of the $Cu_3Au$ disordered alloy

Those are the properties which depend on the V(K) behaviour throughout the whole Brillouin zone. The ordered phase structure is determined by the V(K) minimum coordinates, and here Fermi surface approximation, applied to the $Cu_3Au$ alloy, does not give satisfactory results [14]. Therefore the V(K) calculation was made with the account of the real Fermi surface for the $Cu_3Au$ alloy and also for the alloys of the noble metalls with polyvalent ones $Cu_3Al$, $Ag_3Mg$, $Au_3Cd$. The position of V (k) minimums (see (12)) is determined from the condition

$$|\bar{k}+\bar{K}_n| = 2k_o \qquad (17)$$

where $K_o$ is the radius - vector of the flat section on the Fermi surface. That is, the coordinates of the minimums of V (k) can be obtained geometricaly by means of the construction which is analogous to the Kohn construction for the phonon spectrum S[15]. For the alloys under consideration the sections of the Fermi surface have been calculated by the pseudopotential method with the account of degenerating the states on the Brillouin zone boundaries (Fig. 4.)

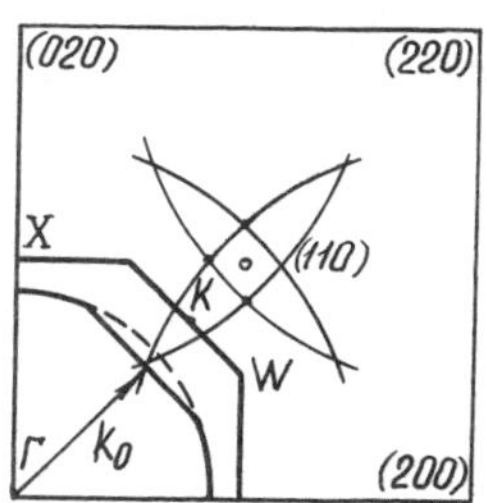

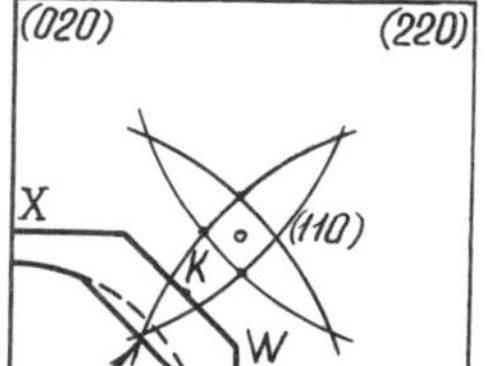

Fig. 4. Fermi surface section of the $Cu_3Au$ disordered alloy

The view of the Fermi surfaces of other alloys is analogous. One can see that there is a flat section on the Fermi surface in the <110> direction. From the Kohn - Moss construction one can obtain the well-known equation [16]:

$$\frac{1}{2M} = \sqrt{\left(\frac{12Z}{\pi}\right)^{3/2} t^2 - 1} - 1 \qquad (18)$$

where $t=K_0/K_{F0}$ is the so-called trancation-factor and M is the antiphase period. The calculated values of $K_0$, t and M are given in table 1.

Table 1

| Alloy | $K_0$ | t | $M_{theor}$ | $M_{exp.}$ |
|---|---|---|---|---|
| $Cu_3Au$ | 0.7409 | 0.9475 | 5.41 | 5 - 6 |
| $Cu_3Al$ | 0.8668 | 0.9688 | 1.20 | 1 |
| $Ag_3Mg$ | 0.8008 | 0.9511 | 1.95 | 1.99 |
| $Au_3Cd$ | 0.8132 | 0.9558 | 1.77 | 1.8-1.9 |

For all the alloys the minimum of V(k) is in the $(1,0,\frac{1}{2M})$ points that corresponds to the experiment (the well-known fourfould splitting of the (110) site in the $Cu_3Au$). Together with M values it indicates that $LI_2(M)$ superstructure is generated in $Cu_3Au$, $DO_{22}$ in $Cu_3Al$, $DO_{23}$ in $Ag_3Mg$ and the mixture $DO_{22}$ and $DO_{23}$ in $Ag_3Cd$.

Thus, the atomic order type and the crystall superstructure type in these alloys are determined by the Fermi surface geometry, namely, by the presence of the flat sections on the Fermi surface in the <110> direction.

Thus, the position of the absolute minimum of $V(\bar{k})$ and so the type of the superstructure formed are defined by the behaviour of the collectivized electron small groups near the Fermi surface sections with a small curvature as well as by the position and the width of d-resonances. That is, the type of long-and short-range order and thus the motive force of the atomic ordering are connected with the local geometry of the Fermi surface and with d-resonance character of electronic states.

In this sense the foregoing quantum theory of the atomic ordering is a common part of the phase transition quantum theory. And the physics of reasons for the order - disorder transition reveals the community with such transitions as the loss of the lattice structural stability at the superconducting transition, the instability in various substances with respect to the generation of the density wave of different nature (concentration density, charge and spin density waves, displacement waves) and so on.

The Fermi surface form deflection from the spherical form influences also the anisotropy and the long - range character of the interatomic interaction in the alloys.

## THE PAIR ORDERING ENERGY. FRIEDEL AND NON FRIEDEL OSCILLATIONS

The pair ordering energy is given by the expression

$$V(\vec{R}) = \frac{2}{N}\sum_{\vec{q}} F(\vec{q})e^{-i\vec{q}\vec{R}} \qquad (19)$$

It is obvious from the (5),(6),(19) expressions that the effective interatomic interaction in the transition metal alloys is formed of the electrostatic interionic interaction and the indirect interaction through the collectivized and s-d - hybridized electronic states. On the whole the pair ordering energy is an anisotropic. The spherical symmetry of V(R) may be in the case of the Fermi surface approximation.

In the noble and transition metal alloys the asymptotic of V(R) is also determined by the singularities in susceptibilities $\chi(\vec{q}), \mathcal{F}_{M}(\vec{q})$: $V(R) \sim A\cos(2k_F R)/R^n +$

$$+ (C_{AB} - C_{AA})\cos\left[\left(\sqrt{2E_d^{(A)} + 2\sqrt{(E_d^{(A)})^2 + (\omega_d^{(A)})^2}}\,R\right]/R^n +$$

$$+ (C_{BA} - C_{BB})\cos\left[\left(\sqrt{2E_d^{(B)} + 2\sqrt{(E_d^{(B)})^2 + (\omega_d^{(B)})^2}}\,R\right]/R^n \qquad (20)$$

The index n depends on the geometry of Fermi surface. For the Fermi sphere of the radius $K_F=K_{Fo}$, n=3 and then the first term is the usual Friedel oscillations $\cos(2K_{F_0}R)/R^3$.

The main difference of the long - range interaction in the transition metal alloys is the appearance of non Friedel oscillations. Firstly, they are slowly damping oscillations of cos (2 PoR)/$R^2$ and cos(2$K_o$R)/R types. These oscillations are due to the screening properties of conduction electrons near the nonspherical Fermi surface sections. For the cylindrical section n=2 and for the flat sections n=1. These contributions occur for the simple metal alloys as well[7]. Thus, the long - - range in the alloy depends essentially on the Fermi surface geometry and becomes considerably stronger in the presence of sections with large principle radii of curvature. Secondly, they are the non Friedel oscillations (the second and third terms in (20)),which are due to singularities in the susceptibility of S-d hybridized states. In the noble metal alloys the width of d-resonance $\omega_d=0$ and $V(R)\sim\cos(2\sqrt{Ed}\,R)/R^n$. Unlike the simple metal alloys in the noble and transition metals alloys as seen from (20) the atomic ordering may be conditioned by one of the element d-subband properties. It is necessary to emphasize that the amplitude of the pair interatomic interaction depends on the long - range order parameter and the pair correlation.

REFERENCES


1. J.A. Moriarty, Phys. Rev., B5, 2066 (1972).
2. E.V. Kozlov, V.M. Dementyev, Red. Izv. VUZov SSSR, Ser.fiz., 12, 63 (1982).
3. E.V. Kozlov, V.N. Emelyanov, V.A. Antonov, Phys. stat. sol. (b), 81, 719.
4. E.V. Kozlov, D.M. Shtern, N.M. Kormin, Red. Izv. VUZov SSSR, Ser. fiz., 9, 22 (1980).
5. E.V. Kozlov, V.M. Dementyev, Izv. VUZov SSSR, Ser. fiz., 6, 30 (1974).
6. S.F. Kolodeznaya, E.V. Kozlov, Izv. VUZov SSSR, Ser. fiz., 3, 118 (1982).
7. A.M. Afanasyev, U. Kagan., Zh., eksper. teor. Fiz., 43, 1457 (1962).

8. L.M. Roth, H.J. Zeger, T.A. Kaplan, Phys. Rev., 149, 519 (1966).

9. V.M. Dementyev, D.M. Shtern, E.V. Kozlov, in: Uporyadochenie atomov i ego vliyanie na sboistva splavov, Kiev, Naukova dumka, 70 (1979).

10. D.M. Shtern, V.M. Dementyev, E.V. Kozlov, ibid, p. 68.

11. E.G. Brovman and U. Kagan, iisp. Fiz. nauk, 112, 369 (1974).

12. S.F. Kolodeznaya, E.V. Kozlov, Phys. stat. sol. (b), 119, 131 (1983).

13. S.F. Kolodeznaya, E.V. Kozlov, in: "Vopr. atom. nauki i tekh", ser."Fiz. rad. povr.i rad. mater." (Kharkov, Kh FTI), 5, 86 (1983).

14. S.F. Kolodeznaya, V.M. Dementyev, E.V. Kozlov, Red, Sh. Izv. VUZov SSSR, Ser. fiz., 10, 84 (1980)

15. S.C. Moss, Phys. Rev. Lett., 22, 1108 (1969).

16. H. Sato, R.S. Toth in "Alloy Behaviour and Effects in Concentrated Solid Solutions - New York; Gordon and Breach, 1965, p. 265.

PART III

# Microstructural Features and Phase Transformation

# MODULATED MICROSTRUCTURES IN β-PHASE ALLOYS

C. M. Wayman
Department of Metallurgy and Mining Engineering and the Materials Research Laboratory, University of Illinois at Urbana-Champaign, 1304 W. Green Street, Urbana, Illinois 61801 U.S.A.

## ABSTRACT

Many ordered β-phase alloys of the B2 or D03 types exhibit unusual diffraction patterns and images when examined in the transmission electron microscope. These effects include "extra" diffraction spots, diffuse scattering, satellites, "tweed microstructures, spinodal decomposition and the formation of incommensurate phases. Examples of these phenomena will be presented and discussed. Although most of the β-alloys also undergo a martensitic transformation at some lower temperature it will be argued that the above effects are not "premartensitic" or precursory in nature. However many of them can be related to certain properties of the β-phase such as elastic anisotropy, a soft phonon mode and electronically driven instabilities.

## INTRODUCTION

Beta-phase alloys are 3/2 Hume-Rothery electron compounds (1.5 electrons/atom) and have many interesting properties. Some of them, e.g., NiAl, show great potential as ordered alloys for high temperature application. These alloys usually order at high temperatures into the B2 or D03 state. In addition, most of them undergo a low temperature thermoelastic martensitic transformation and exhibit the now familiar shape memory effect. These alloys also show anomolous behavior prior to martensitic transformation (sometimes termed premartensitic effects) such as softening of the C' elastic constant, modulated microstructures, diffuse scattering effects, and others.

Many of the peculiarities of the β-phase alloys have emerged since the introduction of transmission electron microscopy. In a classical paper, Delaey, Perkins and Massalski [1] first drew attention to anomalous contrast and diffraction effects in quenched beta-brass type alloys as detected by transmission electron microscopy and diffraction. They described such contrast effects at mottling, striation, cross-hatching, etc., and diffraction effects as diffuse streaking in certain directions, the appearance of additional maxima or their splitting, and slight deviations from cubic symmetry. Some of these effects were attributed, at least in part, to surface rippling from electropolishing, but others were associated with the presence of precipitates or with atomic displacements from ideal b.c.c. packing.

In recent years, my coworkers and I have investigated a number of the β-phase alloys in connection with their martensitic transformation behavior. However, during the process of these investigations we frequently strayed off-course because of numerous peculiarities observed in the "parent" β-phase. Some of these peculiarities are described in the present report and are intended to provide a general insight into the complexities of β-phase alloys. No attempt here is made to present an extensive review of the various phenomena. The examples presented are limited and admittedly mostly from my own research, but should nevertheless prove exemplary of the state of the art concerning the "many faces of β-phase alloys." In particular, I would like to address spinodal decomposition, diffuse electron scattering, tweed microstructures and charge density wave formation in β-phase alloys. In the large, it is argued that the above effects are not premartensitic in the precursor sense. Some of them, however, are "thin foil" effects, e.g., spinodal decomposition.

## Spinodal Decomposition

We have observed thin foils of Cu-Zn [2] and Ag-Cd [3] alloys, both of which show a time-dependent microstructural evolution. That is, no unusual structural features were noted in as-quenched, freshly thinned foils. But "decomposition" in thin foils only of {001} orientation was observed after room temperature aging, the kinetics being more rapid for Ag-Cd alloys than Cu-Zn alloys [3]. Foils of other orientations, e.g., {011} and {111} did not show microstructural features after aging. The decomposition effect is also composition dependent, being observed in a Cu-38.9 at.%Zn alloy, but not in a Cu-47.3 at.%Zn alloy. A typical eample of the as-quenched and decomposed states of (001) foils in a Cu-38.9%Zn alloy are shown in Fig. 1. The basketweave microstructure defined by <100> directions is clearly seen in Fig. 1b. Figure 2 shows an electron diffraction pattern for a Cu-Zn foil after aging at room temperature for 72 hours following quenching and thinning. The satellite reflections, e.g., at (020) and (040), are clearly evident and, moreover, their spacing is seen to be independent of the distance from the origin, (000), as would be expected for spinodal decomposition of a given wavelength. It should also be noted from the appearance of the superlattice reflections in Fig. 2 (indicated by arrows) that B2 order is preserved despite the decomposition. Moreover, for a given alloy (Cu-38.9%Zn) the wavelength of the decomposition depends on the aging temperature, another prominent feature of spinodal decomposition.

The literature is replete with "thin foil" and "surface" effects. Thus, Fig. 3 was obtained, showing a martensite plate (formed initially) within an aged (subsequently) foil. This dark field image, with the martensite out of contrast, clearly shows that the basketweave structure is three-dimensional as can be seen along the martensite-parent interface plane.

In the above work, it was emphasized that the observed decomposition was only observed in {001} foils, which are known to be elastically soft in the <001> direction and hence can promote spinodal decomposition. Although not heretofore found in an ordered alloy system, a theoretical justification for spinodal decomposition in ordered alloys based on a Bragg-Williams model and elastic energy considerations was presented as a sequel to the above work [4]. Other variations in the Cu-Zn spinodal morphology were subsequently published [5]. Figure 4 shows a morphology involving one-dimensional lamellae parallel to [010] and [001] directions. Figure 5 shows a one-dimensional decomposition (within the circular region). Note that the corresponding diffraction pattern, Fig. 5b, shows only one set of sideband reflections, consistent with the appearance of the image, Fig. 5a. Generally speaking, the morphology of the observed spinodal microstructure was found to be dependent upon the foil thickness, ranging from one-dimensional domains in the thinnest regions to the usual basketweave morphology in the thicker regions. However, upon extended aging the spinodal product is replaced by a bainitic structure, Fig. 6, indicating that the metastability of the initial β-phase is increased in the thin foil/spinodal condition.

## Diffuse Electron Scattering

Electron diffraction patterns from β-phase alloys of the B2 or D03 types are considerably more complex than those from a simple ordered b.c.c. structure. The former patterns usually exhibit "extra" reflections and non-

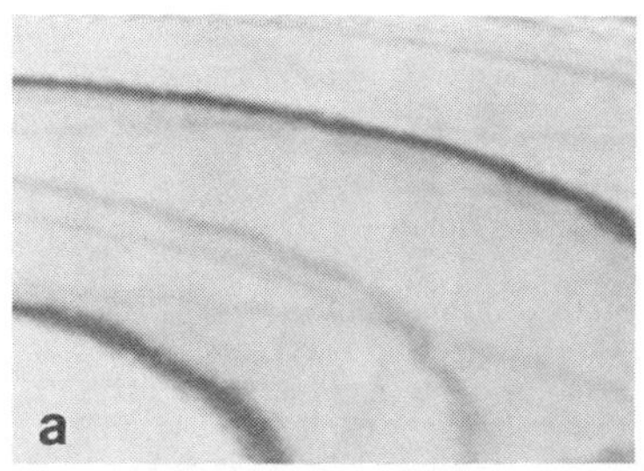


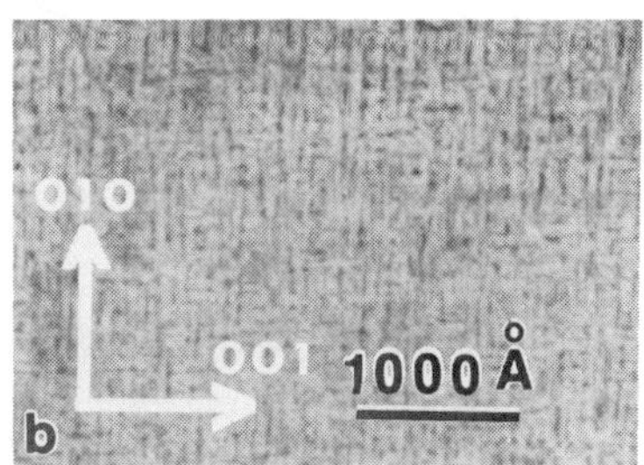


Fig. 1 Transmission electron micrographs (Cu-38.9 at.%Zn) showing (a) the as quenched state and (b) a foil annealed at 55°C for 161 h.

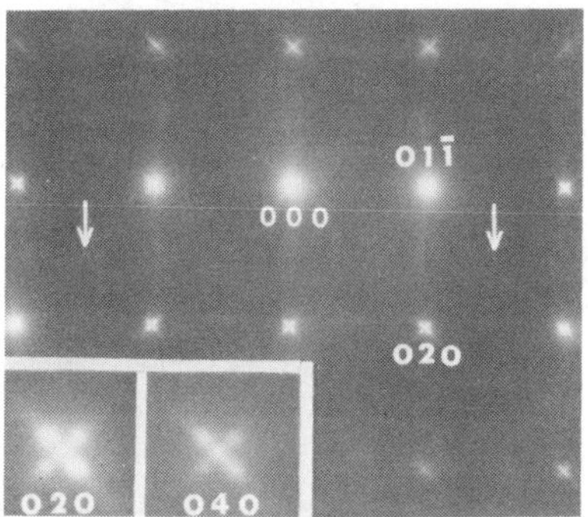


Fig. 2 Electron diffraction pattern (Cu-38.9 at.%Zn) for foil aged at room temperature for 72 h. [100] zone pattern. Arrows indicate superlattice reflections.

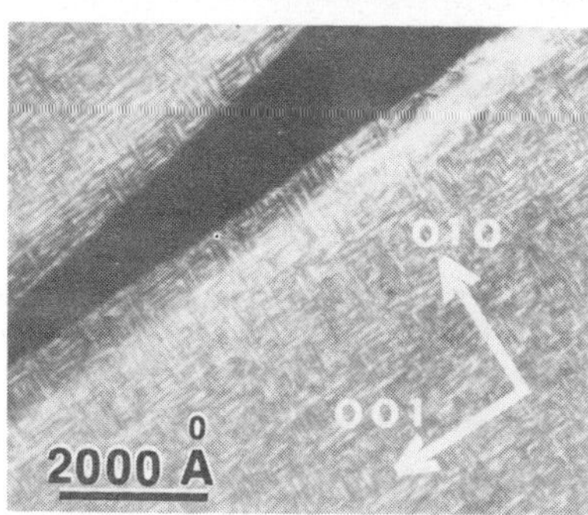


Fig. 3 Transmission electron micrograph (Cu-38.9 at.%Zn) showing martensite and three-dimensional distribution of basketweave structure. Dark field image.

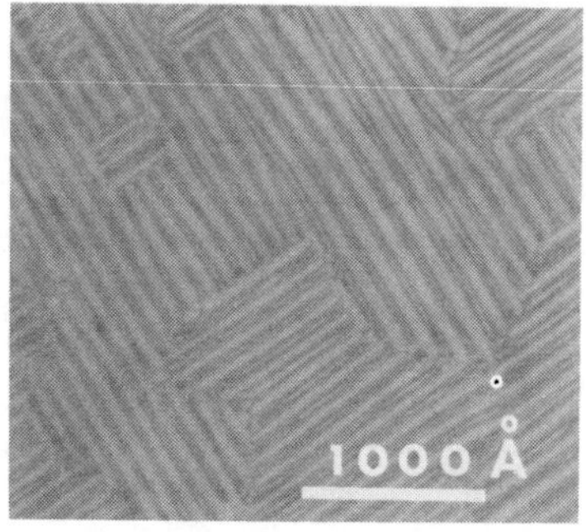


Fig. 4 Decomposed specimen of Cu-38.9Zn alloy showing one-dimensional lamellae parallel to the cube directions [010] and [001].

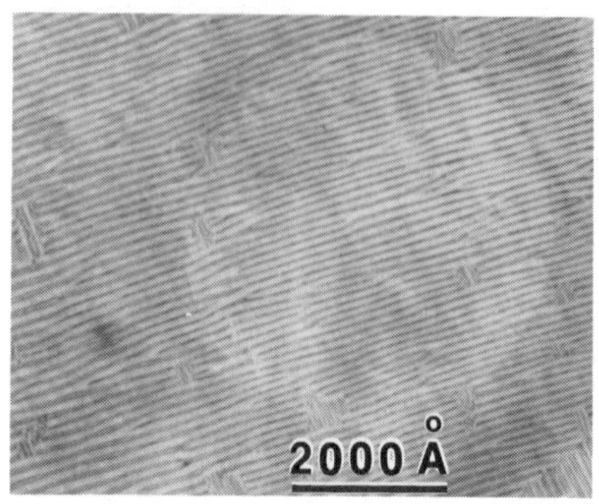


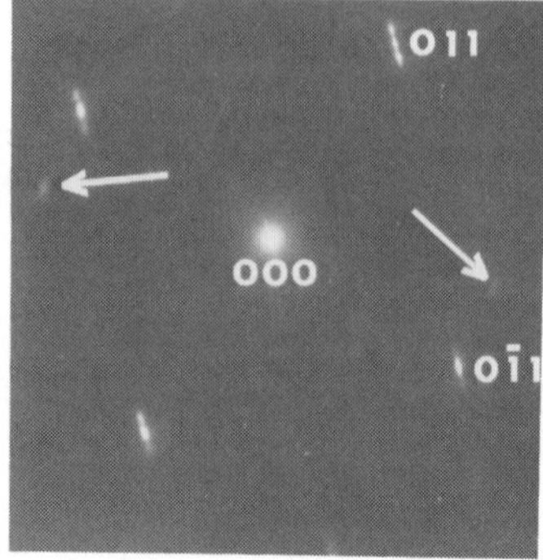


Fig. 5 Lamellae structure showing one-dimensional decomposition and (b) diffraction pattern from circular region.

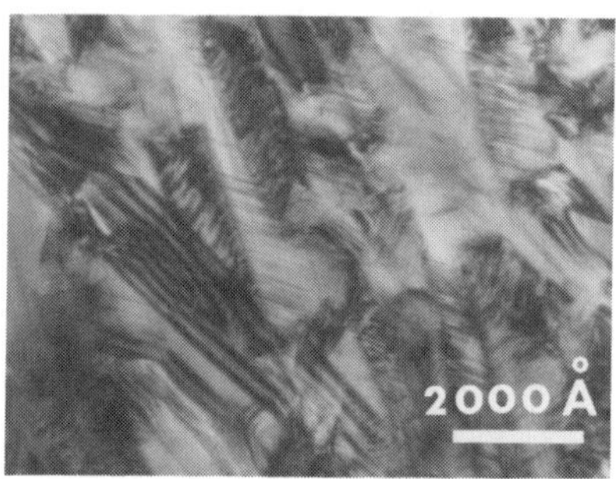


Fig. 6 Thin foil specimen aged at room temperature for 3 months showing bainitic-like microstructure.

radial diffuse streaks (rel rods) in various directions. In an attempt to clarify these diffraction patterns, Otsuka et al. [6,7] studied two β-phase alloys: Cu-39.6 wt.%Zn (B2) and Cu-14.2 wt.%Al-4.3 wt.%Ni (DO3). Both alloys were observed in the β-phase condition after solution treating and quenching to room temperature. Both alloys would transform to martensite below room temperature.

Typical of the diffuse scattering are the $[001]_\beta$ diffraction patterns shown in Fig. 7 for the Cu-Al-Ni alloy and in Fig. 8 for the Cu-Zn alloy. In both cases the streaks were identified as $\langle 110\rangle^*_\beta$ streaks following careful tilting experiments. The streaks were found to obey an extinction rule. For example, in the Cu-Al-Ni alloy $[110]^*_\beta$ streaks are absent in the $[\bar{1}10]_\beta$ and $[\bar{2}21]_\beta$ orientations. Taking a clue from earlier work on Cu-Al-Ni [8] and Cu-Zn [9] alloys that a phonon mode with wave vector $\mathbf{q} \parallel \langle 110\rangle^*_\beta$ and polarization vector $\boldsymbol{\varepsilon}_o \parallel \langle\bar{1}10\rangle$ (i.e., a $TA_2$ phonon) is substantially soft over a wide range of the branch of concern, Otsuka et al. suggested that this phonon mode could explain all the characteristics of the $\langle 110\rangle^*_\beta$ streaks. Accordingly, it was shown that

$$I_{phonon} \propto \frac{|F_g|^2 \cdot |(\mathbf{g}+\mathbf{q}) \cdot \varepsilon_o|^2}{\nu^2} \tag{1}$$

where $\mathbf{g}$ is the reciprocal lattice vector of concern, $F_g$ is the structure factor for $\mathbf{g}$, and the phonon wave vector, polarization vector and frequency are respectively $\mathbf{q}$, $\varepsilon_0$, and $\nu$. If the phonon is a transverse wave for which

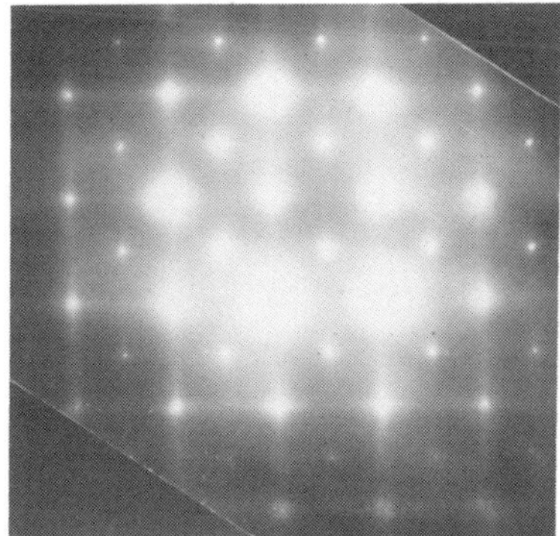

Fig. 7 $[001]_\beta$ diffraction pattern from Cu-Al-Ni alloy showing diffuse scattering.

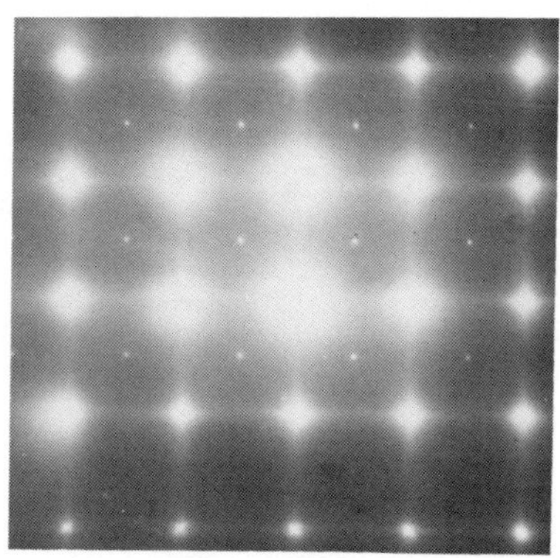

Fig. 8 $[001]_\beta$ diffraction pattern from Cu-Zn alloy showing diffuse scattering.

$\mathbf{q} \cdot \varepsilon_o = 0$, the previous reduces to

$$I_{phonon} \propto \frac{|F_g|^2 \cdot |\mathbf{g} \cdot \varepsilon_o|^2}{\nu^2} \qquad (2)$$

Both of the previous equations indicate that the phonon frequency must be relatively low for phonon scattering to be observable and that the observed streak extinctions for $\mathbf{g} \cdot \varepsilon_o = 0$ are consistent with the soft $TA_2$ phonon mode hypothesis.

Thus the $\langle 110\rangle^*_\beta$ streaks were attributed to the thermal diffuse scattering (TDS) associated with the $\langle 110\rangle^* \langle\bar{1}10\rangle_\beta$ soft phonon mode. The generality of this interpretation was extended to other β-phase alloys. As will be discussed later, TDS is clearly important and now established for β-phase alloys, but other factors should probably be considered in order to more fully tell the story.

Otsuka et al. [6,7] also analyzed "extra" reflections in addition to the diffuse scattering appearing in as-quenched specimens and attributed them to the formation of other phases, e.g., the precipitation of a 2H phase. These will not be discussed here. The interested reader is referred to the original work. Finally, it is noted that Otsuka et al. attributed none of the above effects, i.e., streaking and extra reflections, to premonitory effects insofar as a subsequent martensitic transformation is concerned.

## Tweed Microstructures

Earlier investigations of the martensitic transformation in β-NiAl alloys revealed a type of β-phase contrast which, in retrospect, was very reminiscent of the spinodal-like contrast found in β-brass. However, a more detailed examination [10] of this contrast showed that the modulation (or "tweed") spacing was independent of aging temperature--much unlike the expected situation for a classical spinodal decomposition. Moreover, the striations in NiAl were found to be parallel to traces of $\{110\}_\beta$ planes, not $\{100\}_\beta$ planes. These observations prompted a detailed investigation by Robertson and Wayman [10], who examined β-Ni-Al foils of composition in the vicinity of 63% Ni-Al (at.%).

Tweed microstructures have actually been observed in many alloy systems, and even in nonmetallic solids. Although observed earlier, Tanner [11] was the first to characterize it from studies of Cu-Be alloys (non-β-phase). Typical examples of the tweed structure in NiAl are seen in Fig. 9, a series of dark field images showing that the tweed spacing is markedly dependent on s, the deviation parameter from the exact Bragg condition. Detailed analysis shows that the microstructure shown in Fig. 9 consists of striation whose average direction corresponds to traces of $\{110\}_\beta$ planes. Striations corresponding to several $\{110\}_\beta$ planes are present simultaneously in the same area of the image, and the contrast level for a particular set of striations varies from place to place. These striations show the "best" contrast under two-beam conditions and may be invisible if the two-beam condition is not met. The tweed spacing also depends on the foil orientation. The tweed microstructure gives rise to diffuse intensity in the form of intense rel rods emanating from both fundamental and B2 superlattice reflection spots. Such streaks appear in all six $\langle 110\rangle_\beta$ directions. Examples of the streaking as seen in $\langle 001\rangle$, $\langle 110\rangle$, and $\langle 111\rangle_\beta$ orientations are shown in Fig. 10. The intensity (and extinction) of the streaks appears to be proportional to the value of

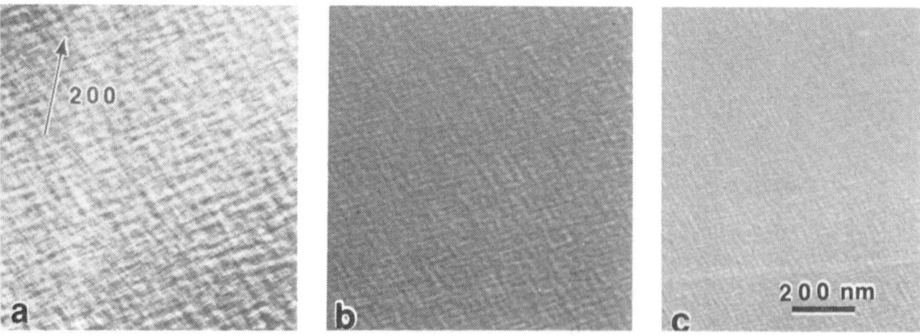


Fig. 9 Tweed microstructure in NiAl. Part of a deviation series of **g** = [002] dark field images with **B** near [100] for (a) s = -0.002 $nm^{-1}$, (b) s = -0.027 $nm^{-1}$ and (c) s = -0.044 $nm^{-1}$.

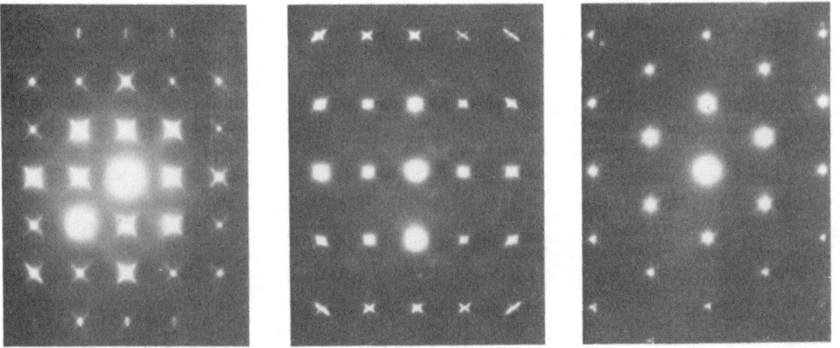

Fig. 10 Tweed streaking as seen in $\langle 001\rangle$, $\langle 110\rangle$ and $\langle 111\rangle_\beta$ orientations.

**g** • ε, where for a <110> <1$\bar{1}$0> "defect" **g** is the diffraction vector, **q** is the propagation vector and **ε** is the polarization vector. With such a **g** • ε visibility condition many of the features of the tweed microstructure can be understood in terms of the same dispersion-surface construction as used for studying stacking fault contrast. In fact, this construction explains the orientation, spacing, and s dependence. The tweed is not localized and appears to occupy the entire volume of the crystal.

The 63%Ni-Al alloy transforms to 3R martensite ($L1_0$ structure) slightly below room temperature and it is interesting to note that the tweed is also found in the martensite, as shown in Fig. 11, where the deviation(s) dependence, orientation dependence, and extinction of the striations or their spacings follows that observed for the parent β-phase. However because of the Bain distortion and symmetry change (the martensite is tetragonal) two of the six $<110>_\beta$ streaks disappear in the martensite. As an additional point, the martensite tweeds (streaks and striations) remained visible at very low temperatures (-120°C, using a TEM cooling stage) and actually became sharper, presumably because of the reduction in intensity of the diffuse background.

The kinematical and extinction characteristics of the Ni-Al tweed streaks resemble closely the phonon streaks described earlier for Cu-Zn and Cu-Al-Ni alloys. However, Otsuka et al. [6,7] claimed no microstructural features corresponding to the streaks. But a reexamination of Cu-Zn and Cu-Al-Ni has actually shown tweed microstructures in these materials with the same characteristics as in Ni-Al. Admittedly the contrast is weaker, but is clearly visible if two-beam conditions are obtained. Hence Robertson and Wayman [10] concluded that TDS cannot be a complete explanation and that it can only be said that the extinctions, etc., indicate a type of {110} <1$\bar{1}$0> shear strain.

Following on the above, Robertson and Wayman made calculations for the 63%Ni-Al alloy and determined the isointensity contours for first-order thermal diffuse scattering (dynamic) and Huang diffuse scattering (static). These results are presented in Fig. 12. Figure 13 shows experimental intensities corresponding to Fig. 12. Clearly the experimental intensity

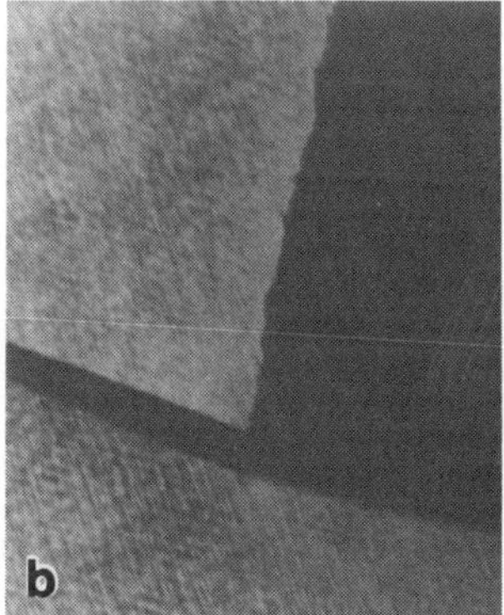


Fig. 11 Bright field images from the same region of a foil showing (a) twins in two 3R martensite variants inclined to the beam and (b) twins in one 3R variant edge-on. Note the similar contrast in the 3R and B2 regions.

profiles more closely match those for the Huang case, suggesting that the tweed displacements are mostly static rather than dynamic in origin.

By a process of elimination, Robertson and Wayman [10] concluded that the perturbing force responible for the tweed contrast in Ni-Al is the random fluctuation of local concentration of the excess Ni atoms (63%Ni-Al) and the degree of B2 order that must necessarily accompany this. Since the Ni sublattice in CsCl Ni-Al is occupied solely by Ni atoms, attention to the Al sublattice is drawn. The second nearest neighbors in the B2 structure are the nearest neighbors on the simple cubic sublattice. For random site occupations and the composition 63%Ni-Al, it was argued that the most probable site symmetry is tetragonal and that significant atomic displacements would be expected around substitutional Ni atoms on the Al sublattice in off-stoichiometric alloys.

Robertson and Wayman concluded that the tweed microstructure in 63%Ni-Al can be explained by <110> <1$\bar{1}$0> static displacment waves (or correlations) of long wavelength via elastic interactions. They claimed that the shapes of diffuse scattering contours around Bragg spots show that a spatially varying tetragonal distortion field can produce the <110> <1$\bar{1}$0> correlations when combined with the high anisotropy of β-NiAl. Tetragonal distortion adjacent to the Ni atom appears to be a reasonable hypothesis.

Other phases in the Ni-Al system such as the γ and $Ni_5Al_3$ phases also show similar tweed effects but do not transform martensitically. It is therefore argued that the tweed effect is not a martensitic precursor, although the softening of the C' elastic constant, sometimes associated with martensitic transformation, is clearly an important factor in observing the tweed per se. Contrary to this view is that expressed by Tanner et al. [12] who proposed that the tweed is a dense array of incomensurate shear strains which are pretransformational in nature.

In clarifying the difference in the physical and possible premonitory interpretation of tweed, it appears important to recognize that the work of Robertson and Wayman [10] emphasizes the importance of static rather than dynamic displacements although <110> <1$\bar{1}$0> shear strain implications are present in either case. As another cautionary note, it should be recalled that many of the martensitic structures formed from β-phase parents have a long period stacking order structure based on "derived" stacking sequences of prior $\{110\}_\beta$ planes. The relation between the $\{110\}_\beta$ tweed plane and the $\{110\}_\beta$ LPSO planes in the martensite may be entirely coincidental. Finally, it should be recalled that tweed has been observed in the NiAl martensite, and that the sharpness of the contrast increased with decreasing temperature below the $M_s$. The argument could be used that tweed in the martensite is simply a precursor effect for the reverse martensite-to-parent transformation. But if this were the case the tweed contrast should be less well-defined at lower temperatures (below $M_s$) which is contrary to the observations.

## Grid Microstructures

In a reexamination of modulatd microstructures in $<001>_\beta$ foils of Cu-Zn and Cu-Al-Ni alloys, Robertson and Wayman [13] reported an image contast which is weaker, but otherwise similar to the tweed contrast found in Ni-Al alloys. Of note is the fact that even as-quenched Cu-Zn specimens showed a tweed contrast. Careful tilting experiments revealed that the satellite spacing is the same in all diffraction patterns and that the behavior is inconsistent with satellites displaced from Bragg spots in <100> directions, the latter being expected if spinodal decomposition had occurred. Accordingly, the terminology "grid microstructure" (rather than spinodal decomposition) was introduced. This issue is currently unsettled and certainly there is no classical description of the grid microstructure. But whatever the case, the observed contrast and its foil normal dependence implies that elastic strain and therefore atomic displacement is important. The resolution of the issue as to grid contrast or spinodal decomposition is further complicated by the fact that calculated satellite intensities for both models are equally good and in agreement with experiment. Complications can also arise from the formation of surface martensite as described in detail for Cu-Zn-Al by Delaey et al. [14].

## Premartensitic Behavior and Charge Density Waves

The intermediate phase TiNi is a well-known shape memory alloy which undergoes a martensitic transformation slightly above room temperature. A puzzling aspect of this material is that it exhibits such "premartensitic" effects immediately above the martensitic start ($M_s$) temperature as: an electrical resistivity anomaly [15,16], streaks in electron diffraction patterns [15,17], and "1/3 spots" seen in both x-ray and electron diffraction patterns [15,18]. Other anomalies in physical or mechanical properties have also been reported. The nature of these so-called "premartensitic" phenomena has long been unclarified.

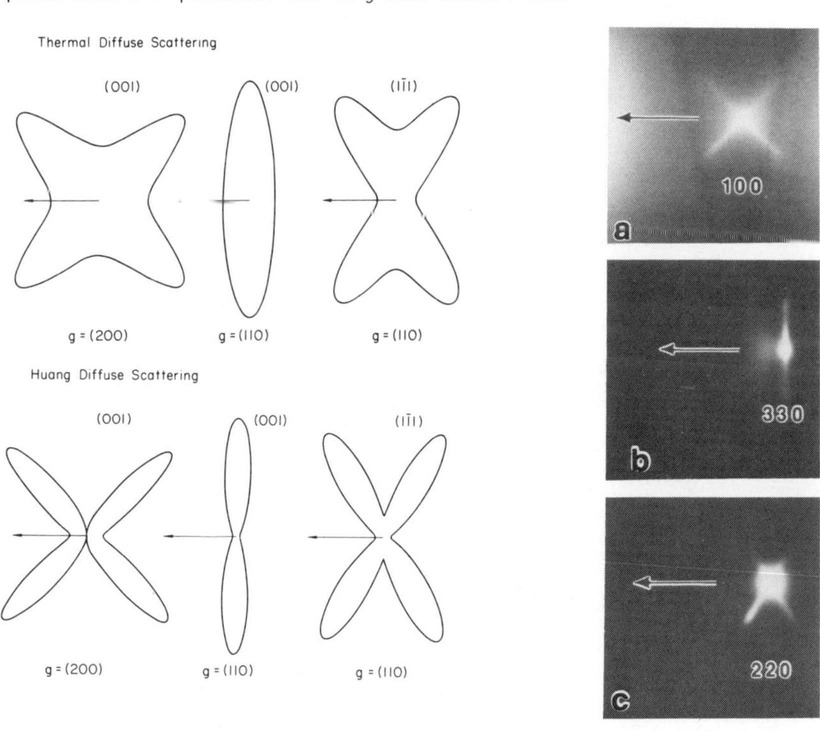


Fig. 12(a) Calculated diffuse scattering isointensity contours and (b) calculated Huang scattering isointensity contours.

Fig. 13 Experimental intensities corresponding to Fig. 12 for 63% Ni-Al. The theoretical isodiffuse contours are the same for all reflections on the systematic row.

The substitution of Fe for Ni in TiNi lowers the temperature at which the martensitic transformation takes place by more than 100°C, and reveals a well-defined "premartensitic" phase. The premartensitic stage of these alloys is found to be comprised of two, rather than one, distinct phases, one of which is a rare example of a charge density wave (CDW) in a three-dimensional metal. Two phase transitions occur, one separating the incommensurate phase from the normal, CsCl high temperature phase, and the other marking the onset of a rhombohedral distortion in which the lattice distorts to accommodate the periodicity of the change density wave. The latter corresponds to an incommensurate-commensurate "lock-in". This paper reports the premartensitic behavior and charge density waves in TiNiFe alloys with Fe substituting up to 3.2 at.% for Ni [19].

Figure 14 shows a plot of electrical resistance vs. temperature for a full thermal cycle between room temperature and liquid nitrogen temperature (-196°C). Upon cooling below room temperature, the resistance is seen to decrease slightly until $T_p \sim 0°C$, the premartensitic start temperature is reached, after which the resistance starts to rise. At the $M_s$ temperature, the resistance starts to decrease until liquid nitrogen temperature, which is approximately the $M_f$ temperature. On heating from liquid nitrogen temperature, the resistance increases up to $A_s$ where the curve changes its slope. At $A_f$, the heating curve crosses the cooling curve and then shows a small hysteresis. The larger thermal hysteresis seen below $A_f$ is typical for a normal martensitic transformation. Preliminary characterization of different phases (I-III) can be inferred from this electrical resistance vs. temperature plot. Above $T_p$, phase I is the high-temperature parent phase. Between $T_p$ and $M_s$, a distinct premartensitic phase II is obvious. Between $M_s$ and $M_f$, the martensitic phase as well as the premartensitic phase coexist. Below $M_f$, the entire crystal would be in the martensitic phase III. At this stage, one might conclude that there are two phase transformations. As will be shown later, additional features show up in electron microscopy at the temperature $T_d$, which is about 12°C lower than $T_p$.

The crystal structure of the high temperature parent phase was examined by electron diffraction at room temperature. The selected area diffraction (SAD) patterns showed the CsCl (B2) type structure. As specimens were cooled below room temperature, they developed superlattice reflections in the vicinity of 0°C ($T_p$). The superlattice reflections were seen to lie along the <110>* reciprocal lattice vectors and to be of the 1/3 (110) type, i.e., they locate at about "1/3" positions of the <110> reciprocal lattice vectors; 1/3 (111) type superlattice reflections were also seen. Typical electron diffraction patterns of the premartensitic phase showing the 1/3

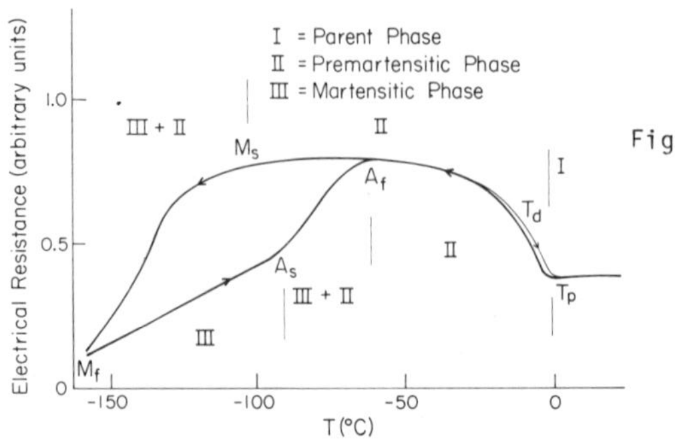


Fig. 14 Electrical resistance as a function of temperature for a full complete thermal cycle: room temperature → liquid nitrogen temperature → room temperature, $Ti_{50}Ni_{47}Fe_3$ alloy.

(110) superlattice reflections in $[001]_{B2}$, $[110]_{B2}$ and $[111]_{B2}$ zones are shown in Fig. 15. The "1/3" superlattice reflections were observed to intensify with decreasing temperatures, as shown in Fig. 16. The intensity change is reversible with temperature and is most obvious between 0°C and -10°C and less so at lower temperatures. The position of the 1/3 (110) superlattice reflections were examined closely by measuring the space between them. The results of such measurements showed that the 1/3 (110) reflections deviated slightly from the exact 1/3 position relative to the parent CsCl structure over a 12°C temperature range.

Antiphase-like microdomains (APD's) ~ 1500 Å in diameter were revealed in dark-field images using the "1/3" reflections. A typical example of these APD's is shown in Fig. 17. These microdomains are similar to antiphase domains observed in ordered alloys. However, no APD's are seen in the dark-field images taken by using $(100)_{B2}$ type superlattice reflections in either the parent phase or the premartensitic regions. It was noted that APD's are temperature independent, in the sense that the domain size and shape do not change with temperature throughout the premartensitic regime.

At about 12°C below $T_R$, needle-like microdomains are observed to form in the $Ti_{50}Ni_{47}Fe_3$ alloy. The temperature corresponding to the formation of the "needle" domains is designated $T_d$. A typical example of the "needle" domains is shown in Fig. 18(a). Figure 18(b) is the corresponding SAD pattern taken from both the needle domains and the matrix in Fig. 18(a).

In addition to 1/3 (110)-type superlattice reflections, 1/3 (111)-type superlattice reflections were also observed in different needle domains as shown in Fig. 19 (a)-(b). Figure 19(b) is the diffraction pattern taken from the encircled region in Fig. 19(a). 1/3 (111) and 1/3 (110)-type superlattice reflections in Fig. 19(b) were proved to be from domain B and domain C respectively.

Below $T_d$, APD's were seen in both the "matrix" and "needle" domains. Figure 20(a) and 20(b) are the bright field and corresponding dark-field images taken using a 1/3 (110) superlattice reflection in the diffraction pattern from the matrix, as shown in Fig. 20(c). In the bright field image, APD's are seen in both the matrix and needle domains, although they are not in good contrast. In the dark-field image, APD's appear in the matrix only because the superlattice reflection used for dark-field imaging was from the matrix, and not from the needle domains. Thus no APD's are seen in the needle domains, which appear dark.

Finally, at still lower temperatures ($M_s$) the martensitic transformation occurs by completely "wiping out" the premartensitic microstructure [20]. The martensite plates grow into the needle domains as if they were not really there at all. No trace of the former needles or APB's was observed in the martensite and there was no indication that the needle domains served as nucleating sites for the martensite.

It was concluded [19,20] that three-dimensional charge density wave (CDW) phenomena and associated phase transitions are involved in the so-called "premartensitic" behavior of TiNiFe alloys and that the first of two premartensitic transitions is associated with the formation of an incommensurate superlattice. The second transition, which occurs at a lower temperature, involves a first order incommensurate-to-commensurate transformation, from a "distorted" CsCl parent to a rhombohedral product, which allows the lattice and the CDW's to "lock-in" and thus be commensurate. These effects were suggested to be separate electronically driven structural changes and not actually precursory effects with respect to the martensitic transformation itself.

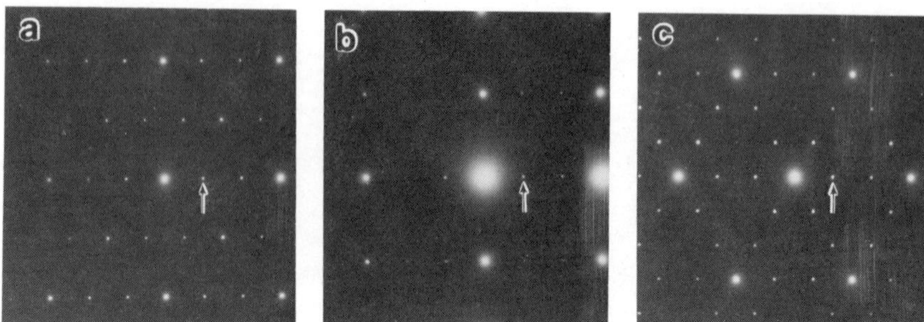


Fig. 15 Typical electron diffraction patterns from the "premartensitic" phase of the $Ti_{50}Ni_{47}Fe_3$ alloy showing 1/3 (110) superlattice reflections. (a) $[001]_{B2}$, (b) $[110]_{B2}$, (c) $[111]_{B2}$ zones.

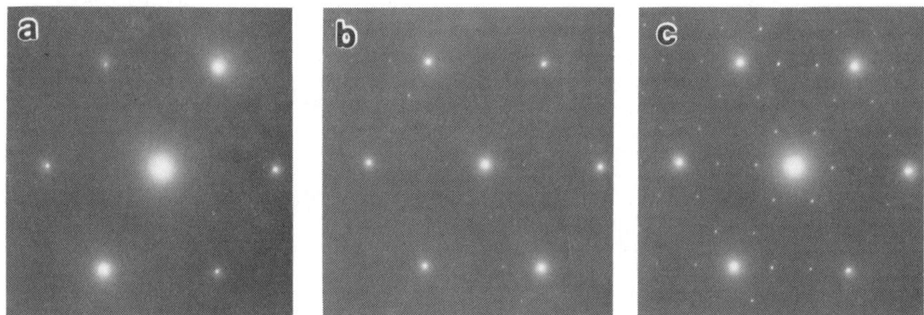


Fig. 16 Electron diffraction patterns showing temperature dependence of 1/3 (110) superlattice reflections in the $[111]_{B2}$ zone. The superlattice reflections appear around 0°C and intensified with decreasing temperature. (a) 4°C, (b) 0°C and (c) -4°C.

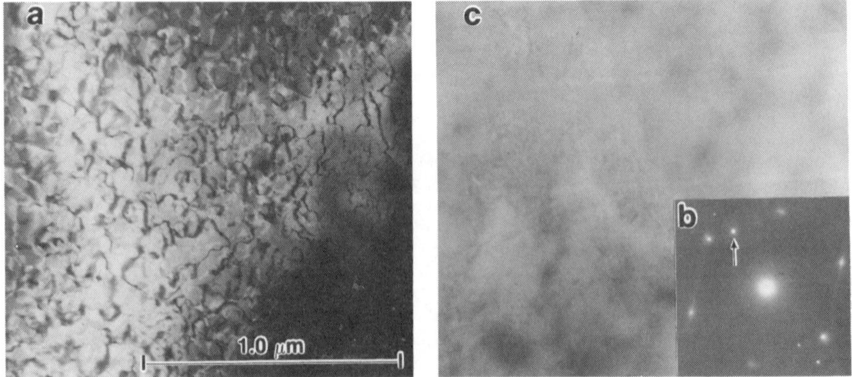


Fig. 17 Typical example showing antiphase-like domains (APD's) in the premartensitic phase of the $Ti_{50}Ni_{47}Fe_2$ alloy. (a) Dark field electron micrograph showing APD's, (b) corresponding selected area diffraction pattern showing the 1/3 (110) superlattice reflection used for the dark field imaging, and (c) bright field imaging.

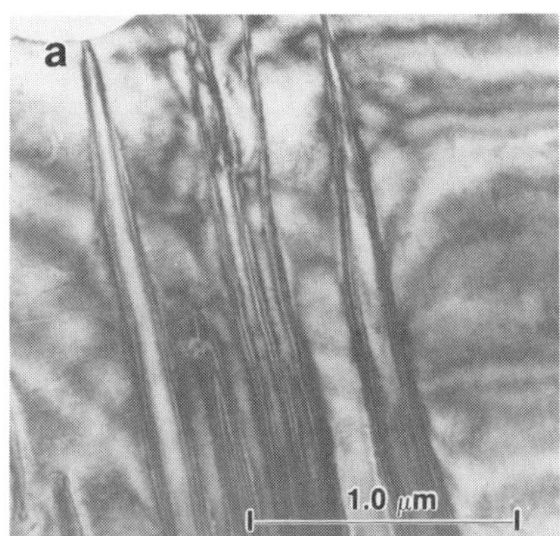

Fig. 18 (a) Transmission electron micrograph showing parallel "needle" domains. (b) Corresponding selected area diffraction pattern taken from both the needle domains and the matrix region. this diffraction pattern is interpreted as a superimposed pattern consisting of the $[001]_H$ and $[201]_H$ zones. The inner "split" spots correspond to the $[20\bar{1}]_H$ zone, while the outer "split spots correspond the the $[001]_H$ zone. Indexing is with respect to hexagonal indices.

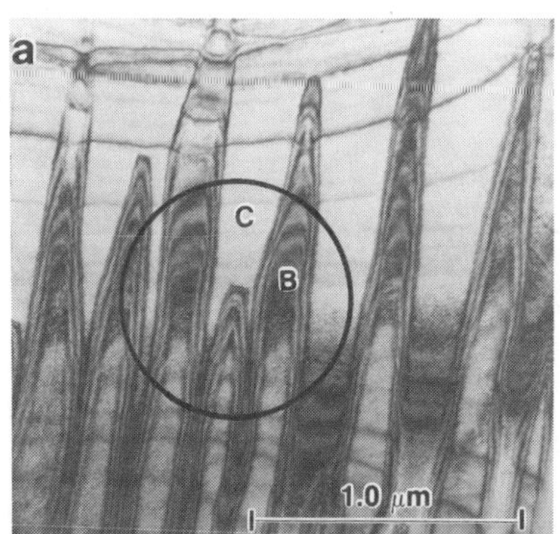


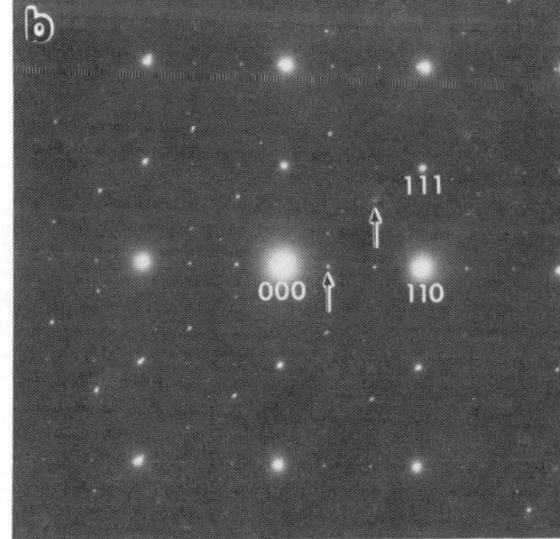


Fig. 19 (a) Electron micrograph showing domains B and C and (b) corresponding diffraction pattern taken from both domains B and C. 1/3 (111) and 1/3 (110)-type superlattice reflections are from domains B and C respectively.

The observations reported here represent a well documented case of a rare but nevertheless simple example of the occurrence of charge density waves in a three-dimensional solid system. The observations are believed to be general in nature and not peculiar to TiNiX alloys. In fact, Sinclair and co-workers [21,22] have made similar observations using a binary TiNi alloy, i.e., "1/3" relfections, domains, etc., which they interpret in terms of displacement waves [22]. The above observations are also consistent with phonon softening as observed by inelastic neutron scattering [23].

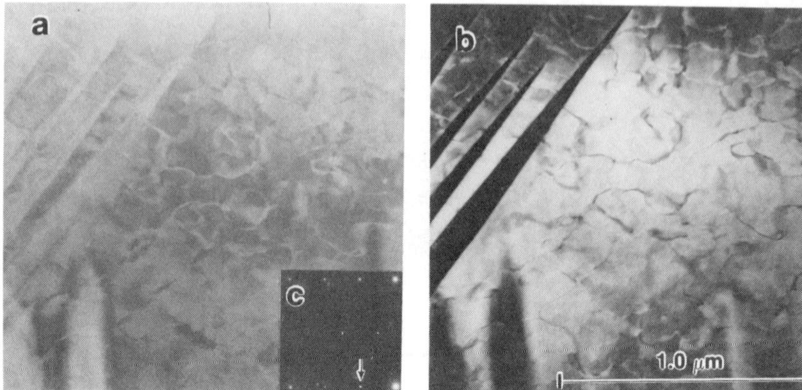


Fig. 20 (a) Bright-field image showing APD's in needle-like domains as well as in the matrix region, (b) Corresponding dark field image taken by using one of the 1/3 (110) superlattice reflections shown in (c).

ACKNOWLEDGEMENTS

The work described herein has been supported by the National Science Foundation through the Materials Research Laboratory at the University of Illinois. Their support is gratefully acknowledged.

REFERENCES

1. L. Delaey, A. J. Perkins and T. B. Massalski, J. Materials Sci. 7, 1197 (1972).
2. H. Kubo and C. M. Wayman, Metall. Trans. A 10A, 633 (1979)
3. H. C. Tong and C. M. Wayman, unpublished research.
4. H. Kubo and C. M. Wayman, Acta. Met. 28, 395 (1980).
5. H. Kubo, I, Cornelis and C. M. Wayman, Acta Met. 28, 405 (1980).
6. K. Otsuka, C. M. Wayman and H. Kubo, Metall. Trans. A 9A, 1075 (1978).
7. K. Otsuka, H. Kubo and C. M. Wayman, Metall. Trans. A 12A, 595 (1981).
8. S. Hoshino, G. Shirane, M. Suezawa and T. Kajitani, Japan J. Appl. Phys. 14, 1233 (1975).
9. S. C. Moss and C. M. Wayman, unpublished research.
10. I. M. Robertson and C. M. Wayman, Phil. Mag. 48, 421, 443, 629 (1983).
11. L. E. Tanner, Phil Mag. 14, 111 (1966).
12. L. E. Tanner, A. R. Pelton and R. Gronsky, J. de Physique, Collque C4, Suppl. No. 12, 43, 169 (1982).
13. I. M. Robertson and C. M. Wayman, Metall. Trans. A 15A, 269 (1984).
14. L. Delaey, G. Van Tendeloo, J. Van Landuyt and Y. Murakami, Proc. Int. Conf. on Martensitic Transformations, ICOMAT-79, Cambridge, MA, 520 (1979).
15. G. D. Sandrock, A. J. Perkins and R. F. Hehemann, Met. Trans. 2, 2769 (1979).
16. C. M. Wayman, I. Cornelis, and K. Shimizu, Scripta Met. 6, 115 (1972).
17. K. Chandra and G. R. Purdy, J. Appl. Phys. 29, 2176 (1968).
18. K. Otsuka, T. Sawamura and K. Shimizu, Phys. Stat. Sol. 5, 457 (1971).
19. C. M. Hwang, M. Meichle, M. B. Salamon and C. M. Wayman, Phil. Mag. 47, 9, 31 (1983).
20. C. M. Hwang, M. B. Salamon and C. M. Wayman, Phil. Mag. 47, 177 (1983).
21. P. Moine, G. M. Michal and R. Sinclair, Acta Met. 30, 109 (1982).
22. G. M. Michal, P. Moine and R. Sinclair, Acta Met. 30, 125 (1982).
23. S. K. Satija, S. M. Shapiro, M. B. Salamon and C. M. Wayman, Phy. Rev. B 29, 6031 (1984).

# KINETICS OF THE SRO-LRO TRANSFORMATION IN $NI_4MO$

E. E. Stansbury, Dept. of Chem., Met. and Poly. Engr.
University of Tennessee
Knoxville, TN 37996

ABSTRACT

The isothermal time-temperature dependence of the transformation from SRO to LRO as measured by resistivity changes is given. Problems of defining an initial state from which the transformation occurs are reviewed with emphasis on the microdomain structure from which the LRO structure develops. Questions of homogeneous transformation relative to heterogeneous transformation by clear nucleation and growth are addressed in terms of experimental evidence and theoretical predictions. Strong evidence for a nucleation and growth mechanism, at least above 750$^{0}$ C, is presented.

## GENERAL FEATURES OF THE SRO-LRO TRANSFORMATION

Classical first order phase transformations typically occur by nucleation and growth leading to C-type time-temperature-transformation curves. Transformation times approach infinity at the equilibrium transformation temperature due to absence of a free energy driving force in this limit; at low temperatures, the transformation times extend due to decreased atom mobility. The result is a maximum transformation rate (minimum time) at some critical degree of undercooling. The trans- formation of the binary $Ni_4Mo$ composition alloy, derived from isothermal resistivity measurements following cooling to the indicated temperatures at about 1$^{0}$ C/s, is represented in Figure 1 [1]. The beginning curve denotes a 0.5 microhm-cm change in the resistivity; the curve to the right represents 75% of the total change (about 20 microhm-cm at 800$^{0}$ C and 40 microhm-cm at 600$^{0}$ C.) There is some indication of a time displacement of the curve near 735$^{0}$ C implying a change in reaction mechanism and hence in morphology but this has not been clearly established. Major differences in the optical morphology are evident in Figures 2a and 2b [1] where the structures following 100 h at 850$^{0}$ C and 385 h at 600$^{0}$ C are shown. At 800$^{0}$ C domains, thin {111} striations and broader "twin-like" {110} striations are visible. At 600$^{0}$ C domains are not visible optically and only faint striations respond to etching after 4500 h. A significant response to aging is grain boundary migration below about 800$^{0}$ C and intergranular cracking above this temperature, both being responses to strains introduced by the volume change of 0.39% on transformation. Some grain boundary migration is visible in the microstructure after 385 h at 600$^{0}$ C.

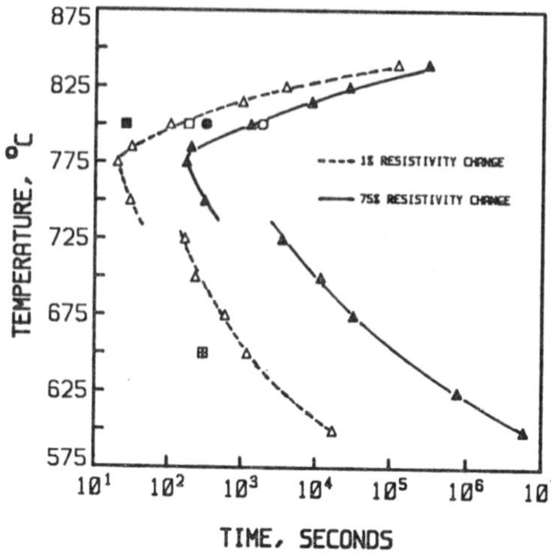


Figure 1
SRO to LRO transformation of $Ni_4Mo$ Alloy
--- 1% Trans.—75% Trans.

Quench and reheat
■ LRO starts
□ SRO gone [2]

● 20% LRO nuclei in SRO
○ LRO nuclei impinge [3]

Initial LRO intensity [4]

The equilibrium transformation occurs at 868° C at which the high temperature short range ordered face centered cubic phase (FCC) transforms on cooling to a body centered tetragonal (BCT) phase with a unit cell containing eight nickel and two molybdenum atoms. The D1a unit cell of

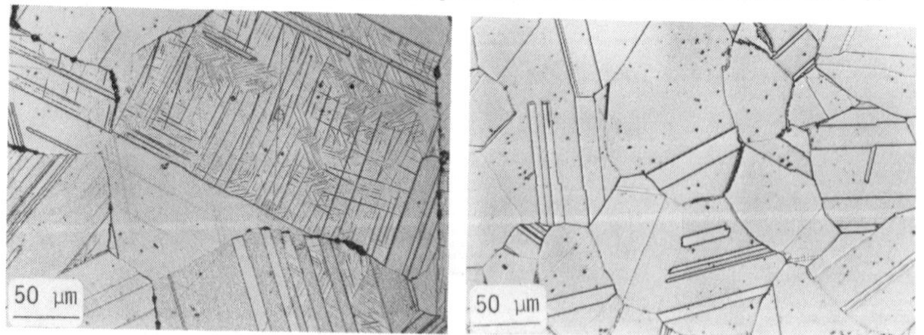


Figure 2. $Ni_4Mo$ structure. (a) 100 h, 800°C, (b) 385 h, 600°C

the long range ordered phase is shown in Figure 3: lattice sites of the initial FCC phase are essentially preserved, there being only a 1.16% contraction in the FCC c-axis and a 0.4% expansion along the a and b axes. The (001) of the ordered phase is parallel to the {100} planes of the FCC parent which allows six possible orientations of the ordered phase from a single FCC orientation, there being two possible orientations of the BCT lattice in any one of the {100} planes. Two of the orientations are shown in Figure 4; two adjacent domains having these relative orientations are said to have an anti-parallel twin (APT) relationship. Types of domain interfaces

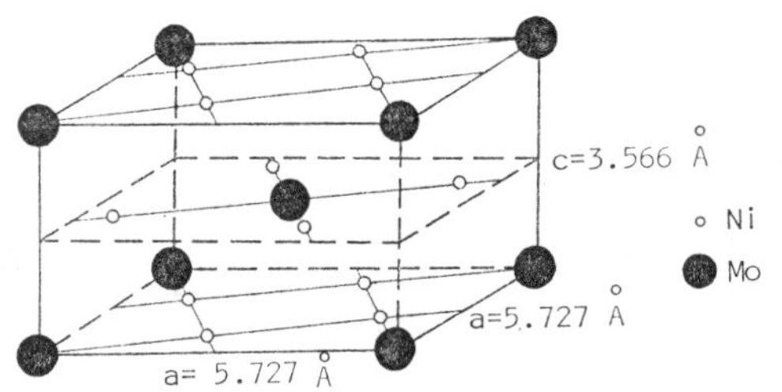


Figure 3

Unit cell of long range ordered $Ni_4Mo$ D1a structure.

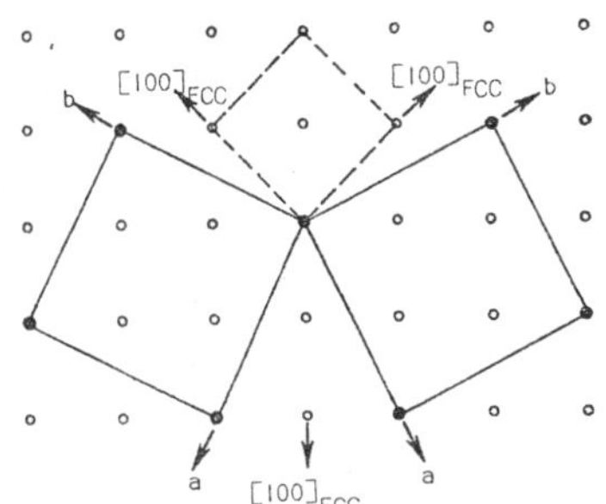


Figure 4

Two variants of LRO D1a lattice generated in FCC lattice. Variants have anti-parallel twin relationship.

have been described by Ruedl et al [5]; in addition to APT boundaries, anti-phase boundaries (APB) result when adjacent regions have the same orientation but are related by translations whose components are parallel to the lattice axes; a third interface, the perpendicular twin boundary (PTB), forms as adjacent regions having c-axes perpendicular and oriented with a common 110 interface.

A feature of the LRO structure pointed out by LeFevre et al [6] and frequently introduced as relevant to the SRO to LRO transformation, is the occupancy of every fifth $\{420\}_{FCC}$ plane by molybdenum atoms with the intervening planes occupied by nickel atoms. This is shown in Figure 5 by the traces of $(420)_{FCC}$ planes in the $(100)_{FCC}$ plane. In addition, $Ni_2Mo$, an intermediate phase sometimes observed in the transformation of nickel-molybdenum alloys [7,8,9], and the structure, identified as $DO_{22}$, can be generated by occupying $\{420\}_{FCC}$ planes with molybdenum atoms at spacings of $3d_{420}$ and $4d_{420}$ respectively. Okamoto and Thomas [4] have pointed out that these structures can occur as a consequence of stacking errors where anti-phase domains meet. Both of these phases however are thermodynamically unstable and are not observed in the final equilibrium state.

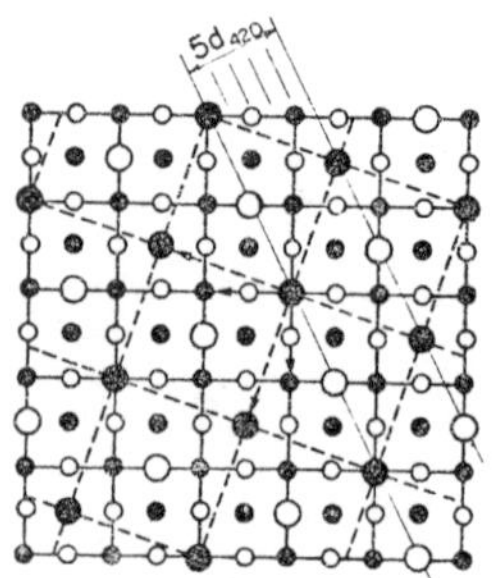


Figure 5

LRO $Ni_4Mo$ lattice (---) generated by placing Mo atoms in planes $5d_{420}$ apart.
○ ● Mo atoms, ○ ● Ni atoms

Spruiell [10] conducted the first definitive research indicating that $Ni_4Mo$ quenched from above the critical temperature was short range ordered. He observed the presence of diffuse scattering at {1 1/2 0} positions in reciprocal space and, from a set of short range order parameters, suggested that local atom arrangements could be similar to those found in the long range ordered $Ni_4Mo$. A significant result of Spruiell's study however, was that the {1 1/2 0} diffuse scattering intensity does not occur at positions of diffracted intensity from the LRO equilibrium superlattice. Further it was shown that on heating to $479^0$ C, the SRO diffuse intensity increased but on reheating to $653^0$ C, the LRO diffraction pattern was observed. Two years later, Ruedl et al [11] published electron diffraction patterns of quenched $Ni_4Mo$ showing diffuse scattering at {1 1/2 0} positions confirming the xray observations of Spruiell.

This diversity of identified structures associated with the SRO/LRO states of the $Ni_4Mo$ alloy has complicated the identification of structural changes under specific conditions which in turn has severly limited the ability to develop quantitative kinetic mechanisms for the transformations as functions of initial states and time. Phenomenologically, beginning and ending states are well established by xray and electron diffraction, and by changes in properties such as electrical resistivity, volume changes and hardness. The structural change limits are clearly seen in the electron diffraction patterns of Figure 6 in which the diffuse SRO intensity is observed at the equivalent {1 1/2 0} positions in Figure 6a [12] and the equivalent 1/5{420} positions of the D1a LRO structure in Figure 6b. These positions are identified in Figure 6c, modified slightly from that presented by Tawancy [9], to include positions for {1 1/2 0} diffraction as well as the positions for the LRO D1a, $DO_{22}$ and $Ni_2Mo$ structures. The {1 1/2 0} positions identifying SRO scattering are also positions for $DO_{22}$ intensity.

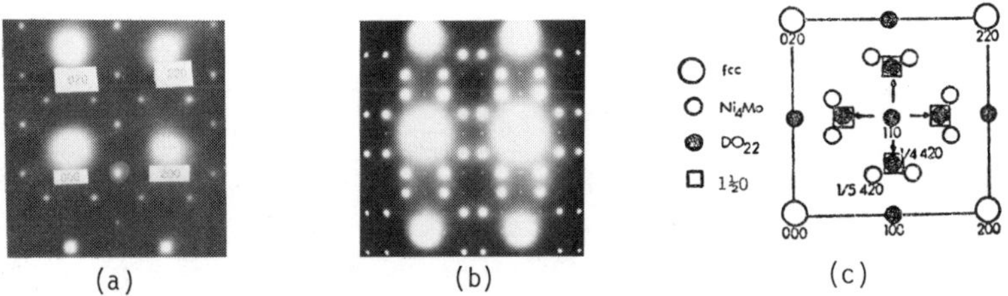


(a) (b) (c)

Figure 6. (001) electron diffraction showing (a){1 1/2 0}SRO [12], (b) 1/5{420} and (c) positions of indicated intensity [9]

However, the $DO_{22}$ structure also has intensity at {100} which should appear if the structure is present. Better discrimination however results by examining other planes in reciprocal space particularly those on which diffracted intensity for one or more of the structures cannot appear. Depending on conditions, the SRO to LRO transformation has been associated with continuous and/or discrete shifts in diffuse intensities between the Dla and {1 1/2 0} positions as well as intensity associated with the $DO_{22}$ and $Ni_2Mo$ structures. It is evident however, that careful experimental technique and critical analysis of the diffraction patterns are necessary to establish the correct sequence of structural changes.

TEMPERATURE DEPENDENCE OF STRUCTURAL CHANGES OF SRO/LRO TRANSORMATION

A recent publication by Banerjee et al [13] reporting observations of effects of electron radiation in the TEM has provided significant new insight into the structure changes and kinetic processes occurring between the LRO and SRO states of $Ni_4Mo$. The initial structures in these studies were obtained by (1) water quenching samples contained in He-filled quartz capsules to give the SRO state and (2) reheating such samples to $802^o$ C for 96 h to produce the LRO state. They report effects of the temperature at which radiation was imposed and time as variables related to the two initial states. Their observations are summarized here to provide a framework for subsequent discussion of current understanding of the structural changes and the kinetics of these changes in $Ni_4Mo$.

Although induced by electron radiation, Bannerjee et al [13] observed complete removal of both LRO and SRO on radiation below 200 K ($-73^o$ C). Recent resistivity measurements at elevated temperatures by Lei [1] and Vasudevan [14] indicate that only near the melting temperature does the temperature coefficient of resistivity become positive implying approach to random atom distribution. Also they note that cold working which initially decreases resistivity due to mechanical dispersion of SRO must exceed 68% reduction before the resistivity decrease is reversed suggesting that deformation in excess of this amount is necessary to disperse the SRO. However, quenching from near the melting point to retain a random state has not been attempted and probably would not prevent SRO from forming on cooling because of the high atom mobility, enhanced by high vacancy concentration, at the high temperatures. Banerjee et al [13] observe that radiation and thermal exposure in the temperature range $200<T<550$ K ($-73<T<277^{\circ}$ C) leads to three conclusions: (1) SRO forms only from a more highly disordered state, (2) in the presence of LRO, SRO does not form and (3) LRO does not nucleate from SRO. In the temperature range $550<T<720$ K ($277<T<447^{\circ}$ C), observation of a "continuous shift of diffraction intensity from SRO to LRO both being linked by diffuse intensity throughout the transition" led to conclusions that no D1a microdomains were present and that the alloy existed in a single phase transitional state. At 770 K ($497^{\circ}$ C), initial SRO disappears with formation of LRO of S = 0.85 and at T>820 K ($547^{\circ}$ C) SRO intensity disappears discontinuously as LRO appears with no intensity between the diffraction spots. Thus at these higher temperatures, LRO nucleates discretely and a mechanism involving pre-existing LRO microdomains is not involved.

Banerjee et al [13] relate their observations to arguments over the past twenty years on the interpretation of experimental measurements and on proposed mechanisms for transformations between LRO and SRO. They are basically addressing four major questions relating to the structure and kinetics of the transformation in $Ni_4Mo$. Specifically, (1) the (1 1/2 0) diffuse intensity be correlated to a microdomain structure, (2) if so, are the microdomains direct precursors of the LRO structure, (3) does the transformation occur homogeneously or heterogeneously by definable nucleation and growth and (4) what kinetic mechanisms are applicable to the changes. A large literature has developed addressing these questions.

## THE SRO STATE OF $Ni_4Mo$

As reviewed earlier, the presence of the {1 1/2 0} diffuse intensity and its interpretation as small regions in which the atomic arrangement is quite similar to that found in the long-range ordered $Ni_4Mo$ was first presented by Spruiell [10]. He also observed that the SRO diffuse intensity increased on thermal treatment at $479^{\circ}$ C. Ruedl et al [5,11] who confirmed the location of the diffuse SRO intensity observed that dark field imaging from the SRO diffraction produced "dotted" images which they suggested could be " very small ordered domains" and that "the transition between long and short range order would be continuous and just a matter of domain size." Both the {1 1/2 0} intensity and the dark field images have received general confirmation as a basic characteristic of the SRO state. Chakravarti et al [15] and Snyder and Brooks [12] used the term "mottled structure" and more recently in a very detailed investigation, Chevalier and Stobbs [16] have used the term "speckled." Field ion microscopy has also been used to study the as-quenched state and stages of ordering. With respect to providing insight into the as-quenched state, Chakravarti et al [15] did not observe ordered domains in a disordered matrix, Yamamoto et al [17] did not attempt a specific interpretation of their field ion micrographs following direct quench, but Okamoto and Thomas [4] suggested that the microdomains that they observed by field ion microscopy possessed an imperfect $DO_{22}$ structure. The early conclusion therefore was that the SRO state of $Ni_4Mo$ was one of microdomains of the D1a structure and/or the related structures of $nd_{420}$ stacking and that the subsequent transformation was the simultaneous growth of the domains and their perfection in terms of increased long range order parameter, S.

In a series of papers, Thomas and co-workers [4,18,19,20,21,22,23,24] studied the structure of the quenched state and changes during early stages of transformation. They examined in detail how a distribution of diffuse intensity relates to the {420} stacking sequences of nickel and molybdenum atoms which result in the $Ni_2Mo$, $Ni_4Mo$ and $DO_{22}$ structures. This led to their proposal of a multi-microdomain model of the dispersion of these structures in a disordered matrix. They recognize that order may be imperfect within these structures and that subsequent transformation involves both increase in perfection and final conversion to the stable D1a structure of $Ni_4Mo$. In this view, the $Ni_2Mo$ and $DO_{22}$ structures are intermediate unstable structures; the {1 1/2 0} intensity is attributed to non-conservative antiphase boundaries formed between the microdomains as a result of adjacent domains "impinging" with molybdenum atoms at every $3d_{420}$ or $4d_{420}$ rather than the equilibrium $5d_{420}$ sequence. Thomas and Goringe [25] however, in discussing current concepts of the short range ordered state and its analysis by transmission electron microscopy, indicate that SRO in $Ni_4Mo$ may be more complex than this although the microdomain concept appears to be applicable to some other ordering alloys.

Charkravarti et al [26] measured diffuse intensity in a larger volume of reciprocal space and concluded that overlap from adjacent {1 1/2 0} intensity gave rise to local intensity that could be interpreted as $Ni_2Mo$ and D1a in the quenched state. From SRO parameters derived from x-ray data, they construct a computer-based distribution of nickel and molybdenum atoms and although groups of atoms tending to have the property of no Mo-Mo nearest neighbors were observed in the computer generated array , they conclude "that there is no evidence for large domains of near perfect LRO to grow at the expense of the SRO matrix." The observation of de Ridder et al [27] that {1 1/2 0} imaging is not observed using material directly quenched, whereas encapsulated samples did allow imaging is an example of the sensitivity of the state of the $Ni_4Mo$ alloy to cooling rate. Similarly, Van Tendeloo [2] observed that severe quenching was required to restirict diffracted intensity to {1 1/2 0} positions.

It is evident that questions of retention of the SRO state as a function of cooling rate give rise to controversy over the structure of the quenched state and hence the state from which the transformation to LRO initiates regardless of path, i.e., continuous cooling or reheating, or isothermal.

Some of this uncertainty was removed by Chevalier and Stobbs [28,29,16] who reported measurements of diffuse intensity in several planes of reciprocal space and observations of the SRO structure using dark field images generated from the SRO {1 1/2 0} diffuse intensity. They ascribe a unique microdomain structure to the SRO state which precludes the presence of regions of D1a, $DO_{22}$ or $Ni_2Mo$ and, furthermore, that the speckle pattern observed by them and others [11,12,15] on dark field imaging is characteristic of this unique SRO state and does not indicate the presence of microdomains of the other structures. It is significant to note that Chevalier and Stobbs used the fastest cooling rate, $500^{\circ}$ C/s, which may be a contributing factor to their conclusions in difference to previous research. Additional evidence for a unique SRO structure is based on the observed increase of SRO diffuse intensity on reheating samples to temperatures below $500^{\circ}$ C [10,30,13,16] and the increase in resistivity observed prior to the decrease associated with LRO [1,14,31]. If LRO microdomains were associated with the SRO state, then these should immediately serve as nuclei for conversion to LRO and the observed intensification of SRO should not be observed.

Although discussed in greater detail subsequently, it is important to point out here that theories of SRO not encompassing a domain concept were developed in the 1930's by Landau and Lifshitz [32] and carried forward by de Fontaine [33] and more recently by Chevalier and Stobbs [16,29]. A somewhat different approach was taken by Clapp and Moss [34,35,36] using a statistical treatment involving pairwise atom interaction potentials and by de Ridder et al [27] using cluster models. These theories account for the {1 1/2 0} diffuse scattering and, according to Stobbs and Chevalier [29], lead to a SRO structure for the $Ni_4Mo$ composition alloy as unbounded regions of continously variable composition with no superlattice structure. The important conclusion is that subsequent transformation whether on continuous cooling or heating, or isothermally, occurs from a limiting "as-quenched" state in the sense of Chavalier and Stobbs [16] or from some pre-nucleated state containing "nuclei" of D1a, $DO_{22}$ and/or $Ni_2Mo$ depending on the exact pre-transformation time-temperature history.

## THE SRO TO LRO TRANSFORMATION BY NUCLEATION AND GROWTH

There is strong experimental evidence that the SRO to LRO transformation occurs by nucleation and growth above approximately 700-750$^{o}$ C but homogeneously at lower temperatures. Exceptions to this are the field ion microscopy observations of Yamamoto et al [17] who concluded that continuous ordering ocurred at 800$^{o}$ C with no stage of LRO domains distributed in a SRO matrix but rather the gradual growth of contiguous domains of the D1a structure with accompanying increase in the long range order parameter.

The evidence of a nucleation and growth mechanism at the higher temperatures is quite positive. On aging at 800 and 750$^{o}$ C, Chevalier and Stobbs [16] observed nucleation of distinct D1a domains with a "pronounced tetragonal prism shape." Similar nuclei in the SRO matrix were reported much earlier by Saburi et al [3] and by others including Lei [1]. Lei examined samples transformed for one hour at 850$^{o}$ C which is a higher temperature than for previous investigations and, in addition to examining both single LRO domains in a SRO matrix using TEM, he developed etching procedures which allowed observations of individual domains optically. Analysis of single domains such as shown in Figure 7 led to proposing the three dimensional model shown in Figure 8. Lei concludes that the domains are "pillow" shaped" with $\langle 001\rangle_{FCC}//\langle 001\rangle_{BCT}$ for the extended top and bottom interfaces and with the appropriate $\{210\}_{FCC}//\{110\}_{BCT}$ as side faces. In Figure 8, the c-axis of the LRO structure is perpendicular to the larger profile of the domain; the two sides and the cross section are parallel to {120} of the matrix which was shown to be in the SRO state. These domains, formed at 850$^{o}$ C, are about 2000x2000x1000 nm in size which compares to 20-80 nm observed by Chevalier

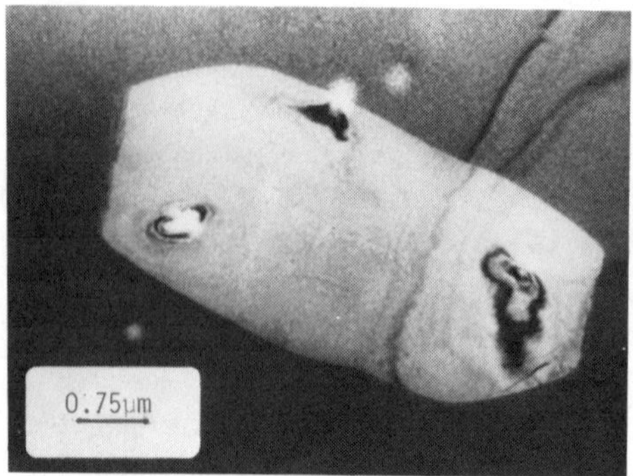


Figure 7

Single LRO domain of $Ni_4Mo$ in SRO matrix. Quenched and reheated 1 h, $850_0$C. [1]

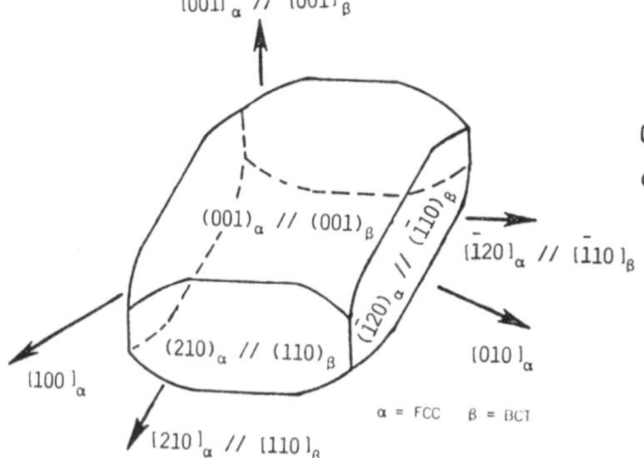


Figure 8

Crystallography of single domain of LRO $Ni_4Mo$. [1]

and Stobbs [16] after 294 s at $800^0$ C, 30 nm after one hour and 350 nm after 186 hours at $775^0$ C observed by Snyder and Brooks [12] and 10 nm after 15 hours at $700^0$ C reported by Chakravarti et al [15]. Saburi et al [3] were the first to observe nucleated LRO domains in TEM reporting a domain size of the order of 100 nm after 5 min at $800^0$ C. It is significant that in all of these cases, the domains are remarkably uniform in size. Chevalier and Stobbs [16], in their studies of aging as a function of time at $800^0$ C, note that the initial stages corresponding to domain sizes <5 nm were not detected and conclude that nucleation is followed by rapid growth.

Chevalier and Stobbs [16] and Lei [1] point out that the morphology of the individual domains corresponds to habit planes of the sides of the domains of {420} type and hence related to the LRO structure as derivable from the occupancy of every fifth {420} plane by molybdenum atoms. In view

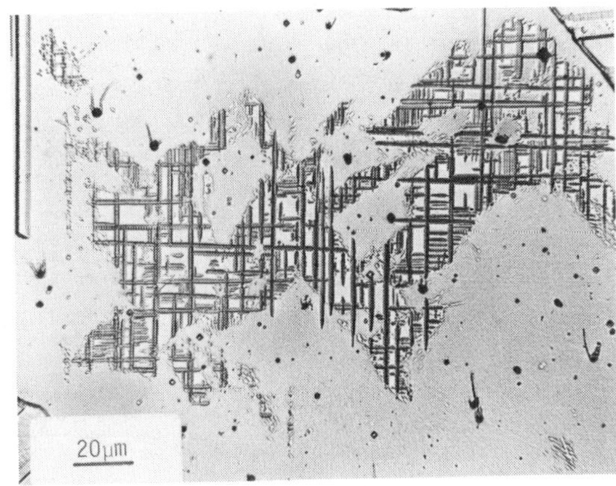


Figure 9

"Walls" of LRO formed by agglomeration of LRO domains in SRO matrix. Quenched and reheated 1 h at $850_{0}$C. [1]

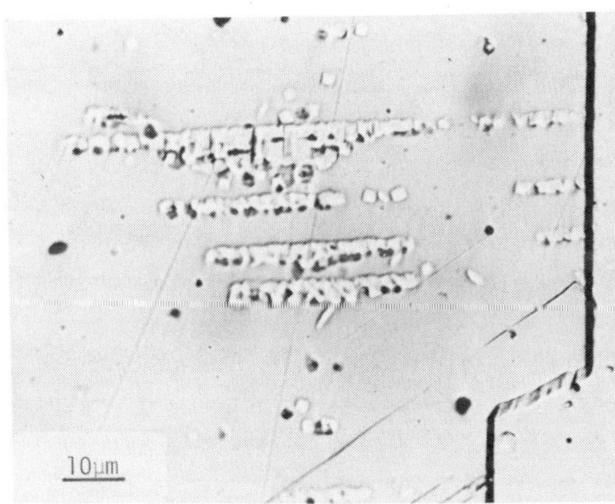


Figure 10

Single, stringers and agglomeration of LRO in SRO matrix. Quenched and reheated 1 h, $850_{0}$C

of the 1.16% contraction along the c-axis and 0.4% expansion along the a- and b-axes, a first order conclusion is that domains should grow faster in the (001) plane as proposed by Lei [1] in Figure 8.

Lei [1] has observed that at $850^{0}$ C individual domains, strings of domains and agglomerated domains will respond to etching relative to the SRO matrix such that the progress of the transformation can be followed with optical microscopy. These features are shown in Figure 9. Generally, the "wall" forms along {100} planes which is confirmed by spreading parallel to the edges of etch pits which develop {100} plane faces. The walls thicken by nucleation and growth of domains over a distance of several domain thicknesses along the wall faces, i.e. isolated domains are observed near the wall faces. The intervening space subsequently transforms allowing the "wall" face to grow. Individual domains and stringers of domains some of which have started to agglomerate are shown in Figure 10. Within the agglomerated regions, individual domain boundaries are still resolvable to some extent. The strings are composed mostly of pairs of square domains, many in the antiparallel twin orientation corresponding to the two variants of Figure 3. In addition to the square domains, elongated domains are observed both along the strings and isolated in the SRO matrix. These are domains with c-axes in the plane of the surface and hence the thin dimension is observed. The thin striations, characteristic of etching the LRO phase and pointed out with reference to Figure 2, are already evident in both Figures 9 and 10.

The mechanism of the transformation at lower temperatures, below 750-725$^{0}$ C is less certain. Chakravarti et al [15] did not observe LRO domains growing in a SRO matrix at 700$^{0}$ C using field ion microscopy but rather concluded that a structure of contiguous domains of D1a structure emerges with continuing perfection as indicated by the increase in the long range order parameter, S. Similar conclusions were drawn by Okamoto and Thomas [19] using the same experimental method on samples aged at 750$^{0}$ C. Some evidence for a change in mechanism around 750$^{0}$ C is evident from the isothermal resistivity measurement of Lei [1] which led to the transformation curve shown as Figure 1. Although the data are somewhat limited, the beginning of transformation curve shows an offset in time between 725 and 750$^{0}$ C suggesting a change in mechanism. However Banerjee et al [13] conclude that the transformation occurs by nucleation and growth as low as 550$^{0}$ C and Stobbs and Chevalier [29] imply that there is no strong reason to preclude this mechanism at lower temperatures.

## CORRELATION WITH THEORETICAL PREDICTIONS

A shift in the mechanism of transformation from distinct nucleation and growth to relatively homogeneous ordering has a basis in the theory developed by de Fontaine [33]. De Fontaine, working in the framework of early theory by Landau and Lifshitz [32] ,Clapp and Moss [34,35,36] and Khachaturyan [37], establishes the k-space configurational energy as a function of atom pair interaction parameters. He then shows that this leads to a statistical expression for the distribution of atoms in the lattice expressed as waves representing the variation of concentration of atoms in different directions in the lattice. These in turn give rise to characteristic diffraction effects which can be represented in the reciprocal lattice for the particular structure. On applying the theory to the {420} structures and confining pair interactions to the third coordination shell, de Fontaine shows that three significant conclusions follow relating to the SRO to LRO transformation in $Ni_4Mo$. First, SRO diffuse intensity is predicted for the {1 1/2 0} position as observed in rapidly quenched $Ni_4Mo$. Second, of particular significance to the time-temperature dependence of the transformation of $Ni_4Mo$ is the conclusion "that all ordering reactions of the 420 series must occur at equilibrium by nucleation and growth." Third, de Fonatine develops the concept of a spinodal ordering temperature above which, at most, concentration waves of small amplitude, but sufficient to give {1 1/2 0} diffraction, exist; below the spinodal ordering temperature, concentration waves can amplify and produce a distinct SRO state. Thus quenching and reheating below the spinodal ordering temperature can lead to intensification of SRO diffraction as has been observed. Since the spinodal ordering temperature is below the critical temperature for LRO, between these temperatures tranformation must occur by nucleation and growth. At lower temperatures however, the concentration wavelength changes with time and the diffuse intensity maxima change fom SRO to LRO positions. A mechanism is the increasing accumulation of molybdenum atoms into every fifth {420} plane. Also non-uniform stacking accompanied by compositional variations could give rise to local regions of the $DO_{22}$ and $Ni_2Mo$ structure observed at least at transformation temperatures in the 600-750$^o$ C range [1,22].

Chevalier and Stobbs [29,16], building on de Fontaine's work, propose that attenuation of the concentration wave by an exponential decay leads to a "static concentration wave packet." They argue that such wave packets can form on all {420} planes in difference to D1a microdomains which are bounded by specific pairs of {420} planes, and that conceptually packets on different planes can interpenetrate. The consequence is an unique SRO structure consisting of unbounded regions of continuouosly varying composition. This structure accounts for the single {1 1/2 0} diffraction and correlates with the "speckled" structure observed on imaging the SRO intensity. The conclusion is that micro-domains as bounded small regions of D1a, $DO_{22}$ or $Ni_2Mo$ do not constitute the SRO state.

## SUMMARY

Current experimental and theoretical investigations of the $Ni_4Mo$ composition alloy lead to the following description of the SRO state and its transformation to LRO.

1. Resistivity measurements indicate presence of SRO from near the melting temperature (1380° C) down to the long-range ordering temperature (868° C), increasing with decrease in temperature. The degree of SRO continues to increase at lower temperatures prior to formation of LRO. SRO has been destroyed by low temperature electron radiation, possibly by severe deformation and by heating to near the melting point.
2. There is a theoretical basis for defining a spinodal ordering temperature (<868° C) below which the SRO structure is more sharply defined than at higher temperatures.
3. The SRO state is a unique structure described statistically by compositional waves in <210> directions. The structure accounts for the observed {1 1/2 0} diffraction and is characterized by "speckled" imaging from the SRO intensity. A SRO state consisting of or containing microdomains of the D1a, $DO_{22}$ or $Ni_2Mo$ structures is doubtful. This is supported by TEM and resistivity measurements which indicate intensification of SRO prior to formation of LRO.
4. In the temperature range 868<T<750° C, the SRO to LRO transformation occurs by nucleation and growth of LRO domains in a SRO matrix. This is confirmed by TEM and at 850° C by optical microscopy.
5. Below about 750° C, transformation by nucleation and growth of LRO regions into a SRO maatrix is less certain. Rather molybdenum atoms diffuse into {420} planes in contiguous regions forming microdomains of predominantly D1a structure but with regions of $DO_{22}$ and $Ni_2Mo$ (Mo atoms occupying planes at $5d_{420}$, $4d_{420}$ and $3d_{420}$). The $DO_{22}$ and $Ni_2Mo$ regions are accounted for as D1a antiphase boundaries across which the D022 or $Ni_2Mo$ sequence of Mo-containing {420} planes exist. With time, D1a domains grow decreasing the relative amount of the other two structures.
6. Subsequent domain growth develops a morphology governed by the interdomain boundary energies of antiphase, antiparallel twin and perpendicular twin boundaries.


## REFERENCES


1. T. S. Lei, Ph.D. Dissertation, University of Tennessee., (1979).
2. G. Van Tendeloo, Mater. Sci. Eng. 26, 209-220 (1976).
3. T. Saburi, K. Komatsu and S. Nenno, Phil. Mag. 20, 1091-1094 (1969).
4. P. R. Okamoto and G. Thomas, Acta Met. 19, 825 (1971).
5. E. Ruedl, P. Delavignette and S. Amelinckx, Mat. Res. Bull. 2, 1045-1054 (1967).
6. B. G. LeFevre, A. G. Guy and R. W. Gould, Met. Trans. 242, 788 (1968).
7. T. Saburi, K. Komatsu, M. Yamamoto, S. Nenno and Y. Mizutani, Trans. AIME 245, 2348-2349 (1969).
8. S. K. Das, P. R. Okamoto, P. M. J. Fischer and G. Thomas, Acta Met. 21, 913 (1973).
9. H. M. Tawancy, Scripta Met. 18, 343-346 (1984).
10. J. E. Spruiell and E. E. Stansbury, J. Phys. Chem. Sol. 26, 811 (1965).
11. E. Ruedl, P. Delavignette and S. Amelinckx, Phys. Stat. Sol. 28, 305 (1968).

12. W. B. Snyder and C. R. Brooks, In B. H. Kear, C. T. Sims, N. S. Stoloff and J. H. Westbrook (eds): Ordered Alloys - Structural Applications and Physical Metallurgy. Lake George, NY, Claitor's, pp. 275-296, (1970).
13. S. Banerjee, K. Urban and M. Wilkens, Acta Met. 32, 299-311 (1984).
14. K. Vasudevan, The Effect of 0-4 Weight Percent Chromium on Transformations of Alpha Phase Ni-Mo-Cr Alloys with Nickel: Molybdenum Ratio Equal to 4:1. Knoxville, TN: M.S. Thesis Univ. of Tenn. (1982).
15. B. Charkravarti, E. A. Starke and B. G. LeFevre, Jnl. Mats. Sci. 5, 394 (1970).
16. J. P. Chevalier and W. M. Stobbs, Acta Met. 27, 1197-1217 (1979).
17. M. Yamamoto, S. Nenno, M. Futamoto and S. Nakanura, Japanese Jnl. of Applied Physics 11, 437-444 (1972).
18. P. R. Okamoto and G. Thomas, Mat. Res. Bull. 6, 45-50 (1971).
19. P. Okamoto and G. Thomas, In B. H. Kear, C. T. Sims, N. S. Stoloff and J. H. Westbrook (eds): Ordered Alloys - Structural Applications and Physical Metallurgy. Baton Rouge, Claitor's, pp. 297-299, (1970).
20. M. N. Chandrasekharaiah, S. Ranganathan, P. R. Okamoto and G. Thomas, Mat. Res. Bull. 7, 13-18 (1972).
21. S. K. Das, P. R. Okamoto, P. M. J. Fischer and G. Thomas, Acta Met. 21, 913 (1973).
22. S. K. Das and G. Thomas, Phys. Stat. Sol. 21, 177 (1974).
23. R. Sinclair, D. S. Gelles and G. Thomas, Proc. Annu. Meet. Electron Microsc. Soc. Amer. 34, 600-601 (1976).
24. G. Thomas and R. Sinclair, Acta Met. 25, 231 (1977).
25. G. Thomas and M. J. Goringe, Transmission Electron Microscopy of Materials. New York: John Wiley (1979).
26. B. Chakravarti, E. A. Starke, C. J. Sparks and R. O. Williams, J. Phys. Chem. Sol. 35, 1317-1326 (1974).
27. R. De Ridder, G. Van Tendeloo and S. Amelinckx, Acta Cryst. 32, 216-224 (1976).
28. J. P. A. A. Chevalier and W. M. Stobbs, Acta Met. 24, 535 (1976).
29. W. M. Stobbs and J. P. A. A. Chevalier, Acta Met. 26, 233 (1978).
30. L. A. Nesbit and D. E. Laughlin, Acta Met. 26, 815 (1978).
31. T. S. Lei, K. Vasudevan and E. E. Stansbury, Proceedings:High-temperature Ordered Intermetallic Alloys. Pittsburgh: Materials Research Society (1984).
32. L. D. Landau and E. M. Lifshitz, Statistical Physics. Reading, MA: Addison-Wesley (1958).
33. D. De Fontaine, Acta Met. 23, 553-571 (1975).
34. P. C. Clapp and S. C. Moss, Phys. Rev. 142, 418 (1966).
35. P. C. Clapp and S. C. Moss, Phys. Rev. 171, 764 (1968).
36. P. C. Clapp and S. C. Moss, Phys. Rev. 171, 754 (1968).
37. A. G. Khachaturyan, Prog. Mat. Sci. 22, 1-150 (1978).

# THE DISORDERING PROCESS IN AN $L1_2$ ORDERED ALLOY*

J.A. HORTON, A. DASGUPTA, AND C.T. LIU
Metals and Ceramics Division, Oak Ridge National Laboratory,
P. O. Box X, Oak Ridge, TN 37831

## ABSTRACT

The antiphase boundary (APB) structure and the disordering process have been studied by transmission electron microscopy (TEM) in a long-range-ordered alloy with $L1_2$ structure and composition $(Ni_{70}Fe_{30})_3(V_{58}Al_{40}Ti_2)$. With an increase in temperature from a single-phase field into a two-phase field, disordering occurs both homogeneously within ordered domains and heterogeneously along APBs. Disordering continues by a ripening process with the disordered islands shrinking while the disordered bands on the prior APBs grow. The APBs are partially aligned on {100} while the material is single phase. As disordering proceeds, the degree of alignment of the disordered bands on {100} increases until complete alignment results. A correlation of the APB structure with room temperature mechanical properties indicates no dramatic change at the transition to the two-phase structure.

## INTRODUCTION

The study of the disordering reactions in the long-range-ordered alloy, $(Ni_{70}Fe_{30})_3(V_{58}Al_{40}Ti_2)$, designated as LRO-55, is part of a larger program investigating ordered alloys originally based on $(Fe, Co, Ni)_3V$ compositions [1–3]. In the present alloy, aluminum was substituted for some of the vanadium in order to enhance atomic ordering and to stabilize the $L1_2$ over the $DO_{22}$ structure. Titanium was added to improve the high-temperature ductility. The LRO-55 undergoes a phase transformation from a single-phase (ordered $L1_2$) to a two-phase (ordered $L1_2$ plus disordered fcc) field as the temperature is increased above the critical ordering temperature, $T_c$ ($\cong$750°C). The purpose of this study was to investigate the disordering process in the alloy, with an emphasis on the role of APBs on the order-disorder transformation.

## EXPERIMENTAL PROCEDURE

The alloy was prepared by arc melting and casting followed by hot rolling at 1100°C and then cold rolling to a thickness of 1 mm (for further details see Refs. 1 and 2). The specimens were vacuum encapsulated in quartz, homogenized for 2 h at 1050°C, and then water quenched with the capsule left intact. A subsequent stepwise heat treatment was employed to yield a highly ordered structure. The encapsulated specimens were annealed 5 h at 800°C, furnace cooled to 700°C, held 24 h at 700°C, then air cooled to room temperature. The material at this point was the normal starting material for subsequent aging studies. Material with this heat treatment was also used in mechanical property studies [2]. Aging experiments were performed for times of 5 min to 10 d at temperatures ranging from 650 to 965°C and were always followed by a water quench. Instead of the stepwise heat treatment described above, an alternate initial heat treatment was employed consisting of a

*Research sponsored by the Office of Energy Systems Research, Division of Energy Conservation and Utilization Technologies (ECUT), U.S. Department of Energy under Contract DE-AC05-84OR21400 with the Martin Marietta Energy Systems, Inc.

30 min anneal in air at 1050°C followed by a direct water quench. The microstructures resulting from this heat treatment and subsequent aging did not differ from those of the stepwise heat treated specimens with similar final anneals.

Specimens for TEM were prepared by electrodischarge machining 3-mm-diam disks followed by mechanically grinding to a thickness of 0.3 mm and electropolishing in a Struers Tenupol with an electrolyte of 1 part sulfuric acid and 7 parts methanol at a temperature of −15°C. Specimens were examined with a JEOL 100 CX and with a Philips EM400T equipped with a field emission gun and an EDAX energy dispersive x-ray spectrometer.

## RESULTS AND DISCUSSION

The microstructure of LRO-55 after the initial stepwise heat treatment described above consisted of a fully ordered ($L1_2$) structure with partially aligned APBs with an average domain diameter of about 100 nm [Fig. 1(a)]. Aging for 24 h at 650°C [Fig. 1(b)] resulted only in a slight increase in domain size as compared to the initial microstructure. A more substantial increase in the domain size was observed after aging at 750°C. Aging at 800°C resulted in a completely aligned APB structure with bands of disordered material along the APBs. This transition occurred between 750 and 800°C (referred to as the transition temperature in the balance of this paper) and corresponds to the entry into a two-phase field composed of ordered $L1_2$ and disordered fcc phases. This transition is illustrated in Fig. 1 which shows a series of micrographs of specimens aged from 650 to 925°C after the same initial heat treatment. For aging temperatures above the transition temperature, the disordered bands continued to form and grow,

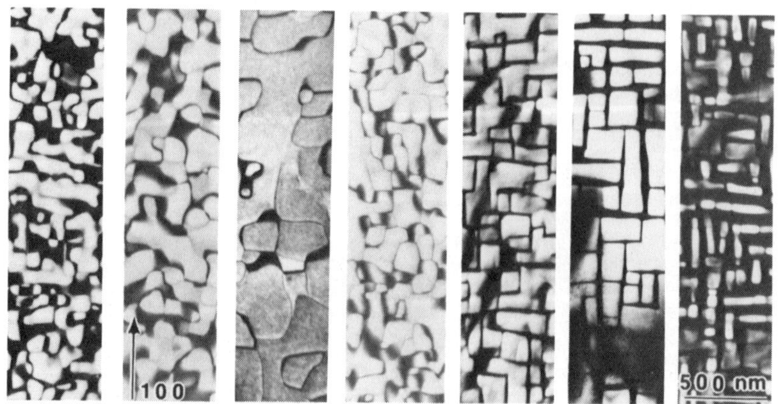


Figure 1. Superlattice dark field TEM micrographs of a series of LRO-55 specimens aged at the times and temperatures indicated above each micrograph. Figure 1(a) shows the microstructure after the initial stepwise heat treatment. In these images APBs have a dark, fringe type contrast in (a-d), better illustrated on higher magnification micrographs; the apparent width results from their inclination to the beam direction, whereas in (e-g) the ordered areas appear bright and disordered regions are dark. The electron beam direction, Z, of the micrographs was within 5 deg of an [001] pole to enable the observation of alignments along {100}.

with the amount of the disordered regions increasing with aging temperature and with some domain coalescence. The highest aging temperature employed (925°C for 1 h), as shown in Fig. 1(g), resulted in a cuboidal $\gamma'$ ($L1_2$ phase) precipitate structure in a disordered matrix.

Room-temperature mechanical properties were measured for each of the specimens whose microstructures are illustrated in Fig. 1. Room temperature tests were used to study the effect of these microstructures on mechanical properties without the complication of the variation of mechanical properties with test temperature. The yield stress of this alloy as well as many other ordered alloys, shows an increase with increasing test temperature and peaks around their critical ordering temperature [2,3]. In this study, the yield stress was measured and is plotted as a function of annealing temperature [Fig. 2(a)]. The yield stress was independent of the annealing temperature to 750°C, indicating that it is insensitive to either the domain size or APB morphology. Also, the yield stress increased linearly with annealing temperature above 750°C, apparently due to the disordering process and the formation of two phases, as indicated in Fig. 1(e)–(f). The result is further demonstrated by comparison of microstructure and yield stress of the specimens annealed at 800°C for 5 min and 24 h, respectively. The increase in annealing time causes a change in microstructure from partially aligned APB structure to the completely aligned disordered band structure; however, both

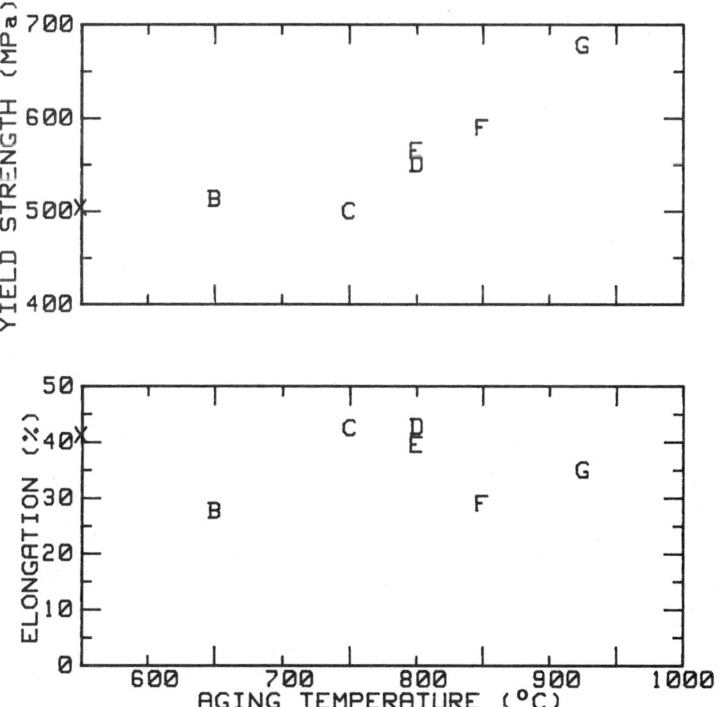


Figure 2. Graphs of room temperature yield stress and elongation versus aging temperature for the seven conditions illustrated in Fig. 1. The x's indicate the properties of the specimen with the initial heat treatment [Fig. 1(a)].

specimens showed virtually the same yield stress [Fig. 2(a)]. The alloy exhibited little change in ductility upon annealing, with perhaps a slight trend of decreasing ductility with increasing annealing temperature [Fig. 2(b)].

## Aging Below the Transition Temperature

Aging of LRO-55 for times up to 10 d at 750°C resulted in domain growth with no apparent increase in the degree of APB alignment as shown in Fig. 3. However, after the 10-d anneal [Fig. 3(c)] a substantial amount of curved APB exists. At the present time, a satisfactory method is not available for estimating the degree of alignment, especially when the domain size is small. Currently not enough data are available to calculate the activation energy for the growth of APBs.

The initial stepwise heat treatment contained an annealing step of 5 h at 800°C, which is above the transition temperature. In order to determine the effect of this stepwise heat treatment on the microstructure, the following comparison experiment was performed. Specimens were annealed for 30 min at 1050°C and directly water quenched to produce a disordered (as ascertained by TEM) structure. This material was then vacuum encapsulated, annealed for 10 d at 750°C, and then water quenched. The resulting microstructure (Fig. 4) was similar to that of the specimen which had the normal stepwise heat treatment plus a 10-d anneal at 750°C [Fig. 3(c)].

Determination of the presence of any disordering along APBs is important especially in a material which does eventually disorder on the APBs. Residual images of the APBs were produced by using $g \cdot R=0$ conditions in dark field. In this type of analysis, a series of micrographs was taken with appropriate <110> and <100> diffracting vectors. Residual images were obtained with at least two different diffracting vectors for each type of APB. These analyses suggest that no disordering was present along the APBs in specimens aged at 750°C or below and indicated that the APBs had fault vectors along <110>. Results are summarized in the micrographs shown in Fig. 5 for a specimen aged 10 d at 750°C. The APBs with a residual image marked A have a fault vector R = a/2[110], B have R = a/2[011], and C have R = a/2[101].

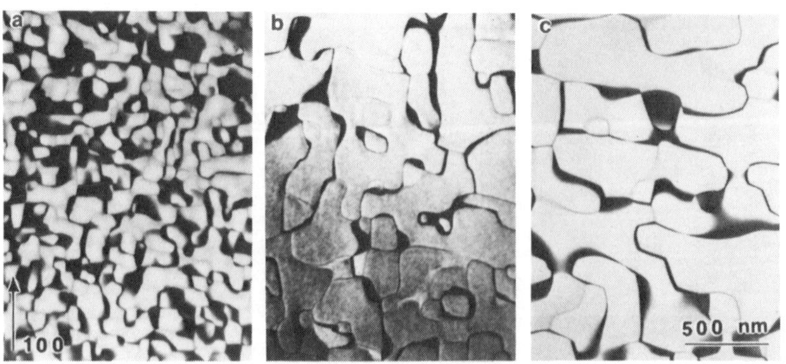


Figure 3. Specimens aged at 750°C for 1 h (a), 1 d (b), and 10 d (c) after the same initial stepwise heat treatment. Z near [001].

Figure 4. TEM Micrograph of a specimen aged 10 d at 750°C after an initial direct quench from 1050°C. This heat treatment resulted in an ordered structure similar to that of Fig. 3(c). Z near [001].

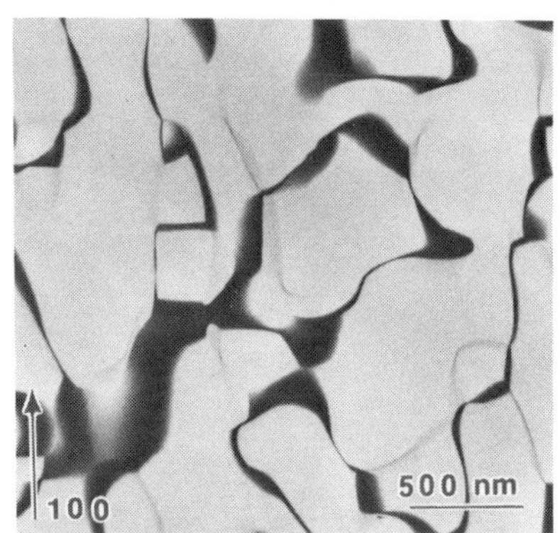


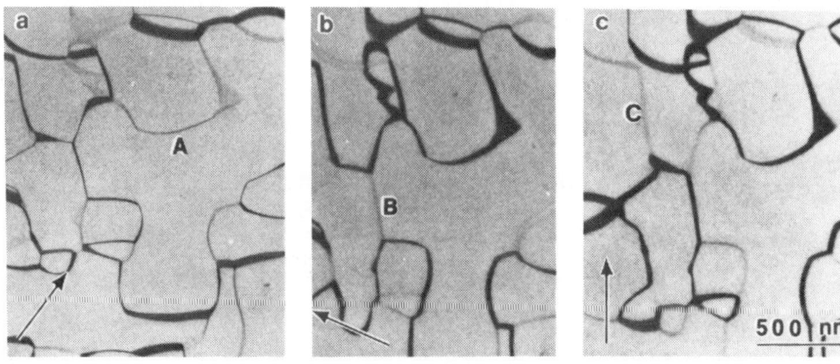


Figure 5. Series of TEM micrographs showing part of an APB fault vector analyses. The presence of residual images suggests that no disordering was present along the APB. The APBs with a residual image marked A have a fault vector R = a/2[110], B have R = a/2[011], and C have R = a/2[101]. In micrograph (a) Z = [001] and the diffraction vector (marked by the arrows) g = $[1\bar{1}0]$, in (b) Z = [111] and g = $[01\bar{1}]$, and in (c) Z = [111] and g = $[10\bar{1}]$.

## Aging Above the Transition Temperature

A short anneal above the transition temperature initiates the disordering process. As shown in Fig. 6, a 5 min anneal at 850°C results in both homogeneous disordering within domains and heterogeneous disordering along the APBs. Stereo imaging of these disordered islands indicated that they were evenly distributed through the foil thickness. A zone denuded of disordered islands is present adjacent to the disordered bands. The disordered bands exhibited a higher degree of {100} alignment than did the APBs in any specimen annealed below the transition temperature. However, some curvature of the disordered bands is still present. Longer anneals at 850°C resulted in a complete alignment of the disordered bands along {100} as shown in Fig. 7. After 20 min at 850°C [Fig. 7(b)], many of the disordered islands have disappeared while the disordered bands have thickened. Little coalescence of the domains appears to have occurred. After 1 h at 850°C [Fig. 7(c)], nearly all of the disordered islands have dissolved while the bands have continued to thicken in an apparent ripening process. Between 1 h and

Figure 6. Aging 5 min at 850°C resulted in initial disordering both homogeneously and along partially aligned APBs. Z near [001].

Figure 7. Specimens of LRO-55 aged at 850°C after the initial stepwise ordering heat treatment for (a) 5 min, (b) 20 min, (c) 1 h, (d) 1 d, and (e) 10 d showing the disordering process. Z near [001].

10 d [Fig. 7(c–e)] at 850°C the thickening of the disordered bands continues although now coalescence of the domains occurs. For coalescence to occur, some of the disordered bands must shrink, possibly resulting in an APB since the disordered bands still separate areas which were out of phase. A $g \cdot R = 0$ analysis, as discussed previously, of the narrow bands shown in Fig. 7(d) and (e) was performed, but no residual images could be produced. This suggests that after the disordered band shrinks to an APB, the APB probably migrates so quickly that too few remain to be readily found. The shrinking of some bands while other bands grow is also an apparent ripening process. The narrow, presumably shrinking bands intersect the growing bands

at a wider point along the length of the larger disordered bands. This process appears to be diffusional in nature possibly along the interfaces. As noted before, the growth of the disordered bands indicates the specimen is in a two-phase region on the phase diagram with associated composition changes.

In order to measure compositional differences between the ordered and disordered regions, specimens were annealed for 30 min at 965°C (after the alternate initial heat treatment) to produce ordered regions more widely separated than those shown so far (Fig. 8). Since this specimen was initially in the disordered state, annealing has allowed the ordered regions to precipitate. The resulting precipitate morphology is similar to that of the specimen shown in Fig. 1(g). It is probable that subsequent annealing of the specimens with the standard heat treatment would result in a similar precipitate morphology. Energy dispersive spectroscopy (EDS) analysis of the specimen shown in Fig. 8 showed that the ordered regions were enriched by 5 at. % in both Al and Ni and depleted by 5 at. % in both V and Fe as compared to the adjacent disordered matrix. The composition of these ordered precipitates is approaching $Ni_3Al$.

The transition between the aligned disordered bands and partially aligned APBs was found to be completely reversible, indicating that homogenization occurred. The specimen from Fig. 1(f) with an aligned disordered band structure was re-annealed at a temperature of 750°C which is below the transition temperature. This resulted in a partially aligned APB structure similar to that shown in Fig. 3(c).

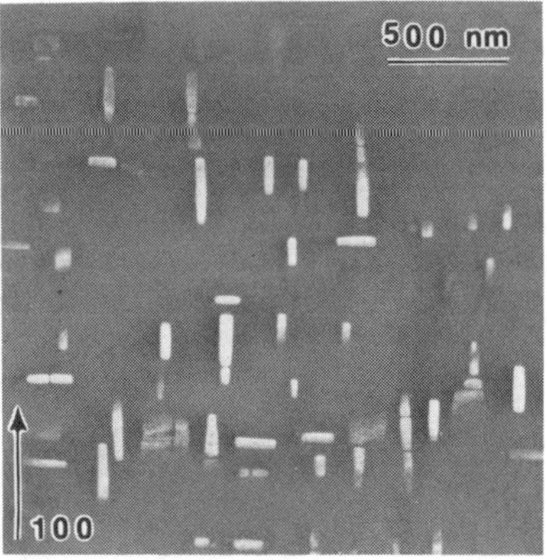


Figure 8. Specimen of LRO-55 aged 30 min at 965°C after an initial direct water quench shows distinct ordered precipitates in a disordered matrix. Z near [001].

## Correlation of Microstructure with Differential Scanning Calorimetry (DSC)

In addition to the transition described thus far, precipitates with a B2 structure were present after the initial stepwise heat treatment. These precipitates had compositions near that of NiAl, averaged about 1 μm in diameter constituted a low volume fraction, and did not dissolve until after the transition to a disordered band structure [4]. DSC curves showed three endothermic peaks for this alloy [1]. DasGupta et al. [1] have shown that for this alloy system both the onset of disordering and the culmination of disordering into a completely disordered structure result in two endothermic peaks. Apparently in LRO-55 an exothermic peak corresponding to the dissolution of

B2 precipitates is added to the first endothermic peak, resulting in what appears to be three separate endothermic peaks.

## SUMMARY

The APBs in LRO-55 have a tendency to align on {100} planes suggesting that the APB energy is anisotropic. After the transition the disordered bands are completely aligned, suggesting that the interfacial energy between the ordered and disordered phases is even more anisotropic than the APB energy. In this system, it does not appear that an increase in aging temperature or time at temperature results in a greater degree of APB alignment, although this is not certain since it is experimentally difficult to determine the degree of alignment when the domain size is small. It also appears that the aging treatment above the transition temperature does not affect the degree of alignment after an anneal below the transition temperature [Fig. 4 versus 3(c)]. From the results presented here, it is still not clear whether a substantial realignment of the APBs occurs at the onset of disordering or whether the disordered bands start and propagate along {100}. In $Cu_2ZnNi$ the APB anisotropy has been found to be a function of temperature [5]. The APB energy becomes anisotropic only near $T_c$. Also similar to LRO-55 is the addition of 3% Ni to $Cu_3Au$ which changes aligned APBs to curved APBs at equilibrium at room temperature [6]. $Cu_3Au$ with {100} APBs has been shown to be near a boundary of stability between $L1_2$ and $DO_{22}$ [7] because of second nearest neighbor pair interaction considerations. An alloy with a lower aluminum content than LRO-55 has the $DO_{22}$ structure [1]. Both $Cu_3Au$ and LRO-55 have similar APB alignment and are near compositions that have the $DO_{22}$ structure. The composition change during the disordering in LRO-55 further complicates the situation. The nucleation of the disordered regions both heterogeneously along the APBs and homogeneously within the disordered islands while the bands grow is an interesting aspect of the study presented here.

## ACKNOWLEDGMENTS

The authors wish to thank H. Inouye and P. Angelini for reviewing this manuscript, M. Winsbro for doing the heat treatments, and K. Russell and W. Smith for preparing the TEM disks and F. A. Scarboro for typing the manuscript.

## REFERENCES

1. A. DasGupta, J. A. Horton, and C. T. Liu, paper in present proceedings.

2. C. T. Liu, Int. Met. Reviews **29**(3), 168 (1984).

3. C. T. Liu, paper in present proceedings.

4. J. A. Horton, A. Das Gupta, and C. T. Liu in: Proceedings of the 42nd Annual Meeting of the Electron Microscopy Society of America, ed., G. W. Bailey (San Francisco Press, San Francisco 1984) p. 486.

5. G.J.L. Van der Wegen, P. M. Bronsveld, and J.Th.M. De Hosson, Met. Trans. A **12A**,2125 (December 1981).

6. M. Yodogawa, D. Wee, Y. Oya, and T. Suzuki, Scripta Met. **14**, 849–854 (1980).

7. M. Morinaga, Acta Met. **25**, 957–962 (1977).

# THE DISORDERING PHASE TRANSFORMATION IN $(Ni_{70}Fe_{30})_3(V_{98-x}Al_xTi_2)$ ALLOYS WITH $0 \leq x \leq 80$*

A. DASGUPTA and J. A. HORTON

Metals and Ceramics Division, Oak Ridge National Laboratory, P.O. Box X, Oak Ridge, TN 37831

ABSTRACT

The sequence of disordering transformation processes in the $A_3B$ type alloy series $(Ni_{70}Fe_{30})_3(V_{98-x}Al_xTi_2)$, currently under development for high-temperature structural applications, was studied by differential scanning calorimetry (DSC), x-ray diffraction, optical microscopy, and transmission electron microscopy (TEM). Results of DSC show that in all alloys there are two endothermic stages of phase transformation from the ordered to the disordered state. With increasing x, the disordering transition temperature, $T_c$, reaches a maximum ~1000°C at $x \approx 50$, and then decreases. Interrupted heating, followed by water quenching, was used to characterize the crystal structure and the microstructure of the intermediate phases. For the x = 20 alloy, TEM observations showed ordered regions of $DO_{22}$ phase in a matrix of disordered fcc (Al) phase at intermediate temperatures. The ordered domains transformed morphologically into cuboid like regions at higher temperatures. From a combined study by all the techniques, we conclude that in alloys with x between 0 and 20, the sequence of phase transformations upon heating is: $DO_{22} \rightarrow DO_{22} + A1 \rightarrow A1$, whereas in alloys with $x > 40$, the major sequence is $L1_2 + B2 \rightarrow L1_2 + A1 \rightarrow A1$.

## I. INTRODUCTION

In recent years, considerable interest has developed at Oak Ridge National Laboratory and elsewhere in ductile long-range ordered (LRO) intermetallic compounds based on $Co_3V$. While $Co_3V$ itself is known to have a hexagonal hP24 structure and is brittle, macroalloying the compound with Fe and Ni, replacing Co, has been used to stabilize the $L1_2$ cubic ordered structure by judiciously controlling the electron concentration (e/a ratio) in the alloy. The LRO compounds, thus produced, have been found to exhibit considerable ductility. A recent review of the physical metallurgy and mechanical properties of such ductile ordered $(Fe,Co,Ni)_3V$ alloys is given in [1].

The alloy series $(Ni_{70}Fe_{30})_3(V_{98-x}Al_xTi_2)$, investigated in the present work, is related to the alloy system $(Fe,Co,Ni)_3V$. With the increase of the index x, Al atoms progressively replace the V atoms in the composition. A small fixed percentage of titanium is added in this alloy system to improve the high-temperature ductility [1]. The purpose of the present work is to determine the course of phase transformations in these alloys as they are heated all the way to the disordered state. Indeed, it is found that these alloys undergo a complicated sequence of phase transformations upon heating. The existence of such phase transformations was first reported in a preliminary way by DasGupta et al. [2].

---

*Research sponsored by the Office of Energy Systems Research, Division of Energy Conservation and Utilization Technologies (ECUT), U.S. Department of Energy Under Contract DE-AC05-84OR21400 with the Martin Marietta Energy Systems, Inc.

Subsequently, some of these observations were supplemented by electron microscopic investigations by Horton et al. [3]. The study of these phase transformations is of considerable interest not only because they relate to the mechanical properties of the alloys concerned but also to a number of their physical and chemical properties such as oxidation and corrosion resistances at high temperatures.

Differential scanning calorimetry (DSC), x-ray diffraction, optical microscopy and electron microscopy have been performed in a complementary manner in the present studies of phase transformations. The details of sample preparation as well as of the above measurements are given in Section II. The results and discussion are presented in Section III. Finally, a summary of the conclusions is given in Section IV.

## II. EXPERIMENTAL PROCEDURES

### Sample Preparation

Weighed amounts of the pure constituent metals Ni, Fe, V, Al and Ti were melted together and arc-cast. All together, seven different alloy compositions were prepared corresponding to x = 0, 10, 20, 40, 50, 60, and 80 in the alloy series represented by $(Ni_{70}Fe_{30})_3(V_{98-x}Al_xTi_2)$. The alloy samples were fabricated into sheets by hot rolling the ingots between molybdenum cover sheets at 1100°C and then cold rolling at room temperature. Rectangular coupons, approximately 1 mm thick and 3 cm × 2 cm in size, were prepared from this material. They were homogenized in vacuum at 1100°C for 30 min, and then given a sequential ordering heat treatment for 5 h at 800°C, 17 h at 700°C and 24 h at 600°C in vacuum before being slowly furnace-cooled down to room temperature. Small portions of these coupons were then subjected to DSC, x-ray diffraction, optical microscopy and electron microscopy for the phase-transformation investigations.

### Differential Scanning Calorimetry (DSC)

To record the heat effects associated with the phase transformations and to determine the order-disorder transition temperatures for various alloys, a Mettler differential scanning calorimeter, type 2000 C, was used. Typically, samples, about 0.2 g in weight, were heated at a constant rate of 5°C/min in argon atmosphere from room temperature to 1200°C, the upper temperature limit of the equipment. No attempt was made to calculate the enthalpies, $\Delta H$, quantitatively. Such calculations are complicated by the fact that often only a part of a two phase alloy undergoes the disordering transformation, and apportioning the heat effect is not possible. In this sense, the present DSC measurements were performed as in differential thermal analysis (DTA). However, for purposes of determining phase transformation temperatures, this procedure proved to be adequate.

### Optical Microscopy

Together with the usual bright field optical microscopy, studies were also made in polarized light. The latter procedure has the advantage that the optical activity of non-cubic phases, such as of the $DO_{22}$ structure, lets it be easily distinguished from the cubic phases, such as of the $L1_2$ structure which are optically inactive. Considerable difficulty was experienced due to lack of contrast in bright field optical microscopy in discerning the $DO_{22}$ from the $L1_2$ regions in the microstructure. This problem was overcome with the application of polarized light microscopy. To identify the phases, the results of optical microscopy were corroborated with the results of x-ray diffraction and electron microscopy.

## X-ray Diffraction

The x-ray diffraction studies were made using a Philips APD 3600 diffractometer with crystal monochromated Cu Kα radiation. Crystal structures and the corresponding lattice parameters were determined from the x-ray data, and then compared with the results of DSC measurements, optical microscopy and electron microscopy.

## Electron Microscopy

To study the phases formed at different temperatures, the DSC-scans were used as a guideline. Appropriate temperatures were selected for each composition corresponding to its DSC scan. Samples were heated to these temperatures and water quenched. The quenched samples thus obtained were examined by transmission electron microscopy (TEM). Besides TEM, energy dispersive analysis was carried out with a Philips EM 400 T/FEG unit to determine the composition of the ordered and other regions in the microstructure. A description of the TEM and EDS techniques is given in more detail in [4].

## III. RESULTS AND DISCUSSION

Figure 1 shows typical DSC scans for the alloys with x = 0, 10, 20, and 40 in the alloy series $(Ni_{70}Fe_{30})_3(V_{98-x}Al_xTi_2)$. The rate of heating was 5°C/min in these measurements. It was seen that for x = 0 there are two endothermic effects, whereas for x > 0 there are three endothermic effects in the DSC scans of the alloys. Plots for x > 40 are not shown in Fig. 1. They were found to be qualitatively very similar to the plot for x = 40, and displayed three endothermic effects in their respective thermograms.

Calling the temperature above which a sample is fully disordered, as evidenced from the thermograms, as $T_c$ (see Fig. 1), the plot of $T_c$ vs x is shown in Fig. 2. Interestingly, with increasing x the transition temperature $T_c$ increases first and then decreases. It reaches a maximum at a temperature ~1000°C at x ≈ 50. Subsequent TEM studies indicated that for the alloys with x ≥ 40, the phase transformations were more complicated than for lower x, and involved the presence of an additional NiAl phase with B2 structure not encountered in the latter. These are generally also the compositions, where $T_c(x)$ is past its maximum value (Fig. 2). In the present paper, we would restrict ourselves primarily to the study of the low-x alloys. A detailed TEM study of the x = 40 alloy can be found in [4].

X-ray diffraction was used to examine the crystal structure of the alloys both in the ordered state at room temperature and in the samples water quenched from various temperatures, which were suitably selected from the corresponding DSC-curves (Fig. 1). These results showed that at room temperature the structure of the alloys with x = 0, 10, and 20 was tetragonal $DO_{22}$, whereas for higher x it was cubic $L1_2$. For the $DO_{22}$ phase, the c/a ratio was found to vary between 2.02 and 2.04. The lattice parameters were found to be: c ≈ 0.726 nm and a ≈ 0.357 nm. For the $L1_2$ phase, the lattice parameter $a_o$ was found to be ~0.359 nm. From the water-quenched samples, it was found that in alloys with x = 0, 10, and 20 the sequence of structural transformations upon heating is : $DO_{22} \rightarrow DO_{22} + A1 \rightarrow A1$, whereas in alloys with $x \geqslant 40$ the major sequence of the transformations was found to be : $L1_2 + B2 \rightarrow L1_2 + A1 \rightarrow A1$. Indeed, the first of the above two stages in (2) is more complicated and is a subject more completely treated in an accompanying paper [4].

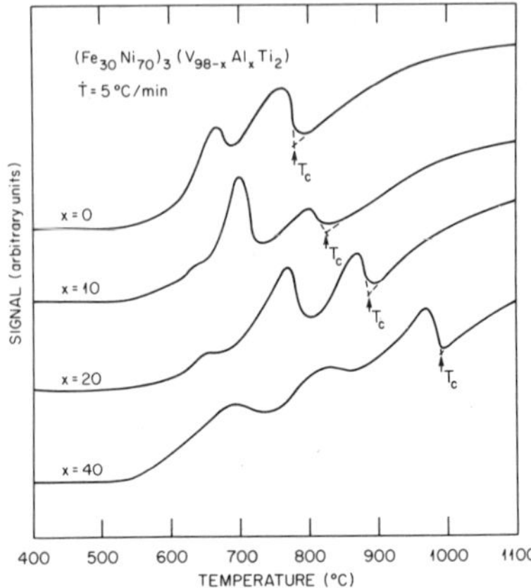


Fig. 1. DSC-traces of the ordered alloys in the system $(Ni_{70}Fe_{30})_3(V_{98-x}Al_xTi_2)$. The heating rate, $\dot{T}$ = 5°C/min. The positive signal direction is endothermic.

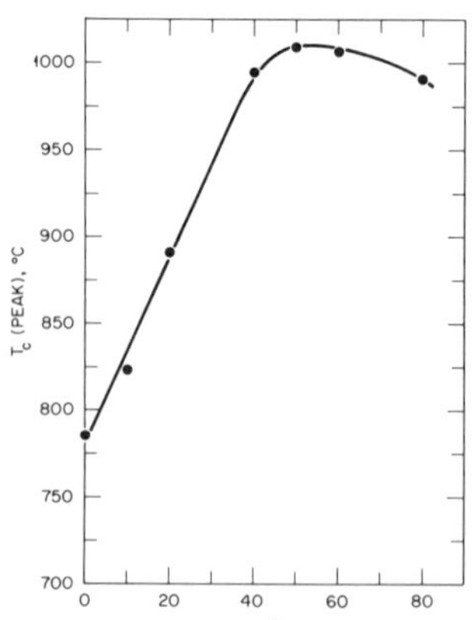


Fig. 2. The critical transition temperature, $T_c$, as a function of the composition x in the $(Ni_{70}Fe_{30})_3(V_{98-x}Al_xTi_2)$ alloys.

In alloys with $x \geqslant 10$, the first of the three endothermic heat effects in the DSC scan (Fig. 1) on the low temperature side was found not to correspond to any phase transformation as revealed by x-ray diffraction, optical- or electron-microscopy. It is thought that this heat effect represents ferromagnetic to paramagnetic transition in these alloys, which is consistent with the fact that at room temperature ferromagnetism was not found in the x = 0 alloy, but it existed and increased in intensity with increasing aluminum content in the other alloys. The peak temperature of this heat effect was found to increase systematically to higher values with increase in the aluminum content. The two remaining endothermic heat effects in the DSC-thermograms were found to represent two distinct stages in the phase transformation process. These stages along with their temperature ranges are summarized in Table I.

TABLE I. Stage of Phase Transformation in Various $(Fe_{30}Ni_{70})_3(V_{98-x}Al_xTi_2)$ Alloys

| x | Transformation | Temperature stage 1 | Range stage 2 |
|---|---|---|---|
| 0 | $DO_{22} \rightarrow DO_{22} + A1$ | 615–692 | — |
| | $DO_{22} + A1 \rightarrow A1$ | — | 692–795 |
| 10 | $DO_{22} \rightarrow DO_{22} + A1$ | 660–725 | — |
| | $DO_{22} + A1 \rightarrow A1$ | — | 725–823 |
| 20 | $DO_{22} \rightarrow DO_{22} + A1$ | 684–800 | — |
| | $DO_{22} + A1 \rightarrow A1$ | — | 800–890 |
| 40 | $L1_2 + B2 \rightarrow L1_2 + A1$ | 749–870 | — |
| | $L1_2 + A1 \rightarrow A1$ | — | 870–994 |

The results of optical microscopy were found to be in good agreement with the result of DSC and x-ray diffraction. In general, the use of polarized light was found necessary to distinguish the $DO_{22}$ from the cubic $L1_2$ and A1 phases. As an example, a sequence of bright field and dark field (polarized light) micrograph pairs is shown in Fig. 3 for the x = 20 alloy, quenched from different temperatures. The observed microstructures clearly agree with the sequence of phase transformations found from DSC and x-ray diffraction for this alloy.

The results of the TEM studies were also found to be in general agreement with the results obtained with other techniques used in this study. As an illustration, Fig. 4 shows a sequence of TEM micrographs for the x = 20 alloy. The homogeneous $DO_{22}$ phase witnessed in Fig. 4a at room temperature transforms to ordereded tetragonal domains of Fig. 4b at 770°C with a two-phase ($DO_{22}$ + A1) structure, and then to the structure of Fig. 4c at 860°C, where the ordered $DO_{22}$ phase is still crystollographically oriented but much diminished in volume fraction. This sequence attests to the evolutionary scheme of transformations $DO_{22} \rightarrow DO_{22} + A1 \rightarrow A1$ previously stated for this alloy in this paper.

Finally, the sequence of phase transformations in the LRO-alloys must have one-to-one correspondence with their mechanical as well as chemical properties. As an illustration, for the x = 20 alloy the DSC - trace is shown superimposed in Fig. 5 on the plots of yield strength and % elongation as functions of temperature. Even though the mechanical-properties data are scanty, there seems to be a good correlation between these data and the DSC trace. Not only at $T_c$ the yield strength sharply drops and the elongation sharply increases with temperature, which is to be expected as the last vestiges of the $DO_{22}$ phase disappear at this temperature, but also the whole process of decrease in the strength and increase in ductility occurs in the same overall temperature range of stages 1 and 2 (~700–900°C) as evidenced by the DSC thermogram. Even the sharp decreases in the yield strength are found to occur roughly between 700–800°C and from 800–900°C, which agree with the temperature ranges of stages 1 and 2 for this alloy respectively.

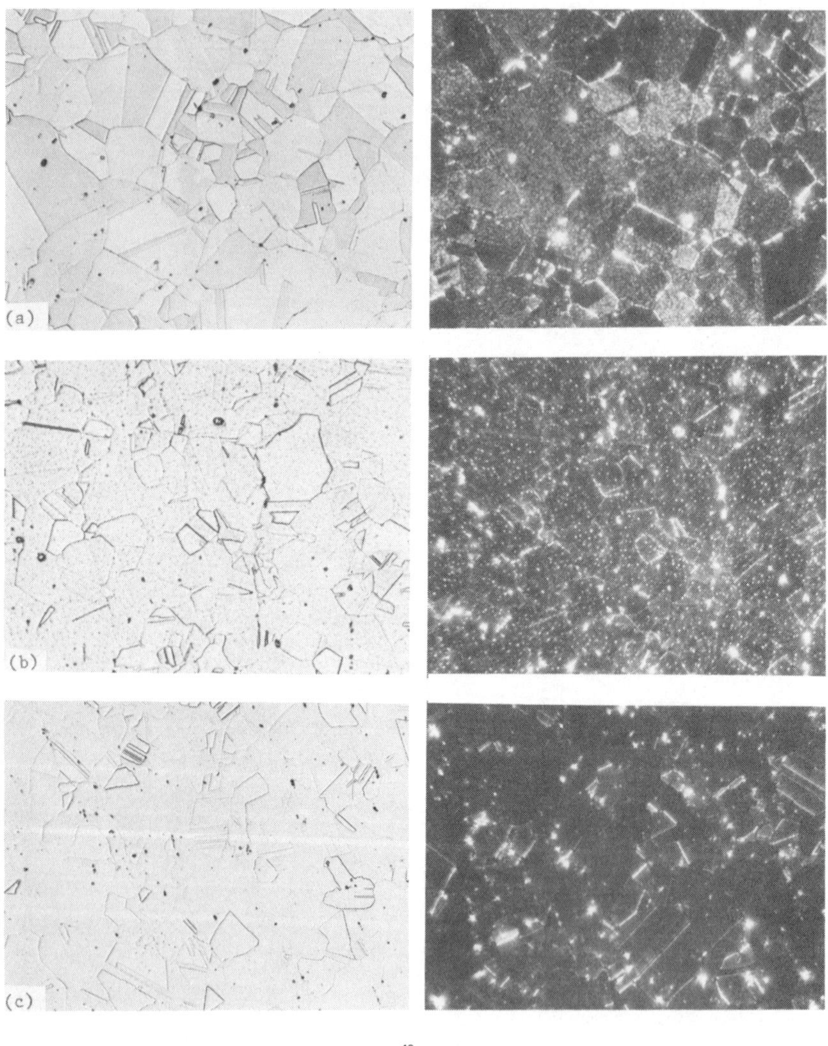


Fig. 3. (a) Optical micrographs of x = 20 alloy, quenched from 720°C, in normal and polarized light. It is a two phase microstructure mostly consisting of the optically active $DO_{22}$ phase. Much of the dark part in the polarized light micrograph above was seen to be optically active at other orientations. Also active in the above micrograph are inclusions of impurities and grain boundaries. (b) Optical micrographs of x = 20 alloy quenched from 850°C, in normal and polarized light. The microstructure consists of two phases $DO_{22}$ and Al (dominant part). (c) Optical micrographs of x = 20 alloy, quenched from 1050°C, in normal and polarized light. The microstructure consists of homogeneous Al phase. The optically active areas consist of extraneous inclusions and grain boundaries.

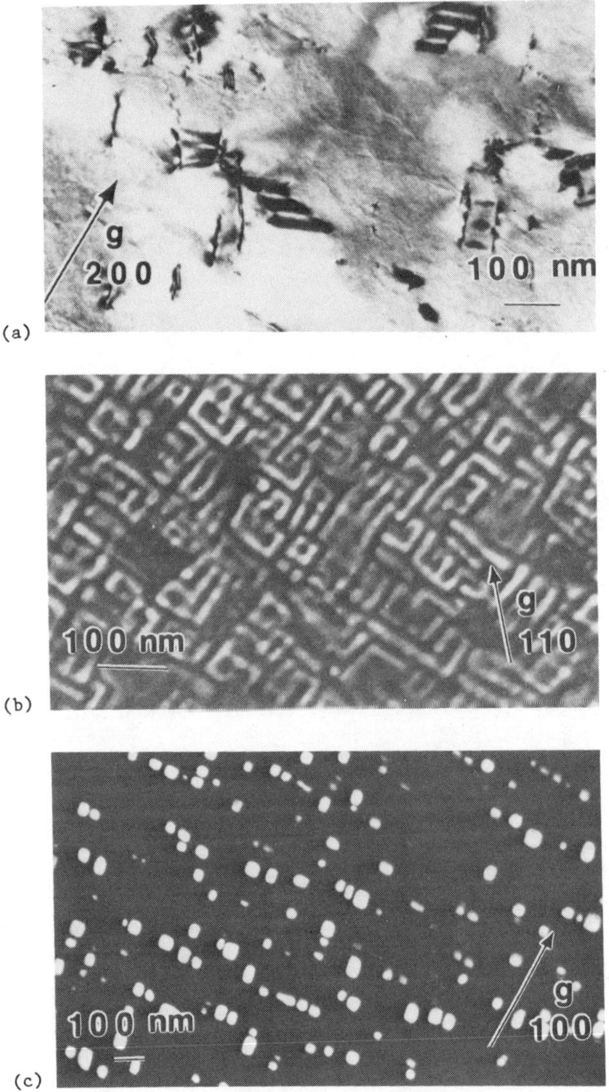


Fig. 4. (a) Bright field TEM micrograph of the x = 20 alloy after the ordering heat treatment at room temperature. The microstructure consists of a single $DO_{22}$ phase. (b) Dark field TEM micrograph of the x = 20 alloy after quenching from 770°C. The microstructure consists of a two-phase mixture. The bright phase is ordered $DO_{22}$ and the dark matrix is disordered fcc. (c) Dark field TEM micrograph of the x = 20 alloy after quenching from 860°C. The ordered tetragonal regions of $DO_{22}$ can be seen in a dark disordered fcc matrix.

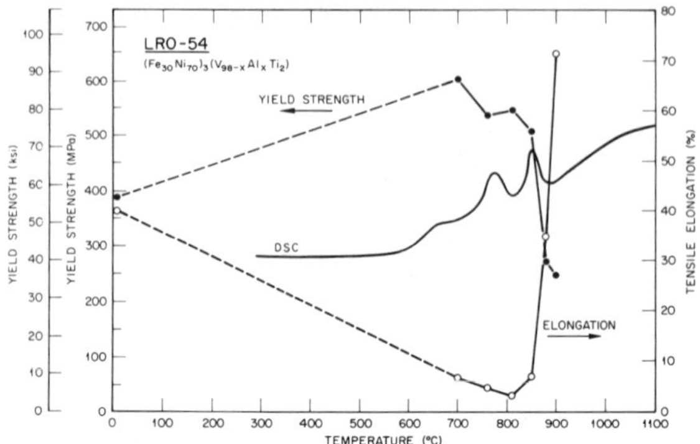


Fig. 5. A superimposition of the DSC scan on the plots of elongation and yield strength data as functions of temperature for the x = 20 alloy. Good correlation can be seen between the DSC and yield strength plots.

## IV. CONCLUSIONS

In the alloy system represented by $(Ni_{70}Fe_{30})_3(V_{98-x}Al_xTi_2)$, the sequence of phase transformations upon heating was established from a combined study of DSC, x-ray, optical and electron microscopy to be: for $x \leq 20$: $DO_{22} \rightarrow DO_{22} + A1 \rightarrow A1$, and for $x \geq 40$: $L1_2 + B2 \rightarrow L1_2 + A1 \rightarrow A1$. In the latter sequence, only the major stages are found here. Clearly, the first of these stages would, in reality, involve multiple steps, which are treated in detail in an accompanying paper [4].

## ACKNOWLEDGMENTS

The authors would like to thank C. T. Liu for preparation and fabrication of the alloy samples used in this work and to provide tensile properties data for the x = 20 alloy. Assistance by O. B. Cavin in gathering x-ray diffraction data is gratefully acknowledged. The authors are also thankful to D. M. Kroeger, H. Inouye, and C. T. Liu for carefully reading the manuscript. We also thank Gwendolyn Sims for preparation of the final manuscript.

## REFERENCES

1. C. T. Liu, Intern. Met. Rev. 29, 168 (1984).
2. A. DasGupta and C. T. Liu, J. Metals 34, No. 12, 87 (1982).
3. J. A. Horton, A. DasGupta, and C. T. Liu, Proc. 42nd Annual Mtg., Electron Microscopic Soc. of America, 486 (1984).
4. J. A. Horton, A. DasGupta, and C. T. Liu, paper in the present proceedings.

# MICROSTRUCTURES IN RAPIDLY SOLIDIFIED $L1_2$-TYPE Ni-Al-Cr ALLOYS

S. C. HUANG, K. M. CHANG, E. L. HALL AND R. P. LAFORCE
General Electric Corporate Research and Development, PO Box 8, Schenectady, NY 12301

## ABSTRACT

A microstructural study of the melt spun alloys $Ni_{75}Al_{25}$, $Ni_{72.5}Al_{22.5}Cr_5$ and $Ni_{67.5}Al_{20}Cr_{12.5}$ was carried out. Variations in grain morphology, phase separation and ordering structure were observed. The results are discussed in terms of alloy stoichiometry, Cr effect on phase reaction, microsegregation of Al and the cooling rate.

## INTRODUCTION

The polycrystalline compound $Ni_3Al$ has regained certain research interest since Aoki and Isumi first made it ductile by boron microalloying [1-4]. The ductilization effect is attributable to boron fitting into the lattice cavities at the grain boundaries and increasing the cohesive strength between grains [5,6]. Subsequently, boron was also found to strengthen $Ni_3Al$ by interstitial solid solution hardening [6,7]. Recently, some ternary $Ni_3Al$-base alloys, though containing no boron, were found to be ductile and strong after rapid solidification [8]. The latter finding was attributed to the low degree of ordering and suppression of phase separation in the rapidly solidified material. This paper reports on the microstructures observed in melt-spun $Ni_3Al$ alloys containing 0, 5 and 12.5 at.% Cr. The results on the solidification morphology, phases, microsegregation and the ordering structure are discussed. The present microstructural characterization should form the basis for understanding the ribbon mechanical behavior which is now under investigation.

## EXPERIMENTAL

Alloys of $Ni_{75}Al_{25}$, $Ni_{72.5}Al_{22.5}Cr_5$ and $Ni_{67.5}Al_{20}Cr_{12.5}$ were rapidly solidified by melt-spinning in vacuum. The ribbons were approximately 35 µm thick. The solidification front velocity in the ribbon was estimated to be greater than 10 cm/s [9], which is equivalent to a cooling rate greater than $5x10^5$ K/s. Sections of the ribbons were studied by optical metallography. Specimens for analytical electron microscopy (AEM) investigation were prepared by electropolishing using a solution of 20% perchloric acid in methanol at -55°C. In all cases, unless specified, the electropolishing was carried out from both surfaces of the ribbon so that the in-plane microstructure in the ribbon midsection was studied. Microscopy was performed using a Hitachi H-600-1 AEM operating at 100 kV. This instrument was equipped with an energy dispersive X-ray (EDX) spectrometer for microchemical analysis.

## RESULTS

### $Ni_{75}Al_{25}$

This compound was difficult to melt-spin into ribbon. To improve its ribbon formability, ~0.1 wt.% boron was added. The small boron addition did not alter the solidification or ordering reactions of the compound.

The longitudinal-section microstructure of the melt-spun boron-doped $Ni_{75}Al_{25}$ ribbon is illustrated in Figure 1. The ribbon shows an equiaxed structure consisting of uniform grains ~5 μm in diameter. No second-phase particles can be seen in the grain interiors or at grain boundaries. A TEM study confirmed the single-phase structure and showed the matrix to be the ordered, $L1_2$-type, γ' phase, Figure 2. The grain interiors contained only dislocations.

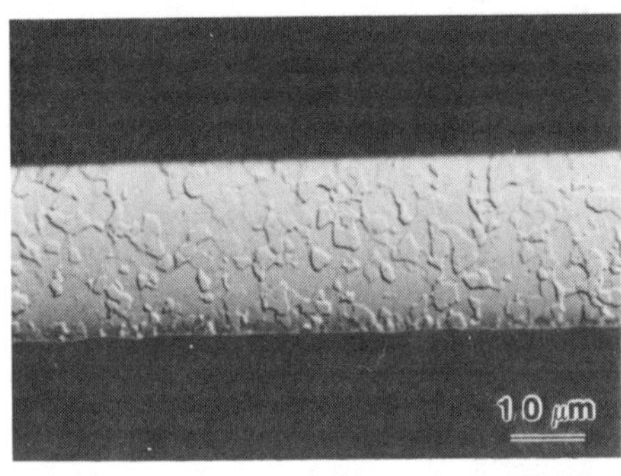


Figure 1. A longitudinal-section light micrograph of melt-spun $Ni_{75}Al_{25}$ containing 0.5 at.% boron.

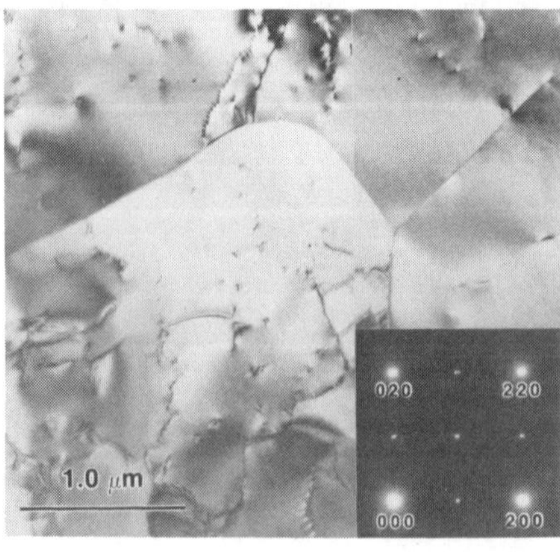


Figure 2. A transmission electron micrograph of the melt-spun $Ni_{75}Al_{25}$ containing 0.5 at.% boron. Inset: Diffraction pattern from γ' grain, [001] orientation, showing γ' superlattice spots.

### $Ni_{72.5}Al_{22.5}Cr_5$

A longitudinal section of this ribbon shows the microstructure of columnar grains extending from the contact to the noncontact surface of the ribbon, Figure 3. The grains have diameters of 2-3 μm.

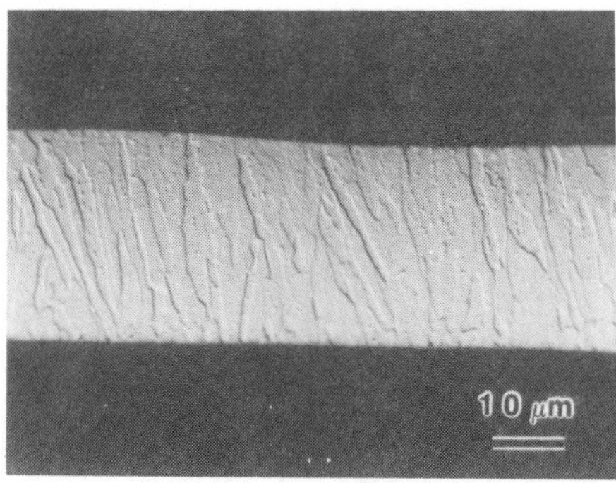


Figure 3. A longitudinal-section light micrograph of the $Ni_{72.5}Al_{22.5}Cr_5$ ribbon. The ribbon surface in contact with the casting wheel is shown at bottom. The left-hand end of the ribbon was cast first.

The TEM micrograph in Figure 4a shows grain boundary particles which are barely noticeable in Figure 3. The particles are ~0.5 μm in size and elongated along the grain boundary. They were identified as the martensitic, NiAl, β phase. Figure 4a also shows antiphase domains of 0.1-1.0 μm in grain interiors. The domains are defined by the swirl-like, non-oriented antiphase boundaries, as further illustrated in Figure 4b. The above microstructure is similar to that reported for melt spun $Ni_{70}Al_{25}Cr_5$ [8], but with less variation in the antiphase domain size.

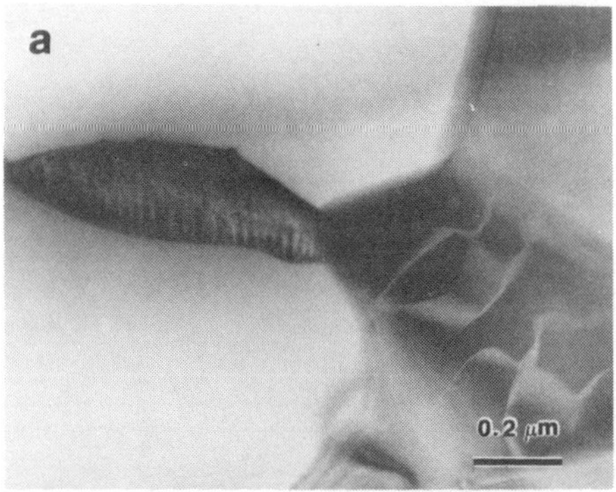


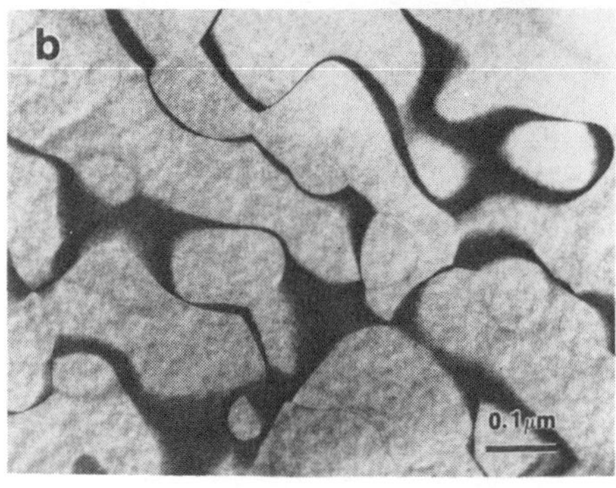


Figure 4. (a) Bright-field TEM image of melt-spun $Ni_{72.5}Al_{22.5}Cr_5$, and (b) Dark-field TEM image taken using γ' reflection, showing detail of the intragranular antiphase domains. Foil normal in (b) near [100].

## $Ni_{67.5}Al_{20}Cr_{12.5}$

The metallographic structure of this ribbon is similar to Figure 3, but the columnar grain diameter is reduced to 1-2 μm.

Cross sections of the columnar grains (in the ribbon plane) were studied by TEM on a specimen prepared by electropolishing from both surfaces of the ribbon. The grains observed in this ribbon midsection specimen generally have orientations close to [100] which is the preferred growth direction for cubic crystals. The dark-field micrograph in Figure 5a, which was formed by using a γ' diffraction spot, shows the intragranular γ' particles. The boundaries between γ' particles appear in some cases to be antiphase boundaries; in other cases, a thin layer of γ may be present. Three areas differing in the γ' size and morphology can be noted. In area A around the center of the grain, the precipitates are ~0.05 μm in diameter and are faceted on the cube planes. In the outer area of B, they have reduced diameters of ~0.02 μm. In area C, a single layer of γ' particles has formed at the grain periphery. These γ' particles are ~0.05 μm in size, and have interfaces perpendicular to the grain boundary. Finally, the grain is enclosed by a continuous grain boundary phase, which is shown in the dark-field micrograph of Figure 5b. This phase (area D) has been identified as the same martensitic β phase seen in the 5-Cr ribbon.

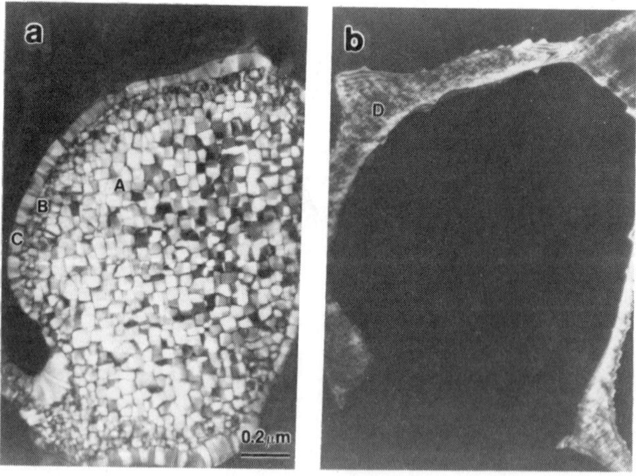


Figure 5. Dark field TEM microstructure in the midsection of the $Ni_{67.5}Al_{20}Cr_{12.5}$ ribbon using (a) a γ' diffraction spot and (b) a β diffraction spot. Foil normal near [100].

The chemistry of each region has been analyzed using energy dispersive X-ray spectrometry, Table I. Compared with the nominal composition, area A is lean in Al, area B is still leaner in Al but rich in Cr, and area C is lean in Cr. The β phase is deficient of Cr and enriched with Al.

TABLE I. MICROCHEMISTRY ANALYSES OF REGIONS IN FIGURE 5 (at.%)

| elements | nominal composition | overall analysis (N) | area A | area B | area C | area D (β) |
|---|---|---|---|---|---|---|
| Ni | 67.5 | 68.2 | 71.5 | 68.9 | 71.7 | 62.8 |
| Al | 20.0 | 19.2 | 16.6 | 14.2 | 19.5 | 30.0 |
| Cr | 12.5 | 12.6 | 11.9 | 16.9 | 8.8 | 7.2 |

A specimen showing the ribbon microstructure near the wheel contact surface, prepared by electropolishing only from the ribbon surface not in contact with the wheel, was also studied. In general, grains of a variety of orientations were seen in this specimen, contrary to the [100] grains seen in the midsection specimen (Figure 5). The grain boundary β phase tended to be in discontinuous segments. More interestingly, intragranular β particles were observed (Figure 6), and they were found to have a composition (Ni-28.8Al-6.86Cr) similar to the grain boundary β. These intragranular particles are believed to be located at the triple-points of the boundaries of cells formed inside of the grain. Also, the grain interior showed either a structure similar to Figure 5, consisting of cuboidal γ' precipitates, or a structure comprised of more irregular γ' domains separated by antiphase boundaries (Figures 6a and 6b).

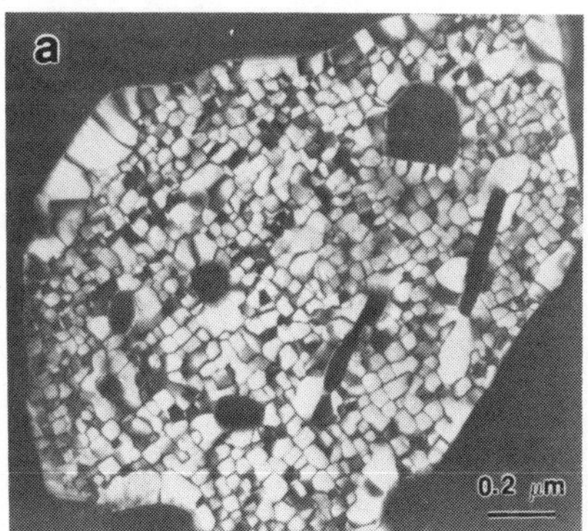


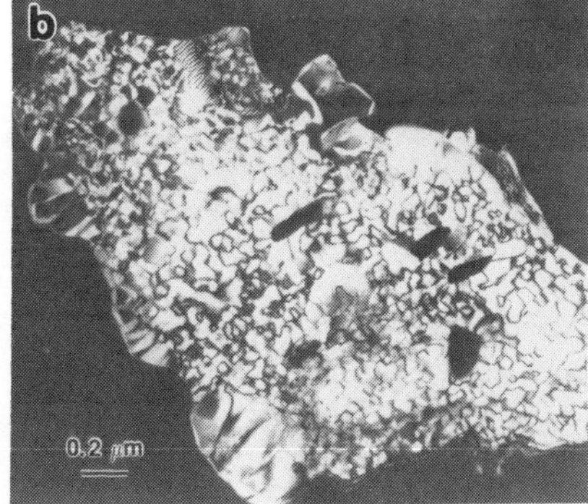


Figure 6. TEM micrographs from the contact surface of the $Ni_{67.5}Al_{20}Cr_{12.5}$ ribbon, showing a structure consisting of (a) cuboidal γ' precipitates and (b) irregular γ' domains separated by antiphase boundaries.

## DISCUSSION

Figures 1 and 2 show that a single-phase material of ordered γ' was produced by melt spinning the boron-doped $Ni_{75}Al_{25}$ compound. Apparently, the ribbon solidification occurred at such high rates that the segregation of Al and B was suppressed to such an extent that the formation of β and boride phases was avoided. Figure 1 also shows an equiaxed structure. Since no nucleation sites such as second-phase particles can be observed in grain interiors, the equiaxed grains are believed to develop not by individual nucleation but by recrystallization from the as-solidified structure. Recrystallization can occur readily in a single-phase material during second-stage cooling of the ribbon following solidification.

Koch et al. [3] studied the microstucture of rapidly solidified $Ni_{76}Al_{24}$ (with and without boron doping) prepared by the arc-hammer technique. The results showed nonuniform antiphase domains in grain interiors, contrary to the present observation. They associated the domain size variation with the segregation of Al, as measured by energy dispersive X-ray spectroscopy and convergent beam electron diffraction techniques. The discrepancies in microstructure and Al partitioning between our work and that of Koch et al. may be related to the difference in the cooling rate.

The coarse equiaxed structure in Figure 1 is in sharp contrast to the fine columnar structure observed in the Cr-containing ribbons, Figure 3. The columnar structure is believed to develop by the directional growth of grains from the contact to the noncontact surface of the ribbon. The grains are inclined toward the leading end of the ribbon, indicating that the solidification occurred under the influence of the melt counter convection within the melt puddle [9]. The martensitic NiAl phase seen in Figures 4a and 5b is formed at the grain boundaries due to Cr and Al segregation. These second-phase particles prohibit the recrystallization and thus substantially retain the initial solidification structure.

Cr thus promotes the β formation, Figures 2, 4a and 5b. This is because Cr alters the phase reaction when added to the Ni-Al alloys, as shown in the ternary Ni-Al-Cr solidification field diagram [10], Figure 7. In this partial concentration triangle, the light lines denote the three-phase invariant lines involving the liquid phase, and the dashed lines the solubility boundaries of the solidified phases. The figure shows no two-phase equilibrium existing between γ and β in the binary Ni-Al system. As Cr is added, a quasi-peritectic reaction L + γ' → γ + β takes place in a four-phase invariant plane. Further Cr additions lead to the eutectic reaction L → γ + β.

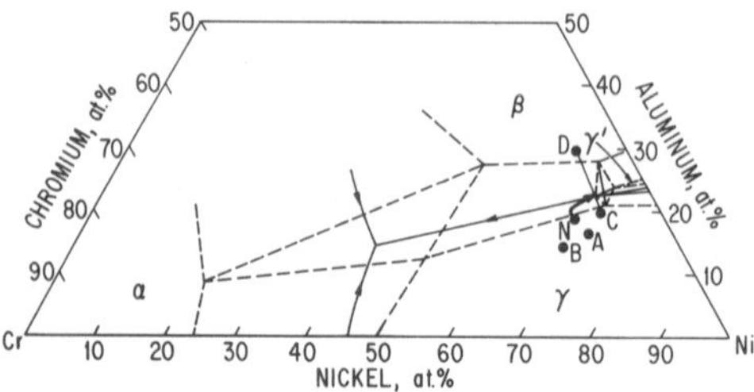


Figure 7. The segregation path (solid line) of the melt spun $Ni_{67.5}Al_{20}Cr_{12.5}$ alloy as shown in the Ni-Al-Cr solidification field diagram [10]. The microchemistry measurements in regions A to D in Figure 5 are from Table I.

The process for the phase separation and the relevant microsegregation in the 12.5-Cr ribbon are further characterized by the microchemistry measurements in regions along the grain radii, Table I. The microsegregation analysis suggests that the solidification occurs first by the formation of an Al-lean core region (area A in Figure 5a), which results in Al enrichment in the liquid phase. It then involves the formation of the Cr-rich region (area B), causing Cr depletion from the liquid. Here the Cr exhibits a reverse segregation (partition coefficient greater than 1), which is not predicted by the Ni-Cr or Ni-Al-Cr phase diagram [10]. Since the microstructural transition from area A to B in Figure 5a is quite abrupt, the Al and Cr segregation appears to be in two distinct steps. Based on these results, the probable segregation route of the liquid is shown in Figure 7 by the dark line, which starts at the nominal composition (N) and moves toward the high Al composition before turning to the low-Cr direction.

It can be seen in Figure 7 that the liquid phase approaches the eutectic trough near the four-phase invariant plane, due to the Al and the inverse Cr segregation during solidification of the grain interior region. The remnant liquid will thus produce $\gamma$ and $\beta$ phases (in the grain boundary region), through the eutectic reaction $L \rightarrow \gamma + \beta$. The measured microchemistries of the grain-periphery $\gamma'$ (area C in Figure 5a) and the grain boundary $\beta$ (area D in Figure 5b) indicate that they correspond to the eutectic products of the remnant liquid, with only slight deviations from the compositions suggested in the solidification field diagram, Figure 7. The "divorced" $\gamma$ and $\beta$ might also result from the four-phase reaction $L + \gamma' \rightarrow \gamma + \beta$, but the observed microstructure demonstrated no evidence of the quasi-peritectic reaction.

By comparing Figures 2, 4 and 5 again, the ordering structure is substantially controlled by the matrix chemistry which in turn is affected by the degree of solute segregation. The $Ni_{75}Al_{25}$ ribbon shows an unsegregated, uniform, ordered structure, Figure 2. The 5-Cr ribbon shows the ordered structure of antiphase domains, indicating that the matrix is only slightly substoichiometric, Figure 4. The 12.5-Cr ribbon shows what may be a $\gamma/\gamma'$ two phase structure, Figure 5a, caused by substantial Al depletion, Table I. The antiphase domain structure as well as the $\gamma/\gamma'$-type structure is observed at the contact surface of the high-Cr ribbon, Figures 6a and 6b. This might be related to the reduced Al segregation in a high cooling rate region. On the other hand, the irregular structure in Figure 6b might result from a non-[100] orientation of the grain investigated. Further microchemistry and diffraction studies are needed to determine the cooling rate and grain orientation effects.

## CONCLUSIONS

The microstructure of melt-spun $Ni_3Al(Cr)$ ribbons has been studied. It was found that the basic $Ni_{75}Al_{25}$ compound could be rapidly solidified into a uniform, single-phase, ordered, $\gamma'$ structure. In the $Ni_{72.5}Al_{22.5}Cr_5$ ribbon the Al-rich $\beta$-phase particles formed at grain boundaries, while the matrix phase which became slightly lean in Al contained non-oriented antiphase domains. In the $Ni_{67.5}Al_{20}Cr_{12.5}$ ribbon, extensive grain boundary $\beta$ phase was found, resulting from the eutectic reaction $L \rightarrow \gamma + \beta$. The grain interiors in this ribbon showed the ordered structure of cuboidal $\gamma'$ precipitates, or occasionally near the ribbon contact surface, irregular antiphase domains. The variation in the ordering structure is related to the matrix stoichiometry, which in turn is affected by the degree of Al and Cr segregation.

## ACKNOWLEDGMENTS

The authors are grateful to R. J. Zabala for his technical support and to L. A. Johnson for his technical suggestions.

## REFERENCES

[1] K. Aoki and O. Isumi, J. Japan Inst. Metals, 43 (1979), p. 1190.

[2] C.T. Liu and C.C. Koch, Technical Aspects of Critical Materials Use by the Steel Industries, Vol. IIB, published by National Bureau of Standards, Washington, DC, 1983, p. 42-1.

[3] C.C. Koch, J.A. Horton, C.T. Liu, O.B. Cavin and J.O. Scarbrough, Rapid Solidification Processing, Principles and Technologies III, edited by R. Mehrabian, published by National Bureau of Standards, Washington, DC, 1983, p. 264.

[4] A.I. Taub, S.C. Huang and K.M. Chang, Met. Trans., 15A (1984), p. 399.

[5] R. P. Messmer and C. L. Briant, Acta Met., 30 (1982), p. 457.

[6] S. C. Huang, A. I. Taub, K. M. Chang, C. L. Briant and E. L. Hall, Proceedings of the Fourth International Conference on Rapidly Quenched Metals, September 3-7, 1984, Wurzurg, Germany.

[7] S.C. Huang, A.I. Taub and K.M. Chang, Acta Met., in press.

[8] A. Inoue, H. Tomioka and T. Masumoto, Met. Trans., 14A (1983), p. 1367.

[9] S.C. Huang and R.P. Laforce, Proceedings of the Materials Research Society Symposium on Metastable Materials, November 15-17, 1983, Boston.

[10] A. Taylor and R.W. Floyd, J. Inst. Metals, 81 (1952-3), p. 451.

# STRUCTURAL UNIT MODELS FOR HIGH ANGLE SYMMETRICAL TILT BOUNDARIES IN ORDERED COMPOUNDS WITH THE $L1_2$ STRUCTURE

D. FARKAS
Department of Materials Engineering, VPI & SU, Blacksburg, VA 24061.

## ABSTRACT

Hard sphere models were used to determine densest configurations in symmetrical [100] and [110] tilt boundaries in compounds with the $L1_2$ structure. The minimum allowed interatomic distances used in these models were estimated from interatomic potentials and the structures of the intermetallic phases in the binary system. The structural unit model is used to analyze the possible ground states for ordering.

Two different cases were analyzed corresponding to compounds with "soft" potentials (i.e. $Cu_3Au$) and "hard" potentials (i.e. $Ni_3Al$). For the $Cu_3Au$ type the grain boundary structures obtained were similar to those reported by other investigators for pure fcc metals. Several boundaries were found to be a "two phase" structure, differing in composition and ordering state. This leads to a certain degree of clustering in the boundaries. The contribution of clustering to the grain boundary energy is calculated in a point approximation based on the first coordination shell.

For compounds of the $Ni_3Al$ type the structures that are densest were found to be generally different from the low energy configurations of boundaries in pure fcc metals and $Cu_3Au$. These configurations preserve order, but are much less dense. The possibility of grain boundary "phases" that are not present in other fcc materials may constitute an explanation for the extreme GB weakness observed in $Ni_3Al$ and other $L1_2$ compounds with high ordering energy.

## INTRODUCTION

There is growing interest in the development of ordered alloys as new structural materials for high temperature use. For example, it has been shown that microalloying can improve ductility in intermetallic alloys based on $Ni_3Al$. The study of grain boundary configurations in compounds with the $L1_2$ structure is therefore of particular interest. In the present work hard sphere models were used to obtain configurations of symmetrical tilt boundaries in two ordered compounds with the $L1_2$ structure, $Cu_3Au$ and $Ni_3Al$. The model includes the consideration of the different atomic sizes as well as the possible translational states of the boundary. The structural unit model was then used to analyze the possible ordering states present in these boundaries.

## THE MODEL

It is now well established that the type of interatomic potential assumed for the calculation of grain boundary structures strongly influences the results. Wolf (1,2) showed that the unrelaxed and relaxed grain boundary energies in twist boundaries in Al depends on the choice of potential. Furthermore, the directions of relaxation were also potential dependent. Even the basic structure of the unrelaxed boundary may change, as pointed out by Frost (3). It seems that the hard spheres models can give reliable starting configurations for relaxation studies in most cases. However, results can be improved by analyzing the interatomic potentials and determining a minimum allowed distance between atoms to be used in the geometrical models. This consideration applies specially to distances between unlike atoms. It is well established (4) that the minimum of the potential in this case need not be at an intermediate position between those of the components of the bond. It should also be pointed out that the analysis of the distances of closest approach found in the different intermetallic phases that may be present in the binary system will also help to determine the minimum distance that can be expected in the grain boundary.

In the present study two different cases were considered as representative of soft and hard potentials. In the soft potential case the distances of closest approach allowed are similar for all bonds. The grain boundary configurations in this case are expected to be similar to those obtained by maximizing order in the structures that have been reported for pure fcc metals. $Cu_3Au$ is an example of a system where the basic structural units of fcc metals are expected to be stable. Note that in this system both $Cu_3Au$ and $Au_3Cu$ are stable with the $L1_2$ structure and similar lattice parameters.

The case of hard interatomic potentials was also studied using geometrical models where the allowed distances of closest approach were obtained from an analysis of the interatomic potentials and the available crystallographic data for the system. The example chosen is $Ni_3Al$. Studies of grain boundaries in Al (5,6) using pseudopotentials show that Al behaves as hard spheres and the configurations obtained by geometrical methods (3) agree with those of computer simulations. In $Ni_3Al$ the nearest neighbor distance is 14% shorter than the equilibrium distance in pure Al. This indicates that for this material Al-Al bonds at the nearest neighbor distance of $Ni_3Al$ may not be present in the grain boundary structure. This is supported by analysis of the other (non-coherent) phases present in the Ni-Al system. In all these structures the Al-Al distance does not deviate more than 1% from the equilibrium distance in pure Al. Furthermore, Ellner et al (7) found a correlation between heat of formation of aluminide phases and Al-Al distances in the structure. Therefore it was assumed in the present work that the Al-Al bond length cannot be smaller than the nearest neighbor distance in pure Al. In constructing GB structures it is also necessary to estimate the position and depth of the minimum in the Ni-Al potential. The position of this minimum is not intermediate between that of Al and Ni but is even smaller than that of Ni. This was reported in experimental studies of Ni-Al liquids by Enderby (8). This is due to the heteroelectronic nature of the Ni-Al bond as Machlin (4) reported in pairwise studies of intermetallic phases using Lennard-Jones potentials. Based on the preceding analysis the minimum Al-Al interatomic distances allowed in the construction of the $Ni_3Al$ boundaries were longer than the nearest neighbor interatomic distance in the bulk (0.28 nm).

The procedure for generating and analyzing grain boundary configurations is as follows:

<u>Step 1:</u> Identify the "favored" boundaries in a series and the structural units that conform them in pure fcc metals (9). If different types of structural units are possible they are considered separately.

<u>Step 2:</u> Determine the densest configurations using the method of Frost and the minimum allowed distances discussed above. Translations perpendicular to the tilt axis were allowed, both perpendicular and parallel to the boundary plane.

Step 3: For the favored boundaries assume coincident sites that are compatible with order in both crystals. This choice can in some cases be taken in two different ways (corresponding to the first coincident site considered as an A or B atom). For most boundaries this will result in two different configurations.
Step 4: Identify the different ordering configurations that occur in the structural units.
Step 5: Determine the constitution of intermediate boundaries in terms of the structural units of step 1.

RESULTS

The first series considered is that of symmetrical [100] tilt boundaries. The favored boundaries in this family are the $\Sigma = 5$, $\theta = 53.1°$ and $\theta = 36.9$, (120) and (130) respectively, measured from $[100]_1$ to $[100]_2$. The $\Sigma = 5$, (130) boundary has been studied extensively in aluminum (3,5,6). Two types of structures have been reported for pure metals. Hashimoto et al. (10) calculated the energies of both types in Al and found that type 1 is stable. In the case of ordered compounds with the $L1_2$ structure the ordering results in different types of structural units for boundaries of type 1, as shown in Fig. 1. In analyzing type 1 boundaries it can be noticed that the structural unit D in the case of the $L1_2$ structure can be ordered in 4 different configurations, labeled DD'D"D". The units D and D' originate in taking the coincidence sites in this boundary as B atoms. The boundary is formed by the sequence DD' (here called α (130)). If the coincidence sites are B atoms the structural units D', D" result and form β (130).

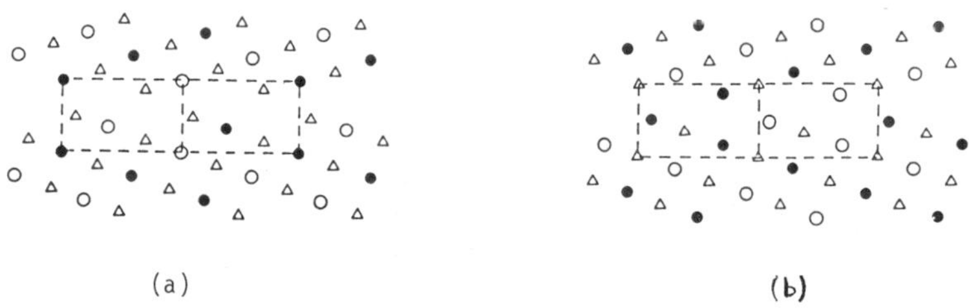

Fig. 1. Structure of the [100] (310) boundary in $Cu_3Au$
(a) α Structure DD' (b) β Structure D"D"'

It is also evident from the structures that α and β (130) are boundaries of different compositions. A schematic free energy composition diagram representing this situation is presented in Fig. 2. In this construction equilibrium between the two GB phases is given by the common tangent and equilibrium between GB phases and the bulk is given by the parallel tangent condition, as shown by Hillert (11). The two grain boundary structures are in equilibrium with each other, as indicated by the common tangent between their free energy curves. This common tangent is parallel to the tangent to the bulk free energy curve at 25% B. The distance between the parallel tangents is the grain boundary energy, that is the free energy necessary to form one mole of grain boundary from the bulk material. Note that both boundaries in this scheme have the same grain boundary energy.

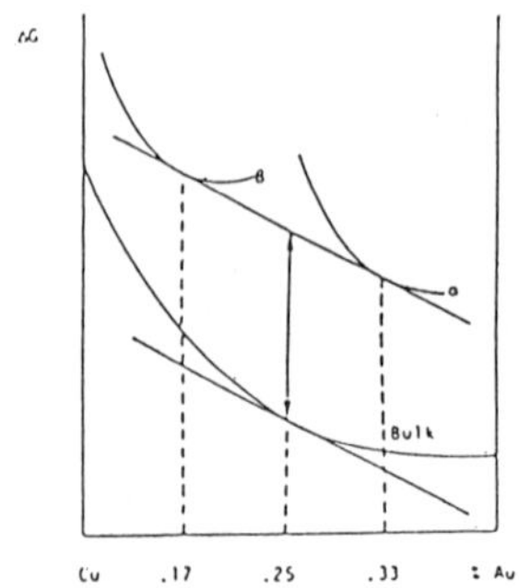


Fig. 2. Free energy-composition diagram for α and β (130) boundary in $Cu_3Au$.

Table 1 shows the average number q of A-B bonds per atom in each unit (12). The α phase of the 130 boundary is B rich and therefore presents a value of q larger than that of the perfect lattice. On the other hand β (130) has a much lower value. The q average for both α and β is lower than that of the perfect lattice (clustering effect).

Table 1

Structure and Composition of [100] Favored Boundaries in $Cu_3Au$ Type Compounds

| Structural Unit | Boundary | Composition | | q | |
|---|---|---|---|---|---|
| A<br>A' | perfect<br>lattice | 0<br>0.5 | 0.25 | 4<br>8 | 6 |
| A"<br>A''' | perfect<br>lattice | 0<br>0.5 | 0.25 | 4<br>8 | 6 |
| B<br>B' | α (120) | 0.5<br>0.167 | 0.333 | 7.5<br>5.5 | 6.5 |
| B"<br>B''' | β (120) | 0<br>0.333 | 0.167 | 3.5<br>6.7 | 5.1 |
| E<br>E' | β' (120)<br>4 layers | 0.375<br>0.125 | 0.25 | 7.8<br>4.4 | 6.1 |
| E'''<br>E" | α' (120)<br>4 layers | 0.5<br>0 | 0.25 | 7.4<br>2.9 | 5.05 |
| I<br>I' | β' (120)<br>2 layers | 0.5<br>0 | 0.25 | 9<br>3.5 | 6.25 |
| I"<br>I''' | α' (120)<br>2 layers | 0.25<br>0.25 | 0.25 | 7.5<br>3 | 5.25 |
| D<br>D' | α (130)<br>4 layers | 0.10<br>0.30 | 0.20 | 3.9<br>5.5 | 4.7 |
| D"<br>D''' | β (130)<br>4 layers | 0<br>0.60 | 0.30 | 5<br>7.8 | 6.4 |
| G<br>G' | α (130)<br>2 layers | 0<br>0.333 | 0.167 | 3.33<br>6 | 4.67 |
| G"<br>G''' | β (130)<br>2 layers | 0.5<br>0.167 | 0.333 | 4.67<br>7.67 | 6.17 |

The disordering contribution to the energy of the GB will be:

$$\Gamma_{disord} = (q-q_0)\varepsilon N/Area$$

where q is the number of AB nearest neighbor bonds in the first coordination shell, N is the number of atoms in the repeat unit, and Area is the area of the repeat unit in the GB plane.

$$\varepsilon = V_{AB} - (V_{AA}+V_{BB})/2$$

For values typical of $Cu_3Au$ this gives $\Gamma_{disord} = 0.15$ J/m. Hashimoto et al. (10) calculated the difference in energy between type 1 and type 2 boundaries to be 0.3J/m. This suggests that type 1 will also be stable in $Cu_3Au$, as in pure metals.

The structure of type 1, however, cannot be stable for $Ni_3Al$, since it contains Al-Al bonds of length equal to the nearest neighbor distance in the $Ni_3Al$ structure. For $Ni_3Al$ a structure of type 2 was found to be densest. It is shown in Fig. 3. This type 2 boundary is characterized by structural units HH'. This is the only ordering structure possible for this boundary and its composition is equal to that of the bulk. This boundary does not destroy the ordering of the perfect lattice and it will be stabilized for ordered compounds of high ordering energy. Note that this boundary does not contain Al-Al nearest neighbor bonds. However, it is a low density boundary.

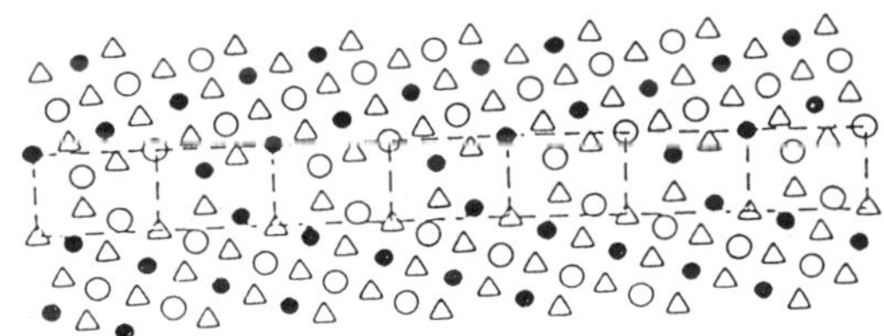

Fig. 3. $\Sigma = 5$ [100] (130) boundary in $Ni_3Al$. Al-Al bonds present with length of 2.8 Å. HH' Structure.

The other $\Sigma = 5$ boundary (210), $\theta = 53.1°$ has also been investigated in detail (3,5,6,13). Two different structures have been found in computer simulations. Structure a has been obtained by Smith et al. (13) for aluminum and Wetzel et al. (14) for several fcc metals. Sutton and Vitek (15) report structure b for Cu. The difference in these two structures is due to the fact that different interatomic potentials were used. (Lennard-Jones type potentials yield structure a for Al, Cu, Ni and Au (14)). The splines potential of Crocker (15) yields type b for Cu. Hard sphere models (3,13) yield configuration a for all fcc metals. The results of the present work are also similar to structure a for $Cu_3Au$ (Fig. 4a,b). It should be noted that both structures are similar. They both yield dense configurations, and when considered for ordered $Cu_3Au$ type compounds result in two-phase boundaries, containing Au-Au nearest neighbor bonds. Both of these structural units can then be classified as type 1 structures. The effect of ordering on grain boundary energy is somewhat smaller in the (210) case than in the (310) boundary.

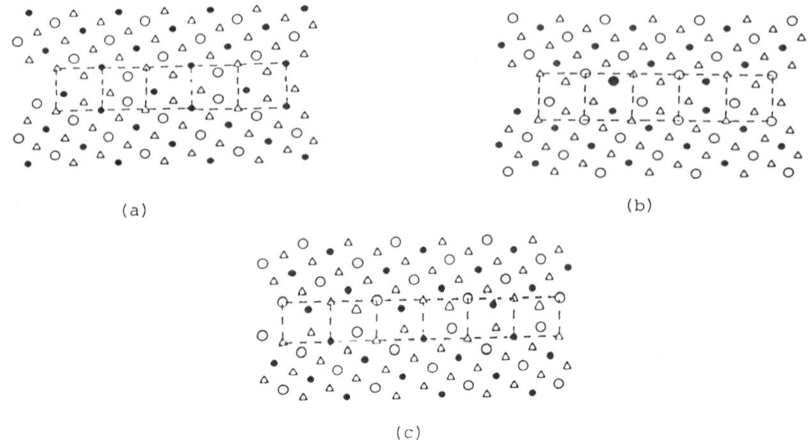

Fig. 4. Structure of the [100] (210) boundary -- (a) α Structure $Cu_3Au$ EE', (b) β Structure $Cu_3Au$ E"E"', and (c) $Ni_3Al$.

Structures of type 2 have not been reported previously for the (210) boundary. In the present work a type 2 configuration was obtained using hard sphere models for $Ni_3Al$. This is shown in Fig. 4c. As in the (130) case this boundary does not destroy order and is less dense. The intermediate boundaries in the [100] family are listed in Table 2 for boundaries in $Cu_3Au$ type compounds. The constitution of these intermediate boundaries in terms of structural units is indicated in the last column. The unit A appears in some of these boundaries and is characteristic of the perfect lattice (9). One of the intermediate boundaries (520) is shown in Fig. 5. Table 2 also shows the constitution of intermediate boundaries in $Ni_3Al$. Note that these results are very similar to those of $Cu_3Au$ if the units corresponding to Type 1 boundaries are replaced by the corresponding units of type 2 boundaries. The unit A of the perfect lattice is replaced by the unit Z of the perfect lattice with a translational boundary.

Grain boundary structures were also determined for boundaries in the [110] series. The results for this series of boundaries differ from the previous series in that the Σ = 3 (111) and Σ = 9 boundaries were found to have the same structure for both types of $L1_2$ compounds analyzed. The other boundaries analyzed show very similar behavior to the [100] series.

Table 2

Structure of [100] Boundaries in $Cu_3Au$ Type Compounds

| Boundary | Σ | θ | Pure Metals Structure | $L1_2$ Structures | |
|---|---|---|---|---|---|
| Perfect lattice | 1 | 90° | AA | AA' | A"A''' |
| (310) | 5 | 36.87° | DD | DD' | D"D''' |
| | | | BB | BB' | B"B''' |
| (210) | 5 | 53.08° | EE | EE' | E"E''' |
| | | | AB | AB"A'''B' | A'BA"B''' |
| (320) | 13 | 67.38% | AE | AE"A'''E' | A'EA"E''' |
| | | | ABB | AB"B'''A'BB'AB"B''' | |
| (530) | 17 | 61.93° | AEE | AE"E'''A'EE'AE"E''' | |
| | | | BAABA | B'AA'BA"B'''A'AB"A''' | |
| (750) | 37 | 71.08° | EAAEA | EAA'EA"E'''A'AE"A''' | |
| | | | AAAB | A'''A"A'''BA'AA'B''' | A"A'''A"B"AA'AB' |
| (540) | 41 | 77.32° | AAAE | A'''A"A'''EA'AA'E''' | A"A'''A"E"AA'AE' |
| | | | DB | D'B'''D"BD' | DB'D'''B" |
| (520) | 29 | 43.60 | DE | D'E'''D"ED' | DE'D'''E" |

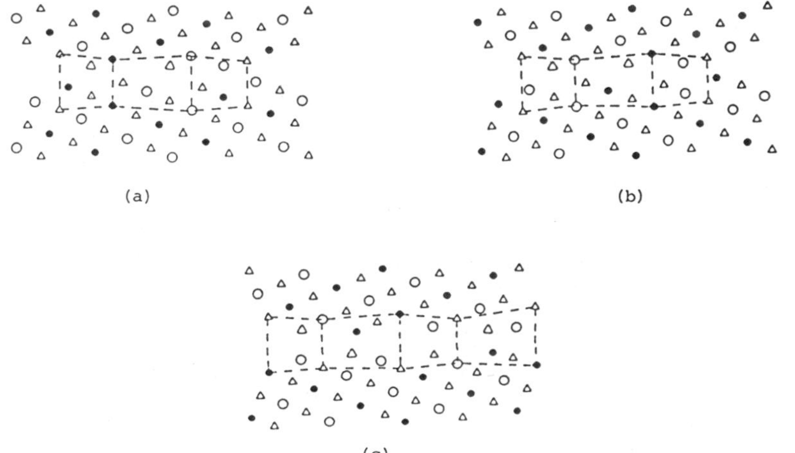

(a) (b) (c)

Fig. 5. Structure of the [100] (520) boundary -- (a) α Structure $Cu_3Au$, (b) β Structure $Cu_3Au$, and (c) $Ni_3Al$.

## CONCLUSIONS

The first conclusion of the present analysis is that the grain boundary structure in ordered alloys may be a two phase structure. The two grain boundary phases differ in composition but have the same grain boundary energy.

Secondly it was shown that the grain boundary structures that are densest for pure fcc metals are also densest for the $L1_2$ structure if the interatomic potentials are "soft", i.e. the equilibrium distances are similar for all types of bonds. This is not the case for compounds with "hard" interatomic potentials where the equilibrium bond lengths differ for the different types of bonds. $Ni_3Al$ is an example of this. Grain boundary structures for this compound were found to be different from those found in pure metals. The new structures mantain order but are much less dense. This may constitute an explanation for the intrinsic grain boundary weakness observed in $Ni_3Al$, not observed in $Cu_3Au$.

## REFERENCES


1. D. Wolf, Acta Met. 32, 245 (1984).
2. D. Wolf, Acta Met. 5, 735 (1984)
3. H. J. Frost and F. Spaepen, J. Phys. (Paris), C6, 74 (1982).
4. J. Shao and E. S. Machlin, J. Phys. Chem. Solids, 44, 289 (1983).
5. D. A. Smith, V. Vitek and R. C. Pond, Acta Met. 25, 475 (1977).
6. V. Vitek, A. P. Sutton, D. A. Smith and R. C. Pond, in "Grain Boundary Structure and Kinetics", ASM Conference, 115 (1980).
7. M. Ellner, U. Kattner and B. Predel, Journal of the Less - Common Metals, 87, 305 (1982)
8. J. E. Enderby and W. S. Howells, in "Interatomic Potentials and the Simulation of Lattice Defects", P. C. Gehlen, J. R. Beeler, Jr. and R. I. Jaffe, 217 (1971).
9. A. P. Sutton and V. Vitek, Phil. Trans. R. Lond. A., 309, 1 (1983)
10. M. Hashimoto, Y. Ishida, R. Yamamoto and M. Doyama, Acta Met. 29, 617 (1981).
11. M. Hillert, in "Lectures on the Theories of Phase Transformations", Hubert Aaronson, 1 (1975).
12. T. Takasugi and O. Izumi, Acta Met. 31, 187 (1983).



### Acknowledgements

This work was supported by DOE, ECUT program under ORNL subcontract No: 19X-89678V.

DYNAMICS OF LAVES PHASE INTERSECTIONS IN A CONCENTRATION GRADIENT

C. W. ALLEN
Department of Metallurgical Engineering and Materials Science, University of Notre Dame, Notre Dame, IN 46556*

## ABSTRACT

In the approximate interval 850-1150C, the Laves phase $TiCr_2$ exhibits several composition dependent structures, ranging from two layer hexagonal (2H), denoted C14, to face-centered cubic, denoted C15, for Ti-deficient and Ti-excess material, respectively. Intermediate stacking variants such a dihexagonal (4H), denoted C36, may also exist. The Laves phase band in a Ti-Cr microdiffusion couple is examined in cross-section TEM at ambient temperature, after a 1000 C anneal. The band consists of very faulty hexagonal material, at least in part reflecting rapid shear transformation on cooling; in addition the cubic phase is present, in some instances within the same grain as the hexagonal phase. The cubic/hexagonal interfaces are approximately parallel rather than perpendicular to the bulk diffusion direction. Due to a difference in Ti and Cr fluxes, there is a gradual transformation of cubic to hexagonal by a shear mechanism. A model for the mobility of the hexagonal/cubic interface in a concentration gradient is proposed.

## INTRODUCTION

Interest in the structures and microstructures developed in the reaction of metal-metal and metal-semiconductor systems has considerably increased in recent years largely because of the important role of such reactions in microelectronic systems. In such reactions it is often found that the sequence of intermediate phases characterizing the equilibrium phase diagram is violated by the absence of one or more of these phases and sometimes even by the appearance of metastable phases in preference to those of higher thermodynamic stability. This paper treats reaction of titanium and chromium, a system with a single intermetallic, the Laves phase $TiCr_2$, which possesses the C14 (2H) structure at elevated temperatures and C15(3C)at lower temperatures. As indicated in the equilibrium phase diagram of Fig. 1, the former transforms to the latter on cooling in the temperature range of approximately 850 to 1150 C, depending on phase composition. In addition to these two equilibrium phases, however, several metastable structures commonly appear in the transformation process including C36(4H) and $6H_1$ [2]. The purpose of this study is to determine the characteristics of the Laves phase band developed at 1000 C, across which the solute concentration variation spans both equilibrium structures.

## SHEAR TRANSFORMATIONS IN $TiCr_2$

$TiCr_2$ is a topologically close packed material in which the transformation at constant composition, 2H→3C, on cooling is martensitic in mechanism. Because the change involves only lattice shear without detectable dilatation

*This work was performed in the Division of Materials Science and Technology, Argonne National Laboratory, and sponsored in part by the Division of Basic Energy Sciences, U.S. Department of Energy.

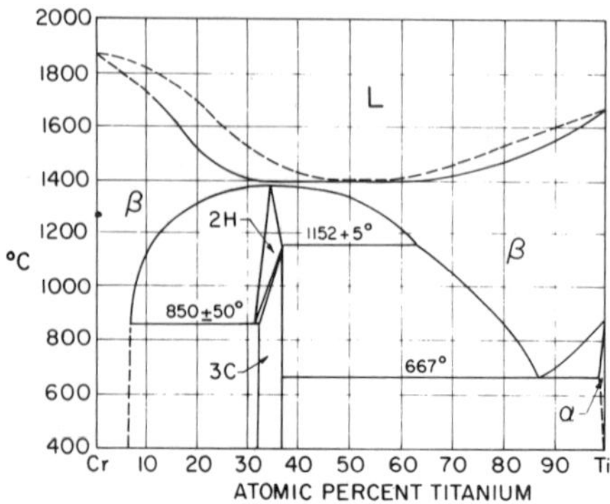


Fig. 1. Equilibrium phase diagram for Cr-Ti, after Farrar and Margolin [1].

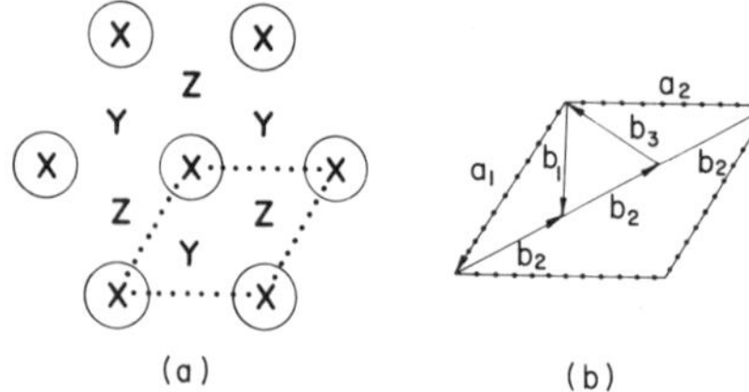


Fig. 2. Basic layer positions with respect to hexagonal or ortho-hexagonal basal plane. Unit cell is outlined in (a) and enlarged in (b). In (b) three Burgers vectors for Shockley-type partial dislocations are shown.

and the close packedness is preserved in the process, the transformation of this type is called simply a shear transformation. The resultant structures depend on cooling rate, and only for an exceptionally slow rate does the cubic phase alone result. It is convenient to describe the transformations in terms of changes in stacking sequence of close packed layers, the individual layers composed of a thin slab about 4Å thick in this case rather than a single close packed plane (e.g.,as in pure Co). If the possible layer positions are denoted X, Y and Z, as indicated in Fig. 2a (rather than A, B. and C, as is customary for Co) then the 2H structure is represented by the stacking sequence ...XYXYXY... and the 3C structure, by ...XYZXYZXYZ...; the underlined symbols represent the extent of a unit cell in [0001] for 2H and in <111> for 3C. In the case of $TiCr_2$ the transformation sequence on cooling is observed to proceed with intermediate structures appearing, the common sequence of events being 2H→4H→$6H_1$→3C, with thin regions of other modifications sometimes appearing as well [3]. In terms of changes in stacking involved in this shear transformation sequence,

...XYXYXY...→...XYXYXZ...→...XYXZYZ...→...XYZXYZ... .

Furthermore the transformation 2H→4H has been observed to proceed by the glide of Shockley-type partial dislocation dipoles [4-6] as predicted [3].

All of these transformations result from glide of partial dislocations, so that where interfaces exist between two structures, the interfaces may be

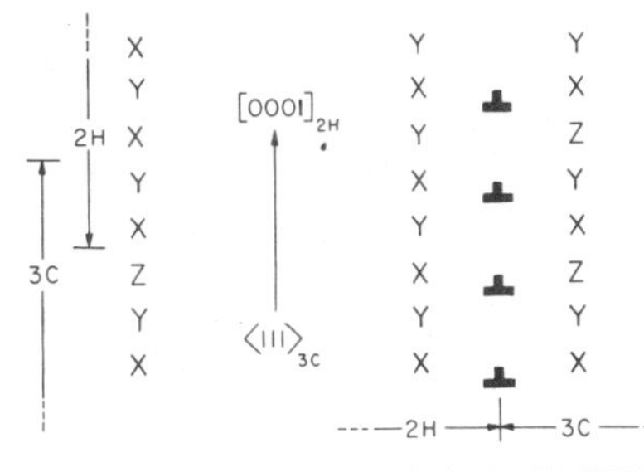


Fig. 3. Two possible interfaces separating shear transformation domains. In (a) interface is horizontal and coherent; in (b) it is vertical and semi-coherent, an array of Shockley partial dislocations.

either completely coherent (parallel to the dislocation glide planes) or semi-coherent (made up of partial dislocation arrays) as depicted in Fig. 3. The nature of these transformations is not very different in the Ti-Cr intermetallic over its solubility range, except that the transformation on cooling occurs more readily in Cr-deficient material, owing to the higher critical temperatures which promote increased dislocation mobility. In the experiment described in this paper, the effect of a composition gradient on the transformations is investigated. However, because the observations are not made _in situ_, the incomplete transformation of the hexagonal phase(s) on cooling, which is very difficult to completely suppress, is superimposed.

## EXPERIMENTAL PROCEDURE AND RESULTS

A diffusion couple was produced by incompletely dissolving beads of chromium with titanium in the liquid state during arc melting. The resultant button was sliced and sections containing the Cr and the Ti-rich β phase were annealed in vacuum at 1000 C for 24 hours and water quenched. TEM cross-section samples were prepared by slicing and core drilling of discs containing the reaction zone near their center, followed by mechanical dimpling and ion milling to produce thin area within the Laves phase band. The procedure has had only limited success because of differential ion milling rates, the Laves phase material being significantly more resistant to sputtering than either the Cr-rich or Ti-rich β regions on either side of it in the diffusion couple.

The TEM micrograph in Fig. 4a shows an area within the Laves phase band in which the C15 cubic phase and highly faulted hexagonal material share a coherent boundary $(111)_{cubic}//(0001)_{hex}$. Fig. 4b shows a somewhat more complicated region of hexagonal and cubic material, separated by a relatively high angle boundary (approximately horizontal); a low angle boundary within the hexagonal material is also present at the left. In Fig. 4b the region at the bottom of the predominantly hexagonal grain is cubic. Within this cubic region the faults in contrast are sheared, indicating that probably this region is in the early stage of conversion to an hexagonal structure. The position and orientation of both of these areas (Figs. 4a and 4b) in relation to the interphase boundaries is shown schematically in Fig. 4c.

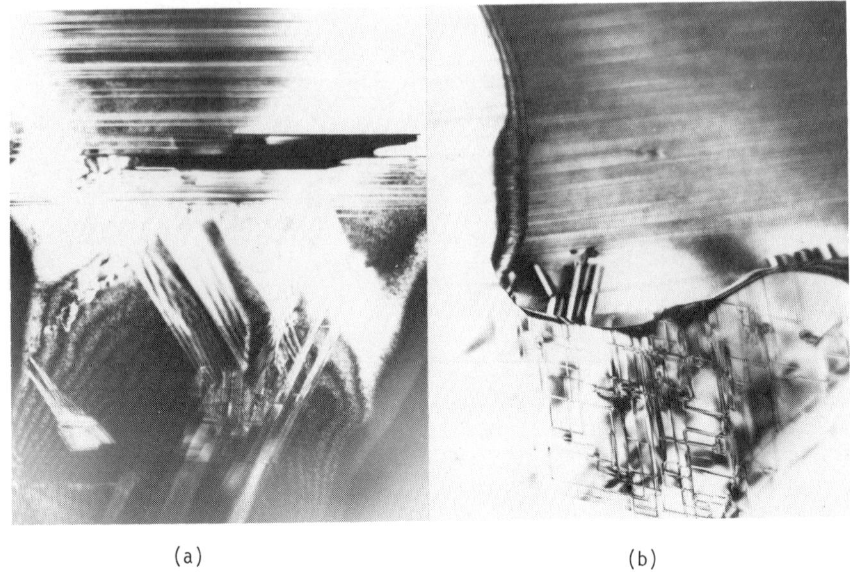

(a) (b)

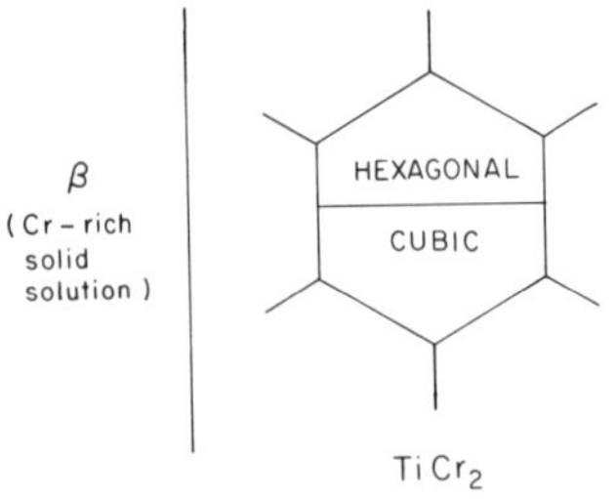


(c)

Fig. 4. TEM bright field micrographs (a) and (b) of hexagonal/cubic interface region in Cr-Ti diffusion couple Laves phase band. (c) Relation of areas in (a) and (b) to Cr-rich β zone. Area in (a) is 0.8 μm wide; (b) 1.2 μm wide.

DISCUSSION

The prediction of (macroscopic) thermodynamic arguments for the character of a Ti-Cr type diffusion couple, based on the assumption of local quasi-equilibrium, is that the interdiffusion zone should consist of parallel single phase bands which lie perpendicular to the diffusion direction. In this way the chemical potential for each of the elemental components will be continuous across the entire diffusion zone. On a fine scale, however, it is evident from Figs. 4a and 4b that the kinetics of the formation and motion of the 2H/3C interface create a morphological situation locally which is quite different from the thermodynamic prediction. In fact the actual interface is approximately parallel rather than perpendicular to the probable concentration gradient in the grain (Fig. 4c). This is surprising because several of the {111} of the cubic phase are oriented much more nearly normal to the flux direction than the one in Fig. 4. It is expected that Cr diffuses faster than Ti (Cr is smaller and the solubility range of the intermetallic extends mainly toward Cr-deficiency, suggesting excess concentration of Cr-vacancies) so that locally there must be a gradual conversion of cubic→hexagonal; that is, the coherent interfaces in the central parts of Figs. 4a and 4b must be migrating downward. It remains now to be shown that the observed orientation of these interfaces is indeed that for which the driving force for the transformation appears to the dislocations responsible for the transformation to be greatest. This is a situation which is really exclusive to the case wherein a shear transformation occurs in a gradient, a case which has not been addressed previously.

The situations in Fig. 4 are shown schematically in Fig. 5. The symbol ⊥ represents a Shockley partial, which, gliding from left to right, produces translation of the material above its glide plane so that the layer designations within the translated material permute cyclically (X becomes Y; Y,Z; Z,X) by convention [3]. Each of the dislocations of Fig. 5 may have Burgers vector $b_1$, $b_2$ or $b_3$ (Fig. 2b) and may be edge, screw or mixed in character.

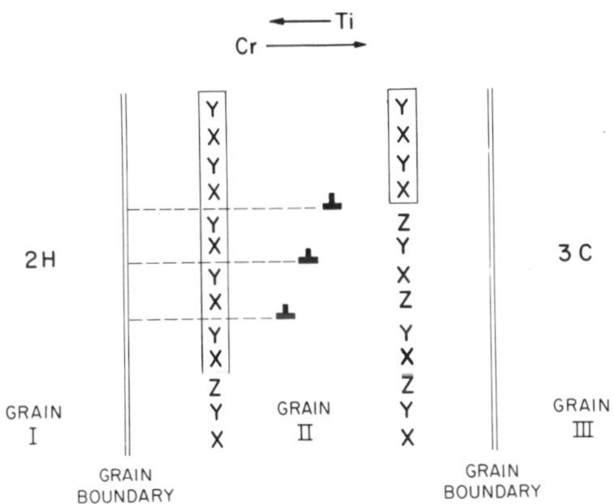


Fig. 5. Model for 3C→2H transformation in which grain boundary acts as partial dislocation source. Dashed lines are intrinsic stacking faults.

In Fig. 5 the transformation is assumed not to involve intermediate structures, but this is of no importance to the following argument. The cubic phase is least stable to the left and most stable to the right so that a dislocation emerging, for instance, from the boundary between grains I and II trails an intrinsic stacking fault which at first has very negative specific stacking fault energy with respect to the cubic phase. As the dislocation glides to the right, the portion of fault adjacent to it has progressively less negative energy per unit area which is the transformational glide force per unit length acting on the dislocation. In the absence of other forces, the dislocation would continue to glide until it just reaches the region where the cubic phase is stable and the stacking fault energy becomes positive. If the grain size were sufficiently large in the diffusion direction (i.e., there were no grain boundary to the right into which the dislocations could disappear), one might thereby expect these transformational dislocations to form a semicoherent interface, such as that shown for 2H|3C in Fig. 3b, oriented approximately as predicted by the thermodynamic argument and slowly gliding as a unit as diffusion proceeds. In the present case, the grain size is not large so that such an interface possibly does not have an opportunity to build up. Rather individual dislocations or small groups as indicated in Fig. 5 form (emerge from the boundary between grains I and II) and glide away to a boundary at the right, generating an increasing amount of stacking fault which eventually organizes to constitute the ultimate 2H structure. Unfortunately the details of this process are not known even for the case of transformation of 3C in chemically homogeneous Laves phase.

Fig. 5 also illustrates the fact that each of the Shockley partials generates two layers of the 2H(C14) structure when the dislocation glide planes are separated by two layers. If all the dislocations were to have the same Burgers vector (e.g., $b_2$ in Fig. 2b) a net macroscopic shear of 0.35 would result, which would be appropriate to a deformation-induced transformation (mechanical martensite). In the case of the diffusion-induced transformation, however, one expects the net shear to be zero. From a geometrical viewpoint the transformation is in no way inhibited by such a requirement but rather the dislocations involved should be equally distributed among those with Burgers vectors $b_1$, $b_2$ and $b_3$ (Fig. 2b). In the mechanically-induced transformation $|\bar{\Sigma} b_i|$ is maximized; in the diffusion-induced case, it is minimized. In fact, this is a quite general rule for the transformation of the cubic phase to any of the other possible structures by a shear mechanism.

The planar faults in the cubic region of Fig. 4b have been analyzed to be crystallographic shear planes on {111} [7] whose density, however, is much too small to accommodate substantial off-stoichiometry. Further these faults do not appear to play any role in the inter-Laves phase transformations, owing to the close relationship of all the Laves phase structures which are simply stacking sequence variants.

## CONCLUSIONS

A mechanism for occurrence of a shear transformation in a concentration gradient has not been proposed previously. Observations of the diffusion-induced shear transformation in $TiCr_2$ suggest that the transformation proceeds by the repetitive nucleation of Shockley partial dislocations, probably in grain boundaries, in regions of the Laves phase where the driving force for transformation is large. In the case of a shear transformation this situation produces structural interfaces which are locally more nearly parallel than normal to the flux as thermodynamic arguments would predict. It is not clear from the present ex situ observations whether the resultant hexagonal structure is 2H or some combination of intermediate hexagonal structures, as is common in the cooling transformation of a single composition of $TiCr_2$.

## REFERENCES

[1] P. A. Farrar and H. Margolin, Trans. Met. Soc. AIME, 227, 1342-1345(1972).
[2] C. W. Allen, P. Delavignette and S. Amelinckx, Phys. Stat. Sol. (a), 9, 237-246 (1972).
[3] C. W. Allen and K. C. Liao, ibid, 74, 673-681 (1982).
[4] K. C. Liao and C. W. Allen, Solid-Solid Phase Transformations, 1493-1497 (1982).
[5] C. W. Allen and K. C. Liao, Proc. Int. Conf. on Martensitic Transformations (ICOMAT), 124-129 (1979).
[6] H. Kolar, High Resolution Electron Microscopy of Phase Transformations in Laves Phase $TiCr_2$, Masters Thesis, Arizona State University (1982).
[7] C. W. Allen and K. C. Liao, "On Crystallographic Shear in Ti-Cr Laves Phases." In preparation for submission to Phys. Stat. Sol. (1985).

# INFLUENCE OF PHOSPHOROUS ON THE KINETICS OF ORDERING IN $Ni_2Cr$

LAWRENCE P. LEHMAN AND T. H. KOSEL
University of Notre Dame, Department of Metallurgical Engineering and Materials Science, Notre Dame, IN 46556

ABSTRACT

The isothermal ordering kinetics of stoichiometric $Ni_2Cr$ was investigated in specimens aged at 500°C and containing 3 different phosphorous impurity concentrations; 0.004, 0.032, and 0.142 atomic percent P. In every case, higher P specimens were found to transform more rapidly than lower P content specimens, with as much as an order of magnitude difference in transformation times. Low and medium P specimens transformed via a continuous ordering process. In contrast, the highest P specimens transformed by an unusual process in which colonies of ordered domains were formed which consumed grains through the advance of an order-disorder interface. Colonies tens of microns in size were typically observed while individual ordered domains were only a few tens of nm in size.

## INTRODUCTION

Phosphorous is an impurity suspected of leading to premature failure in nickel based superalloys due to embrittlement [1]. $Ni_2Cr$ is a model material for many of these nickel based superalloys, which experience the same sluggish long range order (LRO) transformation.

On cooling below the critical temperature of 590°C, $Ni_2Cr$ transforms from a disordered FCC structure to a body centered orthorhombic $D_{2h}^{25}$-Immm superlattice [3,4] by a continuous ordering process [2]. The order-disorder transformation in this material has been observed by several authors [1, 3-6]; changes in hardness [4], volume [4,6], and conductivity [4,5] are commonly tested properties. X-ray diffraction is a convenient technique to measure the lattice contraction during the LRO transformation using the fundamental (FCC equivalent) lines of the diffraction pattern [6].

Experiments were carried out to measure the rate of LRO transformation in $Ni_2Cr$ as a function of its bulk P content; changes in the FCC equivalent lattice parameter were measured by x-ray diffraction and the morphology of the ordered phase was observed by optical and electron metallography. Interrupted isothermal aging at 500°C was used to follow changes in the material properties of individual specimens as a function of time. In addition, transmission electron microscopy (TEM) was used to establish the size and shape of domains after LRO was complete.

## EXPERIMENTAL

### Materials

$Ni_2Cr$ alloys with three different phosphorous levels were prepared by the Cabot Corporation by making P additions to a single high purity vacuum melted heat of $Ni_2Cr$. The phosphorous levels of the low, medium and high P alloys were 0.004, 0.032, and 0.142 atomic percent respectively.

The cast slabs were alternately cold rolled, annealed in an inert gas atmosphere at 1150°C, and water quenched, until various final thicknesses between 0.4 and 3.2 mm were attained.

Mat. Res. Soc. Symp. Proc. Vol. 39. ©1985 Materials Research Society

## Specimen Preparation

In order to ensure that the material was fully recrystallized and annealed, all specimens were prepared from material which was initially heat treated at 900°C for 2 hours and then quenched.

TEM specimens were electrochemically thinned from 0.4 mm thick sheet material after 400 hours aging at 500°C.

Two sets of specimens were prepared from 3.2 mm thick material for x-ray diffraction experiments; one set was used with the 900°C pretreatment while the other set received an additional 600°C pretreatment of four hours duration. The 600°C pretreatment, just above the critical ordering temperature, was intended to induce an advanced state of short range order (SRO) in these specimens for comparison with the results from the 900°C pretreated specimens. Both sets were then aged for times up to 500 hours in appropriate intervals.

Metallography was performed on the same specimens used for the x-ray experiments. The surfaces were polished and chemically etched after each aging interval, to bring out the surface details and to remove damaged material before the x-ray measurements.

# RESULTS

## Transmission Electron Microscopy

Dark field TEM micrographs taken from a single ordered superlattice diffraction spot image ordered domains from only one of the six variants of the ordered phase. After 400 hours aging, dark field photomicrographs revealed that the average domain sizes of the low and medium P specimens were nearly identical, approximately 10 nm (Figure 1a), while the average domain size in the high P specimen was considerably larger, approximately 50 nm (Figure 1b). Both micrographs in Figure 1 show a grain boundary in the upper right hand corner. Notice that in Figure 1a the domain size is uniform

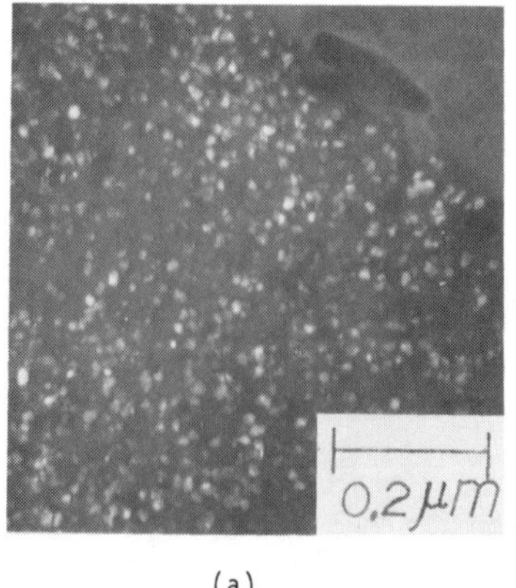


(a)

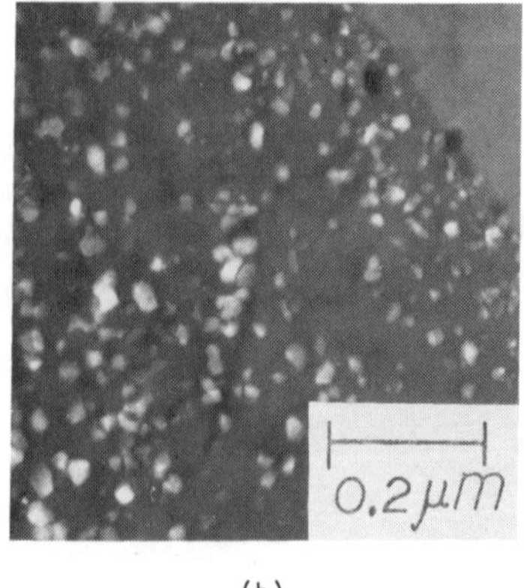


(b)

Figure 1. Dark field images showing the size and morphology of ordered domains within the material. A grain boundary is visible in the upper right corner of each micrograph. a) Medium P specimen with a uniform size distribution of domains typical of both low and medium P specimens. b) High P specimen with a larger domain size inside the grain and smaller domains near the grain boundary.

throughout, even up to the grain boundary, while in Figure 1b the domain size decreases near the grain boundary. Although diffuse scattering effects should have been present at short aging times due to SRO, these were not observable due to the nearly equal atomic scattering factors of Ni and Cr, and for the same reason superlattice reflections were not observed in the x-ray diffraction experiments.

## X-ray Diffraction

The data from the x-ray diffraction measurements of the FCC equivalent lattice parameter are shown in Figures 2a and 2b. For either pretreatment, a

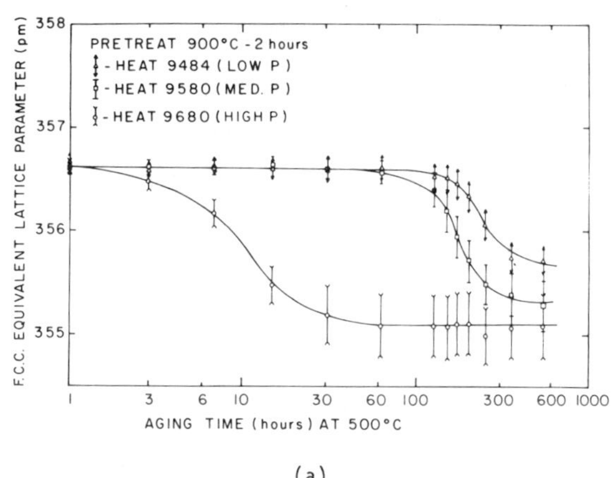


(a)

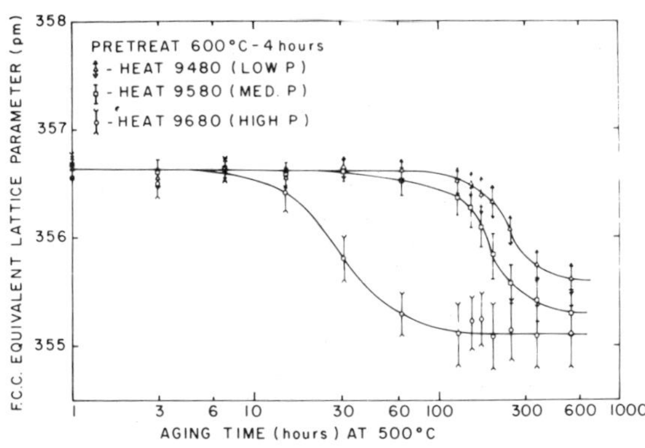


(b)

Figure 2. The progress of the order-disorder phase transformation as seen through the lattice contraction. a) Material quenched from 900°C. b) Material quenched from 600°C.

higher bulk P content was always found to lead to a more rapid order transformation, and the highest P specimens transformed in only about one tenth the time required by the others. For both pretreatments, the initial lattice parameter for all specimens was 356.6 pm, and specimens with the same P content experienced the same ultimate lattice contraction to the same absolute final lattice parameter; the low, medium and high P alloys contracted 0.27%, 0.36% and 0.43% respectively.

In the 900°C pretreatment (Figure 2a), the low, medium and high P specimens reached the inflection point of the sigmoidal transformation curve in 240, 170 and 10 hours respectively while in the 600°C pretreatment (Figure 2b) they required 240, 185, and 30 hours respectively.

During the sigmoidal segment of the transformation curves of the low and medium P specimens, fundamental diffraction lines slowly shifted from their disordered line positions to their ordered line positions without splitting or significantly changing their shape. However, for both pretreatments of the high P specimens, the diffraction lines split into distinct ordered and disordered components during the LRO transformation. This can be seen in Figure 3, where the corrected data represents monochromatic Cu $K_{\alpha 1}$ radiation after the background was subtracted and a Rachinger correction applied. Thus, the high P specimens experienced a clear two phase stage during the transformation while the low and medium P specimens did not.

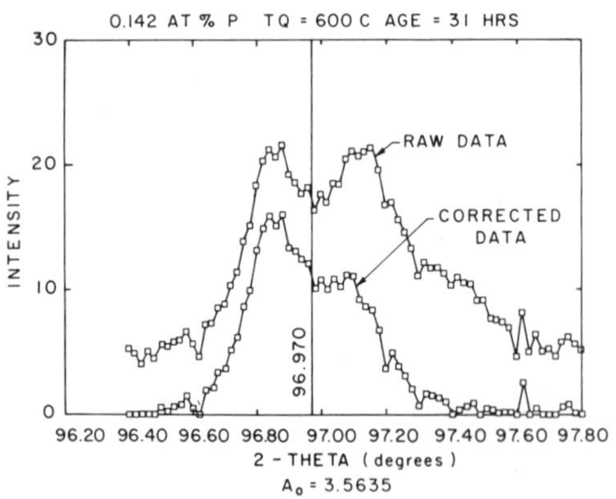


Figure 3. X-ray diffraction line scan from the high P material pretreated at 600°C, aged at 500°C for 31 hours. The material is 53% transformed. The corrected data shows characteristic peaks from material with both the disordered and ordered lattice parameters.

## Metallography

During the LRO transformation, the low and medium P specimens exhibited only small scale surface undulations upon etching (Figure 4). Before and after the sigmoidal portion of the transformation curve, these were not observed, and the surfaces of grain interiors were smooth and grain boundaries were clearly evident. In contrast, high P specimens exhibited evolution of a distinct two phase structure. An example of this two phase structure can be seen in Figure 5, in which the material had achieved a 50% transformed condition. The ordered regions of the crystal are more resistant to the chemical etch, and are seen here in raised relief reaching sizes up to 50 microns. Since the TEM results show that the average domain size does not

Figure 4. SEM micrograph showing the surface undulations within grains of medium P specimen pretreated at 900°C and aged 172 hours; specimen is 51% transformed.

Figure 5. SEM micrograph showing the high P specimen pretreated at 600°C and aged at 500°C for 31 hours, resulting in 53% transformation. Note the ordered phase in raised surface relief.

exceed approximately 50 nm, the ordered regions must represent colonies of ordered domains. The colonies nucleated inside the grains and expanded through the advance of an order-disorder interface which has an irregular contour. In the proximity of grain boundaries and twin boundaries, the advancing colonies stopped, and these interface regions transformed much later in the aging cycle. Occasionally a colony appeared to penetrate a twin boundary, usually at a single point, and then expanded within the twin from that point; however, twins were usually observed to nucleate separate colonies.

DISCUSSION

The following results for the low and medium P heats tend to suggest that a continuous ordering mechanism was active in these specimens: i) The fundamental x-ray lines slowly shifted their positions without splitting or significantly changing their shape as the lattice contracted during ordering. ii) TEM showed a relatively uniform domain size within the grains even up to the grain boundaries. iii) Metallography showed no evidence of any feature which might be construed as an order-disorder interface. The continuous ordering process was reflected in the emergence and disappearance during the transformation of small scale surface undulations within grains, which may be due to local variations in the degree of order. Since the smallest dimensions of most of the raised areas in Figure 4 are about two microns, if they were ordered regions surrounded by disordered material their sizes would not give rise to measurable x-ray line broadening. It can therefore be concluded that the surface undulations were not due to a two-phase structure of ordered and disordered material, since no splitting of the fundamental x-ray lines occurred for the low and medium P specimens. Furthermore, the observation of a gradual shift in line position is not consistent with the presence of such a two-phase structure.

In contrast, the high P specimens exhibited evidence of a two phase structure in both the x-ray scans and the metallography. However, while SEM micrographs showed that throughout the transformation a definite interface existed between the ordered and the disordered regions of the crystal, they offer no direct evidence of the atomic processes occurring at the colony interface.

The following theory is proposed to describe the growth of colonies of domains by movement of an order-disorder interface in the highest P specimens, and the resultant increase in transformation rate. At the beginning of the 500°C aging cycle, subcritical nuclei form throughout the material. At the same time a few supercritical nuclei are formed and begin to grow. As the size of one of these ordered domains increases and the degree of LRO within the domain increases, it begins to experience a lattice contraction. Since the domain is completely coherent with the matrix, the material immediately outside the domain begins to experience a radially tensile and circumferentially compressive stress. The resulting coherency strains begin to interact with neighboring subcritical nuclei, whose own coherency strains create an effective energy barrier to their growth. When the coherency strain field of the growing supercritical domain begins to overlap those of the subcritical nuclei in the surrounding lattice, the strain energies of the nearby ordered nuclei are reduced, and they therefore become supercritical. As they begin to grow, APB's are formed between the first generation and the second generation domains. In this way the order-disorder interface is passed on to the second generation domains and further growth of the first generation domains is arrested by the APB's. At grain boundaries (possibly at twin boundaries as well) P has diffused from the bulk into the nearby interface leaving behind P depleted zones. When the order-disorder interface encounters these zones, the transformation process changes to the slower continuous ordering mechanism appropriate to the lower local P concentrations. Thus the regions near grain boundaries remain untransformed until continuous ordering occurs, producing smaller domains.

An increased P concentration evidently increases the the driving force for the LRO reaction in $Ni_2Cr$, resulting in an increased transformation rate in the continuous ordering process, and in the decreased critical LRO nucleus size for nucleation and growth which leads to the rapid colony growth process. The fact that the fractional lattice contraction upon ordering increases with P content supports the suggestion that P increases the driving force for the LRO reaction. This may be due to an effect of P on the Fermi energy, which would be expected to differ in the ordered and disordered phases, especially in view of the attendant volume difference. Low and medium P specimens were transformed by the continuous ordering process before a critical size LRO nucleus could be formed to initiate nucleation and growth. In the high P specimens, since the critical size LRO nucleus was sufficiently small, supercritical nuclei formed early in the aging cycle and initiated the growth of colonies as described above before the continuous ordering process could take place. This explains the dramatic increase in the reaction rate in the high P specimens, since an alternate, faster transformation mechanism occurs. It is believed that in the presence of the low P concentrations near grain boundaries, the critical nucleus size for nucleation of fully ordered domains is too large to result in domain nucleation even with the aid of the stress field of the nearby domains. Therefore the continuous mechanism operates near the grain boundaries.

## CONCLUSIONS

The ordering transformation has been observed in $Ni_2Cr$ specimens with three different P concentrations and two heat treatments. Specimens with the low and medium P bulk concentrations, 0.004 and 0.032 at.% P, were found to transform via the continuous ordering process. Higher P concentrations were always found to lead to larger domains and more rapid transformation for a given heat treatment.

High P specimens (0.142 at.% P) were found to transform much faster than their lower P counterparts, by as much as an order of magnitude. X-ray diffraction showed that the material develops a two phase structure during the transformation, with characteristic peaks from both the ordered and the disordered material appearing simultaneously when the material is in transition. Metallographic examination also revealed a two phase structure during the same period. Ordered regions nucleated within the bulk and grew outwards toward the grain boundaries. The ordered regions were often observed to be tens of microns in extent, while the ordered domain size seldom exceeded 50nm. The growing ordered regions are actually colonies of ordered domains, which grow by adding new members at the order-disorder interface. Colony growth appears to represent an intermediate level of coordinated ordering behavior between the continuous ordering mechanism and the homogeneous nucleation and growth mechanism.

Since P leads to increased ordering kinetics, and the ordered phase is quite brittle even in the absence of H [1], P appears to be detrimental to H embrittlement resistance.

## ACKNOWLEDGEMENTS

This work was partially supported by Cabot Corporation, the Wyman Gordon Foundation, and the Office of Advanced Studies, University of Notre Dame, and the authors are thankful for this support. The authors would also like to thank Drs. C. W. Allen and G. C. Kuczynski for many helpful discussions.

## REFERENCES

1. J. A. Kargol, M. V. Zeller, R. Asfahani and T. M. Parrill, Applications of Surface Science 15, 129 (1983)

2. L. E. Tanner, Acta Met. 20, 1197 (1972)

3. Ye. Z. Vintakin and G. G. Urushadze, Physics of Metals and Metallography 27, 132 (1969)

4. M. Hirabayashi, M. Koiwa, K. Tanaka, T. Tadaki, T. Saburi, S. Nenno and H. Nishiyama, Trans., Jap. Inst. of Metals 10, 365 (1969)

5. R. J. Taunt and B. Ralph, Phys. Stat. Solidi A 29, 431 (1975)

6. L. Karmazin, Mat. Sci. and Engineering 54, 247 (1982)

## $DO_3$-DOMAIN STRUCTURES IN $Fe_3Al$-X ALLOYS

MADAN G. MENDIRATTA* AND HARRY A. LIPSITT**
*Universal Energy Systems, Inc., 4401 Dayton-Xenia Rd., Dayton, OH 45432
**Materials Laboratory, AFWAL/MLLM, Wright-Patterson AFB, OH 45433

### ABSTRACT

The influence of a number ternary additions in $Fe_3Al$ upon the $DO_3$ domain structures and $DO_3 \rightarrow B2$ transformation temperature, $T_c$, has been investigated. It was found that additions in (atomic percent) of 5 Ti, 5 Cr, 6 Mn, 12 Mn, 10 Ni, 3 Mo, 6 Mo, 3 Si, and 5 Si raise $T_c$. In the Fe-20 Al-5 Ti alloy not only is $T_c$ raised significantly ($\geq$ 750°C), but the $DO_3$ antiphase domain boundaries with fault vector $\frac{1}{2}a_o'$<100> become highly crystallographic and are parallel to {100} planes.

### INTRODUCTION

The alloy Fe-25 at.% Al ($Fe_3Al$) has an ordered cubic $DO_3$ crystal structure at temperatures below ∿ 540°C, above which it transforms to the defect FeAl phase with an ordered cubic B2 crystal structure. In the $DO_3$ phase field, depending upon ordering heat treatments, two types of domains coexist: large domains with a characteristic anti-phase domain boundary (APB) vector $\frac{1}{4}a_o'$<111> and within these large domains, small domains with a characteristic APB fault vector of $\frac{1}{2}a_o'$<100>, $a_o'$ being the lattice parameter of the $DO_3$ phase which is twice as large as $a_o$, the lattice parameter of the B2 phase. In the B2 phase field only one type of domain with an APB fault vector of $\frac{1}{2}a_o'$<111> exists. In the binary Fe-Al alloys these APBs are highly isotropic, i.e., they are smoothly curving with no preference to any specific crystallographic planes.

The influence of ternary elements in $Fe_3Al$ upon the $DO_3 \rightleftarrows B2$ transformation temperature, $T_c$, and upon the domain structure is not well established. We are aware of only one paper [1] in which the effect of substituted elements, Ti, Mn, Ni, Cu, and Ge in $Fe_3Al$ has been investigated by electron microscopy and Mössbauer spectroscopy. The major finding is that these elements substitute selectively on specific sublattice sites. The present investigation was undertaken to investigate the effect of a large number of ternary additions upon the evolution of the $DO_3$ domain microstructure and upon the $DO_3 \rightarrow B2$ transformation temperature. Previous studies have shown that the low $T_c$ of 540°C has an adverse effect upon the high temperature tensile strength of $Fe_3Al$, therefore, it was of some importance to determine which ternary elements significantly increase $T_c$.

## EXPERIMENTAL PROCEDURES

A number of transition metal and some non-transition metal ternary alloying additions were investigated. The transition metal additions were substituted for Fe with the formula Fe-25Al-X (all compositions in atomic percent) where X = 1 Ti, 5 Ti, 1 Cr, 5 Cr, 6 Mn, 12 Mn, 3 Ni, 10 Ni, 2 Nb, 5 Nb, 3 Mo, 6 Mo, 1 Ta, and 5 Ta. The non-transition metal alloys were: Fe-23 Al-5 Cu, Fe-22 Al-3 Si, and Fe-20 Al-5 Si. The alloys were prepared by hot extrusion from powders manufactured by a plasma rotating electrode process.* The hot extrusions were carried out** at ∿ 1120°C with an extrusion ratio of 16:1. Wet chemical analysis indicated the alloy compositions were very close to the nominal compositions.

Specimens machined from the extruded alloys were held at 1120°C for 1 hr. in an inert atmosphere, water quenched, and then subjected to ordering heat treatments at temperatures from 600-750°C followed by water quenching. The domain structures were examined by thin foil transmission electron microscopy. The different superlattice reflections required to generate contrast from the $\frac{1}{4}a_o'\langle 111\rangle$ and $\frac{1}{2}a_o'\langle 100\rangle$ APBs have been calculated by Marcinkowski and Brown [2]. Whether $T_c$ was higher or lower than a given ordering temperature was inferred from the size of the $\frac{1}{2}a_o'\langle 100\rangle/DO_3$ domains, if the ordering temperature was below or above $T_c$ then large or ultrafine quenched-in $DO_3$ domains were expected to be present, respectively.

## RESULTS AND DISCUSSION

Since a large number of alloys were investigated, only a few examples of the domain microstructures will be presented. All the results are summarized in Table I. Figure 1 shows the microstructures for the Fe-25 Al-6 Mo alloy. It shows dark-field electron micrographs taken with $\bar{g} = (\bar{1}11)$ for specimens quenched from 600, 650, and 700°C. Also included is the $[1\bar{1}0]$-zone-axis diffraction pattern for a specimen quenched from 600°C. Similar diffraction patterns were observed for the other two ordering temperatures, with no significant difference being found in the intensities of the superlattice reflections $(\bar{1}11)$ and (002). Smoothly curving $DO_3$ domains are visible for 600 and 650°C treatments, the domains being slightly larger for 650°C than for 600°C; however, the domains are extremely fine in the 700°C quenched specimens. These results indicate that up to 650°C the alloy is in the $DO_3$ phase field and that at 700°C the $DO_3$ phase boundary is exceeded. Thus, $T_c$ must fall somewhere between 650 and 700°C. Figures 2(a) and 2(b) are the (111) dark-field electron micrographs for the two alloys

*Nuclear Metals, Inc.
**Hot extrusions were performed by TRW, Inc.

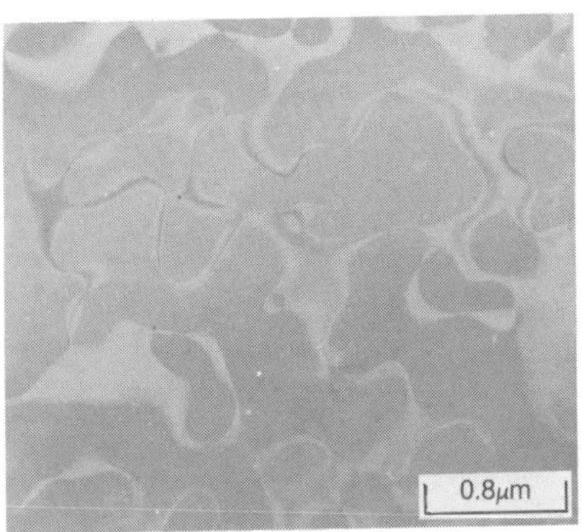


(a) 600°C/1 hr. → wq. $\bar{g}$ = ($\bar{1}$11)

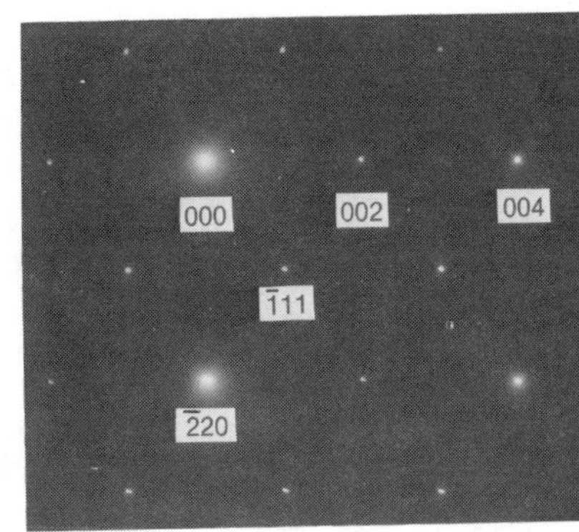


(b) [1$\bar{1}$0] ZONE AXIS

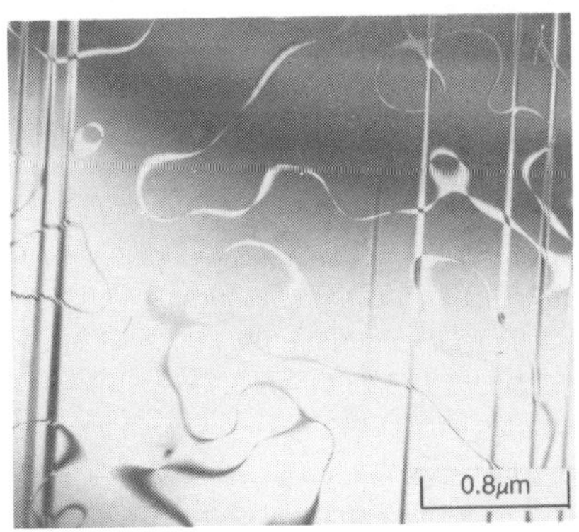


(c) 650°C/1 hr. → wq. $\bar{g}$ = ($\bar{1}$11)

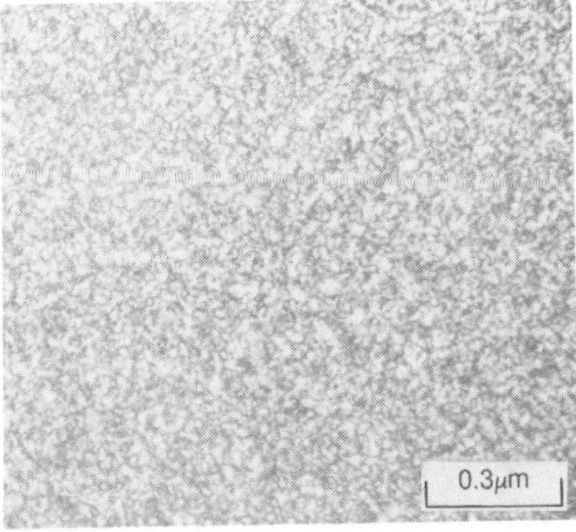


(d) 700°C/1 hr. → wq. $\bar{g}$ = ($\bar{1}$11)

Figure 1. Dark-Field Electron Micrographs and Diffraction Pattern Showing Variation of the $DO_3$ Domain Size with Temperature in Fe-25 Al-6 Mo.

TABLE I. Summary of $T_c$ Data ($T_c$ for $Fe_3Al \cong 540°C$)

| ALLOY | $T_c$(°C) | REMARKS | ALLOY | $T_c$(°C) | REMARKS |
|---|---|---|---|---|---|
| $Ti_1$ | < 600 | | $Cr_1$ | < 600 | |
| $Ti_5$ | ≧ 750 | Anisotropic APBs | $Mo_3$ | 600-650 | |
| $Cr_5$ | 600-650 | | $Mo_6$ | 650-700 | |
| $Mn_6$ | 600-700 | | $Ta_1$ | < 600 | |
| $Mn_{12}$ | 600-650 | | $Ta_5$ | < 600 | Two Phases |
| $Ni_3$ | < 600 | | $Cu_5$ | < 600 | Two Phases |
| $Ni_{10}$ | 600-650 | | $Si_3$ | 650-700 | |
| $Nb_2$ | < 600 | Two Phases | $Si_5$ | 700-725 | |
| $Nb_5$ | < 600 | Two Phases | | | |

Fe-25 Al-1 Ti and Fe-25 Al-5Ti, both quenched from 600°C. The former alloy shows quenched-in very fine domains indicating that $T_c$ is below 600°C, while the latter shows large $DO_3$ domains indicating that $T_c$ is above 600°C; it can be seen that the APBs are crystallographically anisotropic. Thus, replacement of Fe with 5 at.% Ti in $Fe_3Al$ not only raises $T_c$, but also changes the nature of the APBs; more results on this alloy will be presented later in this paper.

The type of data included in the above examples were generated for all the alloys investigated. A summary of the temperature range in which $T_c$ falls is given in Table I. Except for the 5 Ti alloy, all other alloys exhibited highly isotropic APBs. The alloys containing 2 and 5 Nb, 5 Ta, and 5 Cu exhibited a two phase microstructure, a solid solution $DO_3$ matrix, and a ternary element-rich second phase. The solubility limits and the characterization of the second phases in these alloys have been determined [3], however, those results are outside the scope of this paper. It can be seen from Table I that a large number of ternary additions in $Fe_3Al$ increase $T_c$ above 600°C.

Since the addition of 5 at.% Ti produced highly crystallographic APBs, it was decided to perform a number of heat treatments on this alloy to examine the evolution of the APB microstructure. Therefore, thin foils corresponding to the 1120°C/1 hr. → water quench condition and to the subsequent heat-treated conditions from 500-750°C were examined in the TEM. In the 1120°C quenched condition the microstructure consisted of extremely fine $DO_3$ domains and much larger, highly isotropic B2 domains. For the 550°C/24 hr. treatment, somewhat isotropic $\frac{1}{2}a_o'\langle100\rangle$ APBs, Fig. 3(a), enclosed by larger quenched-in $\frac{1}{2}a_o\langle111\rangle$ APBs, Fig. 3(b), were observed.

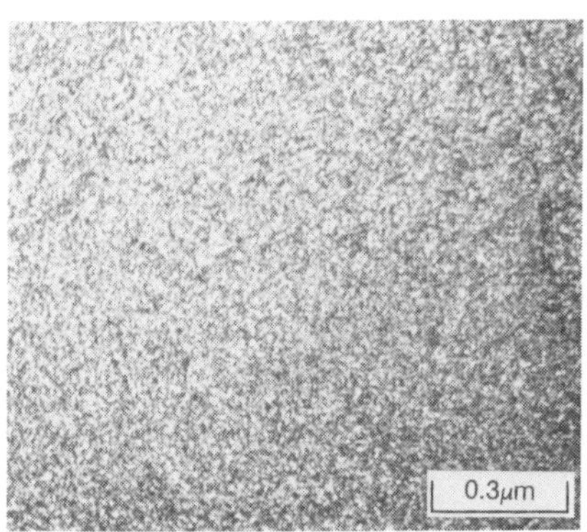


(a) Fe-25 Al-1 Ti

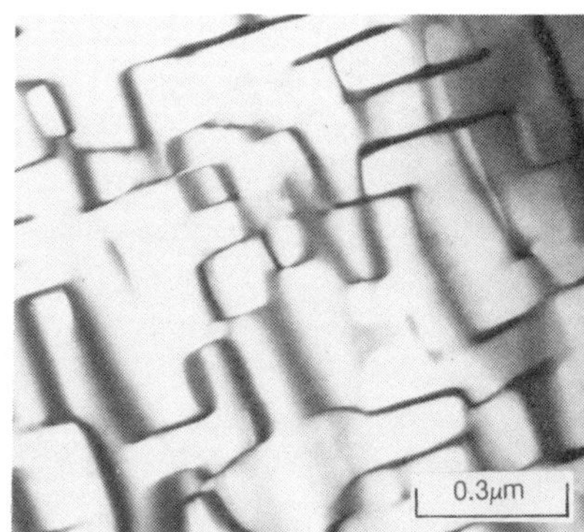


(b) Fe-25 Al-5 Ti

Figure 2. The $DO_3$ - Domain Microstructure in the Two Ti Alloys. Dark field, $\bar{g}$ = ($\bar{1}$11). Both alloys were heat treated at 600°C for 1 hr. → water quench after water quenching from 1120°C.

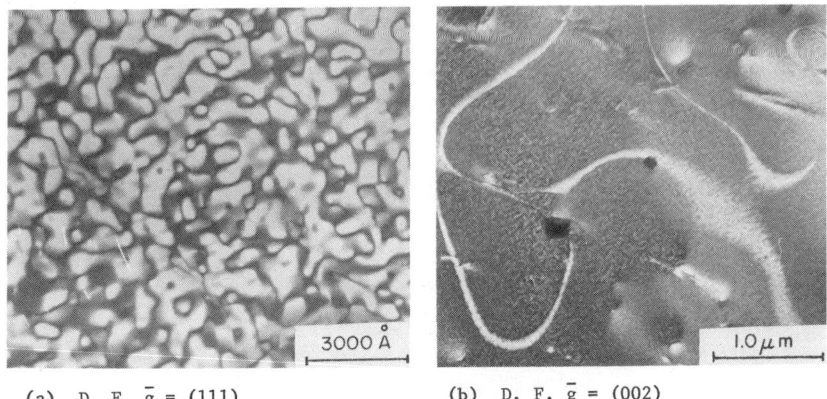


(a) D. F. $\bar{g}$ = (111) (b) D. F. $\bar{g}$ = (002)

Figure 3. Dark-Field Electron Micrographs Showing Isotropic APBs in Fe-25 Al-5 Ti Alloy Heat-Treated at 550°C for 24 hr. After Water Quenching from 1120°C.

With increasing temperature, the $\frac{1}{2}a_o'$<100> APBs became anisotropic rather quickly, e.g., at 650°C/15 min., Fig. 4(a), the APBs exhibit a degree of anisotropy and after 1 hr. at 650°C, Fig. 4(b), they became totally anisotropic. Trace analysis revealed that these APBs lie on {100} planes. These microstructural features persisted up to 750°C indicating that addition of 5 at.% Ti stabilizes the $DO_3$ phase at least up to 750°C. Figure 5 shows the APBs corresponding to a heat treatment of 700°C/1 hr. It can be seen that portions of large $\frac{1}{2}a_o$<111> APBs, [Fig. 5(b)], have begun to align with the cube planes.

The results of this study have shown that although a number of ternary additions increase $T_c$, the greatest increase is provided by 5 Ti ($\geqq$ 750°C). Also, of the alloys investigated the Ti addition is unique in that it produces highly anisotropic $\frac{1}{2}a_o'$<100> APBs. The increase in $T_c$ implies an increase in the second nearest neighbor interaction energy [4].

Using zeroth order Bragg-Williams and pairwise interaction approximations, Marcinkowski and Brown [2] have calculated the APB energies, $\gamma_{hkl}$, for APBs lying on different crystallographic planes for $Fe_3Al$. The maximum and minimum energy corresponded to {111} and {100} planes, respectively, with values of 68 and 40 ergs/cm$^2$. It was argued that the ratio of $\gamma_{111}/\gamma_{100} \simeq 1.73$ was too small to induce APB anisotropy. Following the same calculation procedures and taking $T_c \simeq$ 750°C for the Fe-25 Al-5 Ti alloy, the calculated values of $\gamma_{111}$ and $\gamma_{100}$ are 96.4 and 55.7 ergs/cm$^2$, the ratio $\gamma_{111}/\gamma_{100}$ being exactly the same as for $Fe_3Al$. The reason for this is that under the approximations used, this ratio is a function only of Miller indices and is independent of $V_2$, the second nearest neighbor interaction energy. Based upon this ratio, then, the Ti containing alloy should not exhibit APB anisotropy, contrary to the present results. The absolute difference ($\gamma_{111} - \gamma_{100}$) for the Ti alloy is 40.73 ergs/cm$^2$ which is significantly higher than that for $Fe_3Al$ which is 28 ergs/cm$^2$. It is possible that this higher difference dictates the observed APB behavior. It is also possible that the ternary alloying addition increases the APB energy anisotropy: theory to compute this has not yet been developed.

## CONCLUSIONS

The conclusions of this study are mainly that a number of transition-metal as well as Si additions increase the $DO_3 \rightarrow B2$ transformation temperature in $Fe_3Al$, the greatest increase produced by addition of 5 at.% Ti. Of all of the alloys studied, the 5% Ti alloy produces highly crystallographic

(a) 650°C/15 min. (b) 650°C/1 hr.

Figure 4. Dark-Field Electron Micrographs Exhibiting Transition in the Nature of the $DO_3$ APBs from Somewhat Isotropic to Totally Anisotropic APBs in Fe-25 Al-5 Ti Alloy.

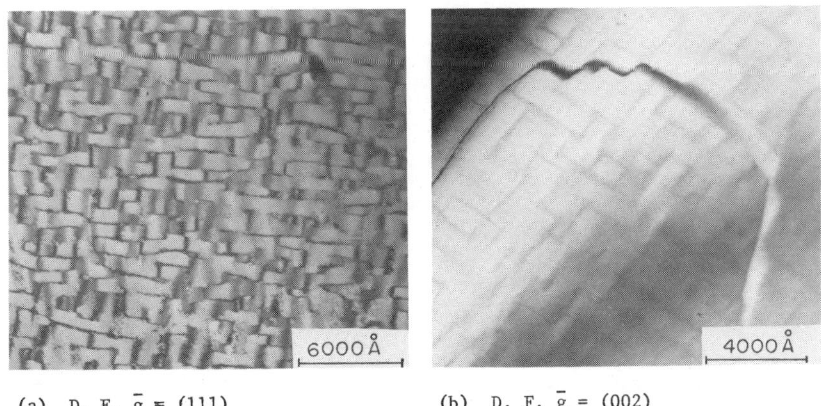


(a) D. F. $\bar{g}$ = (111) (b) D. F. $\bar{g}$ = (002)

Figure 5. The Structure of APBs for a Heat Treatment Corresponding to 700°C/1 hr. → Water Quench in Fe-25 Al-5 Ti Alloy.

$DO_3$ APBs, lying on {100} planes. This anisotropy cannot be satisfactorily explained based upon Bragg-Williams approximations and pairwise interaction energies taking into account only up to second nearest neighbors.

REFERENCES

1. G. Athanassiadis, G. LeCaer, J. Foct, and L. Rimlinger, Phys. Stat. Sol. (a), 40, 425 (1977).
2. M. J. Marcinkowski and N. Brown, J. Appl. Phys. 33, 537 (1962).
3. M. G. Mendiratta, Universal Energy Systems, Inc., Dayton, OH, Unpublished research.
4. P. S. Rudman, Acta. Metall. 8, 321 (1962).

# RESISTIVITY OF SHORT-RANGE AND LONG-RANGE ORDER CHANGES IN $Ni_4Mo$

T. S. Lei*, K. Vasudevan** and E. E. Stansbury**
* National Taiwan Institute of Technology, Taipei, Taiwan.
** Dept. of Materials Science and Engineering, University of Tennessee, Knoxville, TN 37916.

## ABSTRACT

Short-range order (SRO) and long-range order (LRO) reactions in $Ni_4Mo$ as a function of room temperature deformation and thermal treatments between -269 and 1300°C are reported. Resistivity results show that SRO is dispersed by heating near 1300°C and by cold working to 68% reduction of area (RA). The resistivity decreases to a minimum at 68% RA after which an increase is attributed to an increase in defect density.

The resistivity as a function of variations in the time-temperature history is reported. Cooling at 60°C/min results in the maximum "frozen in" SRO with resistivity of 144 μohm cm at 25°C; the resistivity of the LRO state obtained on slow cooling is 48 μohm cm. At -269°C, the resistivities are 139 and 8 μohm cm for the SRO and LRO states respectively. Correlations relate kinetics and microstructures to resistivity changes during heating, cooling and isothermal SRO-LRO transformations. Preliminary results of reheating fully ordered samples to near 868°C and quenching are reported. Results are discussed with relevance to existing theories.

## INTRODUCTION

On cooling below 868°C, the critical ordering temperature, $Ni_4Mo$ transforms from a SRO face-centered cubic structure to a LRO body-centered tetragonal structure with c/a = 0.623 [1,2]. Although the transition from SRO to LRO on isothermal aging below the critical temperature are well established [3-17], information on structural changes on continuous heating and cooling is limited. Apart from the data of Grube and Schlect [18] and Baer [19], who used electrical resistivity measurements to study phase changes in a Ni-20 at.% Mo and a Ni-17 at.% Mo alloy, limited information exists on understanding the resistivity and structural changes associated with SRO and LRO on continuous heating and cooling.

This study was undertaken to completely characterize the electrical resistivity behavior associated with SRO and LRO changes in $Ni_4Mo$ and to correlate these observations with corresponding microstructural changes. Also, preliminary results of reheating fully ordered samples to just below 868°C and quenching are reported. The above results are compared with those of other systems and qualitative comparisons are made with existing theories of the resistivity of alloys exhibiting SRO and LRO.

## EXPERIMENTAL METHODS

The $Ni_4Mo$ alloy was prepared by induction remelting of a Ni-Mo master alloy, with nickel corrections, and casting into water-cooled copper molds to produce ingots 0.75 inches in diameter and 2.5 inches long. The chemical analysis indicated a Ni:Mo atom ratio of 4.032:1. The ingots were homogenized at 1100°C for 100 hours, slow cooled to room temperature, reheated to 950°C for 1 hour and then water-quenched to retain the fcc phase. Subsequent swaging of the ingots with intermediate anneals between 950 and 1000°C produced rods 0.125 inches in diameter. Four inch long samples were cut from the swaged rods for resistivity measurements.

The technique used for measuring the electrical resistivity at room temperature has been described elsewhere [17]. For low temperature measurements, a dry ice and methanol mixture, liquid nitrogen, and liquid helium were used. A computer-controlled resistivity apparatus was designed for measurements at elevated temperatures over wide ranges of heating and cooling rates, or isothermally.

Samples for electron microscopy were prepared in a Struers "Tenupol" jet thinning apparatus. An electrolyte consisting of 660 ml $H_2SO_4$, 25 ml HCl and 340 ml distilled water was used with a voltage of 16 to 20 volts, current of 0.4 to 0.6 amperes and a temperature of 4$^o$C.

## RESULTS

### Effect of Cold Working on Water-quenched $Ni_4Mo$

Figure 1 shows that the resistivity of $Ni_4Mo$ water-quenched from 950$^o$C decreases to a minimum at 68% RA and then increases on further reduction.

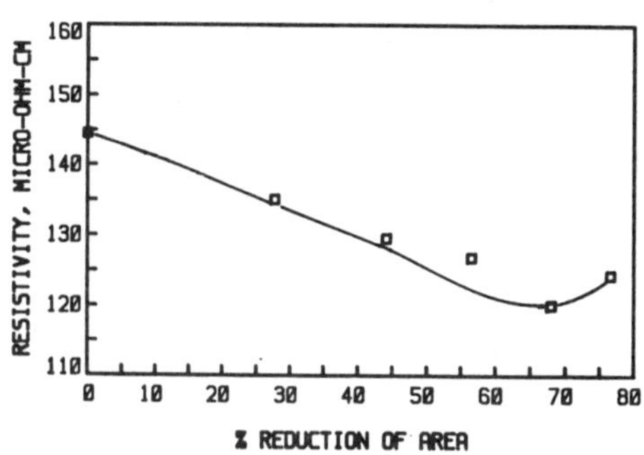


Figure 1. Variation of resistivity with cold work for as-quenched alloys.

### Resistivity Changes on Continuous Heating and Cooling

The resistivity of $Ni_4Mo$ as a function of continuous heating, cooling and isothermal holds is shown in Figure 2, where H, C and I denote heating, cooling and isothermal runs, the numbers 1, 2 ..., the sequence of runs.

In Figure 2, curve 1H shows that heating the initially 77% RA cold swaged alloy, increases the resistivity linearly to about 300$^o$C due to increasing thermal scattering of electrons, followed by a non-linear rapid increase to about 650$^o$C due to progressive return of the SRO destroyed by plastic deformation. The resistivity rapidly decreases between 650 and 800$^o$C due to the onset of LRO, and then increases near the LRO-SRO critical temperature of 868$^o$C due to a decrease in the degree of LRO. Above 868$^o$C, LRO to SRO transformation occurs and the resistivity decreases with increasing temperature with the slope of the resistivity-temperature curve becoming zero near 1300$^o$C, indicating that the SRO is dispersed on approaching the melting temperature. This decrease in resistivity with increasing temperature results in a negative temperature coefficient of resistivity.

On cooling at 60$^o$C/min from 1300$^o$C (curve 2C), the resistivity initially increases to about 151 micro-ohm-cm at 650$^o$C due to SRO. Below 650$^o$C, the SRO is effectively "frozen-in" due to inadequate atom mobility and the resistivity decreases to 144 and 139 micro-ohm-cm at 25 and -269$^o$C respectively. In the following heating curve 3H, the resistivity coincides with the previous curve 2C to 300$^o$C, and then has a higher value to 650$^o$C due to additional SRO which occurs because of adequate diffusion. Above 650$^o$C, the behavior is similar to that of curve 1H.

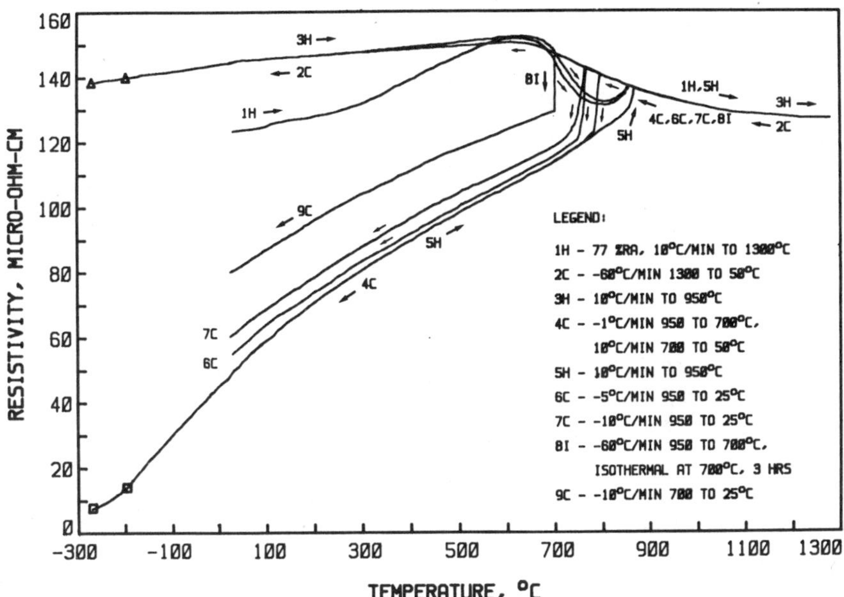


Figure 2. Variation of resistivity with temperature for $Ni_4Mo$ under different indicated conditions.

Curve 4C results from slow cooling at rates of $1^oC$/min from 950 to $700^oC$ and $10^oC$/min from 700 to $50^oC$ respectively. The resistivity increases initially on cooling to about $800^oC$ and then rapidly drops in the temperature range 800 to $775^oC$ with the onset of LRO. This cooling rate is sufficiently slow to allow complete transformation to the LRO state. The resistivity values at 25 and $-269^oC$ are 48 and 8 micro-ohm-cm respectively which contrasts markedly with the corresponding SRO resistivity values of 144 and 139 micro-ohm-cm and the cold worked state value of 124 micro-ohm-cm at $25^oC$.

The resistivity values on reheating the fully long range ordered $Ni_4Mo$ alloy (curve 5H), are essentially similar to $775^oC$. Beyond $775^oC$, the resistivity increases with a greater slope to $850^oC$, followed by a rapid increase near the critical temperature of $868^oC$. The value of this temperature is in exact agreement with the results of Guthrie and Stansbury [2]. Above $868^oC$, the behavior is similar to curves 1H and 3H. However, notably, the experimental transformation temperature itself is dependent on the initial state of order; a mixture of SRO and LRO transforms to SRO at a lower temperature (curves 1H and 3H) when compared to an initially complete LRO state (curve 5H). This result agrees with the recent radiation experiments of Banerjee et al [15]. Another noteworthy aspect of curve 5H is the decrease in slope in the temperature range -150 to $600^oC$.

An increase in cooling rate results in an increase in the LRO resistivity at $25^oC$ to 54.5 and 60.5 micro-ohm-cm for cooling rates of 5 (curve 6C) and $10^oC$/min (curve 7C) respectively. Microhardness measurements showed a corresponding decrease with values of 494, 464 and 434 DPN for cooling rates of 1, 5 and $10^oC$/min respectively.

Curve 8I results for a sample held for 5 minutes at $950^oC$, rapidly cooled to $700^oC$ and held for 3 hours. The change from 149 to 129 micro-ohm-cm is due to LRO. The sample was then cooled to $25^oC$ (curve 9C) which resulted in a value of 80 micro-ohm-cm, indicating a partially ordered state.

The decrease in resistivity on annealing at 700°C is actually preceded by a small increase as seen in Figure 3 which shows the change in resistivity with time.

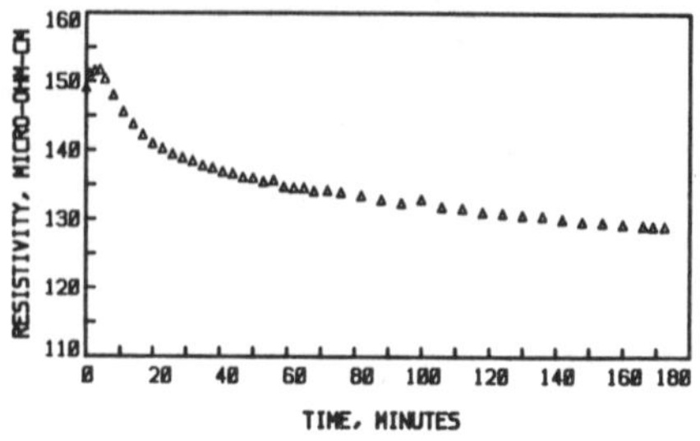


Figure 3. Variation of resistivity with time at 700°C.

## Effect of Reheating and Rapid Quenching on Resistivity

To examine the changes in structure and resistivity near the ordering temperature, fully ordered samples (cooled at 1°C/min) were reheated to 800, 825, 850, and 860°C for 1/2 hour and quenched in iced brine. With an increase in reheat temperature, the resistivity increases relative to that of the fully ordered value of 48 micro-ohm-cm, being 51.9, 57.3, 58.6, and 62.54 micro-ohm-cm respectively for the aforementioned reheat temperatures.

## Electron Microscopy Observations

The electron microscopy results are shown in Figures 4 to 7. All dark field micrographs were obtained using 1/5{420} superlattice spots. A dark field micrograph of $Ni_4Mo$ cooled at a rate of 1°C/min from 950°C is shown in Figure 4a. The corresponding [122] diffraction pattern (Figure 4b) shows superlattice spots from two variants of anti-parallel twin related domains first reported by Ruedl et al [6]. No {11/20} SRO spots are observed thus indicating a fully ordered structure. The resulting domain size was 187 nm. With an increase in cooling rate, the domain size decreases to 80 and 36 nm for cooling rates of 5 (Figure 5a) and 10°C/min (Figure 5b) respectively. Only superlattice spots were observed and {11/20} SRO spots were absent in the [112] and [001] diffraction patterns corresponding to these samples.

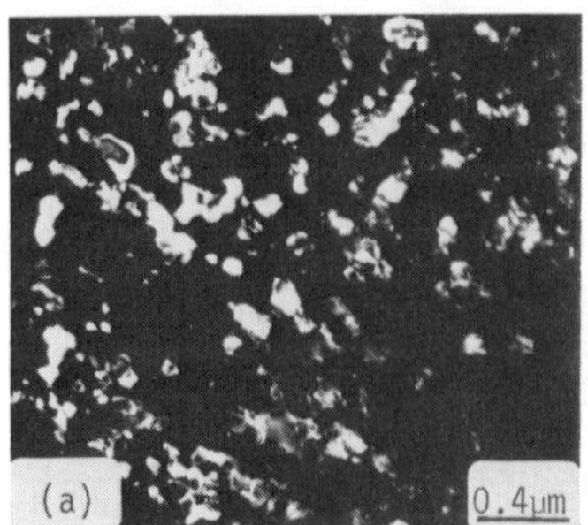


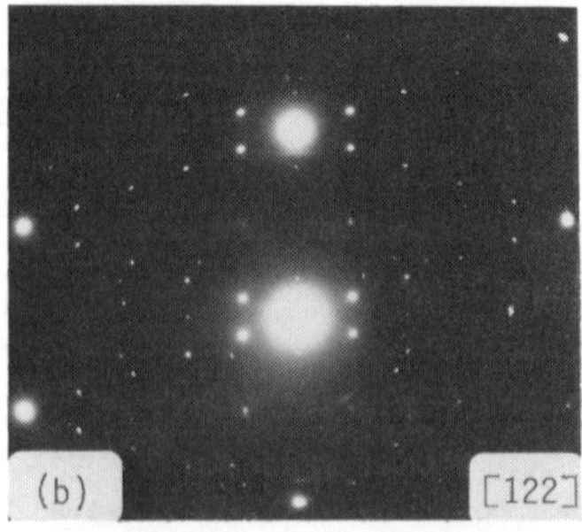


Figure 4. $Ni_4Mo$ cooled at 1°C/min from 950°C. a) Dark field micrograph from superlattice spot; b) [122] SADP.

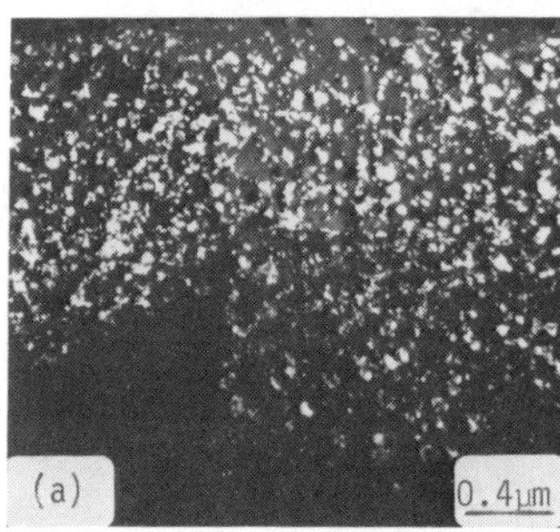


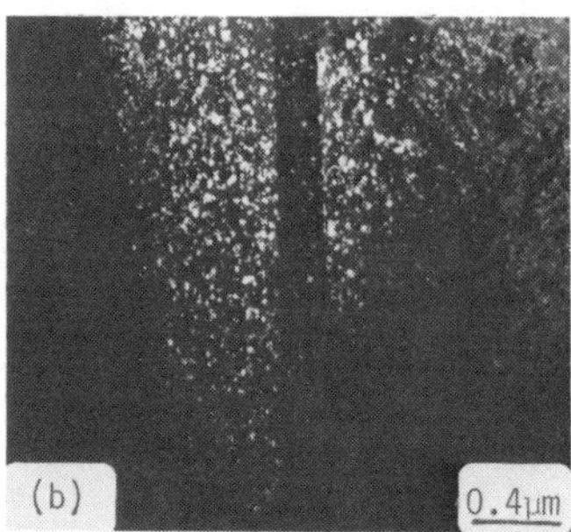


Figure 5. Dark field micrographs from superlattice spot for $Ni_4Mo$ cooled from 950$^oC$ at a) 5$^oC$/min; b) 10$^oC$/min.

A bright field micrograph of the fully ordered sample reheated to 800$^oC$ for 1/2 hour and quenched in iced brine is shown in Figure 6. One striking feature is the presence of rather large plate-like regions at the grain boundaries. Whether these regions are $Ni_4Mo$ or $Ni_3Mo$ is being investigated. A bright field micrograph and the associated [001] diffraction pattern of the same sample from another area are shown in Figures 7a and 7b. A dark field image from a superlattice spot shown in Figure 7c reveals both domain coarsening from the initial size of 187 nm (Figure 4a) to 245 nm and domain alignment along certain crystallographic directions. Increase in the reheat temperature results in further domain coarsening to 261 and 365 nm at 850 and 860$^oC$ (Figure 7d) respectively. The dark field micrograph in Figure 7d was obtained with the objective aperture positioned around two neighboring superlattice spots from both variants of the ordered phase (beam direction [112]). The micrograph in Figure 7d illustrates good contrast between anti-parallel twin-related domains as well as the boundaries themselves. Also, crystallographic alignment of domain interfaces becomes pronounced with increase in reheat temperature.

Figure 6. Bright field micrograph of fully ordered alloy reheated to 800$^oC$, 1/2 hour, quenched in iced brine.

## DISCUSSION

### SRO Resistivity

The decrease in resistivity with cold work and subsequent increase on reheating observed in this work, and previously by others [18-20], was first reported in several Ni-base alloys by Thomas [21] and called the "K-State" effect. Many other systems behave similarly [19,22-31]. It is generally

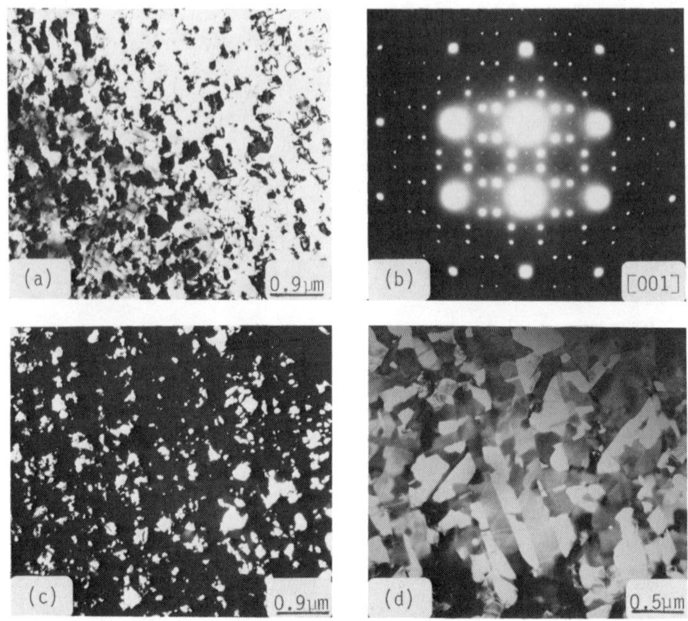


Figure 7. a) Bright field micrograph; b) [001] SADP; c) Dark field micrograph from superlattice spot for fully ordered alloy reheated to 800°C, 1/2 hour, quenched in iced brine; d) Dark field micrograph from superlattice spot for fully ordered alloy reheated to 860°C, 1/2 hour, quenched in iced brine.

agreed [20] that in $Ni_4Mo$, the decrease in resistivity with cold work is due to a decrease in the degree of SRO. To our knowledege, there is no published account of the slight increase in resistivity for deformation in excess of 68% reduction. This increase is attributed to an increase in defect concentration overcompensating for the decrease in resistivity due to decrease in the degree of SRO.

The explanation of the resistivity anomalies vary and depend on the particular alloy. Several theories have been advanced to explain this behavior on the basis of changes in the electronic configuration as a result of lattice distortions [32,33]. Since most alloys evincing this behavior contain at least one transition element, significance has been frequently attached to both the presence of an unfilled d-shell [21,31,34], and the degree of occupation and density of d-bands [35,36]. Others have explained the increase in resistivity by proposing SRO microdomains [37], or at least nonrandom atom configurations providing more effective electron scattering than for the random solid solution. In the case of $Ni_4Mo$, the microdomain description of SRO [9-11] may not be valid and the static concentration wave packet model is most likely to be applicable [12-14]; the small increase in resistivity that is observed initially on isothermal annealing at 700°C (Figure 3) suggests that the SRO attains a critical wavelength before LRO nucleats, confirming the wave packet model. If indeed the SRO state bore a structural similarity to the LRO structure, the increase would not be expected. These explanations of the resistivity anomalies are complicated by the fact that similar to $Ni_4Mo$, SRO increases resistivity in many systems [19,22-31], but decreases it in others [38-40].

Many theories relating SRO and resistivity exist [41-43]. Gibson [41] has pointed out that SRO is expected to decrease the resistivity in solid solution where component atoms differ in valence when the Fermi surface passes close to a Brillouin zone boundary for the random lattice, but increase resistivity when the Fermi surface is close to a Brillouin zone boundary for the superlattice. However, recent calculations using CPA by Nicholson [44] indicate that in $Ni_4Mo$ and perhaps in many systems evincing a high resistivity, the Fermi surface is virtually nonexistent. Although this would explain the high resistivity, the reason for the negative temperature coefficient of resistivity is unclear.

Rossiter and Wells [42] obtained an expression relating the resistivity to $\alpha_i$, the Cowley short range order parameter, $C_i$, the concentration of atoms in the $i^{th}$ shell, an integral $Y_i$, and n, the number of conduction electrons. They showed that the sign of $d\rho/dT$ depended on the sign of $\alpha_i$, the sign and magnitude of $Y_i$, and n. For $n = 1$, $\alpha_1 < 0$ (for SRO), and $Y_1 < 0$, the theory predicts a negative $d\rho/dT$. Our experimental results on $Ni_4Mo$ qualitatively agree with this theory as can be seen from Figure 8, where $d\rho/dT$ (obtained graphically from curve 2C in Figure 2) is plotted against temperature. In Figure 8, $d\rho/dT$ is zero near 1300°C and becomes progressively more negative with a maximum at about 750°C. Following this, the slope becomes rapidly less negative in a narrow temperature range, and then changes sign near 600°C, whereupon the amount of SRO is "frozen-in" due to insufficient atom mobility.

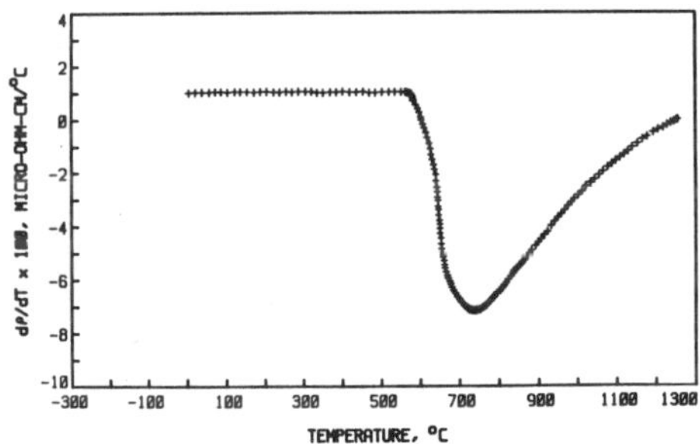


Figure 8. SRO $d\rho/dT$ versus temperature.

The occurrence of negative temperature coefficient of resistivity has received some attention [36,45,46] with formulation of the well known "Mooji Rule", i.e. independent of structure or electronic configuration, $d\rho/dT$ decreases with an increase in resistivity in several disordered, as well as short-range ordered transition metal alloys. Mooji [45] has attributed this to the very small electron mean free path existing in these alloys. Though this may be true in $Ni_4Mo$, it appears that the explanation for the negative temperature coefficient of resistivity more probably relates to the nature of interactions between the degree of SRO and the density of d-states and how these change with temperature. Thus, if the SRO results in an increase in the number of unfilled d-shells, then a decrease in the degree of SRO (increase in temperature) can be visualized as decreasing the density of unfilled d-shells and therefore decreasing the s-d electron scattering and the resistivity.

## LRO Resistivity and Structure

Although there are few systems in which LRO leads to an increase in resistivity [47,48], similar to $Ni_4Mo$, most systems [19,24,27,49], show a decrease. Several theories have been formulated relating LRO and resistivity [50-54]. Rossiter [53,54] derived a simple expression combining the effects of LRO on the relaxation time to give an overall dependence of the resistivity on LRO as follows:

$$\rho(S,T) = \rho_o(0) \frac{(1 - S^2)}{(1 - A S^2)} + \frac{BT}{n_{eff}} \qquad (1)$$

where, S is the Bragg-Williams LRO parameter, B is a constant, A is a parameter whose sign and magnitude depends on the relative positions of the Fermi surface and the superlattice Brillouin zone boundary, $n_{eff}$ is the effective number of conduction electrons, and $\rho_o(0)$ is the residual resistivity for S = 0. The first term in the expression represents the contribution to the resistivity from S and the second term represents the phononic contribution which varies with temperature as follows:

$$\rho_p(T) = T \frac{d\rho}{dT} = \frac{BT}{n_{eff}} \qquad (2)$$

Thus, $d\rho/dT$ varies inversely with $n_{eff}$, which can increase or decrease with temperature depending on the alloy. In his analysis on quenched results of resistivity of FeCo [54], $d\rho/dT$ increased with increase in temperature due to a decrease in $n_{eff}$. An at-temperature analysis on $Ni_4Mo$ using graphically determined values of $d\rho/dT$ obtained from curve 5H in Figure 2 is shown plotted versus temperature in Figure 9. In the temperature range -150 to 600°C, $d\rho/dT$ decreases which is opposite to his results. In fact, Rossiter [53] has pointed out that the linear relationship in equation 2 may not be valid for transition metal alloys owing to the dependence of s-d electron scattering on temperature. Therefore, it appears that in long range ordered $Ni_4Mo$, s-d scattering (density of d-states) decreases with temperature, leading to an increase in $n_{eff}$. As a result, $d\rho/dT$ will decrease, resulting in a relative decrease in resistivity as deduced from equation 1. Hence, the LRO resistivity behavior qualitatively agrees with the theory of Rossiter [53,54]. Above 600°C, the $d\rho/dT$ increases (Figure 9) due to the greater influence of a decreasing degree of LRO as the order-disorder temperature of 868°C is approached.

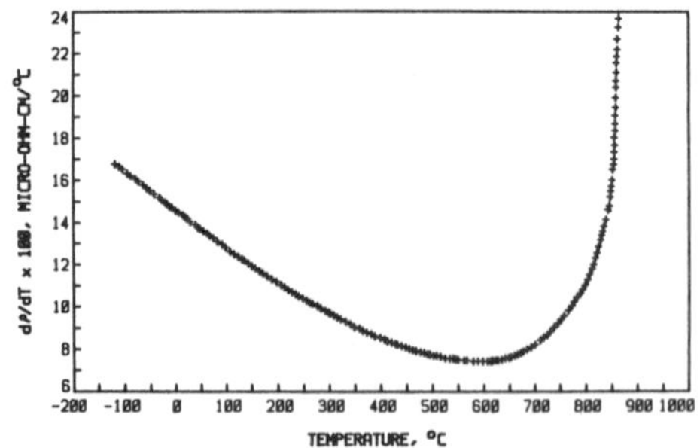


Figure 9. LRO $d\rho/dT$ versus temperature.

The increase in room temperature resistivity with an increase in cooling rate from 1 to 10°C/min cannot be entirely accounted for by scattering due to smaller-sized domains observed in Figures 4a, 5a, and 5b. This is because the domain sizes are large compared to typical values of the electron mean free path, and also it has been observed [17] that samples with very large domain sizes had resistivity values at 25°C identical to that of the sample cooled at 1°C/min. Thus, the major factor increasing the resistivity appears to be the decreasing degree of LRO with cooling rate.

There is some controversy as to the exact orientation of aligned domains in reheated samples, with reports of alignment both along {110} fcc planes [8] and {420} fcc planes [14,55]. It appears that both types of orientations can be present depending on heat treatment and whether these exist on continuous cooling and reheating are being investigated. Finally, the first order phase change in $Ni_4Mo$, indicated by the discontinuous change in resistivity on continuous heating and cooling (curves 5H and 4C in Figure 2), permits quenching-in from near the ordering temperature, specific degrees of LRO as is evident from quenched-in increases in resistivity values.

## CONCLUSIONS

This investigation has resulted in complete characterization of the resistivity changes associated with SRO and LRO in $Ni_4Mo$ and having the following salient features:

1. SRO resistivity decreases with an increase in cold work to 68% reduction of area after which an increase is attributed to the predominant influence of increasing defect concentration. Fast cooling from above the LRO-SRO transition temperature results in retention of SRO with resistivities of 144 and 139 micro-ohm-cm at 25 and -269°C respectivelv. The corresponding resistivities of the fully LRO state obtained by slow cooling are 48 and 8 micro-ohm-cm respectively.
2. Increase in cooling rate within the range permitting LRO, results in a decrease in domain size and an increase in room temperature resistivity. This increase is attributed to the primary influence of a decrease in the degree of LRO with cooling rate.
3. The negative temperature coefficient of resistivity that is observed in the SRO state above the critical temperature is probably due to the interaction between the change in the degree of SRO with temperature and s-d electron scattering, decrease in the degree of SRO (increase in temperature) decreases such scattering and leads to a decrease in resistivity.
4. A decrease in slope of the LRO resistivity-temperature curve in the temperature range -150 to 600°C is attributed to the temperature dependence of s-d electron scattering; increase in temperature decreases such scattering, thereby increasing $n_{eff}$ and decreasing the slope.

## REFERENCES


1. D. Harker, J. Chem. Phys. 12, 315 (1944).
2. P. V. Guthrie and E. E. Stansbury, Report to Oak Ridge National Laboratory, ORNL-3078 (1961).
3. J. E. Spruiell and E. E. Stansbury, J. Phys. Chem. Sol. 26, 811 (1965).
4. F. W. Ling and E. A. Starke, Acta Met. 19, 759 (1971).
5. B. Chakravarti, E. A. Starke, C. J. Sparks and R. O. Williams, J. Phys. Chem. Sol. 35, 1317-1326 (1974).
6. E. Ruedl, P. Delavignette and S. Amelinckx, Phys. Stat. Sol. 28, 305 (1968).
7. W. B. Snyder and C. R. Brooks In: Ordered Alloys - Structural Applications and Physical Metallurgy, B. H. Kear, C. T. Sims, N. S. Stoloff and J. H. Westbrook (eds), Baton Rouge, Claitor's, p. 275 (1970).
8. T. Saburi, E. Kanai and S. Nenno, J. of Less Common Metals 37, 59 (1974).
9. P. R. Okamoto and G. Thomas, Acta Met. 19, 825 (1971).
10. S. K. Das, P. R. Okamoto, P. M. J. Fischer and G. Thomas, Acta Met. 21, 913 (1973).
11. G. Thomas and R. Sinclair, Acta Met. 25, 231 (1977).
12. J. P. A. A. Chevalier and W. M. Stobbs. Acta Met. 24, 535 (1976).
13. W. M. Stobbs and J. P. A. A. Chevalier, Acta Met. 26, 233 (1978).
14. J. P. A. A. Chevalier and W. M. Stobbs, Acta Met. 27, 1197 (1979).
15. S. Banerjee, K. Urban and M. Wilkens, Acta Met. 32, 299 (1984).
16. R. S. Irani, F. W. Ling and R. W. Cahn, Metallography 6, 141 (1973).

17. E. E. Stansbury, K. Vasudevan and T. S. Lei, Proceedings: 17$^{th}$ Annual Technical Meeting of the International Metallographic Society, Philadelphia (1984).
18. G. Grube and H. Schlect, Ztschr. Elektrochem. 44, 413 (1938).
19. H. G. Baer, Z. Metallkde. 56, 79 (1965).
20. B. G. LeFevre, A. G. Guy and R. W. Gould, Met. Trans. 242, 788 (1968).
21. H. Thomas, Z. Physik 129, 219 (1951).
22. A. C. Damask, J. Phys. Chem. 1, 23 (1956).
23. B. M. Korevaar, Acta Met. 9, 297 (1961).
24. A. van den Beukel A. P. Matthijsen, J. M. J. Ritzen and R. den Buurman, Acta. Met. 16, 435 (1968).
25. B. G. Livshits and R. A. Rymashevskii, Phys. Met. Metallog. 13 (2), 41 (1962).
26. E. A. Starke Jr, V. Gerold and A. G. Guy, Acta Met. 13, 957 (1965).
27. E. Torfs, L. Stals, J. van Landuyt, P. Delavignette and S. Amelinckx, Phys. Stat. Sol. (a) 22, 45 (1974).
28. J. van der Wekken, R. Taggart and D. H. Polonis, Metal Science J. 5, 219 (1971).
29. R. J. Willey J. Mater. Sci. 13, 871 (1978).
30. Y. N. Babanova, I. B. Klyuyeva, L. S. Chemerinskaya and F. A. Siderenko, Phys. Met. Metallog. 49 (3), 184 (1980).
31. V. S. Meskin, R. I. Sergienko and L. A. Potapov, Phys. Met Metallog. 13, 113 (1962).
32. W. H. Aarts and A. S. Houston-Macmillan, Acta Met. 5, 525 (1957).
33. H. J. Logie, J. Jackson, J. C. Anderson and F. R. N. Nabarro, Acta Met. 9, 707 (1961).
34. H. G. Muller and P. Muth, Z. Metallk. 50, 217 (1959).
35. F. Brouers and M. Brauwers, J. Phys. Letters 36, L-17 (1975).
36. R. P. Vasilyeva, N. N. Narkulov, V. P. Fadin and Y. N. Arkhipov, Phys. Met. Metallog. 48 (3), 671 (1979).
37. W. Chen and M. E. Nicholson, Acta Met. 12, 687 (1964).
38. A. C. Damask, J. Appl. Phys. 27, 610 (1956).
39. M. K. Sardar and K. P. Gupta, Scripta Met. 14, 835 (1980).
40. P. Meisterle and W. Pfeiler, Acta Met. 31, 1543 (1983).
41. J. B. Gibson, J. Phys. Chem. Solids 1, 27 (1956).
42. P. L. Rossiter and P. Wells, J. Phys. C: Solid St. Phys. 4, 354 (1971).
43. P. L. Rossiter, J. Phys. F: Metal Phys. 7 (3), 407 (1977).
44. D. Nicholson, Private communication, ORNL, Oak Ridge (1984).
45. J. H. Mooji, Phys. Stat. Sol. (a) 17, 521 (1973).
46. J. Richter and W. Schiller, Phys. Stat. Sol. (b) 92, 511 (1979).
47. Y. V. Pal'guyev, A. A. Kuranov, P. N. Syutkin and F. A. Siderenko, Phys. Met. Metallog. 41 (6), 72 (1976).
48. V. A. Phillips, Acta Met. 9, 976 (1961).
49. M. J. Marcinkowski and R. Smoluchowski, J. Phys. Chem. Solids 26, 185 (1965).
50. M. Brauwers, J. Giner, J. Van der Rest and F. Brouers, Solid State Comm. 17, 229 (1975).
51. V. Christoph, J. Richter and W. Schiller, Phys. Stat. Sol (b) 100, 585 (1980).
52. P. L. Rossiter, J. Phys. F: Metal Phys. 9, 891 (1979).
53. P. L. Rossiter, J. Phys. F: Metal Phys. 10, 459 (1980).
54. P. L. Rossiter, J. Phys. F: Metal Phys. 11, 615 (1981).
55. T. S. Lei, Ph.D. Dissertation, University of Tennessee (1979).

# HEAVY ION INDUCED DAMAGE AND DISORDER IN INTERMETALLIC COMPOUNDS WITH THE B2 STRUCTURE

JOHN L. BRIMHALL AND HOMER E. KISSINGER
Pacific Northwest Laboratory, P. O. Box 999, Richland, WA 99352

## ABSTRACT

Although the B2-type intermetallic compounds, NiAl, FeAl, NiTi and FeTi all maintained a high degree of order during irradiation, metastable phase formation was observed. Dislocation loops and network developed in NiAl and FeAl during irradiation and a nickel rich, fcc phase, most likely disordered $Ni_3Al$, formed in NiAl after high doses. The build-up of constitutional vacancies (excess vacancies on Ni sites) and subsequent elemental partitioning of Ni and Al can explain the formation of a nickel rich phase in NiAl. No metastable phases formed in FeAl. Dislocation loop structures did not develop in NiTi or FeTi, rather a transformation to an amorphous phase occurred at relatively low doses. The amorphous transformation results when the free energy of the crystal lattice containing a high defect concentration exceeds the free energy of the amorphous configuration.

## INTRODUCTION

This work has demonstrated that a wide range of microstructural changes can be produced in compounds with the B2 structure by irradiating with heavy metal ions. These changes include less order in the lattice, dislocation loops, metastable crystalline phases as well as complete transformation to an amorphous phase. This report represents part of a larger study to determine the tendency toward a radiation induced amorphous transformation in intermetallic compounds of varying crystalline complexity.[1] In this regard, four compounds with the B2 structure, which represents a relatively simple crystal structure, were investigated.

The four compounds NiAl, FeAl, NiTi, and FeTi were all bombarded with high energy nickel ions at ambient temperatures and to dose levels up to $1x10^{16}$ ions/cm². The microstructural changes induced by the irradiation were subsequently examined by transmission electron microscopy. The observations showed some similarity in behavior as well as some striking differences among the various compounds. The results were rationalized using current principles of phase stability in an irradiation environment.

## EXPERIMENTAL PROCEDURES

Most of the alloys were prepared by arc melting the constituents in a vacuum furnace. The compositions were 50%A-50%B within a spread of ±1%. The NiAl alloy was supplied by Professor T. E. Mitchell of Case Western Reserve University. Thin slices were cut from the cast material and TEM-size specimens extracted by punching or cutting. The specimens were annealed at ∿1000°C for 24 hours to promote homogenization and furnace cooled. The surfaces were subsequently highly polished prior to ion irradiation.

All the alloys were irradiated with 2.5 MeV $Ni^+$ ions at ambient temperatures. The ion current was kept low to minimize beam heating, but a temperature rise as great as 50°C still occurred. The dose ranged from

$5x10^{14}$ ions/cm² to $1.2x10^{16}$ ions/cm². This corresponds to damage levels of 0.1 to 5 dpa in the near surface region using the E-DEP1 code to calculate dpa.[2] The maximum calculated penetration of the ions was about 0.5 μm. The concentration of deposited nickel at 0.5 μm and at the highest dose was < 0.5%.

After bombardment, the specimens were selectively electropolished to analyze the near surface region using electron microscopy. Approximately 0.3 μm was subsequently removed from the surface and the alloy re-examined in the electron microscope. The dose in dpa at 0.3 μm is approximately twice that of the near surface region.

## RESULTS

The ordered aluminides developed expected dislocation loops and a dislocation network, but remained highly ordered after high doses. There was evidence of a radiation induced phase transformation in the NiAl alloy. Conventional microstructure in terms of dislocation loops and dislocation network did not develop in the titanides. Rather, an amorphous structure formed at low doses. The alloys did not fully disorder before transforming to an amorphous state.

### a. NiAl and FeAl

Ion bombardment of NiAl produced small black spots which developed into dislocation loops and a dislocation network with increase in dose, fig. 1a, b. The degree of order remained high even after a dose of 1 dpa as shown by the electron diffraction pattern in fig. 1b. A quantitative estimation of the actual degree of order was not attempted.

At the dose of 5 dpa, the diffraction patterns showed evidence of a second phase (fig. 2). In all cases, the extra spots could be indexed as a fcc structure with a lattice parameter near 3.58 ±.02 A°. Other orientations of NiAl were studied and in all cases, the new phase could be indexed as an fcc structure. This likely corresponds to a nickel rich γ phase containing at least 25% aluminum. The ordered $Ni_3Al$ ($L1_2$) phase is

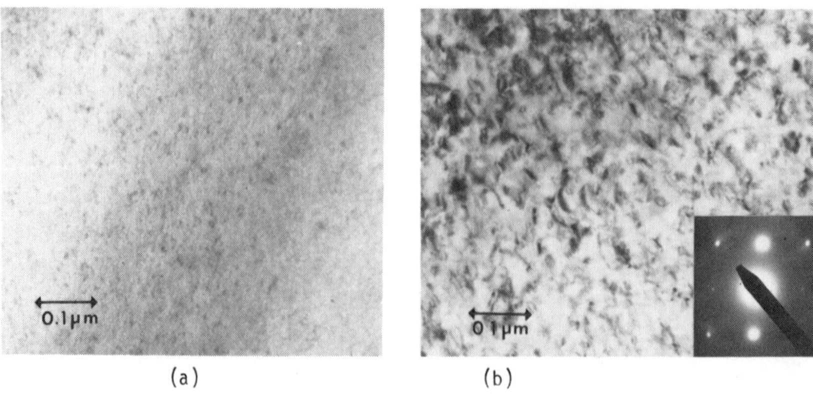


Fig. 1. Microstructure of NiAl bombarded with 2.5 MeV $Ni^+$ ions (a) 0.2 dpa, (b) 1.0 dpa.

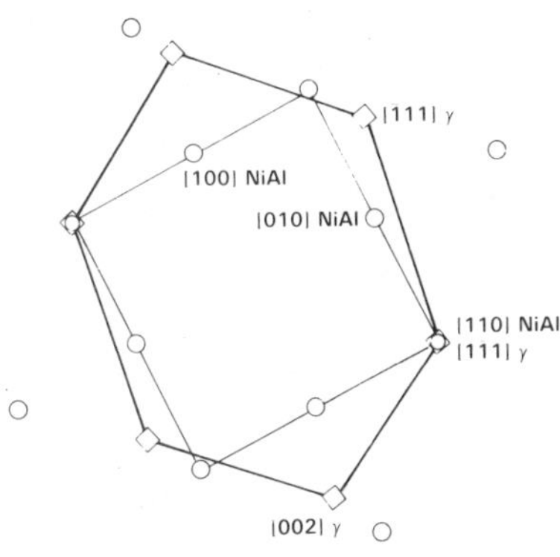


Fig. 2. Selected area diffraction pattern of NiAl bombarded with 2.5 MeV $Ni^+$ ions to 5-10 dpa. Schematic diagram illustrating the orientation relationship of the two phases.

expected at this composition and it is possible that the superlattice spots in $Ni_3Al$ are invisible in these diffraction patterns. However, due to irradiation, the $Ni_3Al$ phase would disorder to some extent so the phase may actually be closer to a fcc nickel-aluminum solid solution and will be referred to as γ phase. The orientation relationship of the new phase to NiAl phase is also illustrated in fig. 2. The (001) plane in NiAl is nearly parallel to the $(1\bar{1}0)$ plane in the γ phase and the [110] NiAl direction is coincident with the $[1\bar{1}1]$ direction in the γ phase. There was considerable double diffraction which can account for the extraneous spots.

The radiation induced phase was imaged in dark field, fig. 3, using the diffraction spot from the γ phase. The phase appeared web-like in nature in the dark field and two dimensional when viewed in stereo. In bright field, the phase was difficult to resolve as the contrast was only slightly different from the matrix. By comparing carefully the bright and dark field micrographs, the phase could be delineated in bright field.

This new phase was only identified at the highest dose examined for the NiAl. No indication of a second phase was evident in the diffraction patterns of specimens irradiated at lower dose levels. The phase was present in the irradiated alloy both in the near surface region, as well as in a region ∿ 0.3 μm from the surface. The amount of new phase appeared to be greater at the greater depth from the surface. This is consistent with a greater damage energy and hence dpa level at a depth of 0.3 μm.

In contrast to the NiAl, no second phase formation was observed in the FeAl alloy. A dislocation network did develop in the FeAl alloy as the radiation dose increased, in a manner similar to NiAl. Fig. 4 illustrates the microstructure after a dose of 5 dpa. A high degree of order was still present in the alloy as shown by the diffraction pattern.

b. NiTi and FeTi

The NiTi alloy did not develop a dislocation microstructure during ion bombardment as did the aluminides. Amorphous regions formed at very

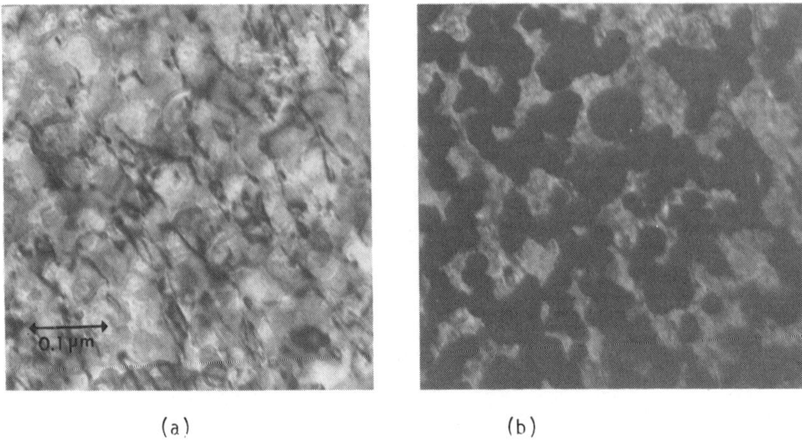


(a) (b)

Fig. 3. (a) Bright field (b) dark field of NiAl bombarded to ∿10 dpa with 5 MeV $Ni^+$ ions. Dark field imaged with second phase diffraction spot.

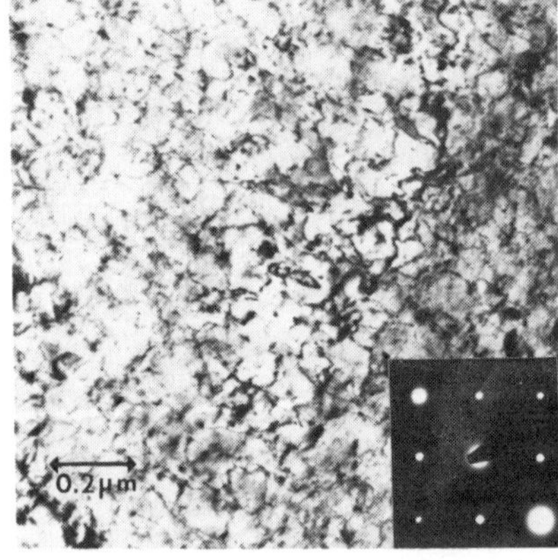


Fig. 4. Microstructure of FeAl bombarded with 2.5 MeV $Ni^+$ ions to 5 dpa.

low dose levels as indicated in fig. 5. This dose level of 0.02 dpa is somewhat less than that dose at which small loops or defect clusters were observed in the NiAl structure. Increasing the irradiation dose increased the amount of amorphous material until the structure was completely amorphous at a dose ≤ 0.1 dpa. The dose dependence of this phenomenon has been previously reported [3].

Even when the material was ≥ 50% amorphous, the remaining crystalline matrix in the NiTi was still ordered. The selected area diffraction pattern in fig. 5 from a partially transformed structure still shows the superlattice spots for the NiTi B2 structure. An estimate of the degree of order was made by comparing the intensities of the diffraction spots. The order parameter S was defined as the ratio of $I_{(200)}/I_{(100)}$ in the irradiated material divided by the same ratio in the unirradiated alloy. The value of S is therefore assumed to be unity in the unirradiated alloy. A plot of S is shown in fig. 7 as a function of irradiation dose. There is an indication of a decrease in the ordering with dose but the degree of order is still greater than 0.5 at the highest dose observed.

Irradiation of FeTi produced microstructures very similar to that in NiTi. An amorphous phase developed at low dose levels and increased in volume as the irradiation dose increased. A partially transformed structure is shown in fig. 8. The remaining crystalline material also showed order, but the degree of order was not quantitatively estimated. The dose dependence of the amorphous transformation was also not quantitatively determined although the transformation was essentially complete at doses <1 dpa.

Fig. 5. Microstructure of NiTi bombarded with 2.5 MeV $Ni^+$ ions to 0.02 dpa. Small contrast features are amorphous regions.

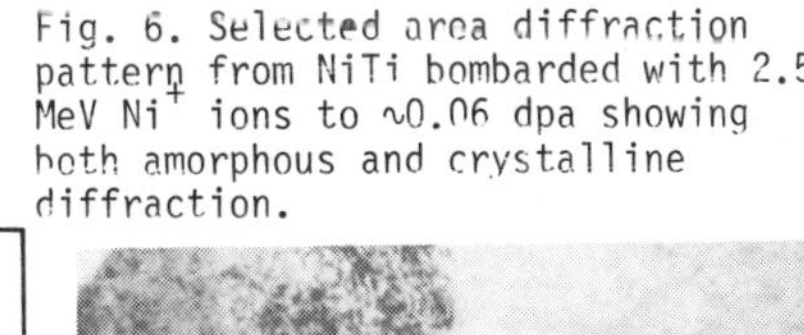

Fig. 6. Selected area diffraction pattern from NiTi bombarded with 2.5 MeV $Ni^+$ ions to ∿0.06 dpa showing both amorphous and crystalline diffraction.

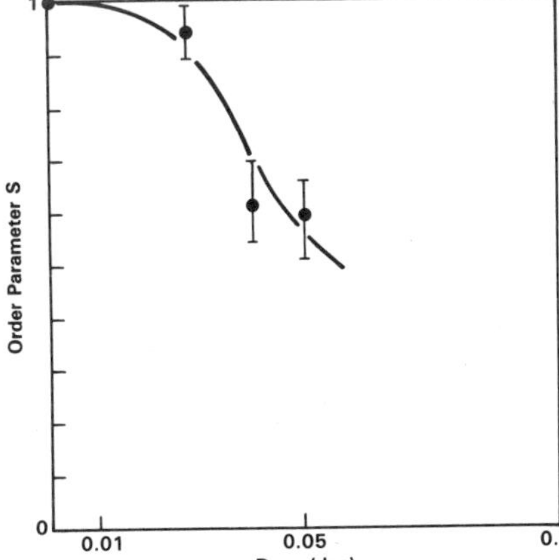


Fig. 7. The decrease in order in NiTi as a function of irradiation dose in NiTi. At the highest dose, the material was ∿50% amorphous.

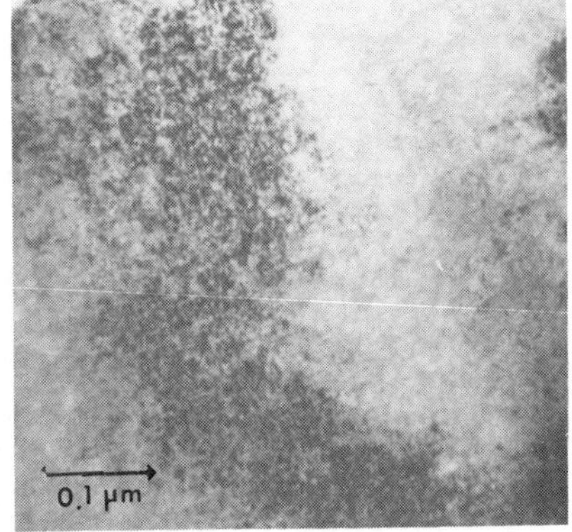


Fig. 8. Micrograph of FeTi bombarded with 2.5 MeV $Ni^+$ ions to ∿0.05 dpa showing a partially amorphous structure.

DISCUSSION

The resistance to radiation induced disorder in all these B2 type alloys is compatible with other observations using different types of irradiating particles. The metastable phase formed in NiAl can be related to the unique point defect behavior in NiAl which allows for excess vacancies on the nickel sublattice. The observations are also compatible with other results which show radiation induced formation of a NiAl phase from $Ni_2Al_3$. The amorphous transformation in the titanide alloys results from the necessity to relieve the high free energy of the point defects. The narrow range of stoichiometry and inability of these alloys to form a dislocation structure are important factors in the transformation to an amorphous structure.

The high degree of order in these B2 type alloys after a relatively high irradiation dose is compatible with other observations of these alloys. Liu et.al. found a strong resistance to disordering in NiAl during electron irradiation [4]. Riviere et.al. also observed a high degree of order after ion irradiation in a Fe-40% Al alloy [5,6]. Neutron and electron irradiation of this latter alloy also did not produce full disorder [7,8]. More disorder could be produced at low temperatures in these alloys, but not complete disorder [4]. However, an equiatomic Fe-Al alloy apparently does fully disorder if irradiated at low enough temperature [9,10]. There is also evidence that NiTi disorders during electron irradiation before becoming amorphous [11].

These results imply a high ordering energy for these B2 type alloys. The high ordering energy is apparently characteristic of this particular structure. Other simply ordered phases such as the $L1_2$ disorder readily during irradiation [12].

Radiation induced phase changes, as observed in NiAl, have also been observed in other intermetallic compounds [13-16]. Liou and Wilkes have given examples of how the phase diagrams can be altered when the free energies of the various phases increase due to irradiation [17]. The combination of disordered regions and a high defect concentration will increase the free energy of a particular phase. The free energy increase may be sufficient to induce the nucleation of a new phase and thereby lower the total free energy of the system.

The formation of a Ni rich fcc $\gamma$ phase from NiAl can also be rationalized with the unique point defect properties in NiAl alloys. The B2 phase can exist in NiAl in a compositional range from 45% to 60% nickel. In high nickel alloys, Ni atoms will occupy some of the Al sites. In aluminum rich alloys, however, there is an excess of Ni vacancies to account for the non-stoichiometry. The large Al atoms do not want to go into the small Ni sites. During irradiation, vacant Al sites can therefore be filled by either Ni or Al interstitial atoms, but the vacant Ni sites would not be filled equally with Al atoms. This slight bias in the recombination behavior will eventually produce an excess of vacancies on the Ni sites which is compensated by excess Ni atoms in the lattice even though the alloy had equiatomic composition initially. As a result of statistical fluctuations, some regions can acquire a sufficient nickel concentration to actually nucleate a new phase.

The ordered fcc phase $Ni_3$ Al which is the nearest equilibrium phase richer in nickel can be derived from the NiAl phase if the Al atoms on every second (110) plane are replaced by Ni atoms[18]. Such a transformation would produce the orientation relationship shown in fig. 2. Also, the fact that the $\gamma$ phase within one grain showed only one orientation relationship with NiAl implies a crystallographic evolution of this type. However, the $Ni_3Al$ phase may also disorder completely during irradiation [1,4]. The actual phase can then be considered an fcc solid solution phase or perhaps partially ordered $Ni^3Al$ with a composition near 72% Ni.

If the remaining NiAl phase reaches the compositional limit of ∿45% Ni, then an elemental balance will show that the Ni-rich phase (∿72% Ni) would occupy a volume of 18%. From a qualitative appearance, this is a reasonable estimate of the amount of phase present. The phase is definitely less than 50% by volume.

Although the remaining NiAl phase is richer in aluminum, it does not transform to an Al rich phase such as $Ni_2Al_3$. Recently, Nastasi et al. [15] have shown that ion irradiation of $Ni_2Al_3$ results in a transformation to NiAl. By proper placement of point defects, the NiAl structure can also be derived from the $Ni_2Al_3$ structure. This observation implies that NiAl is the stable phase relative to $Ni_2Al_3$ in an irradiation environment. There is no driving force for the NiAl to convert to an $Ni_2Al_3$ phase.

A radiation induced phase transformation in FeAl might have been expected based on the similarity in point defect behavior with the NiAl phase. The $Fe_3Al$ phase is not an fcc type phase like $Ni_3Al$, however. A radiation induced metastable phase has been reported in ion bombarded Fe-40%Al [5,6]. The phase was not positively identified, but believed to have an fcc structure with possibly a CuAu type ordering. It is possible that the phase transformation is more difficult for the 50Fe-FOAl composition of the alloy used in our study.

Several factors may account for the amorphous transformation in the titanium compounds. Both NiTi and FeTi show a limited stoichiometric range compared to NiAl or FeAl. Hence, there is no formation of constitutional vacancies in the former alloys so large concentration of vacancies cannot be accommodated. As mentioned above, there is an increase in the free energy of the alloys due to the high point defect creation during irradiation. In NiAl and FeAl, the free energy increase is eventually relieved by the clustering of point defects and the formation of dislocation loops. In NiTi and FeTi, the increase in free energy is relieved by the formation of an amorphous phase. The crystalline structure will relax into an amorphous configuration because the free energy of the highly defected crystalline state is greater than that of an amorphous configuration [1]. Reasons as to why dislocation structures do not form readily in these compounds remain to be explored. A lack of dislocation loop structure has also been reported in ion irradiated $Zr_3Al$ which also became amorphous [19]. It is possible that the defect mobilities are considerably less in the titanium compounds such that the defects cannot cluster sufficiently to nucleate dislocation loops.

## SUMMARY

In B2 type intermetallic compounds, heavy ion bombardment produced distinctly different microstructures in NiAl and FeAl compared to NiTi and FeTi. Irradition produced dislocation microstructures in both NiAl and FeAl as well as a Ni-rich fcc phase in the NiAl compound. The ability of the NiAl to form constitutional vacancies on the Ni sublattice can result in Ni rich regions and eventual phase transformation. Ion irradiation produced an amorphous phase in NiTi and FeTi with no evidence of dislocation loop formation. The limited stoichiometric range in the alloys compared to NiAl or FeAl is believed to be an important factor in this radiation induced amorphous transformation.

## ACKNOWLEDGEMENTS

This work was supported by the Division of Materials Sciences, Office of Basic Energy Sciences, Department of Energy under Contract DE-AC06-76RLO-1830.

## References

1. J. L. Brimhall, H. E. Kissinger and L. A. Charlot, Rad. Eff. 77, 237 (1983).
2. I. Manning and G. Mueller, Comput. Phys. Comm., 1, 85 (1974)
3. J. L. Brimhall, H. E. Kissinger, and A. R. Pelton, Ion. Implantation and Ion Beam Processing of Materials, (Elsevier Pub. Co., 1983).
4. H. C. Liu, C. Kinoshita and T. E. Mitchell, Phase Stability During Irradiations, ed. by J.R. Holland, C. K, Mansur and D. I. Potter, (AIME, NY, 1981) p. 343.
5. J. P. Riviere, J. F. Dinhut and J. Delafond, Nucl. Inst. and Meth., 182/183, 495 (1981).
6. J. P. Riviere, M. O. Ruault, M. Schack, Phil Mag A., 47, 255 (1983).
7. J. P. Riviere and J. Grilhe, Proc. Int. Conf. Fundamental Aspects of Radiation Damage in Metals, ed by M. T. Robinson and F. W. Young, CONF 751006 (NTIS, Springfield, VA) p. 636.
8. J. P. Riviere, Rad. Eff., 33U, 21(1977).
9. T. Mukai, C. Kinoshita and S. Kitajima, Phil Mag A., 47, 255 (1983).
10. J. P. Riviere, Scripta Met., 14, 503 (1980).
11. A.R. Pelton, Proc. of 7th Intl. Conf. on High Voltage Electron Microscopy. ed. by R.M. Fisher, R. Gronsky and K. H. Westmacott, CONF830819 (NTIS, Springfield, VA, 1983) p. 245.
12. E. M. Schulson, J. Nucl. Mat., 83, 239 (1979).
13. C. J. Roussouw, Phys. Stat. Sol.(a), 80, 631 (1983).
14. V. N. Bykov, V. A. Troyan, G. G. Zdorovtseva and V. S. Khaimovich, Phys. Stat. Sol.(a), 32I, 53(1975).
15. M. Nastasi, L. S. Hung, H. H. Johnson, J. M. Mayer and J. M. Williams, "Ion Irradiation Induced Phase Transformation in $Ni_2Al_3$" submitted to J. Appl. Phys.
16. D. I. Potter in Phase Transformation During Irradiation, ed. by F. Nolfi (Appl. Science Pub, N.Y., 1983) p. 213.
17. K. Y. Liou and P. Wilkes, J. Nucl. Mat., 87, 317 (1979).
18. P. Georgopoulos and J. B. Cohea, Acta Met., 29, 1535 (1981).
19. L. M. Howe and M. H. Rainville, Phil. Mag A, 39, 195(1979).

# PART IV

# Mechanical Behavior

## PLASTIC FLOW OF $L1_2$ ORDERED ALLOYS

D. P. POPE AND V. VITEK
University of Pennsylvania, Department of Materials Science and Engineering, 3231 Walnut Street, Philadelphia, PA 19104 USA

### ABSTRACT

The flow stress of many $L1_2$ ordered alloys has a very unusual temperature dependence: the flow stress increases with increasing temperature. This unusual behavior is related to the nature of dislocation dissociation and core structure. The flow stress increase is the result of thermally activated cross slip of $[\bar{1}01](111)$ screw dislocations to the (010) plane which is accompanied by a transformation of the dislocation core from a glissile to a sessile form. Thus dislocations which are mobile on (111) planes become immobile after cross-slip into (010) planes. The dependence of the flow stress on temperature, orientation and sense of the applied uniaxial stress will be discussed in the light of this cross slip model for $Ni_3Al$, $Ni_3Ga$ and for $\gamma/\gamma'$ nickel base superalloys.

The response of $Ni_3Al$ to cyclic plastic strains (plastic strain controlled fatigue) will also be shown to be in accord with the cross slip model. The mean stress in such a test becomes compressive or tensile, depending on the orientation of the sample, even though the net plastic strain is zero after each cycle.

The strengthening of $Ni_3Al$ by ternary additions will also be discussed. It will be shown that ordinary solid solution strengthening models are not applicable but that the cross slip model can also be applied.

Finally, it will be shown that dislocation core simulation studies predict that there should also be a class of $L1_2$ ordered alloys that show a "normal" flow stress-temperature behavior, i.e., the flow stress increases at low temperatures. The results of our studies on $Pt_3Al$ will be used to illustrate this behavior.

### FLOW STRESS

Alloys with the $L1_2$ structure have an $A_3B$ composition and an fcc derivative structure with the majority atoms on the face centered positions and the minority atoms at the corner positions. Since the structure is fcc derivative, one might expect plastic flow to occur in the same manner as in fcc metals, viz., by $[\bar{1}01](111)$ slip, and like fcc materials to show only a slight temperature dependence of the flow stress. This is definitely not the case as has been shown for a large number of alloys having the $L1_2$ structure. There are two kinds of behavior seen in these materials: (i) the so-called anomalous behavior in which the CRSS for $[\bar{1}01](111)$ slip increases dramatically with increasing temperature, and then drops again at still higher temperatures due to the onset of $[\bar{1}10](001)$ slip and (ii) the so-called normal behavior in which the CRSS increases with decreasing temperature, similar to that seen in bcc metals. No $L1_2$ material has shown a flow behavior like that of fcc metals.

The anomalous flow behavior of $L1_2$ alloys was first observed by Westbrook [1] in a study of the hardness of $Ni_3Al$ as a function of temperature. This result was subsequently confirmed by Flinn [2] who measured the flow stress of polycrystalline $Ni_3Al$ as a function of temperature. Flinn showed that the flow stress is insensitive to temperature changes at temperatures below $20^oC$, but between $20^oC$ and $600^oC$ it increases dramatically, almost threefold, then decreases again above $600^oC$. Flinn proposed that the anomalous increase is due to diffusion-controlled changes of the dislocation configuration. He hypothesized that the deformation takes place by the motion of [$\bar{1}$01](111) superdislocations dissociated on the (111) plane into two 1/2[$\bar{1}$01](111) superpartials separated by APB. As the temperature increases, one of the superpartial dislocations climbs such that the APB lies on the (010) plane, and then further motion of the pair on parallel (111) planes can only occur by extending the APB. The driving force for this climb is the anisotropy of the APB energy. On the basis of nearest neighbor violations Flinn showed that the APB energy is a minimum on the (010) plane and near the maximum on the (111) plane. The anisotropy of the APB energy is at the heart of the most successful theories currently being considered. However, diffusive mechanisms are not very plausible since Davies and Stoloff [3] showed that the flow stress of $Ni_3Al$ is not very strain rate dependent. A number of models were proposed to explain the anomalous behavior, some based on intrinsic lattice defects [3,4,5], and others based on changes in long range order [6]; but models based on variations of the Flinn model are currently the most successful.

Later experiments on a number of $L1_2$ alloys showed that: (i) For samples oriented in the standard [001]-[011]-[$\bar{1}$11] unit triangle, slip occurs on the [$\bar{1}$01](111) system at temperatures below the peak in the CRSS vs. temperature curve and on the [$\bar{1}$10](001) system above the peak [7]. (ii) The dislocation structure in samples deformed in the temperature regime of the anomalous increase of the CRSS consists mostly of long straight screw dislocations [8]. (iii) The temperature dependence of the CRSS measured at low offset strains is different from that at larger strains [9,10]. At an offset strain of $10^{-5}$ to $10^{-6}$ the CRSS is nearly temperature-independent. (iv) Schmid's law does not hold for $L1_2$ alloys which show the anomalous flow behavior [11]. (v) The CRSS also depends on the sense of the applied uniaxial stress. The tensile CRSS exceeds the compressive for samples oriented near [001] and the opposite is true for samples oriented on the [011]-[$\bar{1}$11] side of the unit triangle [12].

Takeuchi and Kuramoto [11] proposed a model which successfully explains (i) - (iv) above, but not (v). This model is based on the Flinn model of APB energy anisotropy, the Kear-Wilsdorf [13] cross-slip mechanism originally proposed to explain the high work hardening rate of $L1_2$ ordered alloys and the proposal by Thornton et al. [14] that the Kear-Wilsdorf model is applicable to the temperature dependence of the CRSS. The Kear-Wilsdorf model provides a mechanism for the immobilization of [$\bar{1}$01](111) superdislocations by cross slip from (111) to (010) planes, the driving force for which is the APB energy anisotropy proposed by Flinn. In this model, the leading 1/2[$\bar{1}$01](111) superpartial of a pair of such screw dislocations separated by APB on the (111) plane cross slips onto the (010) plane. The pair is assumed to be mobile on the (111) plane, but immobile on the (010) plane because the 1/2[$\bar{1}$01](111) superpartials are assumed to be dissociated into Shockley partials on the (111) plane. Therefore glide on the (010) plane is impossible, since nonconservative motion of the Shockley partials would be required. Takeuchi and Kuramoto proposed that the cross-slip of segments of the dislocation line provides local pinning points which hinder the motion of the rest of the dislocation. Furthermore, they assumed that the frequency of such cross-slip events is determined by thermal activation, the activation enthalpy for which is reduced by the RSS on [$\bar{1}$01](010), the RSS which drives cross-slip to the (010) plane. They then showed that the increase in the RSS for [$\bar{1}$01](111) slip over the value at low temperatures $\Delta\tau_{pb}$ should obey an equation of the form:

$$\Delta\tau_{pb} = A\ \exp\left[\frac{-H + \tau_{cb}V}{3kT}\right], \tag{1}$$

where A is a constant, p means "primary plane," b means "in the direction of the Burgers vector," H is the activation enthalpy, $\tau_{cb}$ is the RSS on the cube plane in the direction of the Burgers vector, V is the activation volume, and k and T have the usual meaning. Since $\tau_{pb}$ is a function of $\tau_{cb}$ in equation (1), this predicts a breakdown of Schmid's law. The breakdown is predicted to occur as described in (iv) above since $\tau_{cb}$ increases as the uniaxial stress axis is moved away from [001]. Experimental results on the CRSS measured in compression of $Ni_3Ga$ [11], $Ni_3Ge$ [14], and $Ni_3Al$ [15-17] were shown to be in agreement with the Takeuchi and Kuramoto model.

However the early tests of the Takeuchi and Kuramoto model were flawed by the fact that most investigators tested samples of only a few different orientations. Lall et al. [18] tested samples of more orientations and found significant differences with the model, e.g., the relative strengths of samples having nearby orientations is opposite, in some cases, to the predictions of the model. Lall et al. proposed that the differences between the results and the predictions of the model are due to the fact that a "non-glide" stress component which controls the width of the Shockley partial separation is required in equation (1). This stress component which acts on the primary plane in the direction of the edge component of one of the Shockley partials, called $\tau_{pe}$, always extends the partials for a given orientation and sense of the applied stress axis and constricts them for others. When the Shockley partials are constricted cross slip is aided, and the opposite occurs when they are extended. Lall et al. proposed a modification of equation (1) in the form of

$$\Delta\tau_{pb} = A \exp\left[\frac{-H + \tau_{cb}V_1 + \tau_{pe}V_2}{kT}\right], \tag{2}$$

where $V_1$ and $V_2$ are different activation volumes for the two processes. Equation (2) predicts a different CRSS for tensile and compressive measurements, except for those orientations for which $\tau_{pe}$ is zero, that is, on the [012]-[$\bar{1}$13] great circle in the unit triangle. Furthermore, it predicts that the tensile CRSS should exceed the compressive for orientations near [001] and the opposite should occur on the [011]-[$\bar{1}$11] side of the triangle. These predictions were confirmed by Ezz et al. [12] in tests on $Ni_3(Al,Nb)$, with one exception: The orientation at which the tension/compression flow stress asymmetry disappears is not on the [012]-[$\bar{1}$13] great circle, but rather it is located on the [001] side of that great circle. This lack of agreement was later explained by the more refined model of Paidar et al. [19], which was based on the results of dislocation core simulations by Yamaguchi et al. [20] and Paidar et al. [21] and the effects of core width before and <u>after</u> cross-slip on the activation enthalpy as proposed by Friedel [22].

The results of dislocation core simulations in $L1_2$ materials [19,20] showed that there are two possible dislocation core configurations for [$\bar{1}$01](111) superdislocations. For those materials in which the APB on (111) planes is stable the superdislocation dissociates into two 1/2[$\bar{1}$01] superpartials separated by APB and the cores of the superpartials dissociate into a configuration which resembles that of two Shockley partial dislocations separated by complex stacking fault (CSF). The energy of the CSF is sufficiently high, however, that the width of separation of the Shockley partials is only a few interatomic spacings, and therefore the cores of the 1/2[$\bar{1}$01] superpartials are only locally extended, in planar fashion, in the (111) plane. Since the cores of the superpartials are planar on the (111) plane, these dislocations are glissile. If the APB on the (111) plane is unstable, i.e., if the APB energy is high, then the [$\bar{1}$01] superdislocation dissociates in to a 1/3[$\bar{2}$11] and a 1/3[$\bar{1}\bar{1}$2] superpartial dislocation separated by superlattice intrinsic stacking fault (SISF). The cores of the two superpartials are highly nonplanar and are therefore expected to be sessile at low temperatures. Materials in which the dislocations are dissociated in this latter way are expected to show a temperature dependence of the CRSS for [$\bar{1}$01](111) slip which is similar to that of bcc metals. This latter kind of material will be considered later in this section.

If one of the 1/2[$\bar{1}$01] superpartials separated by APB on the (111) plane cross-slips to the (010) plane, the core of the superpartial immediately redissociates on the (111) plane, the (1$\bar{1}$1) plane or on both simultaneously. Consequently, even though the 1/2[$\bar{1}$01] superpartials are not really dissociated into discreet Shockley partial dislocations as assumed by Kear and Wilsdorf [13], the cores of these dislocations dissociate after cross-slip onto the (010) plane in a manner similar to that assumed by Kear and Wilsdorf, and as a result, the dislocations are, indeed, pinned after cross-slip. Also, the cross-slipping dislocation is expected to move only a distance of b/2 on the (010) plane before its core dissociates onto the (1$\bar{1}$1) plane (b = 1/2[$\bar{1}$01]), and the dislocation is pinned. This means that cross-slip onto the (010) plane does not occur by the movement of a bowed segment on the (010) plane, but rather, a double kink mechanism is involved.

The activation enthalpy for the formation of this double kink was determined by Paidar et al. [19] assuming, as did Friedel [22], that the local constriction of the core, the cross-slip and the redissociation all occur simultaneously. In the case of cross-slip to the (010) plane, it must also be assumed that the shift of the core along the (010) plane by b/2 is also a part of this simultaneous process. The critical activation enthalpy for formation of a double kink was shown to be

$$H_c = 2W + b\{4C - [(\gamma_1/\sqrt{3} - \gamma_0 + \tau_{cb}b)\mu b^2/8\pi]^{1/2}\}, \tag{3}$$

where W is the sum of the energies of constrictions on the (111) and (1$\bar{1}$1) planes, C is the self energy of the kink per unit length, $\gamma_1$ and $\gamma_0$ are the APB energies on the (111) and (010) planes respectively, b is |1/2[$\bar{1}$01]|, and $\mu$ is the shear modulus. Successive jumps of b/2 could lead to complex configurations, but it was shown that the initial jump of length b/2 is rate controlling. Equation (3) can be expanded in a Taylor series and if only linear terms in $\tau_{cb}$ are retained, then equation (1) is obtained.

In the above, the core width is a factor only to the extent that the core configuration controls the jump distance on the (010) plane. There is an important additional effect of core width as was shown by Escaig [23]. Escaig showed that

$$W = W_0[1 + \beta_1\tau_{pe} - \beta_2\tau_{se}] \tag{4}$$

where $\tau_{pe}$ and $\tau_{se}$ are the RSS on the edge components of the Shockley partials on the primary and secondary planes, respectively, $\beta_1$ and $\beta_2$ are constants which depend on the width of the unstressed superpartials, and $W_0$ is the energy when $\tau_{pe}$ and $\tau_{se}$ are zero. Both $\tau_{pe}$ and $\tau_{se}$ are taken to be positive when the partials are extended. If equation (4) is inserted into equation (3) and if it is assumed that

$$\Delta\tau_{pb} = A\exp[-H_c/3KT], \tag{5}$$

as in Takeuchi and Kuramoto, then an expression is obtained which contains all the predictions of the Takeuchi and Kuramoto model, but which also contains the effects of core width and makes specific predictions about the orientation dependence of the tension/compression asymmetry. The specific predictions peculiar to the Paidar et al. model are: (i) For orientations near [001] the tensile CRSS should exceed the compressive, (ii) For orientations on a great circle which lies somewhere between [001] and the [012]-[$\bar{1}$13] great circle the compressive and tensile CRSS's should be equal, (iii) For orientations on the [011]-[$\bar{1}$11] side of this great circle, the compressive CRSS should exceed the tensile, and (iv) For orientations near [011], the compressive CRSS should greatly exceed the tensile. These predictions are summarized in Fig. 1. Umakoshi et al. [24] performed experiments on $Ni_3(Al,Ta)$ to check these predictions and found remarkably good agreement. More recent experiments of Ezz [25] on $Ni_3Ga$, see Fig. 2, and on $Ni_3(Al,Nb)$ are also in agreement with the model, as are the results of Heredia [26], see Fig. 3, on a two phase nickel base superalloy containing about 60 vol. per cent of a phase having the $L1_2$ structure in a matrix of an fcc solid solution. Thus it appears that the model of Paidar et al. [19] can very successfully explain the flow behavior of $L1_2$ alloys which show the anomalous flow behavior. The paper by Vitek and Pope which appears in this volume will expand on this theme.

We return now to the flow behavior of $L1_2$ alloys in which the $[\bar{1}01](111)$ superdislocations are dissociated into $1/3\langle 11\bar{2}\rangle$ superpartials separated by SISF. As stated earlier in this section, the cores of these superpartials are expected to be nonplanar, and therefore dislocation motion is only expected to occur by the nucleation of double kinks on the dislocation line, i.e., the CRSS is expected to increase rapidly at low temperatures. Polycrystalline alloys that show such a temperature dependence of the flow stress were only recently reported by Wee et al. [27] and the results were more recently confirmed by Wee et al. [28] on single crystalline $Pt_3Al$. It is interesting to note that $Pt_3Al$ deforms by $[\bar{1}01](111)$ slip or by $[\bar{1}11](001)$ slip, depending on the orientation of the compression axis. Reasons for the appearance of these two slip systems are given in a later paper in this volume by Tichy et al.

## CYCLIC DEFORMATION

There is evidence in the literature that the asymmetry of the flow stress of $Ni_3Al$, as measured in tensile and compression tests, can be considerably enhanced by cyclic deformation. This was first demonstrated by Jablonski and Sargent [29] who showed that the fatigue hardening rates of the tensile and compressive segments of the fatigue cycle of a two phase nickel base superalloy are different at 760°C. They tested [001]-oriented samples and showed that the tensile stress is larger than the compressive stress, leading to a substantially positive mean stress, as measured under total strain control. Ezz and Pope [30] interpreted this result in terms of the asymmetry of the CRSS for $[\bar{1}01](111)$ slip, using the ideas presented in the previous section. Anton [31] subsequently argued that the Jablonski and Sargent results were incorrect due to experimental difficulties with measuring the total strain at 760°C.

In an attempt to resolve this controversy, Ezz [25] measured the rate of cyclic hardening of $Ni_3(Al,Nb)$ single crystals as a function of orientation, temperature and plastic strain. He found that the asymmetry of the hardening rate is definitely observed, but only in a certain range of sample orientation. Furthermore, the sense of the asymmetry was found to be always the same as that observed in monotonic tests. Typical results are shown in Fig. 4. The [001]-oriented sample was cycled at 878K at a plastic strain range of $2 \times 10^{-3}$, the [011] at 573K and $5 \times 10^{-4}$ and the $[\bar{1}11]$ at 583K and $3 \times 10^{-4}$. (The test temperatures were chosen to maximize the tension/compression flow stress asymmetry as measured in monotonic tests.) Note in Fig. 4 that the mean stress of the [001]-oriented sample is only slightly positive, amounting to about 10% of the maximum stress, but the mean stress for the other two samples is much larger and negative, amounting to about 40% of the maximum stress in the case of the [011] oriented sample.

Similar effects have been observed in bcc metals, specifically in $\alpha$-Fe and Mo single crystals [32,33], and the origins of the effect are believed to be the same in both cases: the unusual properties of screw dislocations in both bcc and $L1_2$ materials.

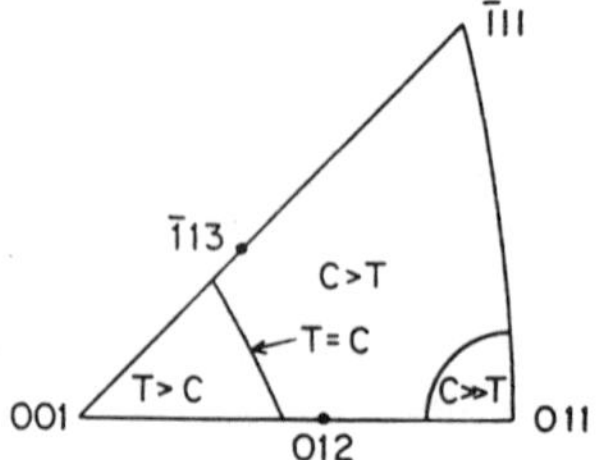


Fig. 1. Summary of the predictions of the Paidar et al. [19] model.

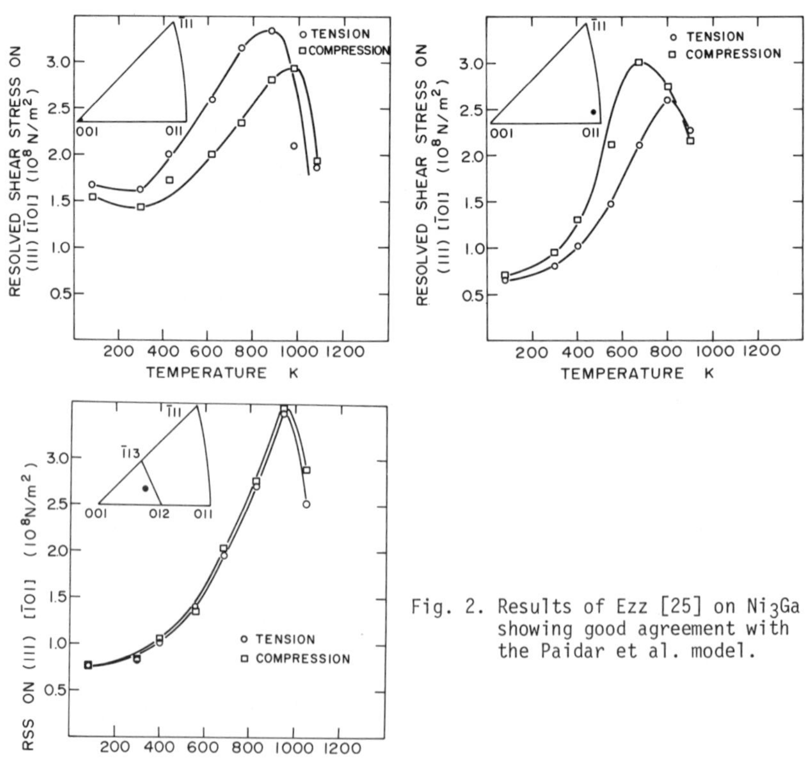


Fig. 2. Results of Ezz [25] on $Ni_3Ga$ showing good agreement with the Paidar et al. model.

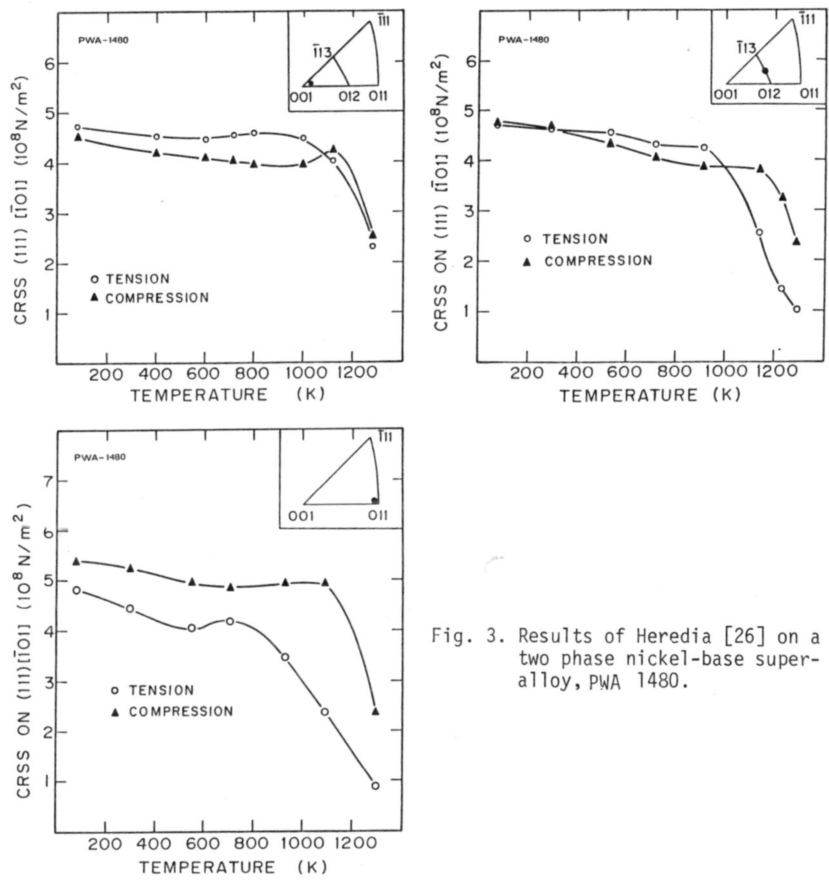


Fig. 3. Results of Heredia [26] on a two phase nickel-base superalloy, PWA 1480.

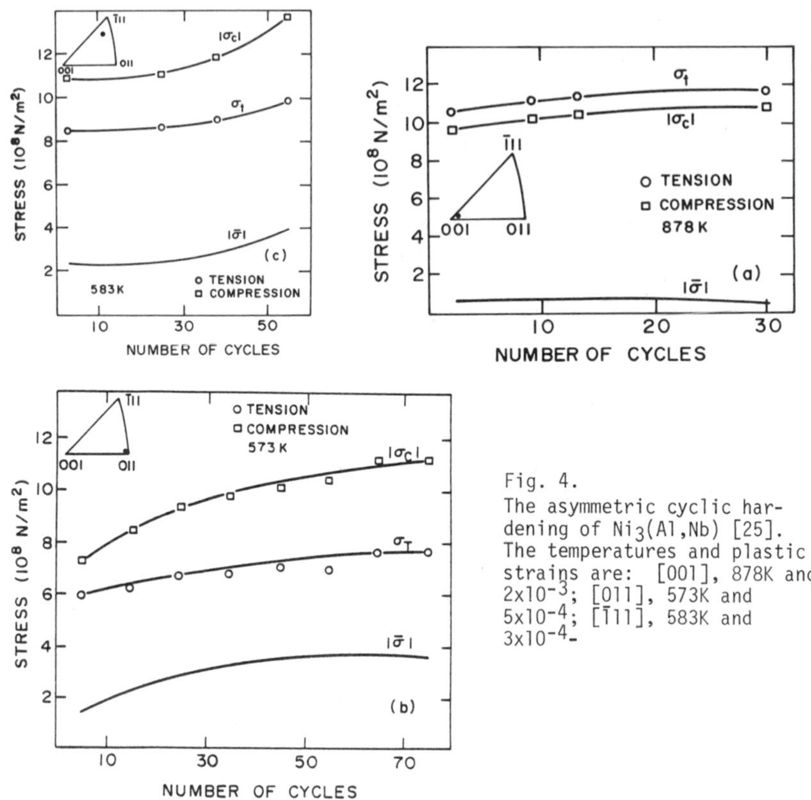


Fig. 4.
The asymmetric cyclic hardening of $Ni_3(Al,Nb)$ [25]. The temperatures and plastic strains are: [001], 878K and $2x10^{-3}$; [011], 573K and $5x10^{-4}$; [$\bar{1}$11], 583K and $3x10^{-4}$.

## COMPOSITIONAL EFFECTS

This complex subject will be only briefly discussed here because it has recently been reviewed in detail [34] and is also the subject of another paper in this volume by Mishima et al. There is considerable interest in this subject since an $Ni_3Al$-based precipitate is the main strengthening element in modern nickel base superalloys and these precipitates usually contain many other elements in addition to Ni and Al. Many different elements can form solid solutions with $Ni_3Al$, substituting for one or both elements. Summaries of the phase diagrams for $Ni_3Al$-X and $Ni_3Ga$-X systems, where X is the ternary addition, have recently been published by Ochiai et al. [35].

Different ternary additions have very different effects on the flow strength of $Ni_3Al$, and in addition, those elements which substitute for Al appear to have the biggest strengthening effect, particularly if the sum of Al and X exceeds 25 at.%. (Off-stoichiometric alloys do not contain excess vacancies, rather, the excess atoms of one type substitute on the sites of the other atom type [36].) The major question which arises form these studies is the following: Is the strengthening of $Ni_3Al$ by ternary additions the result of classical solid solution strengthening mechanisms, e.g., the Fleischer mechanism, or is this strengthening the result of changes in the rate of cross slip due to changes in the various fault energies involved in equations (3) and (4) of this paper? Although the data are not complete, in spite of the fact that a large number of studies have been performed, it appears that the strength changes, at least at elevated temperatures, are more related to changes in the cross-slip rate than to traditional solid solution strengthening mechanisms. This point of view is described most completely in a series of papers from Suzuki's group in Japan [37-39] and in the paper by Mishima et al. in this volume.

## CONCLUSIONS

An increasingly large body of experimental data indicate that the flow properties of $L1_2$ ordered alloys are controlled by the motion of screw dislocations. In alloys which show the anomalous flow stress increase with temperature the CRSS is controlled by the density of pinning points on screw dislocations. These pinning points are produced by cross-slip of dislocations from (111) to (010) planes. Screw dislocations also appear to control the cyclic hardening rate of such alloys and are probably intimately involved in the strengthening produced by ternary additions. Many of these same ideas are applicable to two phase nickel base superalloys [34,40]. There is another group of $L1_2$ alloys which do not exhibit the anomalous flow stress increase with temperature, but in which the screw dislocation core structure is believed to be important. In those alloys the <101> superdislocations are believed to be dissociated into 1/3<112> superpartials with nonplanar cores separated by SISF. The flow behavior of these alloys is very similar to that of bcc metals.

## ACKNOWLEDGMENTS

This work was supported by the National Science Foundation under grant no. DMR79-05556 and by the Office of Naval Research under grant no. 5-21233.

## REFERENCES


1. J. H. Westbrook, Trans. TMS-AIME, 209, 898 (1959).
2. P. A. Flinn, Trans. TMS-AIME, 218, 145 (1960).
3. R. G. Davies and N. S. Stoloff, Trans. TMS-AIME, 233, 714 (1965).
4. T. L. Johnston, A. J. McEvily and A. S. Tetelman, in High Strength Materials (V. F. Zackay, ed.), pp. 363-381, Wiley, New York, 1965.
5. S. M. Copley and B. H. Kear, Trans. AIME, 239, 977 (1967).
6. D. P. Pope, Phil. Mag. 25, 917 (1972).
7. A. E. Staton-Bevan and R. D. Rawlings, Phys. Stat. Solidi(a), 29, 613 1975.
8. B. H. Kear and M. F. Hornbecker, Trans. ASM, 59, 155 (1966).
9. P. H. Thornton, R. G. Davies and T. L. Johnston, Met. Trans., 1A, 207 (1970).
10. R. A. Mulford and D. P. Pope, Acta Met., 21, 1375 (1973).
11. S. Takeuchi and E. Kuramoto, Acta Met., 21, 45 (1973).
12. Salah S. Ezz, D. P. Pope and V. Paidar, Acta Met., 30, 921 (1982).
13. B. H. Kear and H. G. F. Wilsdorf, Trans. TMS-AIME, 224, 382 (1962).
14. H. R. Pak, T. Saburi and S. Nenno, Trans. Jap. Inst. Metals, 18, 617 (1977).
15. T. Saburi, T. Hamona, S. Nenno and H. R. Pak, Jap. J. Appl. Phys., 16, 267 (1977).
16. E. Kuramoto and D. P. Pope, Acta Met., 26, 207 (1978).
17. K. Aoki and O. Izumi, Acta Met., 26, 1257 (1978).

18. C. Lall, S. Chin and D. P. Pope, Met. Trans., 10A, 1323 (1979).
19. V. Paidar, D. P. Pope and V. Vitek, Acta Met., 32, 435 (1984).
20. M. Yamaguchi, V. Paidar, D. P. Pope and V. Vitek, Phil. Mag., 45, 867 (1982).
21. V. Paidar, M. Yamaguchi, D. P. Pope and V. Vitek, Phil. Mag., 45, 883 (1982).
22. J. Friedel, in Dislocations and Mechanical Properties of Crystals, (J. C. Fisher, W. G. Johnston, R. Thomson and T. Vreeland, Jr., ed.) pp. 330-32, John Wiley, New York (1957).
23. B. Escaig, in Dislocation Dynamics, (A. R. Rosenfield, G. T. Hahn, A. L. Bemet, Jr. and R. I. Jaffee, ed.) pp. 655-77, McGraw-Hill, New York (1968).
24. Y. Umakoshi, D. P. Pope and V. Vitek, Acta Met., 32, 449 (1984).
25. Salah S. Ezz, Ph.D. Dissertation, University of Pennsylvania, Philadelphia, PA (1984).
26. F. E. Heredia, M.S. Dissertation, University of Pennsylvania, School of Engineering and Applied Science, Philadelphia, PA (1984).
27. D. M. Wee, O. Noguchi, Y. Oya and T. Suzuki, Trans. JIM, 21, 237 (1980).
28. D. M. Wee, D. P. Pope and V. Vitek, Acta Met., 32, 829 (1984).
29. D. . Jablonski and S. Sargent, Scripta Met., 15, 1003 (1981).
30. Salah S. Ezz and D. P. Pope, Scripta Met., 16, 117 (1982).
31. D. Anton, Scripta Met., 16, 479 (1982).
32. H. Mughrabi and C. Wuthrich, Phil. Mag., 33, 963 (1976).
33. H. Mughrabi, F. Ackermann and K. Herz in Fatigue Mechanisms (J. T. Fong, ed.), ASTM STP 675, p. 69, ASTM, Philadelphia (1979).
34. Salah S. Ezz and D. P. Pope, Int. Metals Rev., 29, 136 (1984).
35. S. Ochiai, Y. Oya and T. Suzuki, Acta Met., 32, 289 (1984).
36. K. Aoki and O. Izumi, Phys. Stat. Sol. (a), 32, 657 (1975).
37. D. M. Wee and T. Suzuki, Trans. Jap. Inst. Met., 20, 634 (1979).
38. D. M. Wee, O. Noguchi, T. Oya and T. Suzuki, Trans Jap. Inst. Met., 21, 237 (1980).
39. T. Suzuki, Y. Oya and S. Ochiai, Met. Trans. 15A, 173 (1984).
40. B. H. Kear and D. P. Pope, in Refractory Elements in Superalloys - Effects and Availability, Proc. USA-Brazil Con. on Superalloys, J. Tien, ed., ASM, 1984.

# THE EFFECTS OF GRAIN SIZE ON THE FLOW AND FRACTURE OF LONG-RANGE ORDERED ALLOYS

ERLAND M. SCHULSON
Thayer School of Engineering
Dartmouth College, Hanover, NH 03755

## INTRODUCTION

Three points usually come to mind when considering the effects of grain size on the mechanical properties of polycrystals: strengthening at low temperatures through grain refinement, strengthening at high temperatures through grain coarsening, and lowering of the ductile to brittle transition temperature in materials such as ferritic steels and zinc through grain refinement. Ordered alloys, as we shall see, exhibit the same effects.

In considering these effects, discussion is limited to short-term behavior. Long-term behavior, although of interest, appears not to have been examined for possible grain size effects. Emphasis is placed upon relatively recent observations and upon the behavior of the aluminides, a class of ordered intermetallic compounds which appear near to exploitation.

The paper is not intended to be a critical review. Rather, it indicates general trends and shows that grain size can have significant effects on the mechanical properties of ordered alloys.

## YIELD STRENGTH

At low temperatures, grain refinement raises the yield strength. The B2 alloys AuZn [1], FeCo [2], FeCo-2V [3], and NiAl [4] and the $L1_2$ alloys $Cu_3Au$ [5], $Ni_3Al$ [6], $Ni_3Fe$ [7], $Ni_3Mn$ [8], and $Zr_3Al$ [9] all exhibit this feature. Partial disordering, whether through heat treatment (FeCo [2], FeCo-2V [3], $Cu_3Au$ [5], $Ni_3Fe$ [6], $Ni_3Mn$ [7]) or through fast neutron irradiation ($Zr_3Al$ [10]) lowers the magnitude of this effect but does not eliminate it. With only one exception, the grain size dependence has been expressed in terms of the Hall-Petch relationship:

$$\sigma_y = \sigma_o + k_y \ d^{-1/2} \tag{1}$$

where $\sigma_y$ s the yield strength, d is the average grain size and $\sigma_o$ and $k_y$ are material parameters which, respectively, measure the resistance of the lattice to dislocation slip and the effectiveness with which grain boundaries impede slip. Table I lists the Hall-Petch parameters at room temperature.

Table I: Hall-Petch parameters for B2 and $L1_2$ ordered alloys at 20° C.

| Structure | Alloy | $\sigma_0$(MPa) | $k_y$(MPa·m$^{1/2}$) | Ref. |
|---|---|---|---|---|
| B2 | FeCo | 50 | 1.0 | 2 |
| | FeCo-2V | 138 | 0.74 | 3 |
| | NiAl | 189 | 0.67 | 4 |
| | AuZn | 20 | <0.1 | 1 |
| $L1_2$ | $Ni_3Fe$ | 108 | 1.3 | 7 |
| | $Ni_3Mn$ | 98 | 1.1 | 8 |
| | $Zr_3Al$ | 90 | 0.76 | 9 |
| | $Cu_3Au$ | 68 | 0.50 | 5 |

The exception is $Ni_3Al$ [11]. In this case, the yield strength of stoichiometric material exhibits a grain size dependence stronger than $d^{-1/2}$; namely:

$$\sigma_y = \sigma_0 + kd^{-(0.80\pm0.05)} \quad (2)$$

where k is a material parameter. That this dependence is not just a better description than $d^{-1/2}$ dependence but is the best description is based upon a regression analysis [11] of all room-temperature data (i.e., 34 points) generated in the author's laboratory, corresponding to grain size ranging from d=2.9 to 1100μm. This analysis revealed that the value of n (in the general expression $\sigma_y = \sigma_0 + kd^{-n}$) which gives the highest regression coefficient, $r_{max}$, is n = 0.80 ± 0.05; $r_{max}$ = 0.992. Correspondingly, $\sigma_0$ = 93 ± 13 MPa and that k = 2080 ± 100 MPa·μm$^{(0.8 \pm 0.05)}$. Figure 1 illustrates this dependence in a plot of $\sigma_y$ versus $d^{-0.75}$, and it shows how well data previously published obey the above description.

Most alloys appear to be "conventional" and can be understood in terms of existing theories of yielding [12,13]. $Ni_3Al$, on the other hand, is non-conventional and for this reason is worth further consideration. In particular, it is significant to note that this material, like other $L1_2$ alloys ($Cu_3Au$ [5], $Ni_3Fe$ [7], $Ni_3Mn$ [8] and $Zr_3Al$ [9,10], yields discontinuously and that the discontinuities in $Ni_3Al$ correspond to the steady but jerky propagation (at stress constant to ± 5 MPa) of Lüders bands along the gauge section of test specimens. It is also significant that the Lüders strain, $\varepsilon_L$, increases with decreasing grain size, as it does for $Zr_3Al$ [9,10], and that $\varepsilon_L$ obeys the relationship:

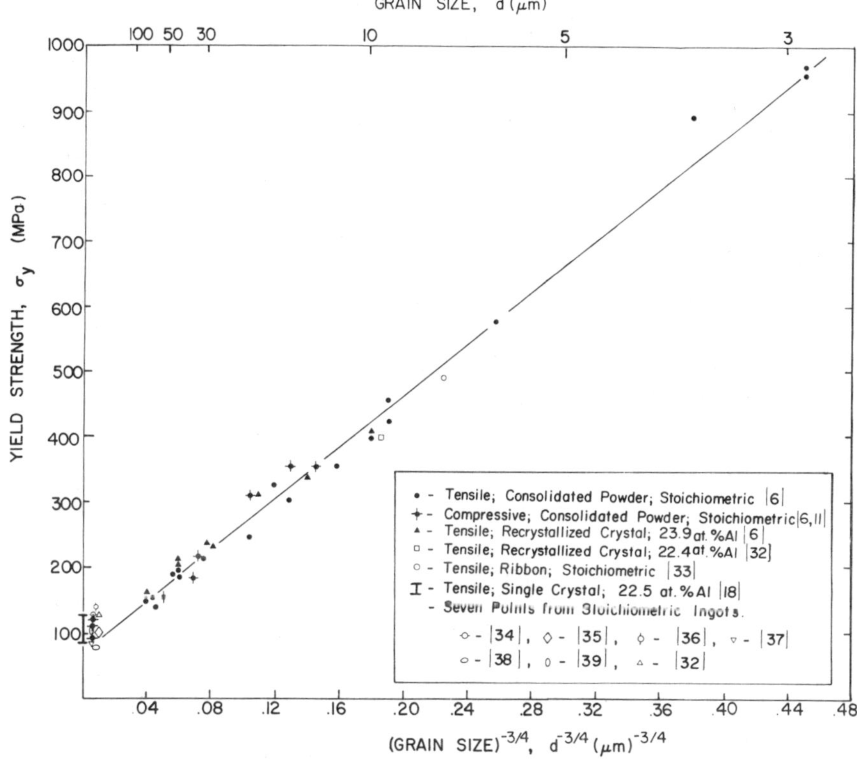


Figure 1 Yield strength versus $(\text{grain size})^{-3/4}$ for $Ni_3Al$ at room temperature.

$$\varepsilon_L = \frac{\lambda}{d^{1/2}} \qquad (3)$$

where $\lambda = 8.4 \times 10^{-2}\ \mu m^{1/2}$ at room temperature. These features are significant because they mean that the yield strength, which was taken as the stress to propagate Lüders bands, corresponds to the stress required to continue plastic flow through a matrix which has strained by an amount $\varepsilon_L$. Since $\varepsilon_L$ increases with decreasing grain size, the attendant strain hardening also increases, thereby raising the yield strength. In more quantitative terms:

$$\sigma_y = \sigma_o + \alpha G b \rho_L^{1/2} \qquad (4)$$

where $\alpha$ is a parameter of order unity, G is the shear modulus and b is the Burgers' vector of the slip dislocations; $\rho_L$ is the dislocation density generated during the Lüders strain and may be expressed as:

$$\rho_L = \frac{m\,\varepsilon_L}{bl} \qquad (5)$$

where l is the average distance by which dislocations move and m is the Taylor orientation parameter. If l is assumed to be proportional to the grain size in lightly strained material (i.e. $l = \beta d$ where $\beta$ is another parameter of order unity), then the combination of Equations 3, 4 and 5 gives the relationship:

$$\sigma_y = \sigma_o + \alpha G \left(\frac{m\,\lambda b}{\beta}\right)^{1/2} d^{-0.75} \qquad (6)$$

This analysis thus gives a grain size dependence in keeping with experiment. Also, the calculated coefficient[11] of $d^{-0.75}$ is 930 $MPa \cdot \mu m^{3/4}$, in rough agreement with experiment.

Turning to behavior at high temperatures, grain-boundary sliding (GBS) may become significant. This process, when not accommodated by dislocation creep or diffusional creep, is evident from intergranular wedge-cracks and from grain offsets. In the $L1_2$ aluminides $Zr_3Al$ [14] and $Ni_3Al$ [6], which remain fully ordered at high temperatures, these features have been observed in specimens strained at a rate of approximately $10^{-4}s^{-1}$ and at temperatures above $\simeq$ 600°C (i.e. $\tilde{<}$ 0.55 $T_m$), suggesting that GBS accounts for the attendant reduction in strength with decreasing grain size, Table II. In the B2 aluminides NiAl [15] (49 at.% Al) and FeAl [16] (39.8 at.% Al), which also remain ordered, this mode of deformation has not been reported for the conditions investigated to date. At 600°C ($\simeq 0.46T_m$) grain size has essentially no effect on the strength of NiAl strained at $10^{-4}s^{-1}$, and at 925°C ($0.75T_m$) grain refinement actually strengthens and not weakens FeAl strained in compression at rates as low as $10^{-7}s^{-1}$. Whether the aluminumlean B2 aluminides are less susceptible to GBS than are the $L1_2$ aluminides remains to be determined as does the possible effect of deviations from stoichiometry.

Table II: Data showing the high-temperature strengthening of $Ni_3Al$ and $Zr_3Al$ through grain coarsening.

| Alloy | T(°C) | $\dot{\varepsilon}(s^{-1})$ | $\sigma_y$(MPa) | d(μm) | Ref. |
|---|---|---|---|---|---|
| $Ni_3Al$ | 750 | $1 \times 10^{-4}$ | 190<br>302<br>336<br>342 | 10<br>15<br>42<br>113 | 6 |
| $Zr_3Al$ | 750 | $3 \times 10^{-4}$ | 280<br>310 | 3.5<br>29 | 9 |
| | 900 | $3 \times 10^{-4}$ | 50<br>280 | 3.5<br>29 | 9 |

The other effect of grain size on the yield strength occurs at intermediate temperatures and relates to the "anomalous " increase in the strength of certain $L1_2$ alloys with increasing temperature. Coarse-grained polycrystals of $Ni_3Al$ [6] and $Zr_3Al$ [9], for instance, exhibit greater thermal hardening than do fine-grained aggregates, and very fine-grained polycrystals ($d \simeq 1\mu m$) exhibit thermal softening, Figure 2. These effects reflect the opposing effects of temperature on the lattice resistance to dislocation glide (which increases with increasing temperature) and on the effectiveness with which grain boundaries impede slip. They imply that the yield strength of a fine-grained polycrystal of grain size $d^*$ will show little thermal sensitivity, where $d^* \simeq [-(\partial k_y/\partial T) / (\partial\sigma_0/\partial T)]^{n'}$ where n' is the absolute value of the reciprocal of the exponent in the $\sigma_y$ vs. d relationship. Moreover, they indicate that caution should be exercised when attempting to extract fundamental information such as activation energies from $\partial\sigma_y/\partial T$ for polycrystals.

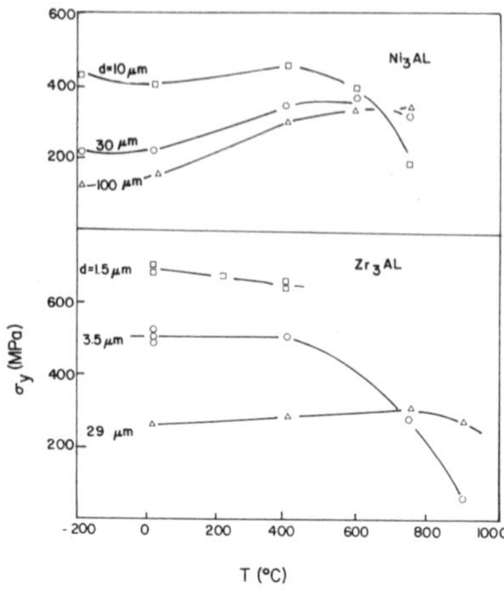


Figure 2 Yield strength versus temperature and grain size for stoichiometric $Ni_3Al$ and for $Zr_3Al$ [9]. Note reduction in slope with decreasing grain size at intermediate temperatures.

FLOW STRESS

Concerning the effect of grain size on the low-temperature flow stress, all studies [1-4,6,7,9,] indicate that the trend noted upon yielding persists upon work hardening; namely, that grain refinement raises the strength. The studies also show that a Hall-Petch type relationship applies:

$$\sigma_f = \sigma_{0,\varepsilon} + k_\varepsilon d^{-1/2} \tag{7}$$

where $\sigma_f$ is the flow stress at a plastic strain $\varepsilon$, $\sigma_{0,\varepsilon}$ is the resistance to slip offered by the lattice and its additional defects (e.g. point defects, dislocations and faults) created during flow, and $k_\varepsilon$ is a measure of the effectiveness with which grain boundaries impede slip when the matrix has been plastically strained. Within this general conformity, however, two types of behavior are noted. In the first (Type-I), $k_\varepsilon$ increases with increasing strain, as found in the B2 alloys AuZn [1], FeCo [2], and NiAl [4]. In the second (Type-II), $k_\varepsilon$ remains constant and equals $k_y$, as seen in the $L1_2$ alloys $Zr_3Al$ [14] and $Ni_3Al$ [6]. Type-I reflects the fact that the work hardening rate decreases with increasing strain and with increasing grain size, while Type-II indicates that the work hardening rate is essentially independent of these parameters.

If one ascribes to dislocation interactions the increment in hardening which accompanies an increment in plastic strain, then behavior of Type-I suggests that throughout the work hardening stage of deformation the average distance moved by dislocations, l, is proportional to the grain size and is thus roughly constant. Type-II, on the other hand, suggests that l decreases with increasing strain in proportion to (dislocation density)$^{-1/2}$. The two types of behavior may reflect the ease with which dislocations cross-slip, type-I implying the higher propensity. Consistent with this view is the fact that the strongly-ordered B2 alloys (e.g. AuZn and NiAl) exhibit rather wavy slip, whereas the $L1_2$ alloys exhibit planar slip owing to differences in dislocation structure between open and close-packed crystals.

## DUCTILITY

The greatest obstacle to the exploitation of the most strongly ordered alloys is brittleness at ambient temperatures. Unlike strength, which is moderate to high and which can be improved through metallurgical methods, and unlike oxidation resistance, which is inherently good due to the formation adherent oxides, ductility under tensile stress is small. In other words, strongly ordered polycrystals are not very tough.

The problem is not always inherent in the crystal lattice. Single crystals of NiAl [17] and of $Ni_3Al$ [18], for instance, are ductile. Rather the problem is often associated with the boundaries which join the crystallites together. In the strongly ordered B2 compounds, such as NiAl, only three independent systems (i.e. of the form {hko} <001>) operate easily, and so grain-to-grain contiguity, which requires at least four active systems [19], is difficult to achieve. In other strongly ordered alloys which do have enough independent slip systems, such as the $L1_2$ alloys which slip on systems of the form {111} <110>, the problem appears to be one of inherently low boundary cohesiveness. Segregants exacerbate the situation [20], but appear not to be the primary cause [21], at least in every case.

Grain size appears to be a factor. Whether boundaries act as barriers to dislocations and thus as sites for the concentration of stress at the heads of dislocation pileups, or whether triple points, owing to elastic anisotropy, act as stress concentrators, the grain size controls the magnitude of the stress intensification. The larger the grains, the higher is the stress concentration factor in the vicinity of the boundary and the lower is the applied stress necessary to nucleate a crack. The other aspect is that the larger the grains, the larger are the cracks that will eventually form, assuming that the crack size is proportional to grain size as it is in ice [22]. Cracks in coarse grained aggregates may be so large that they propagate as soon as they nucleate, thereby preventing macroscopic ductility.

One remedy may be to refine the grains to below a critical size [23,24]. In this case, when cracks form they will be stable. Hardening through plastic flow will then be required to cause propagation, and the aggregate will exhibit some ductility under tension. The smaller are the grains below the critical size, the greater will be the hardening required for propagation and the greater will be the ductility. If the grains (and, hence, the cracks) are very small, then it is possible that "hard" slip modes will become active (e.g. <111> slip in the strongly ordered B2 aluminides) before cracks propagate, in which case extensive ductility may result.

The critical grain size, $d_c$, may be formulated in terms of measurable parameters by assuming that the applied stress to nucleate microcracks is similar to that to propagate slip across grain boundaries (i.e. that cracks nucleate just past the onset of yielding), by invoking the Hall-Petch expression for the yield strength, by invoking linear elastic fracture mechanics, and by assuming that the fracture toughness is independent of grain size. Thus [24], analogous to Cottrell's formulation [23]:

$$d_c = \left( \frac{Y\, K_{Ic} - k_y}{\sigma_o} \right)^2 \tag{8}$$

where $K_{Ic}$ is the plane strain fracture toughness and Y is a geometrical parameter which takes into account the stress state, the constant of proportionally between the crack size and the grain size, the crack shape and the degree to which cracks interact.

Supporting the concept is the observation that the room temperature ductility of FeCo-2V and of FeCo-V-Ni appears to be proportional to $d^{-1/2}$ [25] and the observation [26] that the ductile to brittle transition temperature for FeCo-2V falls from 450°C to -50°C upon reducing the grain size from 100μm to 1μm. Additional support is the observation [27] that the "30° bend transition temperature" in NiAl decreases with decreasing grain size, in proportion to $d^{-1/2}$; the result [40] that melt spun ribbons of $Ti_3\ Nb_5\ Mo_3\ Si_4$, upon annealing, develop an A-15 structure and exhibit some bend ductility, but that upon further annealing the ductility decreases - effects attributed to the development of a very fine grained microstructure and to the subsequent coarsening of this structure; and the observation [41] that the ductile to brittle transition temperature for the A-15 alloy $Nb_3Sn$ under compression decreases by 125°C (i.e. from 1400°C to 1275°C) upon decreasing the grain size from 60μm to 10μm. Detailed support is obtained from a systematic study [4,15] of NiAl (49at.%Al) which showed that at 400°C and $10^{-4}s^{-1}$ the tensile ductility increases from 2 to 3% for aggregates of grains larger than about 20μm to over 40% for d = 10μm with no change in fracture mode, Figure 3. This study also showed that $d_c$ decreased and increased, respectively, upon increasing and decreasing the strain rates and that $d_c$ decreased upon decreasing the temperature, presumably in response to the effects of strain rate and temperature on $\sigma_0$. Upon increasing the temperature to 600°C (Fig.3), the fracture mode changes from a transgranular cleavage/intergranular mixture to the nucleation, growth and coalescence of microvoids in which case grain size has no effect on either ductility or strength. The last point is also evident for the $L1_2$ aluminide $Zr_3Al$ [14] which fractures in a ductile mode at room temperature.

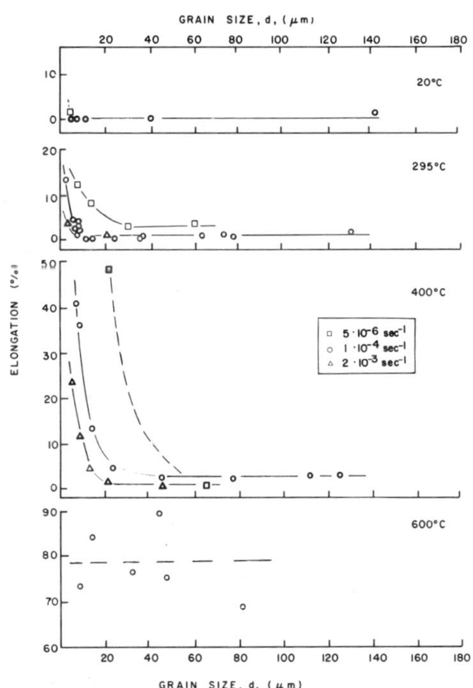


Figure 3 Elongation versus grain size for NiAl (51/49) at temperatures from 20°C to 600°C at strain rates from $5 \times 10^{-6}s^{-1}$ to $2 \times 10^{-3}s^{1}$. Note brittle to ductile transition at grain size which increases with increasing temperature and with decreasing strain rate[4,15].

It should not be concluded, however, that grain refinement to below a critical grain size is the solution to brittleness in every case. For instance, brittleness in $Ni_3Al$ is not ameliorated at any temperature between -196°C and 400°C upon decreasing the grain size to 10μm [6]. In this case, microalloying with boron [21,28,29] is much more effective.

## GRAIN SIZE EFFECT ON THE MICROALLOY PARTITIONING OF MICROALLOYING ELEMENTS?

One final point is considered; namely, the possible effects of grain size on the partitioning of beneficial elements. For instance, small additions ($\lesssim$1at.%) of boron impart extensive ductility to nickel-rich $Ni_3Al$ [28,29]. In this case, boron occupies both interstitial sites[30] as well as intergranular sites [31]. Given the apparent attraction for the latter, it seems possible that in dilute alloys the fraction of the boron atoms which occupy the boundary sites increases with decreasing grain size. One result of such partitioning may be that the strengthening confirmed through boron [30] will be less effective in fine-grained polycrystals than in coarse-grained material. Another may be that polycrystals of a given grain size may exhibit two-stage strengthening upon boron additions, a first stage of low strengthening during which boron favors boundary sites followed by a second stage of high strengthening once the boundary sites have become "saturated". A third possibility is that the "critical boron concentration" necessary to ductilize $Ni_3Al$ may increase with decreasing grain size. Support for the first two of these possibilities is based upon two observations: that an addition of 0.35at.% boron to stoichoimetric $Ni_3Al$ increases the room-temperature yield strength [11] of polycrystals of 42±1μm grain size by 109±11 MPa compared with a smaller increase of 22±22 MPa in an aggregate of 11±1μm grains, Table III; and that a plot of the room-temperature yield strength versus boron concentration in $Ni_3Al$ exhibits positive curvature [30].

Table III: Data showing the room temperature strengthening of $Ni_3Al$ by Boron for two grain sizes. (Taken from ref. 11)

| Grain Size (μm) | $\sigma_y$ (MPa) | | $\Delta\sigma_y$ (MPa) |
|---|---|---|---|
| | $Ni_3Al$ | $Ni_3Al$ + 0.3at.%B | |
| 11±1 | 378±20 | 400±11 | 22±22 |
| 42±1 | 190±5 | 299±10 | 109±11 |

## CONCLUSION

In conclusion, grain size affects the yield strength and the flow stress of long-range ordered alloys. In those $L1_2$ alloys which exhibit "anomalous" thermal strengthening, grain size affects the magnitude of the strengthening. In some cases, it also affects ductility in that the tensile elongation (and the tensile strength) increase with decreased grain size below a critical value. Albeit speculative at this point, grain size may affect the partitioning of microalloying elements, such as boron in $Ni_3Al$.

## ACKNOWLEDGEMENTS

The author acknowledges his students, D.R. Barker, R.B. Graham, D.V. Viens, and T.R. Weihs, from whose work many of the results are taken, and his colleagues I. Baker and H.J. Frost for discussions. The work was supported by the Department of Energy through contract No. DEAC02 - 81ER10907 and grant No. DE-FG02 - 84ER45148 and by the National Aeronautical Space Administration through grant No. NAG - 3-13.

## REFERENCES

1. A.R. Causey and E. Teghtsoonion, Met. Trans. 1 (1970) 1177
2. M.J. Marcinkowski and R.M. Fisher, TMS-AIME 233 (1965) 293
3. K.R. Jordan and N.S. Stoloff, TMS-AIME, 245 (1969) 2027
4. D.R. Barker, M.E. Thesis, Thayer School of Engineering, Dartmouth College, May 1982; see also E.M. Schulson and D.R. Barker, Scripta Met. 17 (1983) 519
5. S.M.L. Sastry, Mater. Sci. Engg. 22 (1976) 237
6. D.V. Viens, M.E. Thesis, Thayer School of Engineering, Dartmouth College, May 1983
7. A.C. Arko and Y.H. Liu, Met. Trans. 2 (1971) 1875
8. T.L. Johnston, R.E. Davies and N.S. Stoloff, Phil. Mag. 12 (1965) 305
9. E.M. Schulson and J.A. Roy, Acta Met. 26 (1978) 29
10. E.M. Schulson, ibid. 26 (1978) 1189
11. E.M. Schulson. D.V. Viens, T.P. Weihs, and I. Baker (to be submitted for publication)
12. R.W. Armstrong, Met. Trans. 1 (1970) 1169
13. J.C.M. Li and Y.T. Chou, Met. Trans. 1 (1970) 1145
14. E.M. Schulson and J.A. Roy, Acta Met. 26 (1978) 15
15. R.B. Graham, M.E. Thesis, Thayer School of Enigneering Dartmouth College, September, 1984
16. J.D. Whittenberger, Mater. Sci. Engg. 57 (1983) 77
17. A. Ball and R.E. Smallman, Acta Met. 14 (1966) 1349
18. S.M. Copley and B.H. Kear, TMS-AIME 239 (1967) 1977
19. J.W. Hutchinson, Met. Trans. A 8A (1977) 1465
20 C.L. White and D.F. Stein, Met. Trans. 9A (1978)13
21. C.T. Liu and J.O. Stiegler, Science 226 (1984) 636
22. D.M. Cole, M.S. Thesis, Dartmouth College, June 1984
23. A.H. Cottrell, TMS-AIME 212 (1958) 192
24. E.M. Schulson, Res Mech. Lett. 1 (1981) 111
25. C.D. Pitt and R.D. Rawlings, Met. Sci. 17 (1983) 261

26. N.S. Stoloff, "Fracture", Vol.VI, ed. H. Liebowitz, New York, Academic Press, (1969) p.1.
27. E.R. Stover and A.U. Seybolt, WADC TR-60-184 Part VII (1966); quoted by J.H. Westbrook in "Ordered Alloys", ed. B.H. Kear, C.T. Sims, N.S. Stoloff and J.H. Westbrook, Claitor's Publ. Div., Baton Rouge, (1970) p.1.
28. K. Aoki and O. Izumi, Nippon Kinzoku Takkaishi 43 (1979) 1190
29. C.T. Liu and C.C. Koch, Proceedings of Workshop on "Trends in Critical Materials Requirements for Steels of the Future; Conservation and Substitution Technology for Chromium", NBSIR-83-2679-2, June (1983)
30. S.C. Huang, A.T. Taub and K.M. Chang, Acta Met. (in press)
31. C.T. Liu, C.L. White and J.A. Horton, Acta Met. (in press)
32. K. Aoki and O. Izumi, Trans. J. 1.M 19 (1978) 203
33. A.I. Taub, S.C. Huang and K.M. Chang, Met. Trans. A 15A (1984) 399
34. E.M. Grala, "Mechanical Properties of Intermetallic Compounds", ed. J.H. Westbrook, John Wiley and Sons, New York, (1960) p. 358
35. R.G. Davies and N.S. Stoloff, TMS-AIME 233 (1965) 714
36. P.H. Thornton, R.G. Davies and T.L. Johnston Met. Trans. 1 (1970) 207
37. J.A. Lopez and E.F. Hancock, Phys. Stat. Sol. (a) 2 (1970) 469
38. O. Noguchi, Y. Oya and T. Suzuki Met. Trans. A. 12A (1981) 1647
39. P.A. Flinn, "Strengthening Mechanisms in Solids," A.S.M. (1962) p.17
40. M.T. Clapp and D. Shi (unpublished results), University of Massachusetts, Amherst, MA., November 1984.
41. J.B. Clark, G.B. Hopple and R.N. Wright, Met. Trans A 14A (1983) 889

DISLOCATION CORE STRUCTURE AND THE ANOMALOUS YIELD BEHAVIOR OF $L1_2$ ORDERED ALLOYS AT ELEVATED TEMPERATURES

V. VITEK AND D. P. POPE
Department of Materials Science and Engineering, University of Pennsylvania, Philadelphia, PA 19104

ABSTRACT

In many $L1_2$ ordered alloys the flow stress increases with increasing temperature and is in this "anomalous" regime strongly dependent on orientation and sense of the applied stress. These dependences can be predicted from the nature of the dissociation and core structure of the 1/2<101> screw superpartials in these alloys. Computer modelling shows that two different configurations, a glissile one on {111} planes and a sessile one on {010} planes, exist and both are described here in detail. The anomalous increase of the flow stress may then be explained by an increasing amount of core transformations from the glissile to sessile forms as the temperature increases. The theoretical model for the immobilization of screw dislocations by this mechanism is then discussed and its validity illustrated by comparison with experimental results on $Ni_3(Al,Ta)$ single crystals.

## 1. Introduction

In earlier developments of dislocation theory core effects were usually not regarded as very important. However, the extensive studies of the plastic behavior of b.c.c. metals in the late sixties and early seventies revealed that it was the core structure of screw dislocations which controlled the basic features of the plastic behavior of these materials (for reviews see e.g. [1,2]). These are, in particular, a sharp increase of the flow stress with decreasing temperature, invalidity of the Schmid's law, very complex slip geometry as well as asymmetries in the slip geometry and flow stress when applying tension and compression, respectively. The core of screw dislocations in these materials is spread spacially, principally into three {110} planes intersecting along a <111> direction (for more details see [3,4]). This core is therefore sessile and it must transform in a complex way prior to motion. This is in contrast, for example, with f.c.c. metals in which dislocations are split into Shockley partials on {111} planes the cores of which are also confined to these planes. This is the reason why in f.c.c. metals {111} planes are practically always the slip planes, the flow stress is insensitive to temperature and no unusual orientation dependences are observed.

However, b.c.c. metals are not the only materials in which core effects are important. Another example is the plastic flow in h.c.p. metals when the slip is not confined to the basal planes. The flow stress then also depends very strongly on temperature (e.g. [5]) and atomistic studies of dislocations indeed show that the screw dislocation cores are spread outside the prismatic and/or pyramidal slip planes (e.g. [6,7]). Furthermore, an anomalous increase of the flow stress with increasing temperature was observed in the cases of pyramidal slip in Mg [8] and prismatic slip in Be [9]. The explanation seems to lie in the formation of special sessile structures of dislocation cores at high temperatures [8,9]. Unusual temperature dependences and orientation dependences of the yield stress and slip geometry have also been observed in various compounds (e.g. B2 and $L2_1$ ordered alloys [10,11]) as well as in non-metallic materials (e.g. spinel [12], sapphire [13], olivine [14]), and explanation can always be sought in the special features of non-planar dislocation cores (for a review see [4]). Hence, in general, if the slip is confined to well defined slip planes and no unusual temperature and orientation dependences of the flow stress are observed the dislocation core effects are negligible. On the other hand, if the flow stress is strongly dependent on temperature and orientation of the tensile/ compressive axis, various slip planes occur for different orientations, and Schmid's law breaks down, then dislocation core effects may be expected to be very important, perhaps even governing the plastic behavior.

In $L1_2$ alloys an anomalous increase of the yield stress with increasing temperature, first observed by Westbrook [15] and Davies and Stoloff [16], is now a commonly known phenomenon. In a number of $L1_2$ alloys, when studied as single crystals, both the temperature corresponding to the peak and the height of the peak are strongly dependent on the orientation of the tensile axis and a pronounced tension compression asymmetry exists (e.g. [17-20] and reviews [21,22]). This is illustrated in Fig. 1 where the temperature dependences of the yield stress in $Ni_3(Al,Ta)$ single crystals [20] are plotted for four different orientations of the tensile/compressive axis. Furthermore, long straight screw dislocations were observed in samples deformed below the peak temperature [23], indicating their low mobility, relative to that of edge dislocations.

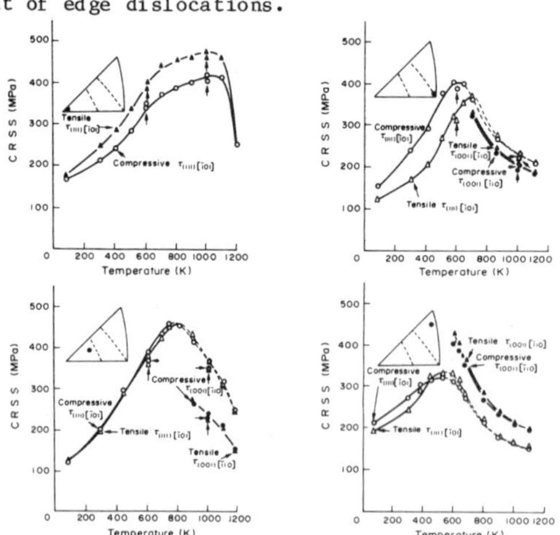


Fig. 1 Temperature dependence of the c.r.s.s. in $Ni_3(Al,Ta)$ for four orientations of the tensile/compressive axis.

All this suggests that dislocation core effects are likely to play an important role in the anomalous yield behavior of $L1_2$ alloys and, perhaps, are responsible for this phenomenon. The first explanation of the anomalous yield behavior, proposed by Thornton et al. [24] and Kear and Wilsdorf [25], indeed, envisaged intrinsic changes in the moving dislocations, namely their cross-slip from $\{111\}$ onto $\{010\}$ planes. However, a full understanding of this unusual plastic behavior, including all the complex orientation dependences, can only be attained if (i) the reason for immobilization of screw dislocations at high temperatures is fully clarified and (ii) the dislocation configurations involved in this process are known. This has recently been achieved by carrying out atomistic studies of screw dislocation cores in $L1_2$ alloys [26,27], and in this paper we first review these results. Although many core configurations discussed here can be described in terms of dislocation splitting the thermally activated processes considered here can occur only if the widths of these dissociations are of the order of lattice spacing and would be impossible otherwise. Hence, we always refer to core spreadings rather than dislocation dissociations. The theory of screw dislocation immobilization based on the atomistic models of the core structure [28] is then described and its validity illustrated by comparing the theoretical predictions with experimental results.

## 2. Stacking-faults and Antiphase Domain Boundaries

The most important "core" phenomenon in any material is dissociation of dislocations into partials separated by a metastable planar fault:stacking fault (SF) or antiphase domain boundary (APB). Since the dislocations in $L1_2$ alloys glide either on $\{111\}$ or $\{010\}$ planes the planar faults on these planes are most important. The situation is relatively simple for $\{010\}$ planes. The only possible metastable fault on the (010) plane is the $1/2[\bar{1}01]$ APB. Its stability is guaranteed by symmetry since there are two intersecting mirror planes ($(\bar{1}01)$ and (101)) for the displacement $1/2[\bar{1}01]$. (For more details on the stability of planar faults see ref. [29]). Hence this fault is metastable in any $L1_2$ alloy regardless of its chemical composition and the $[\bar{1}01]$ superdislocation may always dissociate into two $1/2[\bar{1}01]$ superpartials on this plane.

In the case of (111) planes three different faults are usually considered to be stable: APB ($1/2[\bar{1}01]$), the complex stacking fault (CFS) ($1/6[\bar{2}11]$) and the superlattice intrinsic stacking fault (SISF) ($1/3[\bar{1}\bar{1}2]$). However, in this case only the stability of the SISF is guaranteed on symmetry grounds. The SISF is, therefore, metastable in every $L1_2$ alloy. On the other hand the stability of the APB and CSF is not guaranteed in general. Hence two classes of $L1_2$ alloys may exist [30]. In the first class the APB and the CSF are stable on $\{111\}$ planes. The $[\bar{1}01]$ superdislocation may then dissociate on the (111) plane into two $1/2[\bar{1}01]$ superpartials with APB in between them. Furthermore, each of the $1/2[\bar{1}01]$ superpartials may split in this plane into the Shockley partials of the type 1/6<112> separated by the CSF. In the second class, usually associated with a high ordering energy [30,31], both the APB and CSF are not stable on $\{111\}$ planes. Dislocation dissociations and corresponding core structures encountered in this case are discussed in a separate paper [32].

## 3. Dislocation Core Structures

The core structure of the 1/2[101] superpartial bounding an APB on the (111) plane calculated atomistically using pair-potentials to describe interactions between the atoms, is shown in Fig. 2a using the method of differential displacements, and schematically in Fig. 2b. In Fig. 2a the $[\bar{1}01]$ (screw) component of the relative displacements of neighboring atoms, produced by the superpartial, is drawn on a $(\bar{1}01)$ projection of the lattice as an arrow between the corresponding atoms. The length of each arrow is equal to the magnitude of the displacement and the direction represents the sign of the displacement. Clearly, the core is confined to the plane of APB, the (111) plane. Investigation of the displacements perpendicular to the dislocation line (edge displacements) reveals that the core of the $1/2[\bar{1}01]$ can be described as a narrow splitting into $1/6[\bar{2}11]$ and $1/6[\bar{1}\bar{1}2]$ partials separated by the CSF. The application of a shear stress shows that the lattice friction, the Peierls stress, of this dislocation is very low [27].

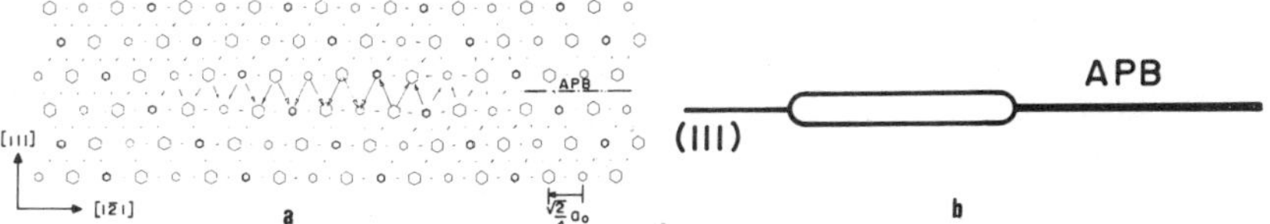


Fig. 2 Core structure of the $1/2[\bar{1}01]$ superpartial bounding an APB on the (111) plane. (a) Differential displacements map (b) Schematic picture of the core spreading.

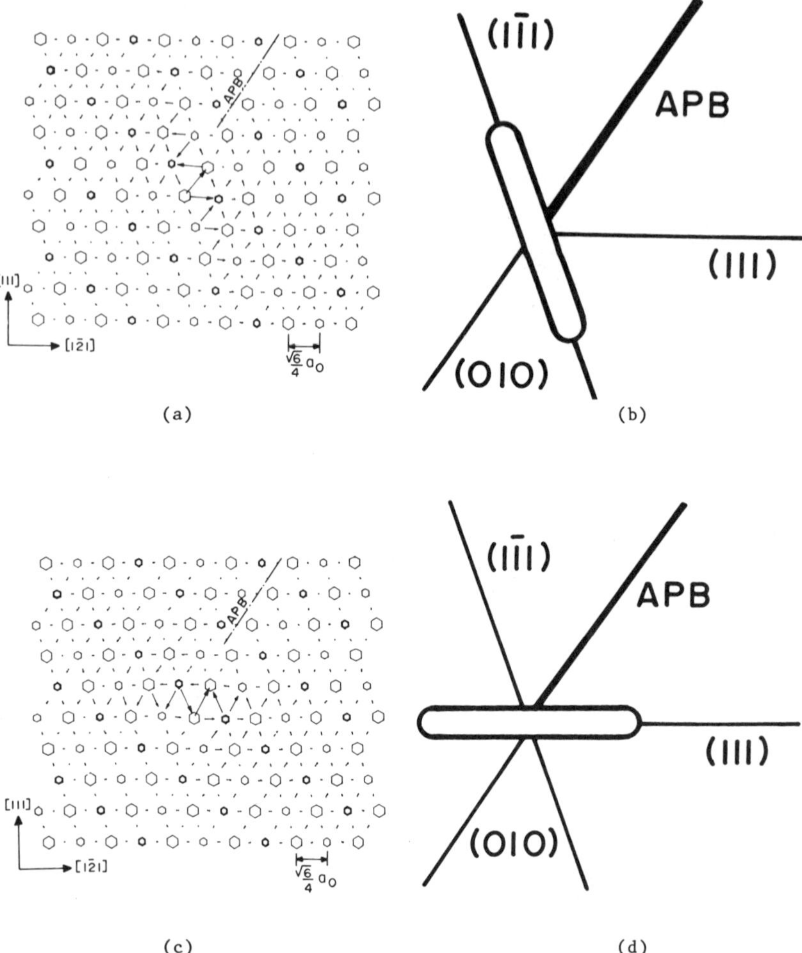


Fig. 3 Two energetically equivalent core structures of the 1/2[$\bar{1}$01] superpartial bounding an APB on the (010) plane. (a) and (c) Differential displacements maps, (b) and (d) Schematic pictures of the core spreading.

Two possible configurations of the core of the 1/2[$\bar{1}$01] dislocation bounding an APB on the (010) plane exist and they are shown in Fig. 3. The differential displacement maps of the [$\bar{1}$01] (screw) components are shown in Fig. 3a and 3c while Figs. 3b and 3d show these core structure schematically. The core is not in this case confined to the plane of the APB but spreads either into the (1$\bar{1}$1) plane (Figs. 3a, b) or (111) plane (Figs. 3c,d). Investigation of the displacements perpendicular to the dislocation line reveals that this spreading can be interpreted as dissociation into 1/6[$\bar{2}\bar{1}$1] and 1/6[$\bar{1}$12] partials for the core spread onto the (1$\bar{1}$1) plane or 1/6[$\bar{2}$11] and 1/6[$\bar{1}\bar{1}$2] partials when the core spreads onto the (111) plane. The spreading of the core outside the plane of APB renders the $[\bar{1}01]$ dislocation sessile. Application of a shear stress in the computer model, indeed, demonstrates that the dislocation starts to move only at a very high stress, in excess of 0.04 G, where G is the shear modulus. It moves along either the (111) or $(1\bar{1}1)$ plane forming a ribbon of APB on this plane [27].

## 4. Physical model for anomalous yield behavior

Since the energy of the APB on {010} plane is always lower than that of APB on {111} plane [33] the splitting of superpartials on {010} planes is always energetically favored. However, as seen in the previous section, dislocations dissociated in this way are sessile and can, therefore, only move with the help of thermal activations. On the other hand, dislocations split on {111} planes move very easily but they are in a state of higher energy. This higher energy dislocation configuration is metastable, i.e. it does not transform spontaneously into the low energy dissociation on {010} planes; thermally activated transformations are, of course, possible. Hence, on loading, dislocations will be primarily generated in {111} planes, though in a metastable form, since they are highly mobile in these planes. This implies that at low temperatures the slip in $L1_2$ alloys will be confined to {111} planes and the yield stress will be temperature independent. As the temperature increases thermal activations will induce cross-slip of screw dislocations into {010} planes initiating thus their transformation into lower energy states. However, such a cross-slip immediately immobilizes the screw dislocations since their cores are always spread into one of the {111} planes (see Fig. 3) and they will then act as obstacles for the other dislocations. The number of these obstacles will increase with increasing temperature which will lead to an increasing yield stress. This is, of course, the model of the anomalous yield behavior proposed in refs. [24,25] and developed phenomenologically by Takeuchi and Kuramoto [34]. In the following we investigate this cross-slip process leading to the formation of obstacles in more detail using our present understanding of the core structure of the dislocations involved.

Since the core of the $1/2[\bar{1}01]$ dislocation is always spread into either the (111) or $(1\bar{1}1)$ plane we postulate, following Friedel [35], that the cross-slipped segment always splits into one of those planes as soon as it starts to form. Let us assume that the superdislocation is originally split in the (111) plane into two $1/2[\bar{1}01]$ superpartials as shown schematically in Fig. 4a, and the applied stress drives the dislocation to the left. When the leading superpartial cross-slips the configuration shown in Fig. 4b ensues since the shortest jump along the (010) plane leads to the dissociation of the core of the cross-slipping superpartial into the $(1\bar{1}1)$ plane. The principal driving force for this cross-slip is the formation of a small ribbon of APB on the (010) plane which is energetically favourable provided $\gamma_{111}^{APB} > \sqrt{3}\,\gamma_{010}^{APB}$, where $\gamma_{111}^{APB}$ and $\gamma_{010}^{APB}$ are the energies per unit area of the APBs on {111} and {010} planes, respectively. Describing the cross-slip jump (Fig. 4a → Fig. 4b) in terms of a double kink formation, the activation enthalpy for this process was derived to be [28]

$$H = W + b\left\{\frac{Gb^2}{8\pi} - \left[\left(\gamma_{111}^{APB}/\sqrt{3} - \gamma_{010}^{APB} + \tau_{010}\, b\right)\frac{Gb^3}{8\pi}\right]^{1/2}\right\} \qquad (1)$$

where b is the magnitude of the Burgers vector of the superpartial and $\tau_{010}$ is the resolved shear stress on the {010} plane in the $[\bar{1}01]$ direction. W is the sum of the energies of constrictions formed on the $a/2[\bar{1}01]$ superpartials in the (111) and $(1\bar{1}1)$ planes.

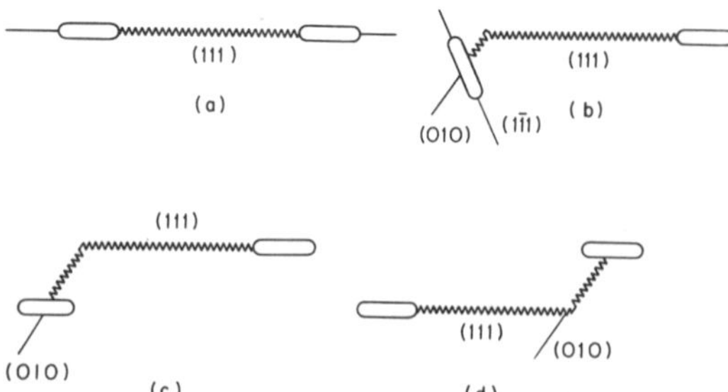


Fig. 4 Schematic picture showing the sequence of the tranformations leading to the immobilization of screw dislocations.

Hence, although we consider the cross-slip from the (111) plane into the (010) plane the constrictions involved are on $\{111\}$ planes as if the cross-slip from one $\{111\}$ plane to another were taking place. This is a very important recognition based on our understanding of the core structure which, as shown below, enables us to explain the observed orientation dependences of the anomalous yield stress. The energy of constrictions, W, is, therefore, the same as in the case of the cross-slip from the (111) plane into the (1$\bar{1}$1) plane in f.c.c. metals and it was derived by Escaig [36] to be

$$W = W_o \ (1 - \alpha_1 \ \tau_{pe} + \alpha_2 \ \tau_{se}) \qquad (2)$$

where, $\alpha_1$ and $\alpha_2$ are positive constants depending on the equilibrium width of splitting and $W_o$ is the energy of the constrictions when no stresses are applied. $\tau_{pe}$ is the shear stress component in the (111) plane acting in the direction of the edge components of the 1/2<112> partials in this plane and $\tau_{se}$ is the shear stress component in the (1$\bar{1}$1) plane acting in the direction of the edge components of the corresponding partials. The stress components $\tau_{pe}$ and $\tau_{se}$ do not exert any total force on the dislocation. The reason why they affect the constriction energy is that they cause either widening or narrowing of the splitting (or dislocation core) in the (111) and (1$\bar{1}$1), respectively, by exerting equal and opposite forces on the partials which have edge components of opposite signs. $\tau_{pe}$ and $\tau_{se}$ are taken to be positive when the applied stress reduces the width of the partials.

The next jump in the cross-slip process discussed will transform the dislocation into the configuration shown schematically in Fig. 4c [28]. The core of the leading superpartial is now again dissociated into the (111) plane, one layer below the original slip plane. It can glide in this plane until the configuration shown in Fig. 4d is reached. The cross-slip of the trailing superpartial into the (010) plane is not, however, energetically favorable because the APB on the (010) plane would now be eliminated and an additional region of APB on the (111) plane formed. Therefore, the dislocation becomes immobilized. It was shown in [28] that the first step in this process controls the rate of formation of these obstacles and, therefore, the yield stress which, following Takeuchi and Kuramoto [34], is proportional to the cube root of the rate at which the obstacles are formed. The activation enthalpy for this process, given by eq. (1), depends not only on the resolved shear stress on the (010) plane but also on the shear stress components $\tau_{pe}$ and $\tau_{se}$ in the (111) and (1$\bar{1}$1) planes, respectively, in the directions perpendicular to the total Burgers vector which appear in the constriction term W (eq. (3)), the effect first noted by Escaig [36] when analyzing cross-slip in f.c.c. metals.

For a given stress, tensile or compressive, the stress components $\tau_{010}$, $\tau_{se}$ and $\tau_{pe}$ are all related by some orientation factors which implies a complex orientation dependence of H and thus of the yield stress in the regime of the anomalous peak. This orientation dependence is certainly different than if H depended only on $\tau_{010}$ as assumed, for example, in [34]. In particular, because H depends on stress components $\tau_{pe}$ and $\tau_{se}$, different rates of obstacle formation and thus different values of yield stress are predicted for tension and compression, respectively, for a given orientation of the tensile (compressive) axis. This explains the tension-compression asymmetry clearly seen in Fig. 1.

A detailed analysis of the orientation dependence of the rate of formation of obstacles and thus of the yield stress in the anomalous regime has been carried out by Paidar et al. [28]. It follows from the dependence of H on orientation that for the tensile axis close to the [001] direction the yield stress should be lower in compression than in tension whilst the opposite is true when the tensile axis is close to the [011] direction. It is seen from Fig. 1 that this prediction is in agreement with measurements on $Ni_3(Al,Ta)$ single crystals. It is also predicted that the tension-compression asymmetry is the most pronounced for the [011] orientation of the tensile axis which is again fully confirmed by experiments (see Fig. 1). Finally, it has been concluded [28] that an orientation must exist in the left part of the standard triangle for which the tension/compression asymmetry disappears. Such an orientation has, indeed, been found as demonstrated in Fig. 1. These comparisons all illustrate that the present theory agrees excellently with the experimental observations and fully explains all the main features of the anomalous yield behavior.

## 5. Conclusions

(i) In $L1_2$ alloys with stable APB on (111) planes the $[\bar{1}01]$ superpartials may dissociate on both (111) and (010) planes. In the former case the core of the superpartials is spread in the plane of APB and is glissile. In the latter case it is not confined to the plane of the APB but it spreads either into the (111) or $(1\bar{1}1)$ plane.

(ii) Splitting into (010) planes is always energetically favored [26] but since dislocations split into (111) planes are glissile they will be generated much more easily by dislocation sources at low and intermediate temperatures. Hence the slip will be confined to (111) planes.

(iii) Thermally activated transformation of the core of screw dislocation on (111) planes into a lower energy configuration will occur at high temperatures. This new configuration corresponds to the formation of a small ribbon of the energetically favored APB on the (010) plane and spreading of the core into the $(1\bar{1}1)$ plane. This is anologous to the cross-slip into (010) planes proposed in refs. [24,25]. After such transformation the screw dislocation becomes sessile and acts as an obstacle to other dislocations moving on (111) planes. This is the likely origin of the anomalous yield behavior of $L1_2$ alloys at high temperatures.

(iv) Although the glissile-sessile transformation is analogous to the "cross-slip" into (010) plane the core of the dislocations spreads after the transformation onto the $(1\bar{1}1)$ plane. Hence, the transformation is affected not only by the shear stresses in (111) and (010) planes but also by the shear stress in the $(1\bar{1}1)$ plane. Futhermore, since displacements perpendicular to the total Burgers vector are present in the core, analogously as in the case of splitting into 1/6<112> partials, the shear stress components in the (111) and $(1\bar{1}1)$ planes perpendicular to the Burgers vector rather than parallel to it are also involved. The reason is that the effect of these components is to increase or decrease the width of the core on (111) and $(1\bar{1}1)$ planes. In a tensile/compression test all these stress components are related through orientation factors and this leads to complex orientation dependences observed in the anomalous regime. In particular the tension-compression asymmetry which arises due to the effect of the shear stress components perpendicular to the total Burgers vector could not be comprehended without understanding that the core spreads into {111} planes even if a ribbon of the APB is formed on a {010} plane.

## Acknowledgements

Support for this research was provided by the National Science Foundation under grant no. DMR79-05556.

## References

1. J. W. Christian, Metall. Trans. A 14, 1237 (1983).
2. M. Meshii, editor, Mechanical Properties of B.C.C. Metals (TMS-AIME, Warrendale, Pennsylvania, 1982).
3. V. Vitek, Crystal Lattice Defects 5, 1 (1974).
4. V. Vitek, Proc. Conf. Dislocations and Properties of Real Materials, The Metal Society London, to be published (1985).
5. I. P. Jones and W. B Hutchinson, Acta Metall. 29, 968 (1981).
6. D. J. Bacon and J. W. Martin, Phil. Mag. A 43, 883 (1981).
7. Y. Minonishi, S. Ishioka, M. Koiwa, S. Morozumi and M. Yamaguchi, Philos. Mag. A 43, 1017 (1983).
8. J. F. Stohr and J.-P. Poirier, Philos. Mag. 25, 1313 (1972).
9. P. Regnier and J. M. Dupouy, Phys. Stat. Sol. 39, 79 (1970).
10. M. Yamaguchi, Mechanical Properties of B.C.C. Metals, edited by M. Meshii (TMS-AIME: Warrendale, Pennsylvania, 1982).
11. Y. Umakoshi, M. Yamaguchi and T. Yamane, Acta Metall. 32, 649 (1984).
12. N. Doukhan, R. Duclos and B. Escaig, J. Phys. Paris 40, 381 (1979).
13. D. S. Phillips and J. L. Cadoz, Philos. Mag. A 46, 583 (1982).
14. J.-P. Poirier and B. Vergobi, Physics of the Earth and Planetary Interiors 16, 370 (1978).
15. J. W. Westbrook, Trans. TMS-AIME 209, 898 (1957).
16. R. G. Davies and N. S. Stoloff, Trans. TMS-AIME 223, 714 (1965).
17. C. Lall, S. Chin and D. P. Pope, Metall. Trans. A 10, 1323 (1979).
18. D. Shah and L. Lin, J. Metals 32, 62 (1980).
19. S. S. Ezz, D. P. Pope and V. Paidar, Acta Metall. 30, 921 (1982).
20. Y. Umakoshi, D. P. Pope and V. Vitek, Acta Metall. 32, 449 (1984).
21. D. P. Pope and S. S. Ezz, Int. Metalls. Rev. 29, 136 (1984).
22. D. P. Pope and V. Vitek, this symposium.
23. B. H. Kear and M. F. Hornbecker, Trans. Am. Soc. Metals 59, 613 (1966).
24. P. H. Thornton, R. G. Davies and T. L. Johnston, Metall. Trans. A 1, 207 (1970).
25. B. H. Kear and H.G.F. Wilsdorf, Trans. TMS-AIME 224, 382 (1962).
26. M. Yamaguchi, V. Paidar, D. P. Pope and V. Vitek, Philos. Mag. 45, 867 (1982).
27. V. Paidar, M. Yamaguchi, D. P. Pope and V. Vitek, Philos. Mag. 45, 883 (1982).
28. V. Paidar, D. P. Pope and V. Vitek, Acta Metall. 32, 435 (1984).
29. M. Yamaguchi, D. P. Pope, V. Vitek and Y. Umakoshi, Philos. Mag. A 43, 1265 (1981).
30. M. Yamaguchi, V. Vitek and D. P. Pope, Philos. Mag. A 43, 1027 (1981).
31. V. Paidar, D. P. Pope and M. Yamaguchi, Scripta Metall. 15, 1029 (1981).
32. G. Tichy, V. Vitek and D. P. Pope, this symposium.
33. P. A. Flinn, Trans. TMS-AIME 218, 145 (1960).
34. S. Takeuchi and E. Kuramoto, Acta Metall. 21, 415 (1973).
35. J. Friedel, Dislocations and Mechanical Properties of Crystals (John Wiley: New York, 1957) p. 330.
36. B. Escaig, J. Phys. Paris 29, 225 (1968).

# DISLOCATION CORE STRUCTURE AND THE NORMAL YIELD BEHAVIOR OF $L1_2$ ORDERED ALLOYS

G. TICHY, V. VITEK and D. P. POPE
Department of Materials Science and Engineering, University of Pennsylvania, Philadelphia, PA 19104

ABSTRACT

A rapid increase of the yield stress with increasing temperature, often observed in $L1_2$ ordered alloys, is commonly called the "anomalous flow behavior". This phenomenon is believed to result from the thermally activated transformation of the core of 1/2<110> screw dislocations from a glissile form to a sessile form at high temperatures. It is shown here that another class of $L1_2$ alloys exists in which these two forms of the screw dislocation core are not available. These are the alloys in which the APB on {111} planes is not stable and the atomistic studies of screw dislocations in such alloys show that their cores are always sessile. The yield stress of these alloys then increases with decreasing temperature and no increase at high temperatures occurs. Such behavior has been observed, for example, in $Pt_3Al$. This "normal" behavior is analogous to that of b.c.c. metals and a theory of the temperature dependence of the yield stress has been developed along the same lines as in the case of b.c.c. metals. Comparison of this theory with measurements on $Pt_3Al$ single crystals shows a good agreement.

## 1. Introduction

A prominent feature of the plastic behavior of many $L1_2$ alloys is the peak in the yield stress vs temperature dependence at high temperatures, first observed by Westbrook [1] and Davies and Stoloff [2]. This phenomenon is commonly called "anomalous flow behavior" and it has been interpreted [3-5] as resulting from the thermally activated transformations of the cores of 1/2<110> screw superpartials from a glissile form to a sessile form. These two forms of the core structure always exist if a metastable antiphase domain boundary (APB) can be present on both {111} and {010} planes. The reason is that the <101> superdislocation may then dissociate into 1/2<101> superpartials on either of these planes and when the splitting is on {111} planes the core has a glissile form while it has a sessile form when the splitting is on {010} planes [3,6,7]. However, while the stability of the APB on {010} planes is guaranteed by symmetry, the APB (and also the complex stacking fault (CSF)) on {111} planes may but need not be metastable [8]. The latter is particularly likely in alloys with very high ordering energies [8,9]. When the APB on (111) planes is unstable the $[\bar{1}01]$ superdislocation can still dissociate on (111) planes but only according to the reaction

$$[\bar{1}01] = 1/3[\bar{1}\bar{1}2] + 1/3[\bar{2}11] \qquad (1)$$

where the 1/3<112> superpartials are separated by a ribbon of superlattice intrinsic stacking fault (SISF). The stability of this fault is guaranteed on symmetry grounds [8] and it can, therefore, exist in any $L1_2$ alloy. Dissociation into two $1/2[\bar{1}01]$ superpartials on the (010) plane is, of course, also possible since the corresponding APB is metastable and the core of these superpartials has a sessile form [6]. Atomistic studies of the core structure of 1/3<112> superpartials suggest that they are also sessile [6,7] and thus that no glissile core configuration exists when the APB is not stable on {111} planes. This is analogous to the situation encountered in b.c.c. metals (see e.g. [10]) and it has been suggested [7] that if such a class of $L1_2$ alloys exists, the anomalous flow stress behavior will be absent in these alloys but the flow stress will increase rapidly with decreasing temperature. We ventured to call such a behavior "normal" since it is not unusual for a material to become harder at low temperatures and softer at high temperatures.

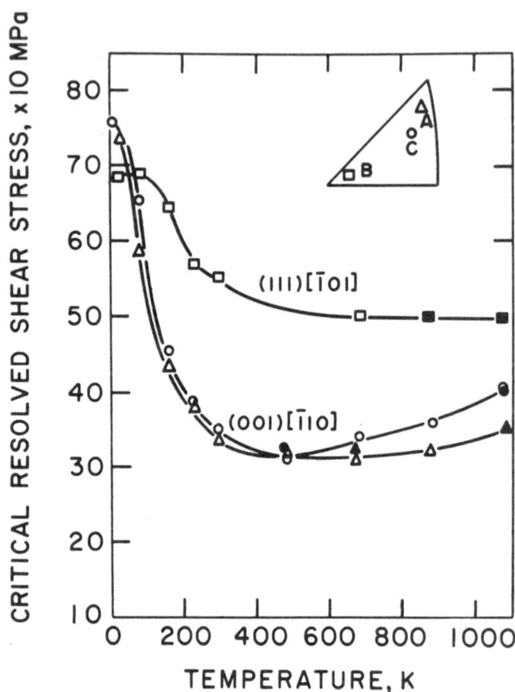


Fig. 1 Temperature dependence of the c.r.s.s. for three different orientations of the compressive axis [15].

Such a normal behavior has, indeed, been found in several platinum based $L1_2$ alloys, e.g. $Pt_3Al$ [11-15]. Results of recent experimental studies of the deformation properties of $Pt_3Al$ single crystals [15] are summarized in Fig. 1 which shows the temperature dependence of the critical resolved shear stress (c.r.s.s.) for three different orientations of the compression axis. The slip planes are either $\{111\}$ or $\{010\}$, depending on the orientation of the compressive axis and while the c.r.s.s. increases with decreasing temperature in both cases, it is higher for $\{111\}$ slip. Such a behavior is typical for the situation when the core structure of dislocations controls the dislocation motion and thus the plastic properties [3]. Hence, in this paper we first show the results of the recent atomistic studies of the cores of superpartials when APB and CSF on $\{111\}$ planes are unstable. We then discuss implications of these results for the understanding of the plastic behavior of this class of $L1_2$ alloys and present a theory of the thermally activated motion of dislocations in these alloys. These results are then compared with the experimental observations shown in Fig. 1.

## 2. Core structure of superpartials

### 2.1 1/2[$\bar{1}$01] superpartials on the (010) plane

The 1/2[$\bar{1}$01] superpartial bounding an APB on the (010) plane has been studied atomistically [17] using pair potentials which lead to mechanically stable $L1_2$ structure but unstable APB on {111} planes, constructed by Yamaguchi et al. [16]. Its core structure is shown in Fig. 2a in terms of differential displacements in the [$\bar{1}$01] (screw) direction and, schematically, in Fig. 2b. (For the meaning of the differential displacements see ref. [3]). The core is not confined to the plane of the APB but spreads symmetrically into the (111) and (1$\bar{1}$1) planes, which intersect along the [$\bar{1}$01] direction. The displacements in the core do not possess any components perpendicular to the dislocation line. The non-planar spreading of the core renders this dislocation sessile and the application of shear stresses in our computer model [17] indeed shows that the dislocation moves only at a very high applied stress, in excess of 0.04 G (G is the shear modulus). The motion is along the (010) plane if the maximum resolved shear stress (m.r.s.s.) plane makes an angle smaller than 30° with this plane, and along a {111} plane otherwise. In the former case the path is always composed of two equal segments of the (111) and (1$\bar{1}$1) planes but it is the resolved shear stress in the (010) plane which determines when movement starts.

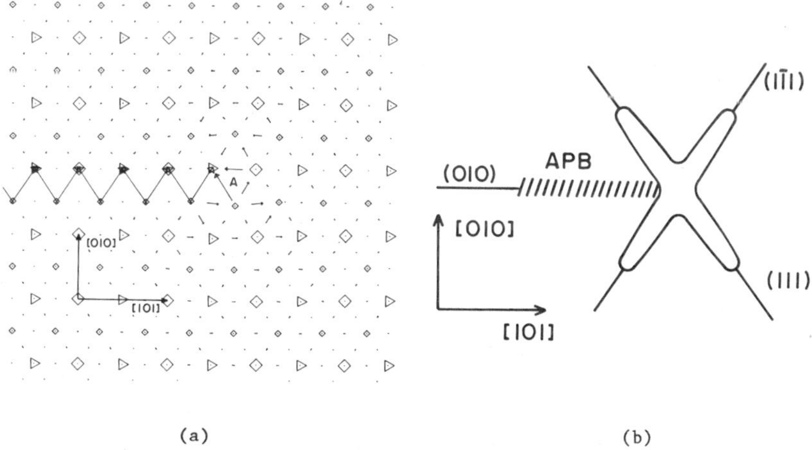


Fig. 2 Core structure of the 1/2[$\bar{1}$01] superpartial bounding on APB on the (010) plane. (a) Differential displacements map. (b) Schematic representation of the core spreading.

### 2.2 1/3[$\bar{2}$11] superpartial on the (111) plane

The atomistic study of the 1/3[$\bar{2}$11] superpartial bounding an SISF on the (111) plane has been made using the same potential as in the previous case. The differential displacement maps showing the [$\bar{1}$01] (screw) and [$\bar{1}$2$\bar{1}$] (edge) components of the lowest energy configuration of the core of this dislocations are shown in Figs. 3a and 3b, respectively. It is seen that while the edge component is confined to the plane of the fault, the screw component is spread into (111) and (1$\bar{1}\bar{1}$) planes intersecting at the point marked A in Fig. 3. This is shown schematically in Fig. 3c. Thus the core of the 1/3[$\bar{2}$11] dislocation appears dissociated into the edge part with the Burgers vector 1/6[$\bar{1}2\bar{1}$], confined to the (111) plane, and to the screw part with the Burgers vector 1/2[$\bar{1}$01], centered in the plane just above the SISF and spread spacially into the two intersecting {111} planes.

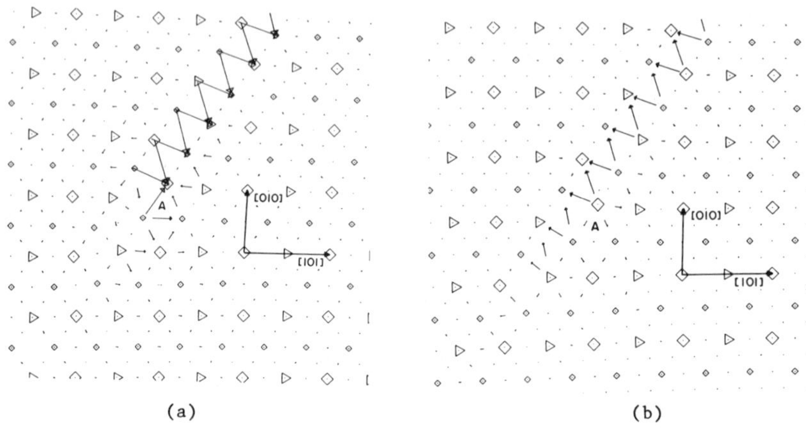


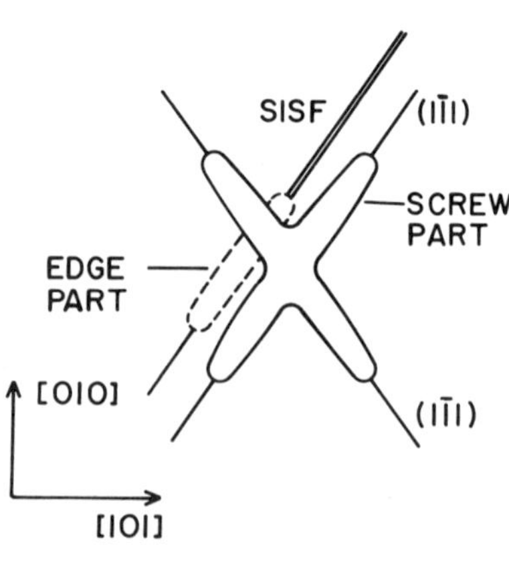


(c)

Fig. 3 Core structure of the 1/3[$\bar{2}$11] superpartial bounding an SISF on the (111) plane. (a) Map of screw components (b) map of edge components. (c) Schematic representation of the core spreading.

The spreading of the screw part is of the same origin as in the case of the $1/2[\bar{1}01]$ superpartials on (010) planes. It again renders the dislocation sessile and the computer study of the motion of this dislocation [17] shows that it only moves at an applied shear stress in the (111) plane in the $[\bar{1}01]$ direction in excess of 0.05 G. The slip plane is either one of the $\{111\}$ planes or the (010) plane, depending on the orientation of the m.r.s.s. plane. An interesting influence of the edge part of the superpartial upon the dislocation motion has been observed when applying simultaneously both the shear stress in the $[\bar{1}01]$ and in the $[\bar{1}2\bar{1}]$ directions. The latter stress component does not exert any total force on the $[\bar{1}01]$ superdislocation but equal and opposite forces on each of the $1/3\langle 112\rangle$ superpartials. When this stress pushes the superpartials together the stress in the $[\bar{1}01]$ direction needed for the dislocation motion is practically unaffected. However, when the superpartials are pulled apart the dislocation will start to move at a lower stress than if there is no shear stress in the $[\bar{1}2\bar{1}]$ direction. In fact if the shear stress in the $[\bar{1}2\bar{1}]$ direction exceeds 0.06 G only a very small shear stress in the $[\bar{1}01]$ direction is needed to move the dislocation, i.e. the dislocation becomes glissile.

## 3. Theory of the temperature and orientation dependence of the flow stress

The principal result of the atomistic studies is that screw dislocations are in this class of $L1_2$ alloys always sessile. Hence they possess a very high lattice friction (Peierls) stress and at finite temperatures their motion can be aided by thermal activation. This is a situation entirely analogous to that of b.c.c. metals and theoretical models of the motion of screw dislocations in b.c.c. metals can be directly employed in our case. In particular we shall assume that the thermally activated motion of screw dislocations occurs via formation of double kinks.

In this model we consider that the dislocation may exist in two different forms. First is the stable sessile form. The second is a glissile form which need not be stable and, in our case, it corresponds to the configuration formed under the effect of the applied stress just before the net movement of the dislocation starts. When in the glissile form the energy of the dislocation is, of course, higher than in the sessile form; in the following we mark this difference $\Delta E$. The thermally activated motion of the dislocation then proceeds as follows [18]: The transformation into the glissile form occurs along a small finite length of the dislocation and this part bows out under the effect of the applied stress. This bowing out cannot, however, proceed further than one lattice period, d, in the slip plane in the direction perpendicular to the dislocation line. The reason is that once this distance has been reached the glissile configuration can transform back to the energetically favored stable sessile form. Using the language common when discussing the dislocation motion through Peierls barriers, the dislocation moved from one Peierls trough to another. A double kink is thus formed consisting of a sessile segment of the dislocation displaced by one lattice period, d, with respect to the original position of the dislocation, and a pair of kinks of opposite signs which are in the glissile form. Constrictions, possessing an energy $E_c$, which correspond to the regions of transition of the sessile to glissile form are present at each junction of the sessile forms with kinks. These kinks then move apart gradually transferring the dislocation into the new position.

It has been shown by Duesbery [18] that the saddle point configuration, corresponding to the maximum of the activation enthalpy, is reached at different stages of this process, depending on the level of the applied stress. For very low stresses the saddle point is reached only after the double kink has been formed and it is determined as in Seeger's treatment [19]. This situation is not important in our case. For intermediate stresses the saddle point configuration is reached once the position of another

Peierls trough has been reached and the transformation into the sessile form becomes energetically favored, i.e. when the transferring segment attains the critical length $l_c = 2E_c/\Delta E$. Following the same arguments as in [18] the activation enthalpy is in this case [17]

$$H = 2E_c (2 - \tau^*) + H_o (1 - \tau^*/4) \tag{1}$$

where $\tau^* = \tau db/\Delta E$, $H_o = 2db \sqrt{G\Delta E}$ and b and G are the Burgers vector of the dislocation and shear modulus, respectively. At high stresses, for $\tau^* \geqslant 1$, the saddle point configuration is achieved before reaching the next Peierls' trough and thus before any transformation to the sessile form occurs. The activation enthalpy is in this case

$$H = 2E_c + 0.67 H_o/\tau^*. \tag{2}$$

$E_c$ depends on $\Delta E$ and according to the calculations of Stroh [20] and Escaig [21]

$$E_c = \alpha\Delta E^{1/2} \tag{3}$$

where $\alpha$ is a constant. Furthermore, at least at a high applied stress, the dislocation core is substantially altered and thus $\Delta E$ must be a function of the applied stress. In fact for the critical stress $\tau_o$ at which the dislocation moves at 0°K $\Delta E$ is equal to zero. This dependence is not known a priori and we shall approximate it here by a linear relationship

$$\Delta E = \Delta E_o(1 - \tau/\tau_o) \tag{4}$$

where $\Delta E_o$ is the difference between the energies of sessile and glissile configurations for $\tau = 0$. A more detailed study of $\Delta E$ vs $\tau$ dependence is given in [17]. Inserting eq. (4) into the expression for $\tau^*$ we find that the high stress saddle point configuration (eq. 2) is achieved for $\tau/\tau_o > (1 + db\tau_o/\Delta E_o)^{-1}$. $db\tau_o$ represents the work done by the stress $\tau_o$ when the dislocation moves the distance d and thus it must be equal or larger than $\Delta E_o$ and therefore $db\tau_o/\Delta E_o \geqslant 1$. Hence for $\tau/\tau_o \geqslant 1/2$ only the high stress saddle point need to be considered.

When considering the thermally activated motion of dislocations the stress $\tau$ entering the expression for the activation enthalpy has to be interpreted carefully. When there are long range internal stresses which determine the level of the flow stress at high temperatures, which we mark $\tau_a$, $\tau$ must be taken as the difference between the applied stress and $\tau_a$. On the other hand if the stress $\tau_a$ originates from an athermal friction mechanism with no internal stresses involved, it is the applied stress which drives the athermally activated process. Of course $\tau$, must always be higher than $\tau_a$ for the dislocation to move. In general a combination of the two cases needs to be considered. However, in ordered alloys the athermal part of the flow stress is usually very high and cannot be explained by internal stresses. Its origin is most likely the presence of extended defects such as APBs or SISF through which the dislocations have to pass and thus the latter interpretation of $\tau$ applies.

It is seen from Fig. 1 that in the case of $Pt_3Al$ single crystals the ratio between $\tau_a$ and the c.r.s.s. at very low temperatures is larger than 0.5. Thus for these stresses only the saddle point configuration with the activation enthalpy given by eq. (2) need to be considered. Using equations (3) and (4) it is

$$H = H_o (1 - \tau/\tau_o)^{1/2} (1 + H_1\frac{1 - \tau/\tau_o}{\tau}) \tag{5}$$

where $H_o = 2\alpha\sqrt{\Delta E_o}$ and $H_1 = 0.67\, G^{1/2}\, \Delta E_o/\alpha$.

Employing the standard rate theory the strain rate can be written as $\dot{\varepsilon} = \dot{\varepsilon}_o \exp(-H/kT)$ where $\dot{\varepsilon}_o$ is a constant and k and T have the usual meaning. Extrapolation to 0°K in Fig. 1 suggests that $\tau_o \approx 750$ MPa for the case of the (010) slip in $Pt_3Al$. Using this value of $\tau_o$ $\tau$ vs $T^* = kT/H_o \ln \dot{\varepsilon}_o/\dot{\varepsilon}$ is plotted in Fig. 4 for several values of $H_1$ (full curves). The atomistic calculations suggest that for the (111) plane the critical stress $\tau_o$ is about 1.3 times higher than for the (010) planes. The broken curves in Fig. 4 show $\tau$ vs $T^*$ dependences for this higher value of $\tau_o$ (1000 MPa). Choosing $\tau_a$ = 350 MPa for the (010) slip and 500 MPa for the (111) slip (straight lines in Fig. 4) the curves corresponding to $H_1$ between 1000 and 1500 MPa are in very good agreement with experimentally measured dependences (Fig. 1) for temperatures above 150°K. However, at lower temperatures the c.r.s.s. for (111) slip becomes about equal to that for (010) slip. This behavior can possibly be explained by the effect of the stress on the $[\bar{1}2\bar{1}]$ (edge) component of the 1/2<112> superpartials which may become important at low temperatures. For the tensile axis close to the [001] direction the shear stress acting in the $[\bar{1}2\bar{1}]$ direction pulls the superpartials apart in the case of compression and as explained in section 2, the motion of the dislocations can then occur at a lower c.r.s.s.

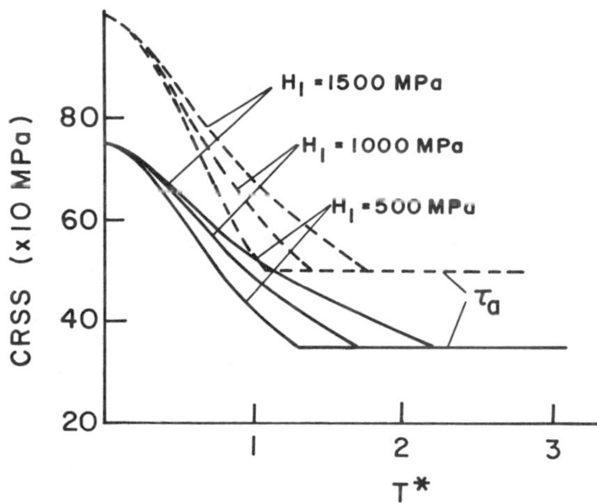


Fig. 4 Calculated temperature dependence of the c.r.s.s. on temperature.

When the activation enthalpies for slip on (010) and (111) planes are known the orientation dependence of the slip plane can also be investigated. Clearly that slip plane will be preferred for which the activation enthalpy is lower. Using the above values for $\tau_o$ and $H_1$ between 1000 and 1500 MPa, the boundary dividing the regions of (111) and (010) slip in the standard triangle has been determined for different stresses larger than $\tau_a$. This boundary always lies between the points B and C shown in the standard triangle in Fig. 1. This is again in a good agreement with the experimental observations.

## 4. Conclusions

(i) Two classes of $L1_2$ alloys exist. In alloys of the first class the APB on {111} planes is metastable and screw dislocations may exist in a glissile form on these planes. The flow stress in these alloys shows the anomalous behavior at high temperatures. In alloys of the second class the APB is not stable on {111} planes and screw dislocations are always sessile. The plastic behavior of these alloys is normal i.e. they are hard at low temperatures and soft at high temperatures.

(ii) In the alloys of the second class the screw superdislocations dissociate either on {010} planes into 1/2<110> superpartials or on {111} planes into 1/3<112> superpartials. Cores of both of these superpartials are sessile and the spacial spreading of their cores which is responsible for their sessility, is of the same origin. In the former case the superpartial itself and in the latter case its screw part spreads into two different {111} planes of the <110> zone.

(iii) The sessile screw dislocations move only at very high applied stresses either on {010} or {111} planes depending on the orientation of the m.r.s.s. plane. In the case of 1/3<112> superpartials a tension-compression asymmetry arises due to the effect of the stress perpendicular to the total Burgers vector of the dislocation on its core structure.

(iv) At finite temperatures the motion of these sessile superpartials can be aided by thermal activations and their motion described using a double kink model, anologously as in the case of screw dislocations in b.c.c. metals. This model describes very well both the temperature dependence of the m.r.s.s. and the orientation dependence of the slip planes observed in $Pt_3Al$ single crystals.

## Acknowledgements

Support for this research was provided by the National Science Foundation under grant No. DMR79-05556.

## References

1. J. W. Westbrook, Trans. TMS-AIME 209, 898 (1957).
2. R. G. Davies and N. S. Stoloff, Trans. TMS-AIME 233, 714 (1965).
3. V. Vitek and D. P. Pope, this symposium.
4. V. Paidar, D. P. Pope and V. Vitek, Acta Metall. 32, 435 (1984).
5. Y. Umakoshi, D. P. Pope and V. Vitek, Acta Metall. 32, 449 (1984).
6. M. Yamaguchi, V. Paidar, D. P. Pope and V. Vitek, Philos. Mag. 45, 867 (1982).
7. V. Paidar, M. Yamaguchi, D. P. Pope and V. Vitek, Philos. Mag. A 45, 883 (1982).
8. M. Yamaguchi, D. P. Pope, V. Vitek and Y. Umakoshi, Philos. Mag. A 43, 1265 (1981).
9. V. Paidar, D. P. Pope and M. Yamaguchi, Scripta Metall. 15, 1029 (1981).
10. J. W. Christian, Metall. Trans. A 14, 1237 (1983).
11. D. M. Wee and T. Suzuki, Trans. Japan Inst. Metals 20, 634 (1979).
12. D. M. Wee, O. Naguchi, Y. Oya and T. Suzuki, Trans. Japan Inst. Metals 21, 237 (1980).
13. M. Yadogawa, D. M. Wee, Y. Oya and T. Suzuki, Scripta Metall. 14, 849 (1980).
14. T. Suzuki, Y. Oya and D. M. Wee, Acta Metall. 28, 301 (1980).
15. D. M. Wee, D. P. Pope and V. Vitek, Acta Metall. 32, 829 (1984).
16. M. Yamaguchi, V. Vitek and D. P. Pope, Philos. Mag. A 43, 1027 (1981).
17. G. Tichy, V. Vitek and D. P. Pope, to be published (1985).
18. M. S. Duesbery, Phil. Mag. 19, 501 (1979).
19. A. Seeger, Phil. Mag. 1, 651 (1956).
20. A. H. Stroh, Proc. Phys. Soc. B 67, 427 (1954).
21. B. Escaig, J. Phys. Paris 29, 225 (1968).

# HIGH TEMPERATURE DUCTILITY MINIMUM IN RAPIDLY SOLIDIFIED $Ni_3Al$ - B

A. I. TAUB, S. C. HUANG and K. M. CHANG
General Electric Corporate Research and Development, PO Box 8, Schenectady, NY 12309

## ABSTRACT

The elevated temperature mechanical properties of rapidly solidified, boron doped $Ni_3Al$ have been investigated. Melt spun ribbon, plasma deposits and HIP'ped powder were tested with the same general results. It was found that the boron doped, rapidly solidified alloy exhibits an increasing flow stress with increasing temperature similar to that reported for conventionally cast $Ni_3Al$. In addition, the rapidly solidified alloys exhibited a severe ductility minimum in the range 500-750C. The ductility minimum is relatively insensitive to both boron and aluminum concentration, but the embrittlement onset temperature changes with processing and testing parameters. In all cases, the onset of the ductility drop coincides with the onset of the decrease of the flow stress. It is postulated that this behavior is due to a change in the flow mechanism at elevated temperature to one that is associated with grain boundary stress concentration.

## INTRODUCTION

The ordered intermetallic compound $Ni_3Al$ exhibits the unusual behavior of an increasing flow stress with increasing temperature [1-3]. This property makes the alloy attractive for high temperature structural applications, but the conventionally cast material exhibits brittle intergranular fracture with zero plastic elongation at ambient temperature. Since single crystals of the compound have been shown to exhibit high ductilities in all crystal orientations even at low temperatures [4], the low ductility of polycrystalline $Ni_3Al$ has been attributed to the inherent weakness of the grain boundaries. Recently, Aoki and Izumi [5,6] found that by introducing small additions of boron to the binary intermetallic, the grain boundary cohesive strength could be improved and that consequently the ambient temperature brittleness could be overcome. The observed effect of the boron addition was quite dramatic. Doping of the binary intermetallic with only 0.5 at. pct. boron resulted in more than 30% plastic strain to failure compared to zero tensile ductility for the unmodified alloy processed under identical conditions. These results were later confirmed by Liu, et. al. [7,8]. More recently, the present authors [9-11] found that when processed by rapid solidification, boron doped $Ni_3Al$ retains its high room temperature ductility and in addition, exhibits a dramatic increase in strength due to boron solid solution strengthening.

In this paper, measurements of the temperature dependence of both the strength and ductility of the rapidly solidified, boron modified alloys will be presented. It is found that the rapidly solidified alloys exhibit an increasing flow stress with increasing temperature similar to that reported for conventionally cast $Ni_3Al$. Further, the rapidly solidified alloys exhibit a severe ductility minimum in the range 550-750C. The effects of deviations from stoichiometry, partial Co substitution for Ni, boron level, heat treatment and testing conditions on the high temperature mechanical properties will be discussed. The majority of the data will be from tensile tests on melt spun ribbon, but the results will be shown to be generic for specimens prepared by other rapid solidification techniques.

## EXPERIMENTAL

The ingots of $Ni_3Al$ with various Ni:Al ratios and boron additions were prepared by vacuum induction melting using high purity Ni, Al and NiB. The ingots were then processed into ribbon form (approximately 6 mm wide x 20-35 μm thick) via melt spinning in vacuum. Details of the melt spinning process for Ni based alloys have been presented elsewhere [12].

Bulk specimens were made from argon atomized powder by both rapid solidification plasma deposition (RSPD) and by hot isostatic pressing (HIP). The plasma specimens were made using an internally fed plasma spray gun operating at low pressure (60 Torr). Powders less than 37 μm in size were deposited on copper plates which were removed after the deposition by chemical dissolution. Sheet tensile specimens 0.15 cm thick, were subsequently machined from the deposit. Details of the plasma deposition process can be found in reference 13. The consolidated powder specimens were prepared by hot isostatically pressing powders sieved to -140 mesh at 1165C for two hours under a pressure of 15 ksi. Details are presented in another paper in these proceedings [14].

Anneals were performed on specimens encapsulated in a quartz tube which had been backfilled with high purity argon. Zirconium foil was used as a gettering agent. Tensile tests for all specimen types were performed in a standard Instron static tensile testing machine at a strain rate of $8.3 \times 10^{-4}$ $s^{-1}$ unless otherwise specified. An argon atmosphere was used for the high temperature tests. Previous work [15] has shown that the yield strengths measured on melt spun ribbon are equal to the values obtained on bulk specimens produced from consolidated powder and plasma deposits as long as the thermal treatments produce specimens with equivalent microstructure. However, because of the thin cross-section of the ribbon the measured ductilities are sensitive to the ribbon surface finish. It has been found that the ductilities measured on ribbon specimens therefore represent only a lower limit of the ductility measured on bulk specimens.

## RESULTS

Figure 1 shows the temperature dependence of the tensile yield strength and plastic strain to fracture of specimens of $Ni_3Al$ doped with 1 at. pct. boron and processed by the indicated methods. The flow stress data from Liu, et. al. [7] for an alloy doped with 0.25 at. pct. boron that was drop cast and subsequently thermomechanically processed is included for comparison. Except for the absolute strength differences among the specimen types that can be attributed to differences in the microstructure caused by process variables and by the differing post-solidification heat treatments [14], the data are very similar and can be used to illustrate the generic features of the mechanical behavior of this system. First, the increase in yield strength with increasing temperature that has been observed in the conventionally processed unmodified alloy is exhibited by the boron-doped alloy. Second, for all the processing techniques, the boron-induced ambient temperature ductility is about 10%. Finally, for all the rapid solidification specimens, the peak in yield strength is found to coincide with the onset of a severe drop in the ductility of the alloy beginning at about 500C and persisting to about 750C. The non-rapidly solidified alloy does not exhibit this ductility minimum [16].

The elevated temperature ductility loss observed in the rapidly solidified specimens is accompanied by a change in the fracture morphology. At low temperatures, the boron modified $Ni_3Al$ exhibits a mixed mode fracture with areas of extensive deformation surrounding patches of intergranular fracture [9]. In the high temperature embrittled regime, the fracture is pure intergranular and indistinguishable from the fracture surface of the brittle, undoped alloy tested at ambient temperature.

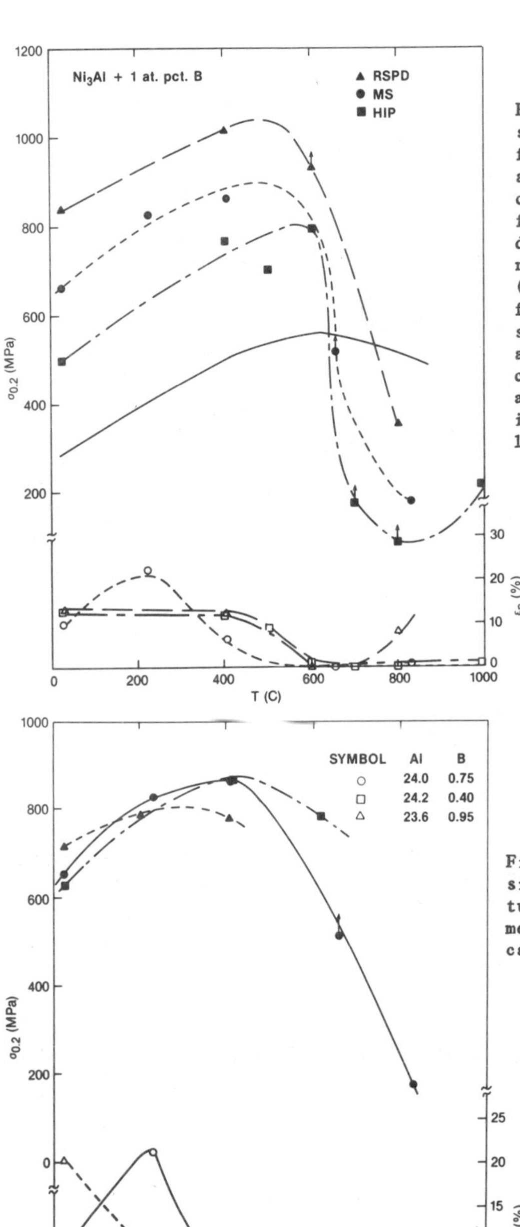


Figure 1. Tensile flow stress and plastic strain to failure of $Ni_3Al$ doped with 1 at. pct. boron. Alloy processed by three rapid solidification processes: plasma deposition (RSPD), melt spinning (MS) and pressed powder (HIP). Arrows indicate failure before 0.2% plastic strain. The flow stress for a thermo-mechanically processed alloy doped with 0.25 at. pct. boron [7] is included for reference (solid line).

Figure 2. Effects of composition on elevated temperature tensile properties of melt spun ribbon in the as-cast condition.

In figure 2, the elevated temperature mechanical properties are shown for melt spun ribbon specimens with the indicated aluminum and boron concentrations as determined by chemical analysis. Previous studies have shown that the boron level and deviations from stoichiometry have a pronounced effect on the room temperature ductility [8,17], and the elevated temperature ductility minimum appears to also vary with the boron and aluminum concentration. However, the data is too limited to draw any significant conclusions.

Figure 3 shows that heat treatment can shift the elevated temperature strength and ductility curves significantly. The data shown are for melt spun ribbons of $(Ni_{80}Co_{20})_3Al$ doped with 1 at. pct. boron and tested in the as-cast condition and after annealing at 1100C for two hours. The data illustrate quite clearly the correlation between the yield strength peak and the onset of the ductility minimum. The as-cast ribbon exhibits a peak in the flow stress almost 100C lower than the annealed material and the ductility minimum begins at a correspondingly lower temperature. At 800C, the ductility of the as-cast ribbon has recovered completely and very large tensile ductilities are obtained.

In figure 4 the variation with strain rate of the elevated temperature strength and ductility are shown. At ambient temperature, the mechanical properties are relatively insensitive to strain rate, but large strain rate effects are found at elevated temperatures. As expected, at any temperature the higher the strain rate, the higher is the flow stress. Further, the higher the strain rate, the higher is the temperature for the onset of the ductility minimum and the higher the temperature at which the ductility begins to recover.

The effects of the various parameters on the elevated temperature mechanical properties as discussed above are summarized in Table I.

## DISCUSSION

The ambient temperature embrittlement problem traditionally associated with $Ni_3Al$ appears to have been solved by the remarkable improvements that are obtained through boron doping of the alloy. However, the potential for the eventual application of this intermetallic as a high temperature structural material now appears to be questioned by the discovery of an elevated temperature ductility loss in the boron-doped intermetallic when formed by melt spinning, plasma deposition, or hot isostatic compaction. The severe ductility minimum starts at about 500C and persists to about 750C, with the exact temperature limits depending on processing and testing conditions. In this temperature range, the ductility drops to nearly zero and the fracture mode reverts to pure intergranular fracture like that observed at low temperature in the undoped alloy.

Examination of the effects of testing conditions and thermal treatment of the alloy on the elevated temperature embrittlement shows one clear correlation, the onset of the ductility minimum corresponds to the onset of the flow stress decrease in the alloy. This correlation has certain implications regarding the ductility loss mechanism.

The anomalous increase in flow stress with temperature observed for this intermetallic compound is generally associated with the cross slip mechanism that operates at low temperatures [18]. At elevated temperature, sufficient thermal energy is available to activate alternate slip planes and the flow stress starts decreasing with increasing temperature. If the change to this alternate slip mode causes some form of grain boundary stress concentration, then premature intergranular fracture could occur.

Alternatively, the flow stress drop may be due to the onset of grain boundary deformation mechanisms (e.g. - grain boundary sliding). The ductility loss that accompanies the flow stress decrease could then be interpreted as an extension of the Bricknell and Woodford model [19,20] which was developed to explain the high temperature ductility loss observed in many nickel base superalloys [21]. They propose that the ductility minimum observed in

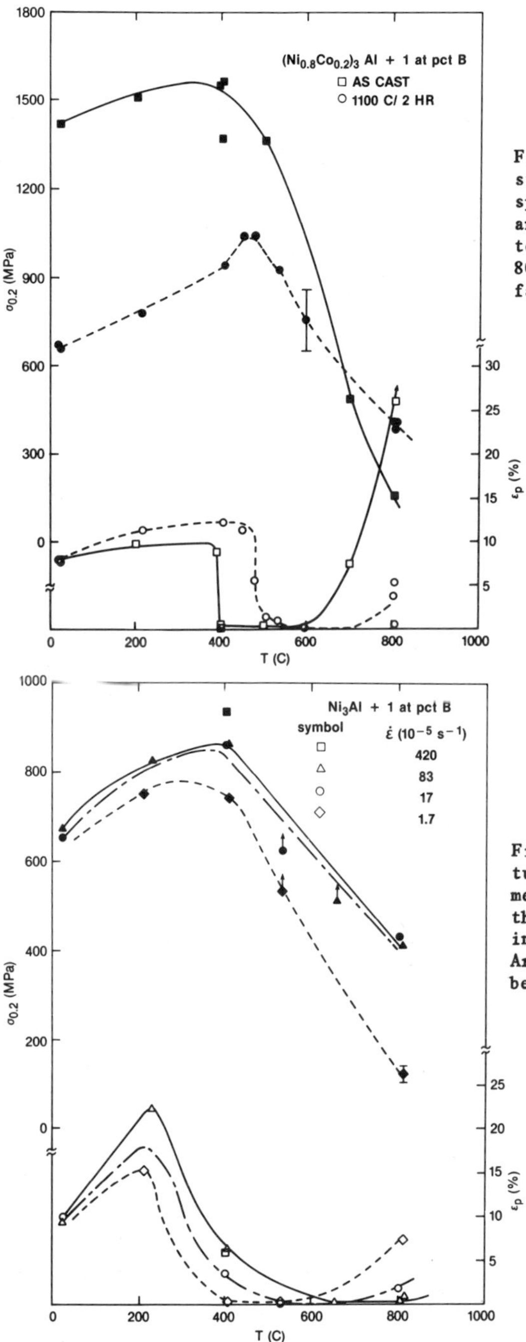


Figure 3. Tensile flow stress and ductility of melt spun ribbon in the as-cast and annealed conditions. The test of the as-cast ribbon at 804C was stopped prior to failure.

Figure 4. Elevated temperature tensile properties of melt spun ribbon tested in the as-cast condition at the indicated strain rates. Arrows indicate failure before 0.2% plastic strain.

TABLE I

VARIATION OF MECHANICAL PROPERTIES OF MELT SPUN $Ni_3Al$+B WITH COMPOSITION AND TESTING CONDITIONS

| Variable | Ambient Temperature Strength | Ambient Temperature Ductility | Ductility Minimum Onset Temperature |
|---|---|---|---|
| Increase boron level doping | Increase | Ductility peak at ~1 at.% | * |
| Increase aluminum level | Small increase | Decreases | * |
| Annealing of as cast material | Decreases | Al rich - Decreases<br>Al poor - No effect | Increases |
| Increase strain rate | No effect | No effect | Increases |

*Data too limited to establish trend.

nickel base alloys is due to the onset of grain boundary sliding and the inability of the grains to accommodate this deformation, leading to high stress concentration at the grain boundary triple points and premature fracture. The observed shift of the flow stress drop to higher temperatures with annealing and increasing strain rate support this hypothesis. The annealing is associated with grain growth (5-10 μm as cast; 25-30 μm annealed) and large grain materials are expected to exhibit grain boundary deformation mechanisms at higher temperatures. Increasing the strain rate also delays the onset of the grain boundary deformation mechanisms to higher temperature.

The ductility loss in the boron doped intermetallic appears to be process dependent. While the high temperature ductility minimum has been observed in specimens produced by three different rapid solidification processes (melt spun ribbon, HIP'ped powder and plasma deposits) it has not been found when the alloy is produced by drop casting followed by thermomechanical processing [16]. Therefore, the ductility loss is probably not an inherent problem with the boron doped alloy but specific to the rapidly solidified material which may contain higher levels of oxygen or other impurities. Higher impurity levels are expected in the rapidly solidified alloy because of the high surface to volume ratio of the melt spun ribbon and atomized powder. For example, the oxygen levels of the specimens used in this work were greater than 100 ppm. These impurities can act to pin the grain boundary triple points thus inhibiting the accommodation of the high stress concentrations and thereby promoting premature intergranular fracture. Sulfur contamination could also account for the ductility loss. It has been shown that low concentrations (< 20 ppm) of sulfur lead to a ductility minimum in pure nickel in the range 600-1000C [22,23] and our sulfur levels are nominally 10 ppm. It is worth noting that the high temperature ductility minimum observed in tensile tests of melt spun [24] and plasma deposited [13] nickel-base superalloys is much more severe than for conventional ingot metallurgy specimens.

## CONCLUSIONS

The elevated temperature mechanical properties of rapidly solidified, boron doped $Ni_3Al$ have been investigated. Melt spun ribbon, plasma deposits and HIP'ped powder were tested with the same general results. It was found that the rapidly solidified alloy exhibits an increasing flow stress with increasing temperature, similar to that reported for conventionally cast $Ni_3Al$. However, the rapidly solidified alloy exhibits a severe ductility minimum in the range 500–750C.

It was found that the ductility minimum is sensitive to both boron and aluminum concentration, and that the embrittlement onset temperature changes with processing and testing parameters. In all cases, the onset of the ductility drop coincides with a peak in flow stress. It is postulated that this behavior is due to a change in the flow mechanism at elevated temperature to one that is associated with grain boundary stress concentration.

## ACKNOWLEDGEMENTS

The authors would like to thank R.J. Zabala and R.P. Laforce for producing the melt spun ribbon, L.A. Wojcik and R.J. Petterson for producing the powder, S. Rutkowski for providing the plasma deposits and C.P. Palmer for performing the tensile tests. They also thank M.R. Jackson, R.H. Bricknell and P.A. Siemers for many helpful discussions. This work was partially supported by the Office of Naval Research.

## REFERENCES

1. R.W. Guard and J.H. Westbrook, Met. Trans. 215 (1959) 807.
2. P.H. Thornton, R.G. Davies and T.L. Johnston, Met. Trans. 1 (1970) 207.
3. O. Noguchi, Y. Oya and T. Suzuki, Met. Trans. 12A (1981) 1647.
4. K. Aoki and O. Izumi, J. Mat. Sci. 14 (1979) 1800.
5. K. Aoki and O. Izumi, Kinzoko 49 (1979) 38.
6. K. Aoki and O. Izumi, J. Japan Inst. Met. 43 (1979) 1190.
7. C.T. Liu, C.L. White, C.C. Koch and E.H. Lee, Proceedings of the Electrochemical Society on High Temperature Materials, ed. Munir Cubicciotti, Vol. 83-7, Electrochemical Society Inc. (1983) p. 32.
8. C.T. Liu, C.L. White and J.A. Horton, Acta Met, in press.
9. A.I. Taub, S.C. Huang and K.M. Chang, Met. Trans. 42A (1984) 399.
10. K.M. Chang, S.C. Huang and A.I. Taub, submitted to MRS Symposia Proceedings: "Rapidly Solidified Metastable Materials", Nov. 1983, Boston, MA.
11. S.C. Huang, A.I. Taub and K.M. Chang, Acta Met., in press.
12. S.C. Huang and A.M. Ritter: "Proc. AIME Symposium Chemistry and Physics of Rapidly Solidified Materials", St. Louis, MO, Oct. 26-27, 1982, B.J. Berkowitz and R.O. Scattergood, Eds., TMS-AIME, Warrendale, PA, 1983, pp. 25-34.
13. M.R. Jackson, J.R. Rairden, J.S. Smith and R.W. Smith, J. Metals 33 (1981) 23.
14. K.M. Chang, A.I. Taub and S.C. Huang, these proceedings.
15. A.I. Taub, S.C. Huang, M.R. Jackson, and E.L. Hall, MRS Symposia Proceedings: "Rapidly Solidified Metastable Materials", Nov. 1983, Boston, MA, in press.
16. C.T. Liu, private communication.
17. A.I. Taub, S.C. Huang and K.M. Chang, Proceedings of 39th Symposium of Mechanical Failures Prevention Group, Gaithersburg, MD, May 1-3, 1984, Cambridge University Press, in press.
18. S.J. Liang and D.P. Pope, Acta Met (1977) 485.
19. D.A. Woodford, Met. Trans. 12A (1981) 299.
20. R.H. Bricknell and D.A. Woodford, Met. Trans. 12A (1981) 425.

21. W.H. Chang: "Superalloys-Processing, Proc. 2nd Intl. Conf. on Superalloys, Sec. V", MCIC-72-10, 1972.
22. C.L. White and D.F. Stein, Met. Trans. 9A (1978) 13.
23. M.G. Lozinskiy, G.M. Volkogon, and N.Z. Pertsovskiy, Russ. Met. 5 (1967) 65.
24. A.I. Taub, unpublished research.

# HYDROGEN EMBRITTLEMENT OF ORDERED ALLOYS

A. K. KURUVILLA and N.S. STOLOFF, Materials Engineering Department, Rensselaer Polytechnic Institute, Troy, NY 12181

ABSTRACT

The susceptibility to hydrogen embrittlement has been investigated in two $L1_2$ superlattices, $(Fe,Ni)_3V$ and boron-modified sub-stoichiometric $Ni_3Al$ and in a $B_2$ superlattice, FeCo-V. The relationship between ordering and hydrogen embrittlement is explained on the basis of characteristic dislocation structure in the ordered lattice and hydrogen-dislocation interactions.

## INTRODUCTION

Long range order is known to change most mechanical properties of solid solution alloys. Many of these changes are attributed to the change in dislocation configuration that accompany ordering in numerous alloy systems. For example, the presence of superlattice dislocations has been found to be responsible for the increase in strain hardening rates whereas segregation of impurities to grain boundaries and restricted cross slip (that result from ordering) contribute to a decrease in fracture ductility and changes in the fracture mode in some ordered alloys. The relationship between long range order and hydrogen embrittlement was established in a previous study where ordering in an $(Fe,Ni)_3V$ type alloy resulted in an enhancement in the susceptibility of the alloy to hydrogen embrittlement[1]. Boron-modified $Ni_3Al$ (an ordered intermetallic compound) also was found to be extremely susceptible to hydrogen embrittlement, although the development of a surface oxide layer prevented such degradation[2]. The present study, which is a continuation of the previous work, was initiated to clearly establish the relationship between ordering and hydrogen embrittlement.

## EXPERIMENTAL PROCEDURE

The alloys used in this study were: i) an $(Fe,Ni)_3V$ type alloy which forms an $L1_2$ superlattice below the transformation temperature, ii) FeCo-V which forms a $B_2$ superlattice below the transformation temperature and iii) boron-modified sub-stoichiometric $Ni_3Al$, that remains ordered up to the melting point and has an $L1_2$ structure. The composition and heat treatment of these alloys is shown in Table I.

The alloys were tensile tested in air or under conditions of simultaneous cathodic charging with hydrogen. In all the tests in hydrogen, charging was done electrolytically at a current density of $50mA/cm^2$ in a solution of IN $H_2SO_4$. 0.05 gm/ℓ of sodium arsenite was added to the solution to serve as a hydrogen recombination poison. The tests were performed at room temperature with a screw driven Instron machine at a cross head speed of 0.0025 cms/min. The $(Fe,Ni)_3V$ alloy (henceforth referred to as LRO-42) and FeCo-V alloy were tested in the fully ordered and disordered condition whereas $Ni_3Al$ was tested in the ordered condition. (This alloy remains ordered to the melting point). An additional test was performed in the case of the FeCo-V alloy, where a partially ordered condition (S=0.4) was tensile tested in air and hydrogen.

The fracture surfaces of these various samples were examined in an AMR 1000 scanning electron microscope to observe fractographic features under different experimental conditions.

 

TABLE I

Composition and Heat Treatment

$(Fe,Ni)_3V$: Fe-39.5w%Ni-22.4%V-0.4%Ti-1000ppm Ce

Disordering: Anneal at 1100°C for 20 mins and water quench

Ordering: 650°C - 24 hrs, 600°C - 24 hrs, 500°C - 48 hrs

FeCo-V: Fe-49.1w%Co-2.03%V

Disordering: 840°C - 1 hr, iced brine quench

Partial ordering: Quench from 690°C

Complete ordering: 550°C - 5 hrs

$Ni_3Al$+B: Ni-24at%Al-0.2%B

Heat treatment: 1000°C - 30 mins

## RESULTS

The results of the tensile tests on LRO-42 are summarized in Table II and Figs. 1a) and 1b). The material yields in air at a considerably lower stress in the ordered condition than in the disordered condition but strain hardening is more rapid, so that the ultimate tensile strength is actually higher for the ordered material. Even so, the alloy in the ordered condition displays a ductility value as high as 25% elongation. Hydrogen charging caused a reduction in less than 10% in the ultimate tensile strength of the disordered material, while the reduction in tensile strength in the case of the ordered material was nearly 60%. Similarly, the reduction in ductility (in hydrogen, as compared to air) was about 50% in the ordered material but only about 20% in the disordered material.

Examination of the fracture surface revealed that the disordered material failed by dimpled rupture in both air and hydrogen as shown in Fig. 2a) and 2b), respectively; ordered material revealed a mixed intergranular and dimpled transgranular fracture in air (Fig. 2c)), but failed in hydrogen by intergranular cracking (Fig. 2d)).

The results of tests on FeCo-V are shown in Table III and Figs. 3a)-c). An increase in the degree of order resulted in a decrease in the yield strength. The ultimate tensile strength, however, increased by about 16% in the case of the partially ordered material (as compared to the disordered material) but decreased drastically with the onset of complete order. The ductility values of disordered and partially ordered material remained more or less constant, but decreased by about 77% with the onset of complete order. Hydrogen charging did not affect the yield stress in any of these conditions, but the ultimate tensile strength and ductility values were significantly reduced. In hydrogen, the UTS decreased by 27%, 45% and 22% and ductility decreased by 84%, 89% and 37% in the disordered, partially ordered and fully ordered conditions, respectively.

TABLE II

Tensile Properties of LRO-42

| Environment | Ordered | | | Disordered | | |
|---|---|---|---|---|---|---|
| | $\sigma_{ys}$ (MPa) | $\sigma_{UTS}$ (MPa) | %El | $\sigma_{ys}$ (MPa) | $\sigma_{UTS}$ (MPa) | %El |
| Air | 344 | 1175 | 25 | 415 | 770 | 38 |
| Hydrogen | 344 | 485 | 13 | 415 | 710 | 30 |

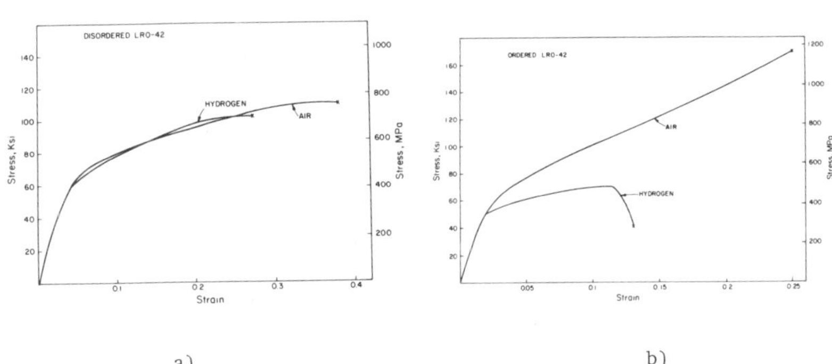


a) b)
Fig. 1 Stress vs. strain: LRO-42

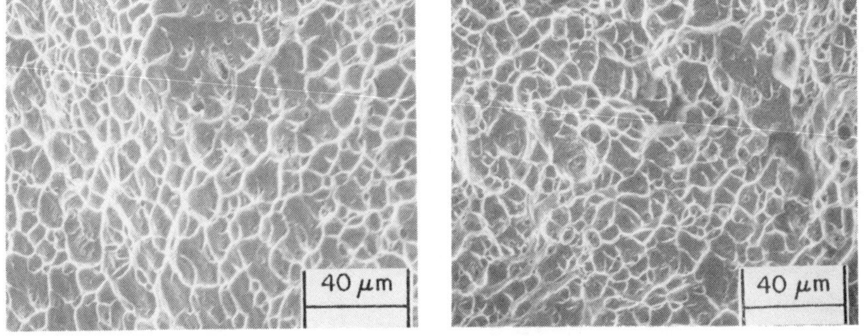


a) Air b) Hydrogen
Fig. 2 Disordered LRO-42, tensile tested

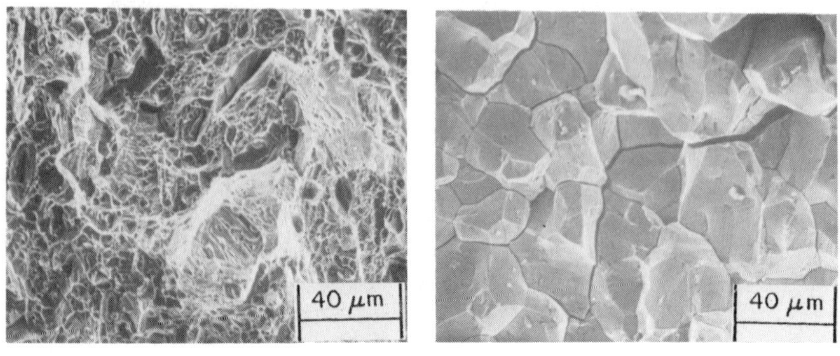


c) Air d) Hydrogen

Fig. 2 Ordered LRO-42, tensile tested

TABLE III

Tensile Properties of FeCo-V

| Environment | Disordered | | | Partially Ordered (S=0.4) | | | Fully Ordered | | |
|---|---|---|---|---|---|---|---|---|---|
| | $\sigma_{ys}$ (MPa) | $\sigma_{UTS}$ (MPa) | %El | $\sigma_{ys}$ (MPa) | $\sigma_{UTS}$ (MPa) | %El | $\sigma_{ys}$ (MPa) | $\sigma_{UTS}$ (MPa) | %El |
| Air | 385 | 842 | 18 | 303 | 980 | 19 | 227 | 552 | 4 |
| Hydrogen | 380 | 615 | 3 | 303 | 537 | 2 | 227 | 434 | 2.5 |

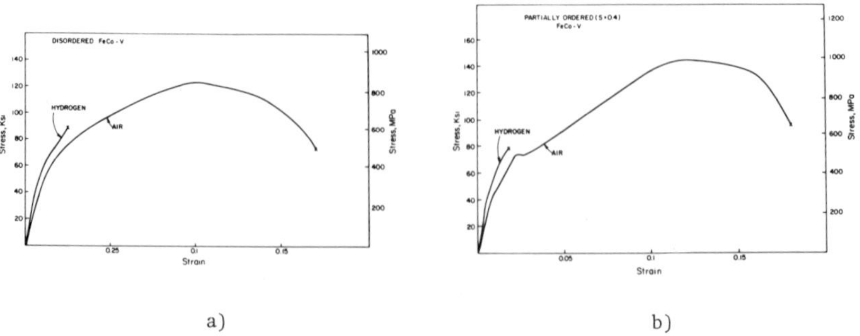


a) b)

Fig. 3 Stress vs. strain: FeCo-V

The fracture surfaces of the disordered and partially ordered material displayed dimpled fracture in air (Figs. 4a) and 4c) whereas, it changed to a brittle cleavage type of fracture in hdyrogen (Figs. 4b) and 4d)). The fully ordered condition displayed brittle fracture features in both air and hydrogen as shown in Figs. 4e) and 4f).

Boron modified $Ni_3Al$ exhibited good ductility and strength when tested in air as shown in Table IV and Fig. 5. In hydrogen, however, the ductility dropped drastically and so did the ultimate tensile strength. The yield strength did not change very much with the hydrogen environment.

Fracture surfaces displayed predominantly ductile transgranular features in air but cracking was entirely intergranular in hydrogen as shown in Figs. 6a) and 6b).

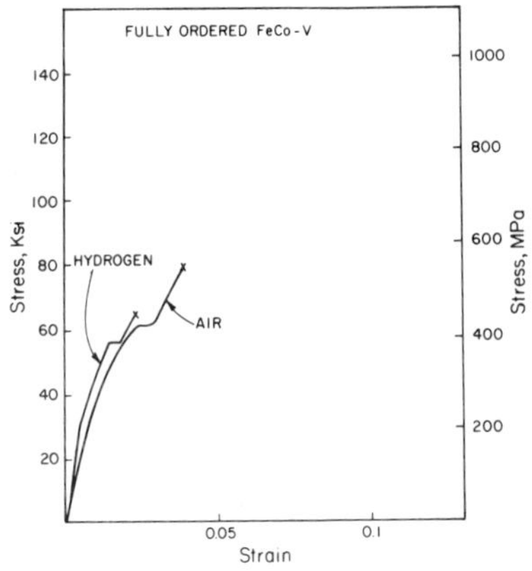


Fig. 3c) Stress vs. strain ordered FeCo-V

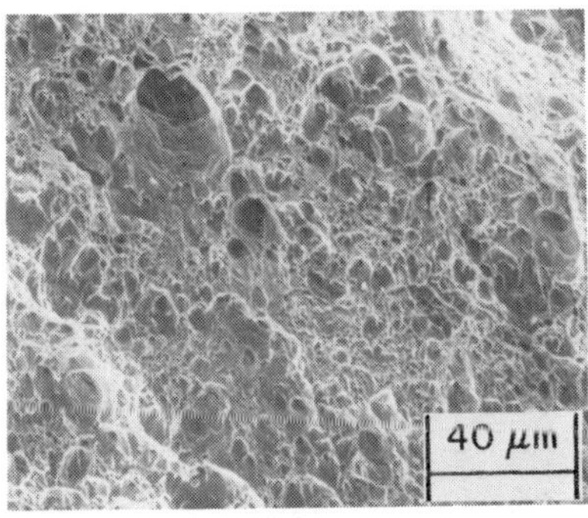


Fig. 4a) Disordered FeCo-V tensile tested in ait.

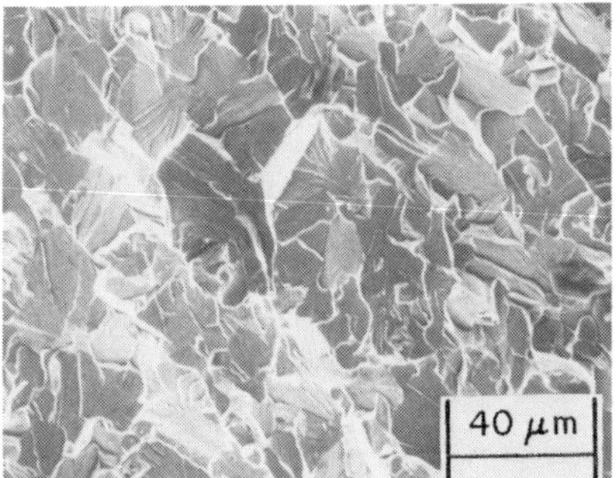


Fig. 4b) Disordered FeCo-V tensile tested in hydrogen

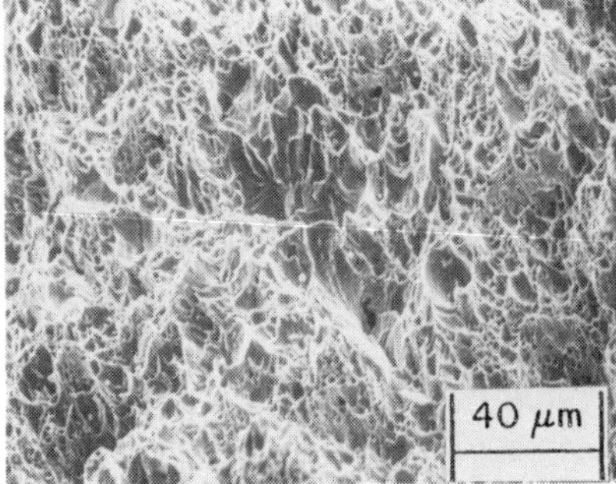


Fig. 4c) Partially ordered FeCo-V, tensile tested in air

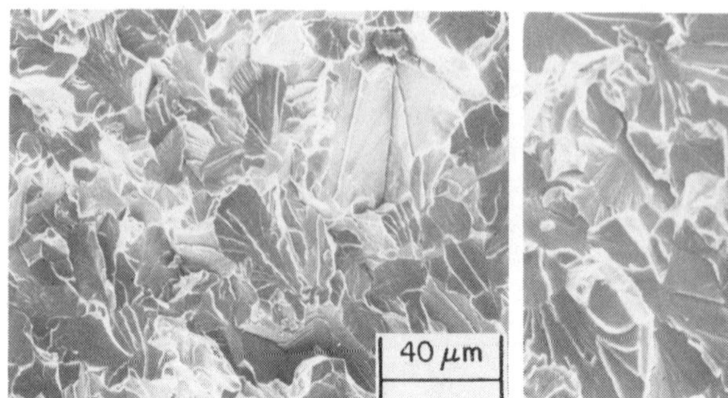


Fig. 4d) Partially ordered FeCo-V, tensile tested in hydrogen

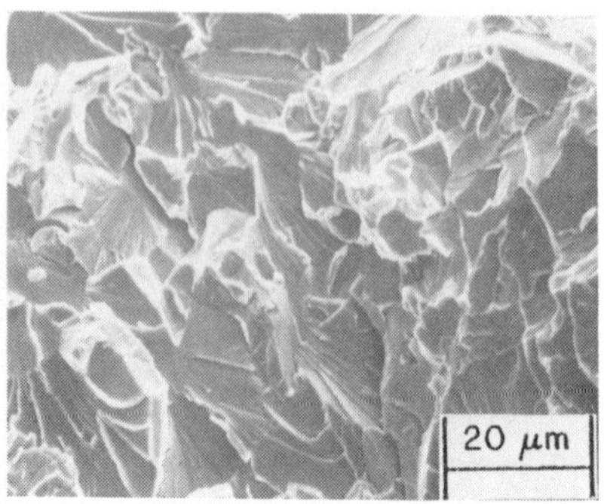


Fig. 4e) Ordered FeCo-V tensile tested in air

Fig. 4f) Ordered FeCo-V tensile tested in hydrogen

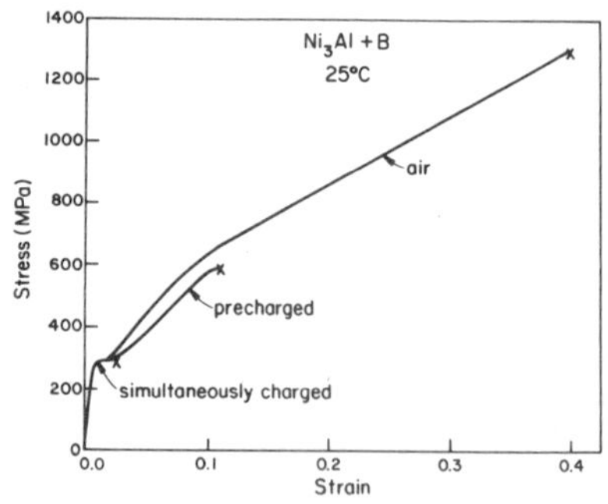


Fig. 5 Stress vs. strain $Ni_3Al$+B

TABLE IV

Tensile Properties of $Ni_3Al$+B

| Environment | $\sigma_{ys}$ (MPa) | UTS (MPa) | %El |
|---|---|---|---|
| Air | 213 | 1296 | 40 |
| Hydrogen | 220 | 289 | ∿0 |

## DISCUSSION

In all these alloys, the introduction of hydrogen to the environment did not change the macroscopic yield stress, which seems to indicate that plastic deformation is a prerequisite for hydrogen induced failure in these materials, both in the ordered and disordered conditions.

Considering first the two $L1_2$ structures, namely, LRO-42 and boron modified $Ni_3Al$, the effect of hydrogen seems to be similar in these two alloys. In LRO-42, ordering promotes an increased susceptibility to hydrogen embrittlement. In both ordered LRO-42 and $Ni_3Al$ (which remains ordered up to the melting point) hydrogen changed the fracture mode from a ductile transgranular-integranular mixture to purely intergranular. The effect of hydrogen on boron modified sub-stoichiometric $Ni_3Al$ is particularly interesting. The boron effect on ductility of $Ni_3Al$ has been well established[3,4,5]. Although no clear reason for this effect has evolved, it has been suggested[14] that boron strengthens grain boundaries by increasing the cohesive forces at the boundary following the arguments forwarded by Messmer and Briant[6]. This could also explain the effect of an optimum boron content that promotes maximum ductility. Below the optimum level, there possibly isn't sufficient boron atoms to strengthen most boundaries and above the optimum level, precipitation of borides or simply the segregation of a large number of interstitial boron atoms to the boundaries, promotes embrittlement. This is substantiated by the observation that the fracture surface of ductile $Ni_3Al$ with the optimum boron content has a reasonable amount of intergranular facets as shown in Fig. 6a). It has also been observed that stoichiometric $Ni_3Al$ cannot be made ductile by boron additions. The fact that stoichiometric $Ni_3Al$ exhibits intergranular failure (even in the presence of boron) and that boron modified sub-stoichiometric $Ni_3Al$ displays a reasonable amount of intergranular fracture leads us to conclude that the grain boundary strengthening effect of boron may only be marginal in that boron doped grain boundaries may not be able to withstand a wide spectrum of loads and temperatures. In fact, support for this hypothesis has been observed in two independent studies. Fatigue work on boron-modified $Ni_3Al$ has shown that the overload region in high cycle fatigue in cyclic crack propagation is purely intergranular in nature[7]. A ductility minimum and severe intergranular failure has been observed in meltspun boron modified $Ni_3Al$ in the temperature range of 600°C-800°C[8].

The effect of hydrogen on dislocation mobility in nickel and iron has been reasonably well established[9,10]. It was suggested by Tabata and Birnbaum[10] that the hydrogen enhanced dislocation mobility in iron was due to a decrease in the stress necessary to move dislocations. Although this may be true, some recent work by Rodriguez and Ficalora[11] has shown that the stresses due to an increase in volume resulting from the formation of a hydrogen anion at a dislocation core in iron are large enough to move dislocations. From either cause, the result is a dislocation pile up and enhanced stresses at grain boundaries. It appears, therefore, that such stresses are sufficient to promote fracture along boron strengthened grain boundaries in $Ni_3Al$. It is interesting to note that boron has been found to eliminate intergranular fracture in nickel in the presence of hydrogen[12]. One of the reasons suggested to explain this behavior was that the boron at the grain boundaries might repulsively interact with metalloids to reduce their segregation. Latanision and Opperhauser[13] have correlated hydrogen embrittlement in nickel with metalloid segregation. No such effect of boron was observed in $Ni_3Al$[2]. It is suggested, therefore, that the principal cause of hydrogen embrittlement in $Ni_3Al$ is the development of high stresses at the grain boundaries which are not strong enough to withstand such stresses, in spite of boron additions.

The behavior of ordered LRO-42 can be explained in a similar way. The long term, low-temperature heat treatment involved in the ordering process

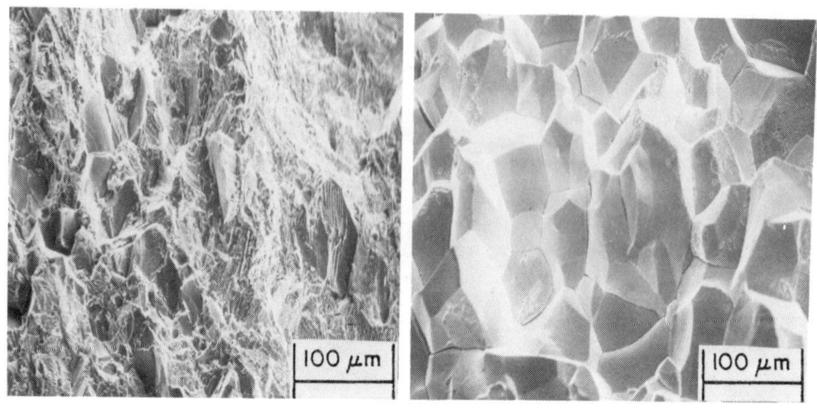


a) Air b) Hydrogen

Fig. 6 $Ni_3Al$+B, tensile tested

could promote the segregation of metalloids to grain boundaries in ordered LRO-42. Such segregation could also affect hydrogen content at grain boundaries; it has been shown by Robertson et al[15] in nickel that the concentration of hydrogen was enhanced at grain boundaries in the presence of sulphur. The weakening of the grain boundaries and development of high stresses at these grain boundaries, resulting from dislocation pile ups and hydrogen-dislocation interaction seems to be the major reason for enhanced susceptibility of LRO-42 to hydrogen embrittlement in the ordered condition as compared to the disordered condition.

However, the ordering process itself could contribute to hydrogen embrittlement. Superlattice dislocations are characteristic of ordered lattices. These dislocations cross slip with difficulty so that accommodation of stress concentrations (eg., at grain boundaries) in the ordered lattice is difficult. Therefore, the build up of stresses at the grain boundaries would be relieved to a much lesser extent in the ordered condition as compared to the disordered condition. This would lead to enhanced susceptibility of an alloy to hydrogen embrittlement in the presence of order in the lattice.

The FeCo-V alloy, which orders into a $B_2$ superlattice, displays different characteristics. The ductile mode of fracture in the disordered and partially ordered condition, changes to a completely brittle (see cleavage facets in Figs. 4b) and 4d)) mode of fracture in the presence of hydrogen. This seems to suggest that the presence of hydrogen leads to a drastic reduction in slip processes such that the material becomes brittle. The fractographic features observed in these conditions, in hydrogen, are similar to those observed in the overload region in a crack growth experiment in vacuum[7]. The brittle features observed in the overload region in crack growth is clearly an indication that the alloy in the disordered condition is strain rate sensitive. With this observation in mind, considering that hydrogen may enhance dislocation motion[9,10], it would be quite reasonable to conclude that the localized effect of hydrogen is similar to that of a high local strain rate, wherein, an attempt to move dislocations rapidly, leads to an increase in flow stress and reduced ductility in the disordered FeCo-V alloy. Another important factor to be considered here is that, unlike the case of fcc alloys, the disordered FeCo-V alloy (which is bcc) is likely to have numerous slip systems that become active in the presence of high stresses, which could result from a hydrogen-dislcoation interaction[11] as mentioned previously.

As for the effect of order in this alloy, increasing the degree of order to S-0.4, without a major change in ductility (Table III) and mode of fracture (compare Figs. 4a) and 4c)) in air, results in an increase in the ultimate tensile strength. Reduced cross slip and the consequent decrease in accommodative slip ahead of a stress concentration seems to be responsible for the enhancement in the susceptibility to hydrogen embrittlement with an increase in the degree of order. In the fully ordered condition, the material is brittle to begin with and hydrogen effects are not manifested to a great extent as compared to the partially ordered alloy.

In terms of a relationship between ordering and hydrogen embrittlement, it is safe to conclude that the onset of order (by virtue of decreased cross slip) enhances the susceptibility of an alloy to hydrogen embrittlement. However, the manifestation of embrittlement could be complicated by factors such as grain boundary segregation, effect of hydrogen on dislocation motion and nature of slip processes that may change with ordering.

## SUMMARY

1. Ordering promotes susceptibility to hydrogen embrittlement.
2. Stresses resulting from hydrogen-dislocation interactions and decreased cross slip cause embrittlement.
3. Boron does not inhibit hydrogen embrittlement of $Ni_3Al$.
4. Embrittlement can be influenced by: grain boundary segregation, hydrogen induced dislocation activity and slip activity.

## ACKNOWLEDGMENT

The authors are grateful to the Office of Energy Systems Research, Division of Energy Conversion and Utilization Technologies (E-Cut) for financial support under Subcontract No. 19X22217C through Martin Marietta Energy Systems, Inc. and to Dr. C.T. Liu at Oak Ridge National Laboratory for helpful discussions.

## REFERENCES

1. A.K. Kuruvilla, S. Ashok and N.S. Stoloff, Proc. Third Int. Congress on Hydrogen and Materials, Paris (Pergamon, 1982), vol. 2, pp. 629-633.

2. A.K. Kuruvilla and N.S. Stoloff, Scripta Met., to be published.

3. K. Aoki and O. Izumi, Nippon Kinzoku Gakkaishi 43, 1190 (1979).

4. C.T. Liu and C.C. Koch, Proc. Conf. on Trends in Critical Materials Requirements for Steels of the Future, Vanderbilt University, TN, Oct. 4-7 (1982).

5. A.I. Taub, S.C. Huang and K.M. Chang, Met. Trans. 15A, 399 (1984).

6. R.P. Messmer and C.L. Briant, Acta Met. 30, 457 (1982).

7. A.K. Kuruvilla and N.S. Stoloff, unpublished.

8. A.I. Taub, private communication, Nov. 1984.

9. J. Eastman, T. Matsumoto, N. Narita, F. Heubaum and H.K. Birnbaum, Hydrogen Effects in Metals (AIME, Warrendale, PA, 1981), pp. 397-409.

10. T. Tabata and H.K. Birnbaum, Scripta Met. 17, 947 (1983).

11. M. Rodriguez and P.J. Ficalora, private communication, June 1984.

12. Y. Ogino and T. Yamasaki, Scrpita Met. 15, 821 (1981).

13. R.M. Latanison and H. Opperhauser, Met. Trans. 5, 483 (1974).

14. C.T. Liu, unpublished.

15. I.M. Robertson, T. Tabata, W. Weis, F. Heubaum and H.K. Birnbaum, Scripta Met. 18, 841 (1984).

# CYCLIC DEFORMATION BEHAVIOR OF $Ni_3Ge$ SINGLE CRYSTALS AT ROOM TEMPERATURE

HAN-RYONG PAK, LEIH-MING HSIUNG* AND MASAHARU KATO**
*Department of Metallurgical and Materials Engineering, New Mexico Institute of Mining and Technology, Socorro, New Mexico 87801
**Department of Materials Science and Engineering, Tokyo Institute of Technology, Yokohama 227, Japan

## ABSTRACT

Fully reversed strain-controlled, tension-compression tests were performed at room temperature to study cyclic-deformation properties of $Ni_3Ge$ single crystals. The cyclic deformation is performed with the stress axis parallel to the [$\bar{1}$ 1.94 2.69] direction at a strain rate of $1.0 \times 10^{-4}$ $s^{-1}$ and at a total-strain amplitude between $5.0 \times 10^{-4}$ and $1.5 \times 10^{-3}$. Cyclic strain-hardening occurs to saturation over the range of the employed strain amplitudes. The cyclic-stress amplitude is higher in compression than in tension at a plastic-strain amplitude larger than $1 \times 10^{-5}$ where screw dislocations are operative. This stress asymmetry becomes more prominent (i.e., the mean stress-amplitude becomes larger in magnitude) as the plastic-strain amplitude becomes larger. At a plastic-strain amplitude less than $1 \times 10^{-5}$ where edge dislocations are operative, the stress amplitude is symmetric. A similar stress asymmetry is also observed for monotonic flow stress. The cyclic hardening is explained by considering an interaction between the screw dislocations.

## INTRODUCTION

A number of investigators have studied the anomalous increase of flow stress (macro-yield stress) with increasing temperature for intermetallic compounds such as $Ni_3Al$-base compounds [1-8], $Ni_3Ga$ [9,10], $Ni_3Ge$ [11,12], $Ni_3Si$ [13] and $Fe_3Ga$ [14]. The most successful model explaining the anomalous behavior is the one proposed by Takeuchi and Kuramoto [9]. This is based on the Kear-Wilsdorf mechanism [16]. In the model, leading superpartial dislocations with a Burgers vector of a/2[$\bar{1}$01] are considered to become immobile by cross slip from (111) to (010). An increase in temperature promotes cross slip in the temperature range where {111}<$\bar{1}$01> slip operates [8,9,12]. The violation of Schmid's law of the critical resolved shear stress for the (111)[$\bar{1}$01] slip is a common feature to the $L1_2$-type compounds showing the strength anomaly [8-11] and is explained by the model, in that the cross slip is aided by the resolved shear stress on the (010) cross-slip plane.

Lall et al [15] recently modified the above model by taking into account cross slip aided by the other stress components which constrict the two Shockley partials on the (111) plane, derived from the a/2[$\bar{1}$01] leading superpartial. According to this model any component of the applied stress which tends to constrict the Shockley partials will promote cross slip and any component which extends them will retard cross slip. Whether the partials are extended or constricted is determined by the direction and the sense of the applied stress. Thus, if measurements of tensile and compressive flow-stresses are made on single crystals having the same orientation, tension/compression asymmetry is expected. When Ezz et al [17] deformed one single crystal of $Ni_3(Al,Nb)$ in tension and then in compression, they found the expected stress asymmetry.

In contrast to such monotonic-deformation behavior of the $L1_2$-type compounds, the fatigue or cyclic deformation behavior has received little attention. Only a few studies have been made during the last two decades

Mat. Res. Soc. Symp. Proc. Vol. 39. © 1985 Materials Research Society

[18-21]. With the above mentioned situation in mind, the main emphasis of the present study was placed on the examination of the cyclic deformation behavior of $Ni_3Ge$ single crystals. This compound was chosen because of its interesting montonic deformation properties such as its large strength anomaly and high stacking-fault energy [10].

## EXPERIMENTAL PROCEDURE

A single crystal was grown by the Bridgman method from an alloy prepared from 99.95% nickel and 99.999% germanium. The single crystal was homogenized for 24 h at $1100^{o}C$ in vacuum and then furnace-cooled. The total weight loss upon melting, growing and homogenizing was less than 0.3% and hence the composition was estimated to be Ni-23.3 at %Ge from the original ratio of components. Specimens for cyclic deformation were cut with a diamond blade into rectangular plates of 7x3x60 mm and the center was filed to a gage size of 3x3x9mm. Electropolishing, strain-annealing and repolishing were done before mechanical testing.

Tension-compression cyclic deformation was conducted at room temperature by using a hydraulic MTS machine with a clip-on extensometer for strain measurement. The total-strain amplitude was controlled between $5.0 \times 10^{-4}$ and $1.5 \times 10^{-3}$. Cyclic tests all began in tension. In all the tests, a triangular-waveform command-signal was used to produce a constant strain-rate of $1 \times 10^{-4}s^{-1}$. The surface was observed during and after cycling with a Nomarski-type optical microscope. Dislocation structure was observed with a Kratos EM-1500 (operated at 1500 keV) at the National Center for Electron Microscopy, Lawrence Berkeley Laboratory, University of California.

The stress axis of the specimens indicated in Figure 1 was $[\bar{1}\ 1.94\ 2.69]$ determined by the back-reflection Laue method. The relevant Schmid factors for the stress axis are listed in Table 1.

Table 1 Schmid factors for the stress axis

| | Tension | Compression |
|---|---|---|
| Primary Octahedral | $(111)[\bar{1}01]$ 0.456 | $(111)[10\bar{1}]$ 0.456 |
| Leading Partial | $(111)[\bar{2}11]$ 0.473 | $(111)[11\bar{2}]$ 0.317 |
| Trailing Partial | $(111)[\bar{1}\bar{1}2]$ 0.317 | $(111)[2\bar{1}\bar{1}]$ 0.473 |
| Edge Component of Leading Partial | $(111)[\bar{1}2\bar{1}]$ 0.156 | $(111)[\bar{1}2\bar{1}]$ -0.156 |
| Edge Component of Trailing Partial | $(111)[1\bar{2}1]$ -0.156 | $(111)[1\bar{2}1]$ 0.156 |
| Cube Cross Slip | $(010)[\bar{1}01]$ 0.422 | $(010)[10\bar{1}]$ 0.422 |

## EXPERIMENTAL RESULTS

### Cyclic stress response

The strain-controlled cyclic formation was monitored by recording hysteresis loops showing the cyclic stress-strain response at different cycles. For the whole range of the strain amplitude employed in this study, the hysteresis loops showed cyclic strain hardening. Figure 1 gives an example of the loops obtained from a specimen cycled at total-strain amplitude of $1.46 \times 10^{-3}$. The magnitude of cyclic hardening per cycle was initially large but declined with cycling, resulting in a change in area of the hysteresis loop. The hardening rate eventually fell almost to zero.

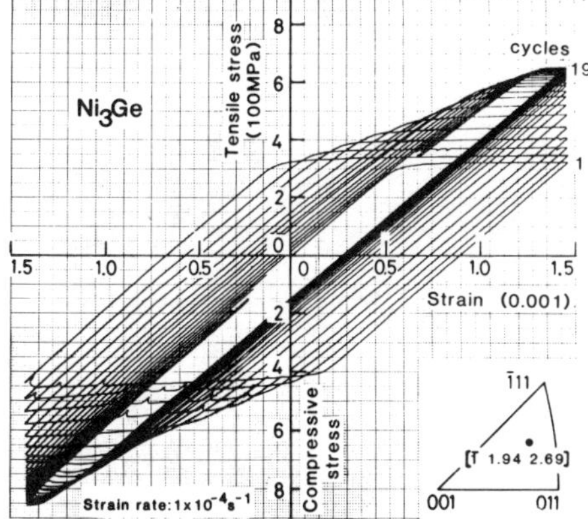


Figure 1. Cyclic strain-hardening of an $Ni_3Ge$ single crystal, cycled at a total-strain amplitude of $1.46 \times 10^{-3}$

This stage is defined as the saturation stage and all the test were continued until the saturation stage was reached.

By plotting the cyclic-stress amplitude (the peak flow-stress at given cycles) against the number of cycles, the hardening behavior was more readily seen. Since most of the hysteresis loops were asymmetric in terms of stress (Fig. 1), the magnitudes of both the cyclic tensile- and compressive- stress amplitudes were plotted in Figure 2. Both of the stress amplitudes increased noticeably as the total-strain amplitude increased. In the case of the smallest total-strain amplitude ($5.0 \times 10^{-4}$), which is slightly larger than the strain necessary for the crystal to begin micro-yielding, the crystal hardened only by 10 MPa in the first few cycles and rapidly went to the saturation stage. As clearly seen in Figure 2, the stress level is higher in compression than in tension. It should be noted that the stress asymmetry becomes more prominent as the total-strain amplitude increases. The behavior of such stress asymmetry is also evident in Figure 3, where cyclic stress-strain curves are shown by plotting the cyclic saturation stress as a function of the total-strain amplitude.

It is known that copper single crystals oriented for single slip have a plateau on the cyclic stress-plastic strain curve. The plateau exists in the range of the

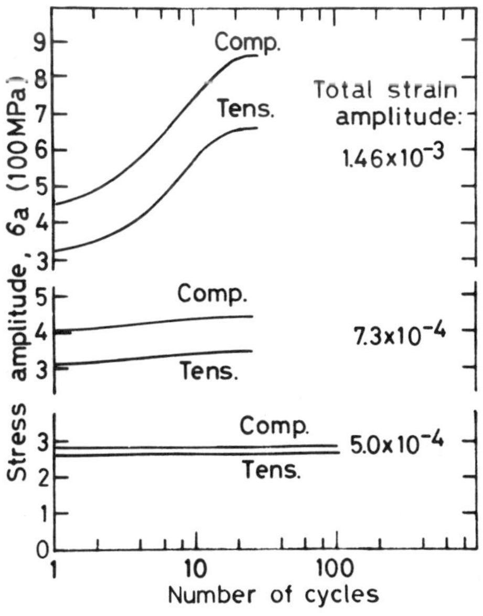


Figure 2. Cyclic hardening of tensile and compressive stress amplitudes for $Ni_3Ge$ single crystals at various strain amplitudes

plastic-strain amplitude between 2.8 x $10^{-5}$ and 3.4 x $10^{-3}$ [22]. Since $Ni_3Ge$ is considered as fcc if atom species are neglected and also since this crystal has a very high stacking fault energy, the existence of such a plateau can can be expected. Figure 4 shows cyclic stress plastic-strain curves, obtained by replotting the data in Figure 3 as the saturation

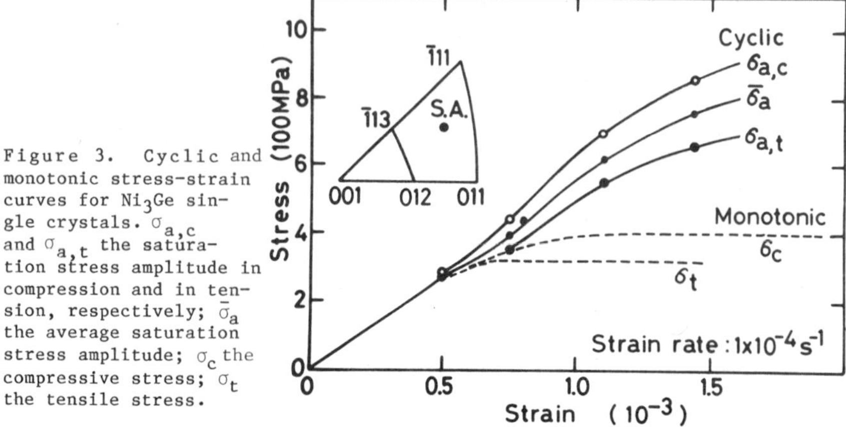


Figure 3. Cyclic and monotonic stress-strain curves for $Ni_3Ge$ single crystals. $\sigma_{a,c}$ and $\sigma_{a,t}$ the saturation stress amplitude in compression and in tension, respectively; $\bar{\sigma}_a$ the average saturation stress amplitude; $\sigma_c$ the compressive stress; $\sigma_t$ the tensile stress.

stress amplitudes against the plastic-strain amplitude measured from the hystersis loop in the saturation stage. Unexpectedly, no such plateau was observable in the present alloy in the same range of the plastic strain amplitude as that for copper. Instead, the saturation-stress amplitude increased monotonically with increasing plastic-strain amplitude.

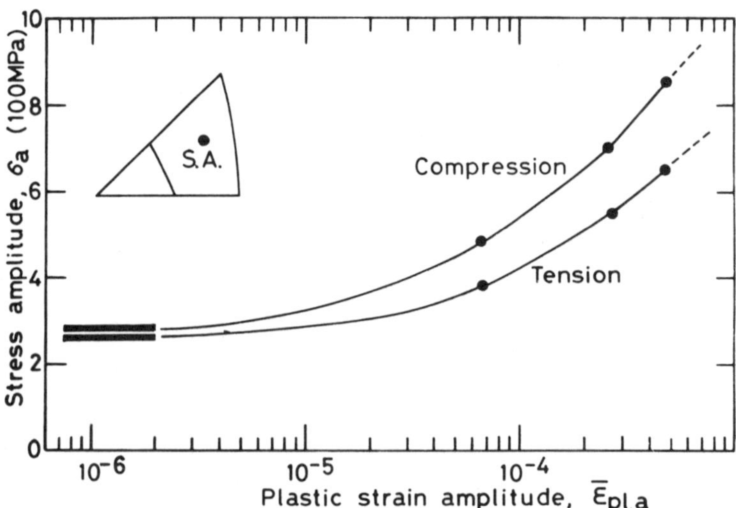


Figure 4. Cyclic stress-strain curves for $Ni_3Ge$ single crystals.

Monotonic stress response

Both monotonic tensile and compressive stress-strain curves were obtained at room temperature with the strain rate of $1 \times 10^{-4} s^{-1}$ for the purpose of comparison to see whether similar stress asymmetry is present. As seen in Figure 3, the compressive flow-stress is higher than the tensile flow-stress in a macro-yielding regime, consistent with the observation in the cyclic deformation shown above. It should be noted, on the other hand, that no stress difference was observed in the micro-yielding regime where plastic strain was less than about $1 \times 10^{-5}$.

Surface observations

Two-surface trace analysis made after cycling showed that plastic deformation took place on the primary slip system $(111)[\bar{1}01]$. The surface observations after reaching saturation revealed that the deformation was fairly uniform for all the employed strain amplitudes. No persistent slip bands were observed on the surface of the specimens cycled to saturation and then electropolished. This observation was consistent with that of dislocation structure, as will be shown later.

Figure 5. Slip bands on one surface of a specimen cycled at a total-strain amplitude of $5.0 \times 10^{-4}$, 22,000 cycles.

DISCUSSION AND CONCLUSION

Monotonic-stress asymmetry

As shown in Figure 3, flow-stress asymmetry was found in the $Ni_3Ge$ single crystals monotonically deformed at room temperature in a macro-yielding regime where screw dislocations are believed to be operative. Ezz et al [17] recently studied monotonic deformation properties of a $Ni_3(Al,Nb)$ alloy and found a similar asymmetry of 0.2% flow stress. It should be noted, however, that in the present study, such stress asymmetry disappeared in a micro-yielding regime corresponding to the plastic strain less than about $1 \times 10^{-5}$ where the motion of edge dislocations are believed to be responsible for plastic deformation [5]. This may naturally be understood since edge dislocations cannot cross slip. Considering common monotonic-deformation properties for the $L1_2$-type compounds, one can expect a similar behavior of micro-yield stress also in $Ni_3(Al,Nb)$ if stress measurements are made in such a low strain regime.

Ezz et al [17] explained the asymmetry of the 0.2% flow stress by the ease of stress-aided cross-slip of screw dislocations from (111) to (010), based upon the model proposed by Lall et al [15]. Here, it should be mentioned, however, that the explanation by Ezz et al [17] was not accurate: in their explanation, such cross slip was considered to be aided by the stress component applied on the edge components of the leading and trailing Shockley partials. Taking the Peach-Koehler relation into account, one can easily find that if such cross slip occurs, cross slip can be aided both by the stress component applied on the edge components and by the stress component on the screw components of those two Shockley partials. That is, the ease of cross slip can be known by comparing the Schmid factors of those Shockley partials. As seen in Table 1, the Schmid factor of tne leading Shockely partial is larger than that of the trailing one in tension while in compression the situation is reversed. Thus, in the present case, the two Shockley partials are extended in tension and constricted in compression, resulting in the fact that the stress level is higher in compression than in tension.

It should be noticed that if one wants to obtain a stress asymmetry in other $L1_2$-type compounds, one must use specimens fully annealed but not subject to any pre-deformation, as in the present study. This is because of strain hardening. A 10-20% increase of flow stress was often seen when specimens were pre-deformed even to 0.15% (see the flow stress in compression of the first hysteresis loop, Figure 1 and the compressive flow stress $\sigma_c$, Figure 3).

## Cyclic strain hardening

In order to understand the cyclic strain hardening, transmission electron-microscope observations were made. Thin foils were prepared from specimens before and after cycling at a total-strain amplitude of $1.0 \times 10^{-3}$. As shown in Figure 6, neither dislocation cell structures nor dislocation ladder structures were observed in the cycled specimen but densely arranged screw dislocations and a lot of dislocation debris were observed. Close observation revealed that screw dislocations were mostly very short compared with those observed in $Ni_3Ge$ specimens deformed about 2% in compression at room temperature [11]. This indicates that cross slip from (111) to (010) frequently occurred during cycling, resulting in the formation of dislocation debris. In the undeformed specimen, on the other hand, only a few as-grown dislocations were observed. Based upon the observations of dislocation structure, a model will be proposed to explain the cyclic strain hardening in $Ni_3Ge$ single crystals cycled at room temperature.

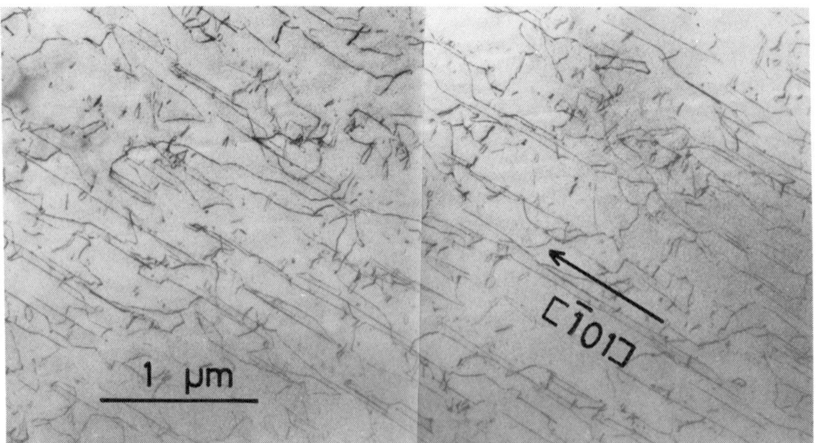


Figure 6. Dislocation structure of a $Ni_3Ge$ single crystal cycled to saturation at a total-strain amplitude of $1 \times 10^{-3}$. Foil surface (111); Kratos EM-1500, operated at 1500 keV.

The observed, cyclic resolved shear stress amplitude, $\tau_a$, is assumed to be composed of the following three terms,

$$\tau_a = \tau_n + \Delta\tau + \Delta\tau(\rho) \quad (1)$$

where $\tau_n$ is the athermal stress required to operate a Frank-Read source of the two $a/2[\bar{1}01]$ superpartial dislocations, $\Delta\tau$ a stress associated with thermally activated cross slip from (111) to (010), and $\Delta\tau(\rho)$ a stress increment which is a function of dislocation density $\rho$. The first two terms are similar to those discussed by Lall et al [15] to explain the critical resolved shear stress of the $L1_2$-type compounds and the third term is introduced in order to take the contribution of strain hardening into account. From the dislocation observations, $\Delta\tau(\rho)$ is considered to be mainly due to the interaction between parallel screw dislocations, which can be expressed in the usual manner as:

$$\Delta\tau(\rho) = \mu b/2\pi r_s \quad (2)$$

where b is the Burgers vector and $r_s$ the distance between two neighboring screw dislocations (or the minimum annihilation distance of screw dislocations), which is a function of screw-dislocation density [23]. From experimental observations (dislocation density and length of screw dislocations), $\Delta\tau(\rho)$ can be estimated to be about 40 MPa for the case of a total-strain amplitude of $1 \times 10^{-3}$ (the details will be reported elsewhere [24]). On the other hand, 100 MPa was experimentally obtained for the shear stress increment at the same strain amplitude (which was the shear-stress difference between the average cyclic-stress amplitude and the average monotonic flow stress), Figure 3. The agreement between the two values is fairly good, and thus it can be concluded that the cyclic strain hardening is probably caused by the interaction between parallel screw dislocations.

## Cyclic-stress asymmetry and mean-stress behavior

The cyclic-stress asymmetry was also found in the $Ni_3Ge$ single crystals. Similar observations were reported in a few other alloys and explained by the following models: the effect of residual stresses due to a difference in thermal condition between Mo-fibers and $Ni_3Al$ [25]; the effect of carbon on the elastic stress-strain relation in tempered 4140 steel [26]; and the effect of image stresses caused by dislocations concentrated at the $\gamma/\gamma'$ interface [27]. None of these can be used to explain the present case. To see the relation between dislocation characteristics and the stress asymmetry, the mean stress (equal to half the saturation-stress amplitude in tension minus half that in compression) was plotted against plastic strain amplitude, Figure 7. The mean stress is seen to be negative and becomes larger in magnitude with increasing plastic-strain amplitude in the strain regime where screw dislocations are operative, while in the low-strain regime where edge dislocations are operative the mean stress is almost zero [28]. Similar observations were made in $Ni_3Al$ by Pope [29] and presented in this conference.

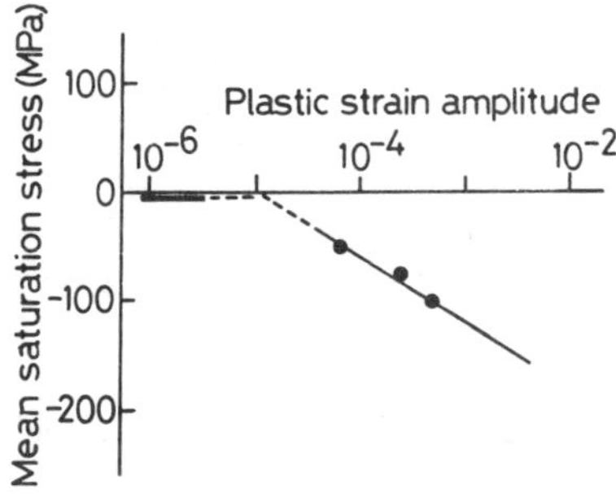


Figure 7. Effect of plastic-strain amplitude on mean saturation-stress.

The symmetry of cyclic-stress amplitude in the micro-strain regime and the negative mean-stress in the macro-strain regime can be explained in a way similar to the monotonic case. However, we can explain neither why the mean stress became larger in magnitude with increasing strain amplitude nor why mean stresses were almost kept constant regardless of number of cycles (Figure 2). For the explanation of these phenomena, further studies on the temperature and orientation dependence of mean stress will be needed.

ACKNOWLEDGEMENTS

The authors are grateful to Professor G. Purcell, Professor A. Miller and Professor O.T. Inal for their support and encouragement. The authors also wish to thank Professor C.C. Koch, Dr. C.T. Liu and Professor N.S. Stoloff for their encouragement. One of the authors (L.H.) would like to thank New Mexico Institute of Mining and Technology for a Research assistantship.

References

1. R.W. Guard and J.H. Westbrook, Trans. Metall. Soc. AIME, 209(1957) 898
2. P.A. Flinn, Trans Metall. Soc. AIME, 218(1960) 145
3. R.G. Davies and N.S. Stoloff, Trans. Metall. Soc. AIME, 233(1965) 714
4. S.M. Copley and B.H. Kear, Trans. Metall. Soc. AIME, 239 (1967) 677
5. R.A. Mulford and D.P. Pope, Acta Metall., 21(1973) 1375
6. A.E. Staton-Bevan and R.D. Rawlings, Phys. Status Solidi, (a) 29(1975) 613
7. K. Aoki and O. Izumi, Phys. Status Solidi, (a) 38(1976)587
8. T. Saburi, T. Hamana, S. Nenno and H.-r. Pak, Jap. J. Appl. Phys., 16 (1977) 267
9. S. Takeuchi and E. Kuramoto, Acta Metall., 21(1973) 415
10. H.-r. Pak, T. Saburi and S. Nenno, Tech. Rep. Osaka Univ., 30(1980) 411
11. H.-r. Pak, T. Saburi and S. Nenno, Trans. Japan Inst. Metals, 18(1977) 617
12. H.-r. Pak, T. Saburi and S. Nenno, Trans. Japan Inst. Metals, 19(1978) 35
13. R. Lowrie, Trans. Metall. Soc. AIME, 194 (1952) 1093
14. D.M. Wee, O. Noguchi, Y. Oya and T. Suzuki, Trans. Japan Inst. Metals, 21 (1980)
15. C. Lall, S. Chin and D.P. Pope, Metall. Trans., 10A (1979) 1323
16. B.H. Kear and H.G.F. Wilsdorf, Trans. Metall. Soc. AIME, 224 (1962) 382
17. S.S. Ezz, D.P. Pope and V. Paidar, Acta Metall., 30 (1982) 921
18. R.C. Boettner, N.S. Stoloff and R.G. Davies, Trans. Metall. Soc. AIME, 236 (1966) 131
19. K.H. Chien, and E.A. Starke, JR, Acta Metall., 23(1975) 1173
20. J.E. Doherty, A.F. Giamei and B.H. Kear, Metall. Trans., 6A(1975) 2195
21. S. Ashok, K. Kain, J.M. Tartaglia and N.S. Stoloff, Metall. Trans., 14A (1983) 1997
22. H. Mughrabi , K. Herz and X. Stark, Acta Metall., 24(1976) 659
23. H. Mughrabi and U. Essmann, Strength of Metals and Alloys, ed. by P. Hassen et al, Pergamon Press (1979) p. 1101
24. H.-r. Pak, L.-M. Hsiung and M. Kato, unpublished work
25. P.R. Bhowal and A.J. McEvily, Metall. Trans., 12A(1981) 1909
26. P.N. Thielen and M.E. Fine, Scripta Metall., 9(1975) 383
27. D.A. Jablonski and S. Sargent, Scripta Metall., 15 (1981) 1003
28. H.-r. Pak, L.-M. Hsiung and M. Kato, J. Metals, 36 No. 7 (1984) p. 38
29. D.P. Pope, appears in this proceedings

# WEAR INDUCED DEFORMATION OF $Ni_3Al$, $(Fe,Co)_3V$, AND $(Fe,Ni)_3V$

B.J. MARQUARDT* AND J.J. WERT**
*General Electric Co., Aircraft Engine Business Group,
1 Neuman Way, Cincinnati, OH 45215
**Mechanical & Materials Engineering, Vanderbilt University
Nashville, Tennessee 37235

## ABSTRACT

Several experimental techniques have been implemented in order to compare the tribological response of $Ni_3Al$, $(Fe,Co)_3V$, and $(Fe,Ni)_3V$ to commercial engineering alloys. Abrasive and adhesive macrowear behavior during sliding contact has been determined. Microstructural data obtained from transmission electron microscopy (TEM) and scanning electron microscopy (SEM) have been correlated with macrowear results. In this correlation, consideration was given to the effects of wear on long-range-order and the presence (or absence) of dislocation subcell structures. The formation of dislocation subcell structures is partially related to the material's stacking fault energy (SFE) which has been the basis of recent wear theory. An experimentally verified wear mechanism based on inhomogeneous slip is defined.

## INTRODUCTION

While the metallurgy of long-range-ordered (LRO) alloys and intermetallic compounds has been studied extensively [1-3], only limited data are available concerning the tribological characteristics of these materials[4-7]. Buckley and Brainard determined the wear resistance of $Cu_3Au$, CuAu, and FeCo in ordered and disordered states and found that the the ordered state possessed a significantly lower wear rate than the corresponding disordered state[4]. Bailey and Sikorski studied $Cu_3Au$, CuAu, and CoPt and concluded that the coefficient of adhesion was lower for the ordered state[5]. Wright and Mikkola have demonstrated that the cavitation erosion rate of ordered $Cu_3Au$ is lower than that observed for disordered $Cu_3Au$[6]. Recent solid particle erosion studies indicate that the erosion rates of ordered $(Fe,Co)_3V$ and $(Fe,Ni)_3V$ are very similar to 316 and 304L austenitic stainless steels[7]. However, more research is essential to obtain a detailed understanding of the wear mechanisms which are operative in LRO alloys and intermetallic compounds.

In recent years, extensive studies have been directed toward the development of LRO alloys and nickel aluminides for possible high temperature applications[8-10]. Alloying techniques have resulted in significant increases in the ductility of these materials. The ductilities of normally brittle cobalt-vanadium, nickel-vanadium, and cobalt-nickel-vanadium hexagonally-ordered alloys were dramatically improved by partially replacing cobalt and nickel with iron[8]. The increase in ductility was attributed to a lower electron-to-atom ratio which stabilized a more ductile cubic-ordered phase. Likewise, it has been demonstrated that small boron additions greatly improve the ductility of polycrystalline $Ni_3Al$[10]. It is believed that the electronic effects of boron increase the cohesive strength of $Ni_3Al$ grain boundaries thus leading to greater ductility of the polycrystalline material[10,11].

In contrast to most conventional alloys, the strengths of ordered $(Fe,Co)_3V$, $(Fe,Ni)_3V$ and $Ni_3Al$ increase with temperature until reaching a critical transition temperature, $T_c$[9,10]. Since these ductile high temperature materials are prime candidates for applications which require wear resistance at elevated temperatures, they were chosen for this investigation. While the long range goal is to understand the high temperature wear characteristics of these materials, the immediate goal was to investigate their structural response under ambient conditions.

While numerous mechanisms have been proposed to explain the details of surface damage and debris production resulting from sliding contact, most wear behavior may be broadly classified as resulting from either adhesive or abrasive wear. A transition from an abrasive to an adhesive wear mechanism may be observed. This transition, which relates to the grit size effect, has been attributed to several different origins which include the ratio of elastic-to-plastic contacts[12], variation of bluntness with particle size[13], and interference between adhesive and abrasive particles[14]. Since the most interesting and material sensitive wear results were observed under adhesive wear conditions, the primary focus of this investigation was placed on adhesive wear mechanisms.

Adhesive wear is characterized by appreciable metal-metal contact and is often attributed to successive formation and rupture of asperity junctions. Based on the observation that flake-like wear debris is often produced by sliding contact, Suh[15,16] has proposed a "delamination theory" of wear based on a dislocation model. The delamination theory proposed by Suh suggests that flake-like wear debris results from cracks initiated by dislocation pile-ups. Through void formation and coalescence, a crack grows parallel to the surface and upon reaching a weak position, shearing occurs and a sheet-like wear particle is produced[16]. Rigney and Glaeser have proposed that flake-like wear debris may also arise from the formation and sequential cracking of cell walls formed during wear induced deformation[17]. Either of these mechanisms may occur in numerous materials under various test conditions.

Stiegler[18] has shown that it is possible to inhibit the formation of dislocation subcells by reducing the SFE. The results of this investigation suggest that subcell formation may be prevented even in the highly stressed region of the wear scar. The presence of inhomogeneous slip along the wear scar would indicate such behavior. In addition, a wear mechanism based on inhomogeneous slip may also account for the production of flake-like wear debris.

## EXPERIMENTAL PROCEDURES

Several variations of pin-on-disk experiments were conducted as a part of this investigation. Weight losses associated with the pin were determined and the resulting surface damage was characterized using SEM. A complete description of the identifications and chemical compositions and a brief summary of the thermal histories of the experimental alloys are presented in Table 1. The engineering alloys included 304L and 316 austenitic stainless steels and AISI 52100 steel. The stainless steels were held at 1025°C for approximately one hour followed by an air cool and the 52100 steel was water quenched from 850°C.

Table I
Composition and History of Materials

| Identification | Composition (wt %) | Thermal History |
|---|---|---|
| LRO 37 | 37.6 Fe 39.5 Ni<br>22.4 V 0.4 Ti | 30 min @ 1100°C<br>1 day @ 600°C<br>2 days @ 530°C |
| LRO 34 | 16.3 Fe 61.1 Co<br>22.2 V 0.4 Ti | 30 min @ 1100°C<br>5 hrs @ 800°C<br>17 hrs @ 700°C |
| IC 15 | 87.3 Ni 0.05 B<br>12.7 Al | 30 min @ 1000°C |

In the first series of experiments, an alloy pin was placed in contact with a 12 inch diameter rotating disk covered with SiC abrasive paper. The test apparatus allowed the pin to precess thus ensuring that it was in contact with fresh abrasive at all times. Prior to testing, each specimen was degreased in acetone and weighed on a precision balance. After sliding approximately 30 meters, the specimen was removed, rinsed in acetone, and reweighed. This procedure was repeated three times for a total sliding distance of 90 meters. Room temperature tests were conducted under normal loads varying from 50 to 500 grams.

Additional wear data for metal-metal contact was obtained from two sources. One set of data corresponds to the testing of pins and disks of the same composition and thermal history. A second set of data was obtained for each alloy pin sliding against a disk of 304L stainless steel. In order to obtain reproducible contact areas, a hemispherical tip was machined on each pin using a radial grinding operation. Before wear testing, the hemispherical tip of each pin was electropolished and each alloy disk was mechanically polished through 0.05 micron alumina. In each metal-metal test, a constant sliding speed of 1.16 meters per minute was employed to minimize frictional heating and its effects. Sliding distances ranged from 29 to 1150 meters under normal loads of 200 and 400 grams.

TEM was utilized for defect analysis of microwear tracks in 304L, LRO 34, LRO 37, and IC 15. Foil preparation was equivalent to that first used by Carpenter[19]. Initially, disks were polished through 600 grit SiC abrasive paper and then electropolished to a smooth surface. After electropolishing, a series of parallel wear tracks were scribed into the surface of the disks using a profilometer equipped with a diamond stylus which measured 12.5 microns in radius. Normal loads used for the production of these microwear tracks were varied from 150 to 300 milligrams and the sliding speed was held constant at 1 centimeter per minute. The microwear tracks were then masked with a lacquer and the specimens were back-thinned to perforation in a submerged jet polisher. After the lacquer was dissolved with acetone, the thin foils were analyzed in JEOL 100C and 100CX microscopes at Oak Ridge National Laboratory.

## RESULTS AND DISCUSSION

A comprehensive list of the wear rates obtained from pins

tested on SiC abrasive paper is presented in Table 2. Each of the LRO alloys and IC 15 demonstrated wear rates similar to those of 316 and 304L austenitic stainless steels. The results indicate that 52100 steel possesses superior abrasive wear resistance. Two general relationships may account for this behavior. When hard second phase particles are large in relation to the penetration depth of the abrasive particles , they present a powerful obstruction to the indenting particles. Hence, the wear resistance of steels increases with carbide content until a fracture mode of the carbides takes over and controls the wear behavior[20]. The 52100 steel used for this investigation contained hard chromium carbides whereas each of the other materials represented a homogeneous single phase. In addition, investigators have shown that wear rates decrease as the hardness of the worn material approaches that of the abrasive[20]. The Knoop hardness value of 52100 steel was 802 as compared to 245 and lower for the other materials.

A typical microwear scar observed in TEM analysis is shown in Figure 1. Selected area diffraction patterns (SADP) obtained from within the scars were observed to be normal spot patterns in all cases. These results differ from previous results reported by Caldwell and coworkers[21] on Cu-Al alloys. Under identical experimental conditions, SADP's from the wear scars of Cu-Al alloys were shown to produce ring patterns indicative of many

Table II
Abrasive Wear Rates (mg/m)

| Load (gms) | LRO 37 | LRO 34 | IC 15 | 52100 | 304L | 316 |
|---|---|---|---|---|---|---|
| 100 | 0.090,0.142* | -- | -- | 0.058,0.070* | .082 | 0.079,0.128* |
| 200 | 0.185,0.197* | .185 | .190 | 0.127,0.153* | .171 | 0.171,0.191* |
| 400 | 0.328 | .288 | .358 | -- | .285 | -- |
| 500 | 0.489,0.549* | .461 | .482 | 0.297,0.371* | .434 | 0.387,0.528* |

* Denotes wear rates on 240 grit SiC paper, all others correspond to 400 grit SiC paper.

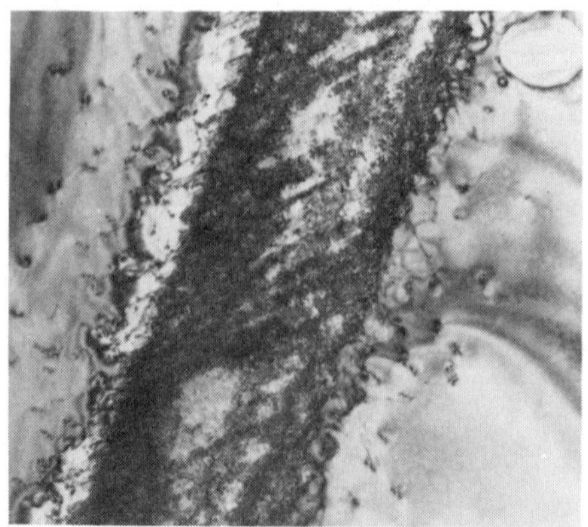

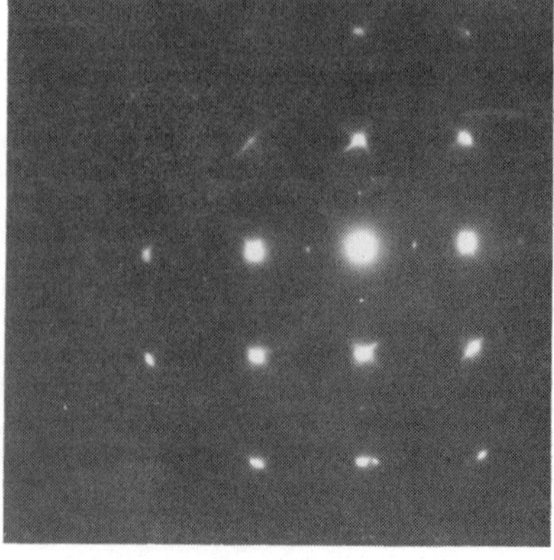

Figure 1 Micro wear scar observed in LRO 34 (A) and the associated SADP which contains superlattice reflections (B).

small crystallites within the scars. In addition, SADP's from the wear scars of each of the LRO materials examined in this investigation contained superlattice reflections. Superlattice reflections indicate that some degree of order is retained in these materials following wear induced deformation.

A number of prominant microstructural features developed during the pin-on-disk wear studies. A serrated structure, as seen in Figure 2a could be distinguished on the leading edge of all hemispherically tipped pins except for 52100 steel. It is believed that this serrated structure is the result of inhomogenous slip along the {111} <110> system. Deformation structures such as these are commonly found in materials with low SFE's.

The data presented in Table 3 which represent metal-metal sliding contact show a large variation of wear rates between materials. A corresponding variation of wear mechanisms is associated with the range of wear data. Detailed microscopic evaluation provided significant information for the interpretation of wear mechanisms.

Unique and significant microstructural information was obtained from the analysis of LRO 34, a cobalt alloy. Other than the serrations shown in Figure 2a, further crystallographic structure was found within the wear scar as revealed by Figure 2b. Detailed analysis of Figure 2b suggests that a lattice rotation aligns the serrated structure perpendicular to the sliding direction. The general structure of the wear scar retains this perpendicular orientation as shown in Figure 3a. In a few isolated cases, the rotation does not occur as indicated by Figure 3b. The apparent explanation for these isolated cases is that the available slip systems are incompatible to the necessary rotation.

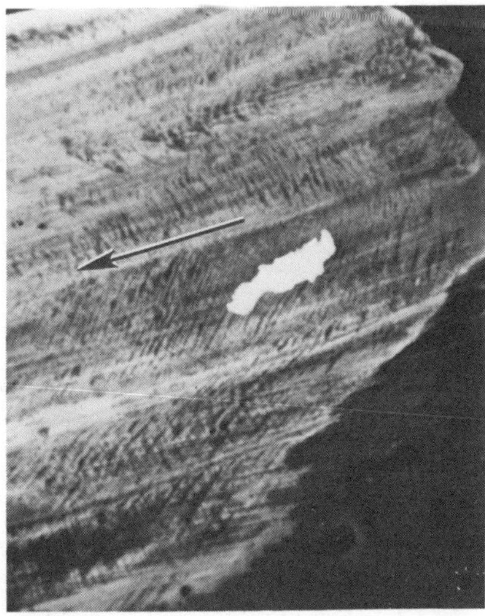

Figure 2 Serrated structure caused by inhomogeneous slip at the leading edge of a wear scar in LRO 34 (A) and a retension of inhomogeneous slip with lattice rotation within the wear scar (B). Arrow indicates sliding direction.

Table III
Weight Loss (gms) of Hemispherically
Tipped Pins Under 400 gm Load

| Sample | 117 Meters | 1150 Meters | 1150 Meters* |
|---|---|---|---|
| LRO 37 | 0.0009 | 0.0207 | 0.0079 |
| LRO 34 | 0.0 | 0.0002 | 0.0004 |
| IC 15 | -- | -- | 0.0115 |
| 52100 | 0.0 | 0.0007 | 0.0003 |
| 304L | 0.0001 | 0.0082 | 0.0082 |
| 316 | 0.0001 | 0.0034 | 0.0058 |

*Denotes tests conducted on 304L disks.

Information gained from TEM analysis has indicated that a subcell structure does not form in microwear tracks of LRO 34. In addition, superlattice reflections have indicated that some degree of order is retained. A microwear track in LRO 34 and its associated SADP are shown in Figure 1. Structures observed by SEM of macrowear specimens indicate similar deformation character is retained under more severe wear environments.

Two factors have a great influence on the deformation of LRO 34. First, a very low SFE greatly inhibits cross-slip mechanisms. Likewise, in order to retain the low energy ordered state, slip becomes more favorable in previously slipped regions where short-range-order has been reduced. As a result, slip tends to be coarse and inhomogeneous in nature. It has previously been shown that inhomogeneous slip leads to crack nucleation and fracture along slip bands with high dislocation densities[22]. On this basis, it is suggested that the wear mechanism which operates in LRO 34 is

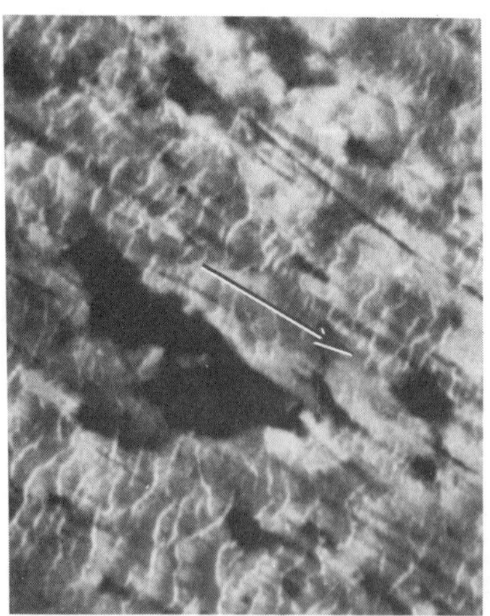

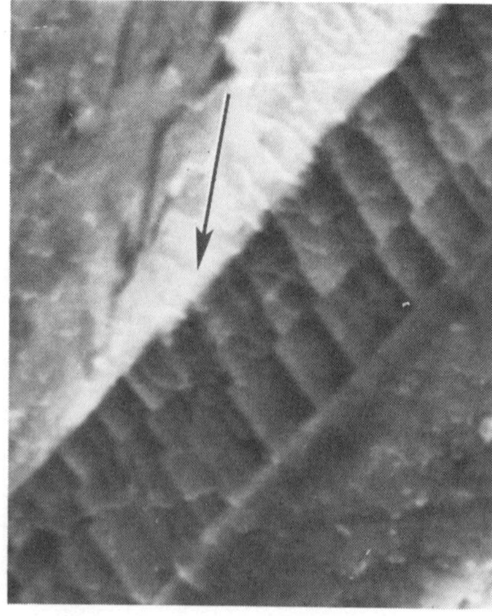

Figure 3 Fracture surfaces in LRO 34 aligned perpendicular to the sliding direction (A) and an isolated case where lattice rotation did not occur (B). Arrow indicates sliding direction.

the fracture and detachment of crystallographic platelets along slip bands of high dislocation density. Low wear rates associated with LRO 34 would suggest that the inhomogeneous slip mechanism is favorable in severe wear environments. Favorable wear rates generally associated with materials possessing low SFE's might also be explained by a similar wear mechanism.

The wear scars found on 52100 steel are generally quite smooth with a few well defined scratches. It is evident that the scratches evolve from carbide particles which are released by the wear process and subsequently act as abrasive particles.

The wear surfaces of 316 and 304L are highly deformed and irregular. At the trailing edge of the wear scar, extrusions are found. The wear process in these stainless steels is complex in nature. A recent investigation has suggested that the wear behavior of austenitic stainless steel is highly dependent upon strain-induced transformations and that a subcell structure forms near the wear surface[23]. Microwear tracks formed in 304L by the single pass of a diamond stylus revealed no indication of a subcell structure or second phase particles. A deformation study has revealed that subcells form in 304 stainless steel, however, the lowest investigation temperature was 1000°F [24]. Localized heating/melting may be responsible for the complex and irregular structure of austenitic stainless steel wear scars.

A highly deformed and fragmented structure covers the majority of the wear surface in IC 15 and LRO 37. In a few isolated cases, signs of inhomogeneous slip may be located along the wear scars of these materials as revealed by Figure 4. However, the randomly fragmented structure which accounts for the majority of the wear surface indicates that a mechanism similar to those described by Suh [16] and Rigney [17] is controlling the wear process. The results from LRO 34 suggest that alloying to lower the SFE of IC 15 and LRO 37 may alter the wear mechanism and improve the wear resistance.

## CONCLUSIONS

Sliding wear tests conducted on SiC abrasive paper indicated that AISI 52100 steel possessed superior wear resistance. This resistance was associated with high microhardness and hard second phase particles.

In conventional pin-on-disk experiments, LRO 34 and 52100 steel demonstrated high wear resistance. A wear mechanism has been proposed for LRO 34 which is based on inhomogeneous slip and fracture along crystallographic planes. In 52100 steel, metal-

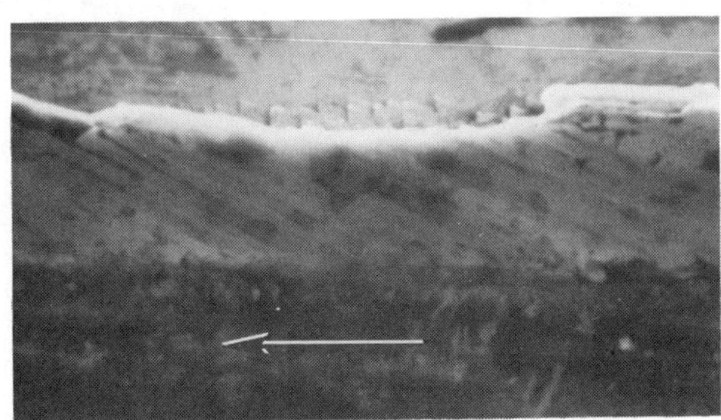

Figure 4 Near the edge of the wear scar, some inhomogeneous slip may be found in IC 15. Arrow indicates sliding direction.

metal contact results in an abrasive wear environment due to the release of carbide particles.

Adhesive wear in austenitic stainless steel is complex in nature and results in highly irregular surface structures.

LRO 37 and IC 15 show relatively poor wear resistance. Alloying to obtain lower SFE's may alter the wear mechanism in these materials and lead to improved wear resistance.

## ACKNOWLEDGEMENTS

This research was sponsored by the Office of Energy Systems Research, Division of Energy Conversion and Utilization Technologies (ECUT), U.S. Department of Energy under contract RC-5329 with the Union Carbide Corporation.

## REFERENCES

1. N.S. Stoloff and R.G. Davies, Prog. Mat. Sci. 13, p.1 (1966).
2. H. Warlimont, ed., Order-Disorder Transformation in Alloys (Springer-Verlag) New York, (1974).
3. J.H. Westbrook, Met. Trans. Vol. 8A, p.1327 (1977).
4. W.A. Brainard and D.H. Buckley, NASA Tech. Memo. Vol. X-52607 (1969).
5. J.A. Bailey and M.E. Sikorski, Wear Vol. 14, p. 181 (1969).
6. R.N. Wright and D.E. Mikkola, Materials Science and Engr. Vol. 26, p. 263 (1976).
7. B.J. Marquardt, D.M. Baker and J.J. Wert, "Erosive Wear of Ductile Ordered Alloys", (in press) Proc. Inter. Conf. on Wear of Matls., April (1985).
8. C.T. Liu and H. Inouye, Met. Trans. Vol. 10A, p.1515 (1979).
9. C.T. Liu and N.S. Stoloff, ORNL Publication Vol. 6032, (1984).
10. C.T. Liu, C.L. White and C.C. Koch, Proc. Symposium High Temp. Matls. Chem., The Electrochemical Soc. Vol. 83-7, p. 32.
11. R.P. Messmer and C.L. Briant, Acta Metall. Vol. 30, p.457 (1982).
12. J. Larsen-Badse, Wear Vol. 11, p. 213 (1968).
13. N.P. Suh, Wear Vol. 55, p. 163 (1979).
14. E. Rabinowicz and A. Mutis, Wear Vol. 8, p. 381 (1965).
15. N.P. Suh, Wear Vol. 25, p. 111 (1973).
16. N.P. Suh, Wear Vol.44, p. 1 (1977).
17. D.A. Rigney and W.A. Glaeser, Wear Vol.46, p. 241 (1978).
18. J.O. Stiegler, in R.E. Reed-Hill (ed.), Phys. Met. Principles, Vol. 2, Van Nostrand, New York, p. 285, (1973).
19. R.W. Carpenter, Proc. 9th Intl. Cong. on Electron Microscopy, p. 588 (1978).
20. M.A. Moore in D.A. Rigney (ed.), Fundamentals of Friction and Wear of Materials ASM, Metals Park, OH (1981).
21. S.G. Caldwell and J.J. Wert, "Surface and Subsurface Behavior of Selected Al-Cu Alloys in Sliding Wear", Journal of Trib., (to be published), ASME-ASLE Lubr. Conf., San Diego, Oct. (1984).
22. A. Gysler, G. Lutjering and V. Gerold, Acta Met. Vol. 22, p. 901 (1974).
23. K.L. Hsu, T.M. Bailey and D.A. Rigney, Wear Vol. 60, p. 13 (1980).
24. R.K. Bhargava, J. Moteff and R.W. Swindeman, Metall. Trans. Vol. 7A, p. 879 (1976).

# EFFECTS OF $DO_3$ TRANSITIONS ON THE YIELD BEHAVIOR OF Fe-Al ALLOYS*

H. INOUYE
Metals and Ceramics Division, Oak Ridge National Laboratory, Oak Ridge, Tennessee 37831

## ABSTRACT

The effect of $DO_3$ transitions on the yield behavior of alloys near the $Fe_3Al$ composition was investigated by tensile tests. Above 400°C, $DO_3$ ($Fe_3Al$) ordered alloys exhibited an anomalous increase in yield strength with temperature with a peak value around 560 to 600°C which was not observed if the $DO_3$ ordered alloys were also aged before testing. This yield behavior is explained by the precipitation of $\alpha$ (disordered solid solution) and the formation of B2 (FeAl) via the sequential reactions: $DO_3 \rightarrow \alpha + DO_3 \rightarrow \alpha + B2$.

## INTRODUCTION

It has been determined that the yield stress, $\sigma_y$, of Fe-Al alloys near the $Fe_3Al$ composition increased with temperature above 400°C to a peak value just below $T_c$, the critical ordering temperature (550°C). Within this same temperature interval, the long-range order parameter, S, decreased from about 0.8 to 0. These alloy properties are shown together in Fig. 1 [1,2,3] as their variation with temperature has been the basis for explaining the yield behavior. The rise in $\sigma_y$ with temperature between about 400°C and $T_c$ has been attributed to superdislocation interactions with the ordered $Fe_3Al$ lattice while the decrease in $\sigma_y$ with temperature was attributed to unit dislocation interactions with the partially ordered lattice. The yield peak was thus explained in terms of the transition from superdislocation to unit dislocation interactions at a critical degree of order [2].

Subsequent to the above studies, the iron-rich end of the Fe-Al system was re-investigated by Oki et al. [4], Swann et al. [5], and Okamoto et al. [6] and their results confirmed the existence of three body-centered cubic phases: a disordered solid solution ($\alpha$), ordered FeAl (B2), ordered $Fe_3Al$ ($DO_3$), and the two phase regions, $\alpha+DO_3$ and $\alpha$+B2 with the phase boundaries shown in Fig. 2. Two versions of the phase diagram are proposed that are in agreement with respect to the shape of the various phase fields, but disagree on the location of the boundaries. Allen and Cahn [7,8] concluded from thermodynamic considerations and critical experiments that both versions of the phase diagram in Fig. 2 were correct; the diagram according to Oki being the metastable one while Okamoto's version was the true equilibrium diagram. This apparent discrepancy, according to Allen and Cahn, arose from the initial generation of large coherency stresses between $\alpha$ and the ordered phases that constricted the two-phase fields and depressed the miscibility gap, then upon relaxation of the stresses by long equilibration anneals, caused the two phase regions to expand to the boundaries found by Okamoto and Beck. The origin of the coherency stresses was not explained by Allen and Cahn but probably

*Research sponsored by the Office of Naval Research under Interagency Agreement DOE No. 40-1241-82, U.S. Department of Energy under Contract No. DE-AC05-84OR21400 with Martin Marietta Energy Systems, Inc.

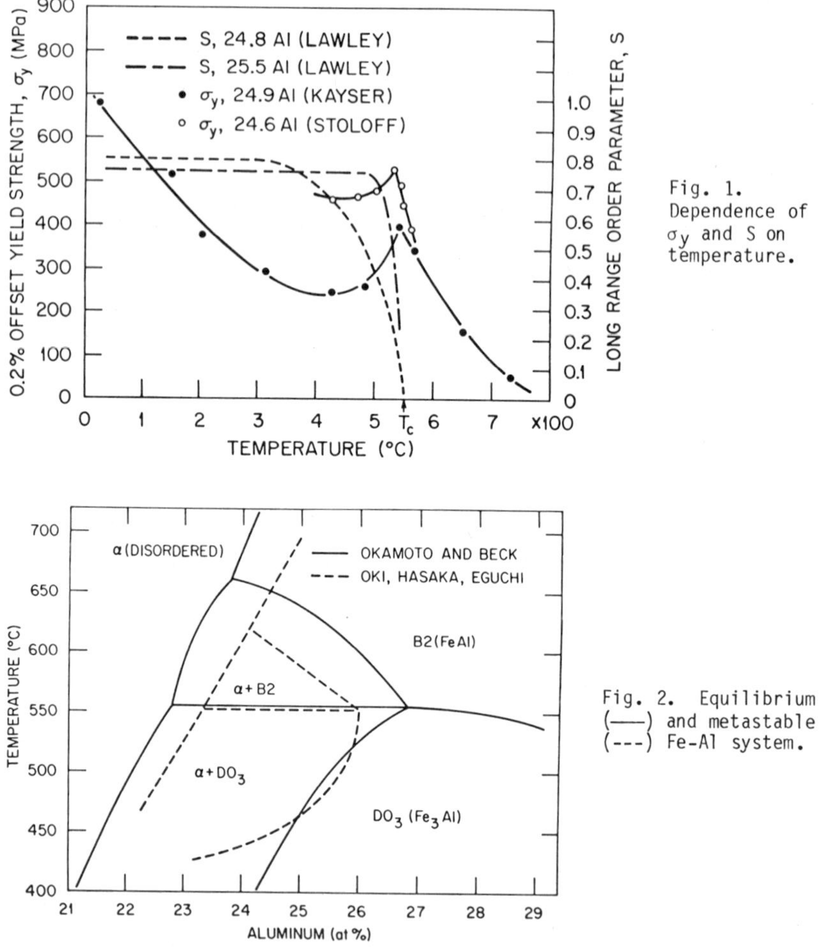


Fig. 1. Dependence of $\sigma_y$ and S on temperature.

Fig. 2. Equilibrium (——) and metastable (---) Fe-Al system.

arose from the differences in the lattice parameters of the phases since $a_0$ of $\alpha$>B2>$DO_3$ [6] and when these phases coexist they caused a considerable amount of line broadening of the X-ray diffraction peaks [4,6,9].

Thus the new versions of the Fe-Al diagram show that Stoloff's furnace cooled specimens were initially $DO_3$ [2] and Kayser's oil-quenched specimens were B2* [1] that could have age hardened when these ordered phases transformed on heating to $\alpha+DO_3$ and $\alpha$+B2. This study shows that this strengthening mechanism could explain the anomalous yield behavior of alloys near the $Fe_3Al$ composition.

---

*Fe-25 Al alloys quenched from the $\alpha$ phase field formed small B2 domains whose formation was not suppressed by quench rates of $5\times10^4$ °C/s [5].

## EXPERIMENTAL

Alloys containing 24 Al and 26 Al (in at. %) were arc melted under 1.3 kPa argon and cast into water-cooled copper molds as 12.7×25.4×127 mm ingots. After a homogenizing anneal in vacuum for 300 min at 1000°C, the ingots were rolled to 0.76-mm-thick sheet starting with an initial rolling temperature of 900°C and finishing at 525°C. Tensile specimens with a gage section of 0.76×3.18×12.70 mm were punched from the sheet with blanking dies and tested under vacuum in an Instron testing machine at a strain rate of $3.3\times10^{-3}\,s^{-1}$ in the $DO_3$ ordered and $DO_3$ ordered plus aged conditions. $DO_3$ ordering was accomplished by a vacuum heat treatment consisting of a 60 min anneal at 850°C, furnace cooling in about 160 min to 400°C, and isothermally annealing for 1000 min at 400°C. The $DO_3$ plus aged condition consisted of isothermal heat treatments of the ordered specimens at 450, 500, and 540°C for times ranging from 5 to 28,800 min. Specimens aged for 60 min or less were done in the vacuum envelope of the Instron testing machine while those aged 500 min or more were done in evacuated quartz capsules, water quenched, and then held for an additional 5 min at the test temperature. Specimens were prepared for analytical electron microscopy (AEM) analysis by spark discharge machining 3 mm disks from 0.7 mm sheet, grinding to a thickness of 0.3 mm, and electropolishing in one part nitric acid to 4 parts methanol in a Struers Tenupol jet polishing unit at –28°C. Analytical electron microscopy was performed in a Philips EM400T equipped with a field emission gun and an EDAX 9100 energy dispersive spectroscopy (EDS) system.

## RESULTS AND DISCUSSION

The dependence of $\sigma_y$ of alloys with 24 Al and 26 Al on aging time at 500°C is compared in Fig. 3 and at 540°C in Fig. 4. In Fig. 3, for an aging temperature of 500°C, the 24 Al alloy had a $\sigma_y$ of about 200 MPa after a 5 min aging treatment then rapidly increased to a peak value near 400 MPa within 60 min. In contrast the 26 Al treatment alloy which also had an initial $\sigma_y$ of about 200 MPa after the 5 min aging decreased steadily with aging time. According to the Fe-Al diagram, $\alpha$ should have precipitated from $DO_3$ in the 24 Al alloy according to the $DO_3\rightarrow\alpha+DO_3$ transition but the alloy with 26 Al should have remained as $DO_3$. A comparable difference of about 200 MPa in $\sigma_y$ attributable to the aluminum content was also measured for alloys aged at 540°C (Fig. 4). The higher aging temperature caused the $\sigma_y$ of the 24 Al alloy to

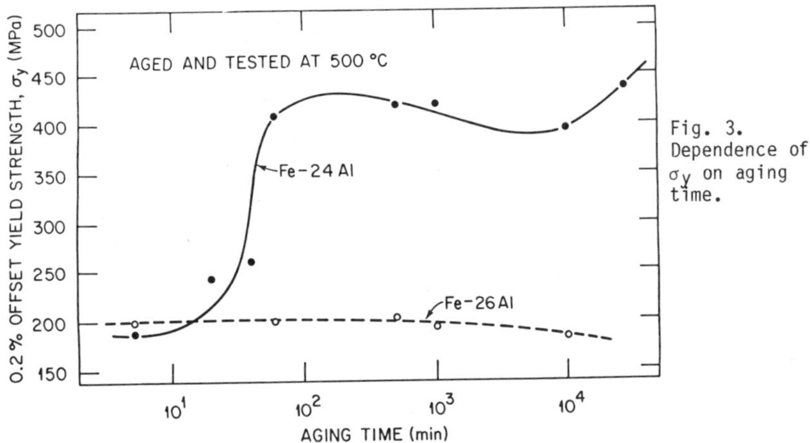


Fig. 3. Dependence of $\sigma_y$ on aging time.

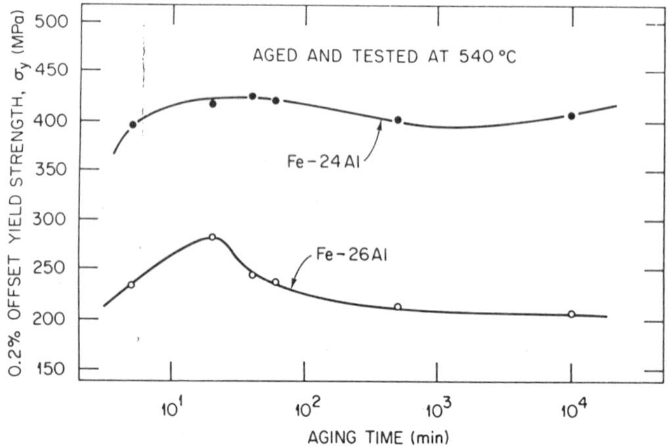


Fig. 4. Dependence of $\sigma_y$ on aging time.

reach its peak value within 5 min of aging while the 26 Al alloy showed a modest peak after 20 min of aging. The phase diagram indicates that the 24 Al alloy should be $\alpha$+$DO_3$ and the 26 Al alloy $\alpha$+$DO_3$ or $DO_3$ at 540°C.

Selected specimens were examined by TEM to confirm the presence of the expected phases and investigate the precipitation process of $\alpha$ from $DO_3$. Figure 5(a) and (b) are micrographs of the 24 Al alloy which was furnace cooled from 850°C then isothermally ordered for 1000 min at 400°C. This specimen which was expected to be single phase $DO_3$ appeared instead to be $\alpha$+$DO_3$. A similar difficulty in identifying the phases present at 400°C was described by Oki [4]. To resolve this discrepancy, a detailed study of a 24.1 Al alloy by Yamanaka showed that the furnace-cooled alloy formed large $DO_3$ domains surrounded by very small particles that could be interpreted as disordered $\alpha$ but were actually aggregates of microdomains having a $DO_3$ structure [10]. Only minor changes in the initial microstructures were apparent in the 24 Al alloy aged for 40 and 60 min at 500°C. Here again we depend on the work of Oki [9] who found that the microdomains of $DO_3$ disappeared within 10 min at 510°C and were replaced by numerous fine particles of $\alpha$ at the edges of the larger $DO_3$ domains. Figure 5(c) and (d) are micrographs showing the $\alpha$+$DO_3$ structure produced when the 24 Al alloy was aged for 500 min at 500°C. A TEM of the 26 Al alloy aged at 540°C was found to be single phase $DO_3$.

The kinetics of the $DO_3 \rightarrow \alpha + DO_3$ transition, measured in terms of $\sigma_y$, was strongly temperature dependent as indicated by the aging curves for the 24 Al alloy in Fig. 6. The broad maxima in $\sigma_y$ observed at the intermediate aging times confirm a prior observation that $DO_3$ ordered alloys with large and small domains transform to their equilibrium microstructure in two steps [9]. Figure 6 also predicts that the $\sigma_y$ of the alloy having an initial $DO_3$ order would increase with temperature simply because the age-hardening rate increased with temperature and its final yield value for a given temperature would depend on the holding time prior to the strength measurement. The results further indicate that if the 24 Al alloy was pre-aged so that it was in its age-hardened state initially, $\sigma_y$ would not be expected to increase with temperature. In accordance with these expectations, $\sigma_y$ of the $DO_3$ ordered 24 Al alloy increased with temperature above 400°C to a peak value around 560°C but when aged for 500 min at 500°C, $\sigma_y$ decreased with temperature similar to conventional disordered alloys, except for a small discontinuity (described below) at 550°C (Fig. 7). For the same reasons, the 26 Al alloy was expected to show only small differences in $\sigma y$ between the ordered and ordered plus aged conditions

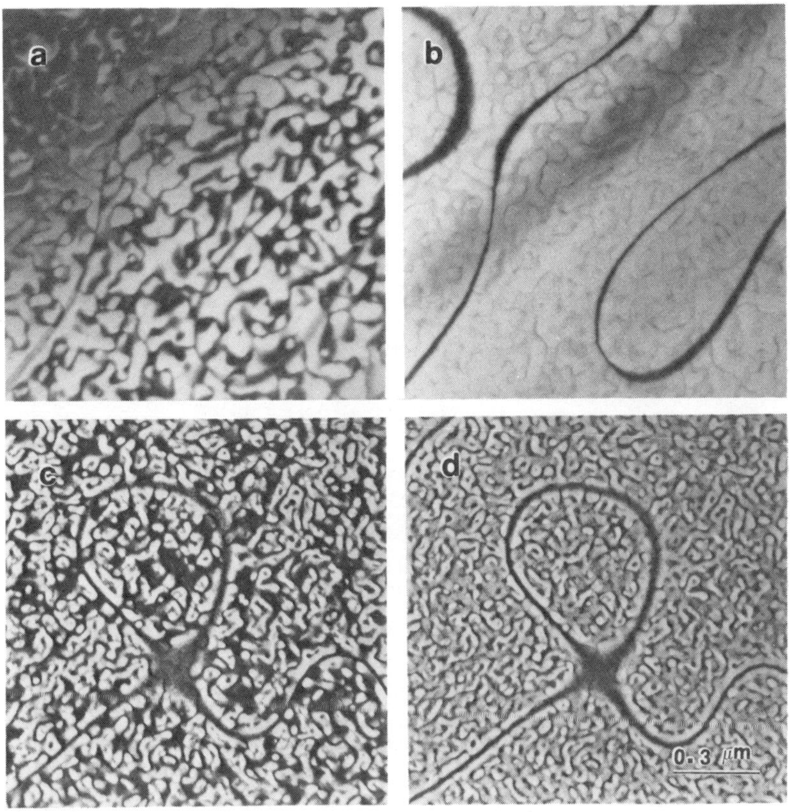

Fig. 5. Domain structure of 24 Al alloy: $DO_3$ ordered for 1000 min at 400°C (a) [111] reflection (b) [222] reflection; $DO_3$ ordered for 1000 mm at 400°C and aged 500 min at 500°C (c) [111] reflection (d) [222] reflection. Dark field micrographs.

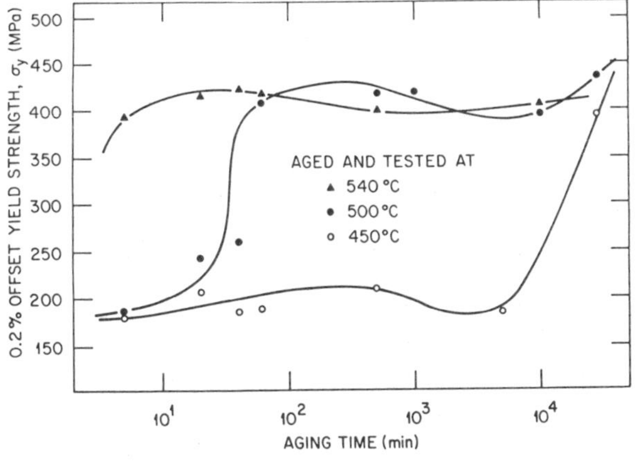


Fig. 6 Dependence of $\sigma_y$ of Fe-24 Al on heat treatment.

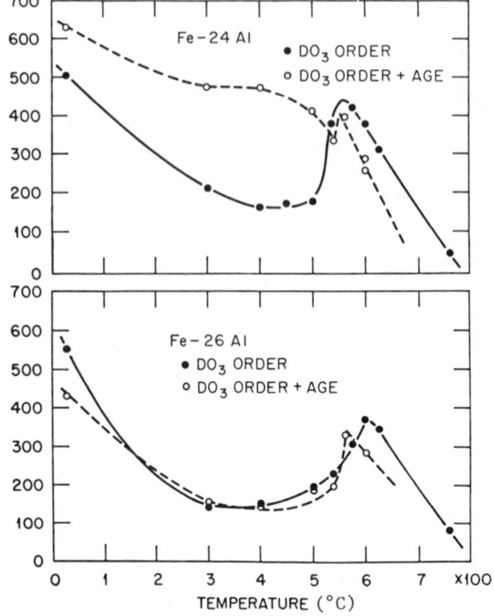


Fig. 7. Effect of 500°C aging heat treatment on $\sigma_y$.

because there was no aging response at 500°C (Fig. 3) and only a small one at 540°C (Fig. 4).

As a final assessment of the anomalous yield behavior, the cause of the $\sigma_y$ peaks require explanation. In the $DO_3$ ordered condition the 24 Al alloy showed a $\sigma_y$ peak near 560°C and the 26 Al alloy gave a peak near 600°C. These data suggest that the $\sigma_y$ peaks were due to $\alpha$ precipitation via the $DO_3 \rightarrow \alpha + B2$ transition in the 24 Al alloy and $DO_3 \rightarrow \alpha + B2$ or $DO_3 \rightarrow B2$ transitions in the 26 Al alloy (see Fig. 2). In the $DO_3$ ordered plus aged condition an abrupt $\sigma_y$ increase was observed at the $\alpha + DO_3/\alpha + B2$ phase boundary (550°C) that was more pronounced in the higher aluminum content alloy (Fig. 7). Because the amount of B2 in the 26 Al alloy is greater than the 24 Al alloy, the $\sigma_y$ discontinuity is believed to be due to the transition of $DO_3$ to B2 via the $\alpha + DO_3 \rightarrow \alpha + B2$ transition. The complete results therefore suggest that the $\sigma_y$ peak for $DO_3$ ordered alloys was caused by age hardening due to $\alpha$ precipitation as well as B2 formation from $DO_3$ and the rapid decrease in $\sigma_y$ above the peak was due to overaging of the $\alpha$+B2 structure and B2 domain growth.

## CONCLUSIONS

Alloys near the $Fe_3Al$ composition were age hardenable above 400°C due to the precipitation of $\alpha$ from the ordered $DO_3$ phase via the $DO_3 \rightarrow \alpha + DO_3$ transition. Consequently, the yield strength of $DO_3$ ordered alloys increased with temperature because kinetics of the age-hardening reaction also increased with temperature. The yield peaks in $DO_3$ ordered alloys occurred at 560°C for the 24 Al alloy and at 600°C for the 26 Al alloy and were caused by the combined effects of $\alpha$ precipitation and B2 formation via the $DO_3 \rightarrow \alpha + B2$ and $DO_3 \rightarrow B2$ transitions.

## ACKNOWLEDGMENTS

Funding for this study was provided by the Office of Naval Research (ONR), with D. E. Polk as technical monitor, and is appreciated. The author acknowledges that C. C. Koch of North Carolina State University and C. T. Liu of Oak Ridge National Laboratory (ORNL) originated the plan to investigate the effects of $Fe_3Al$ stoichiometry on mechanical properties. Thanks are extended to J. A. Horton of the Electron Microscopy Group at ORNL for preparing the micrographs on short notice. Appreciation is acknowledged to C. L. Dowker for preparing the manuscript.

## REFERENCES

[1]. F. X. Kayser, "Iron-Aluminum Systems, Part I, Fundamental Studies and Alloy Development" WADC Technical Report 57-298 (May 1957).

[2]. N. S. Stoloff and R. G. Davies, Acta Metall. **12**, 473 (May 1964).

[3]. A. Lawley, E. A. Vidoz, R. W. Cahn, Acta Metall. **9**, 287 (April 1961).

[4]. K. Oki, M. Hasaka, T. Eguchi, Jap. J. Appl. Phys. **12(10)**, 1522 (October 1973).

[5]. P. W. Swann, W. R. Duff, and R. M. Fisher, Trans. AIME **245**, 851 (April 1964).

[6]. H. Okamoto and P. A. Beck, Metall. Trans. **2**, 569 (February 1971).

[7]. S. A. Allen and J. W. Cahn, Acta Metall. **23**, 1017 (September 1975).

[8]. S. A. Allen and J. W. Cahn, Acta Metall. **24**, 425 (1976).

[9]. K. Oki, H. Sagane, T. Eguchi, Jap. J. Appl. Phys. **13(5)**, 753 (May 1974).

[10]. K. Yamanaka, H. Ino, F. Fugita, J. Jap. Inst. of Metals **35**, 566 (1971).

# CHARACTERISTIC MECHANICAL PROPERTIES AND PHASE STABILITY OF $L1_2$ INTERMETALLIC COMPOUNDS

YOSHINAO MISHIMA, YOSHIHIRO OYA AND TOMOO SUZUKI
Research Laboratory of Precision Machinery and Electronics,
Tokyo Institute of Technology
Nagatsuta, Midori-ku, Yokohama 227, Japan

## ABSTRACT

The present paper deals with the occurence and magnitude of characteristic mechanical anomalies observed in many $L1_2$ intermetallic compounds and their relation to the phase stability of the crystal structure against other geometrically closed packed(GCP) phases. The dislocation configurations in the crystal structure and resultant plastic behavior are in close relation with anti-phase boundary(APB) and stacking fault(SF) energies in the compounds, while the relative magnitude of these energies on various crystallographic planes actually determines the phase stability of the $L1_2$ crystal structure. It is shown here, based on our series of investigations carried out over the past decade, that the general tendencies for particular deformation modes to occur can be systematically interpreted in terms of alloy chemistry of the compounds, including variables such as combination of component species, nonstoichiometry and ternary additions in the compounds, all of which affect the stability of the phase.

## INTRODUCTION

A number of $L1_2$ intermetallic compounds are known to exhibit an anomalous temperature dependence of the yield strength. Actually there are two types of behavior observed in these compounds; 1) the well-known positive temperature dependence of strength as typically observed in $Ni_3Al$(anomalous behavior), and 2) the substantial increase in strength with decreasing temperature occuring below ambient temperature, which is similar to that normally encountered in bcc metals, for example as is observed in $Pt_3Al$. The latter mechanical behavior has become of interest only quite recently, while previous research on the mechanical behavior of $L1_2$ compounds has mainly been concerned with the positive temperature dependence of the strength, known now for several decades. Most of these efforts have been attempted to elucidate the mechanism which gives rise to the mechanical anomaly. The temperature dependence of the flow stress of single crystalline $L1_2$ compounds of various crystallographic orientations subjected to tensile or compressive deformation has been examined and mechanisms involving various types of super-dislocations with different APB and SF energies have been proposed to explain the observed plastic behaviors. Among them the Kear-Wilsdorf mechanism[1,2] most successfully interprets the anomalous positive temperature dependence of the strength. In this mechanism thermally activated cross slip of screw dislocations from {111} principal glide planes to {100} non-glide planes are assumed to be the cause of the strength increase. It is a fact, however, that the magnitude of the anomalous mechanical behavior is different in different compounds and furthermore, there are many $L1_2$ compounds that exhibit only the ordinary negative temperature dependence of strength as seen in fcc metals.

The present authors have attempted to provide a comprehensive understanding of the occurence and the magnitude of such mechanical anomalies in terms of the phase stability of the $L1_2$ phase relative to other GCP phases. By phase stability we mean the relative tendency for that particular intermetallic phase of specific crystal structure to appear in an alloy system as the composition of the alloy or the combination of the components is varied. It is closely related with the physical and electro-chemical nature of the species constituting the intermetallic phases, and thus is a matter of alloy chemistry. We have carried out systematic investigations of the effect of non-stoichiometry and the effect of ternary substitutions in binary $L1_2$ compounds, and the analogy we have drawn is, we believe, one of few examples of a successfull correlation between characteristic plastic behavior and alloy chemistry.

## MECHANICAL ANOMALIES IN $L1_2$ INTERMETALLIC COMPOUNDS

Compressive flow stress-temperature curves showing the temperature dependence of the flow stress of several polycrystalline $L1_2$ intermetallic compounds of stoichiometric $A_3B$ composition are shown in Figs. 1 and 2. In Fig. 1 the positive temperature dependence of the strength in the intermediate test temperature range is shown for several Ni base compounds and $Co_3Ti$, while in Fig. 2, the increase in flow stress with decreasing temperature is shown in several Pt-based compounds, some of which also show a slight positive temperature dependence at intermediate temperature. As to the mechanism which gives rise to these mechanical anomalies, the Kear-Wilsdorf mechanism[1,2] is widely accepted to explain the positive temperature dependence of strength shown in Fig. 1, while for the behavior shown in Fig. 2, a nonplanar dislocation core structure which makes the dislocations increasingly sessile as the temperature is decreased has been proposed by Pope and his co-workers[3-6].

According to the Kear-Wilsdorf mechanism the drag stress increases with increasing temperature due to the formation of sessile segments on the

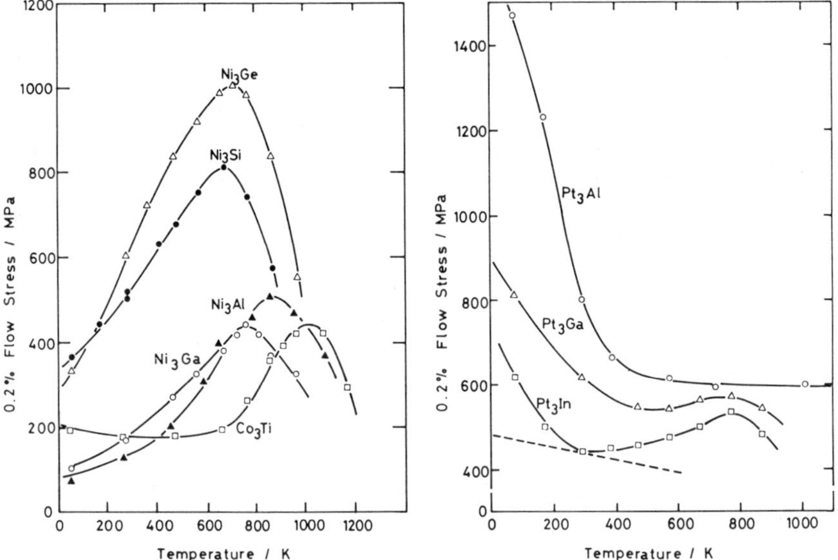


Fig. 1 Temperature dependence of the flow stress for several $L1_2$ compounds exhibiting the anomalous positive temperature dependence of strength[10].

Fig. 2 The unusual negative temperature dependence of flow strength observed in Pt-based $L1_2$ compounds[10].

mobile screw dislocations by thermally activated cross slip from {111} to {100} planes, because in the $L1_2$ crystal structure the normal slip plane and the lowest APB energy plane are not identical. Flinn[7] has shown that the {100} APB energy is the lowest in the $L1_2$ crystal structure. The difference between the {111} and the {100} APB energy is assumed to be the driving force for the cross slip in the Kear-Wilsdorf mechanism. But if the situation is so simple, the magnitude of the positive temperature dependence of strength would be greater for compounds with higher {111} APB energy. It has been shown, however, that a high {111} APB energy is not necessarily responsible for a more pronounced anomalous mechanical behavior because some of the Daltonide- or Berthoride-type compounds[8], in which the ordered structure is stable up to the melting point, show no such anomalous mechanical behavior[9,10].

The discussion hereafter mainly considers the relation between the positive temperature dependence of strength in $L1_2$ compounds and the phase stability of the crystal structure relative to other GCP phases. Thermally activated cross slip must occur in close relation to the anisotropy in APB energy and the resulting configurations of dissociated super-dislocations must be determined essentially by the relative magnitude of APB and SF energies[11,15]. The stability of the $L1_2$ crystal structure relative to other GCP phases is also controlled by these same energies, and they also control the ordered structure and stacking sequence. In a later section, we also discuss the strong negative temperature dependence of strength shown in Fig. 2 in terms of the phase stability of $L1_2$ phase against the $DO_c$($U_3Si$ type) phase.

## PHASE STABILITY OF THE $L1_2$ STRUCTURE AGAINST OTHER GCP PHASES

As stated above, the SF and APB energies on {111} planes are thought to be important factors determining the slip mode and strength of an $L1_2$ compound. These factors obviously affect the phase stability of the $L1_2$ compound with respect to the ordered hcp phase($DO_{19}$) and to the disordered fcc phase (A1). Regarding thermally activated cross slip in $L1_2$ compounds, the value of the APB energy on {100} planes is important, which is closely related to the phase stability of the $L1_2$ phase against the $DO_{22}$, $DO_{23}$ or a long period superlattice phase. These structures are composed of unit-cells of the $L1_2$ structure but have regulary spaced antiphase boundaries on {100} planes spaced at a definite multiple of the unit-cell length.

A family of ordered closed-packed structures is known to form by various types of layers of composition $A_3B$ in various stacking sequences[16]. Several notation systems have been proposed to describe the crystal structures of GCP phases[16,17]. In the following discussion, we shall use the symbols T and R to indicate triangular and rectangular ordering of the atomic layers on close packed planes, as seen in the $L1_2$ and $DO_{22}$ structure, respectively, and the symbols c and h to denote cubic and hexagonal stacking sequences. The relative phase stabilities as influenced by the factors discussed above, of different GCP phases are shown schematically in Fig. 3.

In attempt to understand the relative phase stability of various types of $A_3B$ compounds, two factors have been taken into consideration; 1) the electron concentration criterion[18-21] and 2) the atomic size difference criterion[22-24]. Sinha[18-19] has noted that an increase in hexagonality of the $A_3B$ structure is associated with an increase in electron atom ratio(e/a), defined as the average number of electrons outside the filled shells per atom, whereas, van Vucht pointed out that an increase in hexagonality is associated with an increase in atomic radius ratio of the components, $R_B/R_A$[23,24]. These criteria predict that the crystal structure tends to shift not only in the direction of c → h stacking but also in the direction of T → R layering by an increase in either factor. It should be noted that the crystal structure of a particular phase can be expressed by such notations as cT for $L1_2$, chT for $DO_{24}$, cR for $DO_{22}$ and cTR for $DO_{23}$.

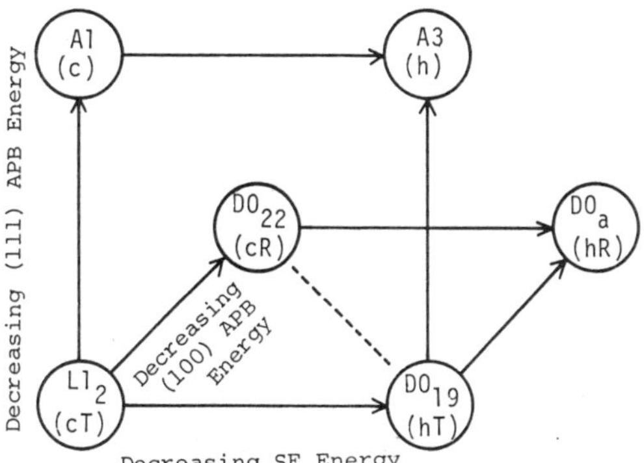


Fig. 3 The relation between phase stability of the $L1_2$ against other GCP phases as influenced by APB and SF energies.

Table 1 summarizes crystal structures of $A_3B$ GCP phases of transition metals compiled from the literature for sixty possible A-B combinations, where A is a transition metal of Group 8B or 8C and B is a transition metal of Group 4, 5, 6 or 7. This table first appeared in ref.[9] but is updated here. Although it is rather busy, the table shows the variation in e/a ratio in the top row and the deviation of the atomic radius ratio from unity for each B element(see the numbers below the B elements), both of which are determined by the combination of A and B elements in the $A_3B$ compounds. The use of Table 1 is based on the premise that structural transitions occur in the sequence cT → hT → cR → hR. When the compounds of 8C metals and the compounds of 8B metals are superimposed, as is done in the table, apparent conflicts with the electron concentration criterion can be seen, as has been pointed out by many workers[18-21]. However, it is observed that the size criterion holds within a group of compounds of a given minority component(B) with the mojority component(A) of the same group in the periodic table. Numerals in the second row from the top indicate $(R_B-R_A)/R_A$, where Goldschmidt's atomic radius for a co-ordination number 12 is employed for the radius of the components. For instance, the hexagonality increases on changing A from Ir to Rh and to Co in an $A_3B$ compound in which B is Nb or Ta. On the other hand, the size effect is not so great in compounds of a given majority component with a minority component of the same group. For instance, the chT structure prevails in $Pt_3Ti$, $Pd_3Ti$ and $Ni_3Ti$. It is to be noted that although the hT structure seems to be absent in a series of compounds of 8C metals, the occurence of chhT or hT structures has been confirmed in ternary compounds by many workers[19-21,23-28]. Some of these results on ternary compounds are indicated by diagonal lines on the diagram, showing pseudo-binary alloys. Along some of these lines transitions from cchT to chhT to hT sequences are seen. Also in this table the $L1_2$ compounds(cT) which exhibit a positive temperature dependence of strength are marked so that the position

Table I Closed packed $A_3B$ structures for alloys of transition metals.

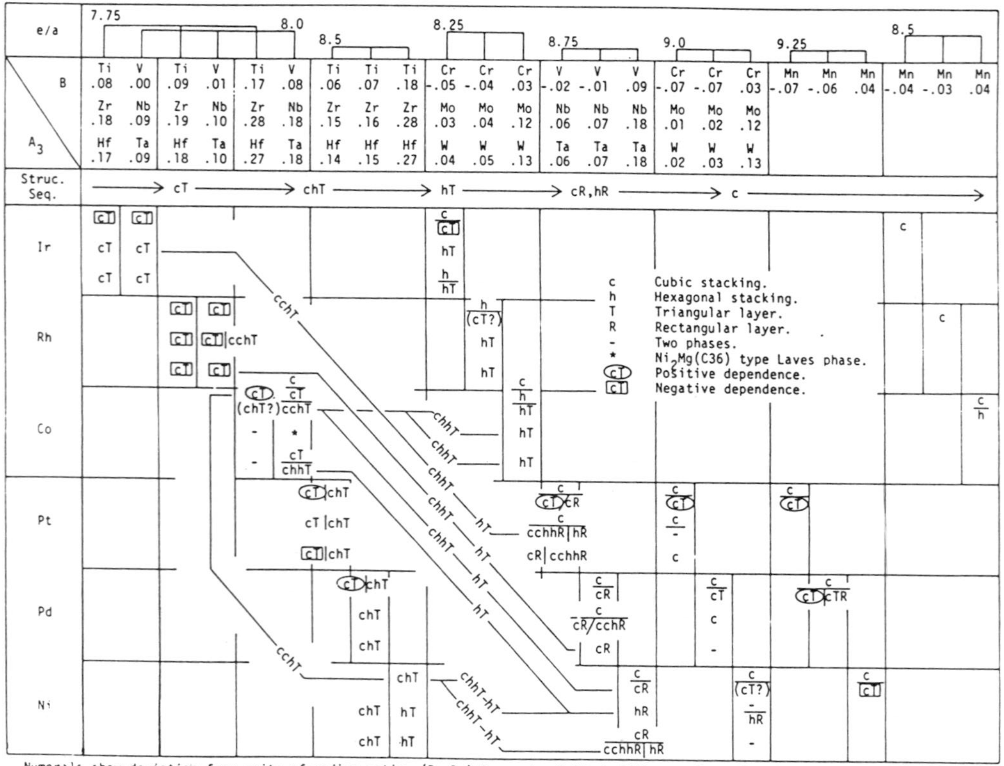

| e/a | 7.75 | | | | | 8.0 | 8.5 | | | 8.25 | | | 8.75 | | | 9.0 | | | 9.25 | | | 8.5 | | |
|---|---|---|---|---|---|---|---|---|---|---|---|---|---|---|---|---|---|---|---|---|---|---|---|---|
| B / $A_3$ | Ti .08<br>Zr .18<br>Hf .17 | V .00<br>Nb .09<br>Ta .09 | Ti .09<br>Zr .19<br>Hf .18 | V .01<br>Nb .10<br>Ta .10 | Ti .17<br>Zr .28<br>Hf .27 | V .08<br>Nb .18<br>Ta .18 | Ti .06<br>Zr .15<br>Hf .14 | Ti .07<br>Zr .16<br>Hf .15 | Ti .18<br>Zr .28<br>Hf .27 | Cr -.05<br>Mo .03<br>W .04 | Cr -.04<br>Mo .04<br>W .05 | Cr .03<br>Mo .12<br>W .13 | V -.02<br>Nb .06<br>Ta .06 | V -.01<br>Nb .07<br>Ta .07 | V .09<br>Nb .18<br>Ta .18 | Cr -.07<br>Mo .01<br>W .02 | Cr -.07<br>Mo .02<br>W .03 | Cr .03<br>Mo .12<br>W .13 | Mn -.07 | Mn -.06 | Mn .04 | Mn -.04 | Mn -.03 | Mn .04 |
| Struc. Seq. | → cT | | | | → chT | | | | → hT | | | | → cR,hR | | | → c | | | | | | | | → |
| Ir | [cT]<br>cT<br>cT | [cT]<br>cT<br>cT | | | | | | | | c / [cT]<br>hT<br>h / hT | | | | | | | | | | | | c | | |
| Rh | | | [cT]<br>[cT]<br>[cT] | [cT]<br>[cT]\|cchT<br>[cT] | | cchT | | | | | h / (cT?)<br>hT<br>hT | | | | | | | | | | | | c | |
| Co | | | | | (cT)<br>(chT?)<br>-<br>- | c / cT / cchT<br>*<br>cT / chhT | | | | chhT<br>chhT | | c / h / hT<br>hT<br>hT | | | | | | | | | | | | c / h |
| Pt | | | | | | | (cT)\|chT<br>cT\|chT<br>[cT]\|chT | | | chhT<br>chhT<br>chhT | hT<br>hT | | c / (cT)cR<br>c / cchhR\|hR<br>cR\|cchhR | | | c / (cT)<br>c / -<br>c | | | c / (cT) | | | | | |
| Pd | | | | | | cchT | | (cT)chT<br>chT<br>chT | | | hT<br>hT | | | c / cR<br>c / cR/cchR<br>cR | | | c / cT<br>c<br>- | | c / (cT)cTR | | | | | |
| Ni | | | | | | | | chT<br>chT | chT<br>hT<br>hT | | chhT-hT<br>chhT-hT | | | | c / cR<br>hR<br>cR / cchhR\|hR | | | c / (cT?)<br>- / hR<br>- | | | c / [cT] | | | |

- c Cubic stacking.
- h Hexagonal stacking.
- T Triangular layer.
- R Rectangular layer.
- \- Two phases.
- * $Ni_2Mg$(C36) type Laves phase.
- (cT) (circled) Positive dependence.
- [cT] (boxed) Negative dependence.

Numerals show deviation from unity of radius ratio, $(R_B-R_A)/R_A$, using Goldschmidt's atomic radius for co-ordination number 12.

of those compounds can be seen relative to other GCP phases. It now becomes evident that $L1_2$ compounds exhibiting the anomalous mechanical behavior are mostly on the boundary of $L1_2$ phase stability. Compounds showing a normal mechanical behavior are isolated in stable regions of the diagram surrounded by other structures, due to unknown incidental effects. For example, the $L1_2$ compounds of iridium with Ti group and V group metals are very stable and do not show any anomalous behavior. It is interesting to note that a strong positive or negative temperature dependence of the strength appears most frequently in compounds based on platinum and in those with titanium. Those $L1_2$(cT) compounds which show the anomalous positive temperature dependence of strength are near regions on the diagram where the $DO_{22}$(cR) phase forms. In these regions a large difference in APB energy on the {111} and {100} planes is expected and thereby provides a large driving force for the Kear-Wilsdorf mechanism. We would like to add that the stability condition for the T and R layers has been given in terms of the ratio of first-neighbor($V_1$) and second-neighbor ($V_2$) bond energies[29-31], where if $V_2/V_1$ is positive the structure $L1_2$(T layer) is preferred while if it is negative, the $DO_{22}$(R layer) is preferred.

## PHASE STABILITY AND THE POSITIVE TEMPERATURE DEPENDENCE OF STRENGTH

In order to describe the magnitude of the positive temperature dependence of strength in $L1_2$ compounds, a method to determine the activation constant for the thermally activated process is first briefly reviewed. The flow stress-temperature curve as shown in Fig. 1 should be divided into two temperature-dependent components. One is the ordinary negative temperature dependence of strength arising from the modulus change with temperature and the other is the positive temperature dependence of strength caused by thermally activated cross slip. Then the flow stress can be written[10,14,32] in the form,

$$\sigma = \sigma_0(1 - BT) + A\exp(-U/RT)$$

where, $\sigma_0$ is the flow stress at 0 K, U is the activation constant, R is the gas constant, T is temperature, and A and B are constants. A bottom temperature $T_b$ is defined by the condition $d\sigma/dT = 0$. These are schematically depicted in Fig. 4. The activation energy, U, can be deduced by an Arrhenius plot of the flow stress after subtracting the term $\sigma_0(1 - BT)$. The smaller is the activation constant, the more extensive is the positive temperature dependence of the strength. In cases where the temperature range from the bottom to the peak in flow stress is too narrow to estimate U, the bottom temperature $T_b$ is a good index to evaluate the activation constant. It has been shown that $T_b$ gradually shifts to higher temperatures at a constant stress level as

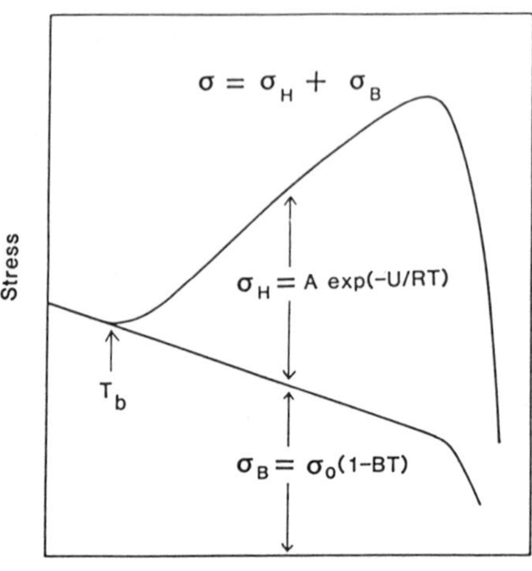


Fig. 4 A schematic representation of flow stress-temperature curve showing anomalous positive temperature dependence of strength.

the activation constant increases[10].

Figure 5 summarizes the relationship between the activation constant and the deviation from unity in atomic radius ratio, $(R_B - R_A)/R_A$, incorporating data for binary nickel-based and platinum-based $A_3B$ alloys having the $L1_2$ crystal structure and for pseudo-binary(ternary) $L1_2$ compounds[10,33]. In the case of pseudo-binary alloys, for example $Pt_3(In, Ga)$, the weighted mean of atomic radius of B-subgroup elements, $R_B$, is derived from the lattice spacing of the alloys, assuming that atomic radius of platinum, $R_A$, is regarded as that of pure fcc metal. In the figure, the reciprocal activation constant is used so as to make the value $U = \infty$ visible. An infinite value of the activation constant means that the $L1_2$ alloy has the ordinary, negative temperature dependence of strength, while a zero activation constant means that the $DO_{22}$ crystal structure is the stable phase. It is successfully shown here that there is a general trend in the relationship between the activation constant, the atomic radius ratio, and the electron concentration. The values of the activation constants of the alloys containing elements of the same B-subgroup pass smoothly through a minimum at about $(R_B - R_A)/R_A = 0.05$. Both curves are quite similar in nature and the increase in the activation constant with increase in radius ratio after the minimum is very drastic. Noting that a small activation constant implies instability relative to the $DO_{22}$ structure, 4B-subgroup elements make the $L1_2$ phase less stable with respect to the $DO_{22}$ phase than do 3B-subgroup elements.

## POSITIVE TEMPERATURE DEPENDENCE OF STRENGTH IN TERNARY $L1_2$ COMPOUNDS

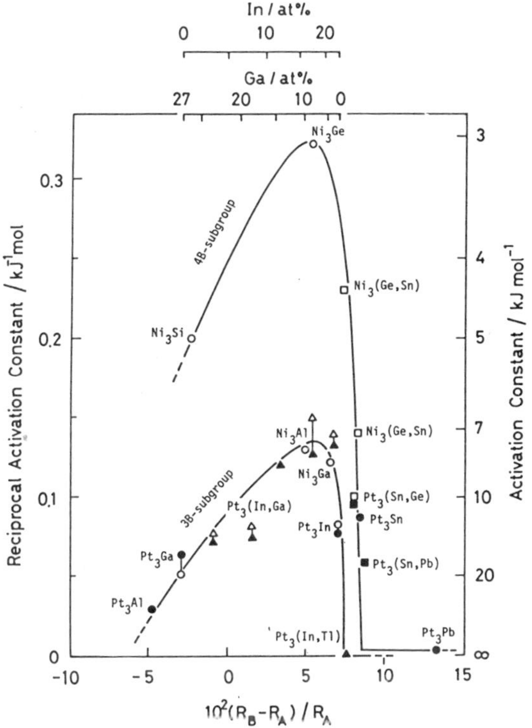


Fig. 5 Activation constant against deviation from unity of the atomic radius ratio for Pt-based and Ni-based $L1_2$ allys with B-subgroup elements. Open marks are the value derived from Arrhenius plot and solid marks from the $T_b$ method[33].

It has already been shown, in Fig. 5, that ternary additions to $L1_2$ compounds can remarkably alter the degree of its positive temperature dependence of strength. For example, it can be seen that only a 0.1 at% addition of Tl to $Pt_3In$ results in a dramatic increase in activation constant resulting in a sudden loss of the mechanical anomaly. In this case Tl additions to the compound would not change the e/a ratio but the atomic radius of Tl is larger than that of In. The present authors have shown elsewhere[32] that $Ni_3Ge$ and $Fe_3Ge$, both of which are $L1_2$ but only $Ni_3Ge$ shows the positive temperature dependence of strength, form continuous $(Ni, Fe)_3Ge$ solid solutions and that the pseudo-binary compound loses the mechanical anomaly at about 30 at%Ge.

Since the findings by Pope[34], $Cu_3Au$ has been recognized as one of the $L1_2$ compounds to exhibit the positive temperature dependence of strength although it is weak compared with that of $Ni_3Al$. This is reasonable to us in view of the presence of a long period superlattice structure nearby the $L1_2$

phase in the Cu-Au system, implying a relatively low phase stability of the $L1_2$ phase. It is well known that the morphology of anti-phase domain(APD) in $Cu_3Au$ is maze-like[35,36] with a strong preference for {100} planes, instead of swirl-like, as is often observed in compounds which do not show the mechanical anomaly. This fact tells us that the {100} APB energy is very low in $Cu_3Au$. We have carried out an investigation of the effect of Ni additions to the compound on the magnitude of the positive temperature dependence of strength. Ni additions are expected to decrease the e/a ratio of the compound if the valence of Ni is taken to be zero and are therefore expected to enhance the phase stability of the $L1_2$ structure. A comparison of the flow stress-temperature curves, normalized by the flow stress at 4 K and Young's modulus, is shown in Fig. 6 between $Cu_3Au$ and $Cu_3Au$-5 at%Ni. It is obvious that the magnitude of the positive temperature dependence of strength, being clearly judged by an increase in $T_b$, is reduced by the Ni addition and concurrently it was shown that the morphology of APD's is swirl-like in a $Cu_3Au$-5 at%Ni alloy[37].

Considerations of the effect of ternary additions on the positive temperature dependence of strength is especially important in $Ni_3Al$, which is one of the major microstructural constituents in Ni-base superalloys. From such a point of view, the present authors have systematically investigated the effect of ternary additions of B-subgroup elements as well as transition metal elements on the substitution behavior, solubility limit, lattice parameter, and mechanical properties of the compound[38-43]. Upon additions of B-subgroup elements, for each of which the valence is strictly defined, the magnitude of the mechanical anomaly can be interpreted in terms of the phase stability concept as affected by electron concentration and the atomic size ratio in

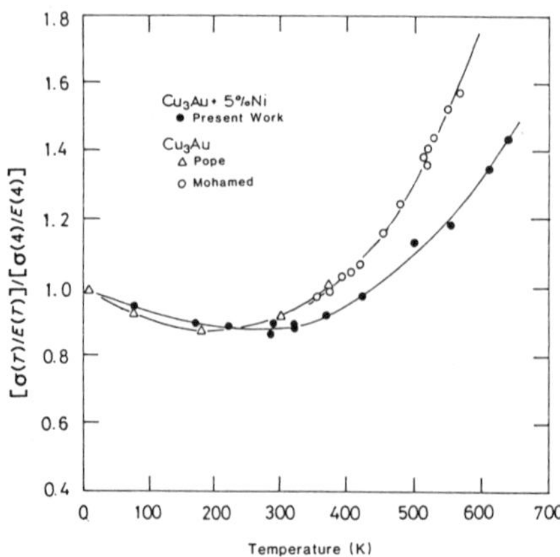


Fig. 6 Temperature dependence of $[\sigma(T)/E(T)]/[\sigma(4)/E(4)]$ in an ordered $Cu_3Au$-5 at%Ni and $Cu_3Au$ alloys[37].

the compounds[44]. For additions of transition metal elements, the atomic size criterion has generally been found to be most useful to interpret the relative magnitude of the mechanical anomaly. In this case the apparent valence of each element could be assigned by an analytical method in light of knowledge of the effect of B-subgroup elements additions[44].

## EFFECT OF DEVIATION FROM STOICHIOMETRY ON THE POSITIVE TEMPERATURE DEPENDENCE OF STRENGTH

Many $L1_2$ compounds are Kurnakov or Berthollide compounds that exist over a range of compositions on either side of stoichiometry. Figure 7 shows that the positive temperature dependence of strength is considerably enhanced by increasing the Al content beyond stoichiometry[45]. This has been shown previously[46-48]. Also as is shown in Fig. 8 the activation constant for the thermally activated cross slip process decreases as the amount of minority element increases and is discontinuous at the stoichiometric composition, not only in $Ni_3Al$ but also in $Ni_3Ga$[45,48].

According to our theory of the relation between the stability of the $L1_2$ phase against other GCP phases and the magnitude of the positive temperature dependence of strength, Fig. 8 implies that an increase in the minority element should reduce the stability of the $L1_2$ phase. Unfortunately, we can not find any evidence for this in the case of the results shown in Fig. 8, however, there are several cases where an increase in the minority element does give rise to a shift from the $L1_2$ to a long period superlattice structure and in which the period, M, of the structure is decreased with further increase in the minority element. Such instances can be found in the Cu-Au, Cu-Pt and Cu-Pd systems and there are more extreme cases, as in the Pt-Sb system, where the $L1_2$(cT) structure appears at $Pt_4Sb$ and the compound $Pt_3Sb$ has the $DO_{23}$(cTR) long period superlattice structure. Sato and Toth[49] have termed such an effect caused by an increase in the minority element(B) in $A_3B$ compounds a "concentration effect". It has also been reported that an increase in minority element often results in a change from the $L1_2$ structure to hexagonal stacking(h) to yield the $DO_{24}$(chT) structure, as is observed in the Pt-Ti, Pt-Zr, Pd-Ti and Rh-Nb systems[50-53]. In view of such evidence, it appears that an increase in the minority element can provide the same effect as increasing the e/a ratio, and thereby reduce the stability of $L1_2$ phases against such GCP phases as the $DO_{22}$ and $DO_{19}$.

It should be noted that $L1_2$ compounds such as $Ni_3Si$ and $Ni_3Ge$, which also exhibit a strong positive temperature dependence of strength, exist only in nonstoichiometric compositions on the Ni-rich side of stoichiometry. The relation between compositional stability range and the magnitude of the positive temperature dependence of strength has been discussed elsewhere in terms

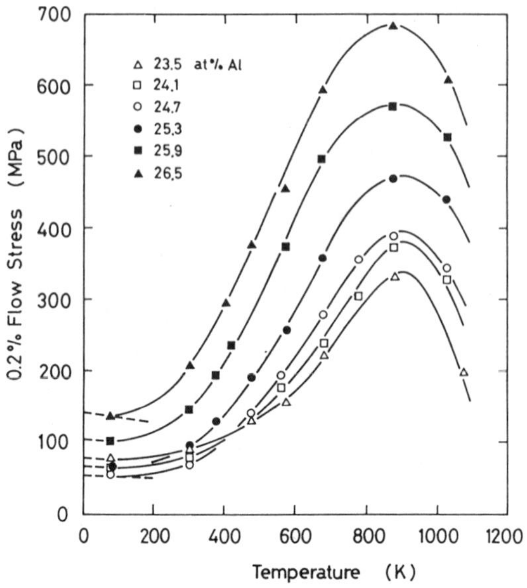


Fig. 7 Temperature dependence of 0.2% flow stress of non-stoichiometric $Ni_3Al$ alloys[45].

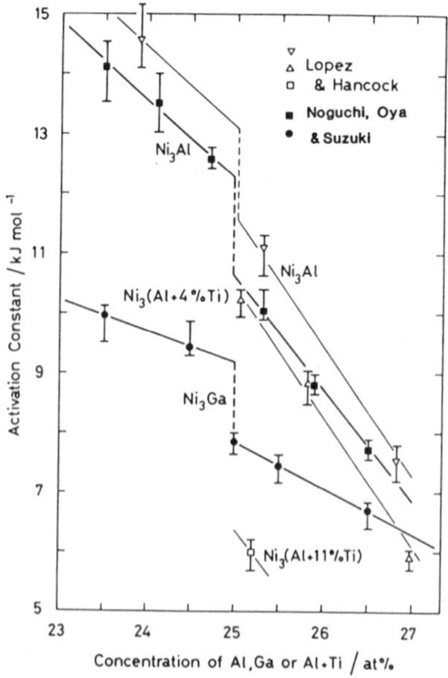


Fig. 8 Effect of deviations from stoichiometry on the activation constant in $Ni_3Al$ and $Ni_3Ga$ alloys[45,48].

of the shape of the free energy curve of the phase in the vicinity of stoichiometry and the bond strength as affected by the valence of the minority component from B-subgroup[54].

## THE LOW TEMPERATURE MECHANICAL BEHAVIOR AND PHASE STABILITY

As has been mentioned at the beginning of this paper, there is an other type of mechanical behavior observed in $L1_2$ compounds that has drawn considerable interest quite recently. A substantial increase in flow stress with decreasing temperature, generally below ambient temperature, is most often observed in Pt-based compounds. Such mechanical behavior is common in bcc metals and alloys but is unusual in $L1_2$ compounds based on the fcc structure. There have been a series of work by Pope and his co-workers in which atomistic studies on dislocation core structures in $L1_2$ ordered alloys and associated Peierls stresses are performed[3-5]. Recently Wee et al have conducted compression tests on single crystalline $Pt_3Al$ and found some supporting evidence for the deformation modes predicted by them[6]. It is proposed that the strength increase at low temperatures is caused by a sessile configuration of the dislocation core structure due to SISF(superlattice intrinsic stacking fault)-type dislocation dissociation on the {111} plane. The other type of dissociation possible on the {111} plane is the APB-type which is glissile and thus does not cause a strength increase at low temperatures.

Now that the behavior is found to occur in association with a particular dislocation configuration determined by the SF and APB energies in the crystal structure, it might be possible again to discuss the occurence and the magnitude of the behavior in terms of the phase stability of $L1_2$ structure against other GCP phases with various SF and APB energies. A series of investigations is now underway by the present authors on various Pt-based compounds on the effect of composition on the magnitude of the mechanical behavior. Preliminary results indicate that such compounds as $Pt_3Al$ and $Pt_3Ga$ exhibiting the low temperature strength increase exist at the stability boundary between $L1_2$ and the $U_3Si$-type $DO_c$ phase. Currently, attempts are being made to correlate the relative phase stability of the two phases in various systems with e/a ratio as well as atomic radius ratio aiming at analytically deducing the relative magnitude of SF and APB energies in the phases. It is then hoped to be able to provide a comprehensive interpretation of the low temperature mechanical behavior of $L1_2$ compounds which can be coupled with the work on dislocation core structure.

## CONCLUDING REMARKS

We have discussed the occurence of particular plastic deformation modes in $L1_2$ intermetallic compounds in terms of the phase stability of $L1_2$ relative to other GCP phases. Such factors as SF and APB energies, which control the configuration of dissociated dislocations in the ordered structure, are actually determined by phase stability as affected by the composition of the compounds and are thus determined by alloy chemistry. The phase stability is essentially governed by the physical and electrochemical nature of the species constituting the phase, and therefore to understand and further to predict the magnitude of the mechanical anomalies approachesare desired in which electronic theory and solid state physics are fully utilized.

References

[1] B.H.Kear and H.G.F.Wilsdorf: Trans. TMS-AIME, 224(1962), 382.
[2] S.Takeuchi and E.Kuramoto: Acta Met., 21(1973), 415.
[3] M.Yamaguchi, V.Vitek and D.P.Pope: Phil.Mag., A43(1981), 1027.
[4] M.Yamaguchi, V.Paider, D.P.Pope and V.Vitek: Phil.Mag., A45(1982), 867.
[5] V.Paider, D.P.Pope and V.Vitek: Phil.Mag., A45(1982), 883.
[6] D.M.Wee, D.P.Pope and V.Vitek: Acta Met., 32(1984), 829.
[7] P.A.Flinn: Trans. TMS-AIME, 218(1960), 145.
[8] I.I.Kornilov: Intermetallic Compounds, Ed. by J.H.Westbrook, John Wiley and Sons Inc., (1967), p.347.
[9] D.M.Wee and T.Suzuki: Trans.Japan Inst.Metals, 20(1979), 634.
[10] D.M.Wee, O.Noguchi, Y.Oya and T.Suzuki: ibid., 21(1980), 237.
[11] M.J.Marcinkowski, N.Brown and R.M.Fischer: Acta Met., 9(1961), 129.
[12] S.Takeuchi, E.Kuramoto, T.Yamamoto and T.Taoka: Japan J.Appl.Phys., 12 (1973), 1486.
[13] K.Suzuki, M.Ichihara and S.Takeuchi: Acta Met., 27(1979), 193.
[14] S.J.Liang and D.P.Pope: Acta Met., 25(1977), 485.
[15] B.H.Kear, J.M.Oblak and A.F.Giami: Met.Trans., 1(1970), 2477.
[16] H.J.Beattie,Jr.: Intermetallic Compounds, Ed. by J.H.Westbrook, John Wiley and Sons Inc., (1967), p.144.
[17] W.B.Pearson: The Crystal Chemistry and Physics of Metals and Alloys, John Wiley and Sons Inc.,New York, (1972), p.1.
[18] A.K.Sinha: Trans. TMS-AIME, 245(1969), 237.
[19] A.K.Sinha: ibid., 245(1969), 911.
[20] V.Sadagopan and B.C.Giessen: J.Metals, 17(1965), 1058.
[21] W.B.Pearson: The Crystal Chemistry and Physics of Metals and Alloys, John Wiley and Sons Inc., New York, (1972), p.87.
[22] A.E.Dwight and P.A.Beck: Trans. TMS-AIME, 215(1959), 976.
[23] J.H.N. van Vucht: J. Less-Common Met., 11(1966), 308.
[24] J.H.N. van Vucht and K.H.J.Buschow: ibid., 10(1965), 98.
[25] B.C.Giessen and N.J.Grant: Acta Crst., 17(1964), 615.
[26] Idem, ibid., 18(1965), 1080.
[27] B.G.Lewis and D.E.G.Williams: J. Less-Common Met., 44(1976), 337.
[28] H.U.Pfeifer, S.Bhan and K.Schbert: ibid., 14(1968), 754.
[29] P.C.Clapp and S.C.Moss: Phys. Review, 171(1968), 754.
[30] S.M.Allen and J.W.Cahn: Acta Met., 20(1972), 423.
[31] M.Kaburagi and J.Kanamori: Prog.Theor.Phys., 54(1975), 30.
[32] T.Suzuki, Y.Oya and D.M.Wee: Acta Met., 28(1980), 301.
[33] T.Suzuki and Y.Oya: J.Mater.Sci., 16(1982), 2737.
[34] D.P.Pope: Phil. Mag., 25(1972), 917.
[35] R.M.Fischer and M.J.Marcinkowski: J.Appl.Phys., 31(1960), 1687.
[36] D.G.Morris, G.T.Brown, R.C.Piller and R.E.Smallman: Acta Met., 24(1976), 21.
[37] A.Fujita, Y.Mishima and T.Suzuki: J.Mater.Sci., 18(1983), 1881.
[38] S.Ochiai, Y.Oya and T.Suzuki: Bull.P.M.E.(T.I.T.), 52(1983), 1.

[39] S.Ochiai, Y.Oya and T.Suzuki: Acta Met., 32(1984), 289.
[40] S.Ochiai and T.Suzuki: Bull.P.M.E.(T.I.T.), 52(1983), 19.
[41] S.Ochiai, Y.Doi and T.Suzuki: ibid., 52(1983), 31.
[42] S.Ochiai and T.Suzuki: ibid., 52(1983), 41.
[43] S.Ochiai, Y.Mishima and T.Suzuki: ibid., 53(1984), 15.
[44] S.Ochiai: Doctor's Thesis, Tokyo Institute of Technology, (1984).
[45] O.Noguchi, Y.Oya and T.Suzuki: Met.Trans., 12A(1981), 1647.
[46] R.D.Rawlings and A.E.Staton-Bevan: J.Mater.Sci., 10(1975), 469.
[47] K.Aoki and O.Izumi: J.Japan Inst. Metals, 39(1975), 1282.
[48] J.A.Lopez and G.F.Hancock: Phs.Status Solidi(a), 2(1970), 469.
[49] H.Sato and R.S.Toth: Alloying Behavior and Effect in Concentrated Solid Solutions, Ed. by T.B.Massalski, Gordon and Breach Science Pub., (1965), p.754.
[50] P.J.Meschter and W.L.Worrell: Met.Trans., 7A(1976), 299.
[51] Idem, ibid., 8A(1977), 505.
[52] E.Raub and E.Roschel: Z.Metallkd., 59(1968), 112.
[53] D.L.Ritter, B.C.Giessen and N.J.Grant: Trans. TMS-AIME, 230(1964), 1250.
[54] T.Suzuki, Y.Oya and S.Ochiai: Met.Trans., 15A(1984), 173.

# CREEP IN ORDERED NICKEL BASE ALLOYS

P. R. STRUTT* AND B. H. KEAR**
* University of Connecticut, Storrs, CT 06268
** Exxon Research and Engineering Co., Annandale, NJ 08801

## ABSTRACT

This paper examines the fundamentals of deformation behavior in ordered γ' ($Ni_3Al$), β (NiAl) and β' ($Ni_2AlTi$) phases, and specific two phase γ/γ' and β/β' alloys. The relative creep strengths of these ordered nickel-base alloys are discussed. Differences in creep behavior are explained in terms of the effects of temperature, crystallographic orientation and alloying on creep deformation mode. In particular, it is shown that trace additions of boron and carbon to γ/γ' alloys, or deviations from stoichiometry in β'-type alloys can exert a strong influence on creep behavior.

## 1. INTRODUCTION

Ordered nickel base alloys are essential materials in many high temperature structural applications. The γ/γ' nickel-base superalloys provide high temperature creep strength, whereas β-NiAl alloys offer resistance to environmental degradation. Thus, in a typical practical situation, the load bearing component is composed of γ/γ' alloy, and its protective coating is β alloy.

Extensive studies have been made of creep in γ/γ' alloys, and some work has been performed on creep of β alloys, and even β/β' alloys, which are phase mixtures of NiAl (β) and $Ni_2AlTi$ (β'). This paper will focus primarily on the latter category of materials, but for the purpose of comparison some highlights of recent work on creep in U-700 crystals (a representative γ/γ' alloy) will be discussed. As will be shown, β/β' alloys have creep strengths comparable with those of the most advanced nickel-base superalloys. Thus, there is a real possibility that β/β' coatings can be developed that will be as strong at high temperatures as the γ/γ' substrate materials. However, the possibility of exploiting β/β' alloys as bulk materials seems less likely, at least until some means can be devised for overcoming the brittleness of these materials.

The first part of the paper will examine fundamental deformation modes in the constituent ordered phases; namely, $Ni_3Al$ (γ' phase), NiAl (β-phase) and $Ni_2AlTi$ (β'-phase) with Heusler structure. This will provide the basis for the subsequent discussion on creep behavior in the two phase γ/γ' and β/β' nickel base alloys.

## 2. DEFORMATION MODES

### 2.1 $Ni_3Al$ (γ'-Phase)

A diagram of the $L1_2$-$Ni_3Al$ unit cell is shown in Fig. 1 (a); aluminum atoms occupy the cube corners of the unit cell and nickel atoms reside at the face centers.

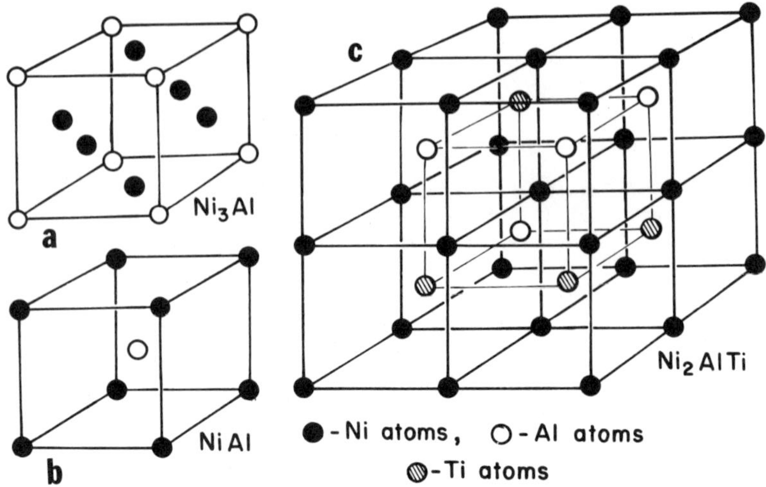


FIG. 1. The unit cells of (a) $Ni_3Al$, (b) NiAl, and (c) $Ni_2AlTi$ (Heusler structure).

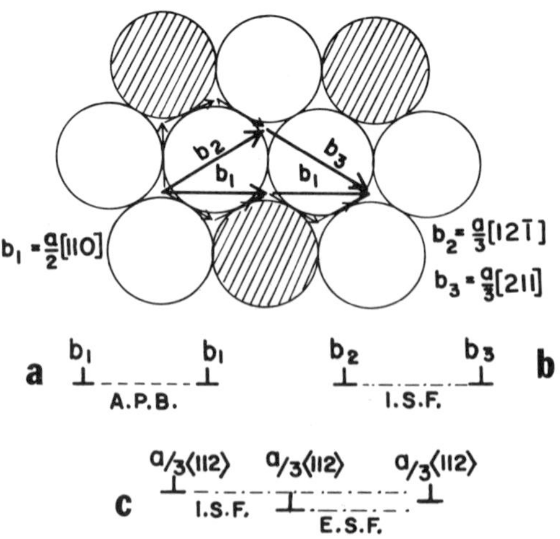


FIG. 2. Schematic diagram of dislocation arrangements in the $L1_2$-type structure; (a) a/2<110> coupled APB pair, (b) a/3<112> coupled intrinsic stacking fault pair, and (c) intrinsic/extrinsic stacking fault pair bounded by a/3<112> partial dislocations.

In the $L1_2$ structure, three types of planar faults can be generated by simple shear displacements on close-packed, ordered {111} planes [1]. These are complex stacking faults, antiphase boundaries and superlattice intrinsic (or extrinsic) stacking faults, corresponding with a/6 <112>, a/2 <110> and a/3 <112> shear displacements, respectively. The corresponding configurations for dissociated superlattice dislocations are shown in Fig. 2. Dissociation (a) represents a pair of superlattice partial dislocations, each with Burgers vector a/2 <110>, which are coupled together by antiphase boundary (APB); in this scheme the leading glide dislocation creates the APB and the trailing dislocation eliminates it. Dislocation (b), Fig. 2, represents a pair of a/3 <112> - type superlattice partials, with net Burgers vector a <110> , which are coupled together by superlattice intrinsic stacking fault (SISF). This configuration is much more relaxed, i.e., widely extended, than the APB type, due to the much lower energy of the SISF fault. Dislocation (c), Fig. 2, represents a superlattice intrinsic/extrinsic stacking fault pair, with net Burgers vector a <112>. The three superlattice partials comprising the fault pair have identical a/3 <112> Burgers vectors, and may exhibit further dissociations into Shockley partials.

Since the energy of an APB with a/2 <110> displacement vector depends sensitively on orientation, the minimum energy configuration for an APB-type superlattice dislocation pair is usually not in the preferred {111} slip plane. For example, an APB-coupled dislocation pair of screw character can reduce its energy by cross slip from {111} into {010}, since the latter is a plane of minimum APB energy. Similarly, dislocation pairs of edge or mixed character can take up lower energy configurations by climb. Such effects exert a profound influence on deformation behavior because they influence dislocation mobility.

### 2.2 NiAl (β-Phase)

A diagram of the $B_2$-NiAl unit cell is shown in Fig. 1 (b); nickel atoms occupy the cube corners of the unit cell and an aluminum atom resides at the cube center.

The $B_2$ type lattice may be regarded as two interpenetrating simple cubic sublattices, one consisting of atoms of one species (B atoms), and the other of atoms of another species (A atoms), see Fig. 3. When the structure is generated from {100} planes, there are alternate layers of A atoms and B atoms. In Fig. 3, B atoms are in the plane of the paper and A atoms in the planes above and below the paper. Obviously glide on {100} <100> involves a simple translation a [100] from $P_1$ and $P_2$, without (i) possible dissociation of the a [100] dislocation into partials, and (ii) formation of APB. In view of (ii), Rachinger and Cottrell [2] predicted glide on {100} <100> to occur in those $B_2$ type alloys with bonding directionality.

The closest-packed plane in the $B_2$ lattice is {110}, see Fig. 3. Hence, in those alloys with covalent bonding (and consequently a low APB energy), it is to be expected that the closest-packed plane is the predominant glide plane. In the {110} plane, A and B atoms are in a 1:1 ratio and, in effect, the $B_2$ lattice is a two-layer repeat structure whose layers are {110} planes. In the correct stacking sequence, B atoms of the second layer are at P sites and, when they are at S sites, an APB is formed in the {100} plane, c.f. Lautenschlager et al [3]. Unit slip in the {110} <111> glide system involves $a[1\bar{1}1]$ translation from one P site to another (e.g. $P_1$ to $P_2$). Lautenschlager et al. [3] considered dissociation of an $a[1\bar{1}1]$ dislocation into partials with Burgers vectors of

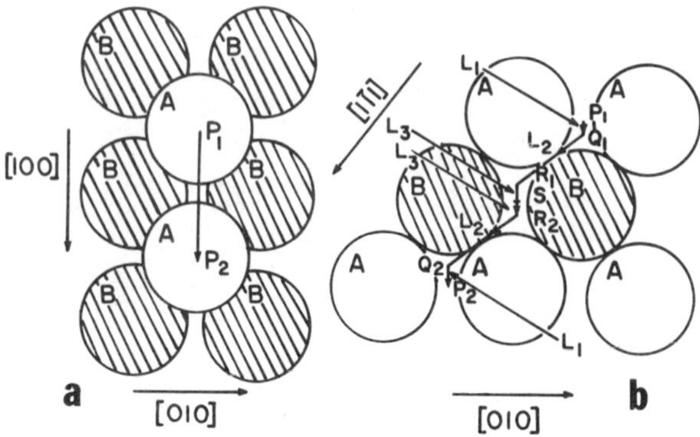


FIG. 3. Basic slip movements in the $B_2$ lattice; (a) view of (001) with [100] translation of an A atom in the second layer from position $P_1$ to $P_2$, (b) sequence of movements involving [1T1] slip of an atom in the second layer from $P_1$ to $P_2$.

magnitude equal to $L_1$, $L_2$, $L_3$ in Fig. 3. It was assumed that Q and R sites are sufficiently close to P and S sites that they correspond to perfect order and antiphase boundary order.

The vertical and lateral displacements ($d_1$ and $d_2$) between two atomic layers that are associated with the movement of a unit or partial dislocation have been calculated by simple geometry by Lautenschlager et al. [3] By considering actual values of $d_1$ and $d_2$ and the decrease in dislocation strain energy by dissociation, the following dissociations may be possible.

$$a\,[010] \rightarrow L_1[101] + L_2[121] + L_2[\bar{1}21] + L_1[\bar{1}0\bar{1}]$$

$$a\,[101] \rightarrow L_1[101] + L_2[121] + 2\,L_3[101] + L_2[1\bar{2}1] + L_1[101]$$

$$a\,[1\bar{1}1] \rightarrow L_1[101] + L_2[1\bar{2}1] + 2\,L_3[101] + L_2[1\bar{2}1] + L_1[101]$$

In the general case the two atomic species are of radii $R_A$ and $R_B$. The values of $L_1$, $L_2$, and $L_3$ depend on $a_o$ and the ratio $R_A/R_B$. Lautenschlager et al. [3] have thus proposed that the value of $R_A/R_B$ is important in determining the glide system in alloys in which bonding directionality is not significant.

From slip line and pole rotation studies, Wasilewski, Butler, and Hanlon [4] and Kanne, Strutt, and Dodd [5] showed slip to occur on the {100} <100> and {110} <100> systems. Pascoe and Newey [6] deduced that slip may also occur on systems of the type {211} <111> and {321} <111>. However, macroscopic slip on {321} or {211} planes may result from pencil glide on two dodecahedral planes. By electron microscopy Ball and Smallman [7] concluded <100> to be the Burgers vector. Since two invisibilities (with different $\bar{g}$-vectors) were not obtained for a specific dislocation set,this determination is not entirely conclusive. Kanne, Strutt, and Dodd [5] and Strutt, Dodd and Rowe [8] have unambiguously determined the <100> Burgers vector for deformation in intermediate and high temperature creep.

## 2.3 $Ni_2AlTi$ (β'-Phase)

A diagram of the $Ni_2AlTi$ unit cell is shown in Fig. 1 (c). It is composed of eight $B_2$ unit cells in which Ni atoms form one sublattice, and Al and Ti atoms form an ordered array on the other sublattice. The small cells constituting the large $Ni_2AlTi$ unit cell are 1.7 pct larger in size than the NiAl unit cell. [9] This difference is sufficiently small that the magnitude of slip vectors in both NiAl and $Ni_2AlTi$ may be expressed in terms of $a_0$ the lattice parameter of NiAl.

As in the case of the $B_2$ type structure, the closest packed planes in the $A_2BC$ or Heusler type structure are of the {110} type. The arrangement of A, B and C atoms in a ($\bar{1}$01) plane is shown in Fig. 4. An instructive way to consider the Heusler phase is to note that it is formed from sets of these planes arranged with a two layer repeat sequence. Possible slip translations between two adjacent ($\bar{1}$01) planes are indicated by the vectors in Fig. 4. For this discussion, however, only one atom (C') in the second atomic layer need be considered, this is of the C atomic specie. In an a[010] slip translation atom, C' is displaced along the [010] vector shown in Fig. 4 into a position in which C atoms are now nearest neighbors. If the entire second atomic layer is considered, the a[010] translation is seen to create both B-B and C-C type wrong bonds; the fault thus formed is termed a sublattice antiphase boundary. Thus, unit slip in <100> type directions must be accomplished by the movement of pairs of a<100> unit disclocations, in which the a<100> dislocations are coupled by a sublattice APB, Fig. 4. It is also evident that slip in <111> type directions is accomplished by the glide of pairs of coupled a<111> dislocations. In this case, however, a further dissociation is possible, since each of the unit a <111> dislocations may dissociate into a pair of a/2<111> dislocations. This dissociation of an a<111> dislocation into a pair of a/2<111> dislocations occurs in $B_2$ structure alloys in which <111> slip is favored. [3] In contrast to <100>, and <111> slip, it is clearly evident from Fig. 4 that an unit a<110> glide translation does not involve the creation of a fault. Therefore a<110> slip is accomplished by the glide of individual a<110> dislocations.

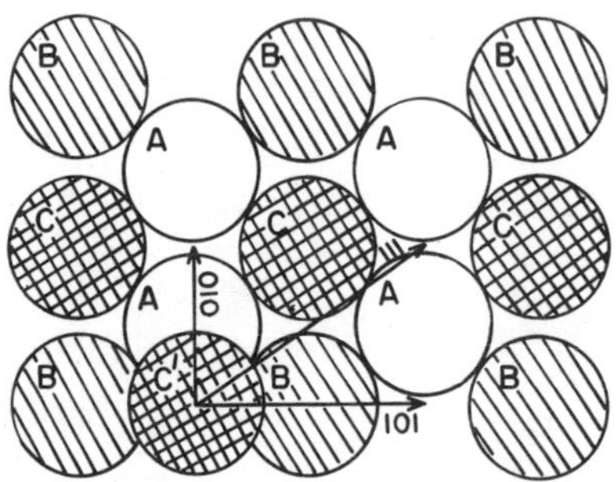


FIG. 4. Basic slip movements in the $A_2BC$ Heusler structure in the (T01) close packed plane; $C_1$ denotes a C atom in the plane immediately above.

## 3. CREEP BEHAVIOR

### 3.1 Two Phase γ/γ' Alloys

The microstructure of a typical γ/γ' nickel-base superalloy contains a high volume fraction of ordered γ' precipitate within a disordered solid solution γ matrix. The two phases are invariably coherent, and frequently exhibit small γ/γ' lattice misfit. A representative superalloy, U700*, contains 35 vol. pct γ' in γ, and exhibits near perfect coherency.

Fig. 5 shows a typical set of creep curves for U700 crystals [10]. Creep curves for [001], [101] and [111] axis crystals are compared at two different stress levels, and at temperatures of 1033°K, 1089°K, 1144°K and 1200°K. The creep curve for the [111] axis crystal at a stress of 655 $MN/m^2$ is omitted, because an almost negligible amount of detectable creep strain occurred at this stress level. Although applied compressive stresses are given in Fig. 5, the basic creep behavior of the [001] and [101] axis crystals is directly comparable since the resolved shear stress for both {111} <110> and {111} <112> shear is the same in each orientation. For the case of the [111] axis crystals the relatively low resolved shear stress for both {111} <110> and {111} <112> slip results in a creep strength significantly higher than that for [001] and [101] axis

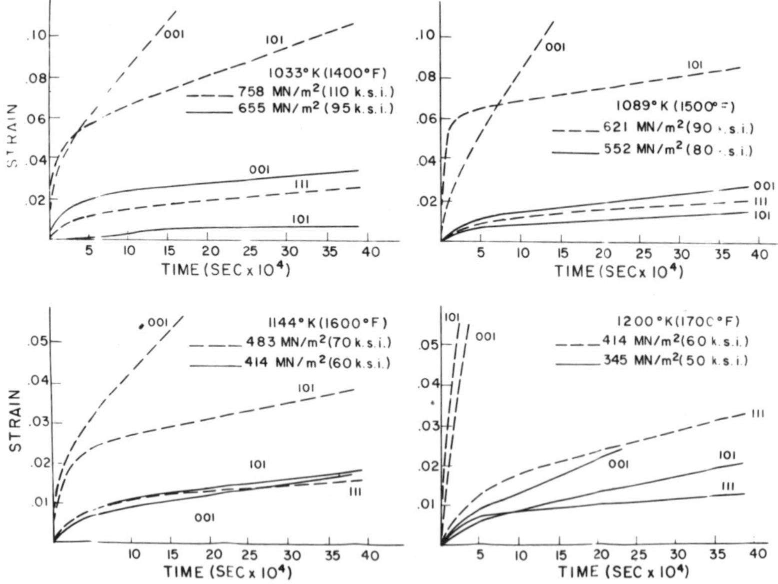


FIG. 5. Creep curves for U700 crystals as a function of temperature, stress and orientation.

*Composition (wt%)

| Ni | Cr | Co | Ti | Al | Mo | B | Zr | C |
|---|---|---|---|---|---|---|---|---|
| Bal | 14.5 | 15.1 | 3.5 | 4.5 | 4.2 | .007 | .001 | .06 |

crystals. In comparing the relative creep behavior of [001] and [101] axis crystals at each temperature, Fig. 5, it is evident that at the higher stress level the [101] axis specimens have primary creep rates that are comparable or higher than that of [001] axis specimens, whereas the corresponding steady state creep rates are lower.

The comparative creep strengths of [001], [101], and [111] axis crystals over the entire stress range (70-827 $MN/m^2$) are represented in a normalized logarithmic plot of $\dot{\varepsilon}_s \exp\Delta H/kT$ vs. resolved shear stress in Fig. 6. Over most of the stress range of interest the <110>-axis crystals are the most creep resistant. A similar behavior has been found in compressive yielding studies carried out on crystals of PWA 1480 by Shah and Duhl [11]. These findings seem to be in agreement with the predictions of the modified cross slip hardening model of Pope et al. [12].

Trace additions of carbon and boron to U700 can influence the steady state creep rate [13]. Fig. 7 shows that an addition of 0.04% boron reduces the creep rate at temperatures below 1144°K by more than an order of magnitude. An anomalous activation energy for creep is also observed when the metalloids are present, which is traceable to an impedance to dislocation motion within the $\gamma$ matrix, or at the $\gamma/\gamma'$ interfaces. Eliminating boron and carbon from U700 results in an apparent activation energy that is consistent with the slowest diffusing specie in the alloy, Table I.

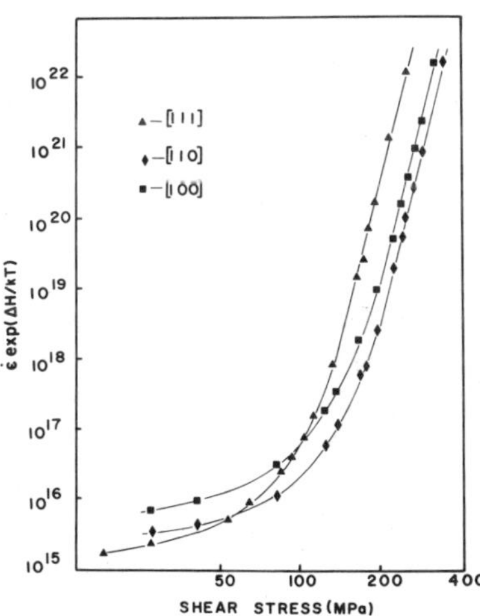


FIG. 6. Normalized creep rate $\dot{\varepsilon}\exp\Delta H/kT$ as a function of shear stress in <100> single crystal compression specimen of U700.

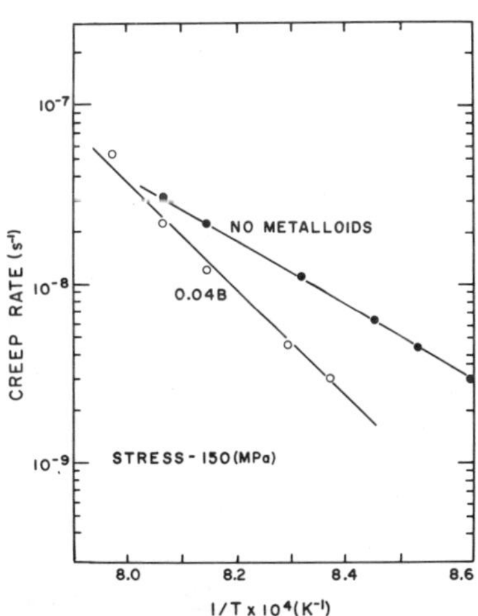


FIG. 7. Arrhenius plots for U700 creep specimens; (a) with 0.04B addition, and (b) without boron.

Table I

Activation Energy for Creep [13]

| Material (at %) | $Q_c$ (kJ $mol^{-1}$) |
|---|---|
| Udimet-700 | |
| (monocrystal 0.04 B + 0.28 C) | 545 |
| (polycrystalline 0.04 B + 0.28 C) | 580 |
| (polycrystalline 0.04 B) | 580 |
| (polycrystalline) | 340 |
| Ni (high purity) | 270 |
| Ni + 0.05 B + 0.02 Zr | 330 |
| Ni + 4.4W | 330 |
| Ni + 4.4W + 0.06 B + 0.2 Zr | 460 |

Using the weak beam contrast technique, small precipitate particles are observed to be decorating the $\gamma/\gamma'$ interfacial dislocations, Fig. 8. The decorated dislocations are observed only in alloys with trace additions of boron, so that the precipitate is probably a refractory metal boride phase. Thus, certain solute atoms in U700 can interact with one another and increase the effective resistance to dislocation motion. It is believed that the temperature dependence of this resistance accounts for the observed anomalously high activation energies for creep in this alloy.

In primary creep of U700 crystals glide occurs on {111} <112> systems, similar to that reported for Mar-M200 crystals. [14] A significant difference, however, is that in U700 this deformation mode occurs at 1200°K, whereas in Mar-M200 it is not observed at temperatures > 1100°K. The specific deformation mode is as depicted in Fig. 9, where the superlattice partials are arranged in the form of intrinsic/extrinsic stacking

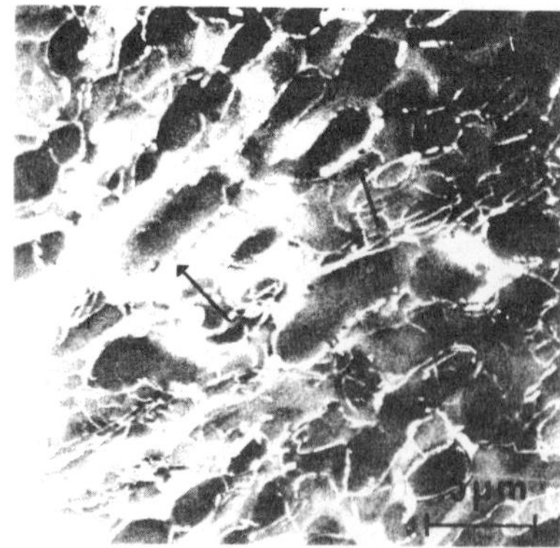

FIG. 8. Weak beam electron micrograph of U700 (with boron) after creep deformation. Micrograph clearly shows interface dislocations, which are decorated with small boride precipitates.

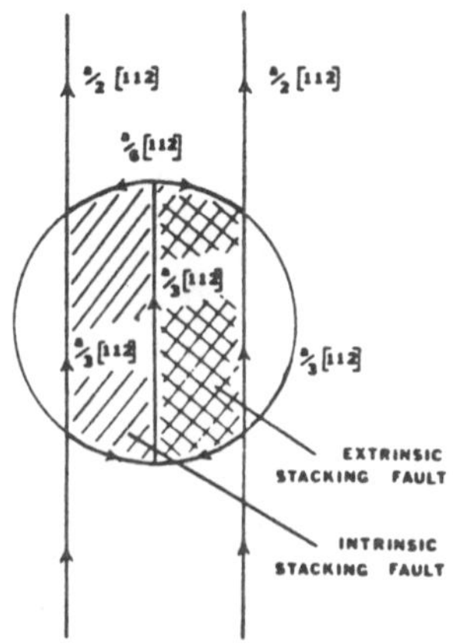


FIG. 9. Schematic diagram showing intrinsic/extrinsic shear of $\gamma'$ particles.

fault pairs. In the early stages of primary creep of U700, electron microscopy reveals a low density of such intrinsic/extrinsic fault pairs that have glided over large distances before encountering obstacles. The obstacles may be interacting a/2<112> dislocation pairs of opposite sign in the same slip system, or interacting a/2<112> dislocation pairs in intersecting slip systems. Fig. 10 (a) shows an example of the latter in an [001] orientation crystal sectioned parallel to the compression axis. The a/2<112> dislocation pairs lie along [$\bar{1}$10] and [110]; the directions in which the {111} slip planes intersect the (001) plane normal to the compression axis. It is apparent in Fig. 10(a) that the partials of the intersecting intrinsic/extrinsic a/2<112> dislocation pairs have interacted. Interactions of this type, which produce a/2<110> dislocation debris, occur more frequently as the dislocation density progressively increases during primary creep. Consequently, as is seen in Fig. 10(b), only a/2<110> dislocation debris is evident. The density of a/2<110> dislocations further increases with increasing strain so that within the steady state region, Fig. 10(c), a/2<110> dislocations form a uniformly dense network, residing for the most part at the $\gamma/\gamma'$ interfaces.

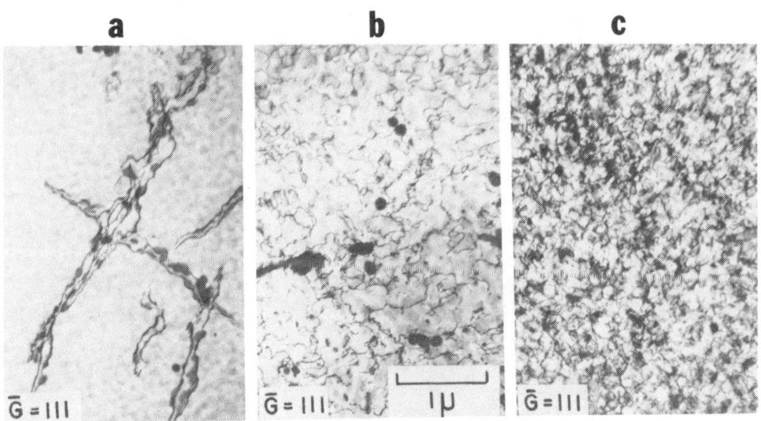


FIG. 10. Progressive changes in dislocation structure in creep of U700; (a) initial stage of primary creep, (b) towards end of primary creep, (c) in steady state creep.

In steady state creep of U700, the effect of increasing (or decreasing) the applied stress is to increase (or decrease) the density of such network dislocations. Strain hardening in steady state creep, therefore, must be a consequence of strong interactions between these network dislocations, and recovery the result of dislocation annihilation within the networks, say by thermally activated cross slip and/or climb processes. Thus, the picture that emerges for steady state creep is that of dislocation rearrangements within the $\gamma/\gamma'$ interfacial networks controlling both the strain hardening and recovery processes.

## 3.2 Two Phase β/β' Alloys

The microstructure of a representative β/β' alloy*, containing about 40 vol. pct β (NiAl) in β' ($Ni_2AlTi$), is shown in Fig. 11. The two phases are readily distinguished by dark field observations using Heusler type superlattice reflections. The lattice misfit between the β and β' phases is accommodated by two orthogonal sets sets of a<100> edge dislocations, Fig. 11. The degree of lattice misfit, as deduced from the spacing of the a<100> interface dislocations, is ~ 1 pct. [15]

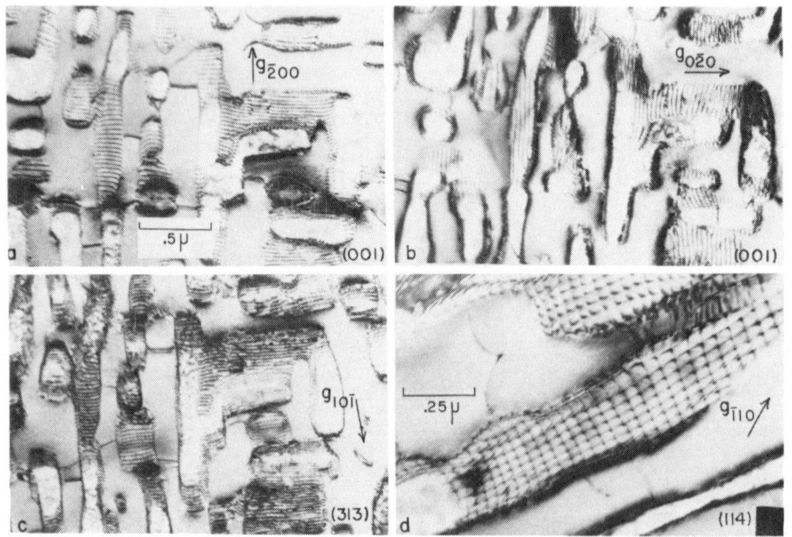


FIG. 11. Transmission electron micrographs of β/β' in the heat treated condition, showing a cross-grid of a<100> interfacial dislocations.

Creep curves obtained at 1323°K for applied stresses of 68.9, 103.4, and 137.9 $MN/m^2$ are shown in Fig. 12. A marked difference between these creep curves and those for NiAl, [8] and $Ni_2AlTi$ [16] is the relatively small or even negligible amount of primary creep.

The activation energy for steady state creep was determined by deforming a series of specimens at the same stress but at different temperatures. Fig. 13 shows an Arrhenius plot of ln $\dot{\varepsilon}_s$ vs 1/T for the β/β' alloy for seven specimens crept at 137.9 $MN/m^2$; the value of ΔH is 280 kJ/mole. This value is in good agreement with that of 290 kJ/mole for self diffusion in nickel, [17] and that of 307 kJ/mole for diffusion of $^{63}Ni$ in stoichiometric NiAl. [18] Arrhenius plots for [001] axis NiAl single crtystals (at 48.3 $MN/m^2$) and polycrystalline $Ni_2AlTi$ (at 68.9 $MN/m^2$) are included in Fig. 13 from data in Refs. 16 and 19. The activation energy for creep in these materials is 314 and 280 kJ/mole. It is of interest to note that the activation energy for creep in β/β' is

*Composition (wt%)

| Ni | Al | Ti |
|---|---|---|
| Bal | 20.5 | 15.7 |

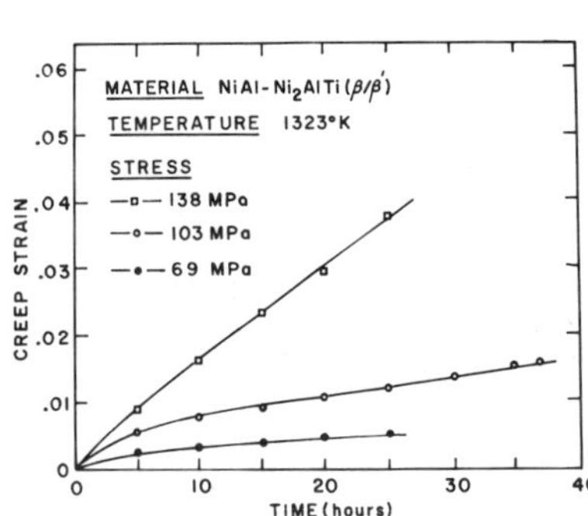


FIG. 12. Creep curves for β/β' alloys at 1323°K.

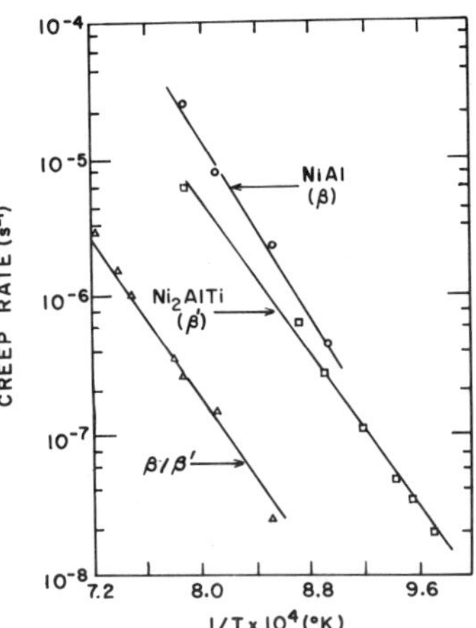


FIG. 13. Arrhenius plots for NiAl (β), $Ni_2AlTi$(β') and a two phase β/β' alloy.

identical with that for creep in single phase β'. In view of these considerations, it seems reasonable that creep in β/β' is diffusion controlled. A comparison of creep strengths of β, β' and β/β' alloys is shown in Fig. 14. Apart from its higher creep strength, a significant feature of the β/β' alloy is its comparatively low stress sensitivity of the creep rate. In the high stress range (275.8 to 620.5 $MN/m^2$), the exponent n in the power law relation is 4.5, whereas values at higher stresses for polycrystalline $Ni_2AlTi$ and [001] axis NiAl single crystals are 8 and 13 respectively.

The effect of titanium concentration on the creep strength at 1275°K for a series of alloys along the NiAl-$Ni_2AlTi$ tie-line is shown in Fig. 15, see [20]. The phase regions along the tie-line were determined from precise lattice parameter measurements. Specimens 2 and 4 have compositions situated slightly within the two phase region, and isolated precipitates or groups of precipitates are revealed by electron microscopy. At compositions between 5 at.% to nearly 12.5 at.% titanium β'-precipitates are distributed within a β-matrix. Conversely, β precipitates are distributed within a β'-matrix at compositions from greater than 12.5 at.% to 20 at.% titanium. In the 12.5 at.% region (specimen 3) a phase mixture is formed with equal amounts of β and β'; there is no discernible precipitate or matrix phase. The dominant feature in Fig. 15 is the remarkably high creep strength of an alloy (specimen 4), which is essentially single phase non-stoichiometric $Ni_2AlTi$; it should be noted that this is for a high value of $T/T_m$, namely 0.75. A transmission electron micrograph from a β/β' alloy specimen crept at 1323°K at an applied stress of 137.9 $MN/m^2$ is shown in Fig. 16. The microstructure shows large semi-coherent β precipitates within the β' matrix. The dislocation arrays at the precipitate-matrix interfaces are more complex than in the underformed heat treated material. Thus, during the creep process glide dislocations have interacted with the pre-existing interfacial networks. A striking feature in Fig. 16 is the existence of paired dislocations (indicated by the symbol a) in the β'phase. These are pairs of a<100> dislocations in climb configurations, such that the coupling antiphase boundary is in a plane

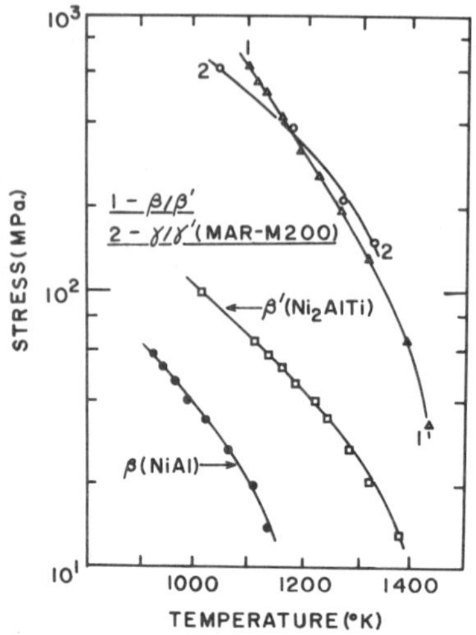


FIG. 14. Creep strength, defined as the stress to maintain a creep rate of $10^{-7}$/sec, vs. temperature for β, β', β/β' and Mar-M200. Data shows that the β/β' alloy is as strong in creep as the γ/γ' nickel-base superalloy Mar-M200.

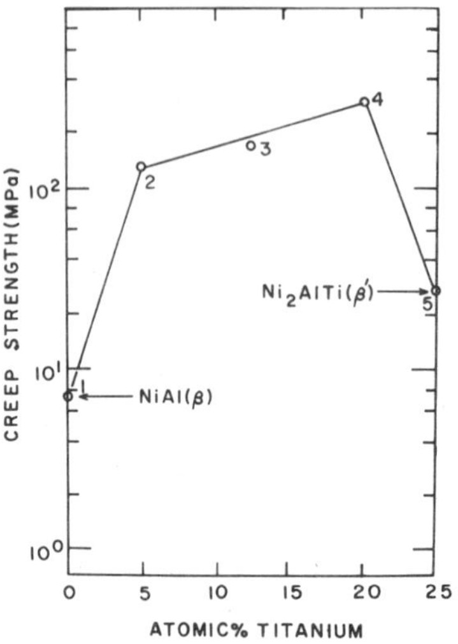

FIG. 15. Creep strength as a function of Ti content for alloys along the NiAl-$Ni_2AlTi$ tie line.

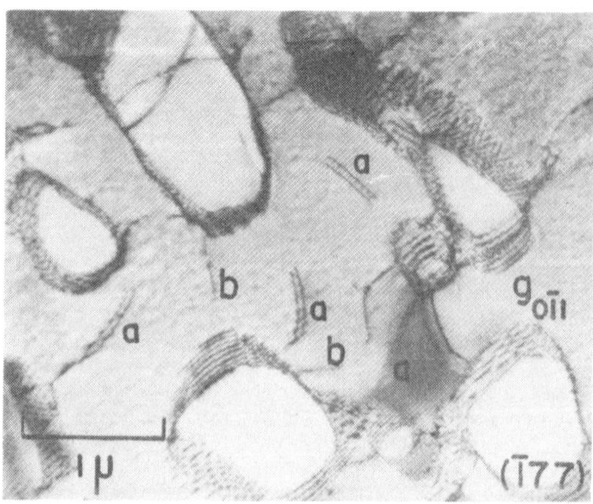


FIG. 16. Dislocation structure in a two phase β/β' alloy after creep, showing both interface dislocations, undissociated a<110> dislocations, and pairs of a<100> dislocations.

perpendicular to the Burgers vector. Such pairs are not observed in single phase β' material; all dislocations in this alloy are of the undissociated a<110> type. In addition to paired a<100> dislocations, undissociated a<110> type dislocations are commonly observed in the β'phase of the β/β' alloy.

## 4. DISCUSSION

A comparison of creep data shows that the β/β' alloy is as strong as the advanced directionally solidified superalloy Mar-M200, Fig. 14. By contrast the creep strength of both single phase NiAl and $Ni_2AlTi$ is inferior to that of the two phase mixture. The constituent γ and γ' phases of Mar-M200 are also much less creep resistant than the normal two phase alloy. These findings are clearly indicative of the importance of interphase interfaces in two phase materials in controlling the high temperature creep strength of ordered alloys. Apparently, such interfaces provide favorable sites for trapping mobile dislocations, and for stabilizing the fine dislocation networks formed during steady state creep. An additional source of strengthening in creep is dislocation-solute interactions, particularly when it takes the form of small precipitates decorating the dislocations, i.e. refractory metal boride precipitates on dislocations residing at γ/γ' interfaces in U700. [13]

In the case of the β/β' alloys based on Ni-Al-Ti, a surprising finding is the high creep strength of alloys 2 and 4, Fig. 15, where the volume fraction of the dispersed phase in the alloys is small. This shows that ternary ordered alloys can exhibit exceptional creep resistance even in the absence of significant amounts of a second phase. Thus, another creep strengthening mechanism must be operative to account for this behavior. Clearly, much more work needs to be done on these complex non-stoichiometric ordered phases.

## REFERENCES


1. B. H. Kear and D. P. Pope, ASM. Conf. Proc. on Refractory Alloying Elements in Superalloys (ASM, Metals Park, Ohio, 1984) p. 135.

2. W. A. Rachinger and A. H. Cottrell, Acta Met., 4, 109 (1956).

3. E. P. Lautenschlager, T. Hughes, and J. O. Brittain, Acta Met 15, 1347 (1967).

4. R. J. Wasilewski, S. R. Butler and J. E. Hanlon, Trans TMS-AIME 239, 1357 (1967).

5. W. R. Kanne, Jr., P. R. Strutt and R. A. Dodd, Trans TMS-AIME, 245, 1259 (1969).

6. R. T. Pascoe and C. W. A. Newey, Met. Sci. J., 2, 138 (1968).

7. A. Ball and R. E. Smallman, Acta Met., 14, 1517 (1966).

8. P. R. Strutt, R. A. Dodd and G. M. Rowe, 2nd., Int. Conf. on the Strength of Metals and Alloys, (ASM, Metals Park, Ohio 1970), p. 1057.

9. A. Taylor and R. W. Floyd, J. Inst. Met., 80, 25 (1952-53).

10. P. R. Strutt, M. Khobaib, R. S. Polvani, and B. H. Kear, 4th Int. Conf. on Strength of Metals and Alloys, (Laboratoire De Physique du Solide, Nancy, France, 1976) p. 314.

11. I. Shah and D N. Duhl, Proc. of Seven Springs Conference on Superalloys, October 1984.

12. C. Lall, C. Chin, and D. P. Pope, Met. Trans. 10A, 1323 (1979).

13. R. S. Polvani, A. W. Ruff and P. R. Strutt, J. Mats. Sci. Letters, 3, 287 (1984).

14. B. H. Kear, J. M. Oblak, J. de Physique, 35, C7-35 (1974).

15. R. S. Polvani, Wen-Shian Tzeng and P. R. Strutt, Trans TMS-AIME, 7A, 33 (1976).

16. P. R. Strutt, R. S. Polvani, and J. C. Ingram, Trans TMS-AIME, 7A, 23 (1976).

17. A. Y. Singaer, Phys. Metals Metallog., 15(1), 100 (1963).

18. G. F. Hancock and B. R. McDonnel, Phys. Stat. Solidi(a) 4, 143 (1971).

19. Y. H. Choo, M. S. Thesis, University of Connecticut, Storrs, Conn.

20. R. S. Polvani, P. R. Strutt and Wen-Shian Tzeng Proc. EMSA, 34th Annual Meeting, Claitor's Publ. Div., Baton Rouge, LA, (1976) p. 595.

# $\gamma/\gamma'$: The Key to Superalloy Behavior

A. F. Giamei, D. D. Pearson and D. L. Anton
United Technologies Research Center, Silver Lane, E. Hartford, CT 06108

Abstract

Gamma prime ($Ni_3Al$), with its unusual temperature dependence of yield strength, is the strengthening phase in Ni-base superalloys. Many facets of $\gamma/\gamma'$ coherent precipitation strengthening have been well studied: coarsening kinetics, volume fraction vs. strength, yield strength vs. temperature, dislocation dynamics, etc.

This contribution deals with several important aspects of two phase behavior in coherent systems (for high temperature application) which are either relatively new or have received inadequate attention. The most desirable $\gamma/\gamma'$ misfit, in sense and magnitude, is dependent on the temperature of application. It must be recognized that the sense and size of misfit are not only a function of alloying, but temperature as well. Sufficient data are presented for various alloying additions to $\gamma$ and $\gamma'$ to allow one to roughly compute the misfit at room temperature or 1000C. The evolution and significance of the $\gamma'$ morphology are discussed. Stability of particle size and shape are related to misfit, temperature and stress. It is demonstrated that for high temperature creep resistance, a fine rafted structure is the most desirable. Finally, some data are presented which demonstrate the dependence of $\gamma'$ strength on stoichiometry as well as temperature.

A portion of this work was sponsored by AFOSR, with A. Rosenstein as contract monitor.

# I. INTRODUCTION

Gamma prime has to be one of the world's most interesting materials. This intermetallic compound with 75 atomic % Ni and 25 atomic % Al is ordered with the $L1_2$ structure. The compound exhibits primarily metallic bonding, is ordered up to the melting point and has reasonable ductility even in the polycrystalline state, particularly if boron is present. It has moderate density and excellent oxidation resistance. The melting point is rather high, approximately 1390C, but the lack of congruent melting makes crystal growth by ordinary techniques impossible at the 3/1 stoichiometric ratio. (This difficulty can be circumvented by alloying with certain species, e.g. Ta.) Gamma prime exhibits significant phase field width at about 4 at. % symmetrically positioned around the 25 at. % Al composition at low to intermediate temperatures. Many of these characteristics are unusual, but the true fascination about this compound is associated with the deformation behavior. The dislocation structures exhibit several different dissociations; two different slip systems are operative and the temperature dependence of the yield strength is truly anomalous. [1-5]. The unique deformation characteristics of gamma prime are largely responsible for the nickel-base superalloys and, in fact, modern gas turbine (jet) engines. Researchers who investigate this material frequently become truly passionate over some of the complexities, such as single crystal mechanical anisotropy or strain rate sensitivity. In fact, as shown in Figure 1, such investigators can be considered to live in a world unto their own.

Gamma prime becomes useful as a constituent in an engineering material when it exists as a coherent or semi-coherent precipitate in gamma Ni or an alloyed solid solution of the fcc Ni phase. In most systems, up to about 70 vol. % can be solutioned at high temperature and reprecipitated as a fine dispersion. A useful dispersion would be 0.25 μm particle size, 0.65 volume fraction and an interparticle spacing of 0.08 μm, for example. Such a dispersion is shown in Figure 2 for the alloy 444: 5Al, 2Ti, 1Nb, 9Cr, 12W, bal. Ni (wt. %). This precipitate is usually brought out by rapid cooling from the solutioned condition followed by isothermal aging and rapid cooling from one or more aging temperatures. This $\gamma'$ is best referred to as aging gamma prime, as distinguished from the typical eutectic or "primary" $\gamma'$. (Actually the eutectic microconstituent is usually the two phase $\gamma/\gamma'$ mixture, at least in the as solidified condition). This eutectic reaction product can normally be diminished in volume fraction by homogenization heat treatment, although in typical cases some remains as a coarse microstructural heterogeneity in commercial alloys. The aging $\gamma'$ can be brought out as coarse, medium or fine in scale, depending on multiple step aging temperatures and times.

Another form of gamma prime in two phase alloys is the so-called cooling $\gamma'$ which forms following the precipitation heat treatments [5]. The fraction of this constituent which is present depends on the temperature dependence of the solubility of $\gamma'$. This cooling (or "hyperfine") gamma prime tends to be important in moderate volume fraction alloys, such as U-700 and in fact is responsible for some very high strength levels. Continuous cooling can lead to convoluted grain boundaries and considerable toughness in polycrystalline materials, Figure 3. It can also lead to some very unusual precipitate morphologies due to the interaction of surface energy and elastic strain energy, Figure 4. Such interaction can lead to the famous ogdoadically diced cubes [6]. There are other forms of $\gamma'$ as well. Isothermal coarsening can lead to rods or plates aligned along the cube direction on a local scale with a reduction in short range elastic strain and system surface energy. This is shown in an early stage in Figure 5. Finally, coarsening under an applied stress can lead to rods or plates aligned perpendicular or parallel to the direction of the applied stress, depending on the crystal orientation and the sense of misfit, where misfit is defined as follows:

$$\delta = \frac{a_{\gamma'} - a_{\gamma}}{a_{\gamma}} \qquad (1)$$

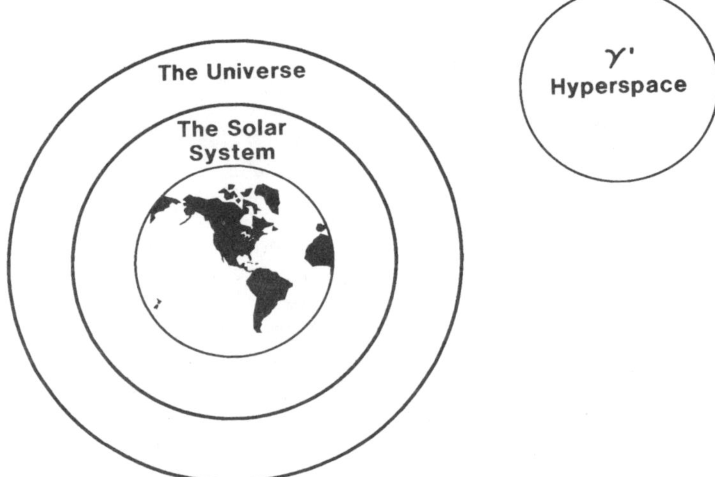


Figure 1. The separate world of gamma prime enthusiasts.

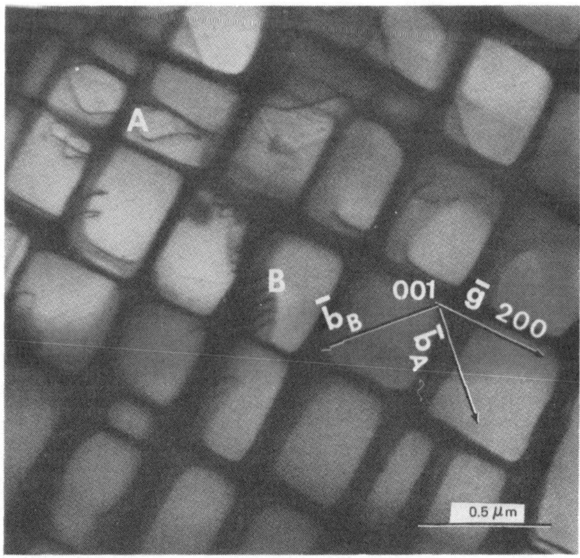


Figure 2. [001] zone axis TEM photograph of carbon free Mar-M200 after conventional heat treatment.

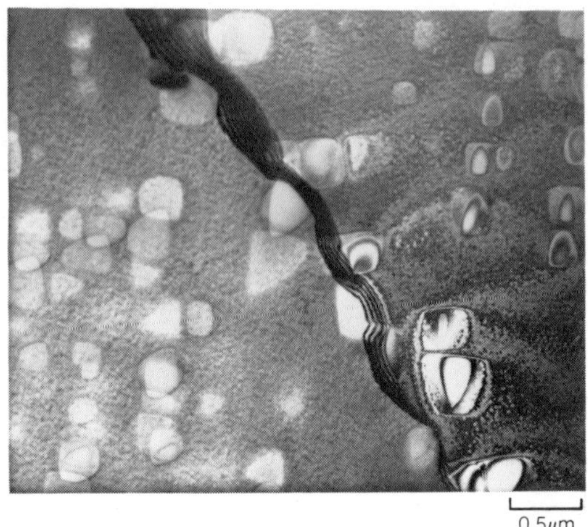


Figure 3. Grain boundary convolution caused by the nucleation and growth of coherent gamma prime in the alloy U-700.

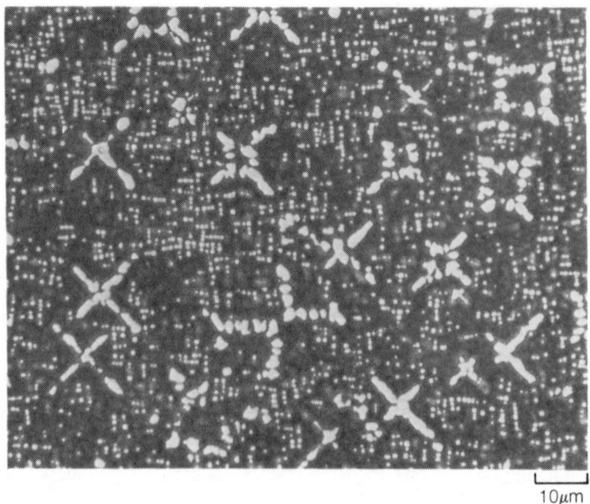


Figure 4. Unusual gamma prime morphology brought about by continuous controlled cooling followed by a flash resolution and a high temperature age in a model Ni-Al-Ta-Re alloy.

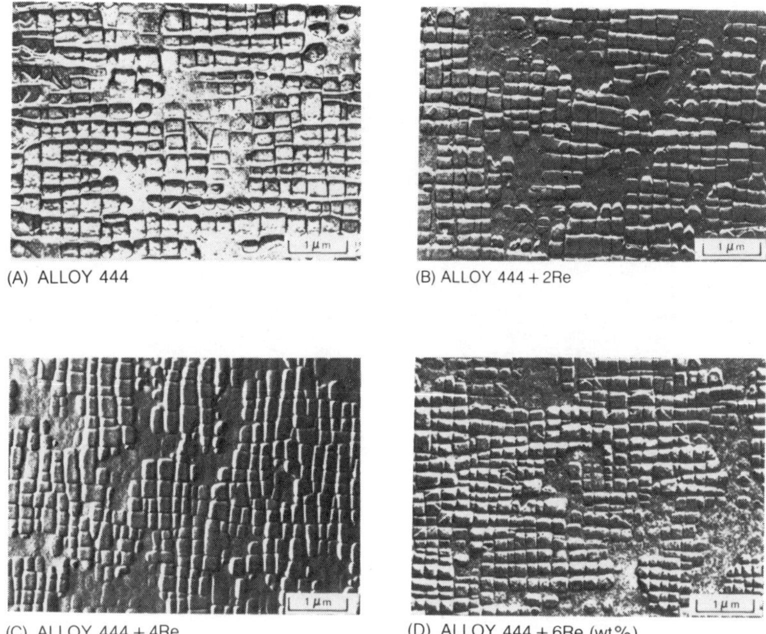


Figure 5. Representative TEM replicas of four alloys aged at 1065C for 2 hr.

A typical stress coarsened microstructure is shown in Figure 6. This type of microstructure is the topic of a number of recent papers [7-10]. One final example of an unusual γ' morphology is the γ/γ' fibrous eutectic which can be generated in the Ni-Al-Ta alloy system, Figure 7.

Alloying elements which are added to superalloys are, in general, not uniformly distributed and can strongly influence the segregation during solidification [11] and/or the partitioning between the two principal phases [12]. They also can have strong effects on properties. Table I summarizes some of the known alloying effects which have particular significance. Some of the commercial alloys have up to 12 major components and the resulting simple microstructures are indeed remarkable.

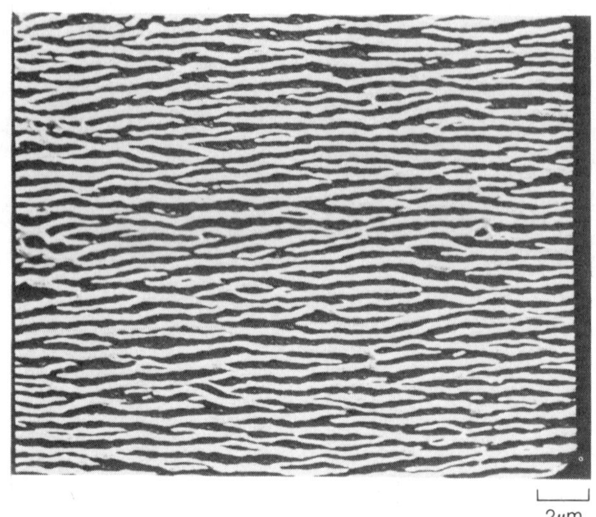


Figure 6. Alloy MMT-143 rafted microstructure following 1.7% creep strain at 1038C/207MPa. Stress axis is vertical.

Table I

Alloying Effects

| Element | $\gamma/\gamma'$ Phase Preference | Principal Effect |
|---|---|---|
| Cr | $\gamma$ | Improves oxidation resistance |
| Co | both | Shifts solvus temperature |
| W | both | Solid solution strengthener |
| Mo | $\gamma$ | Solid solution strengthener |
| Re | $\gamma$ | Solid solution strengthener |
| Ta | $\gamma'$ | $\gamma'$ hardener |
| Ti | $\gamma'$ | $\gamma'$ hardener (and MC carbide former) |
| Nb | $\gamma'$ | $\gamma'$ hardener (and MC carbide former) |
| Hf | $\gamma'$ | $\gamma'$ hardener (and MC carbide former) |
| Zr | neither | grain boundary strengthener |
| C | neither | grain boundary strengthener |

When attempting to relate two phase properties to the behavior of $\gamma'$, it is essential to think about the distribution of elements. It is also important to recognize that due to the significant $\gamma'$ phase field width mentioned above, one is often dealing with sub-stoichiometric or Ni-rich $\gamma'$ in the two phase alloys. This can be a significant effect in terms of mechanical properties, as is shown below.

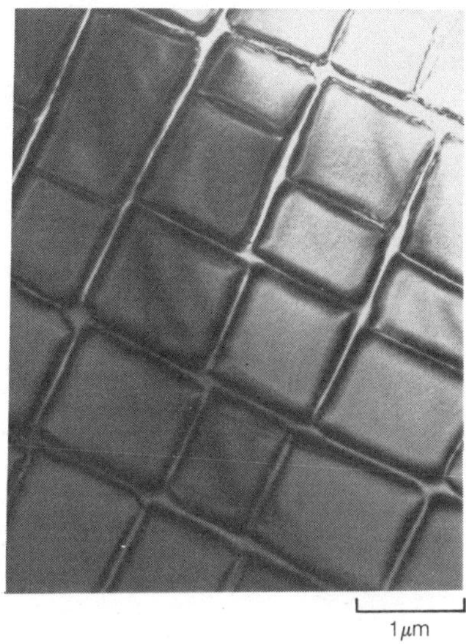


Figure 7. Ni-6.4Al-16Ta (wt%) eutectic TEM [001] micrograph, transverse to growth direction. Gamma prime has rod morphology.

One must also carefully consider the matrix composition. This can be related to matrix stability over long term exposure as well as to creep strength. Stability decreases as the Co, W, Mo, Ta and Cr levels of the matrix increase due to the tendency to form topologically close packed phases within the matrix when it is alloyed beyond its equilibrium solubility limits. Creep strength generally can be correlated with low stacking fault energy (SFE) in coherent fcc systems, since the matrix total dislocations must constrict before entering the particles. It can easily be shown that more energy is required to fully constrict a widely dissociated pair of partial dislocations surrounding a stacking fault. The SFE will also influence the ease of cross-slip by well known mechanisms. As far as the precipitate phase is concerned, alloying is also important relative to several fault energies. Ti, Ta and Nb are thought to increase the APB energy on (111) planes (in this case, a high fault energy is thought to be desirable, see [13]) while Cr and possibly Mo should decrease APB energy. These effects vary as the bonding varies (evidenced by the stability of the relevant binary intermetallic compounds) and as the shape of the gamma prime phase field (and therefore the type of substitution) varies in the appropriate ternary systems. The complex fault energy is probably so high due to first and second nearest neighbor violations, that there is probably no point in trying to influence this through alloying. The superlattice stacking fault energy (intrinsic or extrinsic) is probably generally very low and not involved in deformation mechanisms associated with superalloy yielding or steady state creep over most of the temperature range of interest.

## II. MISFIT

The influence of $\gamma/\gamma'$ misfit has long been recognized as being an important consideration in determining the mechanical properties of nickel-base superalloys. If the $\gamma'$ is precipitated such that misfit dislocations are not formed, the coherency strain contributes to the APB order hardening of the $\gamma'$. This can be important at least up to temperatures of 0.5 Tm where non-diffusional deformation mechanisms predominate. Grose and Ansell [14] have shown conclusively that coherency hardening can effectively increase the yielding behavior in nickel alloys up to 800C. Their study was based on compositional modification of a model Ni-15Cr-Al-Ti-Mo alloy. In their study, they documented the misfit parameter, $\delta$ as in Eq. (1), as a function of temperature by determining the lattice parameters of $\gamma$ and $\gamma'$ using an X-ray diffractometer equipped with a high temperature vacuum stage, employing gold as a standard. In alloys with positive misfit, they showed that the amount of misfit at 20C decreased with increasing temperature and in some alloys, changed sense. At temperatures above 800C, enhanced creep resistance is more desirable than tensile strength and conventional thought is that the misfit should be near zero so that coherency strains do not accelerate particle coarsening [15]. This situation leads to spheroidal particles. Cuboids indicate misfit with coherency strains, tetragonal distortion, and possibly interface dislocations when:

$$\left(\frac{d}{D}\right)_{<hkl>} \sim \delta \qquad (2)$$

where d is the interplanar spacing for a particle size, D, both of which are measured in the same hkl direction. There can be no question that the high temperature creep properties of nickel-based superalloys are dependent on a fine coherent $\gamma'$ precipitate structure [16]. A fine array of $\gamma'$ particles retards creep at high temperature by forcing dislocations to circumvent the $\gamma'$ by climb. Re-bearing superalloys present an interesting case in which negative misfit occurs without necessarily accelerating coarsening. For example, when Mar-M200 is modified with 6 weight % Re at the expense of W, the misfit changes from near zero to -0.3%. The initial particle sizes and coarsening rates are indeed low as indicated in Figure 8. Re partitions almost entirely to the matrix and is thought to retard particle coarsening by requiring the Re to diffuse away from the $\gamma/\gamma'$ interfaces before two neighboring $\gamma'$ particles can impinge. Another interesting alloy with high negative misfit is MMT-143 (Ni-14.3Mo-5.8Al-6Ta by weight %). This alloy has a misfit of -0.8% and coarsens very rapidly until the $\gamma/\gamma'$ interfaces become semicoherent with a high density of misfit dislocations, at which point the rate of coarsening decreases substantially. This alloy exhibits excellent creep resistance with a fairly coarse precipitate size. The loss of coherency during thermal exposure and the effect of this loss on coarsening and creep rate are areas requiring additional study.

The recent interest in negative misfit is that it can be used to develop an extremely fine lamellar $\gamma'$ structure (c.f. Fig. 6), which in turn leads to excellent creep properties since dislocations can no longer climb around the particles and therefore cutting must occur. Rafting of $\gamma'$ at high temperatures is not uncommon in superalloys but it is generally very coarse and develops after many service hours. The trick is to obtain fine rafts in the early stages of creep and this is strictly a function of the fineness of the initial precipitate distribution and the initial coarsening rates. The structure develops best at temperatures near 1000C for both MMT-143 and Re modified Mar-M200. The strong coherency strain driving force to coarsen the $\gamma'$ is modified by the tensile strain to bias the direction of coarsening laterally. The exact mechanism is not fully understood, but probably includes plastic strain as well as the elastic strains considered by Tien and Copley [7] and Pineau [17]. At least it can be stated that rafts decrease the surface energy of the system and that the driving force for this to occur is greater at a

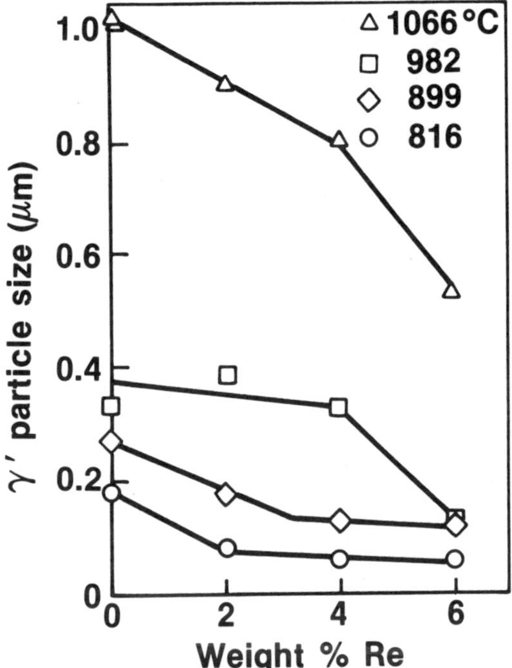


Figure 8. Isochronal coarsening data showing gamma prime particle size vs. Re level for carbon free Mar-M200 aged for 30 hr at the indicated temperatures.

finite negative misfit than at zero misfit. Note that if the particles are coherent, the strain energy can become quite large, but even when they become incoherent (or actually semicoherent), one still has to consider the core energies of the interfacial dislocations, their short range stress fields and typically some residual incoherency. It is known that positive misfit leads to rods rather than plates (in tension) and is therefore not as desirable. The magnitude of the effect can be quite dramatic as shown in Figures 9 and 10. The $\gamma/\gamma'$ eutectic in Figure 10 is significantly refined by the Re addition, which forces cutting rather than bowing or independent dislocation multiplication in the $\gamma'$ phase.

The negative misfit effect on superalloy $\gamma'$ morphology may be much more widespread than commonly realized. Consider the common situation of a zero or slightly positive misfit. As Grose and Ansell have shown, the thermal expansion of $\gamma'$ is less than that of $\gamma$. If we anticipate the temperature dependence of the misfit, a cross-over in the sign of the misfit can frequently be expected. Note that room temperature is a long way from typical operating temperatures of 871-1095C. Other than the work of Grose and Ansel, there is very little data in the literature which can be brought to bear on this subject. We have made some thermal expansion measurements on a variety of Ni-X and $Ni_3(Al,X)$ solid solutions. These were made using a Theta precision dilatometer using Ni as a standard. The linear thermal expansion coefficients derived from this work are listed in Table II.

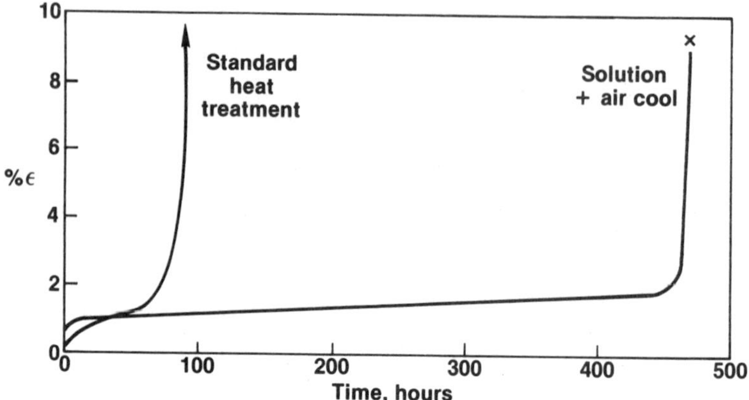


Figure 9. Creep response of [001] single crystal alloy MMT-143 with standard heat treatment, leading to coarse irregular rafts, and solution heat treatment, leading to the formation of finely spaced rafts with uniform spacing; 1038C/207MPa.

Two points can be made when considering the data. First, the thermal expansion of Ni and its solutions is generally greater than that of the $Ni_3Al$ alloys. In particular, Cr has virtually no effect on the thermal expansion of Ni. Second, the heavy refractory elements, Ta, W and Re reduce the thermal expansion of $Ni_3Al$ and Ni substantially on a per atom basis. The alloys of Grose and Ansel contained considerable Cr and none of the heavy refractories. Thus the trend of their data that lower expansion of $\gamma'$ relative to nickel can lead to negative misfit at elevated temperature in alloys which exhibit positive misfit at R.T. is not surprising. However, in modern superalloys, the trend is to reduce Cr and to make increasing use of the refractory elements. The heavier use of refractory metals would tend to reduce the thermal expansivity of the $\gamma$ phase making it more difficult to generate high negative misfits at high temperature from differences in thermal expansion coefficients alone. Figs. 11 and 12 show the lattice parameter variation of $\gamma$ and $\gamma'$ at 20 and 1000°C. Note that W and Mo have a very marked effect on the lattice parameter of Ni but somewhat less on $Ni_3Al$. Re is virtually insoluble in $\gamma'$ and probably behaves similarly to W & Mo in $\gamma$, although we are not aware of any lattice parameter studies. The two alloys previously mentioned, MMT-143 and Re modified MAR-M200, exhibit misfits of -0.8% and -0.3% at R.T. This is due primarily to the high refractory content of the matrix in both phases. The misfit at temperature of MMT-143 has been measured by McKay [18] and shown to be nearly the same as at R.T. This is probably due to the marked reduction of $\gamma$ thermal expansivity due to Mo.

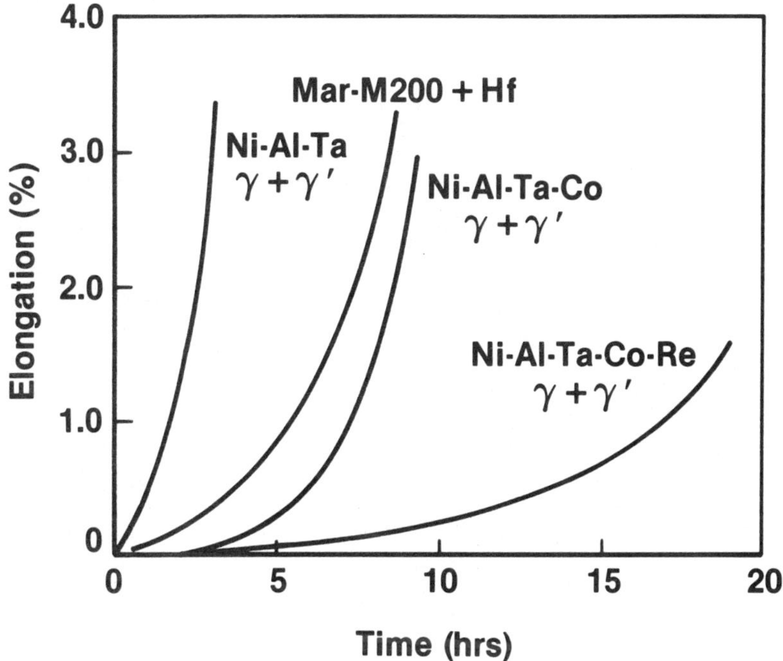


Figure 10. Tensile creep at 982C/276MPa for fibrous γ + γ' eutectic vs. MAR-M200. Co andRe additions refine γ+γ' microstructure.

Table II

Thermal Expansion Behavior of Ni and $Ni_3Al$ Solid Solutions to 1000°C

| Ni-at%X | $\bar{\alpha}$ ppm/°C |
|---|---|
| Ni | 18.4 |
| Ni-10Cr | 18.4 |
| Ni-10W | 16.1 |
| Ni-8Re | 15.8 |
| Ni-20Mo | 16.9 |
| Ni-25Mo | 16.5 |
| $Ni_3(Al,X)$ | |
| $Ni_3Al$ | 16.7 |
| $Ni_{75}Al_{24}Ta_1$ | 16.3 |
| $Ni_{75}Al_{23}Ta_2$ | 16.3 |
| $Ni_{75}Al_{21}Ta_4$ | 15.9 |
| $Ni_{75}Al_{19}Ta_6$ | 15.3 |
| $Ni_{75}Al_{20}Mo_5$ | 14.5 |
| $Ni_{75}Al_{20}Nb_5$ | 15.8 |
| $Ni_{75}Al_{22}W_3$ | 15.8 |
| $Ni_{75}Al_{20}Hf_5$ | 15.9 |

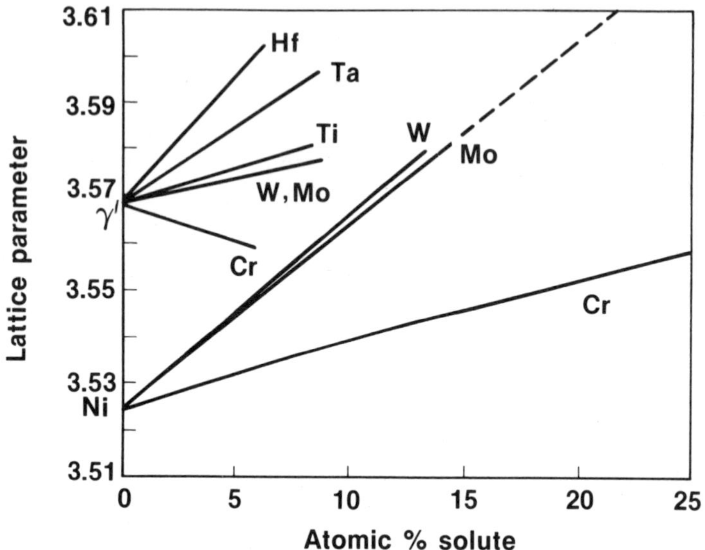


Figure 11. Lattice parameters (Å) of Ni and $Ni_3Al$ solutions at room temperature vs. alloy content

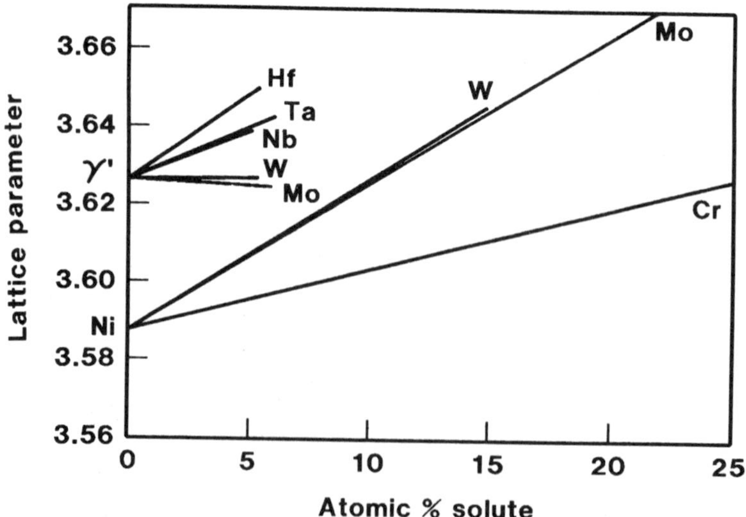


Figure 12. Lattice parameters (Å) of Ni and $Ni_3Al$ solutions at 1000C vs. alloy content.

## III. ALLOYING EFFECTS

As previously mentioned, superalloys have remarkably complex chemistries. The alloy designer is constantly trading some figure of merit such as strength to density ratio with another such as oxidation/corrosion resistance. Furthermore, the partitioning of elements between the γ and γ' phases is not an exact science nor is the γ' that precipitates necessarily of ideal stoichiometry. By this we not only mean that some substitution may be occurring between the Ni substitutional atoms (i.e. Ni, Co, Fe) and the Al substitutional atoms (i.e. Al, Ti, Nb, Ta) but also that the γ' which forms in superalloys will also tend to be rich in Ni. For example, when $Ni_3Al$ is alloyed with 6 atomic % Ta, one of the most potent hardeners of γ', it is possible to vary nickel from 73 to 76.5 atomic %. We have measured the compressive yield strength of [001] oriented single crystals of $Ni_3Al$+6 atomic % Ta and have found a dramatic variation in strength when the Ni stoichiometry is varied. The results are shown in Fig. 13 from which it can be seen that the Ni rich γ' has an approximately 170 MPa lower yield strength than the stoichiometric alloy for temperatures up to the peak temperature. Above 800 C the two alloys are nearly equivalent. Nevertheless, the Ni rich $Ni_3(Al,Ta)$ crystals have a 100°C higher peak temperature and 450 MPa yield strength increment over binary $Ni_3Al$ from room temperature to about 700C.

When considering the benefits that might be derived from negative misfit and the finely spaced rafted γ' microstructure that can be produced, one might ask what is the best approach to further improve creep resistance. One possibility is to harden the γ' phase to further resist penetration of the cutting dislocations. Or it may be that the rate limiting step is the extraction of dislocations at the γ/γ' interfaces into the γ' and that hardening the γ is more important. To partially answer these questions, we prepared three variations of MMT-143 listed in Table III. The series is basically two alloys with 8 and 9 atomic % Mo, which partitions to and hardens predominantly the γ phase; each alloy contains either 2 or 4 atomic % Ta which partitions mostly to the γ'.

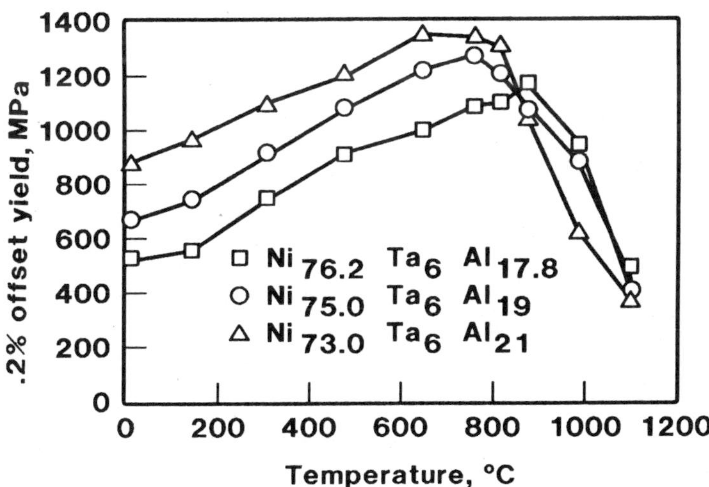


Figure 13. 0.2% offset yield stress vs. temperature for various Ni/Al ratios in gamma prime from Ni-Al-Ta system.

Table III

Alloy Compositions

| Alloy | Atomic % | Weight % |
|---|---|---|
| MMT-143 | Ni-9Mo-13Al-2Ta | Ni-14.3Mo-5.8Al-6Ta |
| 1 | Ni-9Mo-11Al-4Ta | Ni-13.6Mo-4.7Al-11.4Ta |
| 2 | Ni-8Mo-13Al-2Ta | Ni-12.8Mo-5.8Al-6Ta |
| 3 | Ni-8Mo-11Al-4Ta | Ni-12.2Mo-4.7Al-11.5Ta |

The relevant creep properties are shown in Fig. 14. As can be seen, Ta hardens the alloy at each level of Mo, but its effect is not nearly as dramatic as the Mo level. At the higher Mo level, we observe a small but significant amount of δ-NiMo precipitating in the alloy, indicating maximum saturation of the γ/γ' phases. It is possible that at saturation, the cores of the dislocations at the γ/γ' interfaces attract substitutional solutes which add an additional barrier to dislocation penetration into the γ'. Or it may be that at saturation, the γ' is more nearly stoichiometric or perfectly ordered and that viscous creep of the dislocation pairs in the γ' is rate limiting. Higher Mo levels will also force more Ta (and W, if it is present) into the γ'. Further work is required to fully understand the mechanisms and benefits to be derived from high negative misfit and rafted γ' microstructures.

Modern superalloy design and utilization are focussed on single crystal superalloys. It should be remembered, however, that polycrystalline metallurgy (e.g. grain size control and grain boundary strengthening) has played a vital role in superalloy development over the past twenty years. Single crystal superalloys hold many advantages over their isotropic counterparts both industrially as demonstrated in their recent implementation in gas turbine hardware [19] and also scientifically where anisotropic properties of both γ/γ' and single phase γ' Refs. [20-22] have been studied. Growth of single crystal superalloys and single phase pure binary or ternary γ' have been dealt with extensively elsewhere and will not be pursued here. However, a number of processing steps are vital and require some attention. The homogeneity of these crystals after directional solidification has been studied, particularly with respect to interdendritic segregation [23,24]. In most instances it is desirable to minimize this micro-segregation which typically results in Ni and W rich dendrite cores with the balance of the alloying elements segregated between the dendrite arms forming in many instances coarse eutectic phases.

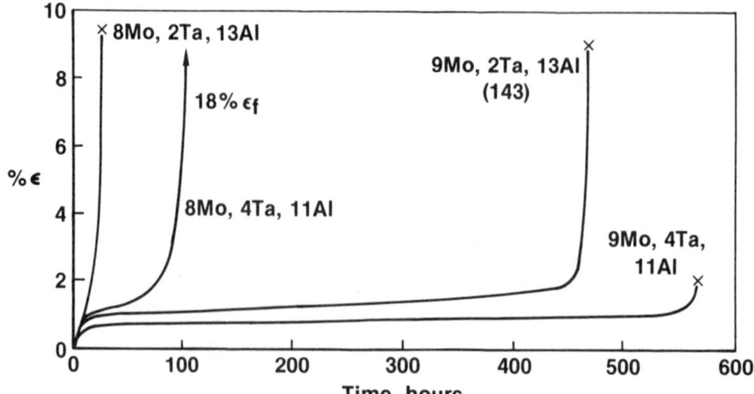


Figure 14. Creep curves for [001] oriented single crystal MMT-143 and three modified alloys of similar chemistry; 1038C/207MPa.

Homogenization at temperatures above the $\gamma'$ solvus for times on the order of 10 hrs is usually sufficient to obtain chemical homogeneity. In specific instances the presence of equilibrium primary $\gamma'$ at the homogenization temperature has been found to be useful in minimizing recrystallization associated with the thermal strains imparted to these crystals in the solidification sequence.

Stoichiometric binary $\gamma'$ single crystals are not readily obtainable from the melt due to the $L+\beta \rightarrow \gamma'$ peritectic reaction. Minor alloying additions of Al-substituting elements (on the order of one or two atomic percent) suppress $\beta$ formation and allow easy growth of crystals. This further enhances the study of single crystal $\gamma'$. (It may be of use to point out that the authors have determined the $L \rightarrow \gamma + \gamma'$ eutectic composition to be at 12.8 wt% Al, as opposed to the published value of 13.3%).

Stress-coarsened $\gamma'$ structures are currently being evaluated for their applicability as gas turbine blade materials. It has been determined that the finer the raft thickness and spacing, the greater the benefit in creep life. Refractory metal alloy additions which retard $\gamma'$ nucleation and growth are beneficial in refining $\gamma'$ both upon quenching and during the "rafting" process. Additions of Re, in particular, slow the $\gamma'$ precipitation and growth kinetics to substantially aid in obtaining fine stress-coarsened microstructures.

The desirability of the presence and magnitude of the $\gamma/\gamma'$ misfit is as yet undetermined but the trend is in the addition of refractory metals which increase the magnitude of negative misfit. If stress-coarsened structures find acceptance, a moderate degree of negative misfit will be necessitated, but not large enough to cause breakdown of the $\gamma'$ rafts. Large negative misfit may accelerate raft generation, destabilize isothermally coarsened $\gamma'$ and even stabilize rafted $\gamma'$.

## IV. Future Trends and Perspectives

Over the past thirty years nickel base superalloy chemistries have changed most markedly in their addition of Al and Al-site filling elements in $\gamma'$. This has lead to increases of $\gamma'$ volume fractions from 25-35 vol.% $\gamma'$ in Nimonic 80 and U-700 type alloys to 65-70% $\gamma'$ in current high pressure turbine blade materials. Recent advances in nickel base superalloy design have emphasized refractory element additions as reflected by the contents of this paper. One may expect these trends to continue, being fueled by the low density and surface stability of $\gamma'$ in spite of the loss of precipitate/matrix interfacial area which is attributed as the primary strengthening agent in current alloys. Research in this area will be centered on the current shortcoming of high temperature strength (> 1000°C) in $\gamma'$. The end result of such efforts may yield materials with equivalent high temperature properties to present alloys with approximately a 15% reduction in density and greatly enhanced surface stability.

## V. ACKNOWLEDGEMENTS

The authors would like to express their appreciation for the assistance of A. Karg of Pratt & Whitney - EDCO in Middletown, CT.

## VI. REFERENCES


1. B. H. Kear, A. F. Giamei and J. M. Oblak, Scripta Met. 4, 567-573 (1970).

2. B. H. Kear, J. M. Oblak and A. F. Giamei, Met. Trans. 1, 2477-2486 (1970).

3. B. H. Kear, A. F. Giamei, G. R. Leverant and J. M. Oblak, Scripta Met. 3, 123-129 (1969).

4. D. P. Pope and S. S. Ezz, International Metals Reviews 29 No. 3, 136-167 (1984).

5. B. H. Kear and D. P. Pope, Proceedings of the Conference on Refractory Alloying Elements in Superalloys - Effects and Availability, ASM (1984).

6. M. Doi and T. Miyazaki, Superalloys 1984, 543-552, M. Gell C. Kortovich, R. Bricknell, W. Kent and J. Radavich, ed., AIME (1984).

7. J. Tien and S. Copley, Met. Trans. 2, 543-553 (1971).

8. D. Pearson, F. Lemkey and B. Kear, Superalloys - 1980, 513-520, J. Tien, S. Wlodek, H. Morrow III, M. Gell and G. Mauer, ed., ASM (1980).

9. R. McKay and L. Ebert, Superalloys - 1984, 135-144, M. Gell C. Kortovich, R. Bricknell, W. Kent and J. Radavich, ed., AIME (1984).

10. T. Kahn, P. Caron and C. Duret, ibid.

11. A. F. Giamei, E. H. Kraft and F. D. Lemkey, New Trends in Materials Processing, 48-97, ASM (1976).

12. O. H. Kriege and J. M. Baris, Trans. ASM, 62 194-200 (1969).

13. S. M. Copley and B. H. Kear, Trans. TMS-AIME, 239, 984-992 (1967).

14. D. A. Grose and G. S. Ansell, Met. Trans. 12A, 1631-1645 (1981).

15. R. Davies and T. Johnston, Ordered Alloys, 447-474, B. Kear, C. Sims, N. Stoloff and J. Westbrook, ed., Claitor's, Baton Rouge, LA (1970).

16. A. F. Giamei, AFOSR Final Report, Contract No. F44620-76-0028 (1979).

17. A. Pineau, Acta Met. 24, 559-564 (1976).

18. R. A. McKay, NASA Technical Memorandum 83698, p. 48 (1984).

19. M. Gell, D. Duhl and A. Giamei, Superalloys - 1980, 205-214, J. Tien, S. Wlodek, H. Morrow III, M. Gell and G. Mauer, ed. ASM (1980).

20. D. M. Shah and D. N. Duhl, Superalloys - 1984, 105-114, M. Gell, C. Kortovich, R. Bricknell, W. Kent and J. Radavich, AIME (1984).

21. C. Lall, S. Chin and D. Pope, Met. Trans. 10A, 1323-1332 (1979).

22. S. S. Ezz, D. P. Pope and V. Paidar, Acta Met. 16, 227-231 (1968).

23. J. J. Montoya-Cruz, R. Kadalbal, T. Z. Kattamis and A. F. Giamei, Met. Trans. 13A, 1153-1159 (1982).

24. R. Kadalbal, J. J. Montoya-Cruz and T. Z. Kattamis, Met. Trans. 11A, 1547-1553 (1980).

# HIGH TEMPERATURE PROPERTIES OF EQUIATOMIC FeAl WITH TERNARY ADDITIONS

R.H. Titran*, K.M. Vedula**, and G.G. Anderson**
*Nasa Lewis Research Center, 21000 Brookpark Road, Cleveland, Ohio 44135
**Case Western Reserve University, Department of Metallurgy and Materials Science, Cleveland, OH 44106

ABSTRACT

The aluminide intermetallic compounds are considered potential structural materials for aerospace applications. The B2 binary aluminide FeAl has a melting point in excess of 1500 K, is of simple cubic structure, exist over a wide range of composition with solubility for third elements and is potentially self-protecting in extreme environments. The B2 FeAl compound has been alloyed with 1 to 5 at % ternary additions of Si, Ti, Zr, Hf, Cr, Ni, Co, Nb, Ta, Mo, W, and Re. The alloys were prepared by blending a third elemental powder with pre-alloyed binary FeAl powder. Consolidation was by hot extrusion at 1250 K.

Annealing studies on the extruded rods showed that the third element addition can be classified into three categories based upon the amount of homogenization and the extent of solid solutioning. Constant strain rate compression tests were performed to determine the flow stress as a function of temperature and composition. The mechanical strength behavior was dependent upon the third element homogenization classification.

## INTRODUCTION

The equiatomic B2 aluminides of Fe, Ni, and Co are considered potential component and structural materials for diverse applications in power systems of aircrafts and automobiles, fossil fuel conversion technology, heat exchangers, and recuperators [1,2] where a high modulus, high tensile strength and high temperature creep strength as well as exceptional oxidation resistance of these intermetallic compounds can improve operation efficiency, reduce component weight, and save strategic elements such as Ta, Nb, and Cr. Knowledge of the mechanical behavior of the FeAl, NiAl, and CoAl intermetallics is quite scant and the effects of composition, temperature, and third element additions are virtually unknown. However, it has been demonstrated that FeAl [3] and NiAl [4] are ductile at elevated temperature, but not very strong while CoAl [5] is strong but not very ductile at elevated temperatures. None of the three intermetallics have exhibited commercially acceptable ductility (>5 percent) at room temperature in tension. Since these B2 binary aluminide systems exist over a range of compositions (Fig. 1) and do possess solubility for third element additions, they offer the potential of producing an acceptable material by alloying. That is, third element additions to induce room temperature ductility, or conversely, third element additions to increase elevated temperature strength whichever is more critical for a particular application.

## PREVIOUS RESEARCH ON BINARY FeAl

The focus of the research effort of the NASA Lewis Research Center on the ordered alloy FeAl, a B2 aluminide, has been alloying with a third element using pre-alloyed binary powders, consolidation by hot extrusion, and then evaluation of the deformation behavior as a function of composition, grain size and temperature. In an effort to understand and realize improvement in mechanical properties of the ternary FeAl compositions, the P/M binary FeAl has been studied extensively at the Lewis Research Center. Clark and Whittenberger [6] studied the thermal expansion of the B2 FeAl in air as a function of composition. Whittenberger [7,8] has shown that for

1600
1400
1200
1000
800
600
400
200
1900
1700
1500
1300
1100
900
700
500
TEMPERATURE, °C
TEMPERATURE, K
Fe2Al5
FeAl2
FeAl3
Fe3Al
MAGN.
TRANSF.
FeAl
0
20
40
60
80
100
ALUMINUM, at. %
20
40
60 80
ALUMINUM, wt %


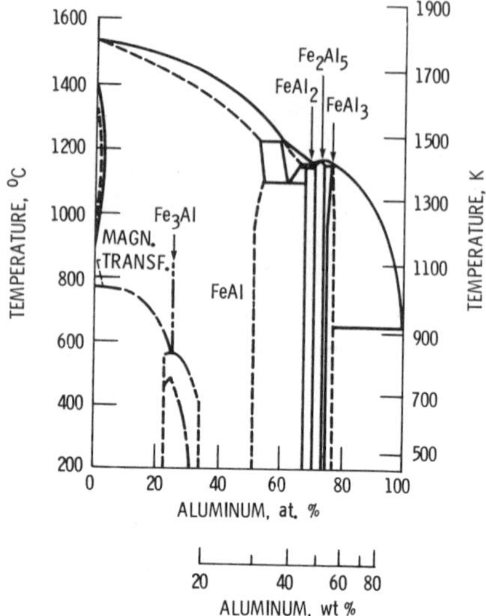

Figure 1. - The iron-aluminum phase diagram showing the large composition range of the B2 phase.

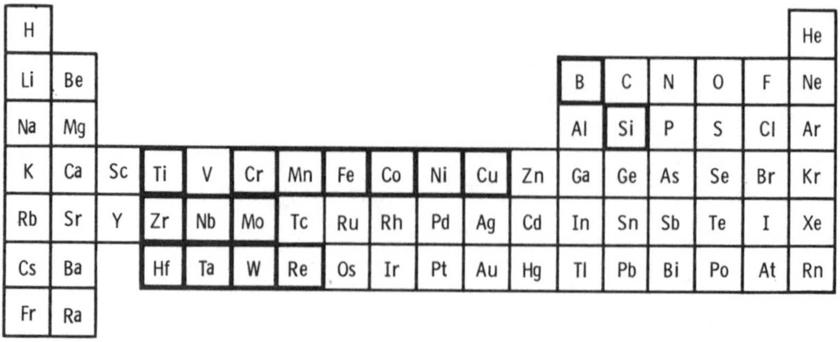

H
He
Li Be
B C N O F Ne
Na Mg
Al Si P S Cl Ar
K Ca Sc Ti V Cr Mn Fe Co Ni Cu Zn Ga Ge As Se Br Kr
Rb Sr Y Zr Nb Mo Tc Ru Rh Pd Ag Cd In Sn Sb Te I Xe
Cs Ba Hf Ta W Re Os Ir Pt Au Hg Tl Pb Bi Po At Rn
Fr Ra


Figure 2. - The periodic table with the elements used for ternary additions in this study outlined.

the Fe-39.8 Al[1] binary, thermomechanically processed by hot extrusion of pre-alloyed powder, the slow plastic deformation behavior was a strong function of grain size; where small grain size resulted in stronger materials. Constant velocity slow compressive plastic flow properties of FeAl materials ranging in composition from about 42 to 49 at % Al were also studied by Whittenberger [9] as a function of grain size and temperature. It was shown that two independent deformation mechanisms with the same apparent activation energy exists, both dependent on grain size which exhibits a stronger influence on strength than composition. In summary, Whittenberger has characterized the binary B2 FeAl material produced by hot extrusion of pre-alloyed powders and modeled its compressive slow plastic deformation mechanism.

## Alloy Preparation

Strengthening of the near equiatomic FeAl (and NiAl) material with elemental ternary additions to the pre-alloyed binary intermetallic powders has been investigated by Vedula and Titran and co-workers [10,11]. A series of ternary FeAl alloys were produced by powder metallurgy techniques. The alloying additions were selected on the basis of past experience in the development of BCC refractory metal alloys and nickel-base superalloys. The selected ternary additions are shown outlined on a Periodic Table in Fig. 2. These alloying additions were added at the 1 to 5 at % level. It was recognized, based on available ternary phase diagrams, that all of the third element additions would not have equivalent solubilities in FeAl. The blended powders were canned in mild steel and extruded at 1250 K at a reduction ratio of 16:1. The nominal chemical compositions of the extruded FeAlX materials are listed in Table I. After hot extrusion, the mild steel cans were leached off in 50:50 nitric acid:water. The 1 and 2 at % ternary alloys were annealed for 100 h at 1425 K. The 5 at % alloy material was heat treated for 175 h at 1525 K in an attempt to maximize the diffusion of the ternary addition into the binary matrix.

# RESULTS

## Microstructural Characterization

Microstructural examination of the as extruded and heat treated materials indicated that they could be classified in three rather broad categories based on the extent to which the 5 at % ternary element had homogenized. Class I FeAlX alloys are single phase materials after homogenization and included the elements Ni, Co, Ti, Si, Mn, and Cr. The solubilities of these elements is consistent with the consideration of atomic size [12] and published phase diagrams. Figure 3 shows a typical single phase structure of FeAl5Ti.

The other elemental 5 at % alloying additions studied showed incomplete solubility even after 175 h of diffusion at 1525 K. In some instances the volume fraction of the second phase appears to have increased during homogenization. This indicates the diffusion of the matrix Fe and/or Al into the elemental powder particle resulted in the formation of either a binary or ternary intermetallic particle in a FeAl matrix with an apparent higher volume fraction of particles than the original extrusion.

The ternary additions which exhibited this type of significant interdiffusion without resulting in a single phase microstructure are classified into a Class II category. Class II elements are Zr, Hf, Nb, Ta, and Re.

---

[1]All compositions are given in atomic percent.

TABLE I. - NOMINAL COMPOSITION IN ATOMIC PERCENT OF CONSTANT STRAIN RATE COMPRESSION TESTED FeAlX MATERIALS

| Specimen | Nominal composition | | |
|---|---|---|---|
| | Fe | Al | X |
| Fe-B1 (B) | 52 | 48 | 0.8 |
| Fe-14 (Si) | 56 | 39 | 5 |
| Fe-22 (Ti) | 56 | 39 | 5 |
| | 50 | 48 | 2 |
| Fe-24 (Cr) | 49 | 46 | 5 |
| Fe-27 (Co) | 49 | 46 | 5 |
| Fe-40 (Zr) | 56 | 39 | 5 |
| | 50 | 48 | 2 |
| Fe-41 (Nb) | 49 | 46 | 5 |
| | 50 | 48 | 2 |
| Fe-42 (Mo) | 49 | 46 | 5 |
| | 49 | 50 | 1 |
| Fe-73 (Ta) | 49 | 46 | 5 |
| | 50 | 48 | 2 |
| Fe-74 (W) | 49 | 46 | 5 |
| Fe-75 (Re) | 49 | 46 | 5 |
| | 49 | 50 | 1 |

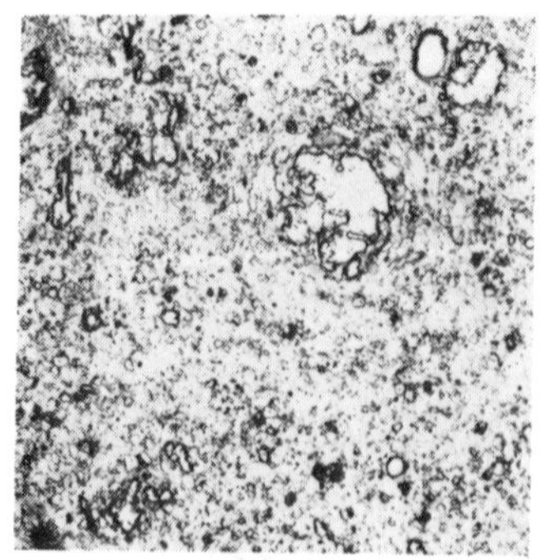

EXTRUDED

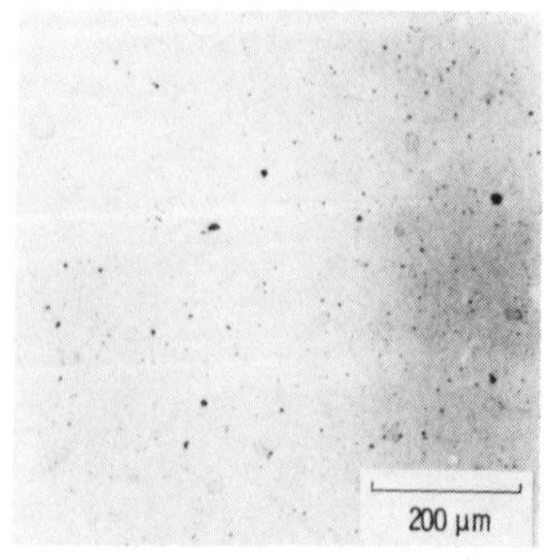


HEAT TREATED

Figure 3. - Photomicrograph showing the solid solutioning of a Class I alloy following the 1425 K - 175 h heat treatment (transverse section).

Figure 4 shows the microstructure of the FeAl5Nb alloy both as extruded and homogenized. It is evident that the volume fraction of second phase particles has increased substantially.

The third category, Class III, consists of those elements which do not exhibit any significant interdiffusion. Mo and W are examples of Class III ternary element additions to FeAl. The original Mo and W powder particles remain essentially as they were for the extruded condition following the homogenization treatment.

The 5 at % ternary alloying additions have been classified into three categories based on the extent of their observed interdiffusion with FeAl as listed in Table II. This classification will be shown to have an important bearing on the slow plastic deformation behavior at elevated temperatures.

## Deformation Behavior

The elevated temperature (1100-1300 K) slow plastic deformation behavior of FeAl with ternary additions was studied by using constant velocity compression tests on right circular cylinders. The stress-axis was parallel to the extrusion axis.

Right cylindrical specimens 5 mm in diameter ranging in length from 10 to 12 mm were machined from decanned and homogenized materials. The extruded bar stock was cut into specimen slugs and both decanned, homogenized, and then centerless ground to 5.0 mm diameter. Specimens 12.5 mm in length were surface ground to length (10 mm minimum) with both ends flat, parallel and true to center axis of the 5 mm cylinder. In general, grinding produced crack-free surfaces.

Elevated temperature compression tests were conducted in air at 1100, 1200, and 1300 K which range from about 0.7 to 0.8 Tm. Testing was accomplished at constant velocities ranging from $2.12 \times 10^{-2}$ to $2.12 \times 10^{-5}$ mm $s^{-1}$ in a universal testing machine where the load was transmitted to the specimen through SiC push rods. Pieces of tungsten foil 25 μm thick were placed between the ends of the test specimen and such rods to prevent welding. Temperature was monitored with type R thermocouples wired to the top and bottom of the test specimens. During testing, the temperature was maintained to within 1 K of the set point, and the difference between the two thermocouples never exceeded 3 K. Specimens were slowly heated to temperature over 2+h while a small stress (about 5 MPa) was applied by operating the test machine in the load cycle mode to retract the expanding push bars and to maintain specimen alignment.

Both stress and strain data were calculated from the load-time charts. Strains were determined either by the offset method [13] when an elastic region could be discerned or by crosshead displacement when little elastic deformation could be seen. In general the calculated strain at the end of a test agreed well (±10 percent) with that determined by the measured change in length; however, for consistency, all calculated strains were normalized to the actual length change. The true stresses were computed on the basis of conservation of volume and uniform deformation.

The 5 at % ternary addition alloys of FeAl were initially compression tested at 1300 K to assist in the ranking of the deformation strength of these compositions. The results of the 1300 K compression tests are shown in Fig. 5. The slopes of the lines which are for clarity only were made parallel to existing FeAl data [7 to 9]. It is noted that the Zr alloys, 48Fe47Al5Zr and 56Fe39Al5Zr have a flow resistance greater than a factor of three times that of the binary FeAl at a comparable grain size, 20 μm. The 49Fe46Al5Nb and 5Mo alloys both appear to have better than a factor of six greater strength than the binary FeAl at a comparable grain size. It is noted that the strongest ternary alloys are those with either a Class II or Class III addition. That is, the volume fraction of second phase appears

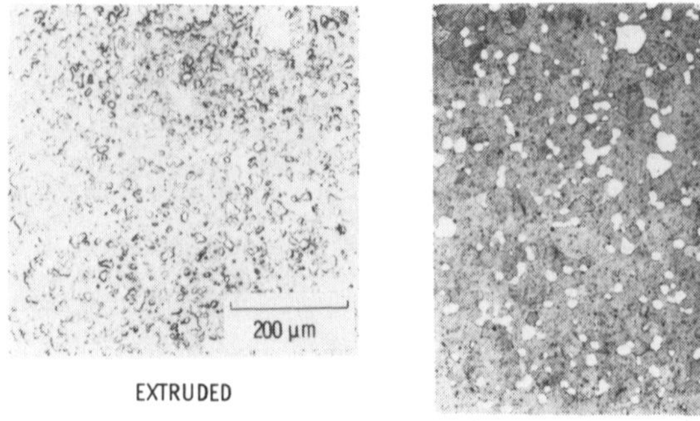


Figure 4. - Photomicrograph showing the increase in volume fraction of Nb rich phase of Class II alloy following the 1425 K - 175 h heat treatment (transverse section).

TABLE II. - CLASSIFICATION OF TERNARY ALLOY ADDITIONS FOR 5 AT % THIRD ELEMENT ADDITIONS TO FeAl AFTER HEAT TREATING FOR 175 h at 1525 K

| Class | Third element addition | Solubility |
|---|---|---|
| I | Fe, Cr, Mn, Co, Ti | Total |
| II | Nb, Ta, Zr, Si, Hf, Re | Partial |
| III | W, Mo | None |

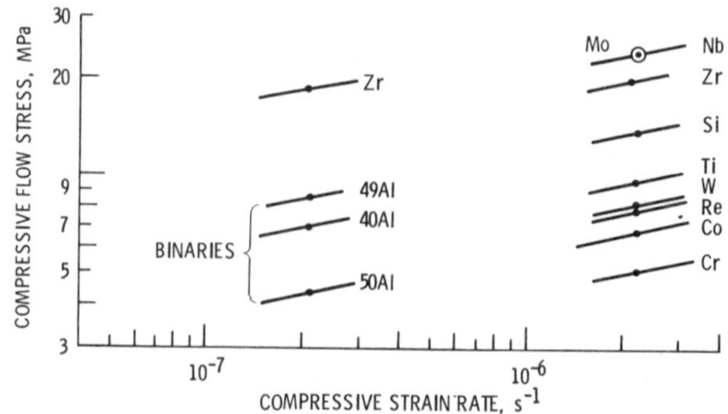


Figure 5. - Comparison of true compressive strain rate against flow stress for various 5 atom percent third element additions to FeAl. Test temperature was 1300 K.

to have increased due to the apparent diffusion of Fe and/or Al into the elemental particle during the homogenization cycle for a Class II, or no noticeable change in volume fraction of second phase for a Class III. Most of the single phase Class I alloys did not exhibit significant improvements in the slow plastic compressive flow strength when compared to the binary alloys. However Ti and Si increased the strength more than the Class III W addition. Based upon the work of Whittenberger [9], we are quite confident the strength increases noted are due to the 5 at % ternary addition and not a grain size effect or Fe:Al ratio effect.

Additional test specimens machined from the extruded bar stock were annealed at 1425 K for 100 h which is sufficient to homogenized the Class I compositions. In addition 0.8 boron alloys were prepared by extrusion and annealed at 1325 K for 100 h. The results of constant strain rate compression tests on the 56Fe39Al5Zr as a function of test temperature is shown in Fig. 6 and is compared to the binary FeAl. At a constant strain rate of $2x10^{-6}$ $s^{-1}$, at 1100 K, the 5Zr addition produces a flow stress greater than 120 MPa compared to only 20 MPa for FeAl. The 5Zr composition shows a 200 K temperature advantage over the FeAl binary. The results of constant strain rate compression tests on the nominal 0.8B addition to the 52Fe48Al binary as a function of test temperature is shown in Fig. 7 along with comparable binary data. At 1100 K and a strain rate of $2x10^{-6}$ $s^{-1}$ the 0.8B alloy has a flow stress of about 80 MPa whereas the FeAl binary has a flow stress of about 20 MPa. The strength improvement of the boron addition is shown to exist to at lease 1300 K, but seems to decrease slightly with increasing test temperature. The results of constant strain rate compressive tests at 1100 K on various ternary FeAl-X alloys is shown in Fig. 8 and reflects a portion of the results of this ongoing screening study.

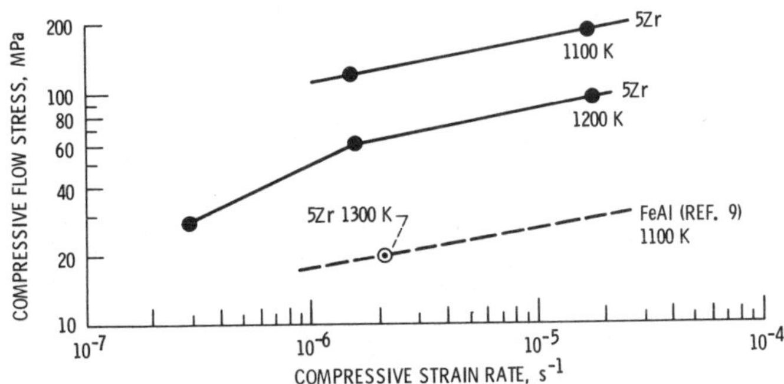


Figure 6. - Compressive flow stress/strain rate behavior for the FeAl-5Zr alloy as a function of temperature.

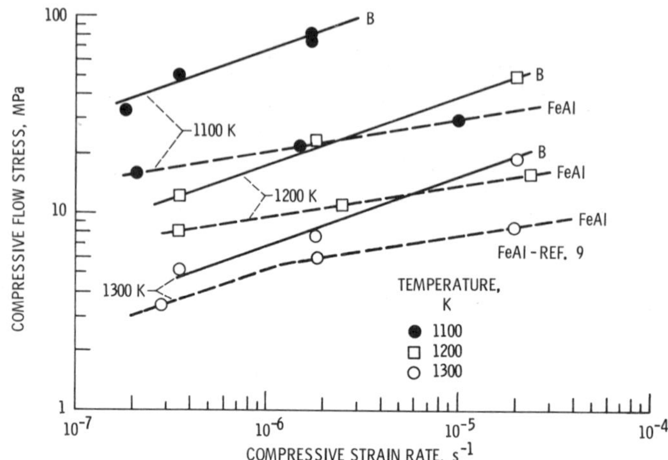


Figure 7. - Compressive flow stress/strain rate behavior of the FeAl-0.8B alloy and comparable FeAl material as a function of temperature.

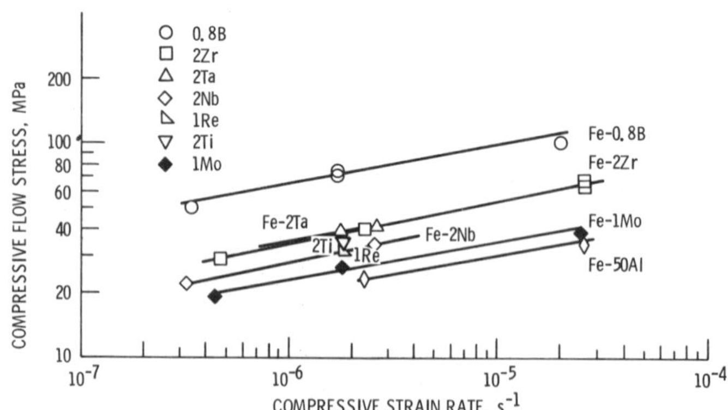


Figure 8. - Comparison of the compressive flow stress/strain rate behavior for several FeAl-X materials at 1100 K. (Numbers before chemical symbols indicate third element concentration in atom percent. Material heat treated for 100 hr at 1425 K prior to testing.)

CONCLUSIONS

The results of constant strain rate compression tests clearly indicate that three of the commonly accepted strengthening mechanisms operate in the B2 FeAl intermetallic system containing ternary additions:

1. Solid solution strengthening - The Class I alloys containing additions of 1 to 5 at % Cr, Ti, Mn, Fe, and Co all exhibit higher flow stress than a comparable FeAl binary at a constant strain rate and a single phase microstructure.

2. Second phase strengthening - The Class II alloys containing additions of 0.8 to 5 at % B, Zr, Ta, Nb, Re, and Hf all exhibited higher flow stress than comparable FeAl tests. It is believed that to some extent, these ternary additions did go into solution in the matrix and reprecipitate possibly as a ternary intermetallic compound which effectively pin dislocations. The Class III ternary additions, Mo and W, which were less effective in strengthening are thought to behave as inert second phase particles which may hinder grain boundary sliding as would the larger particles in the Zr, Ta, Nb, and Hf alloys.

3. Grain boundary strengthening - The 0.8 at % B addition has a remarkable strengthening effect on FeAl. It has been postulated that boron in the $Fe_3Al$ [14] and $Ni_3Al$ [15] systems improves the grain boundary cohesion. The extremely high strength noted for the FeAl0.8B alloy may also reflect a synergistic effect of formation of boride precipitates on dislocation lines along with restriction of grain boundary sliding.

REFERENCES


1. J.R. Stephens, In "COSAM Program Overview," pp. 1-12, NASA TM-83006 (1982).
2. J.D. Whittenberger, in "COSAM Program Overview," pp. 163-174, NASA TM-83006 (1982).
3. G. Sainfort, Mem. Sci. Rev. Metall. 60, 125 (1963).
4. E.M. Schulson, and D.R. Barker, Scr. Metall. 17, 519 (1983).
5. W.D. Nix, in "COSAM Program Overview," pp. 183-190, NASA TM-83006 (1982).
6. R.W. Clark, and J.D. Whittenberger, in "Proceedings of the 8th International Thermal Expansion Symposium," ed. by T.A. Hahan, Plenum Press, NY (1984) pp. 189-196.
7. J.D. Whittenberger, Mater. Sci. Eng. 57, 77 (1983).
8. J.D. Whittenberger, and R.V. Krishnan, Mater. Sci. Lett. 19, 509 (1984).
9. J.D. Whittenberger, Private communication.
10. K. Vedula, G. Anderson, V. Pathare, and I. Aslanidis, in "Proceedings of International Powder Metallurgy Conf.," Toronto (1984).
11. V. Pathare, K. Vedula, and R. Titran, "Proceedings of International Powder Metallurgy Conf.," Toronto (1984).
12. W. Hume-Rothery, Atomic Theory for Students of Metallurgy, the Institute of Metals, London, (1960).
13. J.D. Whittenberger, Metall. Trans. A 10, 1285 (1979).
14. N.S. Stoloff, Int. Met. Rev. 29, (3) 123 (1984).
15. C.T. Liu, C.L. White, C.C. Koch, and H.H. Lee, in High Temperature Materials Chemistry - II ed. by Z.A. Munier, Electrochemical Society Proceedings Volume 83-7 (1983) pp. 32-41.

# THE EFFECT OF SURFACE OXIDE FILMS ON THE MECHANICAL BEHAVIOR OF NiAl

R. D. NOEBE AND R. GIBALA
The University of Michigan, Dept. of Materials and Metallurgical Engineering, Ann Arbor, MI 48109

## ABSTRACT

Thin surface films have been shown to enhance the ductility and decrease the flow stress of several body-centered cubic metals at temperatures $T < 0.2\ T_m$. The origin of this effect lies in the large difference in the intrinsic mobilities of edge and screw dislocations in body-centered cubic crystals. B2 ordered intermetallic alloys, although simple cubic in structure, are based on the body-centered cubic structure and have dislocation core structures, dislocation mobilities and temperature and orientation dependent deformation qualitatively similar to that of bcc metals. This investigation was initiated to examine possible effects of surface films on the mechanical behavior of B2 ordered intermetallic alloys, using oxidized NiAl as the initial material for investigation. Experiments were performed on an impure non-stoichiometric (47.1 at.% Al) single crystal material with an axial orientation near [$\bar{1}$23]. Surface film softening was observed at room temperature in compression at a strain rate of $2 \times 10^{-4}\ s^{-1}$. Flow stresses of the oxide coated crystals were as much as 20% lower than those of identically prepared uncoated crystals. The strains to fracture of coated specimens were larger than those of uncoated specimens and in a few instances the ductility enhancement was as much as four times. Of the several oxides examined, the largest softening effects were found for a thermally deposited delta-$Al_2O_3$ film formed at 1000 °C for 1 hour. The current results for NiAl are compared to results previously obtained for bcc metals. Experiments which could further enhance the film softening effects observed in B2 ordered intermetallic alloys are suggested.

## INTRODUCTION

The B2 ordered intermetallic alloy NiAl is among several ordered alloys which have potential for becoming a widely used high temperature structural alloy [1]. It has excellent oxidation resistance due to formation of a protective $Al_2O_3$ oxide coating [2] and adequate high temperature strength [3]. However, NiAl suffers as an engineering material from its low ductility and tendency toward brittle fracture at ambient temperatures. This limitation should raise some question as to how various surface effects control plasticity in NiAl and other ordered alloys, but there is little in the literature on this subject.

The effect of surface films on the mechanical properties of B2 ordered alloys is of more than just fundamental interest, especially when one considers how surface films can affect the properties of bcc metals. Gibala and co-workers [4-10] have extensively investigated the effects of surface oxide films on the mechanical properties of both single and polycrystalline body-centered cubic refractory metals. Their results have shown that thin surface oxide films on the order of 40-100 nm enhance the ductility and decrease the flow stresses of these metals at $T < 0.2\ T_m$, where $T_m$ is the absolute melting temperature of the metal. This phenomenon has since been referred to as "surface oxide softening" or "surface film softening". Meshii and co-workers have found similar film softening effects for Ni-plated iron single crystals [11-13].

The temperature range over which surface film softening occurs in bcc metals coincides with the well known transition from athermal to thermally activated hardening. Here, the rapid increase in yield stress of these metals with decreasing temperature is associated with the increasing difficulty of screw dislocations to move at lower temperatures as a result of their extended core structures. The mobility of edge dislocations, however, is relatively temperature insensitive. Therefore, the difference in mobility between edge and screw dislocations becomes quite large at low temperatures. Since both screw and edge dislocations must move in order for dislocations to multiply, it is the stress to move screw dislocations which determines the flow stress of the material during normal plastic deformation at low temperatures. On the other hand, if there were a mechanism, such as surface film softening, that produced a large and continuous number of non-screw dislocations, then plastic flow could occur at a lower stress. It is in this large difference in intrinsic mobilities of edge and screw dislocations that surface film softening in bcc metals has its origin.

Although B2 ordered intermetallic alloys are structurally represented in terms of two interpentrating cubic sublattices, they are based on the body-centered cubic structure and exhibit plastic behavior which is very similar to that observed in bcc metals. Both types of metals have a strong temperature and orientation dependence of the yield stress, flow stress and slip geometry, as well as exhibiting a deviation from Schmid's law [14,15]. These similarities in plastic behavior between the bcc metals and bcc-based ordered alloys can be readily explained in terms of dislocation structure and behavior. Materials with bcc-based ordered lattices contain screw dislocations with core structures quite similar to those in bcc metals [16]. Consequently, a large difference in mobility between edge and screw dislocations also exists in these materials, especially NiAl, where at lower temperatures or high strain rates (and, therefore, high flow stresses), the rate controlling process in deformation is overcoming the Peierls barrier of the screw dislocation [17].

The intention of this investigation is to see if similarities in plastic flow behavior of bcc metals and B2 ordered alloys extend to the observation of surface film effects. It would be expected that the B2 materials should also exhibit surface film softening at low homologous temperatures. Consequently, a series of experiments in compression were performed on oxide coated and uncoated samples at room temperature, which is a low "thermal" temperature for NiAl. The results obtained illustrate that surface oxide softening does occur for NiAl at this temperature.

## EXPERIMENTAL

The as-received material was obtained from TRW, Inc., Cleveland, Ohio, through J. K. Doychak of Case Western Reserve University. The NiAl was cast using the selective grain growth technique and resulted in a large grained ingot. A spectrographic chemical analysis of the material was performed by National Spectrographic Laboratories, Cleveland, Ohio and is provided in Table I. It was found that the material was non-stoichiometric (nickel-rich) and contained a significant percentage of residual impurities. Single crystal samples with an axial orientation near $[\bar{1}23]$, determined by Laue x-ray diffraction, were spark machined from the large grained ingot using Cu/W tubes as electrodes. These samples were then centerless ground to a uniform diameter and electropolished in a 2:1 methanol to nitric acid solution in order to remove the disturbed surface layer. Specimens were 4 mm in length, 2 mm in diameter and deformed in compression at an initial strain rate of $2 \times 10^{-4}$ $s^{-1}$ at room temperature.

In order to prepare uncoated and coated samples that were as identical as possible, a single slug of spark machined NiAl was typically centerless ground and electropolished before it was finally cut in half. One specimen was then oxidized and tested while the other was tested unoxidized.

Table I
Chemical Composition of NiAl (at. %)

| Ni | Al | Si | Fe | Cr | Zr | Cu | O | Mg |
|---|---|---|---|---|---|---|---|---|
| 52.10 | 47.13 | 0.52 | 0.09 | 0.05 | 0.03 | 0.03 | 0.03 | 0.02 |

Several different types of oxides and oxide thicknesses were grown thermally on NiAl samples in order to investigate possible softening effects. Thicknesses of these oxides ranged from 40-350 nm and were estimated from kinetic data. Table II gives a brief description of the oxide films investigated.

At 800 °C the predominant oxide phases present are $NiAl_2O_4$, delta-$Al_2O_3$ and gamma-$Al_2O_3$, which form a complex layered oxide coating [18]. A film of $NiAl_2O_4$ or of a solid solution of $NiAl_2O_4$ and gamma-$Al_2O_3$ forms first and has a high epitaxial relationship to the metal. At the surface of this oxide layer grows a layer of delta or gamma-$Al_2O_3$ which thickens with time while the $NiAl_2O_4$ remains incorporated as the inner layer of the scale next to the metal. $NiAl_2O_4$ has a spinel structure, gamma-$Al_2O_3$ has a defect spinel structure, and delta-$Al_2O_3$ is a tetragonal form of gamma-$Al_2O_3$ with a c/a ratio of about 3.

The oxide formed at 1000 °C has not been as definitively determined as the film formed at 800 °C. At short oxidation times ( $\leq$ 8 h) it is thought to be delta-$Al_2O_3$ which at longer oxidation times undergoes a transformation to alpha-$Al_2O_3$ [19]. As at 800 °C the oxide formed at 1000 °C has a high degree of epitaxy with the surface but instead is a complete single phase oxide layer.

Samples oxidized at 1200 °C form a mature alpha-$Al_2O_3$ scale [20] which initially grows as a submicrocrystalline oxide which recrystallizes during its growth to a textured alpha-$Al_2O_3$ film [21].

Table II
Oxide Coatings and Thicknesses

| Oxidation Temperature | Oxidation Time | Total Oxide Thickness (nm) | Predominant Oxide Phases Present | Oxide Structure | Reference |
|---|---|---|---|---|---|
| 800 °C | 0.10 h | 40 | $NiAl_2O_4$ | Spinel | 18 |
| 800 °C | 1 h | 120 | $NiAl_2O_4$,$\delta$-$Al_2O_3$,$\gamma$-$Al_2O_3$ | Spinel | 18 |
| 800 °C | 4 h | 250 | $NiAl_2O_4$,$\delta$-$Al_2O_3$,$\gamma$-$Al_2O_3$ | Spinel | 18 |
| 800 °C | 8 h | 350 | $NiAl_2O_4$,$\delta$-$Al_2O_3$,$\gamma$-$Al_2O_3$ | Spinel | 18 |
| 1000 °C | 0.17 h | 150 | $\delta$-$Al_2O_3$ | Spinel | 19 |
| 1000 °C | 1 h | 350 | $\delta$-$Al_2O_3$ | Spinel | 19 |
| 1200 °C | 0.17 h | 130 | $\alpha$-$Al_2O_3$ | Rhombohedral | 20,21 |
| 1200 °C | 1.25 h | 350 | $\alpha$-$Al_2O_3$ | Rhombohedral | 20,21 |

## RESULTS

Figure 1 gives stress-strain data for pairs of identically prepared specimens of [$\bar{1}$23] orientation, which are either uncoated or oxide coated at 800 °C. In all cases of identically prepared specimens, the oxide coated materials exhibited a lower yield stress and flow stress at the same plastic strain compared to the uncoated samples. The ductility (i.e., the strain to fracture, which is defined as the strain at which the first yield drop associated with a visible crack was observed) was always greater in the oxide coated samples than in the uncoated samples.

For samples oxide coated at 800 °C the average yield stress decreased while the strain to fracture increased for increasing oxide thickness over

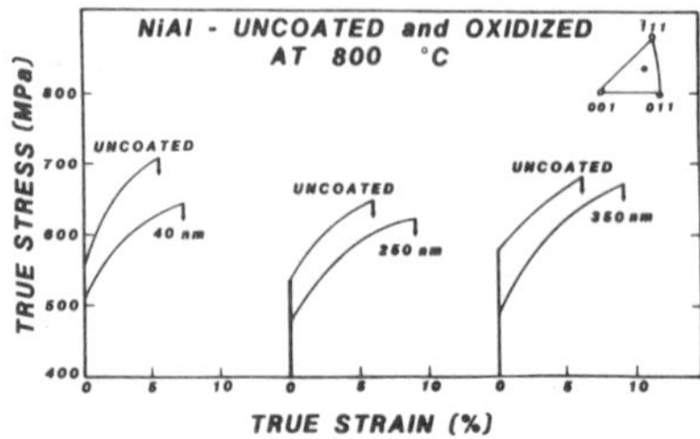


Figure 1: Stress strain data for identically prepared pairs of single crystal NiAl uncoated and oxide coated at 800 °C to various thicknesses. Tests were performed in compression at a strain rate of $2 \times 10^{-4}\ s^{-1}$ at room temperature.

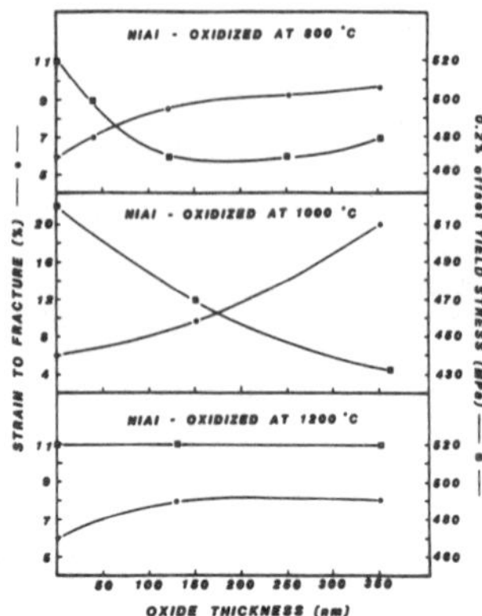


Figure 2: Average mechanical properties of oxide coated NiAl at room temperature as a function of oxide thickness at three oxidation temperatures.

the range of thicknesses tested (Fig. 2). At the larger oxide thicknesses, the yield stress and strain to fracture appear to reach minima and maxima, respectively. Typical results for oxide coated refractory metals have shown that there is an optimum thickness for the surface film softening effect [22]. In bcc metals, the yield stress decreases with increasing thickness to an optimum thickness and then increases with further increase in oxide thickness.

Figure 3 gives stress-strain data for pairs of identically prepared [$\bar{1}$23] specimens which are either uncoated or oxide coated to different thicknesses at 1000 °C. This particular type of oxide exhibited the most pronounced softening effects, primarily at the thicker oxide (350 nm). For this particular oxidation temperature, the average yield stress of the coated material was approximately 20% lower than that of the uncoated material. The average strain to fracture for the oxide coated material was nearly four times that of the uncoated material. The variation in the observed strains to fracture for these materials was also considerable. Strains to fracture of the oxide coated materials ranged from 10% to over 30%, as compared to 5-7% for the uncoated materials. In all other respects the behavior of these oxide coated samples was similar to that of ones oxide coated at 800 °C. Only the magnitude of the effect was different (Fig. 2).

An additional oxidation temperature which was investigated was 1200 °C. Stress-strain data for pairs of uncoated and coated samples of [$\bar{1}$23] orientation are shown in Figure 4. For this particular type of oxide coating, softening was evident for the 130 nm oxide. The flow stresses were lower and the strains to fracture larger for the oxide coated smaples. For the 350 nm oxide the flow stresses for the coated crystals were not always less than those for the uncoated samples. On the average, the yield stress for the uncoated and coated crystals were about the same while the strain to fracture was still improved by the oxide coating (Fig. 2). Overall, the softening effect observed for the alpha-alumina coating was small compared to that seen for the spinel coatings.

## DISCUSSION

There are a number of mechanisms which can result in the softening of bcc metals at low temperatures, where the strength of these materials increases rapidly with decreasing temperature. Examples of these include:

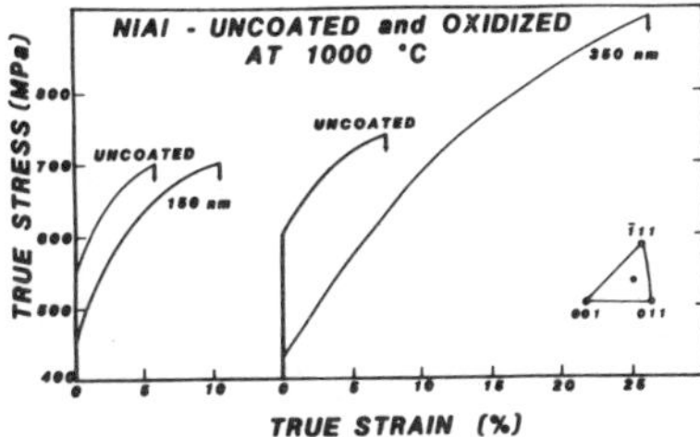


Figure 3: Stress-strain data for identically prepared pairs of single crystal NiAl uncoated and oxide coated at 1000°C to various thicknesses. Tests were performed in compression at a strain rate of $2 \times 10^{-4}\ s^{-1}$ at room temperature.

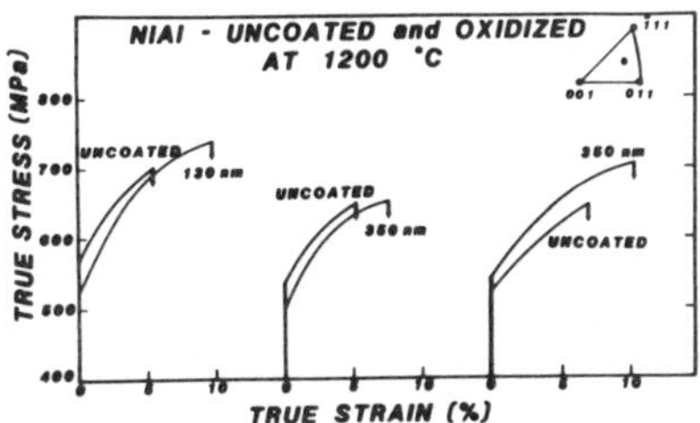


Figure 4: Stress-strain data for identically prepared pairs of single crystal NiAl uncoated and oxide coated at 1200°C to various thicknesses. Tests were performed in compression at a strain rate of $2 \times 10^{-4}\ s^{-1}$ at room temperature.

solid solution softening [14,23,24], irradiation softening [25] and pre-straining at higher (athermal) temperatures [7,26,27]. One mechanism proposed to describe all of these softening mechanisms is based on an increase in the mobile dislocation density, particularly non-screw dislocations. Non-screw dislocations are much more mobile than screw dislocations especially at lower temperatures due to differences in dislocation core structures. Screw dislocations have a nonplanar dislocation core structure which is essentially sessile and must undergo a transformation to a glissile configuration before macroscopic slip can occur on any given plane [28,29]. Non-screw dislocations exist naturally in a glissile configuration so that no transformation of core structure is needed before glide can occur. Consequently, these dislocations glide at much lower stresses than those for screw dislocations [29].

It is interesting to note that direct and indirect evidence for the above softening mechanisms can be found in the literature for NiAl as well. Yang, Dodd and Strutt [30] during the investigation of creep properties of NiAl found that stoichiometric NiAl annealed at 700 to 900 °C was softer and had greater ductility at room temperature than material annealed at 1300 °C. This was at first attributed to small precipitate particles, possibly carbides [30] and/or aluminum oxide particles [31], which are in solution at higher temperatures. The responsible defect was later determined to be silicon impurity atoms which form silicon-vacancy clusters [32]. The particles were responsible for producing prismatic dislocation loops which were then thought to act as dislocation sources, resulting in an abnormally high density of glissile dislocations. This increased number of mobile dislocations was then responsible for the observed softening effect. This was not found to occur, however, for any composition other than stoichiometric. Other non-stoichiometric alloys that were prepared at the same time from similar materials and expected to contain similar impurity levels did not show this effect [33].

Liu and Mitchell [34] have studied the _in-situ_ irradiation of NiAl by high voltage electron microscopy. They found that very high densities of small dislocation loops were formed during irradiation. The character of the loops was identified as pure edge prismatic loops with interstitial nature lying on {001} planes with <001> Burgers vectors. Also, Parthasarathi and Fraser [35] have found that by introducing about 1% compressive deformation in NiAl single crystals it was possible to introduce a high density of vacancy-type edge dislocation loops into the crystal. If these dislocation loops are glissile, then softening might occur when compared to identical but untreated samples. If the loops are sessile then they could act as dislocation sources as in the case described above and softening might again be expected to occur.

The one softening phenomenon that has not been discussed previously is that of surface film softening. As mentioned earlier, this phenomenon has been investigated by Gibala [4-10], Meshii [11-13] and their respective co-workers for bcc metals. Our results indicate that surface film softening also occurs for thermally oxidized samples of single crystal NiAl. The most reasonable explanation for this phenomenon in NiAl is the original one proposed for bcc metals by Sethi and Gibala [4] and which agrees with the above discussion. Briefly, during deformation, the effects of elastic and plastic constraint at interfacial steps between the substrate and the surface oxide act to produce an efficient continuous source of mobile non-screw dislocations (Fig. 5). As a result, the strain rate imposed upon the specimen during deformation can be maintained by the motion of edge dislocations. Therefore, macrostrain occurs by the motion of edge dislocations at a lower stress than the yield stress of the uncoated specimens which must deform by the movement of both edge and screw dislocations.

Even though this is a simplified model it fits the behavior of NiAl quite well. In NiAl slip occurs on systems which do not require disordering by the formation of antiphase domain boundaries (APB) or the production of superdislocations [36]. Deformation typically involves only single dislocations with a <100> Burgers vector [36-41] in which slip can proceed without the creation of APB [36]. Because the dislocation core structures in B2 ordered alloys are quite similar to those in bcc metals [16,17], the dislocation behavior in NiAl should also be quite similar to that in bcc metals. This similarity should extend to surface oxide softening as an operative mechanism in the material.

The specific dislocation generation and motion processes that occur at the oxide-substrate interface are not known in any detail. For bcc metals the dislocation generation mechanisms differ from system to system. For nickel plated iron single crystals, dislocation generation occurs by a film cracking mechanism [12], whereas softening occurs in anodic oxide coated refractory metals without evidence of any film cracking [41]. In this case it is believed that during the deformation of the oxide coated samples the large growth stresses [43,44] and compatibility stresses [45] associated with the oxide/metal composite act to provide an effective source of mobile dislocations at the interface region.

The type of residual stress in the film is also an important factor in film softening. Anodic oxides on molybdenum crystals have a tensile residual stress which would be expected to place a compressive stress on the adjoining substrate. Tottori [46], in experiments on oxide coated crystals of molybdenum, found a much larger softening effect in compression than for samples tested in tension. The reverse case was observed by Talia [47] for anodic coated tungsten crystals. For tungsten the residual stress in the oxide is compressive and a much larger softening effect was seen in tension than in compression.

From consideration of thermal effects, materials with alumina forming scales would be expected to have a large compressive residual stress in the oxide at room temperature [48]. It would be expected, therefore, that an even larger softening effect would be seen in tension for thermally oxidized NiAl as in the

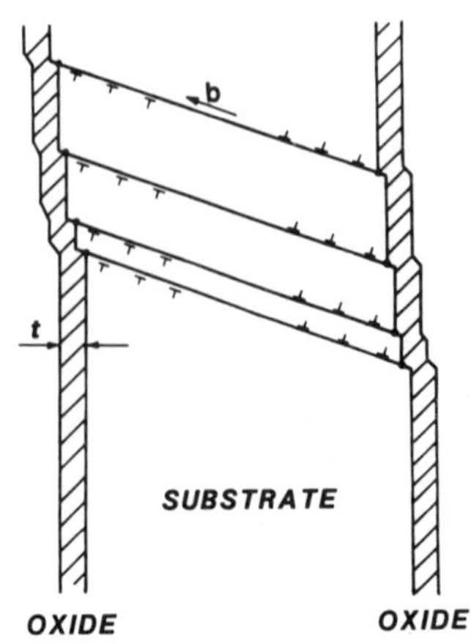

Figure 5: Schematic model for surface oxide softening.

case with tungsten. The magnitude of the thermal stress in the oxide can be estimated by a modified Oxx equation [49]:

$$\sigma = \frac{E \quad \Delta T(\alpha_c - \alpha_m)}{1 - \nu} \qquad (1)$$

where $\sigma$ is the stress in the oxide, which is assumed to be biaxial since $\sigma \simeq \sigma_x \simeq \sigma_y >> \sigma_z$. E, $\nu$, and $\alpha_c$ are the elastic modulus, Poisson ratio, and coefficient of thermal expansion of the oxide, respectively and $\Delta T$ is the temperature difference between the oxidation and ambient temperatures. For alpha-$Al_2O_3$ oxides E is much larger than that for the spinel forming oxides while $\alpha_c$ is about the same for both types of oxides [50]. Therefore, the thermal residual compressive stress in the oxide would be greater in the alpha-$Al_2O_3$ oxide layers and the softening effect might be expected to be less as our results indicate. That the oxide formed at 1000 °C produces a larger effect than the one formed at 800 °C is probably related to the fact that a single oxide layer may be a more efficient source of dislocations during deformation, due to the compatibility stresses involved, than a duplex scale of the type that forms at 800 °C.

Finally, it has been demonstrated by Sethi and Gibala [8] and by Talia and Gibala [47] that the surface softening effect can be enhanced by either abraiding the surface of the specimen or prestraining the specimen at an athermal temperature before applying the oxide coating. The enhancement of the softening effect results from the increased number of dislocation sources present at the surface or an increased number of mobile dislocations available for deformation. Therefore, even though we see that surface oxide softening exists in NiAl, we may still be far from optimizing the effect in this material.

## CONCLUSIONS

The phenomenon of surface oxide softening observed previously in bcc metals also occurs in B2 ordered alloys. In the case of NiAl the largest softening effect was found for a 350 nm delta-$Al_2O_3$ oxide coating deposited at 1000 °C. The softening effect was found to increase with increasing oxide thickness over the range of thicknesses examined. The amount of softening differs with oxide thickness and also with the type of oxide applied. This appears to be related to the stress state at the oxide-substrate interface and the efficiency of this interface to act as a dislocation source. Previous research on bcc metals illustrates that considerable optimization of the softening effect in NiAl is possible.

## ADKNOWLEDGEMENTS

This research was supported by the National Science Foundation, Grant No. DMR 82-03078.

## REFERENCES

1. C.T. Liu and J.O. Steigler, Science 226, 636(1984).
2. H. Hindman and D.P. Whittle, Oxid. Met. 18, 245(1982).
3. J.H. Westbrook, J. Electrochem. Soc. 103, 54(1956).
4. V.K. Sethi and R. Gibala, Scripta Met. 9, 527(1975).
5. V.K. Sethi and R. Gibala, Thin Solid Films 39, 79(1976).
6. V.K. Sethi and R. Gibala, Acta Met. 25, 321(1977).

7. V.K. Sethi and R. Gibala, in Surface Effects in Crystal Plasticity, R.M. Latanision and J.T. Fourie (eds.), p. 599, Noordoff, Int. Pub., Reading, Mass.(1977).
8. V.K. Sethi and R. Gibala, Phil. Mag. 37, 419(1978).
9. J.E. Talia, L. Fernandez and R. Gibala, Scripta Met. 12, 737(1978).
10. T. Tottori, J.E.Talia and R. Gibala, Scripta Met. 14, 1153(1980).
11. K. Kojima, S. Kobayashi and M. Meshii, Scripta Met. 10, 347(1976).
12. S. Kobayashi and M. Meshii, Acta Met. 25, 1515(1977).
13. K. Kojima and M. Meshii, phys. stat. sol. (a) 39, 491(1977).
14. J.W. Christian, Met. Trans. 14A, 1237(1983).
15. M. Yamaguchi, in Mechanical Properties of BCC Metals, M. Meshii (ed.), p. 31, The Metallurgical Society of AIME, Warrendale, Pa. (1982).
16. R.T. Pascoe and C.W.A. Newey, Metal Sci. J. 5, 50(1971).
17. S. Takeuchi, Phil. Mag. 41, 541(1980).
18. J.K. Doychak, M.S. Thesis, Case Western Reserve, Cleveland, Ohio (1984).
19. G.C. Rybicki, NASA Lewis Research Center, Cleveland, Ohio (personal com.).
20. C. Lowell and G. Santoro, NASA TN D-6838 (1972).
21. H.M. Hindman and W.W. Smeltzer, J. Electrom. Soc. 127, 1630(1980).
22. J.E. Talia, L. Fernandez, V.K. Sethi and R. Gibala, Fifth Int. Conf. on Strength of Metals and Alloys, p. 127, Pergamon Press, New York (1979).
23. J.W. Christian, Proc. Second Int. Conf. Strength of Metals and Alloys, p. 31, ASM(1970).
24. W.C. Leslie, Met. Trans. 3, 5(1972).
25. A. Sato and M. Meshii, Scripta Met. 8, 851(1974).
26. W.A. Spitzig and A.S. Keh, Acta Met. 18, 611(1970).
27. A. Sato and M. Meshii, phys stat. sol. (a) 28, 561(1975).
28. M. Yamaguchi and V. Vitek, J. Phys. F: Metal Physics 3, 523(1973).
29. M. Yamaguchi and V. Vitek, J. Phys. F: Metal Physics 3, 537(1973).
30. W. Yang, R.A. Dodd and P.R. Strutt, Met. Trans. 3, 2049(1972).
31. T.C. Tisone, G.W. Marshall and J.O. Brittain, J. Appl. Physics 39, 3714 (1972).
32. R.S. Polvani, P.R. Strutt and B.H. Kear, Proc. 4th Int. Conf. on Strength of Metals and Alloys, Vol. 1, p. 314(1976).
33. W.J. Yang and R.A. Dodd, Met. Sci. J. 7, 41(1973).
34. H.C. Liu and T.E. Mitchell, J. Nuc. Mat. 107, 318(1982).
35. A. Parthasarathi and H.L. Fraser, Phil. Mag. A 50, 89(1984).
36. E.P. Lautenschlager, T.C. Tisone and J.O. Brittain, phys. stat. sol. 20, 443(1967).
37. A. Ball and R.E. Smallman, Acta Met. 14, 1517(1966).
38. R.J. Wasilewski, S.R. Butler and J.E. Hanlon, Trans. Met. Soc. AIME 239, 1351(1967).
39. M.H. Loretto and R.J. Wasilewski, Phil. Mag. 23, 1131(1971).
40. J. Bevk, R.A. Dodd and P.R. Strutt, Met. Trans. 4, 159(1973).
41. H.L. Fraser, R.E. Smallman and M.H. Loretto, Phil. Mag. 28, 651(1973).
42. V.K. Sethi, R. Gibala and A.H. Heuer, Amr. Cer. Soc. Bull. 57, 308(1978).
43. R.E. Pawel and J.J. Campbell, Acta Met. 14, 1827(1966).
44. J.V. Cathcart (ed.), Stress Effects and the Oxidation of Metals, AIME, New York (1975).
45. M.F. Ashby, in Strengthening Mechanisms in Crystals, A. Kelly and R.B. Nicholson (eds.), p. 137, Halsted/Wiley, New York (1971).
46. T. Tottori, M.S. Thesis, Case Western Reserve U., Cleveland, OH (1982).
47. J.E. Talia, Ph.D. Thesis, Case Western Reserve U., Cleveland, OH (1980).
48. Z. Lixin, Y. Yaode, L. Liguang, Z. Qi, Z. Bingshen, in High Temperature Corrosion, R.A. Rapp (ed.), NACE, Houston, Texas (1983).
49. G.D. Oxx, Prod. Eng., Jan. 20 (1958). A.J. Kumnick and L.J. Ebert, NASA Report No. NAG 3-39(1981).
50. W.D. Kingery, H.K. Bowen and D.R. Uhlmann, Introduction to Ceramics, 2nd. ed., John Wiley and Sons, New York(1976).

CREEP BEHAVIOUR OF THE ORDERED INTERMETALLIC (FE,NI)AL PHASE

M. RUDY AND G. SAUTHOFF
Max-Planck-Institut für Eisenforschung Gmbh.,
D-4000 Düsseldorf, Federal Republic of Germany

## Abstract

The deformation behaviour of the ordered intermetallic (Fe,Ni)Al phase with B2 structure is studied as a function of composition and temperature with stress strain tests and creep tests. The alloy composition has been varied with respect to stoichiometry and the Fe/Ni ratio. The creep resistivity depends on the alloy composition in a complex manner: a deviation from the stoichiometric compositions leads to a reduction of the creep resistance, and there is a maximum of the creep resistance for the ternary alloys with about 10% Fe.

## 1. Introduction

Up to now mainly austenitic alloys on the basis of iron, nickel and cobalt have been used for applications above 600°C. Ferritic alloys offer advantages with respect to thermal conductivity, thermal expansion and costs. The intermetallic (Fe,Ni)Al phase is a ferritic alloy with an ordered distribution of atoms, which is considered for high temperature applications [1]. This (Fe,Ni)Al phase has a high oxidation resistance even in atmospheres with a low oxygen potential [2]. The aim of this work is to study the deformation behaviour of the intermetallic (Fe,Ni)Al phase as a function of temperature and composition.

## 2. Alloys

In the system Fe-Ni-Al the B2 structure is obtained in a great range of compositions. The critical temperature of ordering is of the order of the melting temperature. Fig.1 [3] shows the compositions of the alloys which are used in this study.

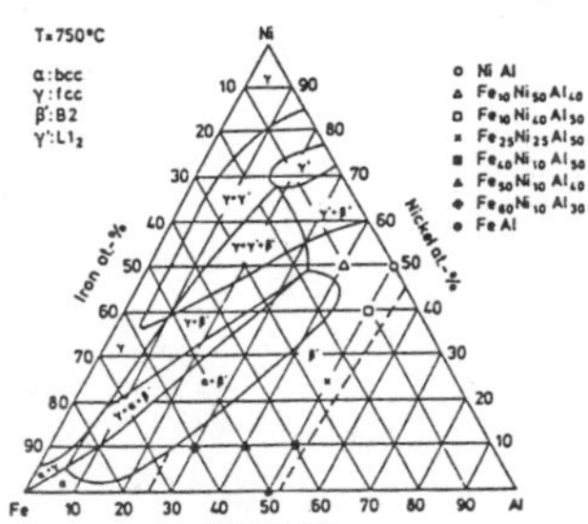

fig.1: Phase diagram at 750°C and composition of the alloys

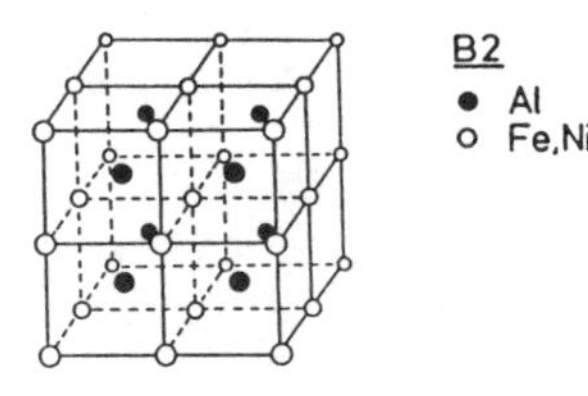


fig.2: Atom distribution in the (Fe,Ni)Al phase

The compositions of the single phase alloys vary with respect to the structural disorder (aluminum contents between 30 at% and 50 at%) and to the Fe/Ni ratio. In the completely ordered B2 structure one sublattice is occupied by Al atoms, whereas Fe and Ni atoms are randomly distributed on the other sublattice (fig.2).

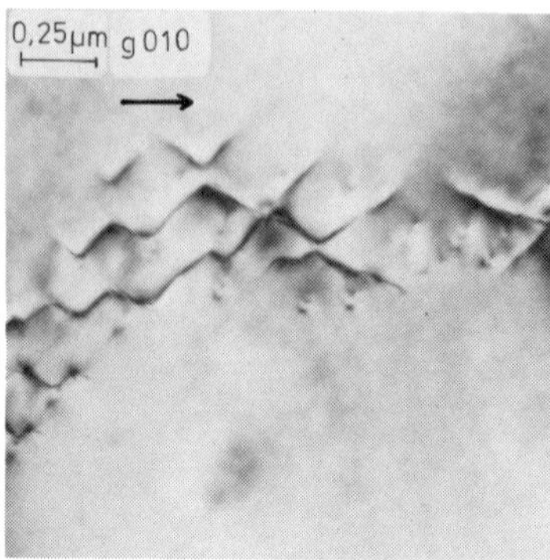


A deviation from the stoichiometric composition leads to Ni atoms on the Al sublattice in the case of an Al deficiency (antistructure atoms) whereas in the case of Al excess vacancies are formed on the Fe/Ni sublattice [4].All alloys were completely ordered,as confirmed by electron micoscropy.Fig.3 is a (100) dark field image of the undeformed alloy Fe10Ni50Al40,which contains 10 at%Fe, 50 at% Ni, 40at% Al. Only subgrain boundaries,but no antiphase boundaries were observed in agreement with [5].

fig.3: TEM micrograph of undeformed Fe10Ni50Al40 with subboundaries

## 3. Experimental procedure

The alloys were prepared in a vacuum induction furnace. Specimens for compression tests were produced by spark erosion. The specimens were solution treated in argon at 1200°C for 24h. The grain size of the studied alloys varies between 20 μm and 3mm.

The mechanical short term behaviour was studied by compression tests (with constant strain rate of $10^{-4}$ $s^{-1}$) with temperatures between 300°C and 1000°C. The dependance of the secondary creep rate on the applied stress at temperatures between 650°C and 1100°C was measured in compression tests with stepwise load increases (with creep rates between $10^{-9}$ $s^{-1}$ and $10^{-5}$ $s^{-1}$).

## 4. Creep Results

### 4.1 Temperature Dependance

With respect to the mechanical short term behaviour fig.4 shows the temperature dependance of the 0.2% proof stress for some Ni-rich alloys. The proof stress is nearly constant up to about 500°C for a given alloy. The steep decrease of the flow stress between 500°C and 800°C is correlated with the transition from brittle to ductile behaviour. For the stoichiometric composition (50% Al) the admixtures of 10% Fe lead to a hardening effect in the whole temperature range. A structural disorder (40% Al) leads to a hardening effect at low temperatures and to a softening at high temperatures.

Fig.5 shows the dependance of the creep resistance (at a given secondary creep rate) on the inverse homologeous temperature for various stoichiometric alloys.

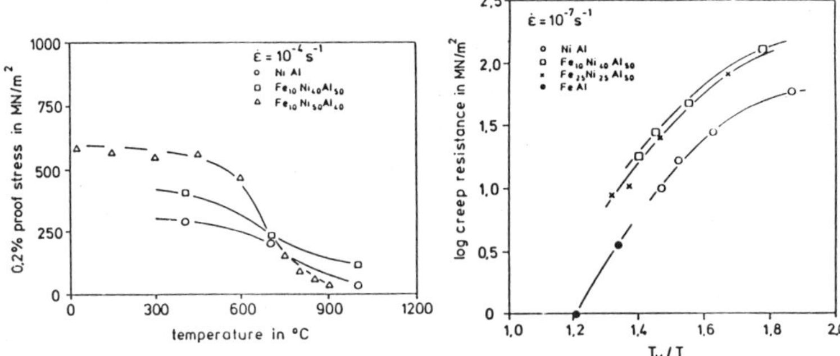


fig.4 : Temperature dependance of the 0.2% proof stress for Ni-rich alloys

fig.5: Temperature dependance of the creep resistance

The data of the binary alloys, as measured in a stepwise loading test are located on a common curve and also the data of the ternary alloys group in a common scatterband.

## 4.2 Stress Dependance

A creep curve of a non-stoichiometric alloy (measured at 750°C in a tensile creep test) is presented in fig.6 (alloy : Fe10Ni50Al40). The secondary stage is reached after about 13% primary strain. In fig.7 the secondary creep rate is plotted for various alloys as a function of the applied stress according to the Dorn equation [6] :

$$(1) \qquad \dot{\varepsilon} = \frac{A\,D\,b\,G}{k\,T}\left(\frac{\sigma}{G}\right)^{n}$$

with $\dot{\varepsilon}$ = secondary creep rate, $\sigma$ = applied stress, D= diffusion coefficient, b = burgers vector, G = shear modulus, A = dimensionless constant, n = stress exponent. The stress exponents as determined from fig.7 vary between 3 and 4.5, and depend on the alloy composition. On one hand the stress dependance of the secondary creep rate is higher for binary alloys (n=4) than for ternary alloys, on the other hand the stress exponent decreases with increasing Fe/Ni ratio.

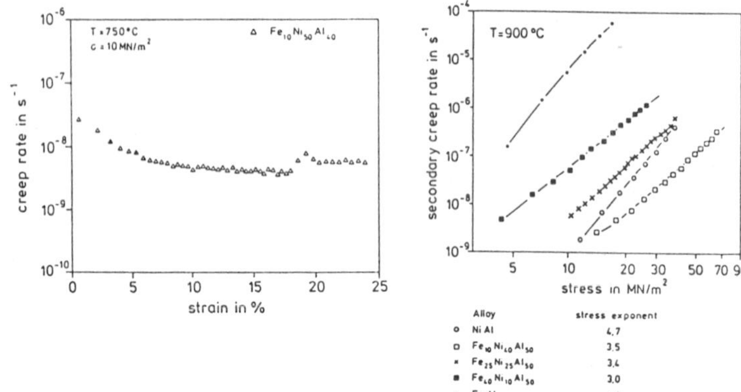


fig.6: strain dependance of the creep rate as measured in a tensile creep test

fig 7.: stress dependance of creep

## 4.3 Composition effects

The creep resistance depends in a sensitive way on the compostion. The influence of the Al content (structural disorder)on the creep resistance at 900°C is visible in fig.8a. The creep resistance decreases with decreasing Al content and it is noticed, that this effect is more pronounced for the Ni-rich alloys than for the Fe-rich alloys (Fe/Ni ratio 1). This means a strong dependance of the creep resistance on the Fe/Ni ratio as demonstrated in fig.8b. In particular the addition of a third component increases the creep resistance compared with the corresponding binary phases FeAl and NiAl. A maximum in the creep resistivity is obtained at all temperatures for alloys with Fe contents of about 10%.

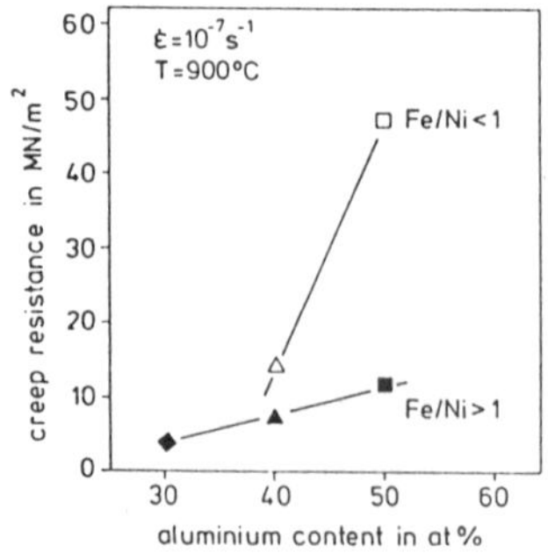


fig.8a: Creep resistance as a function of the Al content

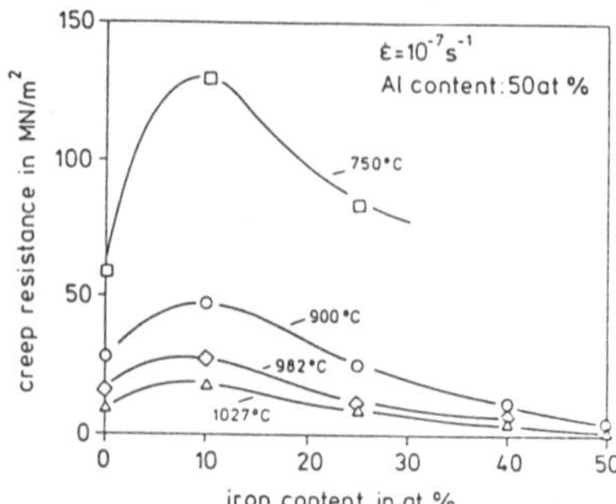


fig.8b: Creep resistance as a function of the Fe content

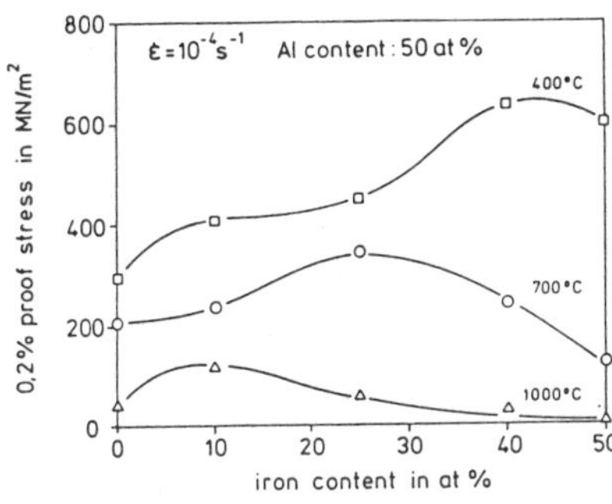


The results of the stress strain tests show a similar behaviour. The dependance of the 0.2% proof stress ($\varepsilon = 10^{-4}\ s^{-1}$) on the composition is presented in fig.9. The strength of the ternary alloy is higher than the strength of the binary phases FeAl and NiAl. With decreasing temperature the strength is displaced to higher contents of Fe. However, it is noted that at low temperatures the Fe-rich alloys show higher strengths than the Ni-rich alloys.

fig.9: 0.2% proof stress as a function of the Fe/Ni ratio for various temperatures

## 4.4 Microstructure after deformation

The microstructure of the (Fe,Ni)Al alloys changes during creep. The grain boundaries in the Ni-rich alloys have a sawtooth like appearance (fig.10). In these alloys a substructure develops during creep (fig.11). Superlattice dislocations ( b=a/2⟨111⟩), which are typical for the deformation of ordered alloys, could not be detected. Only dislocations of the type a⟨100⟩ have been observed within the subgrains.

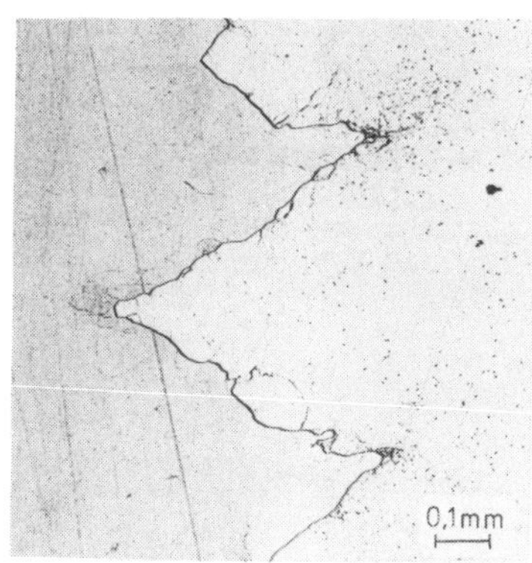


fig.10: Grain boundary in the alloy Fe10Ni40Al50 after deformation at 1027°C

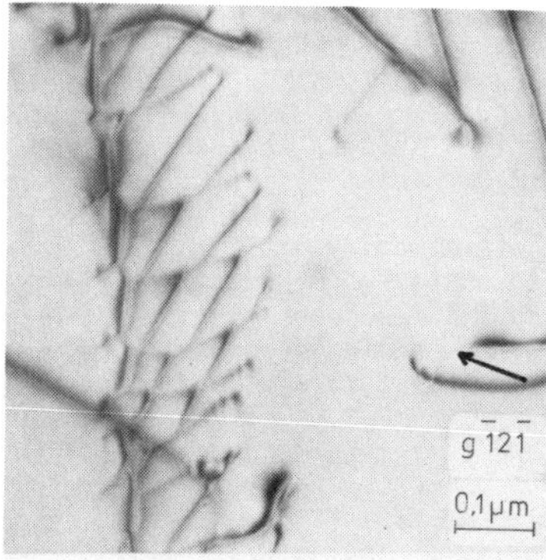


fig.11: Subgrain boundaries in Fe10Ni40Al50 after creep deformation at 900°C (100) dark field image

The development of the microstructure of Fe-rich alloys (Fe40Ni10Al50) differs remarkably from that in Ni-rich alloys. The changes of the grain boundary structure are rather limited, and in the grains only dislocation tangles are observed (fig.12). The type of dislocations could not yet be determined.

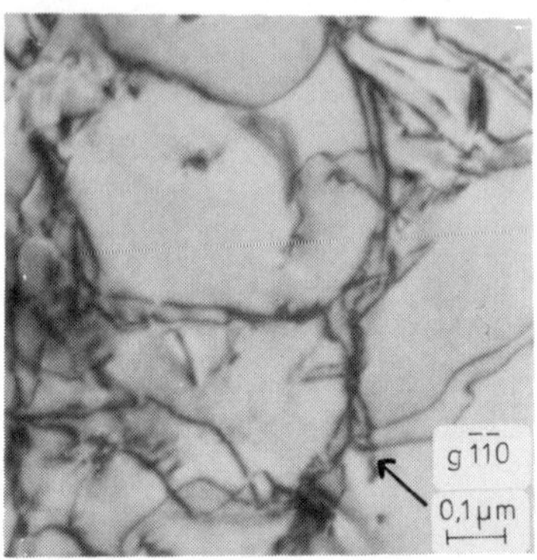


fig.12: Dislocation tangles in Fe40Ni10Al50 after creep deformation at 900°C, (110) light field image

## 5. Discussion

According to fig.4 the reduction of the Al content in the (Fe,Ni)Al phase leads to contrasting effects at low and high temperatures. This was already observed in pure NiAl alloys [7]. Below 50% Al, which is the stoichiometric composition, a decreasing Al content means an increasing number of Ni atoms on Al sites (antistructure atoms), i.e. an increasing degree of disorder. Then the resulting low temperature strengthening may be interpreted as a solid solution strengthening. This is to be checked on the basis of the available theoretical models in a quantitative way. At high temperatures the deformation is controlled by non-conservative dislocation movements, i.e. the rate is determined by diffusion processes. Then the lower high temperature strength of the Al deficient alloy would indicate a higher diffusion coefficient resulting from the presence of the antistructure atoms. A corresponding dependance of the diffusion coefficient on the Al content was indeed observed for NiAl alloys [8].

The temperature dependance of the creep resistivity (fig.5) is expected to be described by the Dorn equation (eq.1). With the respective stress exponents activation energies between 270 kJ/mol (for the Fe-rich alloys) and 350kJ/mol (Ni-rich) are obtained from fig.5. Furthermore the activation energies of the binary alloys FeAl and NiAl are higher than those of the ternary ones. In any case the obtained activation energies are higher than the activation energies of diffusion [8,9] which has to be examined in more detail with respect to the mechanisms of creep and diffusion.

Disordered metal alloys are conventionally classified in the following way [10]: class 1- alloys with stress exponent 3 resulting from viscous dislocation glide without substructure formation, class 2 - pure metals and alloys with stress exponent 4 or 5 resulting from dislocation climb with substructure formation. With respect to the stress exponents the ternary (Fe,Ni)Al would be class 1 and the binary alloys FeAl and NiAl would be class 2. However, this is not consistent with the observed microstructures. Subgrains were observed only in the Ni-rich alloys, whereas the Fe-rich alloys show dislocation tangles.

These observations as well as the observed activation energies may be correlated with the composition effects (fig.8 and 9). On one hand the stoichiometric ternary alloys show an increased strength compared with the binary ones. In this case the composition differences only affect the atom distribution on the (Fe,Ni) sublattice. Therefore this may again be regarded as solid solution hardening which has to be examined on the basis of theoretical models. On the other hand there may be different deformation mechanisms for Fe-rich and Ni-rich alloys since the antiphase energy increases from about 160 $mJ/mm^2$ for FeAl to about 400 $mJ/mm^2$ for NiAl [11]. The APB energy controls the energy - in particular the spacing between the partial dislocations - of the superlattice dislocations [12,13], i.e. an APB energy increase favours the formation of single dislocations of the $a\langle 100\rangle$ type. Indeed $a\langle 100\rangle$ dislocations were observed in the Ni-rich alloys, whereas in the Fe-rich alloys superlattice dislocations and single dislocations are to be expected as reported for FeAl [14]. This is presently studied in more detail.

## Literature


1. N.S. Stoloff, Int. Met. Rev. 29, 123 (1984)
2. A. Rahmel and W. Schwenk in Korrosion und Korrosionsschutz Verlag Chemie, 209 (1979)
3. M.A. Bradley, JISI 168, 233 (1951)
4. W.C. Hagel in Intermetallic Compounds ed. J.H. Westbrook, John Wiley, 380 (1967)
5. N.S. Stoloff and R.G.Davies, Acta Met 12, 473 (1964)
6. B. Ilschner Hochtemperaturplastizität Springer Verlag Berlin, (1973)
7. R.R. Vandervoort, A.K. Mukherjee and J.E. Dorn, Trans. ASM 59, 931 (1966)
8. G.G. Hancock and B.R. Mc.Donell, Phys. Stat. Sol.(a) 4, 143 (1971)
9. W.C. Hagel in Intermetallic Compounds ed. J,H, Westbrook, John Wiley, 383 (1967)
10. O.D. Sherby and P.M. Burke, Progr. Mat. Sci. 13, 325 (1968)
11. G. Inden, unpublished
12. N.S. Stoloff and R.G. Davies, Progr. in Mat. Sci. 13, 1 (1966)
13. A. Ball and R.E. Smallmann, Acta Met. 14, 1517 (1966)
14. Y. Umakoshi and M.Yamaguchi, Phil. Mag. 44, 711 (1981)

# PROPERTY COMPARISON OF MELT-SPUN RIBBONS AND CONSOLIDATED POWDERS OF $Ni_3Al$-B

K-M CHANG, A.I. TAUB AND S.C. HUANG
Metallurgy Laboratory, General Electric Company, Corporate Research and Development Center, P.O. Box 8, Schenectady, New York 12345 U.S.A.

## ABSTRACT

Ductile intermetallic $Ni_3Al$-B alloys have been processed through three rapid solidification techniques: melt spinning, gas atomization, and plasma deposition. Different thermal treatment was required for each rapidly solidified product to form samples for mechanical evaluation. Melt-spun ribbons were tested in ribbon form in the as-cast or annealed conditions, while atomized powders were tested after consolidation by hot isostatic pressing (HIP) or low pressure plasma deposition. Alloy strength, as well as tensile ductility, was found to depend strongly on processing technique, thermal treatment, sample geometry, and most importantly, alloy chemistry. Microstructural and fractographic observations indicate that grain boundary brittleness caused by material processing history plays the major role in determining the alloy's mechanical behavior.

## INTRODUCTION

The positive temperature dependence of the flow stress, high work hardening rate, and excellent oxidation resistance at high temperature have attracted great interest in $Ni_3Al$. Since Guard and Westbrook published their extensive work on this alloy system in 1959 [1], many research efforts have been devoted to investigating the properties of this ordered, $L1_2$ intermetallic material [2]. However, the technology for practical engineering application has been delayed by the brittleness problem; $Ni_3Al$, like other intermetallic alloys, shows no tensile ductility at room temperature. A major breakthrough was achieved by Aoki and Izumi who added a trace amount (0.05 wt%) of boron to polycrystalline $Ni_3Al$ alloys [3]. They reported a tensile elongation of 35% in their $Ni_3Al$-B alloys in contrast to 0% in the undoped alloy.

In spite of successful laboratory results, large scale fabrication and processing of $Ni_3Al$-B type alloys through conventional casting and hot working techniques has encountered serious problems [4]. The difficulties are associated with the intrinsic nature of this type of intermetallic alloy. First, the slow cooling rate of large ingot casting results in dendritic segregation. Since the ductility depends on both stoichiometry and boron level [5], the segregation creates brittle sections in the ingot and difficulties for subsequent processing. Second, the ductility of ordered $Ni_3Al$-B type alloys, like the yield strength, exhibits an abnormal temperature dependence [6]. The tensile elongation decreases monotonically as temperature increases. The ductility level at the regular forging temperature range is enough for engineering applications but far short of that required for ingot conversion practice.

In our work, rapid solidification has been used to overcome the problems outlined above. The rapid cooling eliminates heavy segregation and generates homogeneous chemistry so that the material is ductile throughout, even in the as-cast condition [7]. Therefore, extensive thermo-mechanical processing is not required to homogenize the alloy, and near net shape

TABLE I. Chemical Composition and Processing Technique of Rapidly Solidified $Ni_3Al$-B Alloys

| Atomic Formulation | Composition, wt.% Ni | Al | B | Process |
|---|---|---|---|---|
| $(Ni_{75}\ Al_{25})$-1.0B | 86.54 | 13.26 | 0.20 | -melt spinning<br>-HIP'ed powder |
| $(Ni_{76}\ Al_{24})$-0.75B | 87.19 | 12.65 | 0.16 | -HIP'ed powder<br>-plasma spray |
| $(Ni_{76}\ Al_{24})$-0.5B | 87.16 | 12.73 | 0.11 | -melt spinning |

products can be obtained by thermal treatment only. This paper reports the properties measured from melt-spun ribbons and gas atomized powders of $Ni_3Al$-B type alloys. The powders were consolidated by either hot isostatic pressing (HIP) or low pressure plasma deposition. In addition to the property comparison, the mechanisms related to the differences in mechanical behavior of various rapidly solidified $Ni_3Al$ materials are discussed.

## EXPERIMENTAL

Among the many rapidly solidification techniques being applied to $Ni_3Al$ alloys, three processes, melt spinning, gas atomization, and plasma deposition, were employed for this study. Table I lists the alloy compositions and processes. The starting materials were prepared by vacuum induction melting using high purity Ni, Al, and NiB.

Alloy ingots were remelted in vacuum and melt-spun into ribbons approximately 6 mm wide by 35 μm thick, or gas atomized into powders. Screened powders of -106μm(-140 mesh) were canned and HIP'ed at 1150°C/103MPa for two hours. Plasma deposits were made from powder sieved to -38μm(-400 mesh). Details of each rapid solidification process have been described elsewhere [8,9].

Room temperature tensile tests were performed on the materials processed by the three techniques in both as consolidated and annealed forms. While the melt-spun ribbons were tested without any specimen preparation, the consolidated powders were machined to subsized tensile specimens for testing. The tensile specimens from the plasma deposit were taken along the deposition plane. Metallography and fractography were carried out following standard laboratory procedures. An etching solution of 10 ml $HNO_3$ + 40 ml HCl + 50 ml $H_2O$ was used to reveal the grain structure.

## RESULTS AND DISCUSSION

Fig. 1 shows the cross-sectioned microstructures of melt-spun ribbons and atomized powders of the $Ni_3Al$-B alloys studied. The two major advantages of rapid solidification, homogeneous chemistry and fine grain structure, are revealed in the as-solidified microstructure. Argon atomized powder (Fig. 1b) shows dendritic growth in all particle sizes, but the spacing of secondary dendrite arms is small, between 1 to 2 μm. The cooling rate corresponding to such a dendrite arm spacing is estimated to be on

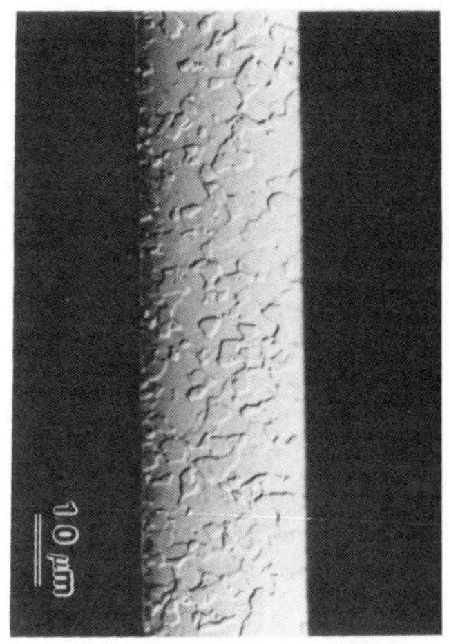

(a) melt-spun ribbon, microcrystalline (b) atomized powder, fine dendrite

Fig. 1 As solidified $Ni_3Al$-B microstructures reflecting the difference in solidification cooling rate.

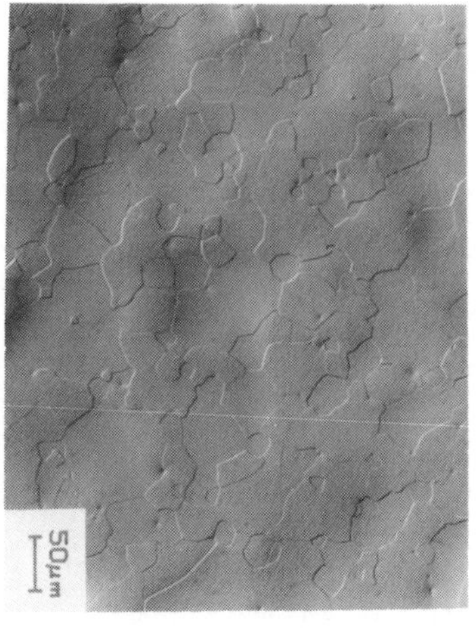


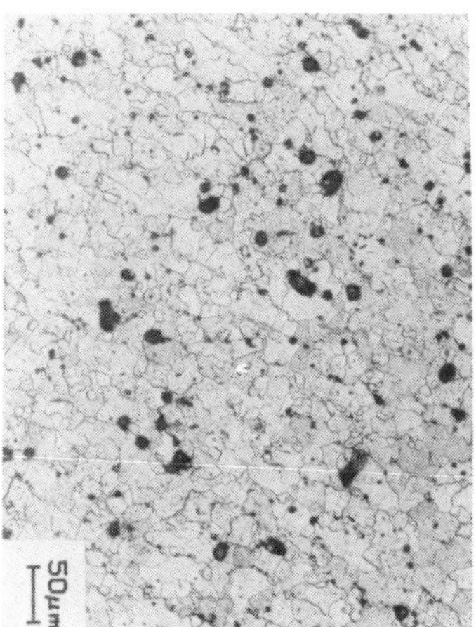

(a) HIP'ed powder, equiaxed (b) plasma deposition, layered

Fig. 2 Grain growth of $Ni_3Al$-B alloys after annealing.

the order of $10^4$ °C/s. Fig. 1a shows that the microcrystalline equiaxed grain structure developed in the ribbon foils completely supresses the dendrite formation. An average grain size of 5 μm distributed uniformly through the ribbon thickness is found with no evidence of chemical segregation. Detailed characterization of ribbon microstructure has been reported previously [10]. A fast cooling rate during solidification, on the order of $10^6$ °C/s, is expected for the melt spinning process accounting for the different microstructure.

All the alloy compositions studied have a single phase, $L1_2$ ordered structure, and high temperature annealing results in significant grain growth. For example, two-hour annealing at 1150°C increases the grain size to 40 μm as shown in Fig. 2a. As a result, annealed ribbons may only consist of 1 to 2 grains through their thickness.

Plasma-spraying $Ni_3Al$-B produces a deposit with an extremely fine microcrystalline structure and a grain size of 2 μm. Many voids are observed in aligned layers parallel to the deposition plane. Further modification of spraying parameters is required to produce void-free deposits. Annealing at 1165°C for two hours eliminates some voids and increases the grain size to 10 μm (Fig. 2b).

The room temperature tensile results are listed in Table II. Every sample studied exhibits some ductility, but variations of two orders of magnitude are observed from one material to the other. In the next two sections, the measured properties from the differently processed $Ni_3Al$-B

TABLE II. Room Temperature Tensile Properties of Rapidly Solidified $Ni_3Al$-B Alloys

| Alloy Composition | Heat Treatment | .2% Yield Strength MPa (ksi) | Tensile Strength MPa (ksi) | Elongation (%) |
|---|---|---|---|---|
| | *** MELT-SPUN RIBBON *** | | | |
| $(Ni_{75}\ Al_{25})$-1.0B | As cast | 731 (106) | 772 (112) | .6 |
| $(Ni_{75}\ Al_{25})$-1.0B | 1100°C/2hrs | 414 ( 60) | 420 ( 61) | .4 |
| $(Ni_{76}\ Al_{24})$-0.5B | As cast | 627 ( 91) | 779 (113) | 10 |
| $(Ni_{76}\ Al_{24})$-0.5B | 1100°C/2hrs | 262 ( 38) | 690 (100) | 23 |
| | *** HIP'ED POWDER *** | | | |
| $(Ni_{75}\ Al_{25})$-1.0B | As HIP | 496 ( 72) | 952 (138) | 13 |
| $(Ni_{75}\ Al_{25})$-1.0B | 1100°C/2hrs | 503 ( 73) | 586 ( 85) | 3 |
| $(Ni_{75}\ Al_{25})$-1.0B | 1000°C/2hrs | 496 ( 72) | 1062 (154) | 17 |
| $(Ni_{76}\ Al_{24})$-0.75B | As HIP | 455 ( 66) [469 ( 68)] | 1311 (193) | 45 |
| $(Ni_{76}\ Al_{24})$-0.75B | 1000°C/1hr | 455 ( 66) [495 ( 72)] | 1324 (192) | 46 |
| | *** PLASMA SPRAY DEPOSITION *** | | | |
| $(Ni_{76}\ Al_{24})$-0.75B | As deposited | 731 (106) [758 (110)] | 1413 (205) | 26 |
| $(Ni_{76}\ Al_{24})$-0.75B | 1165°C/2hrs | 538 ( 78) | 1420 (206) | 34 |

[ ] is the value of upper yield point.

alloys are compared in two respects: process and Ni-Al stoichiometry. The differences will also be correlated with the microstructural observations.

Process

For any given composition, both as-cast ribbons and as-sprayed plasma deposits have a higher yield strength in comparison to as-HIP'ed powders. Since a higher cooling rate occurs during solidification in both former cases, and HIP'ing anneals out the cast structure in the latter case, the Hall-Petch grain size effect is believed to be the major strengthening mechanism to account for the difference in flow stress. Perhaps residual strains induced by the rapid quench also contribute some portion of high strength in the as-solidified materials. The grain size effect on strength can be further reasoned by the annealing results. HIP'ed powders maintain their yield strength in the annealed samples as long as the annealing is performed at a temperature below that used for HIP'ing, to ensure no further grain coarsening. Furthermore, moderate grain growth in annealed plasma deposits decreases the strength to a level about 75 MPa (10 ksi) higher than that of annealed HIP'ed powder; and this strength difference is completely attributable to grain size. However, annealed ribbon foils always show a lower value of yield strength compared to HIP'ed powder, even after the compensation for boron variation [11]. The apparent explanation is that the specific sample geometry of ribbon foils only allows the measurement of minimum value of bulk material properties. In spite of the minor discrepancy of data, the ribbons faithfully reflect the qualitative effects of alloying elements and heat treatments.

Ni-Al Stoichiometry

The boron ductilization effect in $Ni_3Al$ ribbons has been extensively described previously [5]. Maximum room temperature ductility was found to be related to both boron and aluminum concentrations. For the annealed ribbons, substoichiometric aluminum content (Ni:Al = 76:24) has the highest boron solubility and the best tensile ductility. The same observation was found here for powder in both HIP'ed form or plasma deposit. With a substoichiometric aluminum ratio, the as HIP'ed powder exhibits a tensile elongation of 45%, in contrast to 13% for 25 aluminum. The fracture surface of ductile $Ni_3Al$-B alloys reveals a complete transgranular failure mode as seen in Fig. 3a.

Depending on the stoichiometry, high temperature annealing has various effects on alloy ductility. At 25 aluminum, the alloy suffers a serious ductility loss when annealed above 1100°C. The embrittled sample fractures after a small amount of deformation with a completely intergranular fracture surface (Fig. 3b). As the annealing temperature is decreased to 1000°C, the embrittlement effect does not appear; instead, there is a tendency for the ductility to be improved somewhat. Annealing embrittlement does not occur in the alloys with substoichiometric aluminum. The exact mechanism is still under investigation. It is believed to be associated with the redistribution of grain boundary chemistry.

The layered structure with voids in plasma deposits seems to not affect the ductility along the deposition plane. However, on the tensile fracture surface, Fig. 3c, secondary cracks in the deposition plane are observed, with even spacing at a distance of 100 μm. The secondary cracking is caused by the layered structure, and might have a great influence on the ductility in the transverse direction.

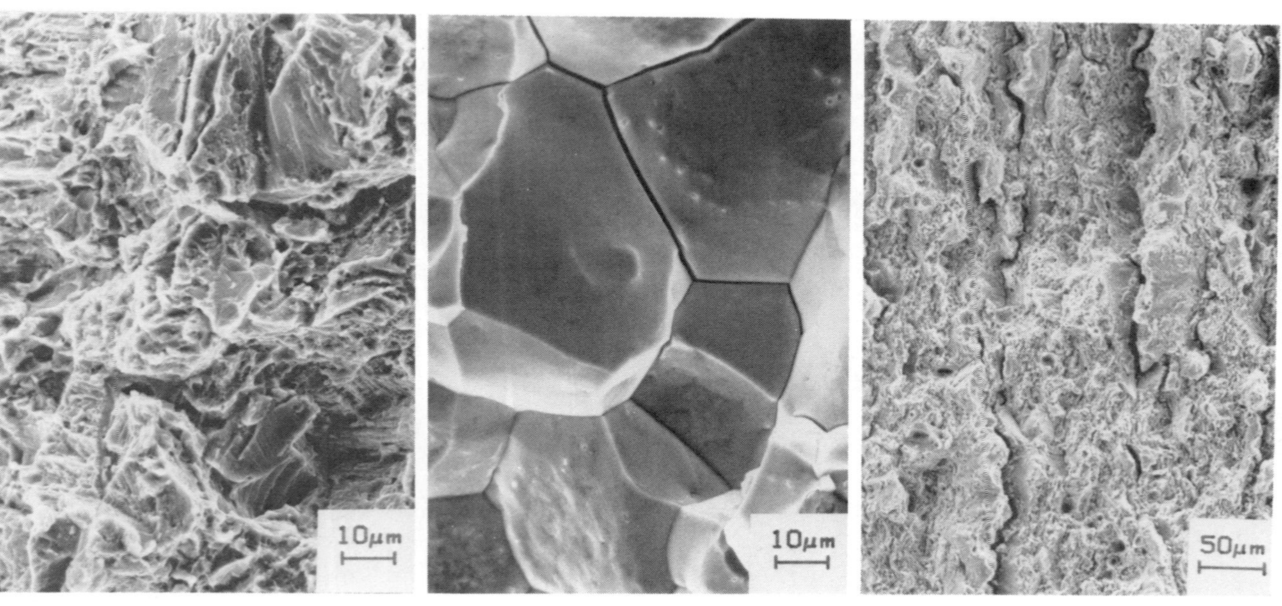


(a) ductile transgranular mode (b) brittle intergranular mode (c) secondary cracking

Fig. 3 Three representative fracture morphologies observed in $Ni_3Al$-B alloys.

Finally, a yield point phenomenon is found during the testing of 24 Al powders as noted in Table II. Solid solution strengthening of interstitial boron atoms [11] is likely to be the explanation, but a more systematic study is needed.

## SUMMARY

1. Ductile intermetallic $Ni_3Al$-B alloys can be successfully fabricated by rapid solidification techniques that provide homogeneous chemistry and uniform fine grain structure.

2. HIP'ed powders provide a direct and commercially feasible route for the utilization of ductile intermetallic alloys. Near net shape HIP'ed powder articles have not only excellent tensile ductility but also ease of manufacturing.

3. Ribbon foils made by melt spinning offer a fast and effective method to screen alloy compositions. Their tensile properties will represent very well the qualitative trend of alloying effects and the low limit value of bulk materials.

4. Plasma spray deposition of $Ni_3Al$-B alloy powder can produce large, thick articles as well as wide, thin coatings, while maintaining the desirable structure of rapid solidification. The deposit exhibits a high yield strength and a good tensile ductility in the deposition plane. Optimization of spraying conditions is needed to avoid the layered structure which may induce premature cracking.

5. The boron ductilization effect in $Ni_3Al$ becomes most pronounced at 24 at.% Al. Annealing embrittlement occurs at temperatures above 1100°C only if the Al ratio is increased to 25%.

## ACKNOWLEDGMENTS

The authors wish to thank C.F. Canestraro, E.H. Hearn, R.P. Laforce, C.P. Palmer, S. Rutkowski, and L.A. Wojcik for their technical support. Helpful discussions with M.G. Benz, R.H. Bricknell and L.A. Johnson are also highly appreciated.

## REFERENCES

1. J.W. Guard and J.H. Westbrook: Trans. TMS-AIME, 1959, vol. 215, p. 807.
2. D.P. Pope and S.S. Ezz: Inter. Metals Rev., 1984, vol. 23, p. 136.
3. K. Aoki and O. Izumi: J. Japan Metal. Soc., 1979, vol. 43, p. 358.
4. D.L. Klarstrom: "$Ni_3Al$ VIM ESR Ingots," presented at ECUT Workshop on "Production and Fabrication of Ordered Intermetallic Alloys," Oak Ridge National Lab., TN, Sept. 25-26, 1984.
5. A.I. Taub, S.C. Huang, and K.M. Chang: "Stoichiometry Effects on the Strengthening and ductilization of $Ni_3Al$ by Boron Modification and Rapid Solidification," in Proc. 39th Symposium of Mechanical Failures Prevention Group, Gaitherburg, MD, May 1-3, 1984.
6. C.T. Liu, et. al.: "Development of Ordered Intermetallic Alloys for High Temperature Use," presented at ECUT Workshop on "Production and Fabrication of Ordered Intermetallic Alloys," Oak Ridge National Lab., TN, Sept. 25-26, 1984.

7. A.I. Taub, S.C. Huang, and K.M. Chang: Met. Trans. A, 1984, vol. 15A, p. 399.
8. S.C. Huang and R.P. Laforce: Mat. Res. Soc. Symp. Proc., Nov. 1983, Boston, MA, vol 28, "Rapidly Solidified Metastable Materials", ed. B.H. Kear and B.C. Giessen, p. 125.
9. M.R. Jackson, J.R. Rairden, J.S. Smith, and R.W. Smith: J. Metals., 1981, vol. 33, no. 11, p. 23.
10. K.M. Chang, S.C. Huang, and A.I. Taub: Mat. Res. Soc. Symp. Proc., Nov. 1983, Boston, MA, vol. 28, "Rapidly Solidified Metastable Materials ed. B.H. Kear and B.C. Giessen.,p. 401.
11. S.C. Huang, A.I. Taub, and K.M. Chang: Acta Met., 1984, vol. 32, p. 1703

# DAMPING AND MODULUS MEASUREMENTS IN B2 TRANSITION METAL ALUMINIDES*

M. R. HARMOUCHE AND A. WOLFENDEN
Texas A&M University, Mechanical Engineering Department, College Station, Texas, 77843

## ABSTRACT

The polycrystalline intermetallic alloys FeAl (50.9 to 58.2% Fe), NiAl (49.2 to 55.9% Ni) and CoAl (48.5 to 52.3% Co) have the B2 structure and are of interest for high temperature applications. We have used the PUCOT (piezoelectric ultrasonic composite oscillator technique) to measure mechanical damping or internal friction ($Q^{-1}$) and Young's modulus (E) as a function of temperature and composition for these materials. The modulus data for six CoAl alloys at temperatures up to 1300 K are presented. Examples are given of the strain amplitude dependence of $Q^{-1}$ for four CoAl alloys. The curves showed the break away phenomenon and are interpreted in terms of a theory dealing with the pinning of dislocation lines and their eventual break away at large strain amplitudes. The dislocation density was calculated to be about $10^8$ $m^{-2}$. For all the compositions (X1) of CoAl studied a single equation could be fitted to the data:

$$E\ (X1,T) = 210.8 + 2.2\ (X1) - 0.08T \qquad (1)$$

where T is in K and E in GPa.

## INTRODUCTION

Intermetallic compounds such as the ordered B2 Co, Fe and Ni-Al alloys have been ignored as potential materials for high temperature service, largely due to the difficulties in their processing and controlling their mechanical properties. However, recent attempts to produce these alloys using hot extrusion techniques have shown that these difficulties could be overcome [1]. The elements of these aluminides are readily available and aluminides offer many additional advantages. They exist over a wide range of composition and possess the ordered cubic B2 crystal structure. They have high melting temperatures and very high solubility for third alloying elements.

Thus, the potential of the CoAl, FeAl and NiAl aluminides as high tempature structural materials is well recognized and efforts to explore their elastic and anelastic behavior are required.

In this paper, Young's modulus and mechanical damping measurements are presented for several polycrystalline CoAl alloys as a function of temperature and composition.

## EXPERIMENTAL PROCEDURES

The experimental procedure used here is the PUCOT (Piezoelectric Ultrasonic Composite Oscillator Technique). The PUCOT experimental arrangement allows us to make simultaneous measurements of Young's modulus, damping and strain amplitude as a function of temperature and is discussed elsewhere [2,3]. The polycrystalline test specimens were cylinders of approximate dimensions 3mm in diameter and 25 to 45 mm in length. Details of their compositions and densities are given in Table 1.

*The research is supported by a grant from NASA Lewis Research Center, Program Manager Dr. J. D. Whittenberger.

Table 1. Composition,[1,2] Density, and Modulus for CoAl.

YOUNG'S MODULUS/GPa

| CoAl Percentage[3] %Co | %Al | Density[4] | T/K 298 | T/K 473 | T/K 638 | T/K 783 | T/K 998 | T/K 1143 | T/K 1303 |
|---|---|---|---|---|---|---|---|---|---|
| 48.49 | 51.31 | 5916 | 293 | 279 | 262 | 255 | 231 | 229 | 209 |
| 49.30 | 50.30 | 5956 | 294 | 281 | 268 | 263 | 241 | 221 | 213 |
| 49.89 | 49.89 | 6000 | 296 | 285 | 270 | 258 | 241 | 228 | 213 |
| 50.49 | 49.30 | 6041 | 298 | (285) | (271) | 261 | 243 | 230 | 216 |
| 51.28 | 48.50 | 6077 | 299 | (286) | (273) | 263 | 242 | 229 | 219 |
| 52.28 | 47.50 | 6127 | 302 | (287) | 274 | 262 | 241 | 231 | 222 |

1. From NASA.
2. All alloys contain nominally: 0.02 C, 0.105 H, 0.004 N, and 0.1 O.
3. Refers to atomic percent.
4. $kg/m^3$

( ) extrapolated or interpolated data.

Table 2. Granato-Luecke Analysis of CoAl Alloys (with Thermal Depinning)

| Alloy Atomic %Co | T/K | Slope of G-L Plot | Intercept of G-L Plot | $U_o$/eV | Dislocation Density $m^{-2}$ | Minor Pinning Length nm |
|---|---|---|---|---|---|---|
| 48.49 | 998 | $-3.54\times10^{-2}$ | 3.9 | 1 | $10^6$ to $10^8$ | 3.6 |
| | | | | 10 | $10^5$ to $10^7$ | 37 |
| 49.89 | 638 | $-3.29\times10^{-2}$ | 76 | 1 | $10^7$ to $10^9$ | 4.9 |
| | | | | 10 | $10^6$ to $10^8$ | 49 |

## RESULTS AND DISCUSSION

Young's modulus has been measured in the composition range 48.49 to 52.28 atomic % Co and in the temperature interval 293 to 1300 K for six CoAl alloys. In addition, mechanical damping has been measured in the strain amplitude interval about $10^{-8}$ to $10^{-4}$ and analyzed in terms of the Granato-Luecke (G-L) theory [4] with thermally assisted depinning effects. The experimental results for the modulus are discussed first, followed by those of the mechanical damping.

### Temperature and Composition Dependence of Young's Modulus for CoAl

Dynamic Young's modulus (E) has been measured as a function of composition (X1) for six CoAl alloys at seven temperatures in the range 293 to 1300 K. The results are given in Table 1. All the modulus data have been pooled to find a general correlation between E and X1 and T. Multiple regression analysis has yielded the following correlation:

$$E\ (X1,T) = 210.8 + 2.2\ (X1) - 0.08\ (T) \tag{2}$$

For this correlation R = 0.996 and the Standard Error of Estimate SE = 2.42 GPa.

Of course, the variation in X1 is small (48.49 to 52.28 at. % Co), making the composition term almost constant. Consequently, the variation in temperature remains the dominant factor in determining E from this correlation.

The modulus is plotted versus X1 in Figure 1 and T in Figure 2. Regression analysis indicates that straight lines fit the data points within the experimental errors. Ideally, these materials should be stiffest at the stoichiometric composition where the melting temperature is the highest. However, the data indicate otherwise. It may well be that the increase in the content of Co contributes a larger increase in the modulus than the decrease expected from deviation from the stoichiometric composition. Another possible explanation could be that the known directional bonding in intermetallic compounds should be a contributor to this deviation. In any case the data points indicate these alloys are very stiff. Even at 1300 K, the modulus for any composition is higher than that for steel at ambient temperature. Future crystallographic study and continued investigation of the modulus should bring about a better understanding of the elastic behavior for these alloys.

### Granato-Luecke Analysis of the Amplitude Dependent Internal Friction in CoAl

Figure 3 shows the strain amplitude dependence of internal friction $Q^{-1}$ in four alloys of CoAl at three temperatures. All the curves are of the form $Q^{-1} = Q_I^{-1} + Q_H^{-1}$, where $Q_I^{-1}$ is the amplitude independent part of the internal friction (horizontal parts of the curves) and $Q_H^{-1}$ is the amplitude dependent part (the parts of the curves that curve upwards). These curves are now analyzed in terms of the Granato-Luecke (G-L) [4] theory of amplitude dependent damping due to dislocation lines and their break away from pinning points at high strain amplitude, with thermally assisted depinning effects. The theory recommends a plot (a G-L plot) of $Q_H^{-1}/\varepsilon$ versus $\varepsilon^{-1}$ to obtain estimates of the minor pinning length of dislocation lines 1 and the total dislocation density $\Lambda$. The slope and intercept of the G-L plot are given by:

$$\text{Slope} = -\ (4/3)\ (U_0/kT)\ (U_0 G/1^3)^{\frac{1}{2}}\ (1/E)$$

$$\text{Intercept} = \Lambda L_n^2\ \nu(3\pi\ kT/2U_0)^{\frac{1}{2}}\ 1^{1.5}\ (2.66E/U_0)^{\frac{1}{2}}/6\pi\omega$$

where $L_n$ = segments of dislocations that break away

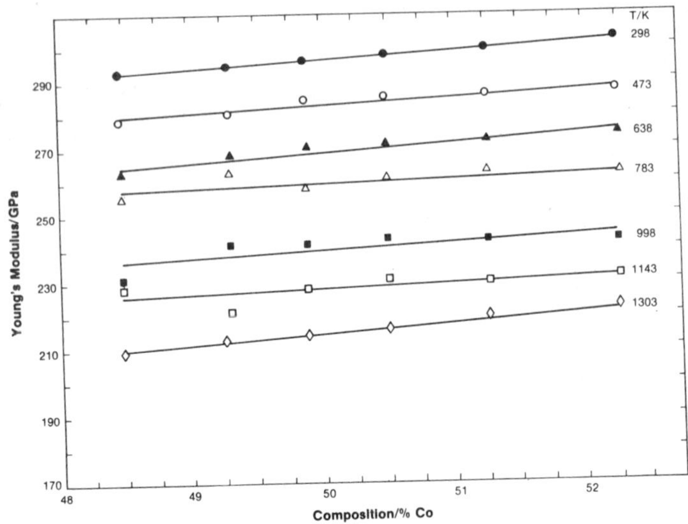


**Fig. 1:** Young's Modulus vs. Composition % Co for CoAl at Constant Temperature.

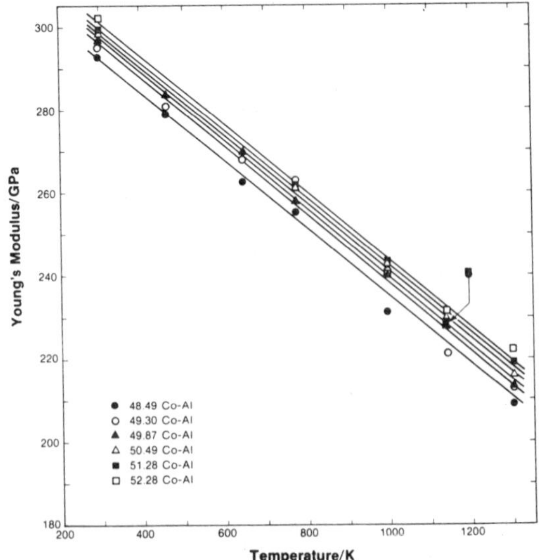


**Fig. 2:** Young's Modulus vs. Temperature for CoAl at Constant Composition.

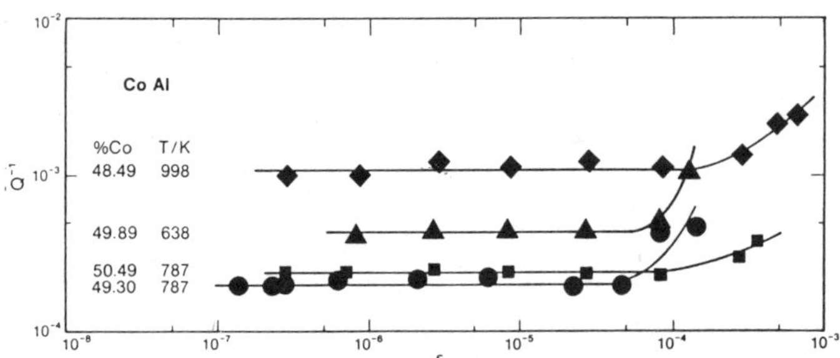


Fig. 3: Amplitude dependence of internal friction for four CoAl alloys.

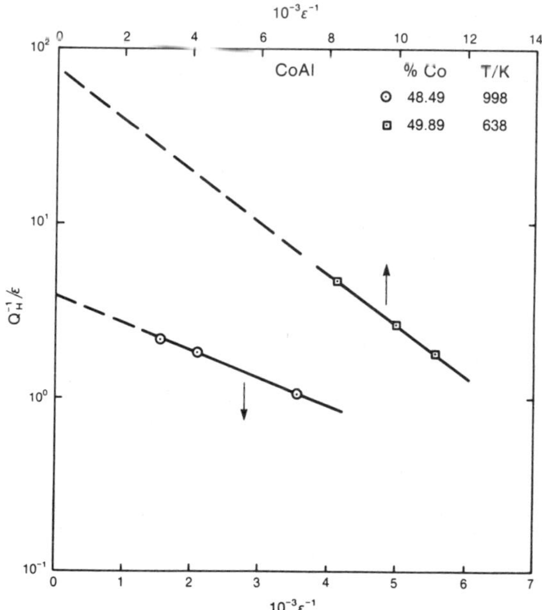


Fig. 4: Granato-Luecke plots for CoAl alloys.

$\nu$ = effective attack frequency (around 10" Hz)

$\omega$ = angular frequency of applied stress

$U_o$= total interaction energy

G = shear modulus (approximately E/2.66)

Two sets of results have been analyzed in this way (those for 48.49% Co at 998 K and 49.89% Co at 638 K) and the G-L plot is shown in Figure 4. The values of E are 232 and 270 Gpa, respectively, with $\omega$ = 5x10$^5$ Hz. From the figure, the measured values of slope and intercept yield from the G-L equations values for 1 and $\Lambda$ with $L_n$ assigned the value of 1 or 0.1 μm. A listing is given in Table 2. Clearly the values of dislocation density are governed only mildly by the choice of $U_o$. Values near 10$^8$ m$^{-2}$ are reasonable for this B2 alloy since low dislocation densities have been observed in FeAl alloys [5]. However, the values of 1 are strongly dependent on the choice of $U_o$, with realistic values of 1 only appearing for $U_o$ in the range 1 to 10 eV. Values of $U_o$ of 10 eV are larger than those usually considered in the G-L theory. However, the defect structure in B2 aluminides is unusual in that large concentrations of vacancies have been deduced to be present by several authors [6,7], and the migration energy of a vacancy in B2 NiAl was found [8] to be over 2 eV. It is likely that the impurities such as C, H, N and O, present to a level of about 0.3 atomic % in the CoAl, could account for some of the pinning points, as a calculation shows that for these impurities the equivalent value of 1 is about 5 nm (pinning points/m$^3$ = 8x10$^{24}$). A more detailed analysis of these data for aluminides and for other (simpler) systems, in terms of the G-L theory for thermal depinning is obviously needed.

## REFERENCES


1. R. W. Clarke, and J. D. Whittenberger, Thermal Expansion 8, T. Han, Editor, Plenum Press, N. Y., 1984, pp. 189-196.

2. J. Marx, Rev. Sci. Instr. 22, 503-509 (1951).

3. W. H. Robinson and A. Edgar, Institute of Electrical and Electronics Engineers Transactions on Sonics and Ultrasonics, SU-21, No. 2, 98-105 (1974).

4. A. Granato and K. Luecke, J. Appl. Phys. 52, 7136-7142, (1981).

5. J. D. Whittenberger, private communication (1985).

6. K. Ho and R. A. Dodd, Scripta Met. 12, 1055 (1978).

7. J. P. Neuman, Y. A. Chang and C. M. Lee, Acta Met. 24, 593-604 (1978).

8. A. Parthasarathi and H. L. Fraser, Phil. Mag. 50, 89-100 (1984).

# PART V

# Alloy Design and Microstructural Control

# TITANIUM ALUMINIDES - AN OVERVIEW

HARRY A. LIPSITT
Air Force Wright Aeronautical Laboratories, Materials Laboratory
Wright-Patterson Air Force Base, Ohio 45433-6533

The object of this paper is to describe a coordinated research and development program which has been pursued by an Air Force-Industry-University team for more than twelve years. The focus of our attention has been on the development, processing, and engine testing of alloys based on intermetallic compounds, specifically on the aluminides of titanium, iron, and nickel. The titanium aluminides, $Ti_3Al$ and TiAl, are the materials with which we have been working the longest and on which development has proceeded the furthest. This Symposium has provided the first opportunity to review the progress of the titanium aluminide development programs sponsored by the Air Force and some of the engine testing efforts undertaken by the engine manufacturers.

In the past twelve years our efforts have progressed from basic studies, as on the mechanisms of the ductile-brittle transitions; to alloy development studies to modify the properties of the compounds; to primary processing studies to learn to cast and forge ingots and to roll sheet; to secondary processing studies to learn to machine, grind, cut, weld, cast, and to superplastically form and diffusion bond; and finally, to studies to put all these technologies together to make real engine hardware, some pieces of which have already been successfully run in aircraft turbine engines.

In this paper the need for such new materials will be discussed, as well as the specific properties of intermetallic compounds that make them especially attractive for use in turbine engines. Next, the general properties of the titanium aluminides will be compared and contrasted with the properties of both conventional titanium alloys and nickel base superalloys. Then, a few of the early published results on the mechanisms of the ductile-brittle transition behavior will be reviewed as a means of indicating the challenges of the alloy development efforts which followed. The studies in which the fabricability of the alloys was assessed will also be reviewed. Finally, some of the actual hardware which has been successfully engine tested will be shown and the fabrication methods used will be described.

 

Now let's see why we need these materials. In designing and operating turbine engines today and in the future, there are two primary problems - both of which demand solutions from the field of materials science. These are the need to operate certain portions of the engine at higher gas and metal temperatures to improve operating efficiency and save fuel. And the need for lighter weight materials to decrease engine weight, engine operating stresses due to heavy rotating components, and to increase the operating life of disks, shafts, and bearing support structure. These latter materials should be less dense than the nickel base superalloys they are intended to replace, but are required to possess roughly the same mechanical properties and oxidation resistance as those materials in current usage.

Intermetallic compounds are particularly suited to these needs because of two properties which derive from the fact that they possess ordered structures. Modulus retention with temperature in these materials is particularly high because of the strong A-B bonding. In addition, a number of high temperature properties which depend on diffusive mechanisms are improved because of the generally high activation energy required for diffusion in ordered alloys. Included in this category are static strength retention, creep/stress-rupture, and fatigue resistance. Finally, in the case of aluminides, the oxidation resistance is particularly good because all of these materials contain a large aluminum reservoir. In many intermetallics the strong A-B bonding also results in low temperature brittleness although the exact mechanism of the ductile-brittle transition seems to be different in every case. It is necessary for the alloy developer to ease the problem of minimal low temperature ductility without destroying the valuable ordered structure which provides the high temperature strength and stiffness.

Table I compares the properties of some example aluminide base materials with those of conventional titanium and nickel base alloys. The higher aluminum content makes the aluminides somewhat less dense than titanium alloys and appreciably less dense than nickel base superalloys. These compounds are also of potential value because they are stiff, that is they have a high elastic modulus. The most important thing about high modulus is the fact that a great many static parts of an engine are sized mainly by the need to minimize elastic deflection under load at the operating temperature. An example of such parts would be the many ring shaped components in an engine. The design stresses on these parts are low, many of them are not in fracture critical positions, and the determinant design parameter is whether or not such parts are stiff enough to retain their shape in the operating environment. Conventional titanium alloys are not stiff enough to be used for many of these applications. Commercial titanium alloys show a modulus that drops rapidly with temperature to a value of about 10 million psi at 540°C. TiAl, on the other hand, shows a higher modulus at 1000°C than titanium does at room temperature [1]. $Ti_3Al$ shows a higher modulus at 815°C than titanium does at room temperature [1].

Table I: Properties of High Temperature Alloys

| | Ti Base | $Ti_3Al$ | TiAl | Superalloys |
|---|---|---|---|---|
| Density ($g/cm^3$) | 4.5 | 4.15-4.7 | 3.76 | 8.3 |
| Young's Modulus ($GN/m^2$) | 110-96 | 145-110 | 176 | 206 |
| Max. Temp. - Creep (°C) | 538 | 815 | 1038 | 1093 |
| Max. Temp. - Oxidation (°C) | 593 | 649 | 1038 | 1093 |
| Ductility - R.T. (%) | ∿ 20 | 2-5 | 1-2 | 3-5 |
| Ductility - Operating (%) | High | 5-8 | 7-12 | 10-20 |

The aluminides have adequate creep strength to much higher temperatures than do conventional titanium alloys. Note that $Ti_3Al$ runs out of oxidation resistance before it runs out of strength. Further alloying or a coating will be needed for operation above about 700°C. TiAl, however, is an alumina former and can serve uncoated to 815°C. The only apparent flaw in the aluminides was that they showed low ductility at room temperature [2,3].

An early investigation [2] showed that many of the superdislocations in TiAl were pinned by an unknown agent. A later study [3] showed that the pinning agent became ineffective at about 700°C and that the ductility of the compound increased very rapidly above that temperature. It was clear that the first step of any alloy development program for this compound must focus on eliminating or reducing the effect of the unknown agent which was causing the high ductile-brittle transition temperature. An early investigation on $Ti_3Al$ [4] showed that the ductility of this compound increased slowly with temperature to 1000°C and electron microscopic examination indicated that the slow increase in ductility was due to the appearance of an increasing density of "c + a" dislocations. It was clear that the first step of an alloy development program for this compound must focus on lowering the temperature at which the "c + a" dislocations appeared in quantity.

The alloy development efforts continue even today, but they have resulted in several compositions having a useful balance of properties. In this same time period, an evaluation was made in which the pay-off which would result from the use of several classes of aluminide components was balanced against the necessary cost and a number of other factors [5]. As a result of this analysis, it was decided to scale up the aluminide technology through a manufacturing process verification program [6,7]. In these studies a number of key technologies were evaluated under pilot plant conditions. The primary technologies evaluated were ingot melting, casting, and forging; sheet rolling, precision casting, superplastic forming and diffusion bonding, isothermal forging, and machining. As a result of this effort, a number of large ingots have been successfully cast, the largest of which weigh 815 Kg. Ingots of $Ti_3Al$ base alloys can be press forged using conventional equipment to form sheet bar, which itself can be rolled on conventional equipment to form sheet. The sheet can be superplastically formed and diffusion bonded to form complex sheet parts. Machining of both materials can be accomplished on conventional equipment using a wide range of feeds and speeds. Forging multiples of $Ti_3Al$ base alloys can be conventionally ring rolled to at least as large as 18 inches diameter. Precision casting of several small parts has been accomplished. In addition to these studies, the engine manufacturers have evaluated many other manufacturing processes necessary to produce turbine engine components.

Figure 1 shows an afterburner divergent nozzle seal being tested in an F100 engine. The part was prepared by superplastic forming and diffusion bonding of sheet (of a $Ti_3Al$ base alloy) to form a one piece unit. Fastening devices were riveted in place. Two such parts were run simultaneously in different positions in the afterburner. Figure 2 shows a precision casting of a low pressure turbine blade for a commercial engine. The blade was cast from a TiAl base alloy. It has not yet been tested. Figure 3 shows a section of the compressor casing for the F100 engine which was fabricated by machining a ring rolled ring of a $Ti_3Al$ base alloy. This part was also run successfully. Figure 4 shows exhaust seals for the F404 engine. The seals were formed superplastically from $Ti_3Al$ base alloy sheet and the lugs were diffusion bonded in place. These parts are currently being engine tested. Finally, Figure 5 shows

Figure 1. Afterburner nozzle seal under test in an F100 engine. Two such seals were run for 65 hours, including 4 hours at intermediate power and 24 minutes at full afterburner.

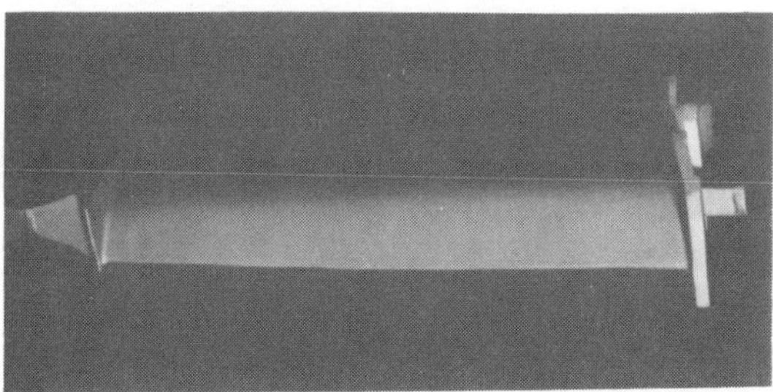

Figure 2. Precision casting of a fifth stage turbine blade, JT9D engine, which is about 25 cm long.

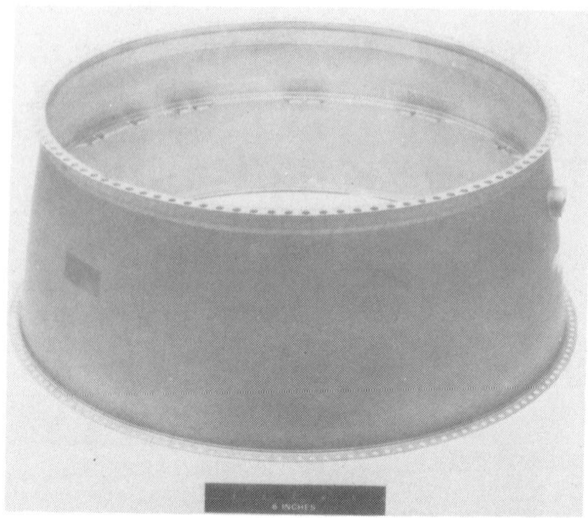

Figure 3. High pressure compressor casing section, F100 engine, run for 65 hours.

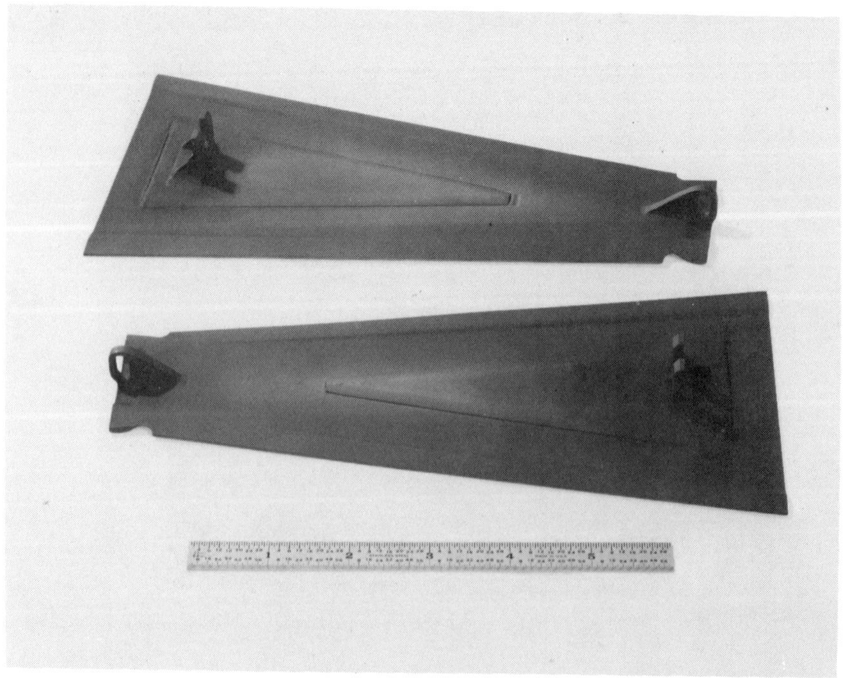

Figure 4. Exhaust seals for F404 engine.

Figure 5. High pressure turbine stator support rings for an experimental engine.

two variable stator support rings. The ring on the left is a $Ti_3Al$ base aluminide, whereas that on the right is a superalloy. These were machined from ring rolled rings. The weight savings is 43%. These and other ring structures are currently under test in a demonstrator engine.

## Conclusions

It should be clear that considerable success has been achieved in this program. The mechanisms of the ductile-brittle behavior of TiAl and $Ti_3Al$ have been determined. Alloying routes have been found to ameliorate this situation and further alloying has resulted in compositions having a useful balance of properties. A wide range of key manufacturing technologies have been evaluated and found to be practical. A number of titanium aluminide alloy parts for turbine engines have been fabricated. Several have already been successfully engine tested; others await testing. It now seems possible that titanium aluminide components may have a place in the turbine engine of the future.

## Acknowledgements

As the opening paragraph of this paper makes clear, the results reported here are based on a long term coordinated team effort spanning twelve years. The author of this paper is but one member of that team. Many of the team members will have their contributions cited in the appended literature review. However, because some of the results reported here are abstracted from reports not afforded unlimited distribution, it is not possible to cite all who have contributed to this program. They are many and their contributions critical. At this time it is appropriate to recognize the contributions of several individuals whose early support was key and should not go unrecognized. Mr. Edward J. Hassell and Colonel Robert Milling, then of the Aerospace Research Laboratories, provided the first administrative and financial support for this effort. More recently, Mr. George Peterson, Dr. Harris Burte, and Mr. Lawrence Hjelm, of the AFWAL Materials Laboratory, have provided continued support for this program.

******

## APPENDIX A - LITERATURE REVIEW

In the past ten years there have been about 15 papers published in the open literature as a result of research on the structure and properties of the titanium aluminides. These papers have not previously been reviewed. This Symposium provides the opportunity to review the status of the science of the titanium aluminides and to indicate those areas where better understanding is needed. The papers rather naturally divide into several areas and it is proposed to divide the review into those same areas.

### Deformation Studies

The deformation substructure of TiAl ($L1_0$ type ordered lattice) contains two types of dislocations [2,8]. One type, the a/2[110], does not disorder the lattice and is free to move as a unit dislocation. The other two types, the a/2[011] and the a/2[101], disorder the $L1_0$ superlattice and therefore would be expected to move in pairs, as superdislocations. Shechtman, Blackburn, and Lipsitt [2] observed these dislocations, however, there was an unexpectedly small proportion of superdislocations observed. Moreover, many of the trailing partials of the superdislocations were shown to be extended such that further movement of the dislocations would have required a continual increase in the dislocation line length. Thus, the superdislocations were effectively pinned by an unknown agent. The observed contrast of many of the dislocation pairs was not consistent with both dislocations being of the same type. Twins of the <112> {111} type play an important role in the deformation of the compound. Lipsitt, Shechtman, and Schafrik [3] studied the deformation substructure of TiAl to 900°C. They showed that the mobility of the a/6[112] partial dislocation controls the plasticity of TiAl. This partial, which is a constituent of the a<011> superdislocation, is essentially immobilized at temperatures below about 630°C by an unknown obstacle. Above about 700°C, the a/6[112] partial is no longer pinned and the activity of the a<011> superdislocations increases rapidly with increasing temperature. The same a/6[112] partial is also the twinning dislocation, so that above 700°C twinning assumes increasing importance as a deformation mode for TiAl. The concomitant result is a ductile-brittle transition near 700°C. The dominant mode of fracture changes from cleavage at room temperature to intergranular above 700°C.

Lipsitt, Shechtman, and Schafrik [4] also studied the relationships between the mechanical behavior, the dislocation activity, and the fracture mode of $Ti_3Al$ as a function of temperature to 900°C, The major mode of deformation in this material was shown to be the motion of a'<11$\bar{2}$0> dislocations on the prism, basal, and pyramidal planes. The highest density of dislocations was on the prism planes because such movement produced no wrong first near neighbors. The incidence of a'/3<11$\bar{2}$3> dislocations remained low up to 900°C, the limit of this study. Twinning was not observed, although microtwins formed at 900°C. Intergranular cracking began above 600°C and increased with increasing temperature, giving rise to an apparent ductile-brittle transition. Some dislocation cross-slipping began above 600°C. Fractographic analysis indicated that below 600°C failure occurred entirely by cleavage. Above 600°C, increasing amounts of ductile rupture occurred, but even at 900°C substantial cleavage existed. No evidence of disordering up to 900°C was observed in the $Ti_3Al$ used in this study.

Yang [9] reported that both "c" and "c + a" dislocations could be identified in $Ti_3Al$ deformed under specific conditions. The "c + a" dislocations were of two varieties, "c + 1/2<a'>" and "2c + < a' >, the presence of which depended on the temperature and strain rate at which the specimen had been deformed. Yang [10] also reported the existence of extended superdislocation networks in annealed and creep deformed $Ti_3Al$-Nb specimens. Such networks had not been observed in the unalloyed compound. It was postulated that the Nb addition caused a reduced APB energy, which in turn allowed the networks to be formed from paired "a" dislocations on basal planes. Extended nodes formed by $a/6\langle 1\bar{1}00\rangle$ partials and extended partials connected with stacking faults were also observed.

In summary, it has been possible to observe examples of a great many types of dislocations in these materials. The close relationship of many of these dislocations to the exact contrast characteristics shown is not consistent with the dislocations being exactly of the types expected in these materials. It will be seen that this lack of clarity leads to later ambiguities in the understanding of the exact deformation modes operating in creep and fatigue.

## Alloying, Structure, and Moduli

Sastry and Lipsitt [11] studied the influence of a small Nb addition on the microstructure and properties of $Ti_3Al$. The results indicated that the ordering transformation in $Ti_3Al$ proceeds by the nucleation and growth of small ordered domains and the subsequent growth of those domains with an activation energy of about 269 KJ/mole. The addition of Nb slowed the domain growth kinetics. The small Nb addition also increased the strength and ductility of a recrystallized martensitic structure. This result was due to the Nb reducing the planarity of slip and increasing the non-basal slip activity. A few years later, Martin, Lipsitt, Nuhfer, and Williams [12] reported on a similar study in some slightly more complex alloys. In a $Ti_3Al$ + Nb + W alloy it was seen that the size of the martensite platelets was reduced by the alloying additions and that the concomitant result was a precipitation of a fine b.c.c. beta phase which may have aided slip dispersal in these alloys. The addition of W to TiAl resulted in a marked strength increase, although there was no accompanying ductility increase. The W addition was shown to refine the microstructure, reduce the slip length, and provide a small degree of solid solution strengthening. Again, some b.c.c. beta phase precipitation was also observed.

Schafrik [1] measured the elastic moduli of TiAl, $Ti_3Al$, and $Ti_3Al$-Nb to 940°C. He showed that the moduli of all the materials decreased slowly and linearly with temperature. The effect of the Nb addition to $Ti_3Al$ was to decrease the room temperature modulus of the alloy by about 10%, but also to slow the rate of modulus decrease with temperature such that the moduli of the two materials was nearly the same at the maximum temperature of the study. Schafrik [13] also showed that it was possible to produce TiAl directly by extruding blended elemental powders and homogenizing the extruded rod. However, the structure and properties of the compound made in this way left much to be desired.

In summary, these results show that the properties of the intermetallic compounds $Ti_3Al$ and TiAl can be altered beneficially by a number of standard metallurgical approaches to achieve an improved balance of strength and ductility. It is obvious that a great deal remains to be learned before it will be possible to tailor the properties of an intermetallic compound to a particular use. However, these data show clearly that the properties of at least some intermetallic compounds are amenable to change and some directions for further work are evident.

## Creep and Fatigue Properties

Mendiratta and Lipsitt [14] studied the steady state creep behavior of $Ti_3Al$ and $Ti_3Al$-Nb. They reported two observations of interest. At temperatures above 700°C, the stress dependence of the steady state creep rate indicated two distinct creep regions: at high stresses the creep was controlled most probably by dislocation climb; at lower stresses a transition regime with a lower stress exponent was obtained. In the high stress and temperature regime the activation energy for creep was considerably increased by the Nb addition. These data are consistent with the reported effects [11,12] of Nb on the thermal stability of $Ti_3Al$. Martin, Mendiratta, and Lipsitt [15] reported on the creep deformation of TiAl and a TiAl-W alloy. No transition region was observed for the range of stress and temperature studied. Again the activation energies for creep were higher than is usual for disordered materials, and the W addition resulted in a further increase in activation energy. The characteristically high creep activation energies in ordered materials are believed to be associated with the slower diffusion rates observed in these materials. It was believed that the activation energy increase caused by the W addition may have been due to both the solid solution and dispersoid effects previously reported for this composition [12].

Sastry and Lipsitt [8,16] studied the fatigue properties (using a stress ratio of R = -1), the development of dislocation substructure during fatigue deformation, and the fatigue fracture modes in TiAl and TiAl-Nb from room temperature to 900°C. The ratio of fatigue strength (at $10^6$ cycles) to ultimate tensile strength was greater than 0.5 at all the temperatures investigated. The fatigue microstructure at low temperature consisted of a high density of a/3[111] faults and dislocation debris of predominantly a/2[110] and a/2[$\bar{1}$10] Burgers vectors with no preferential alignment of dislocations. At high temperatures, a dislocation braid structure consisting of all <110> vectors was observed. As had been reported in tension [4], the fracture mode changed from cleavage at room temperature to intergranular type above 600°C. In $Ti_3Al$ (under conditions of constant plastic strain), the same authors [17] found pronounced fatigue hardening at temperatures up to 700°C resulting from planarity of slip and interaction of "a" dislocations with dislocation debris of "c + a" type. A significant decrease in fatigue hardening, the onset of a fatigue saturation stage, and an increase in cumulative strain for fracture were observed above 700°C. Cyclic deformation in $Ti_3Al$ resulted in a planar dislocation band structure below 700°C, a dislocation braid structure at 700°C, and a recovered structure consisting of knitted dislocation networks at higher temperatures. Kerans [18] also studied the fatigue of $Ti_3Al$ in the same temperature regime (with R = -1), but mainly at lower stresses than did Sastry and Lipsitt. He did not find significant "c" or "c + a" dislocation activity to be associated with the fatigue behavior. An alternate mechanism of failure was proposed since there was also little evidence of cyclic crack propagation observed. It was proposed that as a consequence of the limited number of active slip systems, fracture was controlled by slip incompatibility at grain boundaries. The diversity of the fatigue results reported for $Ti_3Al$ is difficult to understand because the work was all done on precisely the same materials. Clearly, this is an area which bears further investigation.

In summary, the fatigue results on TiAl seem to be consistent with what has been observed under static deformation. The situation with $Ti_3Al$ is not so clear. Successive investigations in the same Laboratory have yielded different results: this dichotomy should be resolved.

## REFERENCES


1. "Dynamic Elastic Moduli of the Titanium Aluminides," Robert E. Schafrik, Met. Trans. 8A, 1003 (1980).

2. "The Plastic Deformation of TiAl," D. Shechtman, M. J. Blackburn, and H. A. Lipsitt, Met. Trans. 5, 1373 (1974).

3. "The Deformation and Fracture of TiAl at Elevated Temperatures," Harry A. Lipsitt, Dan Shechtman, and Robert E. Schafrik, Met. Trans. 6A, 1991 (1975).

4. "The Deformation and Fracture of $Ti_3Al$ at Elevated Temperatures," Harry A. Lipsitt, Dan Shechtman, and Robert E. Schafrik, Met. Trans. 11A, 1369 (1980).

5. "Ti/Al Design/Cost Trade-Off Analysis," J. Holowach and T. K. Redden, General Electric Co., AFAPL-TR-78-74, October 1978.

6. "Private Communication," L. Parsons, AFWAL/MLTM, Wright-Patterson Air Force Base, OH 45433, Contract F33615-76-C-5144, April, 1979.

7. "Private Communication," K. A. Williams, AFWAL/MLTM, Wright-Patterson Air Force Base, OH 45433, Contract F33615-79-C-2091, September, 1984.

8. "Plastic Deformation of TiAl and $Ti_3Al$," S. M. L. Sastry and H. A. Lipsitt, Proceedings of the 4th International Conference on Titanium, Japan, p. 1231, November 1980.

9. "C" Component Dislocation in Deformed $Ti_3Al$," W. J. S. Yang, Met. Trans. 13A, 324 (1982).

10. "Observations of Superdislocation Networks in $Ti_3Al$-Nb," W. J. S. Yang, J. Mats. Sci. Letts. 1, 199 (1982).

11. "Ordering Transformations and Mechanical Properties of $Ti_3Al$ and $Ti_3Al$-Nb Alloys," S. M. L. Sastry and H. A. Lipsitt, Met. Trans. 8A, 1543 (1977).

12. "The Effects of Alloying on the Microstructure and Properties of $Ti_3Al$ and TiAl," P. L. Martin, H. A. Lipsitt, N. T. Nuhfer, and J. C. Williams, Proceedings of the 4th International Conference on Titanium, Japan, p. 1245, November 1980.

13. "Manufacture of TiAl by Extrusion of Blended Elemental Powders," Robert E. Schafrik, Met. Trans. 7B, 713 (1976).

14. "Steady State Creep Behavior of $Ti_3Al$-Base Intermetallics," M. G. Mendiratta and H. A. Lipsitt, J. Mater. Sci. 15, 2985 (1980).

15. "Creep Deformation of TiAl and TiAl + W Alloys," Patrick L. Martin, Madan G. Mendiratta, and Harry A. Lipsitt, Met. Trans. 14A, 2170 (1983).

16. "Fatigue Deformation of TiAl Base Alloys," S. M. L. Sastry and H. A. Lipsitt, Met. Trans. 8A, 299 (1977).

17. "Cyclic Deformation of $Ti_3Al$," S. M. L. Sastry and H. A. Lipsitt, Acta Met. 25, 1279 (1977).

18. "Deformation in $Ti_3Al$ Fatigued at Room and Elevated Temperatures," R. J. Kerans, Met. Trans. 15A, 1721 (1984).

# DESIGN OF DUCTILE POLYCRYSTALLINE $Ni_3Al$ ALLOYS*

C. T. LIU and C. L. WHITE
Metals and Ceramics Division, Oak Ridge National Laboratory, Oak Ridge, Tennessee 37831

## ABSTRACT

This paper provides a comprehensive review of current efforts on design of ductile polycrystalline $Ni_3Al$ alloys. Microalloying has proven to be very effective in alleviating the grain-boundary embrittlement problem. The ductility and fabricability of $Ni_3Al$ (24 at. % Al) are dramatically improved by adding a few hundred parts per million of boron. The beneficial effect of boron is related to its unusual segregation behavior as predicted from the theory of grain-boundary cohesion developed by Rice, based on thermodynamic analyses. Alloy stoichiometry strongly influences grain-boundary chemistry, which, in turn, affects the boundary cohesion and overall ductility of $Ni_3Al$.

The solid-solution hardening of $Ni_3Al$ depends on the substitutional behavior of alloying elements, atomic size misfit, and the degree of non-stoichiometry of the alloy. Hafnium additions are very effective in improving high-temperature properties of ternary $Ni_3Al$ (Al + Hf = 24 at. %) doped with boron. Alloying with $\leq 2\%$ Hf substantially increases the yield stress and raises the peak-strength temperature. In addition, hafnium substantially improves creep properties and oxidation resistance. The $Ni_3Al$ aluminides truly represent a new series of heat resistant materials which do not depend on chromium for oxidation resistance.

## INTRODUCTION

The nickel aluminide, $Ni_3Al$, forms an $L1_2$ ordered crystal structure (Fig. 1) below the peritectic temperature (1395°C) [1]. Because of the long-range ordered crystal structure, it possesses attractive properties for structural use at elevated temperatures. The aluminide shows unusual mechanical behavior, in which the yield stress increases with increasing temperature rather than decreases as commonly observed in conventional alloys [2–4]. It is the most important strengthening constituent ($\gamma'$ phase) of commercial nickel-base superalloys, and is responsible for their high-temperature strength and creep resistance. The $Ni_3Al$-base alloys are resistant to air oxidation as a result of their ability to form adherent oxide surface films that protect the base metal from excessive attack [5,6].

In spite of these attractive properties and the fact that single crystals of $Ni_3Al$ are quite ductile, the development of nickel aluminides for structural use has been frustrated by its propensity for low ductility and brittle intergranular fracture in polycrystalline forms [7–11]. The $Ni_3Al$ materials produced by casting exhibit almost no ductility at ambient temperatures. They failed by brittle intergranular fracture without appreciable plastic deformation within the grains. Considerable effort has been devoted to improving the ductility and fabricability of the aluminides, but significant progress has been achieved only recently, through microalloying. Microalloying involves the addition of minor concentrations (usually in the ppm range) of elements to control grain-boundary structure

*Research sponsored jointly by the Division of Materials Sciences, and the Office of Energy Utilization Research, Energy Conversion and Utilization Technologies (ECUT) Program, U.S. Department of Energy under Contract No. DE-AC05-84OR21400 with Martin Marietta Energy Systems, Inc.

 

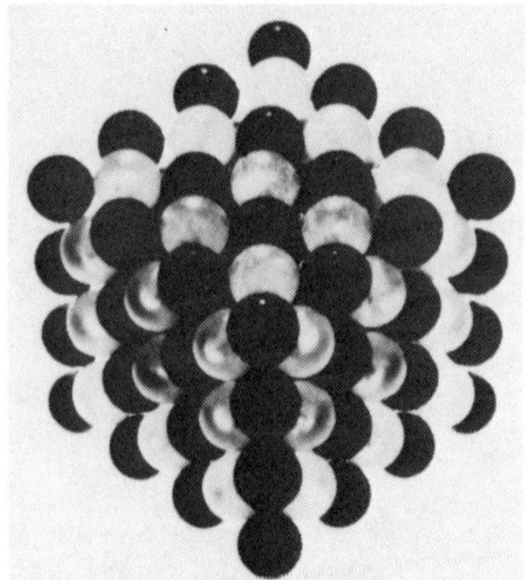

● Al SUBLATTICE SITES
O Ni SUBLATTICE SITES

Fig. 1. Atomic model of $Ni_3Al$ with $L1_2$ ordered structure.

or composition. Aoki and Izumi [12] first discovered that small boron additions substantially improved the ductility of $Ni_3Al$ at room temperature. By control of boron concentration, alloy stoichiometry, and thermomechanical treatment, Liu and Koch [6,13] reported a tensile elongation exceeding 50% — the highest tensile ductility ever achieved by polycrystalline aluminides. The beneficial effect of boron was also confirmed by Taub, Huang, and Chang [14] who tested $Ni_3Al$ foil materials prepared by a melt-spin technique.

Microalloyed $Ni_3Al$ with an adequate engineering ductility serves as a base for the development of a new class of structural materials. Although the homogeneous phase field is limited to a few percent (~3 at. %) around its stoichiometric composition, $Ni_3Al$ has considerable solubility for additional elements [15]. This permits the application of alloying principles to design ductile and strong nickel aluminides. This paper reviews the current efforts on alloy design of $Ni_3Al$-base aluminides through application of physical metallurgical principles. Emphasis is placed on understanding the grain-boundary brittleness phenomenon and boron segregation behavior, and improvement of high-temperature properties through solid-solution effects.

Modern alloy processing techniques have also been used to improve the ductility and fabricability of ordered intermetallic alloys. Rapid solidification via melt-spinning, arc hammer, and rotating disk methods has proved effective in enhancing the mechanical properties of several aluminides, including $Ni_3Al$-B [14,16], $\gamma'$ Ni-Al-X (X = Fe, Mn, Cr, Si, and Co) [17], FeAl [18], and $Fe_3Al$-$TiB_2$ [19]. The merit of splat quenching, which is reviewed separately in this conference [20], is not discussed in detail here, except for comparison and clarity purposes.

## GRAIN-BOUNDARY FRACTURE, SEGREGATION, AND MICROALLOYING

Although single crystals of $Ni_3Al$ are normally quite ductile, polycrystalline $Ni_3Al$ fails intergranularly at both room temperature and at elevated temperatures [13]. Figure 2 shows a scanning electron micrograph of a high purity $Ni_3Al$ that fractured intergranularly at room temperature. The grain-boundary facets are smooth and free of dimpling or slip lines, indicating the brittle nature of the fracture. Figure 3(a) shows a part of an $Ni_3Al$ ingot that had been hot rolled at 1200°C. The ingot was almost pulverized due to extensive grain-boundary cracking as indicated in Figure 3(b).

The first step in design of a more ductile nickel aluminide is to identify the cause of the grain-boundary brittleness. Two distinct factors may contribute to grain-boundary weakness. One factor is intrinsic grain-boundary weakness, and the other is extrinsic weakness resulting from

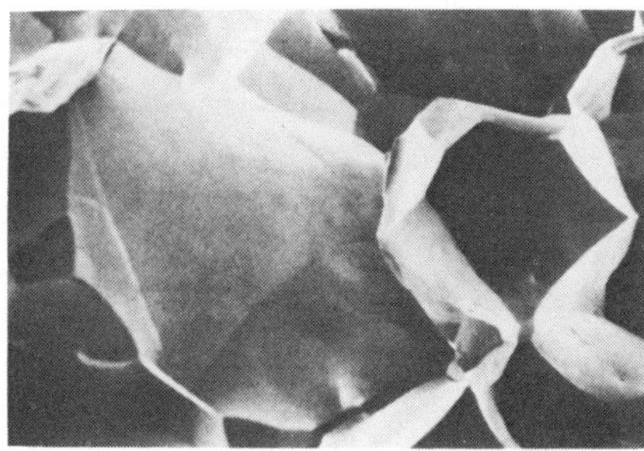

Fig. 2. Brittle grain-boundary fracture produced by bending of $Ni_3Al$ polycrystals at room temperature.

Fig. 3. Cracking of $Ni_3Al$ ingot hot rolled at 1200°C. (a) Fragmented appearance of the ingot, (b) SEM photograph showing the extensive cracking formed along grain boundaries.

segregation of embrittling impurities to the grain boundaries. The propensity of $Ni_3Al$ for intergranular failure led to early speculation that impurity segregation to grain boundaries was primarily responsible for this brittle failure mode [21,22]. Correlation between bulk chemistry and bend ductility indicated that ductility decreased as sulfur content increased in the range 32 to 176 wt ppm [23]. Auger electron spectroscopy (AES) analyses of these alloys indicated that sulfur did indeed segregate strongly to grain boundaries, and the extent of segregation increased with the degree of embrittlement [24].

In spite of the relationship between sulfur segregation and grain-boundary brittleness, attempts to ductilize $Ni_3Al$ by purification have not been very successful. The top Auger spectrum in Fig. 4 is typical of intergranular fracture surfaces of $Ni_3Al$ containing about 1 at. ppm S. These alloys failed completely intergranularly, and the sulfur levels on the fracture surfaces were below the limit of detectability for AES (~0.1 at. % in the top 2–3 atomic layers). This indicates that while certain impurities may segregate strongly to grain boundaries in $Ni_3Al$, and influence their ability to support load and transmit deformation between adjacent grains, the tendency of $Ni_3Al$ to fail intergranularly appears to be intrinsic [25]. Although intergranular failure is most often associated with segregating impurities, intrinsic brittleness similar to that observed in $Ni_3Al$ has been reported for iridium alloys [26] and Fe-Mn steels [27].

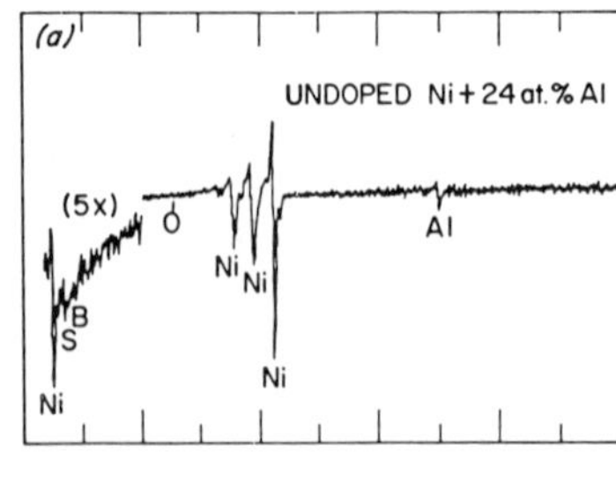


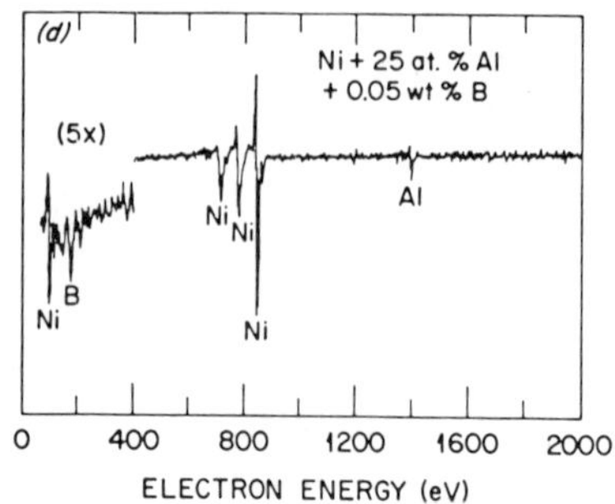


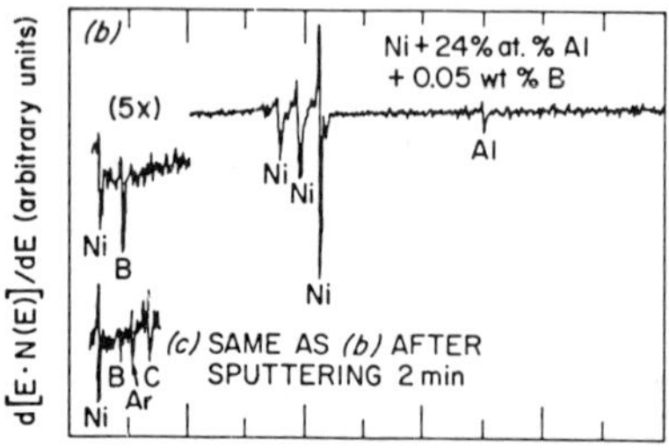


Fig. 4. Auger spectra obtained from intergranular areas on fracture surfaces of undoped and B-doped $Ni_3Al$ specimens. (a) Undoped Ni-24 at. % Al. (b) Ni-24 at. % Al-0.05 wt % B. (c) Same as (b) after sputtered for 2 min. (d) Ni-25 at. % Al-0.05 wt % B.

In order to address both intrinsic and extrinsic factors contributing to grain-boundary failure in $Ni_3Al$, two types of microalloying elements (dopants) are required. Type I dopants are reactive elements such as Ti, Hf, Ce, and Mn that tie up harmful impurities such as sulfur in innocuous precipitates. Such dopants are commonly employed to improve the ductility of conventional alloys whose grain boundaries are embrittled by segregating impurities [28]. Type II dopants are elements that can act as electron donors [29,30] (e.g., B and C) and thereby increase grain-boundary cohesion upon segregating to grain boundaries.

Of the various dopants added to $Ni_3Al$ [6,12,13], boron was found to be the most effective in improving the ductility and fabricability of $Ni_3Al$. Aoki and Izumi [12] first reported the beneficial effect of small boron additions on the ductility and failure mode of $Ni_3Al$, achieving tensile elongations up to 35% (Fig. 5). At about the same time, Liu and co-workers [6,13] found that this beneficial effect of boron is very sensitive to the stoichiometry of the alloy. By careful control of aluminum content and thermomechanical treatment, tensile elongations greater than 50%, and virtually 100% transgranular failure were achieved in Ni-24 at. % Al alloys containing 0.02 to 0.10 wt % B (Fig. 6). With further increases in boron content, the ductility gradually decreases, and the yield stress increases sharply. This striking effect of boron on the ductility of $Ni_3Al$ occurs over a range of compositions where the boron is in solid solution. Optical and transmission electron microscopy indicates that a $Ni_{20}Al_3B_6$ phase begins to precipitate as the boron content is increased above 0.3 wt % [25].

The fracture mode and room-temperature tensile elongation of boron-doped $Ni_3Al$ (24 at. % Al) are not sensitive to grain size (Fig. 7). With an increase in grain size above 100 μm in diameter, the aluminide showed only a moderate drop in ductility (from 50 to 40%), and retained its transgranular fracture mode. These observations suggest that boron is very effective in improving the strength of $Ni_3Al$ grain boundaries, which become sufficiently strong to withstand the high stress concentrations experienced in coarse-grained specimens.

AES studies of partially intergranular fracture surfaces of Ni-24 at. % Al-0.05 wt % B alloys indicate that boron segregates strongly to grain boundaries in these alloys [Fig. 4(b)]. Sputter ion etching

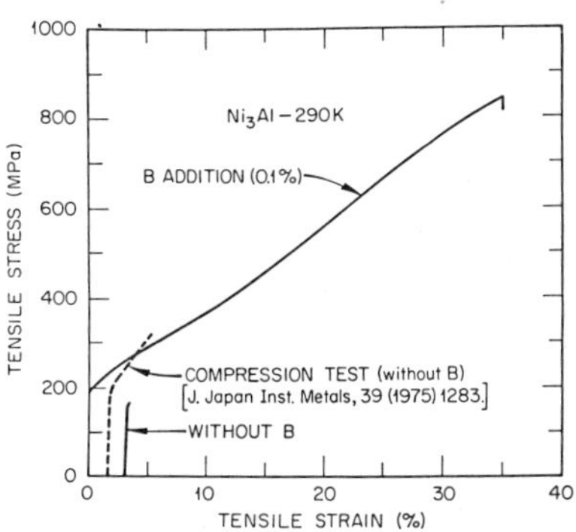


Fig. 5. Room temperature tensile stress-strain curves, showing the effect of boron additions on the elongation of $Ni_3Al$ [12].

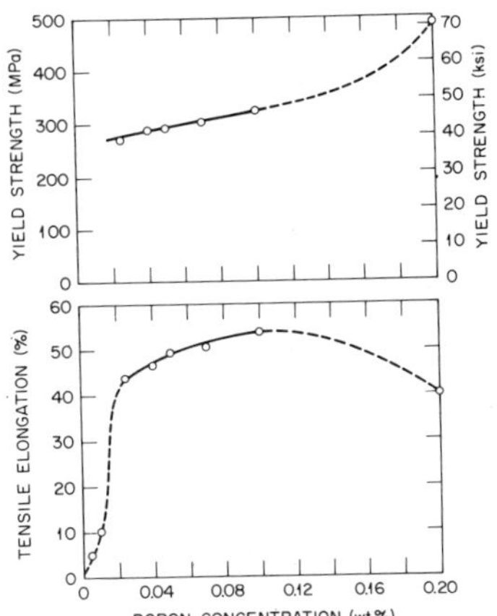


Fig. 6. Effect of boron additions on tensile elongation and yield stress of $Ni_3Al$ (24 at. % Al) [25].

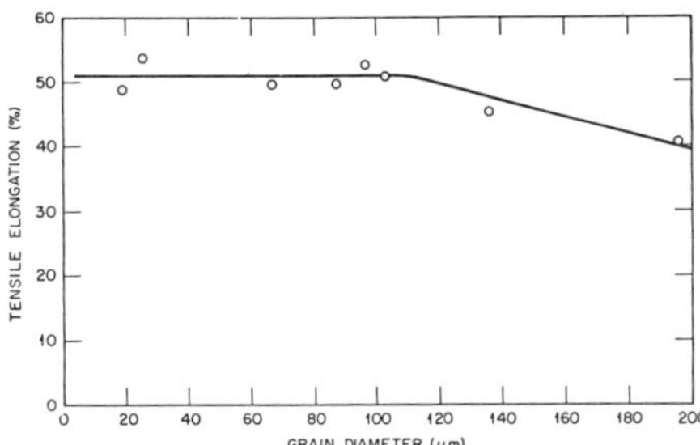


Fig. 7. Effect of grain size on tensile elongation of B-doped $Ni_3Al$ (24 at. % Al).

these fracture surfaces (to remove less than 10 atomic layers) removes all of the boron [Fig. 4(c)], indicating that the boron enriched zone at the grain boundary is only a few atom layers thick.

It appears that the beneficial effect of boron segregation to grain boundaries in $Ni_3Al$ may be explained by the unusual relationship between boron segregation at grain boundaries and at free surfaces [25,31]. Figure 8 shows Auger spectra from the external (polished) surface of Ni-24 at. % Al-0.05 wt % B after sputter etching with argon ions (top spectrum) and after heating that specimen for about 45 min at about 990°C. Comparison of the spectra in Fig. 8 (b) and (c) illustrates the striking difference in segregation behavior at grain boundaries and free surfaces. While boron segregates strongly to grain boundaries and not at all to free surfaces, the converse is true of sulfur. This type of segregation behavior was also observed in a Ni-25.2 at. % Al-0.5% B alloy containing grain-boundary cavities left over from the casting process [25]. Auger spectra from uncavitated portions of a grain-boundary fracture surface showed only boron segregation. Spectra from the cavity surfaces showed intense sulfur segregation, but no boron. Sputter etching and re-analysis of the fracture surface indicated that the sulfur rich zone on the cavity surfaces was only a few atom layers thick.

Interestingly, this unusual type of segregation behavior, and its effect on grain-boundary cohesion, was predicted in a 1976 paper by J. R. Rice [32]. In his paper, Rice developed a classical thermodynamic treatment of segregation effect on grain-boundary cohesion. For grain-boundary separation at low temperatures, where mass transport cannot equilibrate newly formed free surfaces, the fracture surfaces "inherit" a solute excess ($\Gamma^b/2$) equal to half that for the unfractured grain boundary. Rice showed that a segregating solute can either enhance grain-boundary

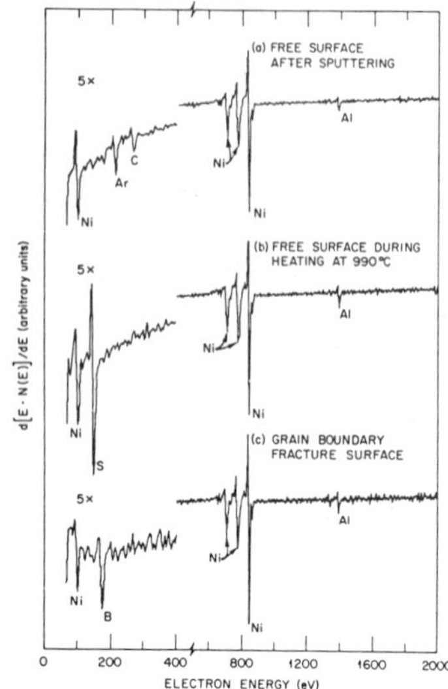


Fig. 8. Auger spectra from an external surface of $Ni_3Al$ containing 24 at. % and 0.05 wt % B, after sputter etching (a), then heating at about 990°C (b). Curve (c) shows an Auger spectrum from an intergranular portion of a fracture surface in this alloy.

cohesion, or degrade it, depending upon whether the inherited solute excess on the intergranular fracture surface is greater or less than the equilibrium free surface excess, $\Gamma^s$. This is illustrated schematically in Fig. 9 where $\Gamma^b$ and $\Gamma^s$ are plotted versus solute chemical potential, $\mu$. Fig. 9(a) represents a case where the solute segregates much more strongly to the free surface than to the grain boundary (like sulfur in $Ni_3Al$ for example). The grain-boundary solute excess of component N, $\Gamma^{b'}$, occurs at a solute chemical potential $\mu_N^b$ $(\Gamma^{b'})$. Only half of $\Gamma^{b'}$ is inherited by each grain-boundary fracture surface, leading to a solute excess much lower than the equilibrium value, $\Gamma^{s'}$, and corresponding to a solute chemical potential, $\mu_N^s$ $(\Gamma^{b'}/2)$, much lower than the actual chemical potential in the alloy. Rice showed that the effect of segregation on cohesive energy, $\phi^*$, could be expressed as:

$$\frac{d\phi^*}{d\Gamma^b} = \mu_N^s\,(\Gamma^b/2) - \mu_N^b\,(\Gamma^b) \qquad (1)$$

In the case where $\Gamma^s >> \Gamma^b$, this leads to a negative value of $d\phi^*/d\Gamma^b$, indicating that segregation should lower the grain-boundary cohesive energy.

Segregation behavior similar to that exhibited by boron in $Ni_3Al$, (i.e. $\Gamma^s << \Gamma^b$) is depicted in Fig. 9(b). Here the inherited solute enrichment, $\Gamma^{b'}/2$, corresponds to a solute chemical potential, $\mu_N^s$ $(\Gamma^{b'}/2)$, that is much greater than $\mu_N^b$ $(\Gamma^{b'})$. This leads to a positive value for $d\phi^*/d\Gamma^b$, indicating that solute segregation should increase grain-boundary cohesion. This appears to be the first direct confirmation of Rice's prediction for beneficial solutes. If the relationship between segregation behavior and its effect on grain-boundary cohesion proves to be general, it could provide a basis for selection of Type II dopants in other alloy systems.

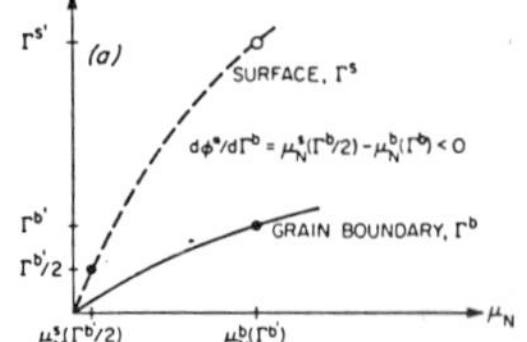


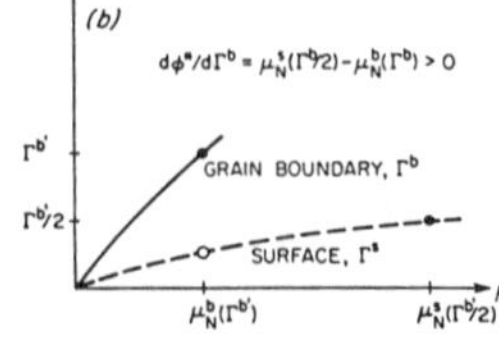


Fig. 9. A schematic drawing showing the effect of solute (N) segregation on the solute chemical potential of free (cavity) surfaces, $\mu_N^s$ and grain boundaries, $\mu_N^b$ for two types of solutes [26]. (a) The solute has a strong tendency to segregate to free surfaces but not to grain boundaries and (b) the solute tends to segregate more strongly to grain boundaries than to free surfaces.

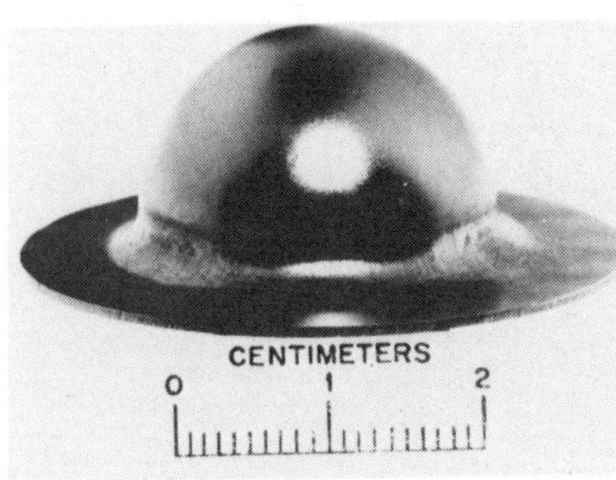


Fig. 10. A cup of B-doped $Ni_3Al$ (24 at. % Al) fabricated by deep drawing at room temperature.

Liu et al. [13] found that no single individual Type I dopant could effectively improve the cold fabricability of $Ni_3Al$. In practice, however, a combination of Type I and II dopants is required to surmount both intrinsic and extrinsic grain-boundary brittleness problems in $Ni_3Al$. In this respect, additions of hafnium and manganese were found to further improve the cold fabricability of boron-doped $Ni_3Al$, perhaps through scavenging sulfur from the grain boundaries. A low level of sulfur (<1 at. %) was detected to segregate to grain boundaries in less pure $Ni_3Al$.

Microalloyed $Ni_3Al$ (24 at. % Al) alloys are best fabricated at ambient temperature rather than at high temperatures. Sheets, foils, rods, and wires have been successfully cold fabricated from small laboratory heats. Figure 10 shows a full cup of $Ni_3Al$ containing 500 wt ppm B, deep-drawn at room temperature. The excellent formability of this microalloyed aluminide is not surprising in view of its good ductility and high work hardening rate.

## STOICHIOMETRY EFFECTS

Alloy stoichiometry has a strong effect on the ductility and fracture behavior of boron-doped $Ni_3Al$ [25]. The boron dopant is most effective in improving the ductility and suppressing intergranular fracture in $Ni_3Al$ containing 24 at. % Al. At higher aluminum concentrations, the ductility drops sharply (see Fig. 11) and the failure mode changes from transgranular, to mixed mode and then to completely intergranular (see Fig. 12).

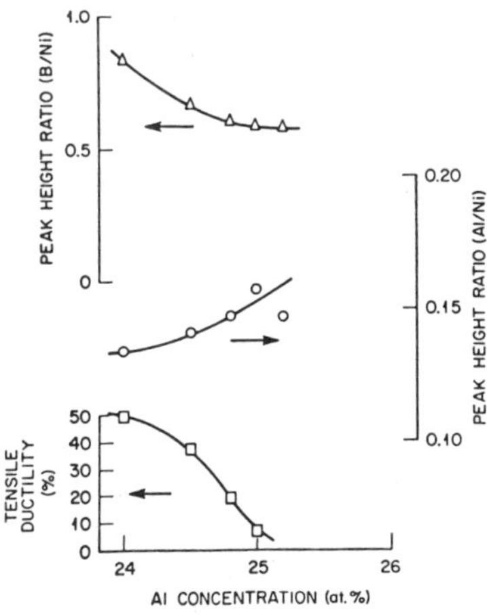


Fig. 11. Effect of stoichiometry on grain-boundary segregation (in terms of peak-height ratio) and room-temperature tensile ductility of B-doped $Ni_3Al$ containing 24 to 25.2 at. % Al.

The Auger spectrum in Fig. 4(d) is from a fresh intergranular fracture surface on a Ni-25 at. % Al-0.05% B alloy. Comparison of this spectrum with the one from the boron-doped alloy containing 24 at. % Al indicates that the increased aluminum content has decreased the intensity of the boron segregation by about a factor of two. The top curve in Fig. 11 shows the ratio of the 180 eV B peak to the 102 eV Ni peak plotted versus aluminum content. This B/Ni peak-height-ratio decreases continuously with increasing aluminum content, while the 1396 eV Al to 848 eV Ni peak-height-ratio (center curve) increases slightly. These results show that ductility and boron segregation depend strongly upon stoichiometry, as well as reinforcing the direct correlation between enhanced grain-boundary cohesion and the boron level at the grain boundaries.

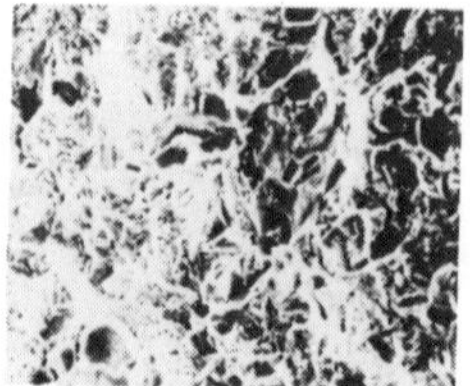

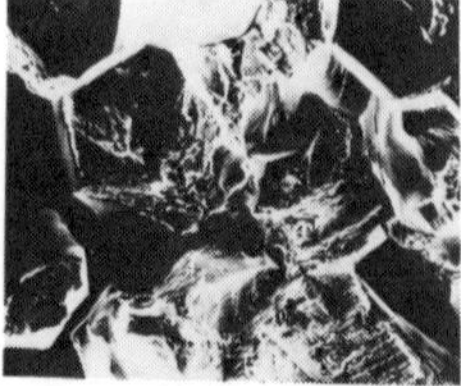

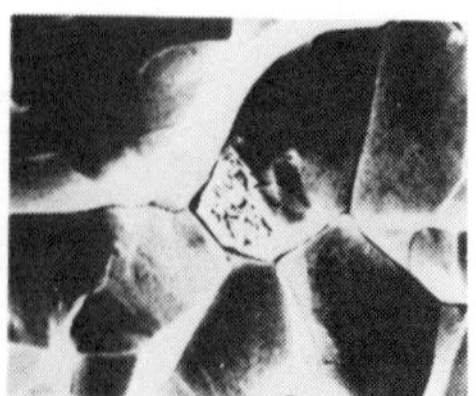

50 μm

(a) 24% Al, ε = 49.4%
TRANSGRANULAR FRACTURE

(b) 24.5 % Al, ε = 37.0%
MIXED FRACTURE MODE

(c) 25.0% Al, ε = 6.0%
INTERGRANULAR FRACTURE

Fig. 12. Effect of stoichiometry on fracture mode of B-doped $Ni_3Al$ at room temperature.

The stoichiometry effect on ductility is also illustrated in Fig. 13 where the ductility is plotted as functions of boron and aluminum concentration for foil samples of $Ni_3Al$ prepared by rapid solidification [33]. These samples start to show considerable ductility when the aluminum level is below about 24.5%. Note that the ductility of the foil materials (~30 μm thick) is considerably lower than that of bulk materials (0.76 mm) prepared from conventionally cast ingots.

In addition to the boron segregation and fracture behavior, alloy stoichiometry is reported to affect the strength and hardness of $Ni_3Al$ and other $L1_2$ alloys [34,35,36]. The deviation from stoichiometry produces an increase in hardness [34] at lower temperatures and a decrease at 800°C, as shown in Fig. 14. The abrupt change in hardness around stoichiometry was suspected to result from excess lattice defects (such as vacancies) introduced to maintain a constant electron density, which is required for phase stability in a localized region. Recent studies [36] of lattice parameter versus aluminum concentration indicated an abrupt change at 25% Al,

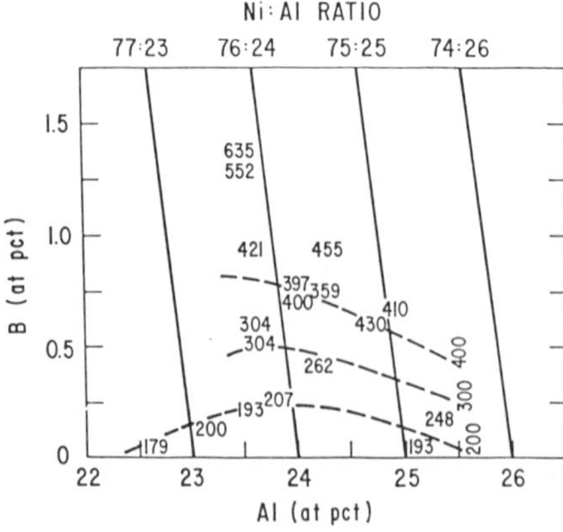


Fig. 13. Plastic strain to fracture (%) of rapidly solidified $Ni_3Al$-B as a function of Al and B concentration. The solid lines indicate constant NiAl ratio (at. %) [33].

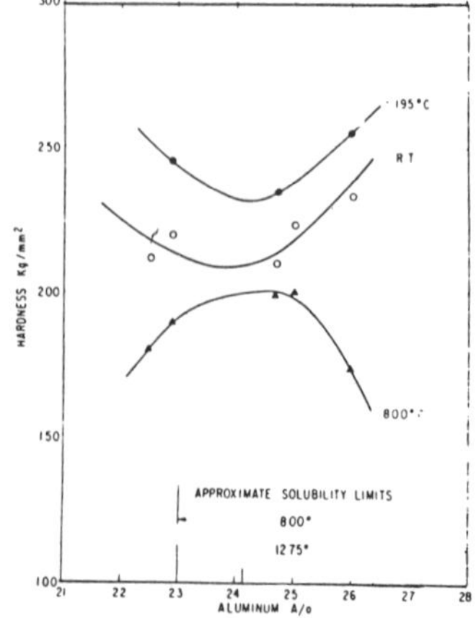


Fig. 14. Effect of stoichiometry on microhardness of $Ni_3Al$ tested at –195°C, room temperature, and 800°C [34].

providing evidence in support of the defect mechanism. Such an abrupt change in lattice parameter was not observed previously [37]. Several investigators [35,36] also reported that deviations from stoichiometry strongly influence the positive temperature dependence of yield strength of $Ni_3Al$. The yield strength showed a more pronounced increase with increasing test temperature for the alloys having higher aluminum concentrations. Noguchi et al. [36] attribute this strengthening effect to the reduction in phase stability of the $L1_2$ structure with respect to the $DO_{22}$ as well as $DO_{19}$ structure.

These observations clearly indicate that the mechanical properties of $Ni_3Al$ are sensitive to the deviations from stoichiometry. A design of the nickel aluminide is thus required to control the stoichiometric composition, and high ductilities can be obtained only by microalloying with boron in $Ni_3Al$ containing less than 24.5 at. % Al.

## SOLID-SOLUTION HARDENING AND MECHANICAL PROPERTIES

$Ni_3Al$ doped with boron serves a base for design of ductile and strong materials for structural uses. The aluminide is capable of being hardened by solid-solution effects because it can dissolve substantial alloying additions without losing the advantage of long-range order. Ochiai et al. [15] recently reconstructed the solubility lobes of ternary $Ni_3Al$ phase ($L1_2$) at 1000°C for various alloying elements. The elements which dissolve substantially in $Ni_3Al$ can be divided into three groups, as shown in Fig. 15. The elements including Si, Ge, Ti, V, Hf, etc. almost exclusively substitute for Al sublattice sites and those of Cu, Co and Pt are for Ni sublattice site. The third group elements such as Fe, Mn, and Cr substitute for both sublattice sites. Guard and Westbrook [34] first suggested that the electronic structure (i.e., the position of elements in the periodic table) rather than atom size factor plays a dominant role on the substitution behavior. The extent of solid solution in $Ni_3Al$, however, is controlled by the atomic size misfit and the difference in the heats of formation between $Ni_3Al$ and $Ni_3X$ [15].

The room temperature solid-solution hardening of $Ni_3Al$ depends on the substitution behavior of alloying elements, atomic size misfit, and the degree of non-stoichiometry of the alloy. Rawlings and Staton-BeVan [35] reviewed the mechanical properties of $Ni_3Al$ and found that it could be effectively hardened only by the elements substituting for Al sites and not by the elements substituting for Ni or for both Ni and Al sites. Figure 16 is a plot of the increment of the room-temperature yield stress ($\sigma_y$) per at. % alloying addition ($\Delta YS/\Delta C$) as a function of atomic size misfit parameter [$\varepsilon = (d_x - d_y)d_y$].* The $\Delta YS/\Delta C$ increases roughly linearly with $\varepsilon$; however, the strengthening is most pronounced for the stoichiometric alloys and Al-rich alloys, and much less pronounced for Ni-rich alloys. Unfortunately polycrystalline alloys with stoichiometric and Al-rich compositions cannot be effectively ductilized by microalloying with boron additions, as described in the foregoing section. Among the solute elements, hafnium, which gives the largest atomic size misfit, strengthens $Ni_3Al$ most effectively at room temperature.

In addition to substitutional solute hardening, Huang, Taub, and Chang [38] recently observed a substantial hardening effect of boron in $Ni_3Al$ prepared by rapid solidification. $Ni_3Al$ can dissolve up to about 1.5 at. % B which tends to occupy the interstitial sites. As shown in Fig. 17, the boron atoms produce the largest lattice dilation ($\Delta\varepsilon$ = 0.0026 per atom percent of boron) and are the most effective room-temperature strengthener in $Ni_3Al$.

---

*$\Delta YS$ is the change in $\sigma_y$ associated with the change in solute content, $\Delta C$. $d_x$ and $d_y$ are the effective atomic diameter of the solute and the atom for which it substitutes, respectively.

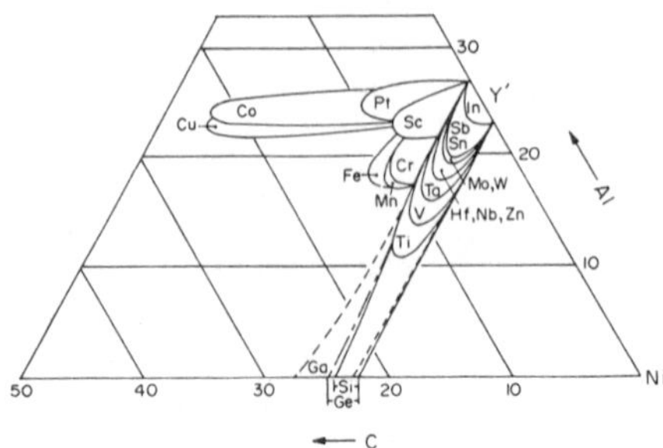


Fig. 15. Semischematic depiction of the solubility lobes of ternary $Ni_3Al$ phase around 1000°C [15].

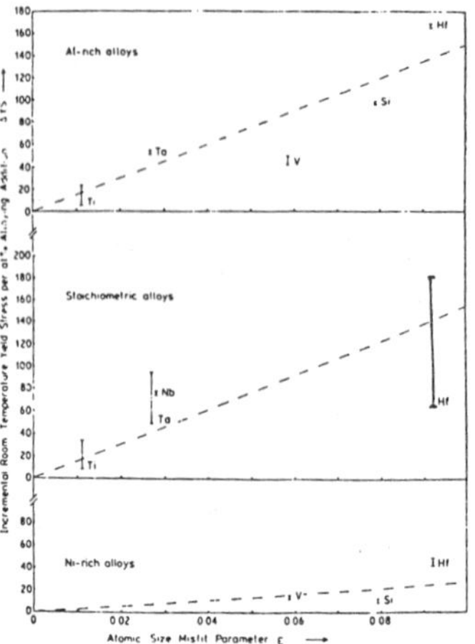


Fig. 16. The room temperature strengthening associated with solute elements which substitute for aluminum in ternary $Ni_3Al$ [35].

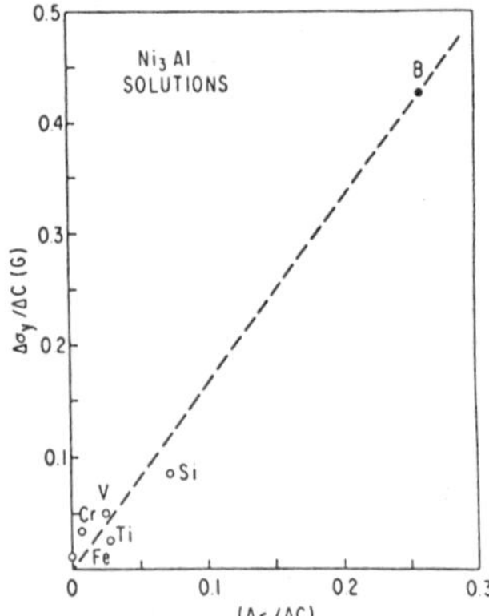


Fig. 17. The correlation of strengthening potency, $\Delta\sigma_y/\Delta C$, to the lattice strain per solute atom fraction, $\Delta\varepsilon/\Delta C$, for boron and a variety of other solute elements in $Ni_3Al$ [38].

In recent studies of solid-solution hardening at elevated temperatures, Liu [39] found that hafnium is also the most effective strengthener when compared with other substitutional solutes in B-doped $Ni_3Al$. Figure 18 is a plot of yield stress ($\sigma_y$) as a function of temperature for B-doped $Ni_3Al$ and Hf-modified nickel aluminide (Ni-22.5 Al-1.5 Hf-0.2 B, at. %) and several commercial solid-solution alloys. The $\sigma_y$ of B-doped $Ni_3Al$ (just like undoped $Ni_3Al$) increases with temperature and peaks around 600°C. The decrease in $\sigma_y$ above 600°C is not a result of thermal disordering because $Ni_3Al$ shows no change in long-range order parameter at temperatures to 1000°C.

Similar increases in $\sigma_y$ with temperature have been observed in other $L1_2$ ordered alloys [2,3]. Of the various models that have been suggested to explain such behavior, the cross slip model first proposed by Kear and Wilsdorf [40] and modified by Takeuchi and Kuramoto [41], most successfully describes the yield behavior of $Ni_3Al$. The model is based on the assumption that the antiphase boundary (APB) energy on {100} planes is lower than on {111} planes. A recent study of the APB morphology of as-quenched and annealed $Ni_3Al$ foils produced by rapid solidification shows alignment of the APBs along {100} planes, and provides the first direct experimental evidence [42] in support of the proposed cross-slip model. The drop in $\sigma_y$ above 600°C is a result of a gradual change in slip systems from {111} to {100} planes, for which there is no hardening in the $L1_2$ structure.

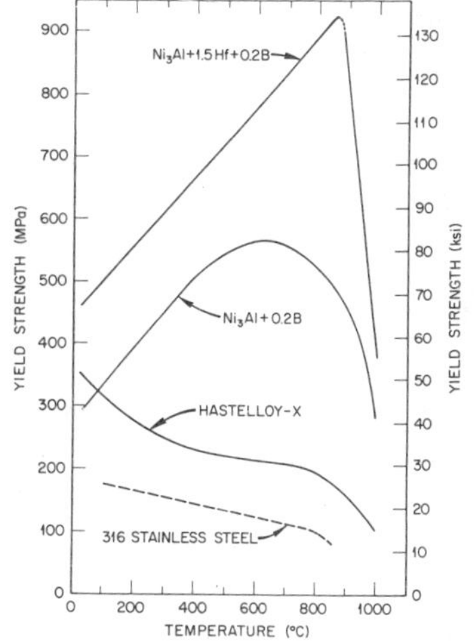


Fig. 18. Yield stress as a function of test temperature for $Ni_3Al$ base aluminide alloys, Hastelloy X, and type 316 stainless steel.

Figure 18 also shows the yield stress characteristics of the aluminide containing 1.5 at. % Hf. Note the hafnium addition substantially enhances the strength level and raises the peak strength temperature to 850°C. The strengthening is entirely a result of solid-solution effects; no precipitates could be detected by TEM. Figure 19 shows that the yield stress increases linearly with hafnium concentration for Ni-rich alloys (Al + Hf = 24 at. %) [39]. This relationship does not fit the prediction from the classic solid-solution model developed based on the elastic interaction between dislocations and symmetrical defects [43] (the model of which predicts $\sigma_y \sim C^{1/2}$ or $C^{2/3}$, where C is the solute concentration). The slope for the 850°C curve in Fig. 19 is higher than that of the room-temperature curve, indicating that hafnium is more effective in strengthening $Ni_3Al$ at elevated temperatures. There are several possibilities that may explain the unusual strengthening effect of hafnium. First, hafnium may increase the critical resolved shear stress (CRSS) more effectively on {100} planes than on {111} planes, and thereby raising the temperature for transition to {100} slip. Recent studies [44] have shown, however, that other solutes (Mo, Ta, Nb, Ti, and W) decrease rather than increase CRSS for (100)[$\bar{1}$10] slip. Second, hafnium may more effectively lower the APB energy on {100} planes and lead to an increase in the rate of cross slip from {111} to {100} planes. Curwick [44] observed that the compressive yield stress of oriented $Ni_3(Al,X)$ single crystals depends strongly on the crystal orientation. For example, the $Ni_3(Al,Ta)$ crystals with <100> orientation exhibit pronounced solid-solution hardening at temperatures to 1000°C whereas the crystals with other orientations (such as <111> and <144>) show no hardening above 400 to 500°C (Fig. 20). This observation suggests a third possibility, namely that the unusual hardening effect of hafnium may simply result from a stronger [100] texture in $Ni_3(Al,Hf)$ alloys.

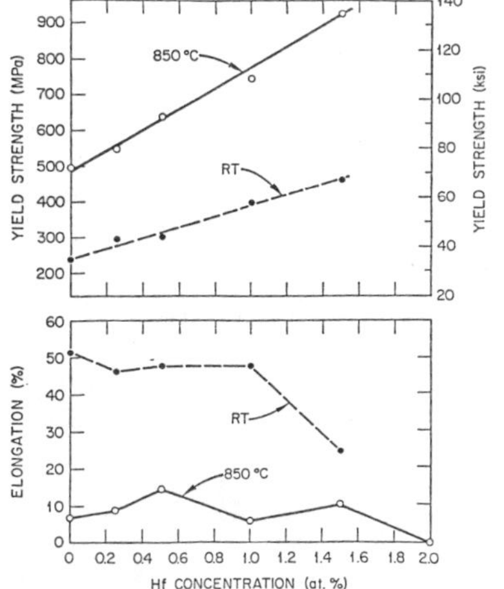


Fig. 19. Effect of Hf concentration on tensile elongation and yield stress of ternary $Ni_3Al$ tested at room temperature and 850°C.

The ductility of $Ni_3Al$ is not sensitive to the hafnium concentrations up to 1% at room temperature, and up to about 1.5% at 850°C [Fig. 19]. Above these levels, the ductility decreases markedly. The aluminides with and without Hf generally exhibit a decrease of ductility with temperature above 600°C and their ductility reaches minimum (5~10%) around 850 to 1000°C. The

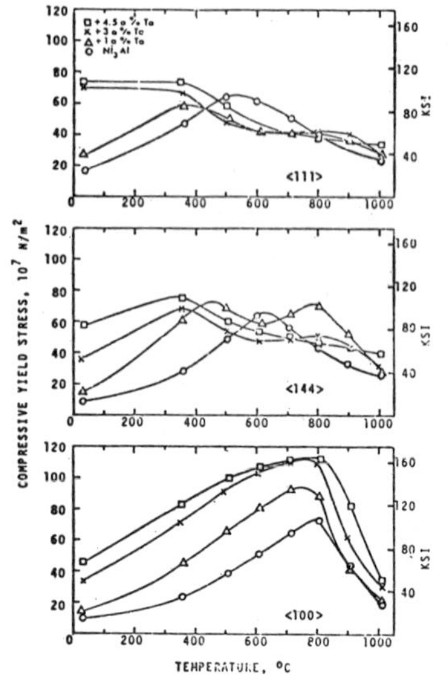


Fig. 20. Effect of Ta on the temperature dependence of the compressive yield stress of oriented $Ni_3(Al,Ta)$ single crystals [44].

decrease in ductility is accompanied by a change in fracture mode from transgranular to intergranular.

Figure 21 compares the density-compensated ultimate tensile strength of advanced aluminides ($Ni_3Al$ alloyed with 0.5 to 1.0% Hf and 0.2% B) with commercial wrought alloys. The comparison clearly indicates the superior strength-to-weight ratio of the aluminides in both well-annealed and 20% cold-worked conditions. The aluminides have a density about 10% less than that of nickel base superalloys, and the lower density could provide considerable advantages in applications involving rotating or moving parts.

Creep properties of $Ni_3Al$ and its alloys have been reported in the temperature range of 530 to 967°C [45–47], but little is known [48] about the mechanisms that control the creep rate, rupture life, and fracture behavior. Hafnium, in addition to increasing the tensile strength, is most effective in improving creep properties of polycrystalline $Ni_3Al$ doped with boron. Limited creep tests indicate that the creep rate of B-doped $Ni_3Al$ (24 at. % Al) depends on grain size, and that creep specimens rupture primarily due to grain-boundary sliding and cavitation at 760°C. As indicated in Table I, alloying with hafnium additions substantially lowers the creep rate and extends the creep life of the aluminides (Al + Hf = 24 at. %).

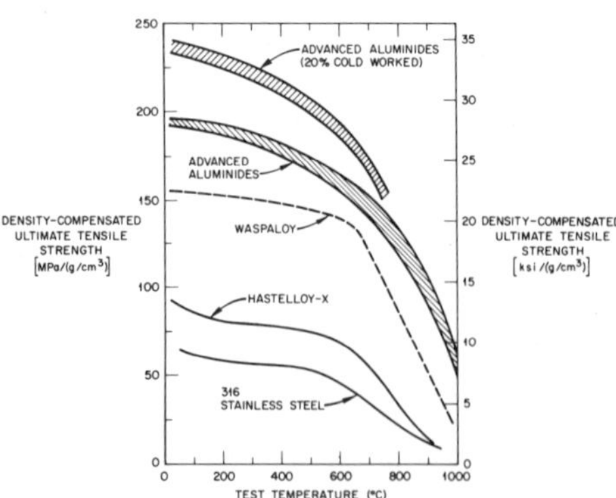


Fig. 21. Comparison of density compensated ultimate tensile strength of nickel aluminides (Ni-23 ~ 1.0% Hf-0.1 B) with commercial wrought alloys.

The beneficial effect of hafnium is not well understood but may be related to slowing the migration of vacancies and gettering harmful impurities segregated to grain boundaries. Sulfur has been detected to segregate strongly to the cavity surface in B-doped $Ni_3Al$. The alloying with hafnium is expected to reduce the extent of sulfur segregated to cavity surfaces and grain boundaries. The creep rates of $Ni_3Al$ alloys modified with $\geq$1% Hf are much lower than those of Hastelloy X and type 316 stainless steel and are comparable to those of Waspaloy (see Table I).

Other properties of B-doped $Ni_3Al$ alloys modified with hafnium additions have been measured, including

Table I. Comparison of creep properties of nickel aluminides with commercial wrought alloys tested at 760°C

| Alloys | Steady-State Creep Rate ($10^{-6}$/h) | Rupture Life or Test Time (h) |
|---|---|---|
| | 138 MPa (20 ksi) | |
| B-doped $Ni_3Al$ (24% Al) | 90 | 350 |
| B-doped $Ni_3Al$ + 0.5 at. % Hf | 4 | >600 |
| Type 316 stainless steel | 8500 | 65 |
| Hastelloy X | 1300 | 200 |
| | 276 MPa (40 ksi) | |
| B-doped $Ni_3Al$ (24% Al) | 500 | 30 |
| B-doped $Ni_3Al$[b] + | | |
| 0.5 at. % Hf | 90 | 500 |
| 1.0 at. % Hf | 40 | >600[a] |
| 1.5 at. % Hf | 20 | >600[a] |
| Waspaloy | | ~ 1000 |

[a]The test was stopped (without rupture) at the time indicated.
[b]Al + Hf = 24 at. %, B = 0.1 at. %.

high-cycle fatigue [49], crack-growth behavior [49], hydrogen embrittlement [50], weldability [51] and oxidation and corrosion resistance [39]. Limited work conducted by Stoloff and Kuruvilla [49] at room temperature indicates that the fatigue and crack growth behaviors of the aluminides are similar to those of $(Fe,Ni)_3(V,Ti)$ alloys having the $L1_2$ structure but are superior to other structural alloys. As shown in Fig. 22, the threshold $\Delta K$ for the ordered alloys is higher than that of commercial alloys by a factor of 2 to 3. The aluminides were embrittled at room temperature by cathodically charging with hydrogen [50]. This embrittlement can be completely eliminated by preoxidation of specimens in air at 900°C. This produced an oxide layer that acts as a barrier to the permeation of hydrogen.

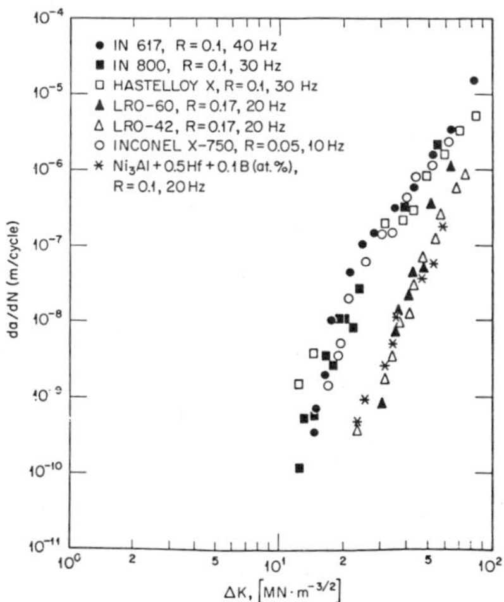


Fig. 22. Comparison of crack-growth rates of nickel aluminides (Ni-23.5 Al-0.5 Hf 0.1 B, at. %) and LRO alloys [$(Fe,Ni)_3(V,Ti)$] with several high temperature alloys tested in air at 25°C [49].

$Ni_3Al$ is basically very resistant to air oxidation; however, it exhibits some degree of spalling at temperatures above 900°C. The spalling is easily suppressed by hafnium additions, which reduce the tendency for void formation and improve the adhesion between the base metal and the oxide film [52]. Oxide films on Hf-modified $Ni_3Al$ alloys showed no indication of spalling or scaling during cyclic oxidation at temperatures up to 1100°C [39]. Moreover, the material remained ductile after oxidation. These $Ni_3Al$ alloys truly represent a new series of high temperature structural materials, which do not depend on chromium, a critical strategic element, for oxidation resistance.

## SUMMARY AND REMARKS

Unmodified polycrystalline $Ni_3Al$ exhibits intrinsic intergranular brittleness which can be further worsened by segregation of harmful impurities such as sulfur. Microalloying has been used to alleviate both intrinsic and extrinsic grain-boundary brittleness in $Ni_3Al$ alloys. The ductility and fabricability of $Ni_3Al$ containing 24 at. % Al are dramatically improved by adding a few hundred parts per million of boron, which tends to segregate strongly to grain boundaries and suppresses the brittle intergranular fracture in $Ni_3Al$. Tensile elongations exceeding 50% at room temperature have been obtained by controlling alloy stoichiometry, boron concentration, and grain size. Additions of reactive elements (e.g., Hf, Mn) further improve the cold fabricability of B-doped $Ni_3Al$. Microalloyed $Ni_3Al$ also exhibits excellent formability, which stems both from its good ductility and from its high work-hardening behavior.

Boron exhibits an unusual segregation behavior in $Ni_3Al$. AES studies indicate that boron has a strong tendency to segregate to the grain boundaries but not to cavity or free surfaces. On the other hand, sulfur, an embrittling impurity, tends to segregate more strongly to free surfaces than to the grain boundaries. The beneficial effect of boron and the embrittling effect of sulfur agree well with predictions from the theory of grain boundary cohesion developed by Rice based on thermodynamic analyses.

Room-temperature ductility and fracture behavior of B-doped $Ni_3Al$ are strongly dependent on deviations from alloy stoichiometry. As the aluminum content of B-doped $Ni_3Al$ is increased from 24 to 25 at. %, the ductility decreases dramatically. Accompanying this decrease is a sharp change in fracture mode from transgranular, through a mixed mode, to intergranular. AES analyses show that the intensity of boron segregation to grain boundaries decreases, and the amount of grain-boundary aluminum increases, with increasing aluminum concentration. These observations indicate that alloy stoichiometry strongly influences grain-boundary chemistry, which, in turn, affects the boundary cohesion and overall ductility of $Ni_3Al$. Alloy stoichiometry also influences the strength and hardness of $Ni_3Al$ at room and elevated temperatures.

The solid-solution hardening of $Ni_3Al$ at room temperature depends on the substitutional behavior of alloying elements, their atomic size misfit, and the degree of nonstoichiometry of the alloy. Ternary $Ni_3Al$ can be effectively hardened only by the elements substituting for aluminum sites and not by the elements substituting for nickel, or for both nickel and aluminum sites. Solute strengthening is most pronounced for the stoichiometric alloys and aluminum-rich alloys, and is much less pronounced for nickel-rich alloys. Among the solute elements, hafnium atoms which substitute for aluminum atoms, and boron atoms which tend to occupy the interstitial sites, both produce large lattice dilations and are the effective strengthener at room temperatures.

Hafnium additions are found to be effective in improving high-temperature properties of ternary $Ni_3Al$ (Al + Hf = 24 at. %) doped with boron. Alloying with less than 2% Hf substantially increases the yield stress and raises the peak-strength temperature from 600 to 850°C. The strengthening is entirely a result of solid solution effects. In fact, hafnium is more effective in strengthening $Ni_3Al$ at elevated than room temperatures. Several possibilities have been suggested to explain the unusual hardening effect of hafnium. Hafnium-modified $Ni_3Al$ alloys are stronger than wrought-commercial alloys including Waspaloy.

The aluminide alloys generally exhibit a decrease of ductility with temperature above 600°C and their ductility reaches minimum (5~10%) around 850 to 1000°C. The decrease is accompanied by a change in fracture mode from transgranular to intergranular.

Other properties of ductile $Ni_3Al$ alloys containing hafnium were determined, including creep properties, high-cycle fatigue, crack growth, hydrogen embrittlement, and oxidation and corrosion resistance. Hafnium additions substantially lower the creep rate and extend the creep life of the aluminides. The room-temperature fatigue and crack growth resistance of the aluminides is superior to commercial structural alloys. The aluminides showed no indication of spalling during cyclic oxidation at temperatures to 1100°C. These intermetallic compounds truly represent a new series of heat-resistant materials which do not depend on chromium for oxidation resistance.

Mechanical and metallurgical properties of $Ni_3Al$ alloys can be further improved by alloy design. For structural applications, we recommend work in the following three areas:

1. High-Temperature Ductility and Fabricability: Boron additions are effective in enhancing the cohesion of $Ni_3Al$ grain boundaries at temperatures below 800°C, but become less effective at higher temperatures. This is indicated by the fact that B-doped $Ni_3Al$ alloys begin to exhibit intergranular fracture and reduced ductility at high temperatures (e.g., 15~30% at 1200°C). Because of limited ductility and poor grain structure in cast ingots, even boron-doped aluminides simply could not be fabricated by hot forging or rolling. Further development is needed to enhance high-temperature grain-boundary cohesion through control of grain-boundary composition and by refining grain structure through precipitation of second-phase particles.
2. Creep Resistance: Polycrystalline $Ni_3Al$ has limited creep resistance possibly because of extensive grain-boundary sliding and cavitation which shorten the creep rupture life. Alloying with hafnium significantly lowers the creep rate and extends the creep life of the aluminides. The creep resistance of Hf-modified aluminides is comparable to that of Waspaloy but not as good as cast superalloys currently in use. Improved understanding of the mechanisms that control the creep rate and rupture processes are required for further improvement in creep properties.
3. Low-Temperature Strength: $Ni_3Al$ alloys have the excellent yield strength at elevated temperatures but their strength is inferior to that of commercial superalloys below 400°C. The use of the aluminides as turbine disk materials requires further improvement of the yield stress at lower temperatures. An effective scheme for increasing the strength appears to superimpose particle hardening upon solid solution (e.g., B, Hf) strengthening, while maintaining careful control of alloy stoichiometry.

REFERENCES


1. M. Hansen, *Constitution of Binary Alloys*, pp. 119, McGraw-Hill Book Company, New York (1958).
2. P. H. Thorton, R. G. Davis and T. L. Johnson, *Metall. Trans.* 1, 207 (1970).
3. P. A. Flinn, *Trans. TMS-AIME* 218, 145 (1960).
4. R. G. Davis and N. S. Stoloff, *Trans. TMS-AIME* 233, 714 (1965).
5. E. A. Aitken, *Intermetallic Compounds,* ed. J. H. Westbrook, pp. 491-515, John Wiley and Sons, Inc., New York (1967).
6. C. T. Liu and C. C. Koch, *Technical Aspects of Critical Materials Used by the Steel Industry, Vol.* IIB, NBSIR 83-2679-2, National Bureau of Standards (1983).
7. E. M. Grala, *Mechanical Properties of Intermetallic Compounds,* ed. J. H. Westbrook, pp. 358, John Wiley and Sons, Inc., New York (1960).
8. R. Moskovic, *J. Mater. Sci.* 13, 1901 (1978).
9. K. Aoki and O. Izumi, *Trans JIM* 19, 203 (1978).
10. A. V. Seybolt and J. H. Westbrook, *Acta Metall.* 12, 449 (1964).
11. K. Aoki and O. Izumi, *Nippon Kinzoku Gakkaishi* 41, 170 (1977).
12. K. Aoki and O. Izumi, *Nippon Kinzoku Gakkaishi* 43, 1190 (1979).
13. C. T. Liu, C. L. White, C. C. Koch and E. H. Lee, *Proc. Symp. High Temperature Materials Chemistry II,* ed. Munir, et al., The Electrochemical Soc., Inc. (1983).

14. A. I. Taub, S. C. Huang, and K. M. Chang, *Metall. Trans. A* **15A**, 399 (1984).
15. S. Ochiai, Y. Oya, and T. Suzuki, *Acta Metall.* **32**, 289–298 (1984).
16. C. C. Koch, J. A. Horton, C. T. Liu, O. B. Cavin, J. O. Scarbrough, in *Rapid Solidification Processing, Principles and Technologies III,* R. Mehrabian, ed., pp. 264–69 (National Bureau of Standards, 1983).
17. A. Inoue, H. Tomioku, and T. Masumoto, *Metall. Trans. A* **14A**, 1367 (1983).
18. M. G. Mendiratta, S. K. Ehlers, D. K. Chatterjee, H. A. Lipsitt, in *Rapid Solidification Processing, Principles and Technologies III,* R. Mehrabian, ed., p. 240 (National Bureau of Standards, 1983).
19. E. R. Slaughter and D. K. Das, in *Rapid Solidification Proceeding: Principles and Technologies II,* Mehrabian, Kear and Cohen, eds., pp. 354–63 (Claitor's Publishing Division, Baton Rouge, 1980).
20. C. C. Koch, *Proc. MRS Symp. High-Temperature Ordered Intermetallic Alloys,* November 26–29, 1984, Boston, MA.
21. A. U. Seybolt and J. H. Westbrook, *Acta Metall.* **12**, 449 (1964).
22. J. H. Westbrook and D. L. Wood, *J. Inst. Metall.* **91**, 174 (1962–1963).
23. B. H. Kear, *Private Communication to C. L. White,* Pratt and Whitney, 1973.
24. C. L. White and D. F. Stein, *Metall. Trans. A* **9A**, 13 (1978).
25. C. T. Liu, C. L. White, J. A. Horton, accepted for publication in Acta Metallurgica, 1984.
26. C. L. White, R. E. Clausing and L. Heatherly, *Metall. Trans. A* **10A**, 683 (1979).
27. S. K. Huang and J. W. Morris, *Metall. Trans. A* **10A**, 545 (1979).
28. D. F. Stein and L. A. Heldt, *Interfacial Segregation,* eds. W. C. Johnson and J. M. Brakely, pp. 239–260, RSM, Metal Park (1977).
29. C. L. Briant and R. P. Messmer, *Phil. Mag.* **B-12**, 569 (1980).
30. R. P. Messmer and C. L. Briant, *Acta Metall.* **30**, 457 (1982).
31. C. L. White, R. A. Padgett, C. T. Liu, S. M. Yalisove, accepted for publication in Scripta Metallurgica, 1984.
32. J. R. Rice, *The Effect of Hydrogen on the Behavior of Metals,* pp. 455–466, AIME, New York (1976).
33. A. I. Taub, S. C. Huang, K. M. Chang, *Private Communication,* May 1984.
34. R. W. Guard and J. H. Westbrook, *Trans. AIME* **215**, 807–814 (1959).
35. R. D. Rawlings, A. Staton-BeVan, *J. Mater. Sci.* **10**, 505–514 (1975).
36. O. Noguchi, Y. Oya and T. Suzuki, *Metall. Trans. A* **12A**, 1647 (1981).
37. K. Aoki and O. Izumi, *Phys. Status Solidi A* **32**, 657 (1975).
38. S. C. Huang, A. I. Taub, and K. M. Chang, submitted to Acta Metallurgica, 1984.
39. C. T. Liu, *Proc. Symp. High-Temperature Alloy Theory and Design,* Bethesda, MD, April 9–11, 1984.
40. B. H. Kear and H. G. Wilsdorf, *Trans. AIME* **224**, 382 (1962).
41. S. Takeuchi and E. Kuramoto, *Acta Metall.* **21**, 415 (1973).
42. J. A. Horton and C. T. Liu, unpublished results, October 1983.
43. R. L. Fleischer, *Acta Metall.* **11**, 203 (1963).
44. L. R. Curwick, Ph.D. Dissertation, University of Minnesota, 1972.
45. P. A. Flinn, *Trans. AIME* **218**, 145–154 (1960).
46. R. K. Ham, R. H. Cook, G. R. Purdy, and G. Willoughby, *J. Mater. Sci.* **6**, 205 (1972).
47. J. R. Nicholls and R. D. Rawlings, *J. Mater. Sci.* **12**, 2456 (1977).
48. D. P. Pope and S. S. Ezz, *Int. Met. Rev.* **29**, 136 (1984).
49. N. S. Stoloff, and A. K. Kuruvilla, *Private Communications,* October 1984.
50. A. K. Kuruvilla and N. S. Stoloff, submitted to publication in Metallurgical Transactions (May 1984).
51. M. L. Santella and S. A. David, *Proc. Symp. High-Temperature Ordered Intermetallic Alloys,* November 26–28, 1984, Boston, MA.
52. D. N. Tsipas, "The Effect of Hf Additions on Oxidation Behavior of $Ni_3Al$," paper presented at the *3rd International Symposium on High-Temperature Corrosion of Metals and Alloys,* Japan, November 18–20, 1982.

# THE B2 ALUMINIDES AS ALTERNATIVE MATERIALS

Joseph R. Stephens
National Aeronautics and Space Administration, Lewis Research Center, Cleveland, Ohio 44135

## ABSTRACT

As part of NASA's Conservation of Strategic Aerospace Materials (COSAM) Program, a research effort is underway to explore the potential of the B2 aluminides as structural material alternatives for the strategic element containing superalloys currently used in gas turbine engines. Emphasis is being place on the equiatomic Fe and Ni aluminides. Although Co is a strategic material, the equiatomic Co aluminide is also being studied to gain a more complete understanding of these fourth period intermetallics. The research effort is a cooperative program involving in-house research at NASA Lewis plus several university grant programs. Research focuses on initial processing techniques such as ingot melting, powder metallurgy, and rapid solidification with and without additional thermomechanical processing; high temperature deformation - primarily compressive creep; compositional effects within the binary B2 aluminides; third-element alloying addition effects on high temperature strength and oxidation resistance; and near room temperature ductility as influenced by processing, alloying, and grain size. This paper will review the various programs now underway and present some of the highlights of research results.

## INTRODUCTION

NASA Lewis Research Center has a research program underway to identify alternate materials to the commonly used nickel-base superalloys utilized in gas turbine engines. This research effort was initially prompted by the cobalt shortage in the 1978 to 1980 time frame along with inflated prices of tantalum and niobium and the near 100 percent import dependence of each of these key elements as well as chromium. These four metals are used as alloying elements in nickel-base superalloys to improve strength (Ta and Nb), corrosion and oxidation resistance (Cr and Ta) and fabricability (Co). In order to minimize the United States aerospace industry's dependence on these four metals, a program was initiated by NASA called Conservation of Strategic Aerospace Materials - COSAM [1]. Two major thrusts of the program involve: (1) strategic element substitution and (2) alternate materials. Research in the first major thrust focused on understanding the role of the strategic elements in nickel-base superalloys and then identifiying non-strategic substitutes [2-5]. This phase of the program has been successful and is now nearing completion. The second major part of the COSAM Program - alternative materials, has focused on the B2 aluminides of Fe, Ni, and Co. Although cobalt was considered a strategic material, it is included in this research in order to gain a better understanding of the properties of this class of materials since the creep strength of CoAl has been reported to be much greater than that of NiAl [6]. The research program has been conducted by in-house investigations at NASA Lewis Research Center in conjunction with several university research programs. This paper will present an overview of the research now underway on the aluminide materials.

Interest in the B2 aluminides stems from a number of reasons. Equiatomic FeAl melts at ∿1600 K which makes it a candidate for intermediate temperature applications in gas turbine engines while stoichiometric CoAl and NiAl melt congruently at ∿1920 K which makes them candidates for higher temperature applications. The binary aluminides exist over a wide range of composition and have large solubilities for substitutional third element additions. Furthermore, the aluminides possess the potential for self-protection in oxidizing and/or corrosive environments, are relatively lightweight, and contain inexpensive readily available elements (Fe, Ni, and Al).

The equiatomic aluminides have the ordered CsCl cubic structure. This is important as it permits use of polycrystalline materials under cyclic temperature service whereas noncubic intermetallics, due to unequal thermal expansion along the principal axes, could be limited to single crystal forms for varying temperature exposures. The CsCl crystal structure is shown in Fig. 1 where possible slip vectors of <111>, <110>, and <100> are indicated for slip or {110} type planes. In general, slip is believed to occur only on the {110} <001> system in B2 intermetallics at low temperatures. If true, then extensive plasticity is not possible in polycrystalline materials because only three independent slip systems exist. Limited room temperature ductility could be a major obstacle to the use of B2 aluminides. Fabricability can also be a problem with the B2 aluminides.

In order to realize the potential of these materials, research efforts have focused on: high temperature deformation behavior of the binary aluminides; alloying to improve high temperature strength; oxidation resistance; near room temperature deformation behavior; and processing including powder metallurgy techniques, conventional casting, and rapid solidification. The bulk of the binary deformation studies are being conducted in-house at NASA Lewis Research Center where emphasis has been place on high temperature creep of powder metallurgy materials with variables such as composition, grain size, and test temperature being explored. In addition, thermal expansion behavior of the binary aluminides has been determined. To augment the in-house program, special emphasis was placed on understanding the mechanisms controlling deformation of the aluminides utilizing transmission electron microscopy techniques primarily through a grant at Stanford University [7]. A second grant at Texas A and M University supports research utilizing the piezoelectric ultrasonic composite oscillator technique (PUCOT) to measure elastic modulii of the B2 intermetallics. There measurements are necessary to support modeling efforts. Detailed results of this study were presented earlier in this symposium by Harmouche and Wolfenden [8].

Alloying to improve high temperature strength is being conducted jointly by R.H. Titran of NASA Lewis and K. Vedula and co-workers from Case Western Reserve University. In a previous session of this symposium the effects of ternary additions to the FeAl intermetallic were described by Titran [9]. Vedula, V. Pathare, and I. Aslanidis [10] of Case Western Reserve University have co-authored a paper to be presented later in this session on alloying of NiAl. The effects of alloying additions on high temperature oxidation resistance and on the ductile-brittle transition temperature are also under investigation.

Studies of the effects of grain size, thermomechanical processing, and alloying on the near room temperature deformation behavior of NiAl were investigated by Schulson of Dartmouth College [11], and presented in an earlier session of this Symposium. In the area of processing, both powder metallurgy techniques and arc melting followed by extrusion have been used to prepare samples of the intermetallics. In addition, rapid solidification processing (RSP) utilizing the melt spinning technique has been explored by D. Gaydosh of NASA Lewis and the microstructure of the melt-spun ribbon characterized by M. Crimp of Case Western Reserve University. A paper describing the results obtained on melt spun ribbons of FeAl and NiAl will be presented by these investigators [12] later in this session. Other processing techniques involving some of the newer melting and fabricating techniques are also being explored at NASA Lewis Research Center.

## HIGH TEMPERATURE PROPERTIES

### Thermal Expansion

Thermal expansion of the aluminides has been measured to provide basic information for (1) thermal stress and oxidation/corrosion modeling and (2) calculation of thermal vacancy concentration. This latter effort requires measurement of the overall thermal expansion ($\Delta l/l_0$) as well as the change in lattice parameter ($\Delta a/a_0$) with temperature [13]. The aluminides have complex point defect structures consisting of antistructure (substitutional) atoms and vacancies [14-16]. In comparison to metals these aluminides can contain enormous numbers of vacant atom sites - approaching 10 percent for Al rich alloys [16-18]. Such large concentrations of vacancies could have a profound influence on mechanical properties.

Typical thermal expansion curves for the B2 aluminides from the work of Clark and Whittenberger [19] are shown in Fig. 2 along with curves for several common classes of materials. The curves illustrate the significantly larger thermal expansion of FeAl in comparison to that of either CoAl or NiAl. The data reveal that the expansion of NiAl is slightly greater than that of CoAl between room temperature and 1273 K. It was also shown that thermal expansion of all three intermetallics varies only slightly with composition. The large thermal expansion of FeAl compared with $Al_2O_3$ could pose problems from the standpoint of spalling of the oxide during cyclic heating and cooling. However, initial oxidation testing suggests that this is not the case.

Attempts to measure lattice parameters of the aluminides as a function of temperature and composition have been made. So far, experimental difficulties have resulted in inconsistent measurements. Steps are being taken to correct these problems.

### Binary Deformation Behavior

During the course of a study on the slow plastic elevated temperature properties of powder metallurgy FeAl the rather unusual flow stress-composition dependence shown in Fig. 3 was observed by Whittenberger [20]. The only plausible explanation for the lower strength of Fe-39.8 Al[1] and the higher strengths of Fe-43.2 Al and Fe-45.7 Al in comparison with other compositions seemed to be the grain size where small grain size resulted in stronger materials. The reason for this range of grain size is unclear since each alloy was extruded under nominally the same conditions. However, there could have been an inadvertent difference in time at temperature prior to extrusion or difference in cooling rates. To further explore this observation in more detail, a study was undertaken to examine the effects of thermomechanical processing on the elevated temperature strength of Fe-39.8 Al. Such treatments affected the grain sizes as summarized below:

---

[1]Unless noted otherwise, all compositions are in atomic percent.

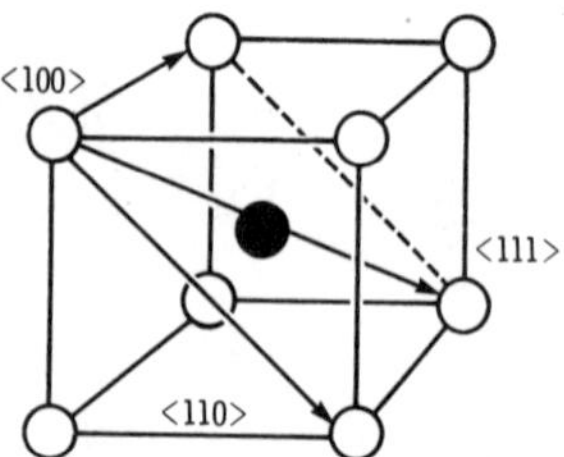


Figure 1. - CsCl (B2) crystal structure and possible <100>, <110>, and <111> slip vectors on {110} planes for the near equiatomic aluminides.

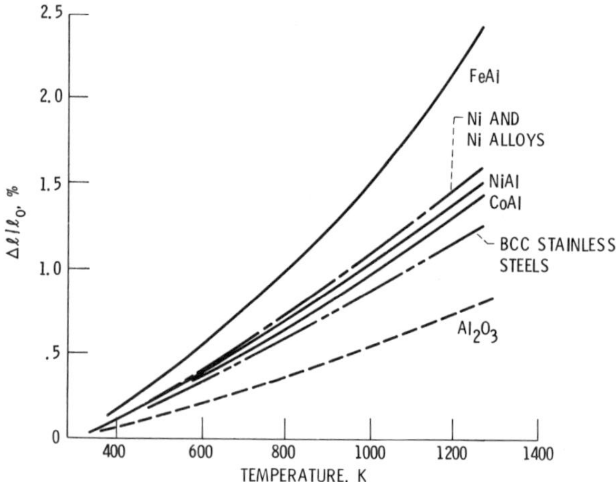


Figure 2. - Comparison of thermal expansion of the B2 aluminides, several common alloys, and alumina.

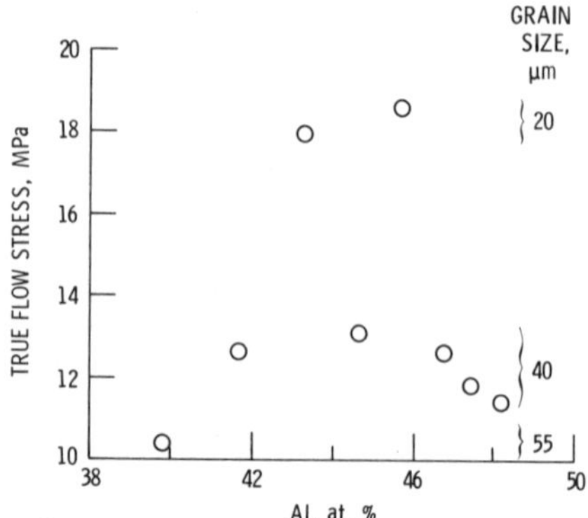


Figure 3. - Effect of aluminum content and grain size on flow stress to produce a strain rate of about $2x10^{-4}$ $s^{-1}$ at 1300 K in FeAl.

| Extrusion parameters | Heat treatment | Grain size, µm |
|---|---|---|
| Extruded at 16:1 at 1200 K | None | 9 |
| Extruded at 16:1 at 1200 K | 16 hr at 1300 K | 11 |
| Compacted at 1505 K and 1420 MPa; extruded at 16:1 at 1505 K | None | 55 |
| Compacted at 1505 K and 1420 MPa; extruded at 16:1 at 1505 K | 16 hr at 1535 K | 185 |

Transmission microscopy and x-ray texture analysis confirmed that all the materials were polycrystalline recrystallized and contained few if any subgrains. All materials were compression tested in air under constant velocity conditions at 1200, 1300, and 1400 K which are homologeous temperatures of about 0.75, 0.81, and 0.87 (based on the measured solidus temperature for Fe-39.8 Al. Typical 1300 K true stress-true strain diagrams are shown in Fig. 4 and are representative of the data for the other material conditions and temperature where the large grain size materials (55 and 185 µm) deformed at an approximately constant stress after ∿1 percent strain while the smaller grain size (9 and 11 µm) aluminides exhibited continuous work hardening.

Flow stress-strain rate behaviors found for the Fe-39.8 Al intermetallic are illustrated in Fig. 5 which shows that at 1200 K the small-grained intermetallics are stronger. However, this advantage is lost at a higher test temperature of 1300 K (Fig. 5(b)).

For materials extruded at 1200 K strain rate-flow stress ($\dot{\varepsilon}$-$\sigma$) can be adequately described by a single stress exponent and activation energy (kJ) where:

$$\dot{\varepsilon} = 1.64\text{x}10^{10}\ \sigma^{3.5}\ \exp\ -\left(\frac{449}{RT}\right) \tag{1}$$

A similar analysis of the material extruded at 1505 K yields

$$\dot{\varepsilon} = C\sigma^{5.4}\ \exp\ -\left(\frac{483}{RT}\right) \tag{2}$$

and C is $9.7\text{x}10^9$ for the extruded material and $1.8\text{x}10^{11}$ for the extruded plus heat treated material.

These results suggest that the 10 µm fine-grained (1200 K extruded materials) intermetallics obey a deformation behavior proportional to a temperature independent stress exponent of 3.5 while larger grained-size material, 55 to 185 µm, obey a stress exponent term of $n = 5.4$. The results further show that decreasing the grain size can effectively strengthen a B2 aluminide to a homologeous temperature of at least 0.75.

The results suggest that dislocation glide mechanism(s) are greatly influenced by grain size. The post test dislocation structure of the compression tested Fe-39.8 Al materials has been studied in detail by Krishnan [21,22], and typical transmission electron photomicrographs are shown in Fig. 6. This figure illustrates that: (1) no subboundaries are formed, (2) the dislocations are straight, and (3) the density, $\rho$, is low. Analysis of individual dislocations to determine character and Burger's vector indicates that all dislocations are of the edge nature with $\bar{b}$ = <101>. This result was confirmed through use of a theoretical dislocation intensity computer program [23]. Further examination of the structure revealed the

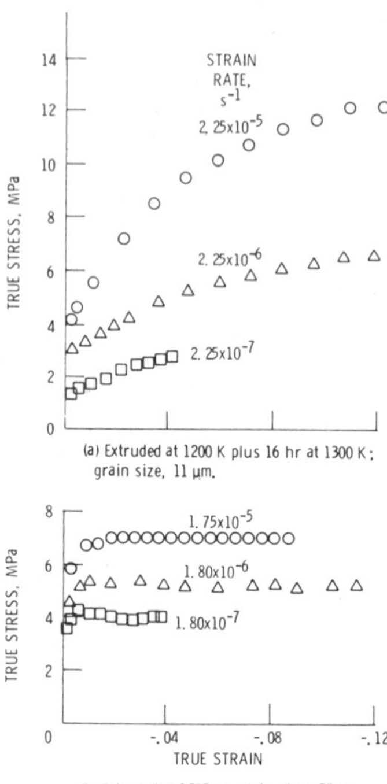


(a) Extruded at 1200 K plus 16 hr at 1300 K; grain size, 11 μm.

(b) Extruded at 1505 K; grain size, 55 μm.

Figure 4. - True stress/true strain compression test curves at 1300 K for Fe-39.8Al materials prepared by extruding at 1200 K plus 16 hr at 1300 K and extruding at 1505 K.

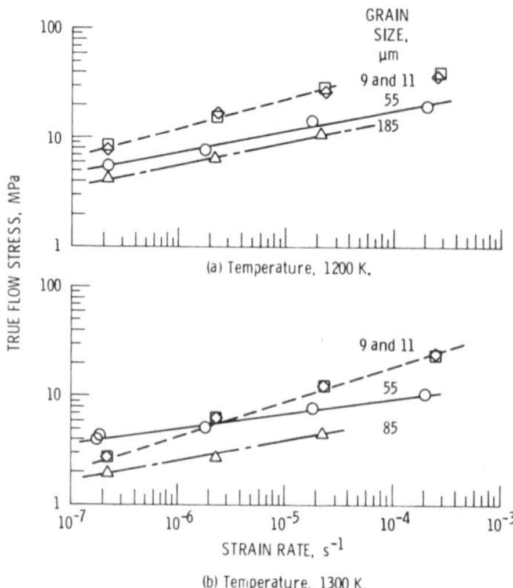


(a) Temperature, 1200 K.

(b) Temperature, 1300 K.

Figure 5. - True flow stress-strain rate curves for Fe-39.8Al materials.

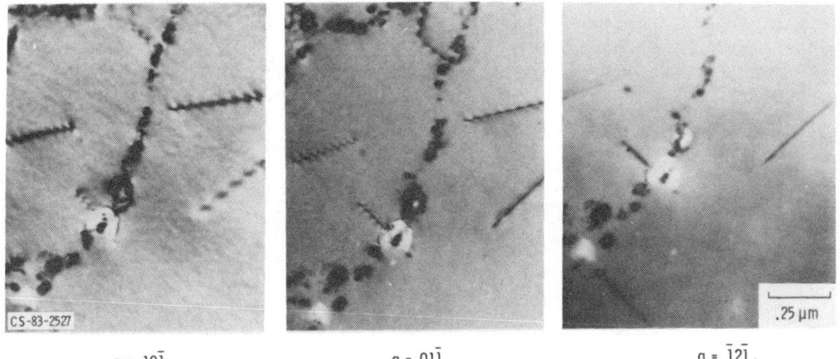


$g = 10\bar{1}$ $g = 01\bar{1}$ $g = \bar{1}2\bar{1}$

Figure 6. - Fe - 39.8 Al tested at 1200 K and $\dot{\varepsilon} \sim 2.1 \times 10^{-7}\ s^{-1}$ to 4.7% strain. Photomicrographs show loss of visibility for $b = 10\bar{1}$ then $g \cdot b$ yields 2, 1, and 0, going from left to right for the noted dislocations.

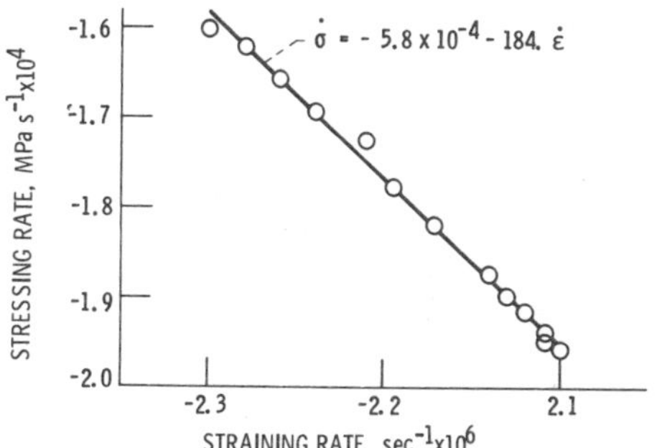


Figure 7. - Typical stressing rate/straining rate curves for as-extruded Fe-39.8Al tested at a constant velocity of $2.12 \times 10^{-5}$ mm $s^{-1}$ at 1200 K.

existence of many dislocation pairs with +b and -b in the near vicinity of one another suggesting that dislocation recovery is governed by a simple annihilation process which produces the observed low dislocation density, $\rho \sim 10^7 cm^{-2}$.

The stress-strain rate data for Fe-39.8 Al have been further analyzed in terms of the Bailey Orowan creep model which envisages high temperature deformation as a competition between recovery and hardening processes. A typical result of this work is shown in Fig. 7 where the slope of the curve is the work hardening coefficient for creep and the intercept is the recovery rate. Since $\dot{\sigma}$ and $\dot{\varepsilon}$ exhibited linear behavior, recovery rates and work hardening coefficients for creep could be readily calculated as functions of strain rate, temperature, and heat-treatment condition. Results for the small grain size Fe-39.8 Al show that recovery rates are dependent upon temperature (Fig. 8), strain rate, and to some extent, upon heat treatment. In contrast, the work hardening coefficient for creep is apparently independent of strain rate at all temperatures and independent of heat treatment at 1300 and 1400 K.

The most puzzling results of this study are the very low hardening coefficients for creep in the intermetallic, approximately 0.0015 of the modulus at 1200 K based on extrapolation of data by Koester and Goedeche [24]. Most results in the literature are greater than or equal to about 0.05 of the modulus of a given material and a similar value was obtained in this study [21] on an eutectic alloy $\gamma/\gamma'$-$\alpha$ thus showing the model is capable of yielding reasonable hardening coefficients. Therefore, deformation in FeAl seems to follow different laws than metals. Evans and Knowles [25] have proposed a model which can account for a low work hardening coefficient for creep in a material which undergoes recovery controlled deformation. Through dislocation glide and climb processes, a three-dimensional network is established, and the maintenance of this network is responsible for creep. In their development it is predicted that hardening coefficient, h, is dependent upon the slip distance, s:

$$h = \alpha\mu/2\ s\rho^{0.5} \tag{3}$$

where $\alpha$ is a constant about equal to one and $\mu$ is the shear modulus. A large value of s would predict a low work hardening coefficient for creep. This seems to be the case for Fe-39.8 Al, as very few barriers to dislocation motion are found even in specimens tested to large strains at relatively high velocities as seen in the transmission electron microscopy photomicrographs (Fig. 6) of as-extruded specimens tested at 1200 and 1300 K.

In addition to the mechanical property results obtained for Fe-39.8 Al, the effects of aluminium content on the elevated temperature slow plastic flow behavior of binary Fe, Co, and Ni aluminides are being investigated. To date only the study of CoAl for aluminum contents ranging form 43.65 to 51.31 percent has been completed. The vast majority of the constant velocity compression tests produced flow curves in which the stress rapidly increased during the first $\sim$1 percent of deformation and then remained constant as was observed for large grain sized FeAl. Typical flow stress-strain rate curves for two CoAl intermetallics are shown in Fig. 9. In this intermetallic Al content is important and this can best be illustrated by Fig. 10 where the stresses required to produce certain rates of deformation are shown as a function of composition. Clearly at faster strain rates the strength peaks near stoichiometry; however this advantage seems to be lessened at lower strain rates and higher temperatures. The effect of grain size on properties of CoAl is uncertain as all materials had a small grain size ($\sim$10 μm) which was little affected by testing.

In terms of the traditional temperature compensated power law behavior, the activation energy and pre-exponential factor for the cobalt aluminides are dependent upon composition with both having their greatest magnitudes at

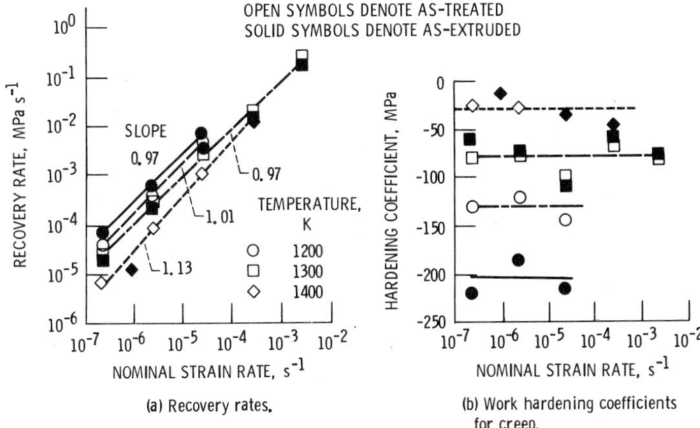


(a) Recovery rates.

(b) Work hardening coefficients for creep.

Figure 8. - Recovery rates and work hardening coefficients for creep as functions of nominal strain rate and temperature for Fe-39.8Al materials.

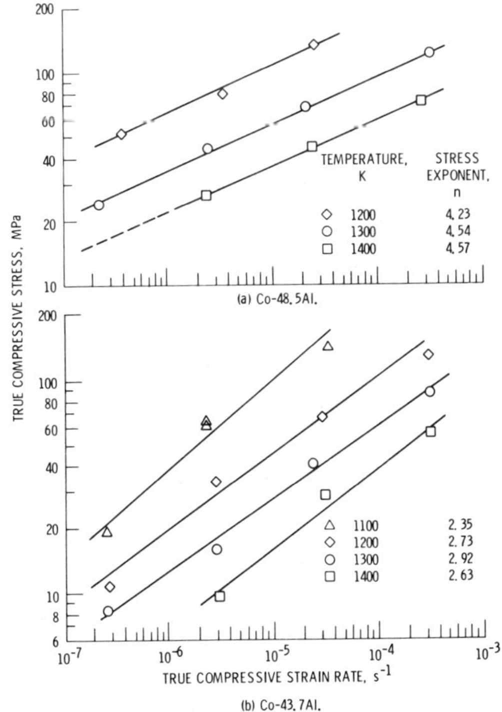


(b) Co-43.7Al.

Figure 9. - Compressive flow stress-strain rate behavior of Co-48.5Al and Co-43.7Al as a function of temperature.

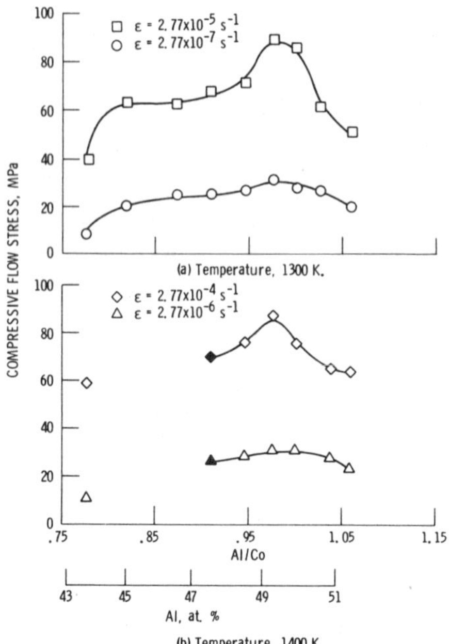


(b) Temperature, 1400 K.

Figure 10. - Flow stresses necessary to produce several strain rates.

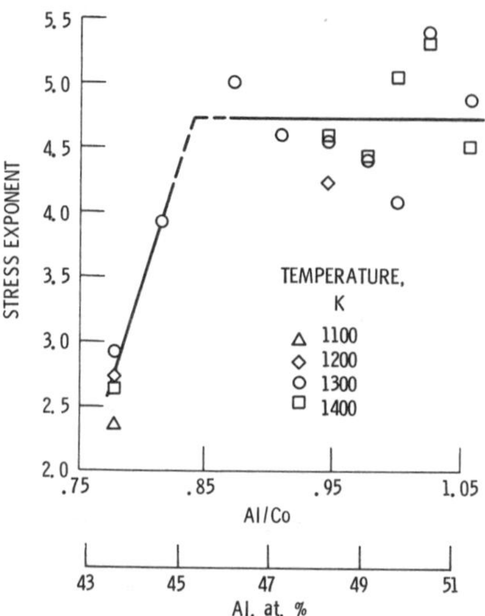


Figure 11. - Stress exponents as a function of composition in binary CoAl materials.

Al/Co $\sim$ 0.975. The stress exponent, on the other hand, seems to be independent of composition for Al/Co < 0.85 where $n \sim 4.7$ as shown in Fig. 11. The highly Al deficient CoAl materials apparently deform by a different mechanism than the other compositions.

The dislocation structures found in both as extruded and as-deformed samples of CoAl have been studied by Nix [7] of Stanford University using transmission electron microscopy. Extensive dislocation networks and very coarse subgrains were found in the as-extruded material plus some isolated dislocations in the deformed material. The Burger's vectors of some of the dislocations have been determined to be <100> and <110>. These dislocations would provide sufficient slip systems for general deformation. The scale of the dislocation substructure is much coarser than one would expect for a metal deformed at the same stress.

## Ternary Alloy Strengthening

Strengthening of the near equiatomic B2 aluminides of Fe and Ni was investigated by Vedula and Titran and co-workers [27,28], via elemental ternary additions to pre-alloyed intermetallic powders. Alloying at the 2 and 5 at % levels were selected on the basis of past experience in superalloy development. Due to the strong effect of boron on low temperature ductility in $Ni_3Al$ [29], boron was added at the 0.1 wt % levels. Microstructural examination of as-extruded and heat-treated materials indicated they could be grouped into three broad categories based on the extent to which the 5 at % of the element diffused into the matrix after long term high temperature annealing for up to 175 hr at temperatures in excess of 0.8 the melting temperature. Class I alloys are single phase after homogenization indicating complete solubility and included the elements Ti, Co, Ni, Fe, Mn, Si, and to a certain extent, Cr and Re. Class II alloying additions were characterized by significant interdiffusion with the matrix, but without resulting in a single phase microstructure. Elements which belong to this category are V, Nb, Zr, Hf, and Ta. Class III elements Mo and W did not exhibit any interdiffusion with the aluminide matrix.

True compressive flow stress-true compressive plastic strain curves for the alloys with ternary additions at the 5 at % level were similar to the binary intermetallics, in general, exhibiting initial work hardening and then leveling off at a constant flow stress for true strains greater than 0.02 to 0.05. For the FeAl alloys Nb and Zr alloy additions produce greater strengthening than other alloying elements. Both Nb and Zr belong to Class II elements, but it is not clear as yet whether this strengthening is due to solid solution effects and/or the effects of second phases present in the microstructure. Results of creep testing at 1300 K are shown in Fig. 12 for NiAl alloys. The lines through the data points are drawn with a slope of 5 which seems to be the general trend for these alloys and the lines allow better comparison of the alloys. Ternary alloys of Ni were tested under constant load conditions rather than constant velocity conditions used in most of the other programs. The data shown in Fig. 12 are primarily for extruded and heat treated alloys, along with some alloys prepared by the Hot Isostatic Pressing (HIP) process. In addition data for two conventional superalloys tested under similar conditions are shown in comparison. The Ta, Nb, and Hf containing alloys are clearly far superior to the other NiAl alloys. These alloys are, in fact, comparable to the conventional superalloys. Once again, as for FeAl alloys, the trend seems to be for Class II alloys to be stronger than the others.

The comparison of the HIPped and hot extruded alloys for Mo and Mn additions along with the binary NiAl shows that the hot extruded material is superior to the HIPped material. The reason for this could be the better bonding in extruded alloys, since the shearing action during extrusion can break up the oxide layers on the individual powder particles. It should

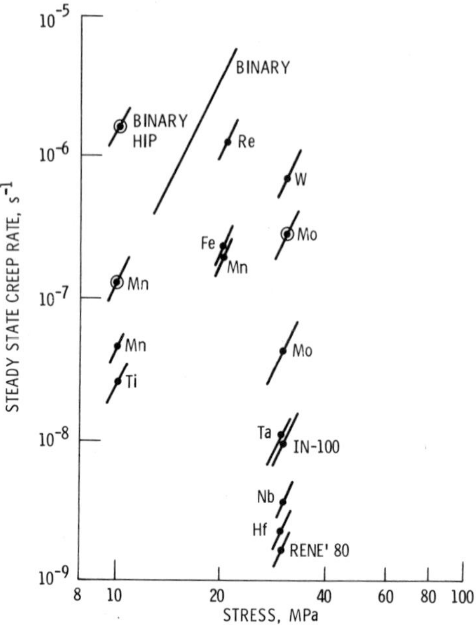


Figure 12. - Compressive creep data for several NiAl alloys tested at 1300 K compared with some conventional superalloys. Lines drawn through data points are expected slopes.

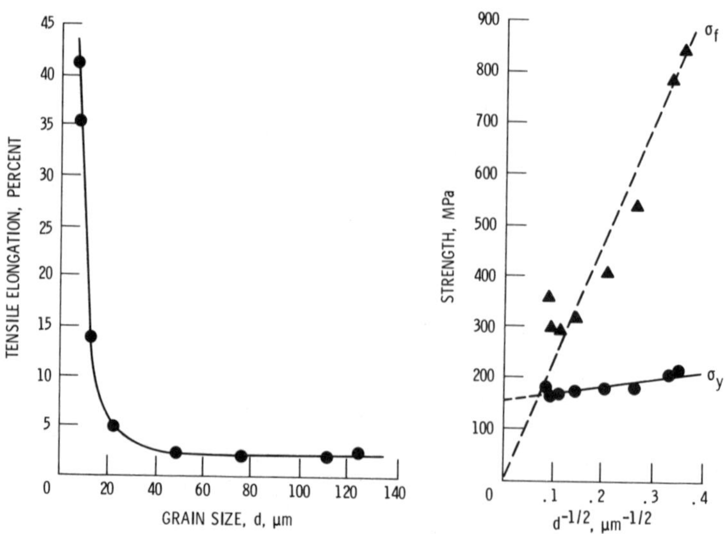


Figure 13. - Effect of grain size, d on tensile elongation, yield strength, $\sigma_y$ and fracture strength, $\sigma_f$ of NiAl at 673 K.

further be noted that the second phase particles in Class II alloys were aligned parallel to the extrusion direction and compression axis which may account for some of the strengthening achieved in these alloys. However, the coarse scale of the second phase particles makes it doubtful that they are the sole factor responsible for the improved strength of these alloys. Further studies are underway to determine the mechanism(s) by which the ternary additions improve the properties of the B2 aluminides.

## LOW TEMPERATURE PROPERTIES

### Grain Size Effects

Schulson [11] described the effects of grain size on the tensile ductility of NiAl near room temperature in an earlier session of this symposium. I would like to call attention to some of his earlier work [30] to show the dramatic effect of grain size on ductility. His work is based on the models of Cottrell [31] and Petch [32] which state that the stress required to nucleate microcracks in coarse-grained materials is more that enough to propagate them. In contrast, for fine-grained materials the stress required to nucleate cracks is less than that required to propagate them. In the first case, coarse-grained materials will fail in a brittle manner while fine-grained material must undergo permanent deformation and work hardening prior to failure. The conclusion is therefore, that a critical grain size should exist for plastic flow. This does indeed appear to be the case for Ni-49 Al at 673 K as illustrated by the results shown in Fig. 13 where the critical grain size is $\sim$20 μm. At lower temperatures the critical grain size becomes even smaller.

### Processing Techniques

Early work on the aluminides in the 1960's indicated that while tensile ductility could be achieved in single crystals at room temperature, polycrystalline materials produced by arc-melting were brittle. Advanced processing techniques give hope that improved low temperature ductility can be achieved in the B2 aluminides. Several of these new techniques are capable of producing very fine grain size material which, as we have just seen, is advantageous for low temperature ductility. One of these techniques is Rapid Solidification Processing (RSP). Gaydosh et al. [33] have investigated melt spinning to improve the ductility of NiAl. Melt spun Ni-50 Al ribbon exhibited a columnar grain structure with an average grain length of 15 μm and an average width of 5 μm. Room temperature bend testing on as melt-spun ribbon indicated an average failure strain of 0.9 percent while ribbons annealed for 1 hr at 1273 K in helium exhibited an average strain to failure of 2.3 percent. Scanning electron microscopy examination of fracture surfaces showed that the as melt-spun ribbon failed by almost entirely intergranular fracture while the heat treated ribbon possessed significant amounts of transgranular cleavage. The fine grain size produced by rapid solidification processing is thought to contribute to the plastic deformation achieved at room temperature, however, since annealing improved the ductility of the melt spun material other factors such as segregation of impurities at grain boundaries must also play a role in the change of fracture mode and increase in ductility.

## CONCLUDING REMARKS

NASA Lewis Research Center through a combined program of in-house research and university grants is investigating the potential of the B2 aluminides as structural materials for high temperature applications in advanced heat engines. The program is aimed at understanding the mechanisms controlling both elevated temperature deformation where the materials may find applications and the near room temperature deformation behavior where limited ductility is a serious deterrent to their eventual use.

Results have shown that grain size plays a major role in controlling the strength of FeAl to homologeous temperatures ∿0.75. A low dislocation density and absence of substructure, along with correlations of creep stressing rates and straining rates, suggest a form of the Bailey-Orowan model describes the deformation of FeAl. The low hardening coefficients, however, indicate that deformation of the aluminide involves a different mechanism(s) than that normally observed for metals. Alloying to improve high temperature strength has shown that best results can be achieved with third element additions that interdiffuse with the matrix of FeAl or NiAl to form a second phase. Nb and Zr are particularly attractive for FeAl, Nb, and Hf in NiAl. Room temperature brittleness especially in NiAl continues to be a problem, but processing techniques that yield a fine grain size suggest that some room temperature ductility can be achieved for the B2 aluminides. A model based on early work of Cottrell and the Hall-Petch relation for yield strength describe the low temperature deformation behavior of NiAl. It was further shown that rapid solidification processing results in measurable room temperature ductility in melt spun NiAl ribbon having a fine grain size. Studies are continuing to further understand and exploit those promising results obtained to date.

ACKNOWLEDGMENTS

The author wishes to express his appreciation to J.D. Whittenberger for many helpful comments in preparing the manuscript. The author gratefully acknowledges the assistance of J.D. Whittenberger, K. Vedula, D.J. Gaydosh, R.H. Titran, and R.V. Krishnan.

## REFERENCES

1. J.R. Stephens, NASA's Activities in the Conservation of Strategic Aerospace Materials, NASA TM-81617 (1980).

2. F.E. Sczezenie, and G.E. Maurer, in "COSAM Program Overview," pp. 21-36, NASA TM-83006 (1982).

3. R.N. Jarret, and J.R. Tien, Metall. Trans. A, 13, 1021 (1982).

4. M.V. Nathal, R.D. Maier, and L.J. Ebert, Metall. Trans. A, 13, 1767 (1982).

5. H.C. Nguyen, B.J. Pletka, and R.W. Heckel in Proceedings of the Conference on High Temperature Alloys: Theory and Design," ed. by J. Stiegler, AIME, New York, (1985). In press

6. L.A. Hocking, P.R. Strutt, and R.A. Dodd J. Inst. Met. 99, 98-101.

7. W.D. Nix in "COSAM" Program Overview: pp. 183-190, NASA TM-83006 (1982).

8. M.R. Harmouche, and A. Wolfenden, in "Proceedings of Materials Research Society, 1984 Fall Meeting," Elsevier Science Publishing Co., New York, (1985). To be published.

9. R.H. Titran, K.M. Vedula, and G.G. Anderson, [Editing Note-Same Info as #8.]

10. K. Vedula, V. Pathare, I. Aslanidis, [Editing Note-Same Info as #8.]

11. E.M. Schulson, [Editing Note-Same Info as #8.]

12. D. Gaydosh, and M. Crimp, [Editing Note-Same Info as #8.]

13. R.O. Simmons, and R.W. Ballutti, Phys. Rev. 125, 862 (1962).

14. N. Ridley, J. Inst. Met. 94, 255 (1966).

15. A.J. Bradley, and A.H. Jay, Proc. Roy. Soc. A, 136 210 (1932).

16. A.J. Bradley, and A Taylor, Proc. Roy. Soc. A, 159 56 (1937).

17. R. Meyer, E. Waschte, and V. Gerold, Z. Metalkd. 67, 97 (1976).

18. D. Paris, P. Lesbats, J. Nucl. Mater. 69-70, 628 (1978).

19. R.W. Clark, and J.D. Whittenberger, in "Proceedings of the 8th International Thermal Expansion Symposium," ed. by T.A. Hahan, Plenum Press, NY (1984), pp. 189-196.

20. J.D. Whittenberger, Mater. Sci. Eng. 57, 77 (1983).

21. J.D. Whittenberger, and R.V. Krishnah, J. Mater. Sci. 19, 509 (1984).

22. R.V. Krishnan, Private Communication.

23. A.K. Head, P. Humblc, L.M. Clarebrough, A.J. Morton, and C.T. Forewood, in "Defects in Crystalline Solids, vol. 7, "American Elsevier Publishing Company, Inc., (1973).

24. W. Koester, and T. Goedeche, Z. Metallkd 73, 111 (1982).

25. H.E. Evans, G. Knowles, in "Creep and Fracture of Engineering Materials and Structures," ed. by B. Wilshire and D.R.J. Owen, Pineridge Press, Swansea (1983), p. 169.

26. J.D. Whittenberger, Mater. Sci. Eng. (1985). To be published.

27. K. Vedula, G. Anderson, V. Pathare, and I. Aslanidis, in "Proceedings of the International Powder Metallurgy Conf.," Toronto. (1984).

28. V. Pathare, K. Vedula, and R.H. Titran, [Editing Note-Same Infor as #27.]

29. C.T. Liu, C.L. White, C.C. Koch, and E.H. Lee, in "High Temperature Materials Chemistry II." ed. by Munir, Electrochemical Society. Proceedings Volume 83-7 (1983), pp. 32-41.

30. E.M. Schulson, and D.R. Barker, Scr. Metall. 17, 519 (1983).

31. A.H. Cottrell, Trans. AIME 212 192 (1958).

32. N.J. Petch, Phil. Mag. 3, 1089 (1958).

33. D.J. Gaydosh, R.W. Jech, and R.H. Titran, J. Mater. Sci. Lett. (1985). To be published.

# A REVIEW OF RAPID SOLIDIFICATION STUDIES OF INTERMETALLIC COMPOUNDS

C. C. KOCH
Materials Engineering Department, North Carolina State University, Raleigh, NC 27695

## ABSTRACT

A review of rapid solidification studies of high temperature ordered intermetallic compounds is presented. Emphasis is on the nickel - and iron - aluminides which are of potential interest as structural materials. The nickel-base aluminides which have been rapidly solidified exhibit changes in grain size, compositional segregation, and degree of long range order (as reflected in APB size and distribution) which markedly affect mechanical properties. Some experiments indicate the formation of a metastable $L1_2$ phase in rapidly solidified Fe-(Ni,Mn)-Al-C alloys, while other work observes only a metastable fcc phase in the same composition range. The metastable phases and/or microstructures in both nickel and iron aluminides are destroyed by annealing at temperatures >750K, with subsequent degradation of mechanical properties. Rapid solidification studies of several other intermetallic compounds are briefly noted.

## INTRODUCTION

Rapid solidification processing has become a major component of materials science and engineering during the last decade. Most large industrial, government, and university laboratories have efforts in this area. Since early interest in this field following Pol Duwez's pioneering work in the early 60's, the field has expanded to over 1000 publications per year at the present time. In recent years most of this research interest has been focused on metallic glasses. However, there is also intense interest in rapidly quenched crystalline alloys such as aluminum alloys and superalloys.

The major effects of rapid solidification are:

1. grain size decreases
2. chemical homogeneity increases
3. extension of solid solubility increases
4. creation of metastable crystal structures
5. creation of metallic glasses

The criterion for rapid solidification is somewhat ill-defined, but usually means the temperature of the liquid is reduced to some value well below the freezing point in a very short time, usually milliseconds. For the rapid solidification to be discussed in this review this means quenching rates of ~ $10^4$ to $10^7$ K/s. The metastability introduced into a material by the non-equilibrium process of rapid solidification can often be attained by other non-equilibrium methods such as vapor quenching (sputtering or evaporation), electrodeposition, or ion bombardment. The emphasis in this review, however, will be on the several methods for rapid quenching from the liquid state.

Almost all classes of materials have been studied by rapid solidification techniques in recent years. In keeping with the theme of this symposium, this review will focus on high-temperature ordered intermetallic alloys that have been subjected to rapid solidification processing. Since these materials have, in general, relatively high melting points, they are not good glass formers, so it is expected that rapid solidification will not produce metallic glasses in such systems. Other non-equilibrium structural and compositional effects may be observed, however. The bulk of this review will be concerned with the nickel- and iron-aluminides although several other intermetallic compounds will be included for completeness.

## RAPID SOLIDIFICATION OF NICKEL ALUMINIDES

Studies of the structure and some properties of rapidly solidified $Ni_3Al$ (with or without small additions of boron and/or titanium) have been carried out by Baker and co-workers (1,2) on rapidly solidified powder, by Koch et al (3) on foils prepared by the arc-hammer method, and by Huang and co-workers (4,5,6) on melt-spun ribbon.

Baker and co-workers (1,2) used powder prepared by the rotary/centrifugal atomization process of the Pratt and Whitney Aircraft, Government Products Division, West Palm Beach Florida. The quench rates varied with droplet size (25 to 80 μm) but were estimated to be greater than $10^5$ K/s. The as-quenched powders were examined by optical microscopy, microhardness measurements, scanning electron microscopy (including analysis by a wavelength dispersive spectrometer), and x-ray diffraction. Powders were subsequently annealed at 800°C. Samples of stoichiometric $Ni_3Al$ and $Ni_3Al$ (24.4 at % Al) containing 1.5 at % Ti and 2.8 at % B were studied.

Stoichiometric $Ni_3Al$ powders exhibited both lamellar and dendritic structures. The relative amounts of lamellar and dendritic structures were independent of particle size and both could be observed in the same particle. Composition orientation within the particles could not be resolved by the wavelength dispersive spectrometry (WDS) technique (detection level 1-2 at.%).

The $Ni_3Al$ powders with Ti and B showed a slightly coarser dendritic structure and no lamellar structure. Again composition variations could not be measured by WDS.

X-ray diffraction patterns revealed broad $Ni_3Al$ lines with weak superlattice lines. The long range order parameter S was calculated to be 0.55 ($Ni_3Al$) and 0.63 ($Ni_3Al$(B,Ti)). Microhardness values were high, 530 VHN and 519 VHN, for the $Ni_3Al$ and $Ni_3Al$ (B,Ti) powders respectively.

After annealing at 800°C most (~ 80%) of the powders developed a grain structure. Some particles contained only a few grains but none were single crystals. Microhardness of the annealed powders dropped to 363 VHN and 413 VHN for $Ni_3Al$ and $Ni_3Al$ (B,Ti) respectively. The hardness impressions in both as-quenched and annealed powders show no evidence of cracking around them indicating some ductility in the material. The long range order parameter increased after annealing as evidenced by the increased intensity of the superlattice x-ray diffraction lines, although values for S were not reported.

Koch et al (3) rapidly solidified foils of $Ni_3Al$ (24 at. % Al) with and without additions of boron (500 and 1000 ppm by weight) in an arc-hammer apparatus. The bend ductility, microhardness, and microstructure of the foils were studied as a function of solidification rate, which was defined by foil thickness. Bend fracture strain, $\varepsilon^f$ versus sample thickness (which is inversely proportional to quench rate) is illustrated in Figure 1. Boron-free $Ni_3Al$ (24 at. % Al) samples exhibited brittle behavior over the entire range of solidification rates used (foil thicknesses 20 to 100 μm). However, boron-containing samples, which were ductile for foil thicknesses greater than 45 μm showed a ductile-to-brittle transition with decreasing thickness in the range of thickness of 30 to 45 μm. The conventionally cast, "bulk", alloys of $Ni_3Al$ (24 at. % Al) without boron were brittle (intergranular failure), while bulk samples of Ni-24 at.% Al with 200 to 2000 ppm by weight boron were ductile. It has been shown by Liu and co-workers (7,8,9,10) that boron segregation to the grain boundaries in substoichiometric $Ni_3Al$ (<25 at. % Al) apparently strengthens them and leads to ductile, at least partial, transgranular fracture. Auger electron spectroscopy (AES) has indicated that boron segregation to grain boundaries in the 24 at. % Al,

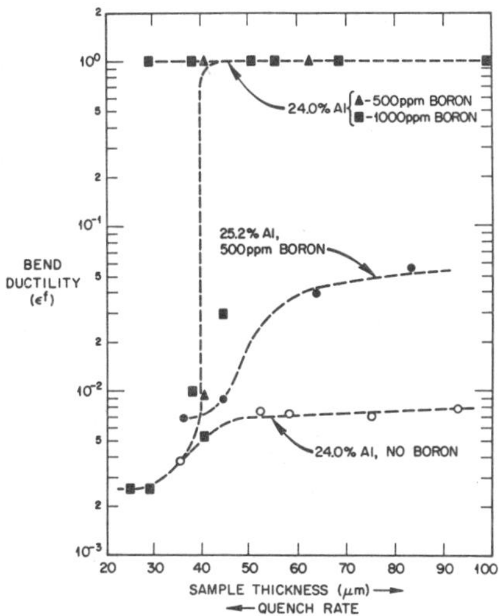


Figure 1. Bend fracture strain, $\varepsilon^f$, versus sample thickness. $Ni_3Al$ with and without boron. (reference 3)

$Ni_3Al$ is approximately 10 at. % boron based on peak-height-ratio of boron to nickel. As the aluminum concentration increases, the boron segregation to the grain boundaries decreases (as measured by AES) as does the ductility. For Al concentrations > 25.0 (i.e. 25.2 at. % Al) the ingots crack during fabrication and the measured peak-height-ratios suggest boron segregation of ∿ 7 at. %. Thus a critical concentration of boron at the $Ni_3Al$ grain boundaries appears to be required to inhibit brittle intergranular fracture. A recent AES study (11) on the brittle rapidly solidified Ni-24 at. % Al - 0.5 at. % B foils of reference (3) reveals a boron segregation at the grain boundaries of ∿ 7 at. %, i.e. similar to that observed in hyperstoichiometric $Ni_3Al$ and below the "critical" value needed for grain boundary strengthening. It is concluded that at the faster quench rates not enough boron can diffuse to the grain boundaries to achieve the critical concentration such that a return to brittle intergranular behavior occurs. At slower quench rates the samples failed in a ductile, transgranular manner. Microhardness was also a function of quench rate as shown in Figure 2. However, the high hardness of the most rapidly quenched foils (∿ 450 VHN) remained approximately constant in the ductile-to-brittle transition range. These values are somewhat lower than those reported by Baker et al. (1,2) for stoichiometric $Ni_3Al$. Not enough data was taken to reveal systematics in hardness versus boron concentration although yield strength has been found to increase with boron concentration in both bulk (10) and melt spun (5) $Ni_3Al$. Transmission electron microscopy revealed a complex anti-phase domain boundary (APB) structure that could not be correlated to the mechanical behavior. Subsequent work by Horton et al. (12) elaborated on the microstructure of rapidly quenched $Ni_3Al$ and $Ni_3Al(B)$ foils. The micrograph of Figure 3 is typical of the rapidly quenched samples. Samples with and without boron additions both exhibited a bimodal size distribution of APB's. APB's are not normally seen in conventionally prepared $Ni_3Al$ because of the large APB energy. The domain structure in these rapidly solidified foils suggests a growth process with elemental partitioning. It was found by x-ray energy dispersive spectroscopy

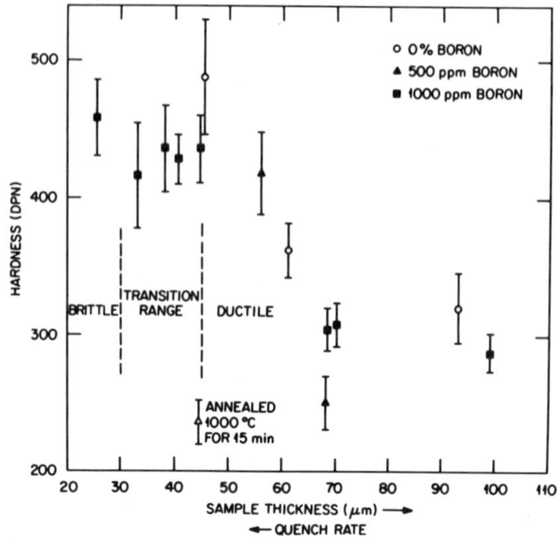


Figure 2. Hardness versus sample thickness. $Ni_3Al$ with and without boron. (reference 3)

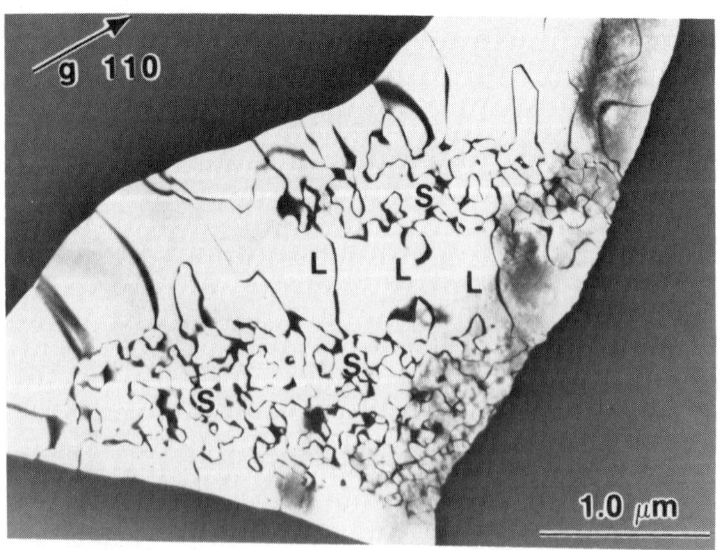


Figure 3. Regions in a specimen of $Ni_3Al$ with 500 wt. ppm boron analyzed by EDS. The large antiphase domain regions (L) were aluminum enriched by ~ 2.5 at. % compared to the small domain regions (S). Foil normal near [112], beam direction near [111]. (reference 12)

(EDS) for the 24 at. % Al alloy that the large domain regions are aluminum rich (25.8 ± .5 at. % Al) and the small domain regions are aluminum poor (23.3 ± 0.1 at. % Al). These results were consistent with lattice parameters measured in the respective regions using convergent beam electron diffraction. The lattice parameter of the large domain regions is 0.3 pm greater than that of the small domain region. From tabulated x-ray data this increase corresponds to a difference of ∿ 4 at. % aluminum between the regions.

Researchers at General Electric Corporate Research and Development (4,5, 6) modified $Ni_3Al$ with boron and rapidly solidified the resulting alloys by melt spinning. Alloys with compositions of $(Ni_{0.75}Al_{0.25})_{100-x}B_x$ containing up to 6 at. % boron were prepared. The alloys were melt spun in vacuum into ribbons approximately 6mm wide by 35 μm thick. Cooling rates were estimated to be greater than 5 x $10^5$ K/s. The structure of the rapidly solidified ribbons was examined by x-ray diffraction, optical microscopy, scanning electron microscopy, and transmission electron microscopy. Mechanical properties were measured by bend tests or by tensile tests.

The microstructure of the ribbons indicated a single-phase structure with equiaxed grains of 5-10 μm diameter for boron concentrations $\lesssim$ 1.5 at. %. The equiaxed grains are believed to develop by recrystallization from the as-solidified structure. This recrystallization can occur during second-stage cooling of the ribbon after the ribbon solidifies. Transmission electron microscopy revealed a subgrain cellular structure with cell diameters of about 0.2 μm. No anti-phase domain boundaries were observed.

The lattice parameters of the melt spun $Ni_3Al$-B alloys increased monotonically with boron concentration as illustrated in Figure 4. The lattice parameters for the higher boron concentration alloys were lower for the top surface of the ribbon compared to those for the wheel-contacting surface of the ribbons-suggesting incomplete solubility at these concentrations for the slower quench that the top surface of the ribbon receives. TEM showed $M_{23}B_6$ borides precipitated at grain and cell boundaries in the boron-rich alloys. It was concluded that a solubility limit of at least 1.5 at. % boron can be attained by melt spinning $Ni_3Al$.

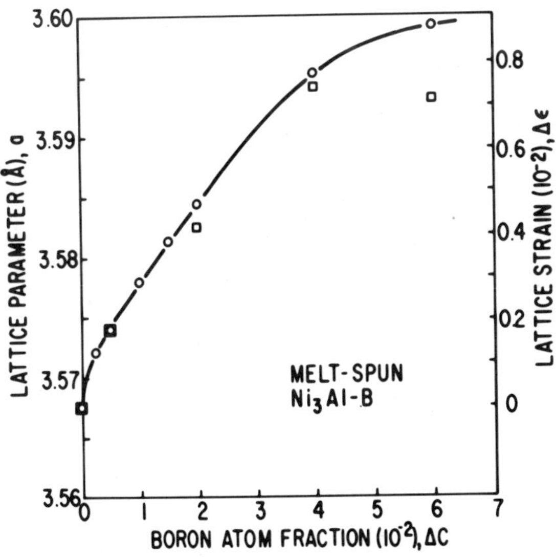


Figure 4. The lattice parameter and lattice strain as a function of the boron concentration in melt spun $Ni_3Al$.

o - wheel contacting surface

□ - top surface (reference 6)

Tensile yield strength and fracture strain as a function of boron concentration is shown in Figure 5. The yield strength exhibits a nearly linear increase with boron concentration, which is presumably associated with the large lattice strain (see Figure 4) boron introduces into $Ni_3Al$ by occupying interstitial lattice positions. As in the previous studies (3,7-11), the increase in fracture strain with boron concentration is believed to be due to segregation of boron atoms to the grain boundaries which increases the bonding strength between grains (13). Auger spectra from fractured surfaces of $Ni_3Al$-1.0 at. % B ribbon showed a segregated grain boundary concentration of 5 at. %. This is illustrated in Figure 6. With increasing boron concentration (> 1.0 at. %), $M_{23}B_6$ boride formation made the ribbons brittle.

The fact that no APB's were observed in the melt spun ribbons of this work in contrast to the rapidly quenched foils made in an arc-hammer (3) may be due to the observation that the melt spun ribbons recrystallized from the as-solidified structure. Since the APB energy is high in $Ni_3Al$, the as-quenched APB structure can be easily removed by annealing, or in the above case, by recrystallization as the quenched ribbon leaves the wheel.

Inoue et al. (14,15) have prepared ternary $L1_2$ structure alloys by rapid solidification of compositions Ni-Al-X (X=Ti, Zr, V, Nb, Cr, Mo, Mn, Fe, Co, Cu, or Si). All the alloys prepared were in the $L1_2$ structure phase field for the respective ternary system. Ribbons approximately 100 μm thick were prepared by melt spinning along with wires 85 to 190 μm in diameter which were prepared by using the in-rotating-water melt spinning technique (16). All the alloys were found to have the single phase $L1_2$ structure after rapid solidification, with grain sizes as small as 2 μm. All samples exhibited an APB structure. However, it was observed that the ductility was markedly different among the alloy systems. The ribbons of Ni-Al-X (X=Cr, Mn, Fe, Co or Si) systems were very ductile and could be bent through 180°. The Ni-Al-X (X=Ti, Zr, V, Nb, or Cu) ribbons, however, were extremely brittle and fractured during a bend test. While the grain sizes for all the Ni-Al-X

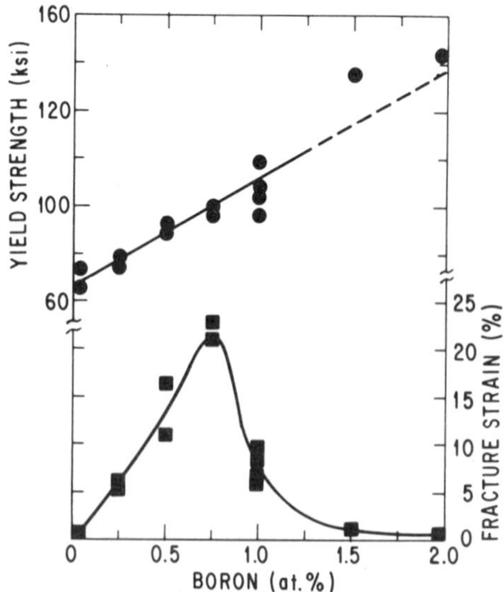


Figure 5. The room temperature yield strength and fracture strain as a function of boron concentration in melt spun $Ni_3Al$. (reference 6)

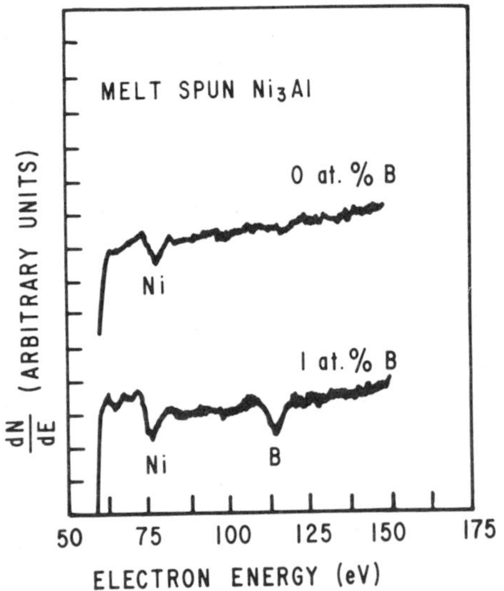


Figure 6. Partial Auger spectra from fracture surfaces in $Ni_3Al$ ribbons doped with 0 and 1 at. % boron. (reference 6).

$L1_2$ structure alloys were similar, the ductility differences could be correlated with the nature of the APB structure. The alloys exhibiting good ductility contained a fine APB domain structure with domain sizes 30 to 55 nm and fairly uniformly distributed throughout a grain. However, the brittle alloys had a much coarser and non-uniform APB structure. The Ni-Al-Si alloys went through a ductile-brittle transition with increasing silicon concentration that clearly revealed the correlation between ductility and APB structure. Alloys with 12 at. % Si showed both good ductility and a fine APB structure. As more Si was added, eg. Ni-6Al-16Si, the APB structure became coarser and non-uniform and the ductility dropped dramatically. As in the work by Koch et al. (3) regions of fine and coarse APB domains were found in a given grain. Selected area diffraction patterns from each region indicated more intense superlattice reflections, therefore a greater degree of long range order, in the regions with coarse APB domains. No measurements of local composition variations were reported. The APB structure was found to be dependent on the average outer electron concentration per atom (e/a) of the X elements, aluminum concentration, and cooling rate. The tendency was for the size of the APB domains to become small and uniform with increasing e/a of X elements, with decreasing aluminum content, and with increasing cooling rate.

Mechanical properties for selected Ni-Al-X alloys which have good ductility are listed in Table I.

Annealing the as-quenched ribbons and wires produced marked decreases in hardness, yield strength, fracture strength, and elongation in the annealing temperature range of 800-1150K. However, no significant change in grain size was observed so the changes in mechanical properties are attributed to structural changes within the grains during annealing. The temperature range where the strengths and elongation decrease significantly on annealing corresponds to that where the APB structure disappears. In this same annealing temperature range the fracture surface morphology of the wires changes from transgranular to intergranular. The authors concluded that the high strengths and good ductility of the rapidly solidified Ni-Al-X alloys were due to the

Table I. Vickers Hardness ($H_V$), Yield Strength ($\sigma_y$), Fracture Strength ($\sigma_f$), Elongation ($\varepsilon_p$), and Average Grain Size of the $\gamma'$ Compounds in Ni-Al-X (X= Cr, Mn, Fe, Co, or Si) Systems Quenched Rapidly from the Melts.

| Alloy(At.Pct) | $H_V$(DPN) | $\sigma_y$(MPa) | $\sigma_f$(MPa) | $\varepsilon_p$(Pct) | Grain Size (μm) |
|---|---|---|---|---|---|
| $Ni_{67.5}Al_{20}Cr_{12.5}$ | 375 | 490 | 810 | ≃ 6 | 3.0 |
| $Ni_{76}Al_{18}Mn_6$ | 250 | 260 | 445 | ≃27 | 2.9 |
| $Ni_{70}Al_{20}Fe_{10}$ | 300 | 390 | 580 | ≃ 4 | 4.9 |
| $Ni_{58}Al_{22}Co_{20}$ | 320 | 420 | 850 | ≃10 | 2.7 |
| $Ni_{78}Al_{10}Si_{12}$ | 445 | 590 | 905 | ≃ 7 | 1.8 |

combined effects of a low degree of long range order containing a high density of APB and the suppression of impurity segregation to the grain boundaries.

## RAPIDLY SOLIDIFIED (Fe, Co, Ni)$_3$ V ALLOYS WITH THE L1$_2$ CRYSTAL STRUCTURE

Alloys of composition Co-16.5 at. % Fe-25 at. %V, Fe-37.5 at.% Ni-24.5 at.% V-0.5 at.% Ti, Fe-37.5 at.% Ni-23.2 at.% V-0.5 at.% Ti-0.4 at.% Si-0.9 at.% Al, and Ni-32.4 at.% Co-25 at.% V were rapidly solidified in an arc hammer apparatus (17). The quenching rate was estimated to be at least $10^6$K/s. The as-quenched samples were approximately 30 μm thick. The structure of the as-quenched samples was fcc. The microstructure consisted of regular hexagonal cells, 300-400 nm in diameter contained in irregularly shaped grains which were approximately 2 ± 1 μm in diameter. The cell boundaries were decorated with MC carbide particles. Selected area diffraction patterns indicated the carbides had a random orientation with the matrix. Small bubbles/voids (∿5-10nm diameter) were observed along the cell boundaries.

After annealing schedules to develop the ordered L1$_2$ structure, superlattice reflections were observed and extensive stacking faults seen. The cell structure remained but the carbide particles at the cell boundaries developed a cube-on-cube orientation relationship with the matrix. The voids at the cell boundaries disappeared on annealing suggesting they were the result of vacancy condensation since gas bubbles of that size would probably grow rather than shrink during annealing. The cellular segregation structure observed in these rapidly solidified L1$_2$ structure materials is similar to that observed after rapid solidification in a number of crystalline iron-based alloys and steels (18).

## RAPID SOLIDIFICATION OF OTHER (NOT L1$_2$) Ni-Al ALLOYS

Chandrasekaran and Mukherjee (19) rapidly solidified alloys of Ni-35 at. % Al by the gun technique. The alloys before quenching had the β-NiAl (CsCl-type structure, B2) which transformed to the L1$_0$ (CuAu type structure) on quenching. The only difference observed between rapidly solidified foils and quenched bulk material was a much finer substructure in the twinned L1$_0$ martensite of the splat-quenched foils.

Duzević (20) vapor-quenched a series of Ni-Al alloys by flash evaporation. The deposition rate was 0.1-0.5 nm/s on a substrate at ∿150°C. Alloys containing 25 at. %, 40 at. %, and 50 at. % Al had amorphous structures in thin films of 60-100 nm thicknesses. No discussion of possible impurity levels were given. Thus, the much faster quench obtainable from the vapor phase than from the liquid state can apparently induce the amorphous structure in Ni-Al alloys.

RAPID SOLIDIFICATION OF IRON ALUMINIDES

The Fe-rich Fe-Al ordered intermetallics (based upon the $DO_3$ crystal structure) are highly oxidation resistant and therefore can be considered as materials for elevated-temperature structural applications. The major problems encountered in their development have been lack of room-temperature ductility and poor workability. This lack of ductility (21) is surprising since the plastic-flow mechanism in $Fe_3Al$ was found to consist of glide and cross-slip of 1/4 $a_o$ <111> dislocations (22); $a_o$ being the $DO_3$ lattice parameter. The predominant slip system, {110} <111>, provides more than five independent slip systems. However, brittle intergranular fracture was observed in polycrystalline $Fe_3Al$. In analogy to the effects of grain size in ductile vs. brittle transitions in polycrystalline aluminides (23,24) it is believed that $Fe_3Al$ might exhibit ductile behavior if prepared with a grain size below some critical value.

Slaughter and Das (25) studied $Fe_3Al$ with the addition of 1.5 mole % $TiB_2$ which had been rapidly solidified by the centrifugal atomization process. The cooling rate was estimated to be $10^5$ to $10^6$ °C/s. The powders were canned and either hot isostatically pressed or forged to 98% of full density. These compacts were then thermomechanically processed. The as-quenched powders and the compacted material were both examined with a scanning transmission electron microscope (STEM). Cellular or dendritic structures were observed in the powders with $TiB_2$ particles approximately 25.0 nm in diameter in the intercellular regions. After thermomechanical treatment a fine uniform dispersion of $TiB_2$ (77 nm average diameter) was observed. The $TiB_2$ dispersion apparently stablized a fine grain or subgrain size of 1-2 μm in the $Fe_3Al$.

This material, and rapidly solidified $Fe_3Al$ without $TiB_2$ dispersions has subsequently been subjected to mechanical testing which revealed a beneficial effect of rapid solidification on the mechanical properties. This work has been sponsored by the Air Force Wright Aeronautical Laboratories and is restricted by special export controls so it is not published in the open literature. Results of this work can be obtained by United States citizens on request to AFWAL (26).

While rapid solidification can apparently provide beneficial effects on the mechanical properties of Fe-Al compounds, the $DO_3$ ($Fe_3Al$) and $B_2$(FeAl) structures, being bcc-related, are inherently less desirable for elevated temperature mechanical properties than the $L1_2$ (fcc-related) structure. It has been reported by Inoue et al. that by rapid solidification non-equilibrium $L1_2$-type structure phases can be obtained in Fe-Al-C (27) and Fe-(Ni,Mn)-Al-C (28) alloys.

Alloys of Fe-Al-C were rapidly solidified by melt-spinning into ribbons approximately 1 mm wide by 30 μm thick (27). The structures of the as-quenched ribbons were examined by transmission electron microscopy and x-ray diffraction. Non-equilibrium phases of fcc (austenite-$\gamma$), $L1_2$ (ordered austenite-$\gamma'$), and hcp (epsilon) were found. The fcc phase was observed in the composition range 2 to 6 wt. % Al and 1.8 to 2.1 wt. % C while the $L1_2$ structure phase was obtained in the range 6 to 12 wt. % Al and 1.7 to 2.1 wt. % C. Although the non-equilibrium fcc phase was found to be ductile (could be bent 180° without fracture), the $L1_2$ structure ribbons were brittle. The fcc alloys had fine subgrains (0.1 to 0.4 μm diameter) and a large dislocation density. The $L1_2$ ribbon TEM microstructure that developed with Al concentrations $\geq$ 6% exhibited a coarsening grain size, disappearance of the subgrain boundaries and the appearance of a very fine modulated contrast accompanied by the superlattice reflections. A rectangular pattern of ordered domains was observed using dark field imaging of superlattice spots ( (100)). The domain size was observed to increase with Al concentration, which suggests that Al enhances the degree of long range order. This observation is consistent with the correlation of domain size (degree of order) and Al concentration in rapidly solidified $Ni_3Al$ (12) and Ni-Al-X (14,15).

A metastable $L1_2$ structure phase was also reported in rapidly solidified

Fe-(Ni,Mn,)-Al-C alloys (28). The formation region was limited to 7-55 wt.% Ni, 7-65 wt.% Mn, 3-9 wt.% Al, and 0.8-2.4 wt.% C. Ribbons approximately 70-200 μm in thickness were prepared by melt spinning. Therefore the quenching rate was presumably lower than for the Fe-Al-C alloys described above (27). The formation range of the phases observed in the Fe-Ni-Al-2.0 %C alloys is illustrated in Figure 7. A similar diagram can be drawn for Fe-Mn-Al-2.0% C alloys. The addition of approximately 7% Ni or Mn to the Fe-Al-C alloys induces a ductile $L1_2$ structure (ribbons can be bent 180° without fracturing) in contrast to the brittle ribbons observed for Fe-Al-C alloys with <7% Ni or Mn. Dark field images from superlattice reflections again reveal a rectangular pattern of ordered domain boundaries. The modulated contrast in the bright field images and the intensity of superlattice spots is much weaker in the Fe-Mn-Al-C than in the Fe-Ni-Al-C $L1_2$ structure alloys. This is reflected in the mechanical properties in that the Fe-Ni base systems possess much higher hardness and tensile strength and smaller elongation than the Fe-Mn base systems even though they both exhibit the $L1_2$ structure. This is attributed to the Fe-Ni-base alloys having much finer grains as well as a more well-developed long-range ordered structure than the Fe-Mn base alloys. The good strength and ductility remain unchanged on tempering (1 hour) until about 750 K. At this temperature the $L1_2$ structure compound decomposes to FeAl and cementite. Additions of Cr, Mo, or Co can improve the thermal stability to about 850K but it is clear that the $L1_2$ structure with its superior mechanical properties can not be maintained at high temperatures.

In contradiction of the above work Han and Choo (29) could not observe any superlattice reflections in rapidly solidified Fe-Ni-Al-C and Fe-Mn-Al-C alloys in the composition ranges where Inoue et al. (27,28) observed the $L1_2$ structures. Han and Choo (29) also used melt spinning to produce rapidly solidified ribbons 30 to 48 μm thick. Instead of the $L1_2$ structure they observed metastable fcc structures which exhibited sidebands around the Bragg peaks. Thus they claim that the structure of these rapidly solidified alloys

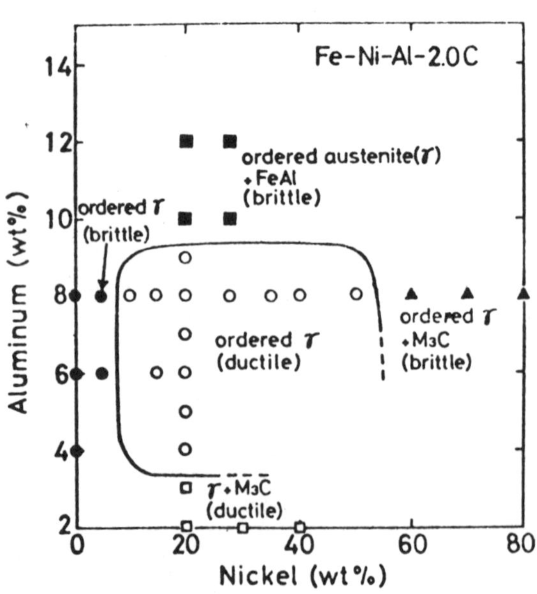


Figure 7. Composition ranges for metastable $L1_2$ structure phase in Fe-Ni-Al-2.0% C alloys which were rapidly solidified. (reference 28)

is a partially decomposed supersaturated fcc solid solution. The differences between these results and those of Inoue et al. (27,28) cannot be resolved from the data presented. It is especially difficult to explain why no superlattice reflections were observed in Han and Choo's experiments in contrast to Inoue et al.'s work. Perhaps impurity levels in the starting materials were different but these were not reported by either of the groups. While Inoue et al. used slower quenching rates (thicker ribbons) than Han and Choo for the Fe-(Ni, Mn)-Al-C alloys both groups used similar quenching rates (ribbon thickness) for their Fe-Al-C alloys. More work is needed to clarify the situation in these alloy systems.

## OTHER RAPIDLY SOLIDIFIED INTERMETALLIC COMPOUNDS

While the emphasis of this review has been on the Ni- and Fe-aluminides, several other intermetallic compounds which have been rapidly solidified will be included for additional information on the range of structural changes possible.

### A-15 Structure Compounds

The metastable A-15 structures in $Nb_3Ge$ (30) and $Nb_3Ga$ (31) were synthesized by rapid solidification (splat quenching) or by rapid quenching from above the solvus, respectively. More recently Lo et al. (32) have rapidly solidified the A-15 compound $Nb_{73}\ Al_{25.5}\ Si_{1.5}$ by melt spinning at quenching rates of $10^5$ to $10^6$ $^oC/s$. The grain size of the A-15 structure phase formed was a function of quench rate with the scale of microstructure (grain size) decreasing from 20 μm at a ribbon thickness of 4 mm down to <0.5 μm for a 25 μm thick ribbon. The fine grain structures result in very high values for the superconducting critical current density ($\sim 1.8 \times 10^6$ $A/cm^2$ in the magnetic field range of 60 to 150 KG).

### Zr-V and Hf-V Laves Phases

Tenhover (33) has rapidly solidified Zr-V and Hf-V alloys by the piston and anvil technique and obtained foils about 30 μm thick. An amorphous phase was found over a fairly broad range of compositions for both the Zr-V and Hf-V systems. Single phase amorphous material was obtained from 38% to 45% V. Above 45% V the foils contained a mixture of $ZrV_2$ (or $HfV_2$) and an amorphous phase. The $ZrV_2$ composition consisted of $ZrV_2$, hcp, and amorphous phases. The Bragg peaks associated with the crystalline phases were unusually broad, indicating small grain sizes. Heat treating Zr-V foils at temperatures greater than 300$^o$C led to the formation of the equilibrium phases of $ZrV_2$ and hcp Zr.

### Cu-Base Shape Memory Alloys

The influence of rapid solidification on the structure and behavior of shape memory alloys in the Cu-Al (34), Cu-Al-Ni (34), and Cu-Zn-Al (35,36) systems has been reported. Rapid solidification processing can be used to obtain fine grained material which display the shape-memory phenomenon due to martensitic transformations between the ordered structures ($\beta_1{}^1 \leftrightarrow \beta_1$ in Cu-Zn-Al, $\beta_1{}^1 \leftrightarrow \gamma_1{}^1$ in Cu-Ni-Al). Rapid solidification provides a technique for grain refinement without the complexity of adding further alloying elements to prevent grain growth.

In melt-spun Cu-Zn-Al (36), evidence for recrystallization after the ribbon leaves the wheel was noted. No solidification structure was observed, nor could the surface solidification pattern be related in scale or morphology to the ambient temperature grain structure.

## The $Zr_2Al$ Compound

The $Zr_2Al$ ($B8_2$ structure) phase was observed to form supersaturated Zr-27 at.% Al alloys by rapid solidification in a piston and anvil apparatus (37). The development of the $Zr_2Al$ phase from the supersaturated bcc $\beta$ phase was inferred from electron microscopy. A model involving spinodal decomposition and an omega-like transformation was proposed to describe the observed microstructure.

## SUMMARY

A review of rapid solidification studies of high temperature ordered intermetallic compounds has emphasized work on the nickel- and iron-base aluminides. No new phases are observed during rapid quenching at "usual" rates ($10^5$-$10^6$ °C/s) in the nickel-base aluminides. However, changes in microstructure, compositional segregation, and degree of long range order as reflected in APB size and distribution are observed and have major effects on mechanical properties (3,6,15). Ternary Ni-Al-X $L1_2$ alloys rapidly solidified (14,15) to a fine, uniform APB structure exhibit excellent strength and ductility. Those ternary $L1_2$ alloys with a coarse APB structure (higher degree of long range order) are brittle. Annealing at 800-1150K degraded the mechanical properties because of increases in the LRO and/or impurity segregation to grain boundaries.

Rapidly solidified Fe-(Ni, Mn)-Al-C alloys exhibited a metastable $L1_2$ structure (27,28) with good mechanical properties. However, annealing at >750K, decomposes the $L1_2$ phase and degrades the properties. In addition, others (29) could not reproduce the $L1_2$ phase in these alloys by similar rapid solidification procedures. More research is required on these Fe-base systems to resolve the controversy.

Some work on Fe-aluminides is restricted and other work on rapid solidification of aluminides is being reported at this symposium. It is expected that much more information on this subject will be forthcoming in the near future.

## ACKNOWLEDGEMENT

The author gratefully acknowledges the support provided by the National Aeronautics and Space Administration (Grant Number NAG8-475) for research on rapid solidification of nickel and iron aluminides.

## REFERENCES

1. I. Baker, F. Ichishita, and E. M. Schulson in Rapidly Solidified Metastable Materials, ed. B. H. Kear & B. C. Giessen, North Holland, (1984) p. 395-400.
2. I. Baker, F. S. Ichishita, V. A. Surprenant, and E. M. Schulson, Metallography, 17, 299-314 (1984).
3. C. C. Koch, J. A. Horton, C. T. Liu, O. B. Cavin, and J. O. Scarbrough, "Rapid Solidification Processing, Principles and Technologies III, ed. R. Mehrahian, National Bureau of Standards, 1983, p. 264.
4. A. I. Taub, S. C. Huang, and K. M. Chang, Metallurgical Transactions A, 15A, 399-402 (1984).

5. S. C. Huang, A. I. Taub, and K. M. Chang, General Electric Co. Technical Information Series, No. 84CRD049, March 1984; also to be published in Acta Metallurgica.
6. S. C. Huang, A. I. Taub, K. M. Chang, C. L. Briant, and E. L. Hall, to be published in the Proceedings of the Fifth International Conference on Rapidly Quenched Metals, Würzburg, Sept. 3-7, 1984.
7. C. T. Liu and C. C. Koch, Technical Aspects of Critical Materials Used by the Steel Industry, Vol. IIB, NBSIR 83-2679-2, National Bureau of Standards (1983).
8. C. T. Liu, C. L. White, C. C. Koch, and E. H. Lee, Proc. Symp. High Temperature Materials Chemistry II, Vol. 83-7, ed. M. Cubicciotti (Electrochemical Society Inc. 1983) p. 32.
9. C. L. White, R. A. Padgett, and C. T. Liu, to be published in Scripta Metallurgica, 1984.
10. C. T. Liu, C. L. White, and J. A. Horton, to be published in Acta Metallurgica.
11. C. C. Koch, C. L. White, R. A. Padgett, and C. T. Liu, submitted to Scripta Metallurgica.
12. J. A. Horton, C. C. Koch, and C. T. Liu, Proceedings of the 41st Annual Meeting of the Electron Microscopy Society of America, Ed. G. W. Bailey, San Francisco Press (1983) p. 248.
13. R. P. Messmer and C. L. Briant, Acta. Metallurgica 30, 457 (1982).
14. A. Inoue, H. Tomioka, and T. Masumoto, Jour. Mat. Sci. Letters 1, 377-80 (1982).
15. A. Inoue, H. Tomioka, and T. Masumoto, Metall. Trans. A, 14A, 1367-77 (1983).
16. T. Masumoto, I. Ohnaka, A. Inoue, and M. Hagiwara, Scripta Met., 15, 293 (1981).
17. E. H. Lee, C. C. Koch, and C. T. Liu in "Rapidly Solidified Amorphous and Crystalline Alloys" ed. by B. H. Kear, B. C. Giessen, and M. Cohen, Elsevier Science Publishing Co. (1982) pp. 375-379.
18. Brian Cantor, in "Rapidly Solidified Amorphous and Crystalline Alloys" ed. by B. H. Kear, B. C. Giessen, and M. Cohen, Elsevier Science Publishing Co. (1982) pp. 317-330.
19. M. Chandrasekaran and K. Mukherjee, Mat. Sci. and Eng. 14, 97-99 (1974).
20. D. Duzević, Sorysta Met. 9, 543-6 (1975).
21. M. J. Marcinkowski, M. E. Taylor, and F. X. Kayser, J. Mater. Sci. 10, 406 (1975).
22. M. J. Marcinkowski and N. Brown, Acta Met. 9, 764 (1961).
23. K. Aoki and O. Izumi, Trans. J. I. M. 19, 203 (1978).
24. E. M. Schulson, Res. Mechanica Letters 1, 111 (1981).
25. E. R. Slaughter and S. K. Das in Rapid Solidification Processing, Principles and Technologies II, ed. R. Mehrabian, B. H. Kear, and M. Cohen, Claitor's Publishing Division, Baton Rouge, Louisiana (1980), pp. 354-363.
26. H. A. Lipsitt, Air Force Wright Aeronautical Laboratories, AFWAL/MLLM, Wright-Patterson Air Force Base, OH 45433, private communication.
27. A. Inoue, T. Minemura, A. Kitamura, and T. Masumoto, Metallurgical Transactions A, 12A, 1041 (1981).
28. A. Inoue, Y. Kojima, T. Minemura, and T. Masumoto, Metallurgical Transactions A, 12A, 1245 (1981).
29. K. H. Han and W. K. Choo, Metallurgical Transactions A, 14A, 973 (1983).
30. B. T. Matthias, T. H. Geballe, R. H. Willens, E. Corenzwit, and G. W. Hull, Jr., Physical Review, 139, 1501 (1965).
31. G. W. Webb, L. J. Vieland, R. E. Miller, and A. Wickland, Solid State Communications, 9, 1769 (1971).
32. K. Lo, J. Bevk, and D. Turnbull, J. Appl. Physics 48, 2597 (1977).
33. M. Tenhover, Appl. Phys. 21, 279 (1980).
34. J. V. Wood, Chemistry and Physics of Rapidly Solidified Materials, Ed. by B. J. Berkowitz and R. O. Scattergood, Met. Soc. of AIME, Warrendale, PA, (1983) pp. 79-94.

35. J. Perkins, J. J. Rayment, B. Cantor, and R. W. Cahn, Scripta Met. 15, 771 (1981).
36. J. V. Wood, Scripta Met. 16, 271 (1982).
37. S. Banerjee and R. W. Cahn, Solid-Solid Phase Transformations, Ed. by H. I. Aaronson, D. E. Laughlin, R. F. Sekerka, and C. M. Wayman, Met. Soc. of AIME, Warrendale, PA, (1982) pp. 1005-1009.

# ALLOYS BASED ON NiAl FOR HIGH TEMPERATURE APPLICATIONS

K. VEDULA*, V. PATHARE*, I. ASLANIDIS* and R. H. TITRAN**
* Department of Metallurgy and Materials Science, Case Western Reserve University, Cleveland, Ohio;** NASA Lewis Research Center, Cleveland, Ohio

## ABSTRACT

This paper presents some of the results obtained in an on-going study of NiAl alloys for potential high temperature applications. Alloys were prepared by powder metallurgy techniques. Flow stress values at slow strain rates and high temperatures were measured. Some promising ternary alloying additions (Hf, Ta and Nb) have been identified. The mechanism of strengthening in alloys containing these additions appears to be a form of particle dislocation interaction. Interesting effects of grain size and stoichiometry in binary alloys are also presented.

## INTRODUCTION

NiAl is one of the interesting intermetallic compounds for potential use as a high-temperature structural material. The ordered structure is believed to provide high-temperature strength and the aluminum content provides the oxidation resistance. This paper presents some of the results of a broader study on the potential for NiAl and FeAl based alloys for high-temperature applications.

Stoichiometric NiAl has an ordered L2o (B2) crystal structure up to its congruent melting point of 1911°K (1638°C). Figure 1 shows the binary Al-Ni phase diagram [1]. The ordered B2 structure is maintained over the composition range from 45.0 to 58.5 at % at room temperature. This provides the potential for improving high temperature strength by solid solution alloying. However, very limited information on the alloying behavior of NiAl is available. A previous publication by these authors [2] contains some information on the potential for solid solubility of several ternary additions. Ti, Mn, Cr, Fe and Cu are found to have considerable solubilities in binary NiAl, whereas Nb, Hf, Ta, Mo, W and Zr are found to have very limited solubilities.

Deviations from stoichiometry on either side are accomodated by defect structures [3,4]. The defects on the nickel-rich side are believed to be antistructure defects (with excess nickel atoms occupying aluminum sites) and those on the aluminum-rich side are believed to be vacancies (with the nickel deficient sites remaining vacant). There is, however, some indication that even the stoichiometric alloy may contain a combination of both these types of defects and that the aluminum rich structure may contain a complex triple defect structure. Hence, within the single phase field of the ordered structure, mechanical properties are sensitive to deviations from stoichiometry [5]. At low temperatures, deviations from stoichiometry result in defect hardening. On the other hand, at high temperatures defects are believed to increase the diffusion processes and hence weaken the alloys.

Previous investigations of B2 aluminides have shown some peculiar grain size effects [6,7,8], in which fine grained material is observed to be stronger than coarse grained material even at temperatures

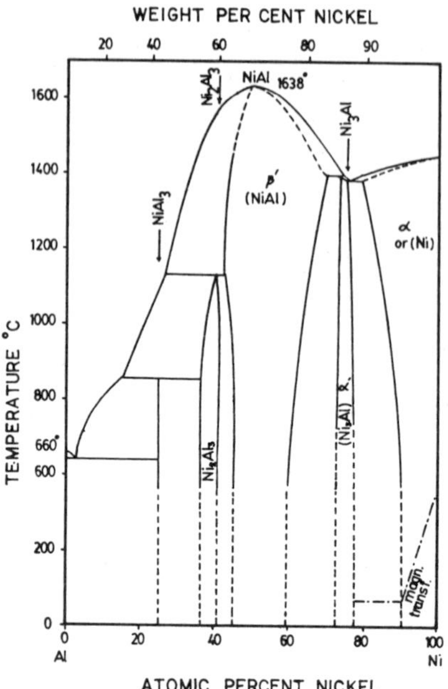


Fig. 1. Binary Al-Ni Phase Diagram [1]

of the order of 0.7 Tm. In particular, this effect has been noted for extruded FeAl alloys [6]. The reasons for this behavior are unknown and must be related to the effects of ordering on grain boundary structure and high temperature deformation mechanisms.

This paper discusses some of the observations of potential strengthening mechanisms in NiAl based alloys. These observations form part of an overall program at NASA Lewis Research Center to develop NiAl alloys for high temperature applications. In particular, the effects of ternary additions, grain size and deviations from stoichiometry will be discussed.

## MATERIALS AND PROCEDURE

Powder metallurgy processing was found to be a convenient method for processing these alloys.

Ternary alloys were prepared from blends of prealloyed NiAl and elemental ternary additions. Consolidation was achieved by either hot extrusion or hot isostatic pressing of canned powder. Alloys were heat treated after consolidation in order to ensure maximum homogeneity. Grain sizes of these ternary alloys were typically in the range of 25 to 50 μm. Specific details of the processing with ternary additions are presented elsewhere [2]. Compositions of alloys reported in the previous publications are presented in Table I. New alloys containing 2 at% Ti, Ta and Nb were later prepared and are discussed in this paper. Compositions of these alloys are presented in Table II.

TABLE I. Gross Chemical Composition of Alloys From Reference 2 (Atomic %)

| Specimen | Ni | Al | Ternary | Fe[1] |
|---|---|---|---|---|
| HIPPED ALLOYS | | | | |
| NiAl | 55.28 | 44.60 | -- | 0.001 |
| NiAl-Ti | 49.89 | 47.18 | 2.16 Ti | 0.330 |
| NiAl-Ta | 51.12 | 46.39 | 1.17 Ta | 1.310 |
| NiAl-Nb | 49.12 | 46.28 | 1.00 Nb | 3.630 |
| NiAl-Mn | 50.03 | 47.22 | 2.52 Mn | 0.002 |
| NiAl-Mo | 49.70 | 47.18 | 2.92 Mo | 0.002 |
| HOT EXTRUDED ALLOYS | | | | |
| NiAl-Fe | 45.19 | 49.54 | 5.27 Fe | |
| NiAl-Mo | 41.84 | 53.73 | 4.43 Mo | |
| NiAl-W | 41.87 | 53.63 | 4.50 W | |
| NiAl-Re | 41.98 | 53.59 | 4.42 Re | |
| NiAl-Mn[2] | 53.00 | 42.00 | 5.00 Mn | |
| NiAl-Hf | 45.32 | 53.63 | 1.05 Hf | |

1 Impurity
2 Intended Composition

TABLE II. Intended Gross Chemical Compositions of Second Alloy Series (Atomic %)

| Specimen | Ni | Al | Ternary |
|---|---|---|---|
| NTi2 | 50.00 | 48.00 | 2.00 Ti |
| NNb2 | 50.00 | 48.00 | 2.00 Nb |
| NTa2 | 50.00 | 48.00 | 2.00 Ta |

Fine grain sizes (< than 10 μm) were obtained in the binary alloys by reextruding at a lower temperature (1073°K) after initial extrusion at a higher temperature (1350°K). This was carried out for binary alloys having compositions of 48 at% Ni, 50 at% Ni and 52 at% Ni. Coarse grains were obtained by annealing extruded specimens in argon for varying lengths of time up to 144 hours at 1625°K and 1723°K. Deviations from stoichiometry were obtained by using different compositions of prealloyed powders.

Microstructures of alloys were observed using optical and electron microscopy.

High temperature testing was accomplished using constant strain rate compression testing of small cylindrical specimens (approximately 5 mm by 10 mm) on an Instron machine in air at temperatures ranging from 1100°K to 1400°K. Specimens were normally tested to about 10% reduction in heights and flow stress was recorded. Load - deformation information was converted to true stress - true strain information neglecting the slight barreling of the specimens. Flow stress remained fairly constant during the plastic deformation. Values of the maximum flow stress and corresponding strain rate were used to compare the high temperature behavior of various specimens.

## RESULTS AND DISCUSSION

The results will be discussed in the light of potential strengthening mechanisms resulting from: (a) alloying additions, (b) grain boundaries and (c) deviations from stoichiometry.

### (A) Alloying Additions

The alloying additions used could be classified on the basis of microstructural evaluation [2]. Extent of homogeneity in the specimens was evaluated by EDAX analyses. The alloys containing Ti, Mn and Fe were essentially shown to be homogeneous single phase alloys with no evidence of undissolved ternary powders. Alloys containing Ta, Nb, Hf and Re showed a significant amount of undissolved ternary powder in spite of a certain amount of dissolution and alloys containing Mo and W showed a negligible amount of dissolution.

The alloying additions which have poor solubility in the matrix, in general, appear to provide more strengthening at high temperatures than those with large solubilities. This trend was clearly indicated in the results presented previously and reproduced in Figure 2 [2]. Compositions of the alloys shown in this figure are tabulated in Table I. The alloys containing Ti, Mn and Fe did not show any significant improvements over binary alloys in compressive creep tests conducted in air at 1300°K. These are single phase alloys and the small amount of strengthening observed is probably due to solid solution effects. The alloys containing Ta, Nb and Hf exhibited very substantial strengthening compared to the binary alloys. These alloying additions only dissolved to a very small extent in the matrix NiAl, as indicated by the presence of undissolved ternary additions even after very long homogenization treatments.

This same trend is observed in the second series of alloys (NTi2, NNb2, NTa2) tested in hot compression at 1300°K. Figure 3 compares these alloys with binary data obtained from [2]. Compositions of this series of alloys are given in Table II. The alloys containing Ta and Nb are much stronger than the binary alloy and the alloy containing Ti.

The strength of the stronger alloys (Ta and Nb containing alloys) appears to be more temperature dependent than that of the binary alloys. This is illustrated in Figure 4 which shows that although the Ta alloy is significantly stronger (about twice as strong at constant strain rate) than the binary at 1400°K, its superiority over the binary alloy is much more at 1200°K (about ten times as strong at constant strain rate). This difference in temperature dependence suggests that whatever the actual mechanism of high temperature deformation in these alloys, the activation energy for the mechanism increases with the alloying addition.

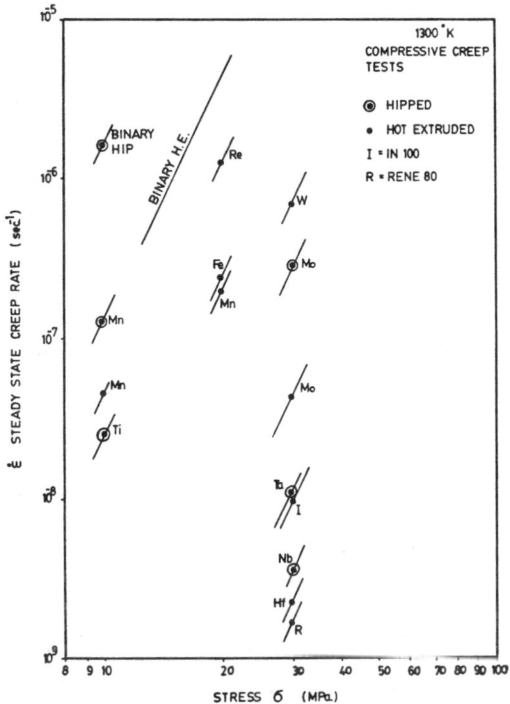


Fig. 2. Comparison of Compressive Creep Behavior of Several Ternary NiAl Alloys and Conventional Superalloys at 1300°K [2]. The ternary alloying addition is indicated at the data point.

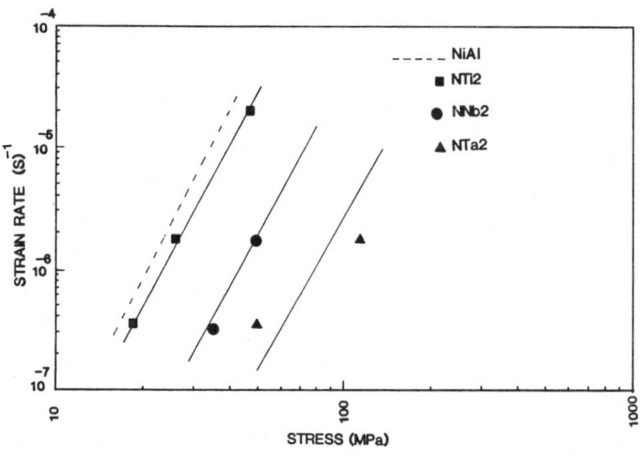


Fig. 3. Comparison of 1300°K Hot Compressive Flow Stress Behavior for additions of Ti, Ta and Nb at 2 at% level.

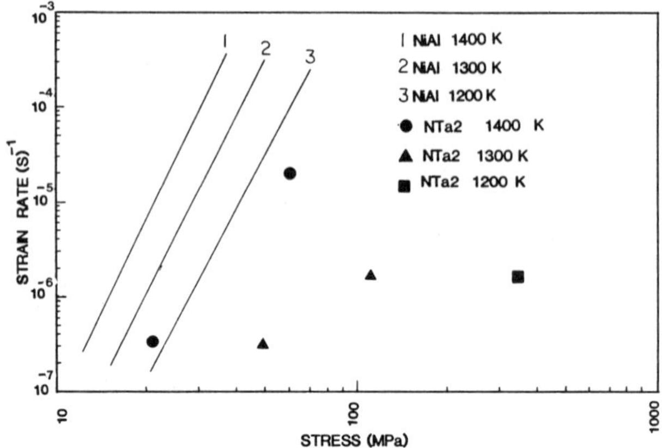


Fig. 4. Comparison of Hot Compresssive Flow Stress Behavior of Binary Stoichiometric NiAl [8] with Ni(50 at%)-Al(48 at%)-Ta(2 at%) at three temperatures.

The mechanism of strengthening for the alloys containing Ta and Nb appears to be some form of interaction between dislocations and fine precipitates which are richer in the alloying addition. This is indicated quite strongly by the TEM microstructure of the deformed Ta-containing alloy illustrated in Figure 5. Typical bowing of dislocations between sub-micron sized particles is obvious. These particles are found to be rich in Ta and are probably a ternary intermetallic of Ni, Ta and Al. These fine precipitates appear to have formed during the slow cooling of the specimen after homogenization treatment, since homogenization in these alloys does result in a certain amount of dissolution of the ternary addition and since the solubility of this ternary element is expected to decrease at lower temperatures. A more detailed TEM investigation is needed in order to clearly understand the nature, the mechanism of formation and role of these precipitates.

## (B) Grain Size and Stoichiometry

A wide range of grain sizes have been obtained using the technique of a second extrusion at a lower temperature. Typical grain sizes obtained by reextrusion followed by annealing are illustrated in Figure 6. (Note the factor of 20 difference in magnifications). Grain sizes obtained for the three alloys studied under different annealing conditions are presented in Table III.

An interesting aspect of the grain growth of these alloys appears to be that it is slow in the beginning but becomes much more rapid after some time resulting in a deviation from the standard power law grain growth kinetics. Figure 7 for grain growth at 1723°K illustrates this type of behavior. This appears to be indicative of some form of abnormal grain growth and, in fact, optical microstructures do reveal a bimodal distribution of grains in the coarse grained specimens.

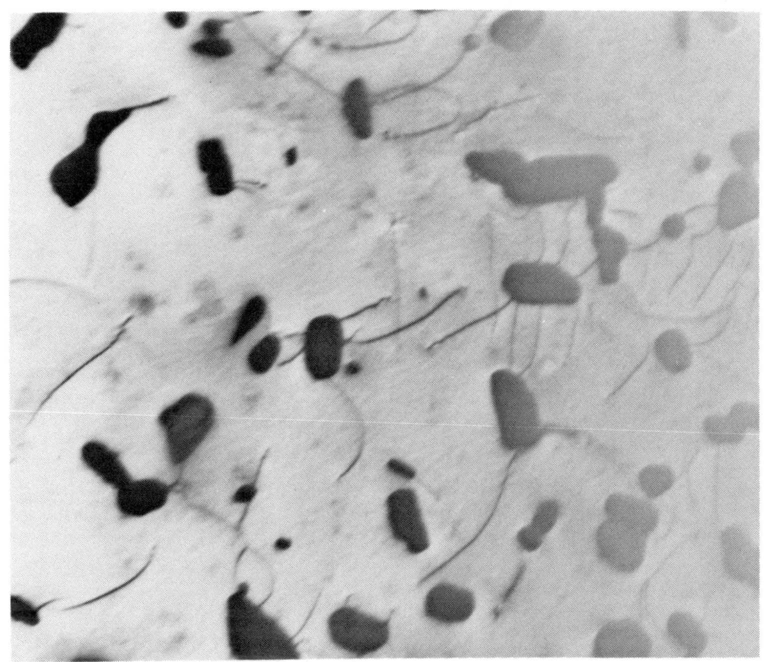

Fig. 5. Transmission Electron Micrograph of NiAl-2-at% Ta specimen deformed in Hot Compression at 1300°K to about 7% strain illustrating dislocation pinning at precipitation as a possible strengthening mechanism.

TABLE III. Grain Size Data for Binary NiAl (Micrometers)

| Composition | Annealing Temperature | Time (hrs) 0 | 0.5 | 3.0 | 10.0 | 144.0 |
|---|---|---|---|---|---|---|
| 48 at% Ni | 1350°C | 3 | 12 | - | 13 | - |
| | 1450°C | 3 | 13 | 15 | 15 | 45 |
| 50 at% Ni | 1350°C | 2 | 15 | - | 19 | 28 |
| | 1450°C | 2 | 17 | 20 | 20 | 48 |
| 52 at% Ni | 1350°C | 5 | 11 | - | 12 | 40 |
| | 1450°C | 5 | 16 | 18 | 20 | 92 |

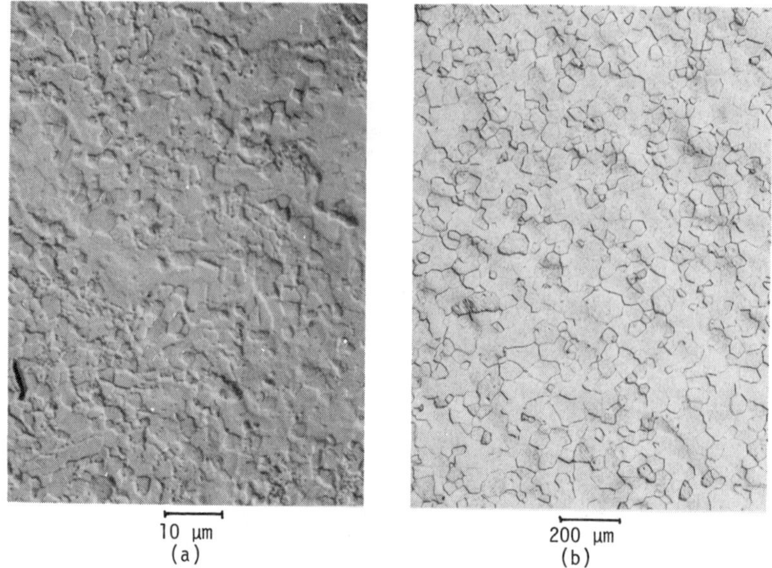


Fig. 6. Comparison of Optical Microstructure of the range of grain sizes of Binary Alloys used in this study. (a) As-Extruded Stoichiometric NiAl (1000X); (b) Stoichiometric NiAl annealed at 1723°K in argon for 144 hrs. (50X).

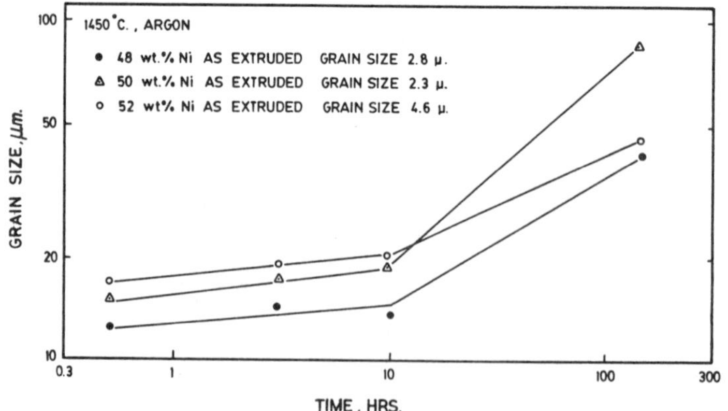


Fig. 7. Grain Growth Kinetics at 1723°K (1450°C) in Argon for Binary NiAl Alloys Prepared for this study.

The hot compressive flow behavior of these alloys under slow strain rates exhibits some very unusual effects of grain size. Figure 8 compares the flow curves of the three compositions of binary NiAl at 1300°K and 3 x $10^{-6}$ $sec^{-1}$. The fine grained (2µm) stoichiometric alloy (50 Ni) is twice as strong as the coarse grained (48 µm) stoichiometric alloy. This is unusual since at high temperatures and slow strain rates metals normally do not exhibit grain boundary strengthening. The explanation for this behavior is not obvious although it seems to agree with some other results of this nature in ordered alloys [6,7,8]. Figure 9 is a typical TEM micrograph of the stoichiometric alloy and shows the presence of subgrain boundaries which are likely to play a key role in the explanation of the observed effects of grain size.

Strengthening due to grain boundaries is, however, observed to a very significant extent only in the stoichiometric alloy. The other alloys containing 48 Ni and 52 Ni do not exhibit the same magnitude of the effect, although the fine grained material is still slightly stronger. Once again, the explanation for this is not clear, although it must be related in some way to the effect of ordering on grain boundary structure and deformation mechanisms. A complication may be introduced by the possibility of grain boundary contamination during the annealing treatment in argon in view of the susceptibility of these alloys to grain boundary pest phenomenon [9].

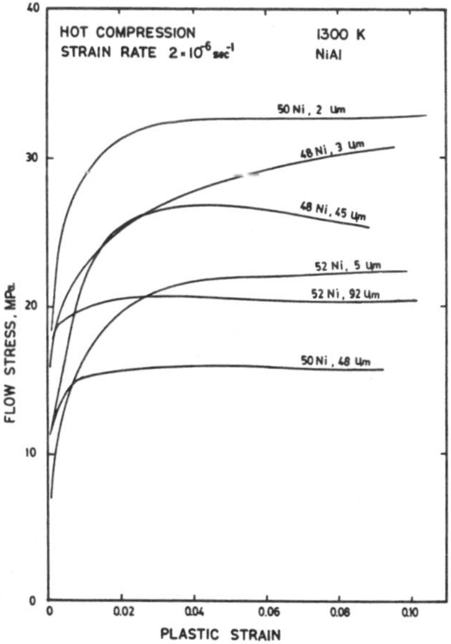


Fig. 8. Effect of Grain Size and Stoichiometry on Hot Compressive Flow Curves of Binary NiAl at 1300°K.

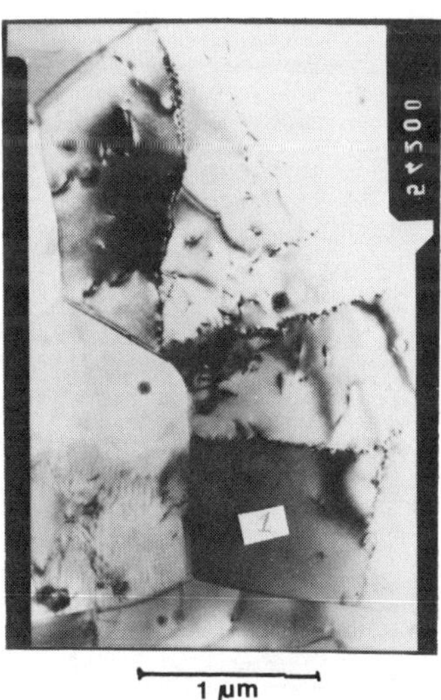


Fig. 9. Transmission Electron Micrograph of As-Extruded Binary Stoichiometric NiAl specimen illustrating the presence of Sub-grain Boundaries.

Some limited results at different temperatures presented in Figure 10 further illustrate the complexity of this behavior. The stoichiometric alloy which exhibits a large grain boundary strengthening effect at1300°K does not exhibit such a dramatic effect of grain boundary strengthening at the lower temperature of 1100°K, although grain boundary strengthening should normally be expected to be more important at lower temperatures. The non-stoichiometric 52 Ni alloy, however, appears to show the generally expected trend of grain boundary weakening at the higher temperature of 1400°K.

These results illustrate the complex interactions between high temperature deformation, defect structures, ordering and grain size which need to be unraveled by carefully controlled studies.

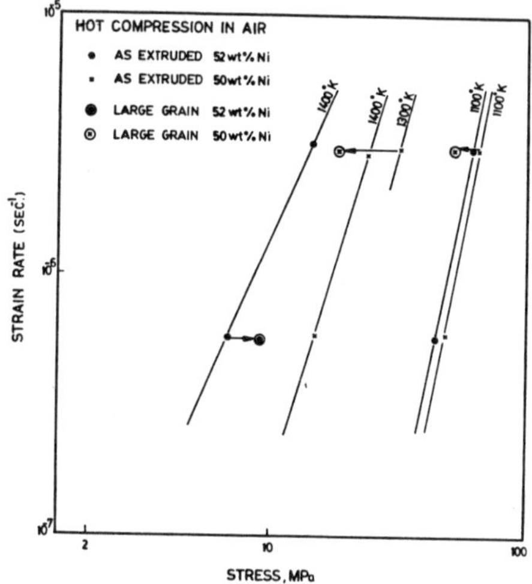


Fig. 10. Hot Compressive Flow Stress Behavior of Binary NiAl Alloys illustrating some of the interesting effects of temperature, grain size and stoichiometry.

## CONCLUSIONS

The results presented in this paper show that a potentially useful strengthening mechanism for high temperature applications of NiAl alloys is by second phase strengthening. In particular, small additions of Nb, Ta and Hf are found to exhibit this kind of strengthening to a significant extent. However, the mechanism of strengthening needs to be more clearly understood and optimized.

Some very interesting effects of grain size and stoichiometry have been observed, which need to be investigated more carefully before any explanations can be offered. In particular, the presence of grain boundary strengthening in the stoichiometric alloy even at fairly high temperatures is very unusual.

## ACKNOWLEDGEMENT

The financial support of NASA Lewis Research Center is greatly appreciated.

REFERENCES


1. Hansen, Handbook of Binary Phase Diagrams.
2. V. M. Pathare, K. M. Vedula and R. H. Titran, to be published in Proceedings of the International Powder Metallurgical Conference, Toronto, June 1984.
3. A. J. Bradley and A. Taylor, Proc. Roy. Soc., 103, (1956), p. 54.
4. M. J. Cooper, Phil. Mag., 8, (1963), p. 805.
5. R. R. Vandervoort, A. K. Mukherjee and J. E. Dorn, Trans. Amer. Soc. Metals, 59, (1966), p. 930.
6. J. D. Whittenberger, Mat. Sci. and Engg., 57, (1983), p. 77.
7. E. M. Grala, "Mechanical Properties of Intermetallic Compounds", ed. J. H. Westbrook, John Wiley & Sons, New York, 1960, p. 358.
8. E. P. Lautenschlager, D. A. Kiewit and J. O. Brittain, Trans. Met. Soc. AIME, 233, (1965), p. 1297.
9. J. H. Westbrook and D. L. Wood, J. Inst. Metals, 91, (1962),p.174.

Ni-BASE INTERMETALLICS FOR HIGH TEMPERATURE ALLOY DESIGN.

P. Nash, Illinois Institute of Technology, Chicago, Illinois 60616

## ABSTRACT

Further development of Ni-base superalloys based on $\gamma$-$\gamma'$ structures offers little hope of substantial improvements in properties for gas turbine applications. However because of the substantial technology base which exists for nickel base alloys there are considerable advantages, in research time and money, in developing new nickel base alloys. Interest in the 1970's in directionally solidified composites indicated some possible new systems to replace $\gamma$-$\gamma'$ alloys. More recently there has been renewed interest in aluminides such as FeAl, CoAl and NiAl. These materials exhibit melting points which are substantially greater than for current $\gamma$-$\gamma'$ alloys. Problems of lack of ductility for these alloys may be resolved by alloying and novel processing routes such as rapid solidification.

In addition to NiAl there are a number of other nickel intermetallics which from limited experimental data are worth considering at least as second phases. Some of these compounds are $Ni_7Hf_2$, $Ni_7Zr_2$, $Ni_5Zr$, $Ni_6AlTa$ and Heusler alloys of the type NiAlx where x is a transition element. The solubilities and phase relationships and where available the mechanical property data in a number of nickel ternary systems will be reviewed in terms of their potential for development of new high temperature alloys.

## INTRODUCTION

This paper summarizes the Ni-rich phase equilibria in Ni-Al-X systems, where X is a refractory element from group IVB, VB or VIB of the periodic table. The NiAl phase will be mentioned but not emphasized, even though it is the most important currently under investigation and is dealt with in a number of papers in this volume. Mechanical property data, where available, will be mentioned to give some indication of strength, but it is the purpose of this work to bring attention to several phases with potential for alloy development without prejudicing the choice of alloy designers. The possible advantages of some of these phases over $\gamma$ or $\gamma'$ are; enhanced oxidation resistance and lower density through the use of higher aluminum contents and increased creep resistance and microstructural stability because of higher solidus temperatures and increased bond strengths.

## DISCUSSION

The ternary phase equilibria in Ni-Al-X systems may be divided into two groups for Ni > 50 at. %:

1. Systems with ternary compounds.
2. Systems without ternary compounds.

The group into which the system falls is controlled by the third element addition as indicated below:

| Group 1 | Group 2 |
|---|---|
| Ti, Zr, Hf<br>V, Nb, Ta | Cr, Mo, W |

In addition on going down groups IVB, VB and VIB in the periodic table there is an increasing tendency for stabilization of binary Ni-X phases. Thus the ternary equilibria exhibit a logical sequence based on their positions in the periodic table relative to nickel. The stabilization of ternary compounds also decreases on going to lower atomic numbers in a periodic group. For example in Group VB the phase $Ni_6AlV$ cannot be produced (3).

The shape of the $Ni_6AlTa$ phase field indicates that nickel and aluminum may substitute for one another on the lattice, whilst little variation in Ta content is allowed. The $Ni_6AlTa$ phase has the hexagonal $DO_{24}$ crystal structure and yet, in a matrix of $\gamma'$, alloys containing this phase exhibit tensile strengths at 1000°C which are greater than that of IN-100 (4,5). The strengthening phase was believed to be $Ni_3Ta$ but it has been shown that this was probably $Ni_6AlTa$ (1). At 1250°C this phase is in equilibrium with $\gamma$, $\gamma'$, and $\beta$ any of which could be used as a matrix. Since the crystal structure of this phase is the same as that of $Ni_3Ti$ one would expect that a series of solid solutions exist from $Ni_6AlTa$ to $Ni_3Ti$ in the quaternary system, which offers the opportunity of manipulating its properties to some extent. All of the systems in group 1 show substantial solubility of the third element on the Al lattice. This permits a certain amount of control over the properties of the $\gamma'$ phase (6,7). These systems also show a solubility of several atomic percent of the 3rd element in $\beta$-NiAl, Table 1. The combination of $\gamma'$ and $\beta$ seems to be a logical development from $\gamma$-$\gamma'$ alloys. In this case one could have $\gamma'$ as either the matrix or as the precipitate. Alloys of the latter type have been produced and hot compression tests on 'as cast' structures reveal strengths in excess of Nimonic 105 up to the limit of the test, 900°C (8).

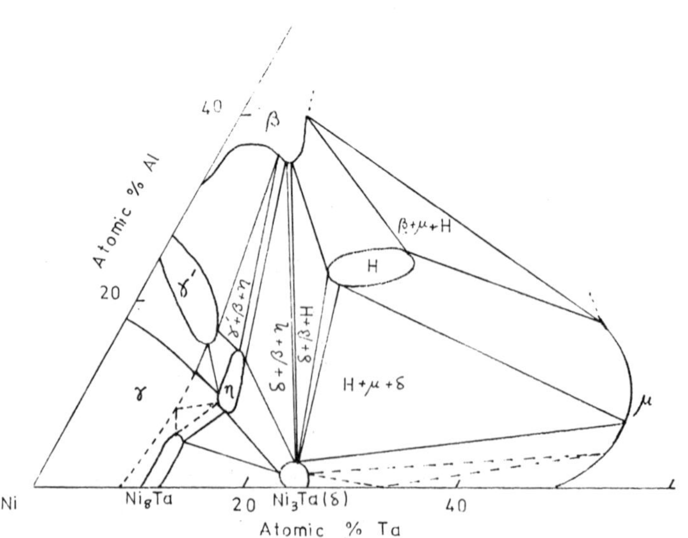


1. 1523 K isothermal section of the Ni-Al-Ta system.[1]

The effects of solute additions to NiAl have been studied to a very limited extent (9,10) but there is some indication that refractory element additions improve the mechanical properties. The martensitic transformation in Ni-rich β may be avoided by alloying additions (1).

Table 1. Solubilities of third elements in γ', β and $Ni_7X_2$

| Element | at. % in γ' | at. % in β | at/ % in $Ni_7X_2$ |
|---|---|---|---|
| Ti | 17 | 11 | - |
| Zr | ~5 | 1.5 | 11 |
| Hf | 9 | >4 | 14 |
| V | 10 | >10 | - |
| Nb | 10 | >5.5 | - |
| Ta | >8 | 5 | - |
| Cr | ~10 | ~10 | - |
| Mo | ~6 | ~1.5 | - |
| W | ~6 | ~0.2 | - |

The Heusler type phase $Ni_2AlX$, H, appears in all of the group 1 systems. This phase is structurally related to the B2-CsCl structure of β but with Al and X elements being ordered on one sublattice. In all but the Ni-Al-Ti system (11) the extent of the H phase field has not been established, however it appears that it tends to extend in the direction of constant nickel content, allowing substitution of Al or X on the same sublattice. High temperature creep of $Ni_2AlTi$ has been studied (12) showing that the creep resistance is ~3 times that of β. Studies have also been made of β-H alloys in the Ni-Al-Ti system with H being the matrix (13) and these showed creep strengths comparable with MAR-M200. The H phases in these systems probably molt congruently at temperatures in excess of 1300°C (1,11) and they form pseudo-binary eutectics with β which generally molt in excess of 1250°C. Solidus data on alloys of this type are not available but will be essential for alloy development.

An unusual effect is noted in the Ni-Al-Hf and also Ni-Al-Zr systems where the phase $Ni_7Hf_2$ extends deep into the ternary, figure 2, Table 1, with substitution of Al for Ni (14). By itself this phase is too dense for turbine applications, but might be useful as a second phase with γ', β or H with which it is in equilibrium. The $Ni_7Hf_2$ phase is congruent melting and forms eutectics with γ', β and H (14). No mechanical property data are available for this phase or alloys containing this phase.

An example of the group 2 systems with Mo as the alloying addition is shown in figure 3 (15). Solubilities of the 3rd element in γ' and β are more restricted but the γ phase field tends to be larger than for group 1. The most important feature of this group is the phase equilibria between the Ni-rich phases and the 3rd element terminal solid solution α. Equilibrium between γ and α can be affected by binary compound formation at low temperatures but γ'-α and β-α equilibria should be free of such effects. The β-α equilibrium generally exhibits a pseudo-binary eutectic. Extended solid solubility may be obtainable by rapid solidification permitting homogeneous structures to be obtained after appropriate heat treatment. Because the β and α structures are related the possibility of spinodal ordering transformation exists.

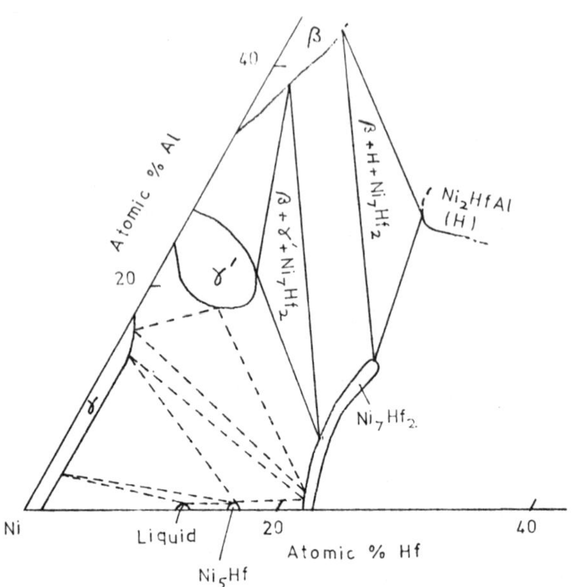


2. 1473 K isothermal section of the Ni-Al-Hf system.[14]

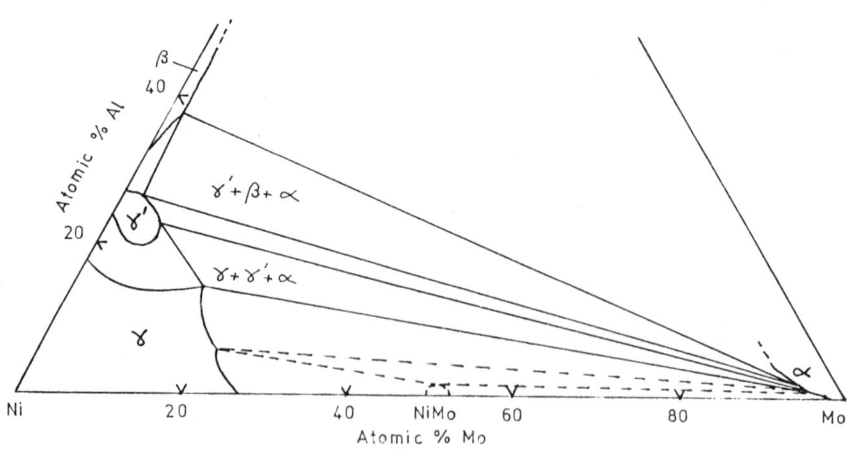


3. 1473 K isothermal section of the Ni-Al-Mo system.[15]

## CONCLUDING REMARKS

There are a number of alternative nickel alloy phase equilibria to the $\gamma$-$\gamma'$ system which may provide the alloy designer with a starting point for an alloy development program. Additional information on solidus and liquidus temperatures needs to be obtained to complete the phase equilibria data which forms the basis of any alloy development program. Problems of lack of ductility for these alloys may be resolved by alloying and novel processing routes such as rapid solidification.

## REFERENCES


1. P. Nash, D.R.F. West, Metal Science 13, 670, 1979.
2. B.C. Giessen, N.J. Grant, Acta Cryst. 18, 1080, 1965.
3. W.W. Liang, R. Standley, P. Nash, M. Skowron, J. Mats. Sci. Letters, 3, 259, 1984.
4. J.C. Hubert, W. Kurz, B. Lux, J. Crystal Growth, 13/14, 757, 1972.
5. F. Mollard, B. Lux, J.C. Hubert, Z. Metallk, 65(7), 461, 1974.
6. R.W. Guard, J.H. Westbrook, Trans. AIME 215, 807, 1979.
7. Y. Oya, S. Ochiai, T. Shinoda, T. Suzuki, Z. Metallk, 75, 53, 1984.
8. D.R.F. West, Private Communication 1978.
9. R.D. Grinthal in 'Mechanical Properties of Intermetallic Compounds' ed. J.H. Westbrook, 1960, p. 337.
10. E.M. Grala, ibid p. 358.
11. P. Nash, W.W. Liang, accepted for publication in Met. Trans.
12. P.R. Strutt, R.S. Polvani, J.C. Ingram, Met. Trans., 7A, 23, 1976.
13. R.S. Polvani, W.S. Tzeng, P.R. Strutt, Met. Trans., 7A, 33, 1976.
14. P. Nash, D.R.F. West, Metal Science 15, 347, 1981.
15. P. Nash, S. Fielding, D.R.F. West, Metal Science 17, 192, 1983.

# RAPIDLY SOLIDIFIED NiAl and FeAl

DARRELL J. GAYDOSH* AND MARTIN A. CRIMP**
* NASA Lewis Research Center, Cleveland, Ohio 44135
** Case Western Reserve University, Cleveland, Ohio 44106

## ABSTRACT

Melt spinning was used to produce rapidly solidified ribbons of the B2 intermetallics NiAl and FeAl. Both Fe-40Al and Fe-45Al possessed some bend ductility in the as spun condition. The bend ductility of Fe-40Al, Fe-45Al, and equiatomic NiAl increased with subsequent heat treatment. Heat treatment at approximately 0.85 $T_m$ resulted in significant grain growth in equiatomic FeAl and in all of the NiAl compositions. Low bend ductility in both FeAl and NiAl generally coincided with intergranular failure, while increased bend ductility was characterized by increasing amounts of transgranular cleavage fracture.

## INTRODUCTION

The B2 crystal structure aluminide intermetallics of Fe and Ni represent potential structural materials for use at elevated temperatures. These materials possess a high melting point, low density, and resistance to oxidation and high temperature deformation. Much work has been done in evaluating the mechanical properties of NiAl and FeAl in the cast [1], cast and wrought [2], cast and extruded [3,4], and P/M and extruded [5] forms. One problem limiting the potential use of these materials is their low room temperature ductility. Application of rapid solidification processing to these materials has already produced improvements in strength and ductility as compared to conventional processing [6,7,8].

Taub, Huang, and Chang [6] applied rapid solidification by melt spinning to $Ni_3Al$ and B modified $Ni_3Al$ and found that rapidly solidified material (5-10 μm grain size) had approximately twice the room temperature yield strength of cast material (200-1000 μm grain size) in both alloys. Although they found that increasing the grain size of the rapidly solidified material from 5-10 μm to 25-30 μm by heat treatment decreased the yield strength significantly in both alloys, they still stated that grain size alone could not explain the difference between rapidly solidified and as cast properties. Improvements in the room temperature ductility and strength of $L1_2$ compounds in Ni-(10-22)Al-X (X=12.5Cr, 6Mn, 10Fe, 20Co, or 12Si) alloys using the melt spinning technique were reported by Inoue, Tomioka, and Masumoto [7]. Although grain refinement was observed, the improvements in ductility and strength were attributed to the suppression of grain boundary segregation, and the low degree of ordered state with a high density of APB's, which was caused by suppression of ordering. Gaydosh, Jech, and Titran [8] also found that melt spun Ni-50Al having a grain size of 5.4 μm exhibited limited plastic deformation in a free bend test at room temperature. The fine grain size was thought to be at least partially responsible for the appearance of ductility.

This paper presents the results of an investigation into the effects of melt spinning on the microstructure and room temperature ductility of the B2 intermetallics NiAl and FeAl. The effects of annealing time and temperature on the melt spun structure and properties are discussed. Ductility of the melt spun ribbons was determined by using a simple bend test. Standard metallographic, X-ray diffraction, and electron microscopy techniques were used to evaluate the structures and fracture surfaces of

Mat. Res. Soc. Symp. Proc. Vol. 39. © 1985 Materials Research Society the melt spun and heat treated ribbons.

## MATERIALS AND PROCEDURES

Free jet melt spinning [9] was used to produce the rapidly solidified material. NiAl and FeAl melt stock was prepared from elemental materials by triple arc-melting in argon. For each run a charge of approximately 13 grams was placed into a dense alumina crucible in a melt spinning apparatus. The charge was heated above its melting point and ejected by pressurized argon onto an uncooled mild steel wheel rotated at a surface speed of 20 meters per second. Melt spinning was performed in a positive pressure argon atmosphere to suppress aluminum vaporization.

The compositions of the rapidly solidified materials are listed in Table I; these compositions were chosen to span the B2 phase fields in each alloy system. Ribbons of each composition were heat treated at two different temperatures for one hour in helium. The FeAl ribbons were treated at 775°C and 1000°C, which represent homologous temperatures of 0.64 to 0.68 and 0.77 to 0.83, respectively, based on the solidus temperatures of the FeAl system. The NiAl ribbons were treated at 1000°C and 1310°C, which represent homologous temperatures of 0.67 to 0.71 and 0.83 to 0.88, respectively, based on the solidus temperatures of the NiAl system.

Optical metallography, X-ray diffraction, scanning electron microscopy (SEM), and transmission electron microscopy (TEM) were used to assess the structure of both the as spun and heat treated ribbon. Grain size was measured using the linear intercept method, where the reported values are the average grain width measured across a longitudinal section, and do not account for the length or depth of the grains. Measurements were made at three different places on each of two different ribbons for every composition. Disks 3 mm in diameter for TEM observation were cut from the ribbons using a Gaton ultrasonic disk cutter. The disks were electropolished to thin foils at -30°C in a 33 percent $HNO_3$ in methanol solution at 12 volts. Finally, room temperature bend testing was used to provide a relative measure of the ductility of the ribbons.

## RESULTS AND DISCUSSION

Rapidly solidified ribbon was produced for every composition except Ni-53Al and Ni-55Al. Melt spinning these compositions resulted in only discontinuous flake and powder product. The melt spun ribbon was 2 to 3 mm wide and ranged from 0.025 to 0.038 mm in thickness. A rough surface and serrated edges resulted from the use of an argon atmosphere. Gas entrapment between the wheel and the ribbon caused some surface roughness on the wheel side of the ribbon. The serrated edges and surface roughness on the ribbon's free side were caused by interaction of the solidifying metal with the gas boundary layer. Because of this surface roughness and the small dimensions of the ribbon, tensile testing was not possible.

TABLE I

Compositions of Melt Spun Ribbons
Al, at.%

| | | | | | |
|---|---|---|---|---|---|
| NiAl | 42 | 46 | 50 | 53 | 55 |
| FeAl | 40 | 45 | 50 | 50.5 | 51 |

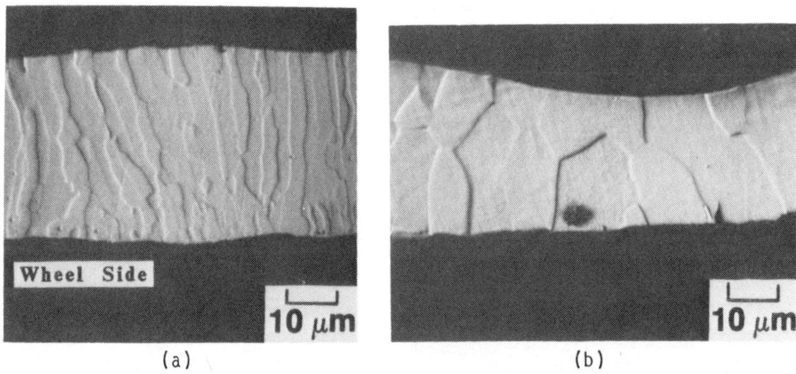


(a) (b)

Figure 1. Microstructure of as spun Ni-50Al.

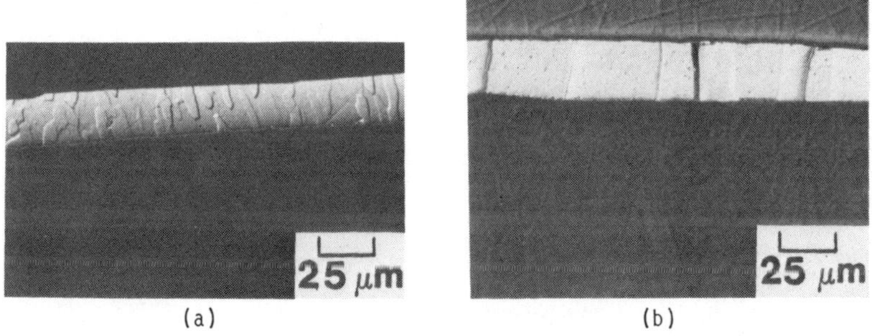


(a) (b)

Figure 2. Microstructure of Fe-50Al (a) in the as spun condition and, (b) treated at 1000°C for 1 hour in He.

Microstructure

Optical metallography revealed that both NiAl and FeAl possessed a columnar grain structure extending outward from the wheel side of the ribbon. A typical photomicrograph of the melt spun microstructure of NiAl is shown in Figure 1(a), while that of FeAl is shown in Figure 2(a). The as spun grain width of the NiAl series ranged from 4 to 6 μm, while that of the FeAl series ranged from 6 to 9 μm. Because the grain length is limited by an approximately constant ribbon thickness, the larger grain width in the FeAl ribbons represents a decrease in the grain aspect ratio in comparison to that of NiAl. In addition, small equiaxed grains were observed on the wheel side of the NiAl ribbons. Both the difference in grain size and the presence of equiaxed grains in NiAl indicate differences between the solidification of NiAl and FeAl during melt spinning.

Figure 1(b) is another photomicrograph of equiatomic NiAl from the same melt spin run as the ribbon shown in Figure 1(a). The change in ribbon thickness, grain size, and grain shape illustrate the major problem with this melt spun material; heterogeneity. Differences in microstructure such as those shown here have been observed within a 2 cm length of ribbon. Statistical analysis of a number of specimens was therefore used to determine the grain size and bend ductility of the materials.

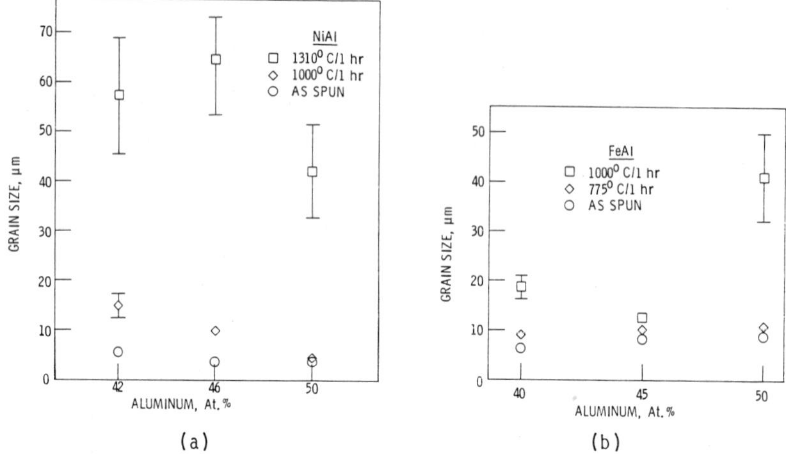


Figure 3. Grain size as a function of heat treatment temperature and composition for melt spun (a) NiAl and, (b) FeAl.

## Grain Size

Grain widths measured for the as spun and heat treated ribbons of NiAl and FeAl are shown in Figure 3. Error bars indicate the 95 percent confidence interval. Where no error bars appear, they are smaller than the symbols. Heat treatment at 1000°C for 1 hour produced some grain growth. Although not shown in the figure, the grain size of equiatomic NiAl grew from 8 to 26 μm when treated at 1000°C for 23 hours, indicating the effect of time on grain growth. Heat treatment at 1310°C for 1 hour resulted in large increases in grain width up to 40 to 60 μm.

Heat treatment of the FeAl series at 775°C resulted in little if any growth, while a 1310°C treatment increased the grain size of Fe-40Al and Fe-50Al. The 45 at.% Al composition, however, had some resistance to grain growth. The 30 μm increase in grain width caused by the 1000°C treatment of equiatomic FeAl is illustrated in Figure 2(b). Note how each grain has grown completely through the thickness of the ribbon. This is typical for grain growths of 30 μm and above in all of the compositions studied. Width increases of about 10 μm resulted in nearly 80 percent of the grains growing through the ribbon thickness.

## Transmission Electron Microscopy

The substructure of the Ni-50Al ribbons observed by TEM consisted primarily of dislocations and precipitates. In the as spun condition (Figure 4(a)), the dislocations occurred throughout the grains with no apparent orientation preference. Precipitates were distributed in an apparently prior subgrain structure and had little obvious effect on the dislocations. Upon annealing, the Ni-50Al ribbon (Figure 4(b)) showed a decrease in dislocation density. Again, the dislocations were randomly oriented, and the precipitates outlined prior subgrain boundaries.

Examination of as-spun Fe-50Al ribbon revealed a relatively large dislocation density, as shown in Figure 4(c). These dislocations were long and straight and assumed three distinct orientations. Precipitates,

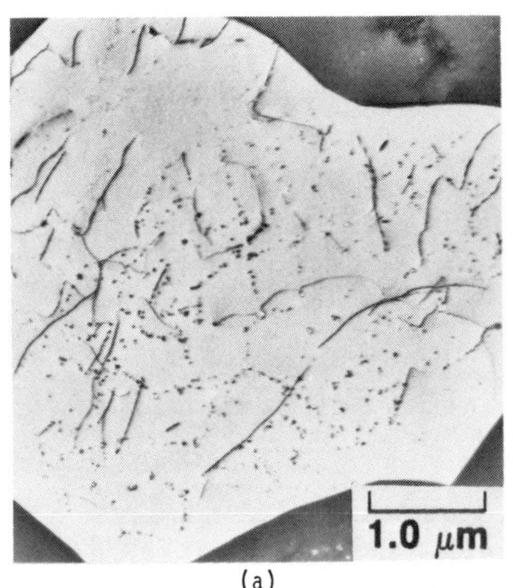

(a)

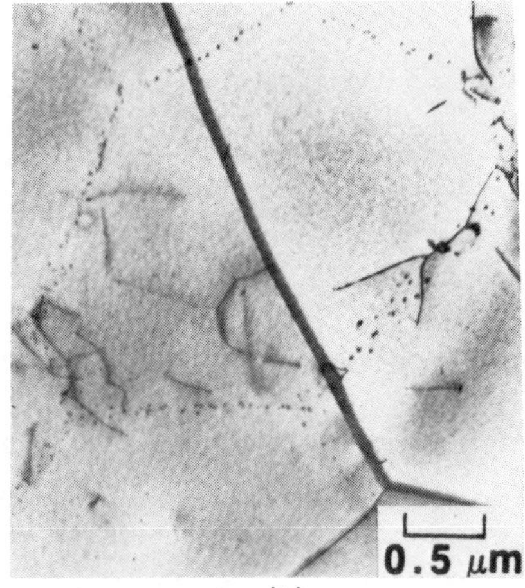

(b)

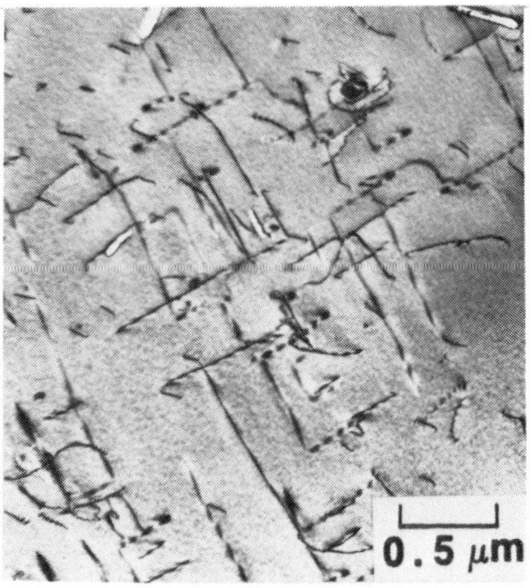

Figure 4. TEM substructure of (a) as spun Ni-50Al, (b) Ni-50Al heat treated at 1000°C for 1 hour in He, and (c) as spun Fe-50Al.

(c)

which are much larger than those in NiAl, occurred randomly in the structure and some dislocation-precipitate interactions were observed. The as spun Fe-40Al and Fe-45Al ribbons displayed similar structures.

Additional work is required to determine the composition, structure, and origin of the precipitates in both NiAl and FeAl.

## X-Ray Diffraction

X-ray diffraction of as spun materials revealed a {100} texture perpendicular to the ribbon surface on the free side of the ribbons, and little or no preferred orientation on the wheel side. All three NiAl compositions exhibited a definite {100} preferred orientation on the free side. Fe-40Al and Fe-50Al had a minor {100} texture, while Fe-45Al possessed little or no preferred orientation. In general, heat treatments which caused grain growth also caused the {100} texture to appear on the wheel side of the ribbon, and heat treatments which had no affect on grain size did not change the original texture. It appears that directional growth occurred in the solidifying ribbons after an initial solidification next to the wheel. These oriented grains then grew through the ribbon when heat treatment caused grain growth.

## Bend Test

Room temperature bend testing of both as spun and heat treated ribbons was conducted by free bending the ribbon around a mandrel of successively smaller diameter until failure occurred. Total strain at failure $e_f$ was calculated from the minimum bend radius r and the ribbon thickness t [10].

$$e_f = \frac{t}{(2r + t)} \qquad (1)$$

For each condition, half of the tests were conducted with the wheel side of the ribbon as the outer bend surface, and the other half with the free side of the ribbon as the outer bend surface. Typically, a total of eight tests were conducted for each condition. It should be emphasized that bending of these ribbons produced large tensile stresses perpendicular to the columnar grain boundaries. This resulted in a more severe test than if the grain boundaries had been randomly oriented.

The results of the bend test for NiAl and FeAl are shown in Figure 5. The 95 percent confidence interval was calculated for each failure strain, and it lies within the symbols for each average value. The primary type of deformation during the bend test as determined by visual observation is also listed in the two figures. For NiAl, as spun failure strains ranged from 0.9 to 1.3 percent. Heat treatment had relatively little affect on the low aluminum content (42 and 46 at.%) compositions, while it increased the failure strain up to 2.25 percent for the equiatomic composition. Also, deformation in the low aluminum content compositions was primarily elastic, while limited plastic deformation was observed for equiatomic NiAl. Heat treatment at 1000°C either increased or had no effect on ductility. However, exposure at 1310°C seemed to lower the observed failure strain in comparison to the 1000°C heat treatment. Because the only effect of heat treatment on ductility occurred at the equiatomic composition, and because grain growth was relatively uniform for NiAl (Figure 3(a)), it is reasoned that grain growth had no direct effect on the measured bend ductility.

The failure strains of FeAl decreased with increasing Al content between 40 and 50 at.% Al and remained constant for hyperstoichiometric Al

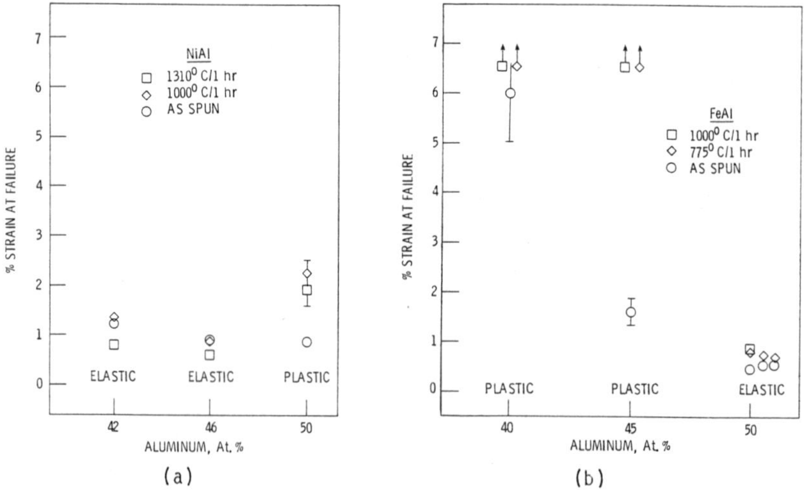


Figure 5. Room temperature bend test results as a function of composition for melt spun (a) NiAl, and (b) FeAl.

levels (Figure 5(b)). The strain to failure for as spun Fe-40Al was quite high (6 percent) and was accompanied by considerable plastic deformation; such ductility was considerably greater than that observed in any NiAl material. The effect of heat treatment with composition was opposite to that in NiAl. The failure strains of the low Al compositions increased dramatically, while the near equiatomic compositions were not affected. In fact, the 40 and 45 Al compositions could be plastically bent 180° and flattened without breaking (This is indicated by the upward arrows in Figure 5(b)), while the near stoichiometric materials failed in a brittle elastic manner. Finally, comparison of Figures 3(b) and 5(b) revealed no direct correlation between grain growth and the measured bend ductility, even though grain growth in FeAl as a function of composition was not uniform.

## Fracture Surfaces

Fracture surfaces were examined for ribbons in the as spun and heat treated conditions. Equiatomic NiAl in the as spun condition exhibited almost entirely intergranular fracture with rare instances of transgranular cleavage fracture. The fracture of ribbon heat treated at 1000°C for one hour was also primarily intergranular, but there was a noticeable increase in the amount of transgranular cleavage. Note that the bend ductility of the heat treated ribbon was twice that of the as spun material. Ni-42Al exhibited a mixture of intergranular and transgranular cleavage fracture in both the as spun and heat treated conditions. Grain growth during heat treatment made any changes in fracture mode hard to quantify for this composition.

Equiatomic FeAl exhibited completely intergranular failure in both the as spun and heat treated conditions. Note that this composition exhibited a low initial value of bend ductility which did not change with heat treatment. Figure 6 shows Fe-40Al in the as spun condition and after heat treatment at 1000°C for one hour. The as spun material possessed a

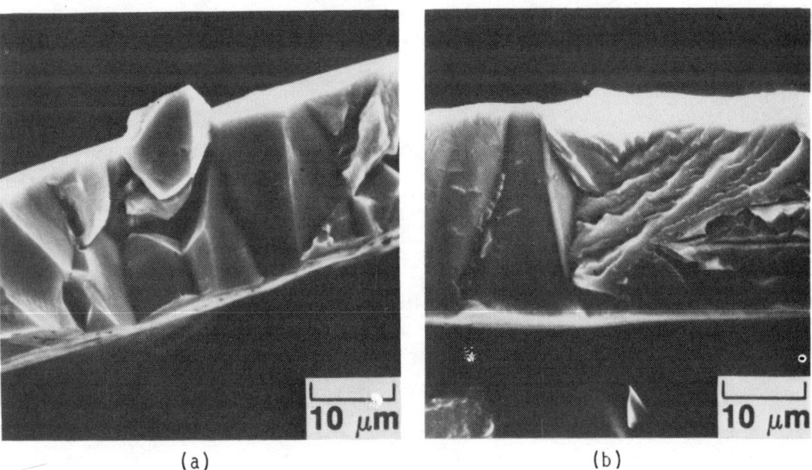


(a) (b)

Figure 6. Scanning electron photomicrograph of the fracture surface of (a) as spun Fe-40Al: $e_f$ = 6%, and (b) Fe-40Al treated at 1000°C for 1 hour in He.

primarily intergranular fracture with some transgranular cleavage; whereas the fracture surface of the heat treated ribbon was primarily transgranular cleavage with some instances of intergranular failure. Although this was the most ductile of the compositions examined, some intergranular failure always occurred in failed specimens.

For both NiAl and FeAl, low ductility was accompanied by primarily intergranular failure, while increases in ductility generally coincided with an increase in transgranular cleavage fracture. Schulson [3] also observed a similar mixture of intergranular and transgranular cleavage fracture in Ni-49Al that exhibited tensile elongations of 14 and 41 percent at 400°C.

One possible explanation for the appearance of ductility and change in fracture mode with heat treatment involves the effects of grain boundary segregation and hardening [11]. These effects could be suppressed by rapid solidification, but reintroduced by subsequent heat treatment. If segregation strengthens the grain boundaries sufficiently, deformation and failure occur through the bulk material instead of along the grain boundaries, and a more ductile failure can result.

## SUMMARY OF RESULTS

Melt spinning of NiAl and FeAl produced a heterogeneous ribbon with some variation in both thickness and microstructure. Both Fe-40Al and Fe-45Al possessed some bend ductility in the as spun condition. The bend ductility of Fe-40Al and Fe-45Al increased dramatically with subsequent heat treatment, while Ni-50Al exhibited some increase in ductility with heat treatment. The measured bend ductility itself was a result of the special situation of producing tensile stresses perpendicular to columnar grain boundaries. This resulted in a more severe test than if the grain boundaries had been randomly oriented. Heat treatment at approximately 0.85 Tm also resulted in significant grain growth in equiatomic FeAl and in all of the NiAl compositions. Low bend ductility in both FeAl and NiAl was generally characterized by essentially complete intergranular failure, while increases in transgranular cleavage fracture occurred when the amount of bend ductility increased.

## REFERENCES


1. A. Ball and R. E. Smallman: Acta Met. 14 (1966) p. 1349.
2. A. Lawley, J. A. Coll, and R. W. Cahn: Trans. Metall. Soc. AIME, 218 (1960) pp. 166-176.
3. A. G. Rozner and R. J. Wasilewski: J. Inst. Metals, 94 (1966), 930-944.
4. E. M. Schulson and D. R. Barker: Scripta Met. 17 (1983) p. 519.
5. J. D. Whittenberger: Mat. Science and Eng., 57 (1983) pp. 77-85.
6. A. I. Taub, S. C. Huang, and K. M. Chang: Met. Trans., 15A (1984) pp. 399-402.
7. A. Inoue, H. Tomioka, and T. Masumoto: Met. Trans. 14A (1983) p. 1367.
8. D. J. Gaydosh, R. W. Jech, and R. H. Titran: Journal of Mat. Science Letters, In Press.
9. R. W. Jech, T. J. Moore, T. K. Glasgow, and N. W. Orth: Journal of Metals, 36 No. 4 (1984) pp. 41-45.
10. G. E. Dieter: "Mechanical Metallurgy, Second Edition," (McGraw-Hill, Inc., 1976) p. 681.
11. R. Bakish, "Intermetallic Compounds," J. H. Westbrook, ed. (John Wiley and Sons, 1967) p. 298.

# THE PRODUCTION AND THERMAL STABILITY OF A REFINED DISPERSION OF $Er_2O_3$ IN $Ti_3Al$ USING RAPID SOLIDIFICATION PROCESSING

D.G. KONITZER AND H.L. FRASER
Dept. of Metallurgy and the Materials Research Laboratory, University of Illinois, 1304 W. Green St., Urbana, IL 61801

## ABSTRACT

The use of rapid solidification processing to produce a refined dispersion of second phase in $Ti_3Al$ is demonstrated. Laser surface melting produces a disordered solid solution of Ti and Al with a supersaturation of Er and O. A refined dispersion of $Er_2O_3$ is formed on heat treatment at 973 K. This oxide is found to be stable at elevated temperatures up to 1173 K.

## INTRODUCTION

The high strength to weight ratio of the intermetallic phase $Ti_3Al$ makes it an attractive material for applications such as aircraft engines [1]. However, the material exhibits little ductility at room temperature [2]. It has been suggested that a possible method of increasing this ductility involves a decrease in the slip length [3], either by refining the grain size or by some transformation, such as quenching to form martensite. The slip length may also be decreased by dispersing a strong second phase in the matrix which would interact with dislocations. This second phase must be stable with respect to coarsening so that the alloy, after prolonged thermal exposure, would still retain its room temperature properties. Recent studies on Ti-Er alloys [4,5] have shown the feasibility of producing a refined stable dispersion of the rare earth oxide, $Er_2O_3$, in Ti by the technique of rapid solidification. In one of the studies [4] the rapid cooling allowed a metastable supersaturation of Er and O to be formed in Ti, and subsequent heat treatment of this structure resulted in precipitation of the rare earth oxide with an average particle size of approximately 20 nm. The stability of this dispersion up to the transus temperature was demonstrated [6] by only slight coarsening when heating below the transus and significant coarsening when heat treated above the transus. The reasoning for this increase in coarsening rate has been suggested to be the increased diffusion rate of O in the BCC structure of β-Ti compared to that in the HCP phase (α-Ti).

For the composition corresponding to $Ti_3Al$, the applicable transformation on heating is the $\alpha_2 \rightarrow \alpha + \alpha_2$ [7] transus and not the $\alpha \rightarrow \beta$ transus as in pure Ti. The transus in $Ti_3Al$ occurs at 1100°C as compared to 873°C in the case of pure Ti. Thus the rare earth oxide dispersions in the close packed inter-metallic compound are expected to exhibit less coarsening at temperatures above 873°C than in Ti-rare earth alloys which have transformed into the β-Ti phase.

In the present paper studies have been undertaken to verify the stability of a rare earth oxide dispersion in $Ti_3Al$. The technique of laser surface melting with subsequent heat treatment has been used to produce the dispersion. The thermal stability of the dispersion will be investigated by prolonged exposure at elevated temperature.

 

## EXPERIMENTAL PROCEDURE

The alloy of composition $Ti_3Al$ + 0.4Er (atomic percent), was prepared by non consumable arc-melting together 99.9% pure Ti, 99.9999% pure Al and 99.9% pure Er, all with respect to metallic impurities. The elements were in the form of pellets and thus the Er to O ratio is expected to be greater than 2:3. O is introduced into the alloy as a major impurity in the Er and a minor impurity in the Ti and Al. The resulting buttons were laser melted as has been described previously (8). The rapidly solidified material was then removed from the button, encapsulated under Ar in quartz tubes, and heat treated. The specimens were then prepared for examination in the transmission electron microscope (TEM) by jet thinning [8].

## RESULTS AND DISCUSSION

The arc melted material consisted of large Er rich precipitates and antiphase boundaries in the equiaxed grains of the ordered matrix, figure 1.

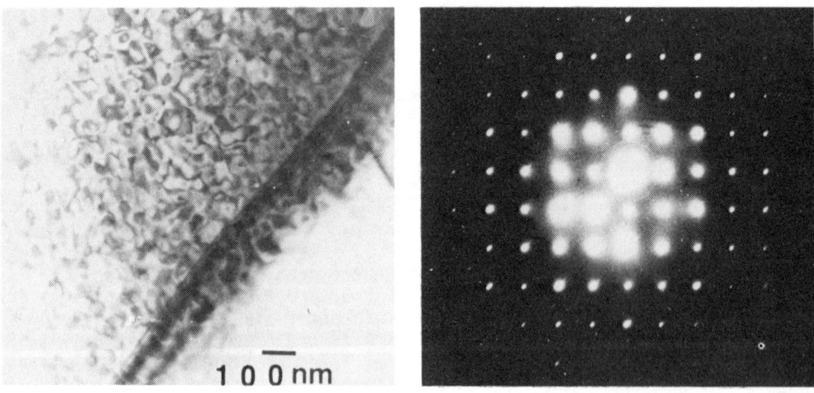


Figure 1. Bright field micrograph and $11\bar{2}0$ zone axis selected area diffraction pattern obtained from the as-arc melted alloy.

Rapid solidification of this material resulted in a microstructure which consisted of disordered martensite laths with no evidence of second phase precipitation, figure 2. Presumably, the matrix in this condition contains a supersaturation of Er which has remained in solution due to the rapid solid state cooling. Heat treatment of this microstructure for 10 h at 973 K caused the precipitation of a fine uniformly dispersed second phase, figure 3. Furthermore this heat treatment produced a fully ordered matrix of equiaxed grains as the martensite laths were consumed during recrystal-Contrary to the case of precipitation of rare earth oxides in binary Ti-Er alloys [9] the precipitates in the present case do not exhibit a well defined orientation relationship as shown by the rings of diffracted intensity in figure 4. The second phase can be identified as the rare

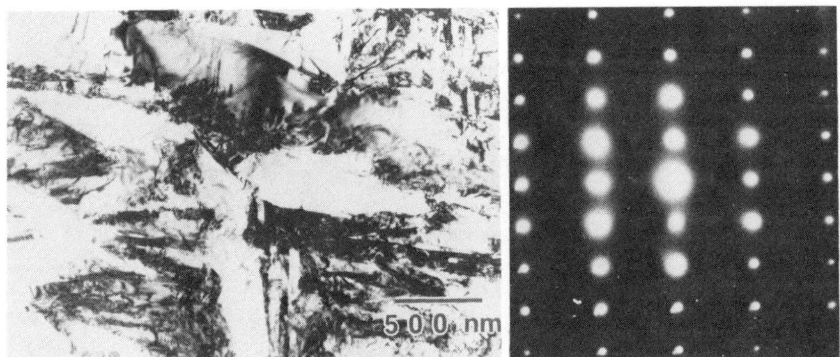


Figure 2. Bright field micrograph and $11\bar{2}0$ zone axis selected area diffraction pattern obtained from the alloy following rapid solidification. Note the absence of superlattice reflections.

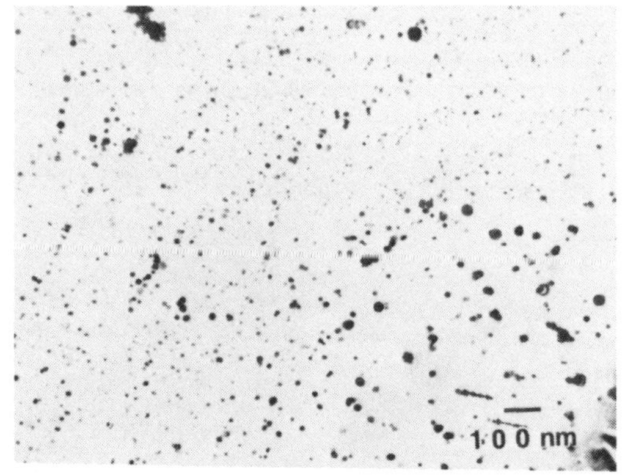


Figure 3. Bright field micrograph of a specimen which has been laser treated and subsequently heat treated for 10 h at 973 K.

earth oxide through an analysis of these rings. Thus, using the single crystal $Ti_3Al$ pattern as a standard, the spacing of the rings is found to be consistent with the spacings for the rare earth oxide, $Er_2O_3$. Also single crystal microdiffraction patterns obtained from a single precipitate are identical to those which have previously been obtained from the rare earth oxide [8].

The thermal stability of the rare earth oxide was investigated by heat treatment of the refined dispersion depicted in figure 3. Thus, specimens of the rapidly solidified alloy were each heat treated for 10 h at 973 K to produce the initial microstructure and then heat treated for 10 hours more at higher temperatures. The dispersion did not coarsen significantly following thermal exposure for 10 h at 1073 K or 10 h at 1173 K, figures 5a and b respectively. This is an important result since heat treatment for 10 hours at 1173 K caused significant coarsening in the binary Ti-Er alloy

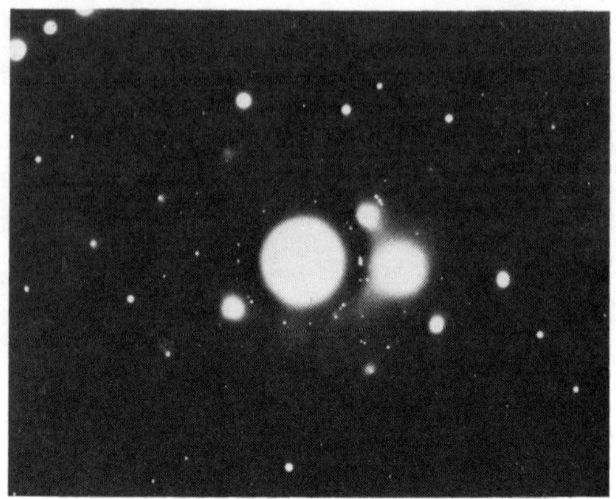

Figure 4. Selected area diffraction pattern, exhibiting diffraction rings, obtained from the same condition shown in figure 3.

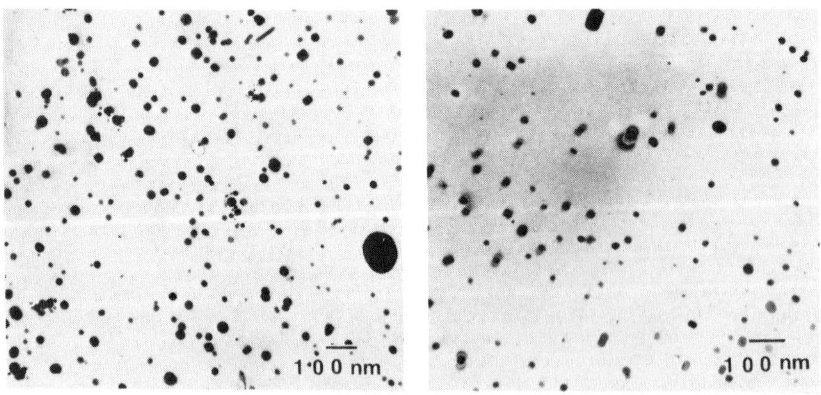


Figure 5. Bright field micrograph of specimens which had been laser treated, heat treated for 10 h at 973 K and subsequently heat treated for 10 h at a) 1073 K and b) 1173 K.

and thus it confirms the importance of the transus temperature on the stability of these oxides. Furthermore since these temperatures are near those employed during the compaction of this compound, it should be possible to produce these refined dispersions in bulk pieces. Heat treatment of the as-rapidly solidified alloy for 10 h at 1273 K does result in significant coarsening of the oxide, resulting in an average size of the precipitates of approximately 60-70 nm, figure 6. As mentioned

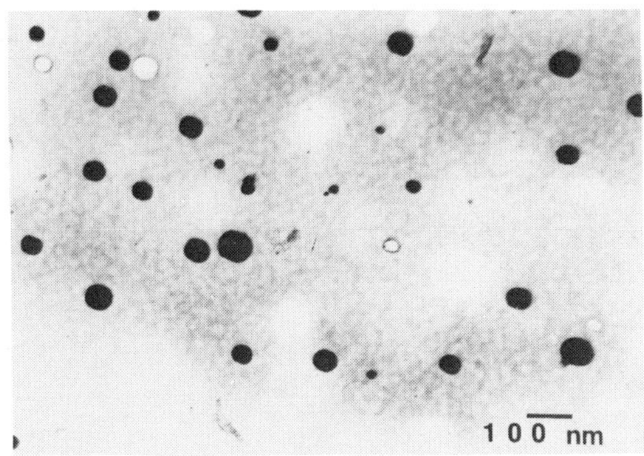


Figure 6. Bright field micrograph of specimen which had been laser treated, heat treated for 10 h at 973 K and subsequently heat treated for 10 h at 1273 K.

previously the Er to O ratio is expected to be greater than 2:3 and thus the rate of O transport is expected to control the rate of coarsening [6]. Although the diffusivity of O in $Ti_3Al$ is not known, it is thought that the increase in this diffusivity with temperature is responsible for the large particle size in this last heat treatment. If the diffusivity of the O in the $Ti_3Al$ is assumed to be similar to the diffusivity of O in the pure α-Ti the approximate size of the second phase can be calculated [6] with respect to time and temperature. A plot of this calculation is shown in figure 7. There is a good correlation between the trends predicted by

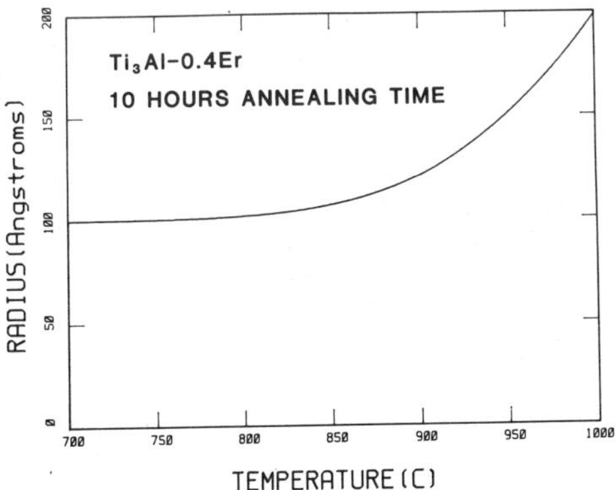


Figure 7. Plot of the predicted particle size following heat treatment for 10 h at a given temperature. Note, initial particle size is 100 angstroms.

this calculation and the experimental results. However it should be pointed out that because of the assumptions made in this calculation, such as the diffusivities of the components, interfacial energy and stability of the oxide, the numbers in figure 7 are only intended to show trends in the variation of particle size with temperature and the sizes are not expected to be exact.

## CONCLUSIONS

The use of laser surface melting is a suitable process for producing a supersaturation of Er in the intermetallic phase $Ti_3Al$. Subsequent heat treatment of this metastable structure results in the production of a fine dispersion of the rare earth oxide. The second phase is resistant to coarsening at elevated temperatures up to 1273 K.

## ACKNOWLEDGEMENTS

This research was supported in part by the US DoE, Division of Materials Sciences, under contract DE AC02 76ER01198. The electron microscopy was performed using the facilities of the Center for the Microanalysis of Materials, operated by the Materials Research Laboratory of the University of Illinois.

## REFERENCES

[1] J.B. McAndrew and H.D. Kessler: J. Metals, Vol. 8, 1956, 1348.

[2] H.A. Lipsitt, D. Shechtman and R.E. Schafrik: Metall. Trans. A., Vol. 11A, 1980, 1369.

[3] A.N. Stroh: Proc. Roy. Soc. London, Vol. A232, 1955, 548.

[4] D.G. Konitzer, B.C. Muddle and H.L. Fraser: Scripta Met., Vol. 17, 1983, 963.

[5] S.M.L. Sastry, P.J. Meschter, and J.E. O'Neal, Metall. Trans. A, Vol. 15A, 1984, 1451.

[6] D.G. Konitzer, R. Kirchheim and H.L. Fraser: in preparation.

[7] R.D. Shull, A.J. McAlister and R. Reno: presented at the 5th International Conf. on Ti.

[8] D.G. Konitzer, B.C. Muddle and H.L. Fraser: Metall. Trans. A., Vol. 14A, 1983, 1979.

[9] D.G. Konitzer, M.H. Loretto and H.L. Fraser: submitted to Acta Met.

PART VI

# Physical and Metallurgical Properties

# THE OXIDATION OF NICKEL-ALUMINUM AND IRON-ALUMINUM ALLOYS*

JOHN V. CATHCART
Metals and Ceramics Division, Oak Ridge National Laboratory, P.O. Box X, Oak Ridge, Tennessee 37831

## ABSTRACT

The high-temperature oxidation behavior of several ordered alloys in the Ni-Al and Fe-Al systems is reviewed with special emphasis on $Ni_3Al$ and NiAl. Ordering influences oxidation through its effect on the activities of the alloy components and by changing the point defect concentration in an alloy. Three categories of Ni-Al alloys are distinguished based on Al content and oxidation behavior. A characteristic feature of the oxidation of high-aluminum Ni-Al and Fe-Al alloys is the formation of voids in the substrate at the oxide-metal interface. The mechanism of void formation and its suppression by minor additions of oxygen-active elements is discussed. A brief description of the effect of preoxidation on the reactions of $Ni_3Al$-base alloys in $SO_2/O_2$ environments is also included.

## INTRODUCTION

Ordered intermetallic alloys have recently attracted attention because of their excellent high-temperature mechanical properties. However, of major importance in the ultimate application of these alloys is the question of their corrosion resistance. This paper reviews the high-temperature oxidation behavior of the nickel aluminides with special attention to the ordered $Ni_3Al$ and NiAl phases and alloys based on them. The effects of minor alloying additives on the oxidation behavior of alumina formers are considered, and oxidation data for the iron aluminides are also presented.

## PHASE RELATIONSHIPS IN THE Ni-Al AND Fe-Al SYSTEMS

Before discussing the oxidation properties of these ordered aluminides, it is useful to consider the phase diagrams for the two systems. As shown in Fig. 1, the ordered $\gamma'$ phase in the Ni-Al system exhibits a relatively narrow range of stability (~4.5 at. % at 1000°C) centered on the stoichiometric $Ni_3Al$ composition [1]. It is fcc with an $L1_2$-type superlattice [2]. The $\gamma'$ phase remains ordered up to its melting point.

The $\beta'$ phase, based on NiAl, is also ordered with a stability range of 19 at. % at 1000°C. It is bcc with a B2-type superlattice isotypic with CsCl. According to Hultgren et al. [2], the order-disorder transformation temperature is "above 800 and below 1200°C." However, Hancock and McDonnell [3] reported marked changes in the diffusivity of Ni in the $\beta'$ phase as a function of composition at temperatures as high as 1350°C. It would appear likely, therefore, that ordering persists at least to 1350°C in this compound.

---

*Research sponsored by the Office of Energy Utilization Research, Energy Conversion and Utilization Technologies (ECUT) Program and the Office of Technical Coordination, Advanced Research and Technology Development (AR&TD) Fossil Energy Materials Program, U.S. Department of Energy, under contract DE-AC05-84OR21400 with Martin Marietta Energy Systems, Inc.

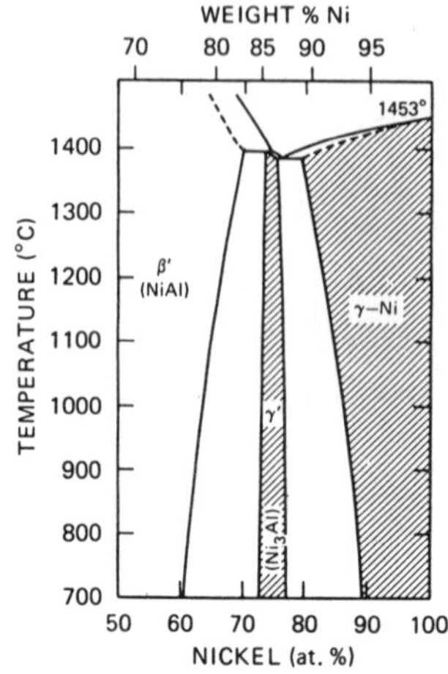


Fig. 1. Partial Ni-Al phase diagram. After [1].

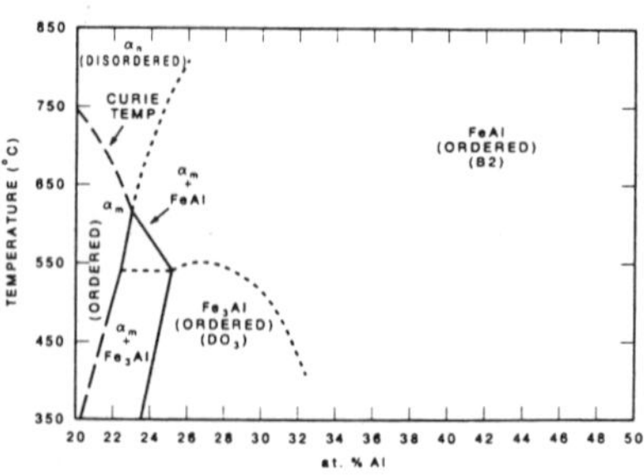


Fig. 2. Partial Fe-Al phase diagram. After [4].

The iron-rich portion of the Fe-Al phase diagram is shown in Fig. 2. The ordered phases include $Fe_3Al$ (bcc with a $DO_3$ type superlattice) and FeAl (bcc with a B2 superlattice). The transition from the $Fe_3Al$ structure to FeAl involves a second order transformation as does the transition from the disordered alpha solid solution to ordered FeAl [4-6]. Note that the ordered FeAl structure persists all the way to stoichiometric FeAl.

## EFFECT OF ORDERING ON OXIDATION BEHAVIOR

The effects of ordering on the oxidation behavior of Ni-Al and Fe-Al intermetallic compounds are, with some exceptions, largely indirect. For example, ordering is always accompanied by large deviations from ideal solid solution behavior, and large changes in the activities of alloy components as a function of composition are to be expected. Under some circumstances these activity changes can, indeed, alter oxidation behavior, but these changes result from altered activities rather than from ordering per se.

Changes in activity with composition can be dramatic as illustrated in Figs. 3 and 4 for the Ni-Al and Fe-Al systems. The activities of both Ni and Al change by more than three orders of magnitude from one phase boundary to the other of the β' phase in the Ni-Al system. From a thermodynamic point of view, this drastic lowering of $a_{Al}$ is not particularly important because in almost all practical cases the activity of oxygen at both the oxide-metal and oxide-gas interfaces will be high enough to make $Al_2O_3$ stable. However, the formation of a complete layer of $Al_2O_3$ at the alloy surface requires a certain minimum flux of Al atoms to that surface, and "... one of the most important aspects of the ordering process is its influence on diffusion ..." [7]. This effect may be expressed in terms of the intrinsic diffusion coefficient, $D_i^I$, for one component of a binary alloy, which is given [8] approximately by

$$D_i^I = D_i^*[1+(\partial \ln \gamma_i/\partial \ln N_i)] \, , \tag{1}$$

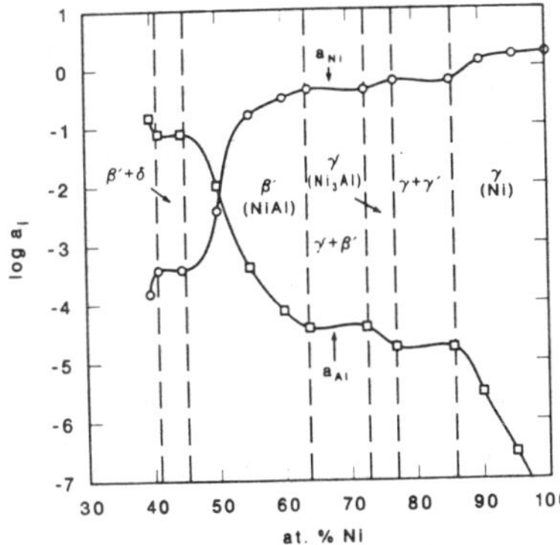


Fig. 3. Calculated activities of Al and Ni in Ni-Al alloys at 1000°C. After [51].

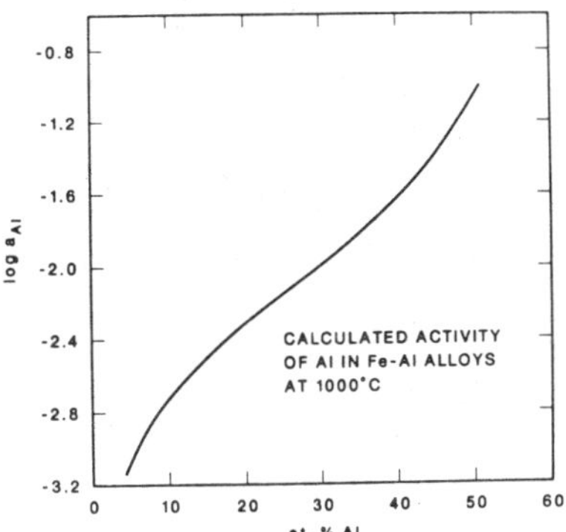


Fig. 4. Calculated activities of Al in Fe-Al alloys at 1000°C. After [51].

where $D_i^*$, $\gamma_i$, and $N_i$ are the tracer diffusivity, the activity coefficient, and the atom fraction, respectively, of the component i. Thus the substantial departure from ideality that is the expected corollary of ordering does influence the flux of Al to the metal oxide interface in a Ni-Al alloy, and, therefore, could be a determining factor in whether internal oxidation occurs or a complete, protective layer of $Al_2O_3$ is formed.

This departure from ideality can have more significant effects on the oxidation of less oxygen-active elements such as Ni. For example, during the oxidation of a Ni-Al alloy in a mixed gas such as $SO_2 + O_2$, one mechanism of attack of Ni by sulfur [9] involves the reaction of Ni with $SO_2$ at the oxide-metal interface:

$$7Ni_{ss} + 2SO_2 = Ni_3S_2 + 4NiO \tag{2}$$

and the equilibrium $SO_2$ pressure for the reaction, expressed in terms of the equilibrium constant, K, and the activity, $a_{Ni_{ss}}$, of Ni in the Ni-Al solid solution is

$$P_{SO_2} = K^{-1/2} a_{Ni_{ss}}{}^{-7/2} \tag{3}$$

Thus the minimum $P_{SO_2}$ pressure required to form $Ni_3S_2$ is a sensitive function of $a_{Ni}$. This phenomenon has been cited as the explanation for the greater resistance to sulfidation of NiAl as compared to $Ni_3Al$ where the Ni activity is much higher (c.f., Fig. 3).

In addition to the effect of activity changes on the diffusivities of components of intermetallics, ordering in these materials can influence diffusion through its effect on the defect concentration in the alloys. In NiAl, for example, an x-ray analysis by Bradley and Taylor [10] indicated that, for Ni-rich compositions, Ni atoms substitute freely on sites in the Al sublattice; however, in Al-rich compositions Al atoms substitute for only about 1% of the vacant sites in the Ni sublattice. At the Al-rich phase boundary, about 16% of the nickel sites are vacant with the vacancies being distributed randomly on the Ni sublattice. Consistent with this result is the observation by Hancock and McDonnell [3] of a minimum in the tracer diffusivity of Ni in NiAl at a composition of approximately 49 at. % Ni (Fig. 5). They found abrupt increases in $D^*_{Ni}$ at both higher and lower concentrations of nickel. These variations in diffusivity are important in determining the microstructure of aluminizing coatings on Ni-base alloys [11] and in fixing the Al gradient established in an alloy as a result of the preferential oxidation of Al.

A final effect relates to the possibility that the preferential oxidation of a component of an ordered phase may result in the formation of a new, perhaps disordered, metal phase at the oxide-metal interface. Such a phenomenon occurs during the oxidation of $Ni_3Al$ where after a period of oxidation a layer of disordered Ni solid solution is formed between the oxide and the base intermetallic (Fig. 6). Thus in oxidizing ordered $Ni_3Al$, one ends up oxidizing a disordered phase containing only a little more than half the Al in the original alloy (14.3 vs 23 at. % at the phase boundaries at 1000°C [2]). The underlying $Ni_3Al$, however, continues to play a role as a source of Al as oxidation continues.

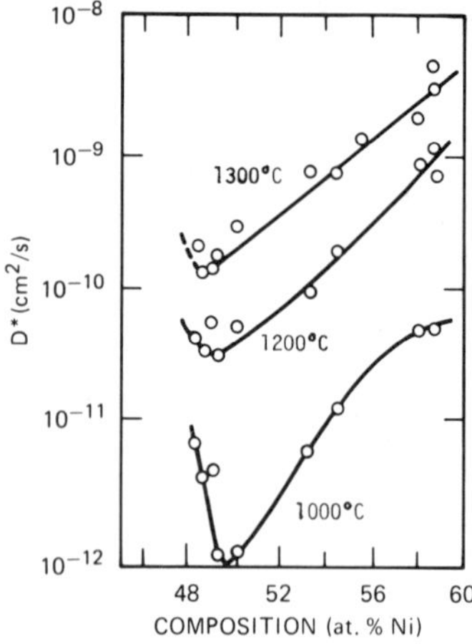


Fig. 5. Tracer diffusivity of Ni in NiAl. After [3].

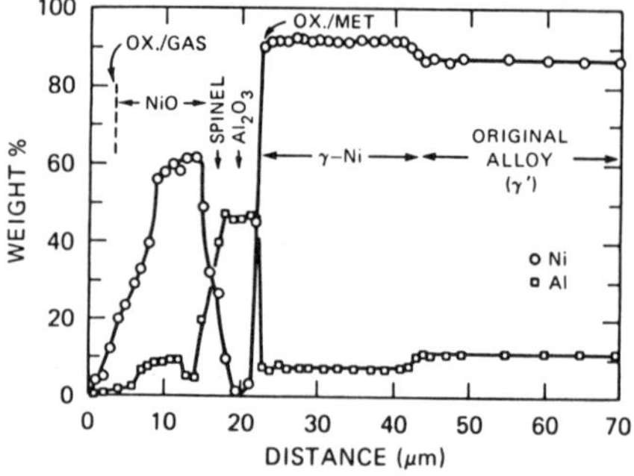


Fig. 6. Microprobe trace across Ni-24Al-0.04 at. % Ce oxidized 1200 h at 1000°C.

## OXIDATION PROPERTIES OF NICKEL ALUMINIDES

Nickel aluminides, especially NiAl, have been used for some time as corrosion resistant coatings on high-temperature alloys, and in recent years a moderate number of studies [12-27] of the oxidation, per se, of these materials have been carried out. One of the more comprehensive studies was that of Pettit [12] who divided Ni-Al alloys into three groups depending on the Al concentration. The boundaries between the three groups vary somewhat with temperature (see Fig. 7), but generally speaking, Group I alloys extend from 0 to about 10 at. % Al. These alloys all undergo internal oxidation, the activity of Al being too low to allow the formation of a complete layer of $Al_2O_3$ at the metal surface. At an oxygen pressure of 0.1 atm, parabolic oxidation was observed with the parabolic oxidation rate constant being an order of magnitude greater than that for pure Ni under the same conditions. This increase in rate was subsequently attributed to an increase in Ni vacancies produced by the solution of $Al_2O_3$ in the layer of NiO that forms at the specimen surface [19].

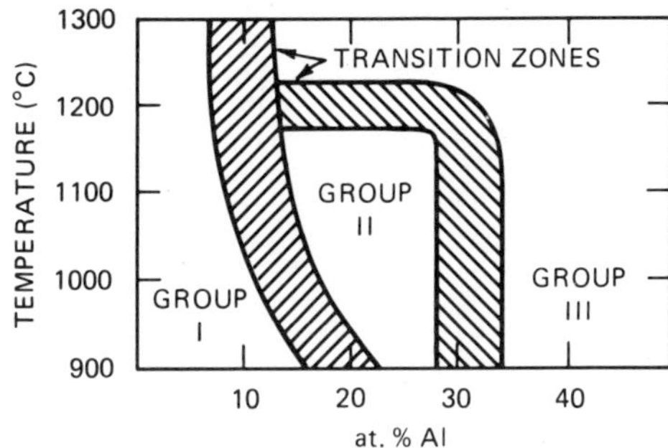


Fig. 7. Categories of Ni-Al alloys. After [12].

Group II alloys (10 to ~30 at. % Al) do contain enough Al to allow the formation of a continuous surface layer of alumina, but for oxidation at 0.1 atm oxygen the maximum flux of Al to the oxide-metal interface is less than that required to satisfy the growth rate of the $Al_2O_3$ scale. Thus during an initial period of slow parabolic oxidation ($Al_2O_3$ scale formation), the concentration of Al at the oxide-metal interface drops below that needed to prevent internal oxidation, and a sharp increase in oxidation rate is observed corresponding to the formation of $NiAl_2O_4$ and NiO.

Group III alloys, which at temperatures below about 1200°C contain at least 30 at. % Al, can all form continuous layers of $Al_2O_3$, and the activity of Al is great enough so that the flux to the oxide-metal interface can always match the growth rate of the oxide. As a consequence, parabolic scale growth is observed with a reaction rate constant some three times smaller than that for nickel.

Pettit also points out that the activities of oxygen in the gas phase and of Al and Ni in the alloy determine the oxides that can be formed. He gives a stability diagram illustrating regions of stability for the alloy-$Al_2O_3$, the alloy-$Al_2O_3$-$NiAl_2O_4$, and the alloy-$NiAl_2O_4$-NiO.

## Low Aluminum Alloys

Pettit's categorization of oxidation behavior is in moderately good agreement with the results of subsequent investigators. For example, the several studies of the oxidation of $\gamma$ solid solution alloys [15, 18, 19, 23, 25] are in agreement as to the predominance of internal oxidation of Al in the temperature range 800 to 1200°C. Hindam and Smeltzer [19] observed the formation of rodlike internal precipitates of $Al_2O_3$ during the oxidation of Ni-Al alloys containing 4 and 12 at. % Al, with the ultimate development on the alloy of a healing layer of $Al_2O_3$ whose formation was triggered by the impingement of the alumina rods on one another. Hindam and Whittle [18] also observed rodlike $Al_2O_3$ precipitates formed at 1200°C in alloys with 4 and 8 at. % Al, although in more dilute alloys the precipitates occurred as polyhedral crystallites and epitaxed platelets. The precipitate morphology was also sensitive to small (0.5 at. %) additions of Ta and Hf, a result suggesting a change in the $Al_2O_3$-matrix interfacial energy induced by the presence of the refractory metal additions. Stott et al. [25] observed internally formed, acicular precipitates in 2 and 8 at. % Al alloys. Oxidation at the oxygen activity furnished by a Ni-NiO pack resulted in large compressive stresses in the surface of the metal; however, stress effects were minor during oxidation at 1 atm oxygen where NiO could also form. Vacancies produced by the oxidation of Ni are apparently injected into the metal where they relieve the compressive stresses associated with internal $Al_2O_3$ formation. Penetration of grain boundaries and sub-grain boundaries in the alloy was also observed [23] and discussed in terms of stress generation and stress relief mechanisms.

Increasing the Al content of an alloy decreases the tendency for internal oxidation although not necessarily the complexity of the oxide scale. Wood and Stott [15] oxidized Ni-14 at. % Al, a composition that is very close to the $\gamma$ phase boundary and observed at 1000°C the internal oxidation of Al and the formation of an external scale consisting of an outer layer of NiO and an inner layer of NiO mixed with $NiAl_2O_4$ particles. At 1200°C, where Al diffusion in the substrate is faster, an external scale of $\alpha$-$Al_2O_3$ containing minimal amounts of Ni was formed, but at both temperatures failure of the scale led to the development of oxide nodules rich in NiO with a "rehealing" of the $Al_2O_3$ layer occurring later in the oxidation process. By contrast the oxide formed in a day on a Ni-23.7 at. % Al ($\gamma$'-$Ni_3Al$) at either 1000 or 1200°C was almost entirely $Al_2O_3$, and only $Al_2O_3$ formed during continued oxidation after the initial oxide failed.

## Oxidation of the $\gamma$' ($Ni_3Al$) Phase

A more extensive study of the oxidation of the $\gamma$' phase was carried out by Kuenzly and Douglass [16]. They oxidized alloys containing 24.3 and 24.9 at. % Al in air at temperatures from 900 to 1200°C. The effect of small (0.3 at. %) Y additions to the base alloy was also studied, and both isothermal and cyclic oxidation kinetics were determined. No spalling was observed with either $Ni_3Al$ or Ni-24Al-0.3 at. % Y during isothermal tests of 50 h duration. Total weight gain in the presence of Y additions increased by 10 to 20%. Cyclic tests lasted up to 11 days with oxide spalling occurring on cooling, although spalling was much worse at 1200°C than at 900°C. At 800°C, the oxide composition on $Ni_3Al$ varied from NiO at the external surface to $NiAl_2O_4$ to $Al_2O_3$ at the oxide-metal interface, but at 1200°C only $Al_2O_3$ could be detected. The presence of Y prevented spalling at both 900 and 1200°C, and at both temperatures after 7 days of oxidation the scales consisted of $Al_2O_3$ covered by an external layer of the spinel. The presence of the spinel was attributed to the trapping of particles of the $Ni_9Y$ intermetallic in the oxide scale at the beginning of oxidation; after oxidation of these particles to NiO and $Y_2O_3$, a solid state reaction between NiO and $Al_2O_3$ apparently occurred.

## Oxidation of the $\beta$' Phase (NiAl)

Hutchings and Loretto [21] measured the oxidation rate in $O_2$ at 1 atm of NiAl as a function of composition at 900°C and observed a minimum in the rate at about 58 at. % Al (see Fig. 8). They attributed this behavior to the way the aluminum activity varies with composition and to the effective Al concentration established in the metal at the oxide-metal interface during oxidation. In a subsequent study Hutchings et al. [20] used a combination of sputtering and Auger spectroscopy to establish the distribution of Al, O, and Ni in the oxide on Ni-Al alloys for specimens oxidized at

900°C; they found an outer layer of $Al_2O_3$ and an inner layer of $NiAl_2O_4$ separated by a thin transition zone. This latter result is not in agreement with the data of Hindam and Smeltzer [30] who found for oxidation in $O_2$ at 1000-1300°C only Al and O in the scale except for a small amount (0.5 wt %) of dissolved Ni. Smialek [17] developed a model for the spalling of the oxide from NiAl, which he attributed, in agreement with the studies mentioned above, to the formation of voids in the metal substrate at the oxide-metal interface (see below).

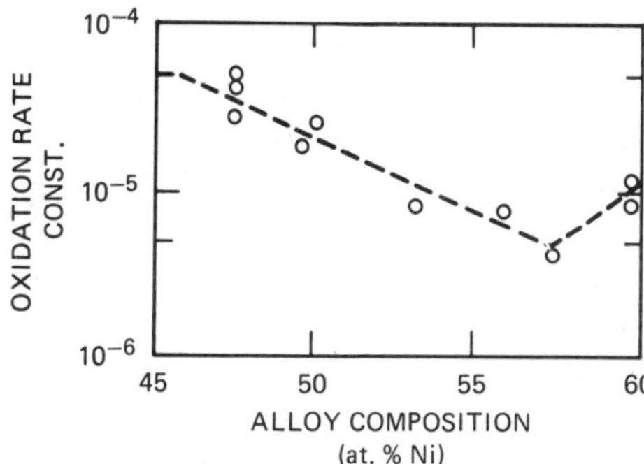


Fig. 8. Variation of oxidation rate of NiAl with composition. The rate constants are expressed in units of $(\mu g/mm^2)^n \cdot s^{-1}$, where n is a constant.

## Oxide Scale Morphology

Various oxide morphologies have been observed in the early stages of oxidation of $Ni_3Al$-base alloys. On electropolished samples, as illustrated in Fig. 9, the oxide scales are highly oriented as indicated by the extensive parallel facets on the surface of the oxide, and the change in the orientation of the facets from one grain of the substrate to another suggests that the scale and the substrate are highly epitaxed. Oxidation rate anisotropy is indicated by the variation in oxide thickness from one grain to another (Fig. 10). More surprising is the variation in the composition of the oxide from grain to grain. The scale in the lower left corner of Fig. 10 is rich in Ni, while the scale over the grain in the upper right corner is almost pure $Al_2O_3$ as indicated by EDS measurements. At higher magnifications such $Al_2O_3$ scales are seen to consist of a mat of $Al_2O_3$ whiskers and platelets; a more highly developed example of these features is shown in Fig. 11.

Smialek and Gibala [24] and Kear et al. [29] investigated transient oxide structures formed on Ni-15Cr-24 at. % Al and Ni-15Cr-12 at. % Al at 1000 and 1200°C and observed that the formation of a continuous layer of $\alpha$-$Al_2O_3$ is preceded by a transient stage of oxidation characterized by a rapid uptake of oxygen and the conversion of surface layers of the alloy to a spinel. Smialek and Gibala also noted the formation of oriented $\alpha$-$(Al,Cr)_2O_3$ or $\gamma$-$Al_2O_3$ subgrains prior to the final development of a layer of randomly oriented $\alpha$-$Al_2O_3$ at the oxide-metal interface. They also found evidence for the deformation of the substrate, and in characterizing the oxide itself they noted the existence of fine porosity in the oxide scale. Kuenzly and Douglass [16] also reported the existence of scale porosity.

As already discussed, the steady state microstructure of these oxide scales is dependent on temperature, alloy composition, and oxygen pressure. At sufficiently high temperatures only $\alpha$-$Al_2O_3$ is formed, although the presence of Y and perhaps other oxygen-active additives can lead to the

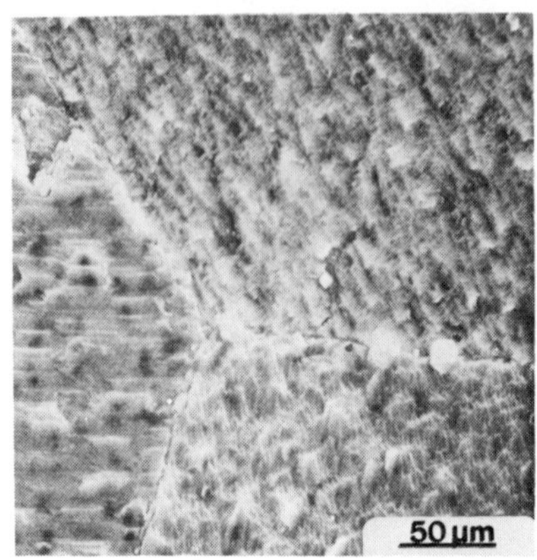


Fig. 9. Surface of Ni-24 at. % Al oxidized 144 h at 900°C in oxygen at 1 atm.

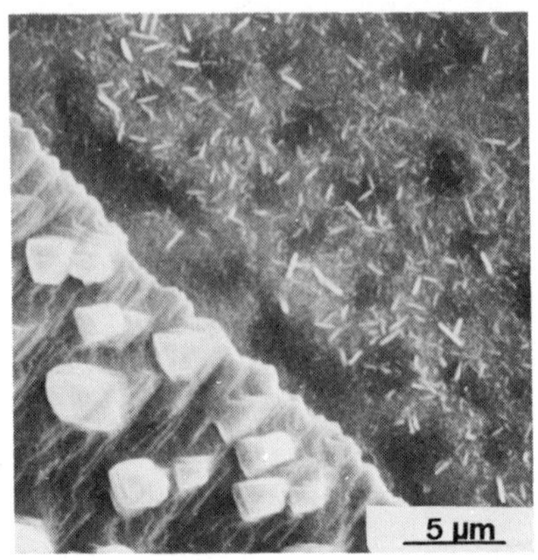


Fig. 10. Surface of Ni-24Al-0.04 at. % Ce oxidized 2 h at 1000°C in oxygen at 1 atm.

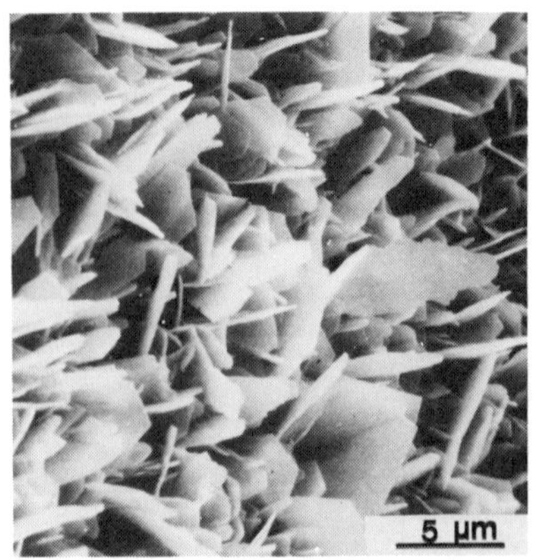


Fig. 11. $Al_2O_3$ platelets on Ni-24 at. % Al oxidized 624 h at 900°C in oxygen at 1 atm.

formation of $NiAl_2O_4$ at the outer surface of the scale. At lower temperatures (e.g., 900°C), both NiO and $NiAl_2O_4$ can form in addition to $\alpha$-$Al_2O_3$ providing the oxygen pressure is such that the former two oxides are stable. Below about 900°C $\gamma$-$Al_2O_3$ is likely to develop rather than the less permeable $\alpha$-$Al_2O_3$.

## Mechanisms of Void Formation

Most authors (e.g., 26, 28, 17, 15) agree with the finding of Kuenzly and Douglass that voids form in the substrate at or near the oxide-metal interface during the oxidation of Al-rich alloys. An example of this phenomenon is given in Fig. 12 for a Ni-24 at. % Al alloy oxidized isothermally for 192 h at 1000°C. Spalling of the scale occurred on cooling. Note the smooth underside of the oxide over the void and the uniformity of the scale thickness over the void. Such a morphology strongly suggests vapor transport of Al across the void, and calculations by Hindam and Smeltzer [30] show that at 1000°C the possible flux of Al available by vapor phase transport considerably exceeds the growth rate of the $Al_2O_3$ scale. As discussed at greater length below, small additions of oxygen-active elements such as Y, Hf, Ce, La, etc. repress the formation of these voids and thus improve scale adherence since the voids are regarded as stress concentrators in the oxide scale [16]. A similar tendency toward void formation in chromia formers has also been observed. Insight into these phenomena requires a review of the mode of transport in alumina and chromia scales.

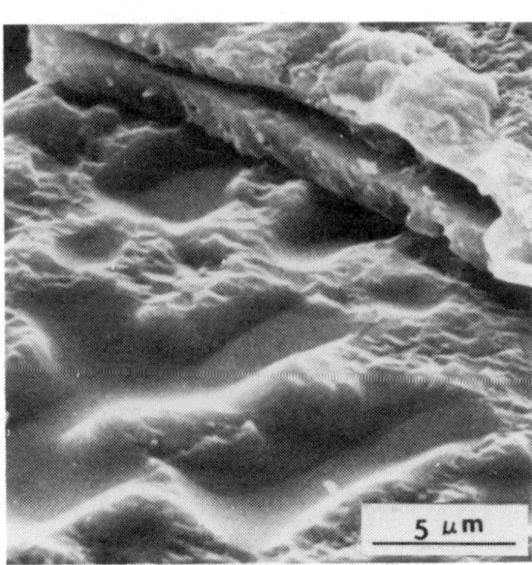


Fig. 12. Voids in substrate at oxide-metal interface of Ni-24 at. % Al oxidized 192 h at 1000°C in oxygen at 1 atm.

It is generally agreed that chromia films grow, at least at temperatures up to 1000 or 1100°C, by a combination of cation and oxygen transport along grain boundaries with cation transport dominating [see, e.g., 31-33]. On the other hand, Hindam and Smeltzer [34] used marker techniques to prove the inward diffusion of oxygen in $Al_2O_3$ scales. Reddy et al. [35] confirmed this result in studies of $^{18}O$ diffusion in alumina scales and showed that diffusion occurred down grain boundaries in the scale. They also found evidence for some diffusion of Al outward as well, although oxygen transport clearly predominates. As will be seen, a major effect of oxygen-active additives is to prevent the outward diffusion of cations through chromia and alumina scales.

Returning to the question of the mechanism of void formation, several authors [e.g., 36] have suggested that void development is related to the occurrence of cation vacancy transport in the scale and the subsequent coalescence of the vacancies at the oxide-metal interface. (Note that interstitial cation diffusion will result in the same effect.) Kuenzly and Douglass [16], however, observed void formation on $Ni_3Al$ under conditions where Ni as well as Al was oxidized (i.e., conditions where Ni diffusion through an $Al_2O_3$ scale occurred) and also, although in smaller numbers, when only $Al_2O_3$ was formed (i.e., in the absence of Ni diffusion). They, therefore, suggested that voids develop because of a Kirkendall-type effect associated with the interdiffusion of Ni and Al near the oxide-metal interface as a result of the preferential oxidation of Al.

## Void Suppression and Improved Scale Adherence

As noted already, small additions of oxygen active elements eliminate or greatly reduce void formation and improve scale adherence. Many theories have been advanced to account for these effects. Stringer [37] proposed that the large rare earth and refractory metal atoms act as strain centers and, hence, as vacancy sinks, causing the vacancies generated by the oxidation process to be distributed throughout the surface region of the substrate rather than precipitating at the oxide-metal interface. This mechanism is seemingly supported by the observation by Tsuzi [38] of voids centered on $Y_2O_3$ particles formed internally during the oxidation of a Ni-24 at. % Cr alloy containing additions of Y.

A second proposal [39] for improved scale adherence involves the oxidation of oxygen-active additives and their incorporation into the oxide in such a way as to form "pegs" that hold the scale to the substrate. However, there are cases where spalling does not occur even though no pegs develop [40] and, certainly, many instances in which very rough $Al_2O_3$ scales spall readily.

The possibility that the oxygen-active elements might concentrate at the oxide-metal interface and form a "graded seal" between the oxide and the substrate was suggested by Pfeiffer [41], and the possibility that oxidation-induced stresses in the scale might be reduced by an enhancement of oxide plasticity was proposed by Francis and Jutson [42].

A recent, alternate explanation was proposed by Kofstad [43] who began by noting what oxidation parameters are changed by the presence of oxygen-active elements in growing $Cr_2O_3$ and $Al_2O_3$ scales. The most striking change is the reversal of the direction of material transport in oxide scales. Stringer et al. [44], for example, reported a reversal from outward diffusion of Cr to the inward diffusion of oxygen in chromia scales formed on Ni-Cr alloys containing $Y_2O_3$ and $CeO_2$. A similar reversal of transport properties of $Cr_2O_3$ scales on Cr containing a dispersion of $ThO_2$ was also observed by Stringer et al. [45]. In effect, although the mechanism is still not understood, the presence of these oxygen-active elements blocks the diffusion of cations in the scale grain boundaries and increases the relative importance of oxygen diffusion.

It is much more difficult to demonstrate this phenomenon in alumina scales where inward oxygen diffusion predominates in any case. However, Reddy et al. [35] did find evidence of a small counter current of Al ions in scale grain boundaries during alumina formation and noted its suppression in the presence of small additions of Zr. Furthermore, the results of Kuenzly and Douglass [16] can be taken as evidence for the suppression of Ni diffusion in $Al_2O_3$ scales by Y if one accepts their explanation for the presence of $NiAl_2O_4$ at the oxide-gas interface of their samples. Thus even in the case of $Al_2O_3$ it is reasonable to argue that oxygen-active elements in the scale prevent cation transport along grain boundaries. The essence of Kofstad's model, then, is that no voids develop at the oxide-metal interface of alumina formers containing Y, Hf, etc. because new oxide grows by the inward diffusion of oxygen and, therefore, no vacancies are generated by the oxidation process. Any vacancies produced by the slowed outward diffusion of cations (e.g., Ni) in the scale grain boundaries or along microcracks could be absorbed at the oxide-metal interface as a part of a stress relief mechanism by which the excess volume of $Al_2O_3$ formed at that interface is accommodated.

This is an interesting mechanism that focuses attention on the effect of minor additives on the properties of grain boundaries (diffusion, mechanical effects, etc.) and as such can explain why very small quantities of oxygen-active elements can have such profound effects on the oxidation properties of alloys. This mechanism does not account for the ability of oxygen-active additives to prevent the occurrence of "Kirkendall-type" voids associated with the interdiffusion of alloy components in the surface regions of the alloy. Thus it does not necessarily invalidate the idea that oxygen-active metals act as strain centers and vacancy sinks in the alloy substrate. Obviously, additional research in this area is needed.

## IRON ALUMINIDES

Iron-aluminum alloys, like Ni-Al, exhibit ordered phases. However, unlike $Ni_3Al$, $Fe_3Al$, and FeAl are bcc, and Fe-Al alloys tend to become very brittle as the aluminum content is increased. Perhaps for this reason the subject of the oxidation of Fe-Al alloys has received considerably less attention in the open literature than has the oxidation of Ni-Al alloys. Tomaszewicz and Wallwork [46] recently reviewed the oxidation of Fe-Al alloys, and for this reason only a brief discussion is given here.

Approximately 15 at. % Al must be added to iron to produce a continuous layer of $Al_2O_3$ during oxidation. Small Cr additions reduce this amount but Ni additions require increased amounts of Al to avoid the formation of nodules of Fe-rich oxides that disrupt the $Al_2O_3$ scale. In analogy to the Ni-Al system, three groups of Fe-Al alloys can be identified. In air at low concentrations of Al, internal oxidation of Al occurs, and $Fe_2O_3$/$Fe_3O_4$ form on the exterior of the alloys. At intermediate Al concentrations, although a transient $Al_2O_3$ layer is formed, it tends to be disrupted by iron oxide nodules, while at high Al concentrations only thin surface layers of $Al_2O_3$ are observed. In a recent paper Tomaszewicz and Wallwork [47] quantified the Al concentrations corresponding to the grouping given above for oxidation at 800°C: 0-4 at. % Al (internal oxidation + $Fe_2O_3$); 4-14 at. % Al ($Al_2O_3$ + $Fe_2O_3$ nodules); and >14 at. % Al (thin $Al_2O_3$ layers only).

In discussing the feasibility of Fe-Al-base alloys as high-temperature materials, Hardwick and Wallwork [48] concluded that binary Fe-Al alloys "... have been shown to be highly oxidation resistant ..." although lacking both room temperature ductility and high-temperature strength. After discussing various ways of remedying these mechanical shortcomings, they correctly point out that the effect of any alloying additions on oxidation properties must be examined.

## REACTIONS OF $Ni_3Al$-BASE ALLOYS IN $SO_2/O_2$ MIXTURES

Extensive studies of M-Cr-Al alloys (where M is Co, V, Ni, or Fe) as coatings for high-temperature alloys have been carried out in a variety of gas mixtures containing both oxygen and sulfur [see, e.g., Refs. 49 and 50]. Such coatings with small additions of oxygen active elements and with Cr contents varying from 18 to 25 at. % and Al from 16 to 20 at. % form dense, protective $Al_2O_3$ scales that resist sulfur attack for long periods at high temperatures.

These coatings are not themselves, of course, suitable for use as structural materials. While their corrosion properties are of great practical importance, they are not ordered alloys, and, therefore, a discussion of their properties is beyond the scope of this paper. Coating results do indicate, however, that it is possible to develop $Al_2O_3$ scales that are essentially impervious to sulfur penetration, and in principle one could hope to duplicate this property in $Al_2O_3$ scales formed on ordered $Ni_3Al$-base alloys.

$Ni_3Al$ itself, even with scale adherence improved through the addition of Hf, is subject to attack in $SO_2/O_2$ atmospheres. Figure 13 shows a Ni-^^Al-0.5Hf-0.2 at. % B sample exposed to the gaseous decomposition products of $CaSO_4$ for 168 h at 871°C. The specimen was heavily attacked, and the small spherical particles on the surface were at temperature a liquid Ni-S solution. However, when a specimen was preoxidized first in wet hydrogen (188 h at 1100°C; $P_{O_2} = 10^{-18}$ atm) and then in pure oxygen at 1 atm for another 188 h at 1100°C, no sign of any sulfur attack was observed in a subsequent test in the $SO_2/O_2$ environment lasting 168 h at 871°C (Fig. 14). (The light colored lines visible in the figure correspond to ridges of oxide formed over substrate grain boundaries during preoxidation.)

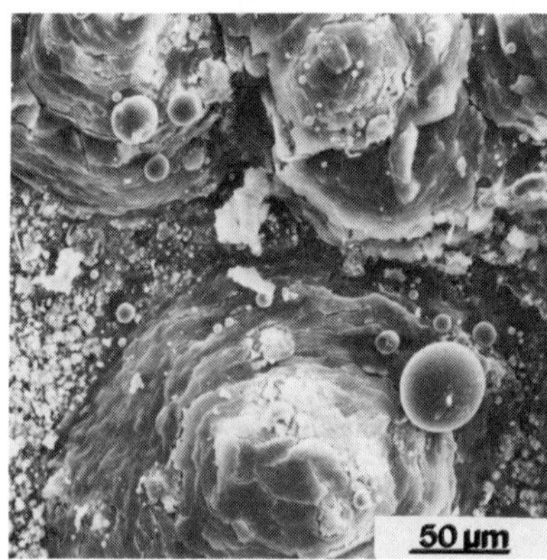


Fig. 13. Surface of Ni-24Al-0.5Hf-0.2 at. % B. exposed 160 h at 871°C to gaseous decomposition products of $CaSO_4$.

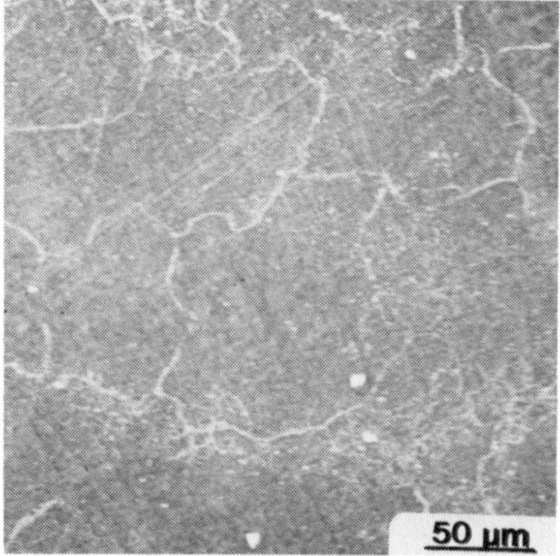


Fig. 14. Ni-24Al-0.5Hf-0.2 at. % B, preoxidized and exposed to $CaSO_4$ decomposition products for 168 h at 871°C.

The effects of various pre-oxidation treatments are summarized in Table 1 where the tabulated weight changes represent weight gain during exposure to the $SO_2/O_2$ environment. As shown in the table, important preoxidation parameters include oxygen pressure and total oxide thickness. Undoubtedly temperature is also significant since $\gamma$-$Al_2O_3$, which is less protective than $\alpha$-$Al_2O_3$, is known to form at temperatures near 900°C and below.

Table I. Effect of preoxidation on sulfidation in capsule tests[a] of Ni-24Al-0.5Hf-0.2 at. % B

| Surface pretreatment | $\Delta W (mg/cm_2)$ |
|---|---|
| None | 8.75 |
| Air; 96 h at 1000°C | 1.3 |
| Wet $H_2$ (D.P. = 0°C); 188 h at 1100°C | 0.32 |
| Double preoxidation[b] | 0.018 |

[a]Exposure in a sealed quartz capsule to the gaseous decomposition products of $CaSO_4$ for 168 h at 871°C.

[b]Oxidation in wet $H_2$ (D.P. = 0°C) for 188 h at 1100°C followed by oxidation in $O_2$ (1 atm) for 188 h at 1100°C.

These results demonstrate that it is possible to grow $Al_2O_3$ scales on $Ni_3Al$ that resist sulfur penetration. As was discussed earlier, transport through $Al_2O_3$ scales occurs via grain boundary diffusion of oxygen. Presumably the greater resistance to sulfur penetration exhibited by the pre-formed $Al_2O_3$ scale is indicative of the fact that this scale has a different microstructure than that of the scale that forms directly in $SO_2/O_2$ mixtures. Thus efforts to produce a $Ni_3Al$-base alloy that is inherently resistant to sulfidation should be directed toward finding additives that appropriately modify the grain boundary properties of $Al_2O_3$.

## SUMMARY

Lowered activity of Ni and/or Al in ordered Ni-Al alloys can affect their oxidation behavior by (1) influencing diffusion processes in the substrate and (2) changing the equilibrium oxidant pressure required to form the oxides/sulfides of Ni. Substrate diffusion may also be affected by changes in alloy defect concentration induced by ordering.

The oxidation behavior of Ni-Al alloys may be divided into three categories based on Al content. Aluminum is internally oxidized at lower concentrations, forms a "transient" surface layer of alumina at intermediate levels, and forms a stable $Al_2O_3$ scale at high concentrations.

In the early stages of oxidation of $Ni_3Al$ the oxide on electropolished samples is highly oriented and epitaxed with the substrate. Different oxide compositions develop on different grains at the oxide gas interface, a result probably related to variations in the degree of crystalline perfection of the underlying $Al_2O_3$ layers.

$Ni_3Al$- and Ni-Al-base alloys characteristically form voids in the substrate at the oxide-metal interface unless small quantities of oxygen-active elements are added to the alloys. The presence of these voids correlates with oxide spallation and poor scale adherence. Void formation has been attributed to the coalescence of vacancies generated at the oxide-metal interface by cation diffusion in the scale or as a consequence of interdiffusion effects in the substrate arising from the preferential oxidation of Al.

Many theories have been advanced to account for the beneficial effects of oxygen-active element additions in increasing scale adherence and repressing void formation. Included among these ideas are the "vacancy sink" model, the occurrence of "pegging," the formation of a "graded seal" between oxide and metal, and enhancement of oxide plasticity. A recent idea emphasizes the avoidance of vacancy generation by the blocking of cation diffusion in grain boundaries by oxygen-active elements.

Fe-Al alloys exhibit excellent high-temperature corrosion resistance, but their mechanical properties are not as attractive as those of the Ni aluminides. A recent review of their oxidation properties is available.

Care must be taken to avoid the formation of the low-melting Ni-$Ni_3S_2$ eutectic when $Ni_3Al$-base alloys are exposed to mixed potentials of oxygen and sulfur. Pre-oxidation under certain conditions creates $Al_2O_3$ scales on $Ni_3Al$ that resist sulfur penetration.

## ACKNOWLEDGMENTS

The authors gratefully acknowledge several useful discussions with Dr. P. Kofstad, and thanks are due to R. S. Crouse for the SEM photographs used in this paper. The efforts of J. H. DeVan in reviewing this manuscript are also appreciated. The assistance of F. M. Foust in the preparation of the manuscript was invaluable.

## RFFERENCES

1. M. Hansen, Constitution of Binary Alloys, McGraw-Hill, New York, (1958), pp. 118-121.
2. R. Hultgren, P. D. Desai, D. T. Hawkins, M. Gleiser, K. K. Kelley, Selected Values of the Thermodynamic Properties of Binary Alloys, American Society for Metals, Metals Park, Ohio (1973), pp. 191-195.
3. G. F. Hancock and B. R. McDonnell, Phys. Stat. Sol. (a) 4 (1971), 143-150.
4. P. R. Swann, W. R. Duff, and R. M. Fisher, Trans. AIME, 245, (1969), 851-853.
5. H. Okamoto and P. A. Beck, Met. Trans. 2 (1971) 569-574.
6. K. Oki, H. Sagane. and T. Eguchi, Japan. J. Appl. Phys., 13 (1974) 753-761.
7. L. N. Larikov, V. V. Geichenko, and V. M. Fal'chenko, Diffusion Processes in Ordered Alloys, (translated from the Russian), Oxonian Press Pvt., Ltd., New Delhi (1981) p. 44.
8. J. R. Manning, Diffusion Kinetics for Atoms in Crystals, D. van Nostrand Co. Inc., Princeton, N. J. (1968), p. 189.
9. P. Singh and N. Birks, Oxidation of Metals 19 (1983) 37-52.
10. A. J. Bradley and A Taylor, Proc. Roy. Soc. (London) A159, 56-72 (1937).
11. G. W. Goward and D. H. Boone, Oxidation of Metals, 3 (1971), 475-495.
12. F. S. Pettit, Trans. Met. Soc. AIME, 239 (1967) 1296.
13. J. S. Wolf and E. B. Evans, Corrosion, 18 (1962) 129t.
14. F. H. Stott and G. C. Wood, Corros. Sci. 17 (1977) 647.
15. G. C. Wood and F. H. Stott, Br. Corros. J. 6 (1971) 247.
16. J. D. Kuenzly and D. L. Douglass, Oxid. of Metals 8 (1974) 139.
17. J. L. Smialek, Met. Trans. A, 9A (1978) 309.
18. H. M. Hindam and D. P. Whittle, J. Mat. Sci. 18 (1983) 1389.
19. H. M. Hindam and W. W. Smeltzer, J. Electrochem. Soc. (1980) 1622.
20. R. Hutchings, M. H. Loretto, and R. E. Smallman, Metal Sci. 15 (1981) 7.
21. R. Hutchings and M. H. Loretto, Metal Sci. 12 (1978) 503.
22. L. A. Panyushin and V. P. Valyev, Protection of Metals (English translation) 12 (1976) 211.
23. Y. Shida, F. H. Stott, B. D. Bastow, D. P. Whittle, and G. C. Wood, Oxid. of Metals 18 (1982) 93.

24. J. L. Smialek and R. Gibala, Met. Trans. A (14A1983) 2143.
25. F. H. Stott, Y. Shida, D. P. Whittle, G. C. Wood, and D. Bastow, Oxid. of Metals 18 (1982) 127.
26. D. N. Tsippas, Proceedings JIMIS Symposium on High Temperature Corrosion, p. 569
27. D. P. Whittle, Y. Shida, G. C. Wood, F. H. Stott, and B. D. Bastow, Phil. Mag. 46 (1982) 93b.
28. J. V. Cathcart, unpublished research.
29. B. H. Kear, F. S. Pettit, D. E. Fornwalt, and L. P. Lamaire, Oxid. of Metals 3 (1971) 557.
30. H. M. Hindam and W. W. Smeltzer, J. Electrochem. Soc. 127 (1980) 1630.
31. G. C. Wood and F. A. Stott, "Development and Growth of Protective $Al_2O_3$ Scales on Alloys" in High Temperature Corrosion, R. A. Rapp, ed., NACE (1983), p. 227.
32. K. P. Lillirud and P. Kofstad, "High Temperature Oxidation of Chromium", ibid., p.155.
33. K. P. Lillirud and P. Kofstad, Oxid. of Metals 17 (1982) 127.
34. H. M. Hindam and W. W. Smeltzer, Oxid. of Metals 14 (1980) 337.
35. K. P. R. Reddy, J. L. Smialek, and A. R. Cooper, Oxid. of Metals 17 (1982) 429.
36. J. K. Tien and W. H. Wood, Scripta Met. 6 (1972) 55.
37. J. Stringer, Met. Rev. 11 (1966) 113.
38. E. Tsuzi, Met. Trans. A, 11A (1980) 1965.
39. E. J. Felton, J. Electrochem. Soc. 108 (1961) 490.
40. J. K. Tien and F. S. Pettit, Met. Trans. 3 (1972) 1587.
41. H. Pfeiffer, Werkst. Korros. 8 (1957) 574.
42. J. M. Francis and J. A. Jutson, Corros. Sci. 8 (1968) 445.
43. P. Kofstad, "High Temperature Corrosion", John Wiley and Sons, to be published.
44. J. Stringer, B. A. Wilcox, and R. I. Jaffe, Oxid. of Metals 5 (1972) 11.
45. J. Stringer, A. Z. Hed, G. R. Wallwork, and B. A. Wilcox, Corros. Sci. 12 (1972) 625.
46. P. Tomaszewicz and G. R. Wallwork, Rev. High Temperature Materials, 4 (1978) 75.
47. P. Tomaszewicz and G. R. Wallwork, Oxid. of Metals 19 (1983) 165.
48. D. Hardwick and G. R. Wallwork, Reviews of High Temperature Materials, 4 (1978) 47.
49. P. R. Clark, C. M. Packer, and R. A. Perkins, "Development of Coatings for Corrosion/Erosion Protection of Internal Components of Coal Gasification Vessels," Vol I. Laser Fused MCrAl, FE-2592-29, March 1981.
50. R. A Perkins, P. R. Clark, and C. M. Packer, ibid., Vol II. Furnace Fused MCrAl Coatings, FE-2592-30, March 1981.
51. A Steiner and K. L. Komarek, Trans. Met. Soc. AIME, 230 (1964) 786.

VOLUME AND GRAIN BOUNDARY DIFFUSION IN $L1_2$ ALLOYS WITH SPECIAL REFERENCE TO $Ni_3Al$ COMPOUNDS

T. C. CHOU AND Y. T. CHOU
Department of Metallurgy and Materials Engineering, Lehigh University, Bethlehem, Pennsylvania 18015

ABSTRACT

Experiment and theory of diffusion in L1(2) alloys are briefly reviewed.* Current work on volume and grain boundary diffusion in Ni(3)Al alloys are presented and discussed.

## 1. INTRODUCTION

In contrast to what is known about diffusion in pure metals and dilute alloys, the knowledge of diffusion in intermetallic compounds is rather sparse [1]. In recent years, there has been a great demand for diffusion data for semiconducting compounds because of rapid advances in thin-film technology [2]. However, the study of diffusion in ordered compounds has not been adequately pursued and most of the available data refer to compounds of B2 structure [1]. The diffusion behavior of compounds with the L1(2) structure is largely unknown.

More recently, extensive research has been conducted on long-range ordered alloys for high temperature usage [3]. A potent candidate in this group is Ni(3)Al, a L1(2)-type compound widely known as the principal strengthening phase ($\gamma$'phase) in nickel-base superalloys [4].

Although Ni(3)Al single crystals are ductile [4,5], the polycrystals are extremely brittle at both low and high temperatures. Recent studies have demonstrated that the ductility of Ni(3)Al polycrystals can be markedly improved by the addition of a small amount ($\sim$ 0.1 at%) of boron [6,7]. In the presence of boron, the grain boundary embrittlement is suppressed. According to Liu and coworkers, the doped compounds exhibit outstanding physical and mechanical properties at elevated temperatures and are superior to conventional alloys, such as Hastelloy-X and 316 stainless steels [3].

The success in developing ductile Ni(3)Al compounds has stimulated intensive research on chemical, physical, and mechanical properties of the ordered alloys. One of these basic investigations is the study of diffusion.

The diffusion data for Ni(3)Al alloys are very limited. In the past, the self-diffusion of Ni in Ni(3)Al compounds has been studied by Larikov et al. [8], Hancock [9], and Bronfin et al. [10], and that of Al in Ni(3)Al by Larikov [11]. Also, the interdiffusion of Al from Ni(3)Al compounds into Ni matrix has been studied by Janssen [12], Janssen and Rieck [13], and Bridge and Maniar [14], and that of Ni from Ni(3)Al into Al matrix by Janssen and Rieck [13]. In view of the demonstrated evidence of the beneficial effect of microalloying on the ductility of Ni(3)Al [3,6,7], it is essential to have a better understanding of the kinetics of solute diffusion in the lattice as well as in the grain boundary of these alloy compounds.

The purpose of this paper is to briefly review the diffusion processes in ordered compounds and to present the results of some recent work on volume and grain boundary diffusion in simple and doped Ni(3)Al alloys.

---

*Hereafter, we write L1(2) for $L1_2$ and Ni(3)Al for $Ni_3Al$.

## 2. THEORY OF DIFFUSION IN ORDERED ALLOYS

In ordered alloys, the effects of the degree of ordering and the defect structures complicate the mechanism of diffusion. This means that the environment of the jumping atoms can be changed merely by varying the temperature or the composition [15]. The diffusion process for such alloys can no longer be a random walk process when the temperature is below the critical temperature [16].

Consider the case of well-ordered binary alloys of stoichiometric composition having a crystallographic structure with two sublattices (such as Ni(3)Al). If the mechanism for atomic migration is by vacancy jumps from a given site to a nearest-neighbor site, one species of atoms (Al) must jump from its own sublattice to an adjacent Ni sublattice by overcoming the ordering energy of the system. However, the other species of atoms (Ni) either can jump from its sublattice to the adjacent Al sublattice by surmounting the ordering energy or can jump in its own sublattice. If the ordering energy is too large, self-diffusion may take place either through next-nearest-neighbor jumps for Al atoms or through nearest-neighbor jumps for Ni atoms. If the concentration or the temperature of the material is changed, the number of lattice defects through which the atomic diffusion takes place can also be changed and, therefore, the possible interplay of several diffusion mechanisms, as well as the correlation of the atomic movements, can be controlled [17].

A theory of diffusion in ordered-disordered fcc $AB_3$ alloy was proposed by Schoijet and Girifalco [16]. Formulas for the ordered-dependent part of the vacancy formation and migration energies have been derived for $Cu_3Au$ alloy, using Kirkwood's second moment approximation [18]. Due to the unsymmetric character of the occupation probabilities for the four sublattices [16] in a fcc alloy, the vacancy formation energy (a product of occupation probabilities) for Au sublattice sites ($\alpha$) is strongly order-dependent and increases with decreasing temperature. The vacancy formation energy in the Cu sublattices ($\beta$) is weakly order-dependent and smaller than the energy for vacancy formation in Au sublattice. Also, the variation of the vacancy formation energies with temperature will affect the interaction of vacancies and the antiphase domains. As an antiphase domain is a region of strong local disordering, a vacancy on a domain wall will find itself in a disordered region of lower energy, if it is an $\alpha$ vacancy, or, in a region of higher energy, if it is a $\beta$ vacancy. In other words, $\alpha$ vacancies will tend to get trapped and $\beta$ vacancies will tend to get away from the domain boundaries. For diffusion normal to a domain boundary, both types of vacancies will become less effective because either they will get trapped at the boundary or the vacancy will act as a barrier to vacancy motion within the three equivalent $\beta$ sublattices. Based on their analysis, the vacancies on the Au sublattice will be the most affected, the order-disorder transition being determined by the diffusion of Au atoms, according to the magnitudes of migration energy and the diffusion activation energy.

In $\beta$-brass, the critical temperature (Tc) for order-disorder transformation occurs at 736°K. It is reasonable to expect a change of diffusion behavior across Tc. Kuper et al. [19], Camagni [20,21], and Bassani et al. [22] performed diffusion measurements of Cu and Zn in $\beta$-brass at the critical temperature and over both sides of the critical temperature. Their results exhibit the same type of temperature dependence, namely a linear Arrhenius plot in the disordered phase and a curved log D - 1/T plot below Tc [15].

In Ni(3)Al intermetallic compounds, no experiment has been made to detect disorder below 1250°C [23]. Aoki and Izumi observed a maximum value of S (the long-range order parameter) at stoichiometric composition and symmetrically decreasing values of S at off-stoichiometric compositions [24]. Unlike the situation in NiAl [25], the substitution of Ni and Al sublattice sites in Ni(3)Al can occur at both Al-rich and Ni-rich off-stoichiometric compositions [3], and has been confirmed experimentally [24].

From the above information, it may be expected that the atomic diffusion paths of Ni and Al in Ni(3)Al are highly selective at all temperatures up to the melting point. The diffusion behavior of Ni and Al, if different, might be attributed largely to the variation of the long-range order parameter resulting from the composition change. The change in diffusion behavior associated with the variation of temperature would be expected to be minor.

## 3. DIFFUSION MECHANISM OF ORDERED ALLOYS

The mechanism that is most effective in explaining diffusion in ordered alloys is the so-called "six-jump cycle" mechanism proposed by Elcock and McCombie [26]. In such a cycle, a single vacancy migrates along a strictly defined path. In the first three migration paths, the atoms are displaced to the wrong sites, but these wrongly-displaced atoms are restored to the right positions in the next three consecutive moves and, consequently, a complete atomic order is maintained. This mechanism has been successfully applied to intermetallic compounds with B2 structure, e.g., AuCd [27,28], AgMg [29,30], AuZn [31], FeCo [32], and CoGa [33]. Based on this mechanism, Elcock [34] calculated the upper and lower bounds for the ratio (G) of the self-diffusion coefficients in a stoichiometric simple cubic binary alloy and in a bcc ordered alloy in the low temperature range ($0 < T/Tc < 0.5$), where Tc is the critical temperature of ordering. In this temperature range it is found that G has a value in the range 0.5-2.0 for simple cubic, stoichiometric ordered alloys, and in the range 0.67-1.5 for bcc ordered alloys, where G is only slightly temperature-dependent in both cases. Experimental evidence was obtained in self-diffusion for bcc Cu-Zn alloy with slightly off-stoichiometric composition [35]. A computer simulation of atomic movements based on the Monte Carlo method [36] indicates that the six-jump cycle actually is the predominant mechanism of the atomic movement in the highly ordered state [17].

It seems likely that such correlated atomic motion is responsible for diffusion in the L1(2) ordered structure. Two possible migration circuits were illustrated by Hancock [9], (see Fig. 1). Such cyclic diffusion mechanism could account for atomic diffusion of both nickel and aluminum. If Al is to diffuse readily in the Ni(3)Al lattice, it must move by such a cyclic mechanism, as any one Al atom has exclusive Ni atoms for the nearest neighbors. Consequently, a nearest-neighbor jump of an aluminum atom always involves disordering. Nickel atoms,on the other hand, have 4 aluminum atoms and 8 nickel atoms as nearest neighbors and could conceivably undergo exchange with any of the 8 near-neighbor nickel-atoms without producing disorder.

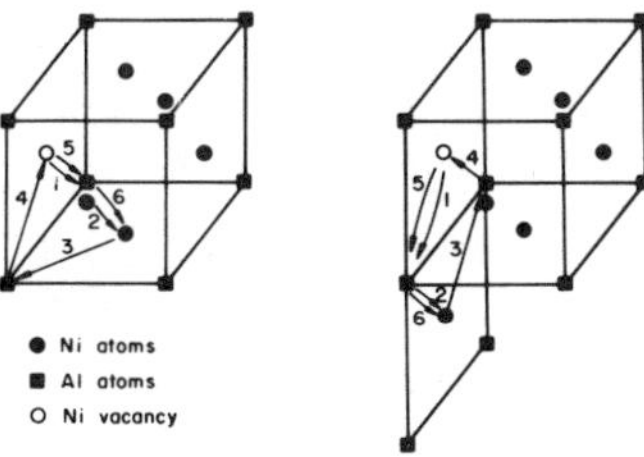


Fig. 1. Two six-jump diffusion cycles in Ni(3)Al.

Although the idea of six-jump cycle mechanism is reasonable if the atomic composition is stoichiometric and the degree of order is high, this mechanism is of limited applicability in practice [17]. Since the system is large, the thermodynamic equilibrium of the system can be obtained without the restoration of equilibrium of each unit process* as assumed. The disordering process of one unit jump at some place can be compensated by a reverse jump at another place in the system. In order to draw more general information from the theoretical calculation, the path-probability method (PPM) was proposed by Kikuchi and Sato [17], which does not include unknown approximations. Furthermore this method fully takes into account the correlation effects (the effect of the movement of other types of atoms which contribute to diffusion) that are essential to the six-jump cycle mechanism of diffusion suggested previously. However, such correlation effects were not considered in the work of Girifalco [37], Schoijet and Girifalco [38], and Cheng et al.[39]. Based on Kikuchi and Sato's treatment, the diffusion coefficients for ternary alloys of any degree of order by vacancy mechanism can be calculated. Also, the isotope diffusion coefficients in binary alloys (as functions of the degree of order) and the composition can be given by regarding the third atomic species as the isotope of the second atomic species.

## 4. GRAIN BOUNDARY DIFFUSION

It is well established that the mean jump frequency of an atom in a grain boundary is much higher than that in the lattice. The diffusivity is therefore higher in the grain boundary than in the grain [40-43]. In the past decade the study of grain boundary diffusion has been increased extensively because of the advance of thin-film technology for fabrication of microelectronic devices where a quantitative understanding of the mass transport at low temperature is important. By measuring the diffusion coefficients D' in various types of grain boundaries, it is possible to learn more about the structure of grain boundaries and how the atoms transport through them. Since the mathematical analyses and the sample preparation for grain boundary diffusion are far more complicated than those for volume diffusion, the progress in the understanding of the grain boundary diffusion is slow. To provide a brief background for grain boundary diffusion, we summarize some basic theoretical analyses and experimental methods for the measurement of D'.

### 4.1 Analysis and Modelling of Grain Boundary Diffusion

Grain boundary diffusion was first quantitatively analyzed by Fisher in 1951 [44]. Fisher's analysis, which is based on a constant concentration source model, is quite initiative but not exact. Subsequently, Whipple [45] solved the problem exactly by using the method of Fourier-Laplace transforms. An asymptotic form of grain boundary concentration was also given by using the method of steepest descent. Suzuoka [46] later treated the same problem by adopting an instantaneous source model which takes into account the variation of the concentration at contact surface with respect to time.

The only atomistic model for grain boundary diffusion is based on pipe diffusion along the dislocation cores [47]. The model is well demonstrated experimentally for low-angle boundaries, but is not applicable to high-angle boundaries where the dislocation cores overlap.

*A unit process is defined as a vacancy jump process starting and ending with the local environment of the vacancy perfectly ordered and giving rise to net displacement of atoms.

Several attempts have been made to extend the dislocation model to high-angle boundaries [48]. For misorientations greater than 15°, the dislocations are assumed to start agglomerating, and, at angles greater than 35°, they join together to form a slab of dislocation cores. The theory implies that the anisotropy of diffusion observed in low-angle boundaries should persist for angles up to 35°. Beyond it, the boundary should be isotropic. This was confirmed by experimental measurements [49-51]. To date, there is no quantitative theory of diffusion in high-angle and special boundaries, such as coincidence-site lattice boundaries [52,53].

### 4.2 Experimental Methods and Techniques

A proper approach to study grain boundary diffusion is to use bicrystals with the boundary normal to the free surface. There are four basic methods that are used to measure the grain boundary diffusivity.

(1) Contour angle method [54,55].
The values of D' can be calculated by measuring the angle $\phi$ between the boundary and the tangent to a contour of constant concentration intersecting at a boundary point.

(2) Penetration depth measurement [54].
The distance, y, from the contact interface to the boundary point at which the concentration of the diffusant vanishes is measured.

(3) Sectioning method [54].
The total amount of diffusant is determined in a number of thin slices cut parallel to the free surface. The integrated concentration, $\overline{c}$, at a depth, y, is determined as a function of y.

(4) Surface accumulation method [55].
The amount of atoms diffusing along the grain boundary from the source surface to the exit surface, where they spread out and accumulate, is measured with respect to time.

There are several techniques commonly used to measure the desired quantities: (a) autoradiography [56-62], (b) electron microprobe analysis [63], (c) radiotracer profiling [64], (d) surface analysis techniques, e.g., SIMS [55,65], AES [55,65,66], (e) other experimental techniques, e.g., macrographic etching technique [67,68], and analytical electron microscopy.

Although a number of review articles on grain boundary diffusion have been published [69-74], most of the available data are for pure metals [47,57-60,63,75-79], alloys [80-82], and ionic crystals [83-85]. To our knowledge, there are no published grain boundary diffusion data for the ordered intermetallic compounds, particularly compounds with L1(2) structure such as Ni(3)Al.

The following are some preliminary results of research on volume and grain boundary diffusion in Ni(3)Al alloys currently being conducted at Lehigh University.

## 5. RECENT EXPERIMENTAL RESULTS

### 5-1. Volume Diffusion

Volume diffusion studies were carried out on diffusion couples composed of pure Ni and Ni(3)Al compounds with different compositions and additives (B and Hf). The diffusion temperature ranges from 800°C to 1212°C. All measured data obey the Arrhenius relationship, i.e., a linear relation for $\ell n\ \overline{D}$ vs. 1/T (see Fig. 2). The results indicate that the diffusion coefficients are insensitive to the composition variations and to the addition of boron or hafnium. However, the activation energies calculated by the least-square fitting show that the diffusion couple of pure Ni and

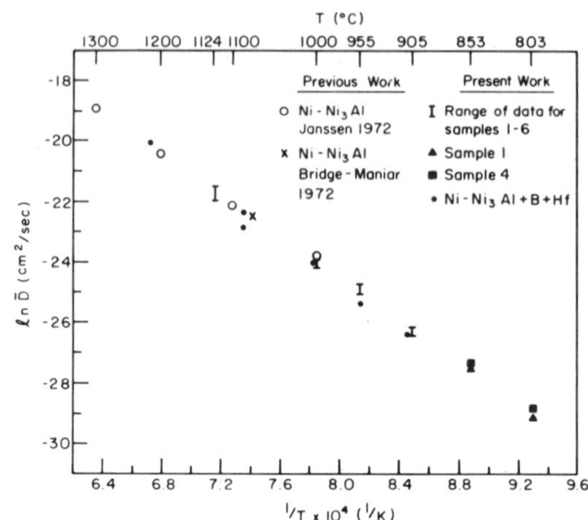


Fig. 2. Plot of $\ell n\,\overline{D}$ vs. 1/T for Ni-$Ni_3Al$ couples with alloying elements. T=1124°C. Sample (1) Ni-Ni24Al, (2) Ni-Ni25Al, (3) Ni-Ni26Al (4) Ni-Ni24Al0.2B, (5) Ni-Ni25Al0.2B, and (6) Ni-Ni26Al0.2B. A number preceding an element refers to its atomic percent.

Ni(3)Al at stoichiometric composition exhibits the highest activation energy (see Table I ).

Based on an assumed quardratic relationship between the layer thickness (x) of Ni-rich solid solution formed in the degraded Ni(3)Al region and the diffusion time (t), $x^2 = Kt$, the penetration constants (K) were measured from metallographic observations. The K values are close to those of the interdiffusion coefficients measured in the temperature range. A comparison between the plots of $\ell n\,\overline{D}$ vs. 1/T and $\ell n$ K vs. 1/T is shown in Fig. 3. It appears that the layer growth of Ni-rich solid solution follows parabolically with respect to diffusion time. This suggests that the layer growth is controlled by volume diffusion.

A linear experimental relationship between $\ell n\,\overline{D}_0$ and Q was also observed as shown in Fig. 4. This relationship can be expressed as $\overline{D}_0 = A \exp(Q/B)$, where $A = 9.0 \times 10^{-12}$ and $B = 2407.3$.

TABLE I. Measured Activation Energies and Preexponential Factors in Volume Diffusion of $Ni_3Al$ Alloys

| Diffusion Couple | Activation Energy,Q (kcal/mol) | Frequency Factor,D ($cm^2$/sec) |
|---|---|---|
| #1 Ni-Ni24Al* | 65.3 ± 3.3 | 5.5 ± 0.3 |
| #2 Ni-Ni25Al | 71.9 ± 3.6 | 86.6 ± 4.3 |
| #3 Ni-Ni26Al | 67.9 ± 3.4 | 17.0 ± 0.9 |
| #4 Ni-Ni24Al0.2B | 65.8 ± 3.3 | 7.0 ± 0.4 |
| #5 Ni-Ni25Al0.2B | 70.5 ± 3.5 | 45.0 ± 2.3 |
| #6 Ni-Ni26Al0.2B | 65.6 ± 3.3 | 5.8 ± 0.3 |
| Ni-Ni23.5Al0.2B0.5Hf | 71.5 ± 3.6 | 57.4 ± 2.9 |

*A number preceding an element refers to its atomic percent.

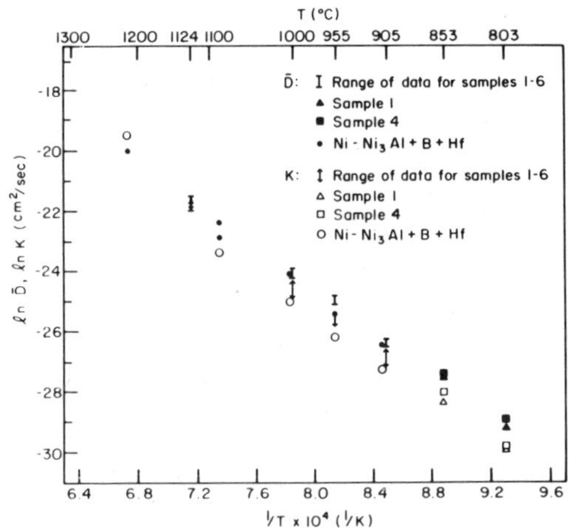


Fig. 3. Comparison between plots of $\ell n\,\overline{D}$ vs. 1/T and $\ell n$ K vs. 1/T.

## 5-2. Grain Boundary Diffusion

In addition to volume diffusion, grain boundary diffusion along certain specific grain boundary (Fig. 5a) was also observed on the metallographically etched sample (etching solution--a mixture of equal parts of $HNO_3$, $H_2SO_4$ and $H_3PO_4$). Since there is no preferential penetration along the coherent twin boundary (Fig. 5b), it is apparent that grain boundary diffusion would occur at high energy grain boundaries. The sharp V-shaped line (Fig. 5c) which intersects the grain boundary was identified to be a constant concentration contour. By measuring the angle, $\phi$, between the grain boundary plane and the tangent of the intersecting V line at the tip (the contour angle method), the grain boundary diffusion coefficient can be calculated using either Fisher's [44] or Whipple's [45] theory. Several grain boundary diffusion coefficients obtained by this method are presented in Table II and III.

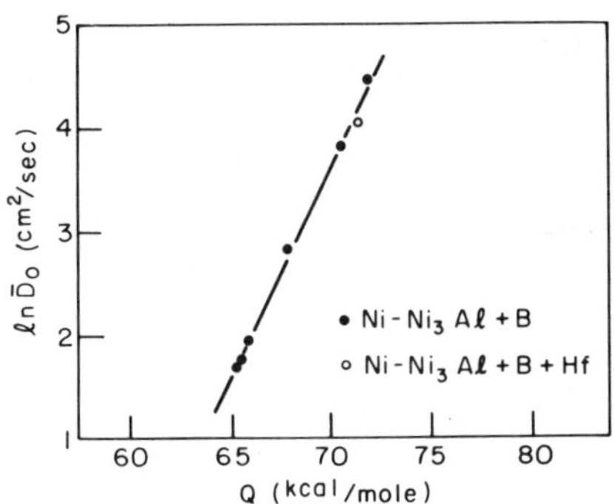


Fig. 4. Plot of $\ell n\,\overline{D}_0$ vs. Q.

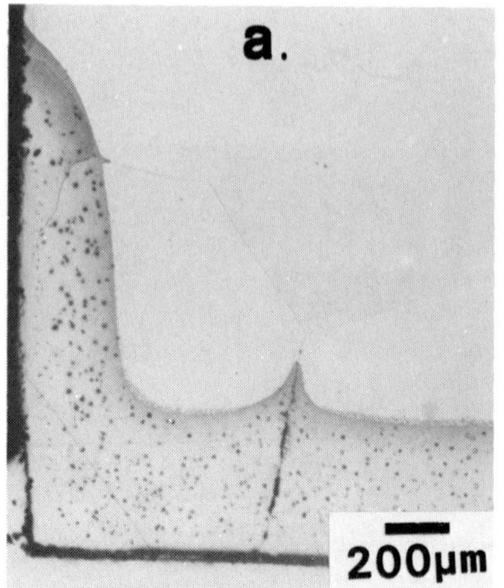


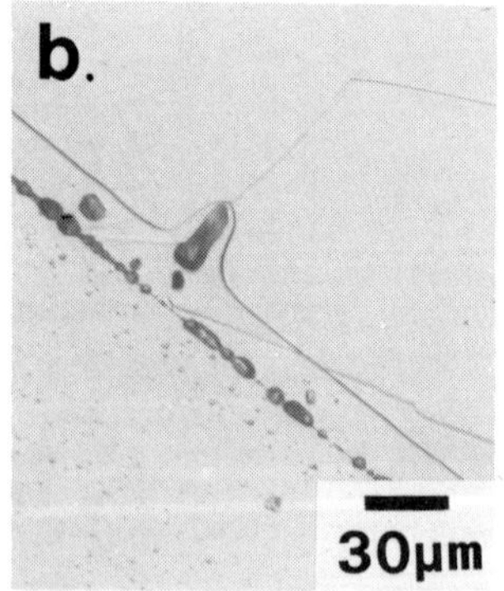


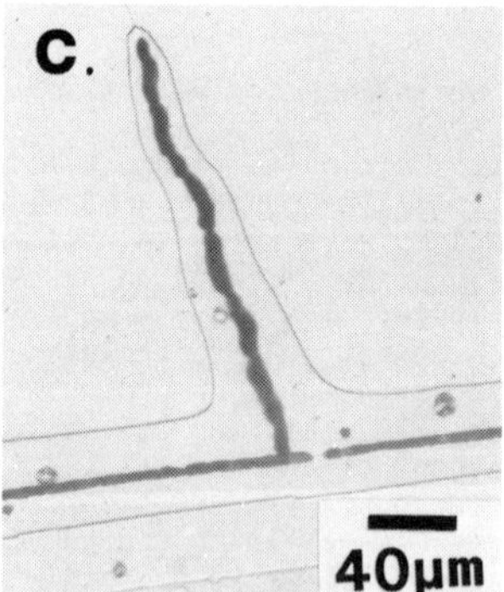


Fig. 5. Micrographs of Ni-$Ni_3Al$+B diffusion couple after diffusion anneal. (a) Relative scale of volume, grain boundary and surface diffusion, T=1212°C; (b) Lack of enhanced diffusion along a coherent twin boundary, T=1000°C; (c) V-shaped isoconcentration contour of a high-angle boundary, T=1000°C.

At 955°, 1000° and 1124°C, the grain boundary diffusivity is about five orders of magnitude higher than the volume diffusivity and the addition of boron (0.2 at.%) decreases the grain boundary diffusivity by about five times. Measurements of grain boundary diffusivity were also made by the sectioning method, using the well known expression [45,54,86],

$$D'\delta = \left(\frac{\partial \ \ell n \ \bar{c}}{\partial \ y^{6/5}}\right)^{-5/3} \left(\frac{4\bar{D}}{t}\right) (0.78)^{5/3}$$

TABLE II. Grain Boundary Diffusion Coefficients D' of Ni-Ni25Al Couples Measured by the Contour Angle Method

| T(°C) | D'δ(cm³/sec) | D'/$\overline{D}$ |
|---|---|---|
| 955 | $9.3 \times 10^{-14}$ (F)<br>$13.8 \times 10^{-14}$ (W) | $1.4 \times 10^{5}$ (F)<br>$2.1 \times 10^{5}$ (W) |
| 1000 | $3.9 \times 10^{-13}$ (F)<br>$3.7 \times 10^{-13}$ (W) | $1.8 \times 10^{5}$ (F)<br>$1.7 \times 10^{5}$ (W) |

F-based on Fisher's analysis
W-based on Whipple's analysis

TABLE III. Grain Boundary Diffusion Coefficients of Ni-Ni(3)Al+B Diffusion Couples. T = 1124°C.

| Couple | | Contour Angle Method | Sectioning Method |
|---|---|---|---|
| Ni-Ni24Al* | D'δ+ | (I) $1.0 \times 10^{-11}$ (F)<br>$1.1 \times 10^{-11}$ (W) | (I) $5.3 \times 10^{-12}$ |
| | | (II) $4.9 \times 10^{-12}$ (F)<br>$4.5 \times 10^{-12}$ (W) | (II) $4.9 \times 10^{-12}$ |
| | | (III) $3.4 \times 10^{-12}$ (F)<br>$4.1 \times 10^{-12}$ (W) | |
| | D'/$\overline{D}$ | (I) $6.0 \times 10^{5}$ (F)<br>$6.1 \times 10^{5}$ (W) | (I) $3.1 \times 10^{5}$ |
| | | (II) $2.9 \times 10^{5}$ (F)<br>$2.6 \times 10^{5}$ (W) | (II) $2.9 \times 10^{5}$ |
| | | (III) $2.0 \times 10^{5}$ (F)<br>$2.4 \times 10^{5}$ (W) | |
| Ni-Ni24Al0.2B | D'δ | $1.1 \times 10^{-12}$ (F)<br>$1.7 \times 10^{-12}$ (W) | $1.7 \times 10^{-12}$ |
| | D'/$\overline{D}$ | $6.2 \times 10^{4}$ (F)<br>$9.8 \times 10^{4}$ (W) | $9.6 \times 10^{4}$ |

*A number preceding an element refers to its atomic percent.
+In units of cm³/sec.
I, II, III represent three different grain boundaries.
F-based on Fisher's analysis, W-based on Whipple's analysis.

where $D'$ = grain boundary diffusivity (cm²/sec)
$\overline{D}$ = volume diffusivity (cm²/sec)
$\overline{c}$ = integrated concentration at a distance y from the interface
t = diffusion annealing time (sec)
$\delta$ = grain boundary thickness (cm)

It is seen that the agreement between the measured grain boundary diffusivity data by using the sectioning method and the angle measurement method is good (see Tables II and III). The measured data also show that the effect of boron doping is measurable.

## 6. DISCUSSION

From the present study, it seems that the interdiffusion coefficients obtained from Ni-Ni(3)Al+B and Ni-Ni(3)Al+B+Hf couples are insensitive to the composition variation in the temperature range studied. However, the calculation of activation energies does indicate that the diffusion couple of pure Ni with stoichiometric Ni(3)Al(B) alloys exhibits the highest total energy for vacancy formation and migration. This is consistent with the previous conclusion [24] that the long-range order parameter S of Ni(3)Al compounds is composition dependent, being maximum at the stoichiometric composition.

The above results differ from those of the Ni self-diffusion in Ni(3)Al alloys [9]. The disagreement can be justified by considering the difference of the environments of Ni and Al atoms experienced during the interdiffusion and the self-diffusion process. For the self-diffusion of Ni in Ni(3)Al ordered compounds, each Ni atom could have 8 different diffusion paths to take without causing any local disorder. However, each Al atom, being exclusively surrounded by 12 Ni atoms, will take only very restricted diffusion paths because any near-neighbor diffusion jump will result in disorder. Under such circumstances, it is reasonable to expect that the self-diffusion of Ni atom in Ni(3)Al will be the least one to be affected by ordering, and, consequently, the compositional variations will only exert very small or negligible influence on the self-diffusion of Ni, as observed in Hancock's study. On the other hand, the self-diffusion of Al atoms in Ni(3)Al would be the one most affected by the ordering. In our study, the composition dependence is related to the behavior of interdiffusion which takes into account not only the diffusion behavior of Ni atoms but also the diffusion behavior of Al atoms. Therefore, the strong ordering dependence, i.e., the composition dependence of diffusion behavior caused by Al atoms, is expected to come into play. This explains the dependence of activation energy on composition in interdiffusion.

The ratio between intrinsic diffusion coefficients of Al and Ni, $D_{Al}/D_{Ni}$, can be calculated by the equation of van Loo [88,89], assuming that during the diffusion process there is no porosity generation and the molar volume does not change appreciably.

Following the analysis of van Loo, it is easy to locate the marker interface. In most of the couples used in our study, the marker interface coincides with the Matano interface. However, in some couples the marker interface moves toward the side of Ni(3)Al component with respect to the Matano interface. In such cases, Al atoms are identified to be the faster diffusion species, i.e., $D_{Al}/D_{Ni} > 1$. This is consistent with Janssen's observation [12] which he concluded from the position of porosities. Assuming that the partial molar volume of Ni is comparable to that of Al at the Kirkendall interface, the intrinsic diffusion coefficient of Al is 2 to 13 times greater than that of Ni over the temperature range studied.

The relationship between $\ell n\overline{D}_0$ and Q is generally recognized to be linear [90-94] and can be represented by $\ell n\overline{D}_0 = a + bQ$. The coefficients a and b for the interdiffusion in Ni-base alloys [92-95] have been reported to be a = - 4.2 to -18.4 and b = 0.09 to 0.27 x $10^{-3}$, with $\bar{D}_0$ in units of $cm^2/sec$ and Q in cal/mole. For the self-diffusion of Ni in Ni(3)Al measured by Hancock [9], a = -30.4 and b = 0.44 x $10^{-3}$. In the present study, a linear relationship between $\ell n\bar{D}_0$ and Q is also obtained (Fig. 4), given by $\ell n\bar{D}_0$ ($cm^2/sec$) = 0.42 x $10^{-3}$ Q(cal/mol)-25.4 with a slope very close to the value obtained from the self-diffusion measurement [9].

The study of Ni-rich solid solution layer growth during diffusion has been carried out by measuring the layer thickness from the marker interface to the phase boundary on the Ni(3)Al side. The values measured on different orientated grains are essentially the same, which suggests that the volume diffusion of Ni into Ni(3)Al is insensitive to the orientation of the grain, as expected in cubic metals [67]. The penetration constants (K), based on the parabolic relationship between the layer thickness and the diffusion time, were also calculated over the the temperature range studied. The nearly-equal values of K and $\bar{D}$ suggest that degradation of Ni(3)Al controls the process of volume diffusion. Similar penetration constants have been calculated based on the layer thickness measured from the marker interface to the pure Ni region. However, no relationship can be correlated between K and $\bar{D}$. This may be partially explained as follows. As the migration of Al atoms in Ni(3)Al compound is highly restricted, for Ni(3)Al to be degraded into Ni-rich solid solution, a major portion of the diffusion time would be spent on vacating the aluminum sites so that the exchanging with Ni atoms might occur. If that is the situation, the values of K and $\bar{D}$ would be compatible.

In the grain boundary studies, the agreement between the data measured by the sectioning method and the contour angle method, suggests that the latter provides a relatively simple and accurate alternative for the measurement of grain boundary diffusion in intermetallic compounds. The earliest experimental evidence of showing preferential grain boundary diffusion was demonstrated by Barnes [67] on the Cu-Ni system. He suggested that the apex angle of the concentration contour at a grain boundary may be used as a measure of the relative rates of diffusion along the grain boundary and through the crystal lattice. However, the V-shaped concentration contours, reported in his study were rather diffuse at the apex and had not been experimentally determined as a set of isoconcentration contours. In our study of Ni(3)Al alloys, the V-shaped concentration contour is sharp, due to the thermodynamic restriction, and has been identified as a constant concentration contour by the electron probe microanalysis. A quantitative treatment for determining the ratio of the actual diffusion coefficients for grain boundary and for volume diffusion has been developed by LeClaire [87] based on Fisher's analysis [44]. However, practical application of this technique is still limited due to the experimental difficulty in revealing sharp concentration contours.

Based on the present study, the addition of 0.2 at.% boron in the Ni(3)Al alloys, does not seem to have any appreciable effect on the interdiffusion coefficients. Similarly, little effect was observed when 0.5 at.% Hf was added to the alloys. On the other hand, a dragging effect, induced by the addition of boron, was observed from the measured grain boundary diffusivities. The grain boundary diffusion coefficients measured from boron-doped Ni(3)Al compound are 3 $\sim$ 5 times smaller than those measured from the boron-free Ni(3)Al compound at 1124°C. This observation should not be considered seriously, however, for the boundaries under consideration may not be of the same structure. At present, it can be concluded that for the diffusion of Ni in Ni(3)Al alloys, the grain boundary diffusion coefficient is about 5 $\sim$ 6 orders of magnitude greater than the interdiffusion coefficients over the temperature range studied.

## Acknowledgments

The authors wish to thank Dr. C. T. Liu and Dr. J. I. Goldstein for helpful discussions. The work was sponsored by the Office of Energy Systems Research, Energy Conversion and Utilization Technologies (ECUT) Program, U.S. Department of Energy, under subcontract 19X-43367C with Martin Marietta Energy Systems, Inc.

REFERENCES

1. N. A. Stolwijk, M. van Gend and H. Bakker, Phil. Mag. A42, 783 (1980).
2. A. Gangulee, P. S. Ho and K. N. Tu, eds. "Lower Temperature Diffusion and Applications to Thin Films," (Elsevier, Lausanne), 1975.
3. C. T. Liu and J. O. Stiegler, Science, 226, 636 (1984).
4. K. Aoki and O. Izumi, Trans. J.I.M., 19, 203 (1978).
5. S. M. Copley and B. H. Kear, Trans. AIME, 239, 977 (1967).
6. K. Aoki and O. Izumi, J.J.I.M., 43, 1190 (1979).
7. C. T. Liu and C. C. Koch, NBSIR, 83-2679-2 (1983).
8. L. N. Larikov, T. K. Yatsenko, L. F. Chernaya and L. M. Kumok, in "Mobility of Atoms in Crystal Lattices," edited by V. N. Svechnikov, (Keter Press, Jerusalem), 1970, pp. 55.
9. G. F. Hancock, Phys. stat. sol. (a), 7, 535 (1971).
10. M. B. Bronfin, G. S. Bulatov and I. A. Drugova, Fiz. Metal. Metalloved., 40, 363 (1975).
11. L. N. Larikov, in "Diffusion Processes in Ordered Alloys," edited by L. N. Larikov, V. V Geichenko and V. M. Fal'chenko, (Oxonian Press, New Delhi), 1981, pp. 118.
12. M. M. P. Janssen, Met. Trans. 4, 1623 (1972).
13. M. M. P. Janssen and G. D. Rieck, Trans. AIME, 239, 1372 (1967).
14. J. E. Bridge, Jr. and G. N. Maniar, Met. Trans., 3, 1005 (1972).
15. L. A. Girifalco, in "Diffusion" (ASM, Metals Park, Ohio), 1973, pp. 185.
16. M. Schoijet and L. A. Girifalco, J. Phys. Chem. Solids, 29, 911 (1968).
17. R. Kikuchi and R. Sato, J. Chem. Phys., 51, 161 (1969).
18. J. Kirkwood, J. Chem. Phys., 6, 70 (1938).
19. A. B. Kuper, J. R. Manning and C. T. Tomizuka, Phys. Rev., 104, 1536 (1956).
20. P. Camagni, Proc. 2nd Intern. Conf. on Atomic Energy, Geneva, 20, 11365 (1958).
21. P. Comagni, Studia Ghisleriana, Ser. IV, Vol. II, 239 (1959).
22. C. Bassani, P. Camagni and S. Pace, Nuovo Cimento, Ser. X, 19, 393 (1961).
23. D. P. Pope and J. L. Garin, J. Appl. Cryst., 10, 14 (1977).
24. K. Aoki and O. Izumi, Phys. Stat. Sol.(a), 32, 657 (1975).
25. R. W. Guard and J. H. Westbrook, Trans. AIME, 215, 807 (1959).
26. E. W. Elcock and C. W. McCombie, Phys. Rev. 109, 605 (1958).
27. H. B. Huntington, N. C. Miller and V. Nerses, Acta Met. 9, 749 (1961).
28. D. Gupta, D. Lazarus and D. S. Lieberman, Phys. Rev., 153, 963 (1967).
29. H. A. Domian and H. I. Aaronson, Trans. AIME, 230, 44 (1964).
30. P. Wynblatt, Acta Met., 15, 1453 (1967).
31. D. Gupta and D. S. Lieberman, Phys. Rev. B 4, 1070 (1971).
32. S. G. Fishman, D. Gupta and D. S. Lieberman, Phys. Rev. B 2, 1451 (1970).
33. A. Bose, G. Frohberg and H. Wever, Phys. Stat. Sol.(a), 52, 509 (1979).
34. E. W. Elcock, Proc. Phys. Soc. (London), 73, 250 (1959).
35. A. B. Kuper, D. Lazarus, J. R. Manning and C. T. Tomizuka, Phys. Rev., 104, 1536 (1956).
36. W. M. Young and E. W. Elcock, Proc. Phys. Soc. (London), 89, 735 (1966).
37. L. A. Girifalco, J. Phys. Chem. Solids, 25, 323 (1964).
38. M. Schoijet and L. A. Girifalco, J. Phys. Chem. Solids, 29 481, 497 (1968).
39. C. Y. Cheng, P. M. Wynblatt and J. E. Dorn, Acta Met., 15 1035, 1045 (1967).

40. R. Smoluchowski, in "Imperfections in Nearly Perfect Crystals," edited by W. Shockley, J. H. Hollomon, R. Maurer, and F. Seitz (John Wiley, New York), 1952, pp. 451.
41. L. Slifkin, D. Lazarus and C. T. Tomizuka, Phys. Rev., 93, 973 (1954).
42. A. Z. Austin and N. A. Richard, J. Appl. Phys., 33, 3569 (1962).
43. T. Suzuoka, J. Phys. Soc. Japan, 19, 839 (1964).
44. J. C. Fisher, J. Appl. Phys., 22, 74 (1951).
45. R. T. P. Whipple, Phil. Mag., 45 1225 (1954).
46. T. Suzuoka, Trans. J.I.M., 2, 25 (1961).
47. D. Turnbull and R. E. Hoffman, Acta Met., 2, 419, (1954).
48. J. C. M. Li, J. Appl. Phys., 32, 525 (1961).
49. R. E. Hoffman, Acta Met., 4, 97 (1956).
50. S. R. L. Couling and R. Smoluchowski, J. Appl. Phys., 25, 1538 (1954).
51. M. Jurisch, Ph.D. Thesis, Freiburg, Sachsen, 1969.
52. V. I. Arkharov and A. A. Pentina, Fiz. Metall. i Metalloved 5, 68 (1957).
53. I. Herbeuval and M. Biscondi, C. R. Acad. Sci. Paris, C273, 1416 (1956).
54. A. D. LeClaire, Brit. J. Appl. Phys., 14, 351 (1963).
55. J. C. M. Hwang and R.W. Balluffi, J. Appl. Phys., 50, 1339 (1979).
56. T. J. Renouf, Phil. Mag., 9, 781 (1964).
57. W. R. Upthegrove and M. J. Sinnott, Trans. Am. Soc. Metals, 50, 1031 (1958).
58. S. Yukawa and M. J. Sinnott, Trans. AIME, 203, 996 (1955).
59. R. F. Canon and J. P. Stark, J. Appl. Phys., 40, 4361 (1969).
60. R. F. Canon and J. P. Stark, J. Appl. Phys., 40, 4366 (1969).
61. A. M. Huntz and P. Lacombe, Can. Met. Quart., 13, 155 (1974).
62. D. Gupta and D. R. Campbell, Phil. Mag. A, 42, 513 (1980).
63. R. Butz, W. Erley and H. Wagner, Phys. Stat. Sol.(a), 7, K5 (1971).
64. Y. Adda and J. Philibert, La Diffusion Dans Les Solides, (Presses Universitaires De France, Paris), 1966, vol. II, pp. 1.
65. W. Gust, M. B. Hintz, A. Lodding, H. Odelius and B. Predel, Acta Met., 30, 75 (1982).
66. D. Gupta and R. T. C. Tsui, Appl. Phys. Lett., 17, 294 (1970).
67. R. N. Barnes, Nature, 166, 1032 (1950).
68. J. Herbeuval, M. Biscondi and C. Goux, Mem. Sci. Rev. Met., 70 39 (1973).
69. R. W. Balluffi, Phys. Stat. Sol., 42, 11 (1970).
70. H. Gleiter and B. Chalmers, Prog. Mater. Sci., 16, 77 (1972).
71. N. A. Gjostein, Short Circuit Diffusion in Diffusion in "Diffusion," (ASM, Metals Park, Ohio), 1973, pp. 241.
72. G. Martin and B. Perraillon, J. de. Physique Colloque, C4, 165 (1975); in "Grain Boundary Structure and Kinetics," (ASM, Metals Park, Ohio), 1979, pp. 239.
73. D. Gupta, D. R. Campbell and P. S. Ho, in "Thin Films-Interdiffusion and Reactions," edited by J. M. Poate, K. N. Tu and J. W. Mayer (John Wiley, New York), 1981, pp. 161.
74. H. Bakker, in "Diffusion in Metals and Alloys," edited by F. J. Kedves and D. L. Beke (Trans Tech Publications, Tihany, Hungary), 1983, pp. 266.
75. A. R. Wazzan, J. Appl. Phys., 36, 3596 (1965).
76. D. Gupta and K. K. Kim, J. Appl. Phys., 51, 2066 (1980).
77. B. Okkerse, Acta Met., 2, 551 (1954).
78. R. E. Hoffman and D. Turnbull, J. Appl. Phys., 22(5), 634 (1951).
79. J. P. Stark and W. R. Upthegrove, Trans. ASM, 59, 479 (1966).
80. D. Gupta, Phil. Mag., 33, 189 (1976).
81. P. Guiraldencq and P. Poye, Mem. Scient. Rev. Met., 10, 715 (1973).
82. A. F. Smith, Met. Sci., 9, 425 (1975).
83. V. N. Kaygorodov, Y. N. Rabovskiy and V. K. Talinskiy, Fiz. Met. Metall., 24, 117 (1967).
84. J. Cabane, J. Chem. Phys., 59, 1165 (1962).

85. V. N. Kaygorodov, Y. N. Rabovskiy and V. K. Talinskiy, Fiz. Metall., 24, 661 (1967).
86. H. S. Levine and C. J. MacCallum, J. Appl. Phys., 31, 595 (1960).
87. A. D. LeClaire, Phil. Mag., 42, 468 (1951).
88. F. J. J. van Loo, Acta Met., 18, 1107 (1970).
89. G. F. Bastin and G. D. Rieck, Met. Trans., 5, 1827 (1974).
90. G. B. Fedorov, Author's Summary of Candidate's Thesis--Moskovskii Mekhanicheskii Institute, 1952.
91. G. B. Fedorov, in "Mobility of Atoms in Crystal Lattices," edited by V. N. Svechnikov, (Keter Press, Jerusalem), 1970, pp. 28.
92. T. Yamamoto, T. Takashima and K. Nishida, Trans. J.I.M., 21, 601 (1980).
93. T. Yamamoto, T. Takashima and K. Nishida, J.J.I.M., 43, 1196 (1979).
94. M. Yokota, R. Harada and H. Mitani, J.J.I.M., 43, 793 (1979).
95. Y. Iijima and K. Hirano, J.J.I.M., 35, 511 (1971).

# HIGH TEMPERATURE OXIDATION OF β-NiAl

JOSEPH K. DOYCHAK, T. E. MITCHELL
Dept. of Metallurgy and Matl's Sci., Case Western Reserve University
Cleveland, OH 44106
J. L. SMIALEK
NASA Lewis Research Center, Cleveland, OH 44135

## ABSTRACT

The oxidation of single crystal β-NiAl has been studied primarily using electron microscopy. Oriented metastable $Al_2O_3$ phases form during transient oxidation at 800°C. Specific orientation relationships exist on all metal orientations studied and are a result of the small mismatch along aligned close-packed directions in the cation sublattices of the metal and oxide. Transformation of the metastable $Al_2O_3$ phases at 1100°C results in an oxide morphology described as the "lacey" structure of α-$Al_2O_3$ scales. This structure results from impingement of oriented patches of α-$Al_2O_3$ as the transformation initiates and moves radially parallel to the surface. Scale growth occurs by diffusion along high angle grain boundaries. A drastic reduction in oxidation rate accompanies the change in oxide morphology.

## INTRODUCTION

The increasing interest in the use of intermetallic materials at high temperatures suggests that aluminides be utilized because of their excellent oxidation resistance. β-NiAl is known to be one of the most oxidation resistant materials in use at high temperatures. For that reason, nickel-aluminide is used as a coating material for components subjected to severe atmospheric oxidizing environments as well as a prospective structural material for such components. The oxidation resistance of these materials is achieved by the formation of an $Al_2O_3$ scale. Because of the high aluminum concentration in the alloys, $Al_2O_3$ is easily formed and maintained at all applicable temperatures [1]. The scales also possess a slow growth rate and are usually quite adherent when oxygen active elements are added to the alloy as dopants or oxide dispersions [2,3].

Common to most alumina formers are the details involving the formation and growth of $Al_2O_3$ scales. Upon exposure to the atmosphere at temperature, oxides of all alloy components form on the surface and grow according to kinetic considerations [4]. Because it is more thermodynamically stable, $Al_2O_3$ forms a healing layer at the metal-oxide interface. Further thickening of the scale involves growth of $Al_2O_3$ by diffusion of mobile species through the oxide layer.

During oxidation, the $Al_2O_3$ scale undergoes a transformation from a transient state to a mature state [5]. In the transient stage, metastable $Al_2O_3$ phases exist within the scale and grow by outward cation diffusion. These phases are based on the spinel structure and include: γ-$Al_2O_3$, a defective spinel; δ-$Al_2O_3$, a tetragonal form of spinel having a triple spinel block as a unit cell; θ-$Al_2O_3$, a monoclinic phase. When the mature stage is reached, the oxide phase is α-$Al_2O_3$ having the corundum structure. The order of transformation with temperature is $\gamma \rightarrow \delta \rightarrow \theta \rightarrow \alpha$ [6]. Prior to formation of α-$Al_2O_3$, cation ordering with slight lattice distortions are

Mat. Res. Soc. Symp. Proc. Vol. 39. 

involved in the phase changes. When the transformation to $\alpha$-$Al_2O_3$ occurs, the anion sublattice restructures into an hexagonal lattice [7]. Associated with this transformation to $\alpha$-$Al_2O_3$ is an approximate 14% decrease in volume [8].

Common to $\beta$-NiAl are the interesting mature oxide morphologies associated with slow growth rates. In this study, the characteristics of $Al_2O_3$ formation and growth using $\beta$-NiAl as a substrate have been determined. Particular attention is paid to the effect of orientation of the substrate on transient oxides and the resultant mature scale morphology.

## EXPERIMENTAL PROCEDURE

A casting of $\beta$-NiAl containing 0.1 wt% Zr was purchased from TRW, Inc. The casting consisted of large(10-20 mm) grains of textured material. Single crystal specimens were obtained from large grains that were oriented using the Laue technique to the desired orientation. Polycrystalline specimens were obtained from buttons of the casting where equiaxed grains were 0.2-0.8 mm in size. Specimens were spark machined to appropriate dimensions for transmission electron microscopy or thermogravimetric oxidation weight gain analysis. The specimens were then mechanically polished through 3 micron diamond and electropolished in a 2:1 methanol:nitric acid solution at 12 V and -20°C.

The single crystal metal orientations studied were (001), (012), (011) and (111). The oxidation of the electropolished specimens was performed in a laboratory furnace in air at 800°C or 1100°C. Oxidation times ranged from 0.1 to 100 hours. All oxidation treatments were performed isothermally. Transmission electron microscopy(TEM) specimens were backthinned and studied in a Philips 400T analytical electron microscope. Thermogravimetric(TGA) studies were performed using a Cahn 1000 electrobalance. Because the oxidation treatments were isothermal, no effect was noticed from the addition of zirconium which improves the oxide adherence according to proposed mechanisms [9].

## RESULTS AND DISCUSSION

Due to the distinct differences between the nature of the transient and mature scales, the results will be presented in two parts. The first section deals with the transient stages at 800°C followed by a treatment of the mature scales formed at 1100°C.

### Transient Oxidation, 800°C

After 0.1 hours of oxidation, nickel-aluminate spinel, $NiAl_2O_4$, was the main oxide phase on all metal orientations. This result was confirmed by electron diffraction using the $\beta$-NiAl substrate as an internal calibration. Also, a high relative nickel concentration in the oxide scale was detected by energy dispersive spectroscopy(EDS). Figure 1 shows diffraction patterns of the oxide scales from all four metal orientations after 0.1 hours oxidation. In some cases, diffraction patterns of the oxide overlapping thin metal regions are shown.

For an (001) substrate orientation(Figure 1a), the oxide is highly

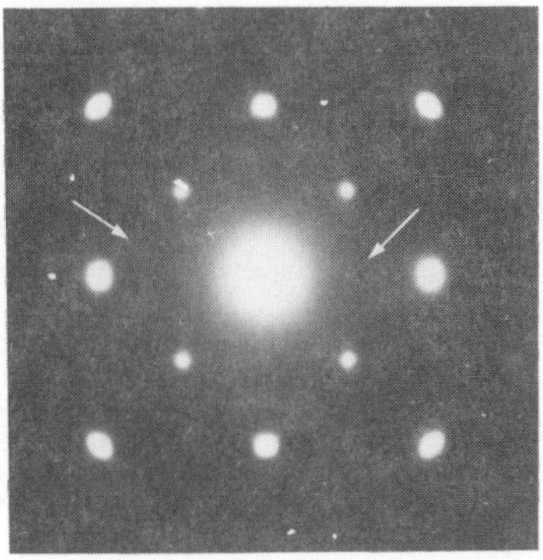

a)

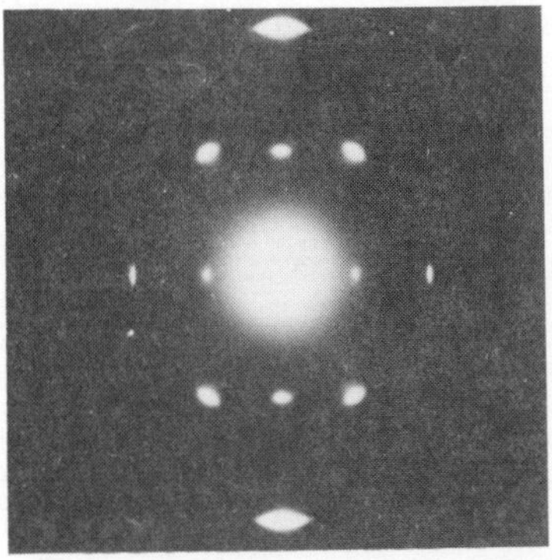

b)

c)

d)

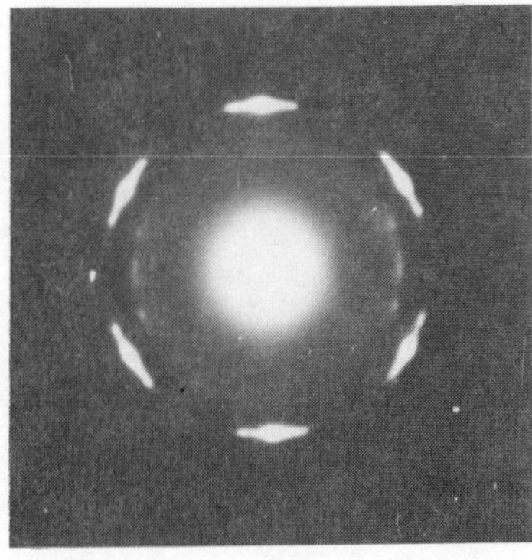

e)

Figure 1
Electron diffraction patterns of oxide scales formed on oriented β-NiAl at 800°C for 0.1 hours of oxidation; a) oxide on an (001) orientation with diffuse intensity indicated by arrows, b) oxide on an (012) metal orientation, c) (011) metal and overlapping oxide showing the N-W and another orientation relationship, d) (011) metal and overlapping oxide showing the K-S orientation relationship, e) oxide on a (111) metal orientation.

textured. The orientation relationship between the metal and oxide is the classical Bain relationship; a common orientation relationship between BCC and FCC materials given by: (001)m || (001)ox, [100]m || [110]ox.

Another prominent feature in this diffraction pattern is the diffuse intensity indicated by arrows. This could arise from another oxide phase, probably a metastable phase of $Al_2O_3$ in its early stages of formation. A higher relative concentration of aluminum was found within the scale as compared to scales formed on other orientations on which only $NiAl_2O_4$ was observed, and suggests strongly that an $Al_2O_3$ phase was present. On (012) metal orientations, the oxide scale is similar in phase and composition to that formed on (001) orientations (Figure 1b). The orientation relationship between the metal and oxide for an (012) substrate orientation is given by: (012)m || $(\bar{1}12)$ox, [100]m || [110]ox.

The oxide that forms on (011) metal orientations after 0.1 hours appears to be only $NiAl_2O_4$. Three different orientation relationships were observed for this condition. Figure 1c shows two orientation relationships between the metal and oxide, one being the Nishyama-Wassermann relationship: (011)m || $(\bar{1}11)$ox, [100]m || [110]ox; and another orientation relationship given by: (011)m || $(\bar{1}11)$ox, $[21\bar{1}]$m || $[01\bar{1}]$ox. Figure 1d shows the third relationship which is the Kurdjumov-Sachs orientation relationship: (011)m || $(\bar{1}11)$ox, $[11\bar{1}]$m || $[01\bar{1}]$ox.

Just as in the case of the similarities between oxides formed on (001) and (012) substrate orientations, the same phase and composition exist for the oxides formed on (011) and (111) metal orientations. However, the metal-oxide orientation relationship on (111) metal faces is slightly more complex. Three variants of the orientation relationship exist mirroring the threefold symmetry of the (111) metal planes (Figure 1e). This orientation relationship is given by: (111)m || near (021)ox, $[1\bar{1}0]$m || [100]ox.

The microstructures of the oxides formed after 0.1 hours at 800°C consist of highly textured subgrains in the range 20-50 nm in size [10]. The distribution of subgrain orientations reaches a maximum at the angular values which are suggestive of the orientation relationships for each condition [11].

With increasing times at 800°C, the strong degree of preferred orientation persists on all metal orientations. The orientation relationships between the metal and oxide remain the same. However, only the Nishyama-Wassermann orientation relationship persists after 1.0 hours oxidation on (011) substrates. The orientation relationships have in common particular parallel directions which minimize the mismatch between cation sublattices in the metal and oxide phases. These directions are the close-packed directions in each phase and involve only a 3% mismatch in atom spacings when properly oriented as in the orientation relationships [12]. This small mismatch accounts for the strong degree of epitaxy which is maintained throughout the transient stages.

After 1.0 hours of oxidation, the relative aluminum concentration in the scale has increased for all metal orientations, suggesting that only Al containing oxides are forming. This indicates that a healing layer of $Al_2O_3$ has formed. The phases that comprise the healing layer are still metastable $Al_2O_3$ phases with the kinetics of formation being a function of metal orientation. On (001) and (012) substrate orientations, $\delta$-$Al_2O_3$ is evident after 1.0 hours and becomes the predominant oxide phase at 10.0 hours of oxidation. The presence of $\delta$-$Al_2O_3$ is indicated by superlattice reflections at one-third positions in the spinel reciprocal lattice in the diffraction pattern of Figure 2a. On (011) orientations, $\delta$-$Al_2O_3$ is evident after 1.0 hours but does not persist. Instead, $\gamma$-$Al_2O_3$ becomes the predominant oxide phase (Figure 2b). On (111) orientations, only $\gamma$-$Al_2O_3$ is observed for these longer oxidation times.

According to findings for bulk alumina, $\gamma$-$Al_2O_3$ is a lower temperature phase than $\delta$-$Al_2O_3$ [6]. However, epitaxial restraints on these growing oxide scales effect the kinetics of metastable $Al_2O_3$ phase development.

Either $\delta$-$Al_2O_3$ formation is enhanced on (001) and (012) orientations or suppressed on (011) and (111) orientations. The latter appears to be the case because at 800°C, $\delta$-$Al_2O_3$ is thermodynamically the more stable phase.

Growth rates of the oxide scales at 800°C were obtained by TGA measurements. In addition to single crystal measurements on (001) and (011) substrate orientations, polycrystalline specimens were also used. The results are shown in Figure 3. As is immediately evident, the range of possible error is too large to determine if any differences in growth rate exist for the different orientations. The large scatter is a result of mechanical instabilities within the electrobalance system. However, the parabolic growth rate values obtained are in the range, kp = 1.2-1.5 X $10^{-4}$ $mg^2/cm^4/hr$ (2.8-14.4 X $10^{-6}$ $kg^2/m^4/s$).

## Mature Scales, 1100°C

At ambient pressures, a transient stage of oxidation always exists prior to the formation of $\alpha$-$Al_2O_3$. The extent of this transient stage depends on many variables, the most important being temperature and alloy composition. Therefore, there will always be a preexisting oxide scale composed of transient phases such as those described in the first section, before any $\alpha$-$Al_2O_3$ forms. At the present time, the exact nature of the transformation that occurs within the scale is not certain. However, the initial and final states are known.

In this study, the initial state was achieved by growing transient scales at 800°C for 100 hours. Specimens were then removed and later placed in a furnace at 1100°C as the growth kinetics were measured. The parabolic weight gain versus time curve is shown in Figure 4 for the 1100°C oxidation. The curve consists of an initial steep portion which levels off after 3-5 hours into a range in which slow parabolic growth kinetics exist. The initial portion corresponds to transient oxidation at 1100°C. Combining the value of the parabolic rate constant at 1100°C for transient oxidation,

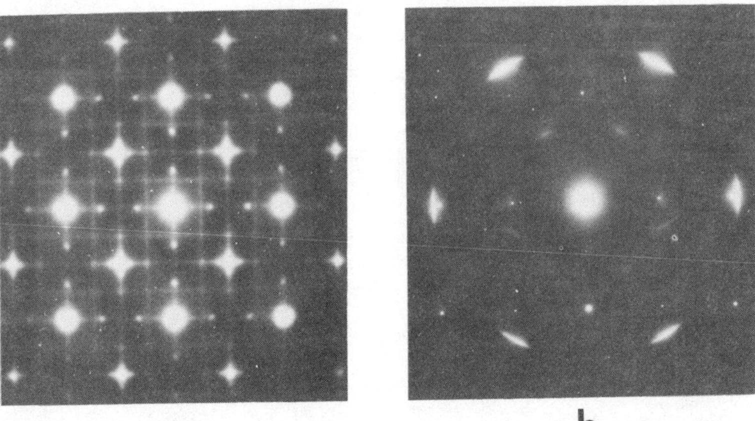


Figure 2 Electron diffraction patterns taken after 10.0 hours of oxidation at 800°C of a) $\delta$-$Al_2O_3$ scale on an (001) metal orientation showing superlattice reflections, b) $\gamma$-$Al_2O_3$ on an (011) metal orientation showing only the N-W orientation relationship.

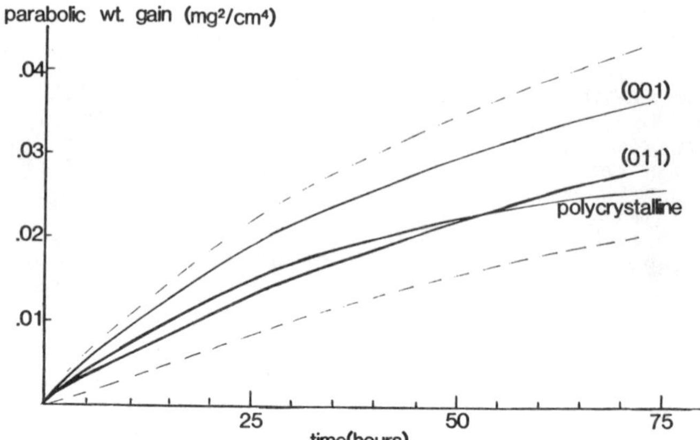


Figure 3 Parabolic oxidation weight gain versus time for 800°C oxidation of β-NiAl. The dashed lines indicate the range of scatter within the data.

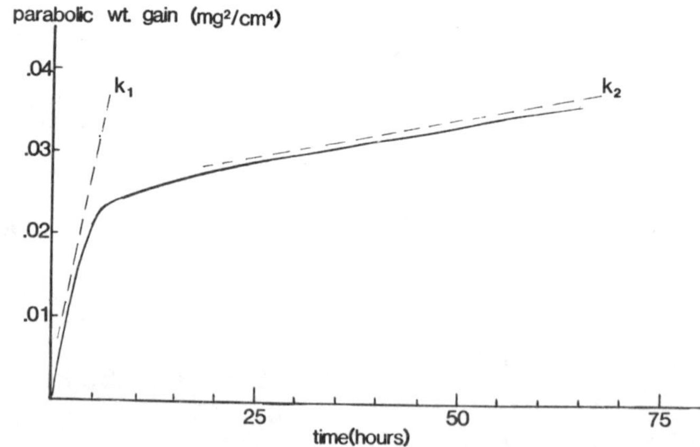


Figure 4 Parabolic oxidation weight gain versus time for polycrystalline β-NiAl at 1100°C. The initial slope, $k_1$, corresponds to the transient stage whereas the final slope, $k_2$, corresponds to the mature oxidation stage.

kp = 6.6 X $10^{-3}$ $mg^2/cm^4/hr$ (1.8 X $10^{-4}$ $kg^2/m^4/s$), with that at 800°C from the previous section, an activation energy of 1.25 X $10^5$ J/mole is obtained for the oxidation of transient scales on β-NiAl. The final portion of the curve in Figure 4 corresponds to the growth rate of α-$Al_2O_3$ at 1100°C. Parabolic kinetics are followed with a rate constant, $kp^2$ = 1.12 X $10^{-4}$ $mg^2/cm^4/hr$ (3.11 X $10^{-6}$ $kg^2/m^4/s$). The transformation to α-$Al_2O_3$ from the transient stages appears to be complete after about five hours. Similar bimodal parabolic growth kinetic curves have been observed in other studies of $Al_2O_3$ scales involving a regime where the transformation from a transient to a mature stage occurs [13-15].

Corresponding with the major change in growth rate is an equally drastic change in oxide morphology. The scale morphology has changed from that of a uniformly fine-grained δ-$Al_2O_3$ scale after 100 hours at 800°C to a largely non-uniform α-$Al_2O_3$ scale after being heated for an additional 70 hours at 1100°C. This structure, termed the "lacey" structure, has been observed in Pt-Al, Ni-Al and Fe-Cr-Al alloys under certain conditions and consists entirely of α-$Al_2O_3$ [16-18]. The microstructure contains numerous features that are best referred to in terms of the contrast differences in the STEM bright field image of Figures 5a,c of an oxide scale extending over the hole of a back-thinned specimen. What is observed are patches of thin, relatively uniform oxide(A), seperated by thicker ridges(B). Two different types of cracks are evident. Some cracks are flanked by thick regions(C) while others are cracks without having thick regions associated with them(D). The uniform patches of oxide consist of subgrains having a radial texture as ahown in the TEM bright field image of Figure 5b. The misorientation between adjacent subgrains as determined by electron diffraction is slight (approximately 1°) while the difference in orientation of subgrains from the center of a patch to a boundary can extend to as much as 7°. Many voids and dislocations are also observed. A thin region runs down the center of each ridge indicating a healed crack. The subgrain size does not change in the ridge region as was suggested in other studies [16,17].

A representative schematic cross-section of this microstructure is shown in Figure 6, a model suggested further by scanning electron microscopy (SEM) images of the metal-oxide and gas-oxide interfaces. A complete description requires the development of this morphology from the onset of the transformation to α-$Al_2O_3$. Nuclei of α-$Al_2O_3$ form at what are the centers of patches in Figure 5a, the misorientation between subgrains being minimal. As the oxide continues to transform, tensile cracks are formed(C) due to stresses resulting from the large decrease in volume associated with the transformation. At once, rapid oxidation occurs to heal the cracks. The healed cracks then act as fast diffusion paths at which oxidation occurs primarily by inward oxygen diffusion with some slight counterdiffusion of aluminum outward. The transformation and subsequent strained growth yields porosity and dislocations amongst the subgrains which extend radially outwards. As two patches meet, a natural high angle grain boundary is formed(B) by large relative misorientations in subgrains, represented by arrows in the schematic of Figure 6. Again, boundary diffusion predominates as evidenced by the observation of ridges at the gas-oxide surface (Figure 5d) but only slight undulations in the underside of the oxide surface. This indicates predominant outward aluminum diffusion for these boundaries. Scale growth proceeds mainly by boundary diffusion along healed cracks and high angle grain boundaries. With increasing time, the ridges expand laterally consuming the thinner patches(grains) of oxide. Oxidation along subgrain boundaries of higher misorientation also occurs as suggested by fine, hair-like dark lines extending radially inward from high angle grain boundaries shown in Figure 5a. These eventually form ridges on the gas-oxide surface in the same manner as high angle grain boundaries. Upon cooling or during specimen preparation, cracks form within the oxide scale (D) and are not associated with the oxidation process.

## CONCLUDING REMARKS

The oxidation of β-NiAl is indicative of many alumina formers. The slow growth rates that accompany the lacey structure are an illustration of the cause/effect relationship between growth mechanisms and oxide microstructure. Boundary diffusion plays the major role in the process of oxidation during the mature stage. Both inward oxygen and outward aluminum

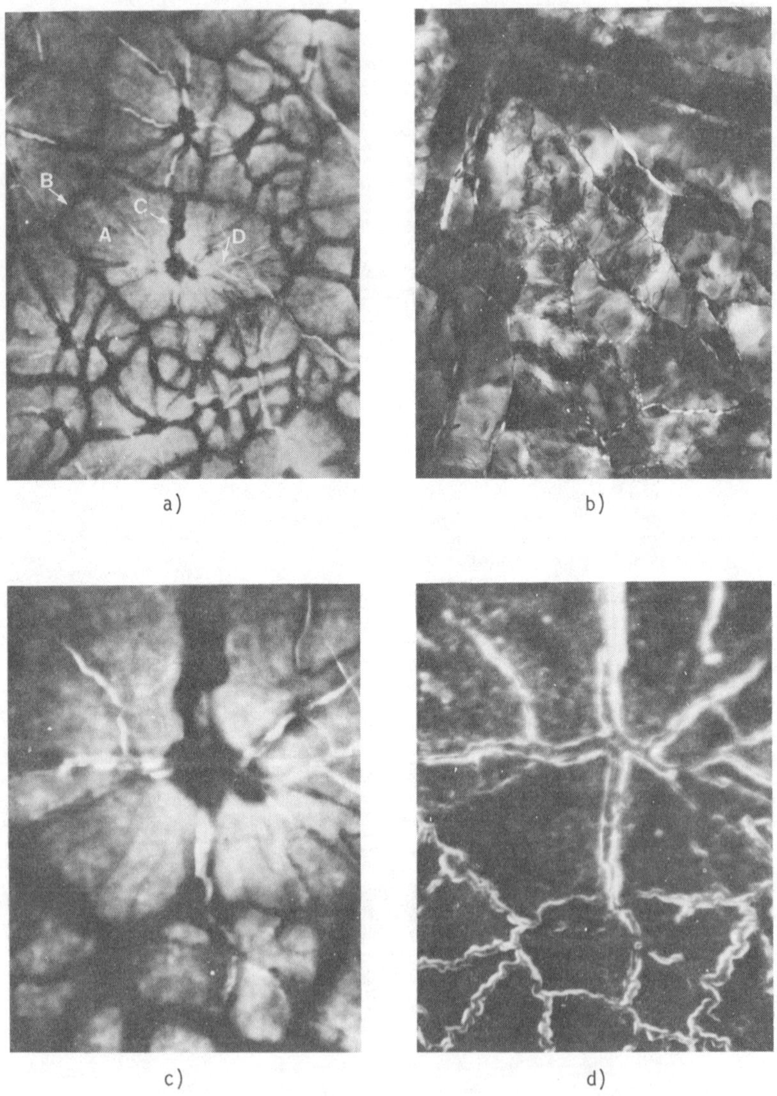


Figure 5 Microstructure of the mature $\alpha$-$Al_2O_3$ scale formed on $\beta$-NiAl at 1100°C; a) STEM bright field image showing patches of oxide and other features described in the text, b) TEM bright field image of a sector of an oxide patch showing subgrains, dislocations and porosity, c) higher magnification image of Figure 5a with d) the corresponding SEM image of the gas-oxide interface.

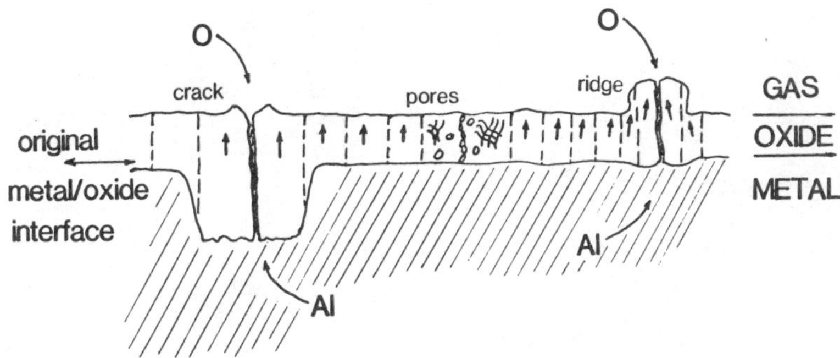


Figure 6 Schematic cross-section of the mature $\alpha$-$Al_2O_3$ scale that forms on $\beta$-NiAl at 1100°C. Nucleation of $\alpha$-$Al_2O_3$ occurs followed by crack formation due to tensile stresses arising from the transformation. The transformation proceeds radially outward until two regions of transformed oxide impinge, the location where ridges are formed. Dashed lines within the scale represent subgrain boundaries, the arrows within each subgrain being an indicator of relative orientation. Inward oxygen and outward aluminum diffusion occurs along healed cracks and high angle grain boundaries.

diffusion are observed to occur but the nature of the boundaries in which each predominate are different.

The transformation to $\alpha$-$Al_2O_3$ and its ensuing morphology are being studied as a function of substrate orientation. The dependence on substrate orientation is readily observed during transient stages and the effects of this prior strong preferred orientation are of interest when developing slower growing oxides which, in this case, should be free of high angle grain boundaries acting as fast diffusion paths. The added dimension of a possible strong substrate or prior transient oxide orientation dependency during growth of $Al_2O_3$ scales ensures continued interest.

ACKNOWLEDGMENT

This work is supported by NASA Grant #NAG 3-498

REFERENCES

1. F.S. Pettit, Trans. Met. Soc. AIME, 239, p. 1296 (1967).

2. E.W.A. Young and J.H.W. de Wit, Surface and Interface Analysis, 5, No. 5, p. 177 (1983).

3. J.K. Tien and F.S. Pettit, Met. Trans., 3, p. 1587 (1972).

4. B.H. Kear, F.S. Pettit, D.E. Fornwalt and L.P. Lemaire, Oxid. Met., 3, p. 557 (1971).

5. J.L. Smialek: "Microstructure of $Al_2O_3$ Scales Formed on NiCrAl Alloys," Ph.D. Thesis, Case Western Reserve University, Cleveland, OH, 1981.

6. W.H. Gitzen: Alumina as a Ceramic Material, American Ceramic Society, Columbus, OH, 1970.

7. E.J.W. Verwey, J. Chem. Phys., 3, p.592 (1935).

8. W.T. Donlon, T.E. Mitchell and A.H. Heuer, J. Mat. Sci., 17, p.1389 (1982).

9. D.P. Whittle and J. Stringer, Phil. Trans. Roy. Soc., A295, p. 309 (1980).

10. J.K. Doychak in: Proceedings of the 42nd Annual Meeting of the Electron Microscopy Society of America, Detroit, MI, 1984, p. 598.

11. J.K. Doychak: "The Transient Oxidation of Single Crystal NiAl + Zr," M.S. Thesis, Case Western Reserve University, Cleveland, OH, 1984.

12. J.K. Doychak, J.L. Smialek and T.E. MItchell in: Proceedings of the 9th International Congress on Metallic Corrosion, Toronto,1984, Vol. 1, p. 35.

13. H.M. Hindam and W.W. Smeltzer, J. Electrochem. Soc., 127, No. 7, p. 1622 (1980).

14. G.C. Rybicki and J.L. Smialek: NASA Lewis Research Center, 1982, to be published.

15. J.S. Sheasby and D.B. Jory, Oxid. Met., 12, No. 6, p. 527 (1978).

16. E.J. Felten and F.S. Pettit, ibid., 10, No. 3, p. 189 (1976).

17. H.M. Hindam and W.W. Smeltzer, J. Electrochem. Soc., 127, No. 7, p. 1630 (1980).

18. S.N. Basu: "Analysis Techniques for Tracer Studies of Oxidation," M.S. Thesis, Case Western Reserve University, Cleveland, OH, 1985.

ORDERING BEHAVIOR AND CORROSION PROPERTIES
OF Ni-Mo AND Ni-Mo-Cr ALLOYS

H. M. TAWANCY, FIRST AUTHOR* AND A. I. ASPHAHANI, NEXT AUTHOR**
* First Author, Cabot Corporation, 1020 West Park Ave., Kokomo, Indiana 46901
** Next Author, Cabot Corporation, 1020 West Park Ave., Kokomo, Indiana 46901

ABSTRACT

Ni-Mo alloys containing 26-28 wt. pct. Mo are known for their excellent corrosion properties in reducing environments. When exposed to temperatures in the range 600°C to 800°C, however, these alloys undergo a long-range ordering reaction from disordered fcc lattice to ordered $Dl_a$ superlattice. This ordering phenomenon has been found to have detrimental effects on corrosion properties. Also, alloys based on the Ni-16Mo-16Cr system undergo a long range ordering reaction from disordered fcc lattice to ordered $Pt_2Mo$-type superlattice in the temperature range of 400° to 600°C. However, in contrast to the case of Ni-Mo alloys, ordering in the Ni-Mo-Cr alloys has some beneficial effects on the aqueous corrosion properties. This difference in behavior between the Ni-Mo alloys and the Ni-Mo-Cr alloys is discussed in-terms of the ordering characteristics determined from transmission electron microscopy and diffraction, X-ray microanalysis in the scanning transmission electron microscope and Auger electron spectroscopy.

INTRODUCTION

Many structural alloys rely on existing alloy theory for mechanical and corrosion properties. For example, it is well known from theory that a fine uniform dispersion of second phase particles is essential for high yield strength.(1) Also, it is known qualitatively that alloys in which dislocations can cross slip with relative ease (i.e., high stacking fault energy) are least susceptible to stress corrosion cracking.(2) Most of the materials requirements dictated by design and from prior service performance in corrosion applications, are satisfied by alloys based on solid solution systems. A relatively new approach in alloy design is to utilize long range ordering as a source of mechanical strength and corrosion resistance.(3,4) Ordered alloys are characterized by high mechanical strength, however, they usually but not always, have poor ductility.(4) Also, ordered alloys can be expected to have better corrosion resistance than disordered alloys because of stronger binding of atoms.

Nickel-base alloys have been of particular importance as corrosion resistant materials because of their unique combinations of mechanical strength, corrosion resistance and fabrication characteristics. Alloying with Mo is essential for corrosion resistance in reducing media and for resistance to localized corrosion attack.(5) Chromium is known to impart corrosion resistnce in oxidizing media. These corrosion characteristics are obtained in alloys provided that the concentration of the alloying elements (e.g., chromium) is above a critical concentration defined as the "reaction limit." There are several commercial alloys based on the Ni-Mo and Ni-Mo-Cr systems. Among these alloys are HASTELLOY® alloy B-2 which contains 26-30 wt. pct. Mo and HASTELLOY alloy C-276 which is based on the Ni-16Mo-16Cr system. Alloy B-2 has an excellent corrosion resistance in reducing media, e.g., HCl at various temperatures and concentrations. It also withstands hydrogen chloride and sulfuric, acetic and phosphoric acids. Alloy C-276 has excellent

® HASTELLOY is a registered trademark of Cabot Corporation

 

resistance to both oxidizing and reducing media and is generally considered the most versatile corrosion resistant alloy available. Also, this alloy offers outstanding resistance to localized corrosion. Both alloys B-2 and C-276 have excellent resistance to stress corrosion cracking; however, this property is sensitive to the particular environment.

A common feature of the Ni-Mo alloy B-2 and the Ni-Mo-Cr alloy C-276 is that they both undergo long-range ordering reactions when exposed to the appropriate temperature (600°-800°C in the case of alloy B-2 and 400°-600°C in the case of alloy C-276). The resultant ordered phases are crystallographically related as shown later. It is interesting, both from a theoretical and from a practical point of view, to understand the corrosion behavior of ordered alloys and its relationship to the microstructure. The objective of this paper is to shed some light on such a relationship.

## EXPERIMENTAL PROCEDURE

All the property and microstructural data presented in this paper was derived from commercial grade materials in sheet form (1-2 mm thick). Table 1 shows the nominal chemical compositions of alloys B-2 and C-276. Samples from each alloy were given the appropriate heat treatment to induce long range ordering. This consisted of up to 1000 hours of exposure at temperatures in the range of 600° to 800°C in the case of alloy B-2. Ordering in alloy B-2 occurs with rapid kinetics at temperatures in the neighborhood of 700°C, however, morphological changes which significantly affect the properties continue to occur with extended exposure. This is why samples were exposed for up to 1000 hours at a temperature. In the case of alloy C-276, the order-disorder temperature is below 600°C and ordering proceeds with very sluggish kinetics in annealed material. After the extended exposure time required to induce long range ordering, precipitation of Mo-rich mu phase also occurs which complicates the interpretation of the effect of ordering on the corrosion behavior.[6] This problem could be overcome by cold working the material prior to the ordering heat treatment. Cold working accelerates the ordering reaction and can cause that reaction to be completed prior to the precipitation of mu-phase. Therefore, samples from alloy C-276 were cold reduced by 60 pct. and then exposed for 100 hours at 500°C.

The corrosion rates were calculated from weight loss measurements made on 25.4 x 25.4 x 1.0 mm specimens after 24-hour immersion in different acqueous environments. Stress corrosion cracking tests were conducted on flat specimens (45 x 4.5 x 1 mm) which were strained as C-shape to a final length of 39 mm and then held at a constant plastic strain (about seven pct.) in slotted holders having 50 x 20 x 6 mm overall dimensions. The specimens were stressed transverse to the original rolling direction.

Thin foil transmission electron microscopy and diffraction were used to characterize the ordered microstructure. Microchemical analyses were conducted in the scanning transmission electron microscope and by Auger electron spectroscopy.

Table I
Nominal Chemical Compositions of Alloys
B-2 and C-276 (wt. pct.)

| Alloy | Ni | Mo | Cr | Fe | W | Co | Mn | Si | C |
|---|---|---|---|---|---|---|---|---|---|
| B-2 | Bal | 26-30 | 1.0* | 2.0* | - | 1.0* | 1.0* | 0.10* | 0.10* |
| C-276 | Bal | 15-17 | 14.5-16.5 | 4-7 | 3-4.5 | 2.5* | 1.0* | 0.08* | 0.01* |

* Maximum

EXPERIMENTAL RESULTS AND DISCUSSION

Properties of Alloys B-2 and C-276 in the Disordered State

The heat treatment of alloy B-2 consists of annealing at 1065°C followed by water quenching. Alloy C-276 is typically heat treated at 1120°C followed by rapid air cooling. Such heat treatments produce optimum corrosion properties. The resulting microstructure in each case is essentially a single phase (disordered fcc solid solution). Typical corrosion rates and room temperature tensile properties in the disordered state are given in Table 2. As can be seen alloy B-2 is more resistant to reducing media (e.g. HCl) because of its high Mo content and the absence of significant amounts of Cr. In the sensitized condition, the boiling reducing acid corrosion test (HCl) detects Mo-depleted zones while the oxidizing acid test (ferric sulfate solution) highlights the corrosion of Cr-depleted and/or Mo-rich regions.

Table II
Characteristic Properties of Alloys B-2 and C-276
In the Annealed Condition (Disordered)

| Alloy | Corrosion Rates, 10% Boiling HCl | MM/Year Boiling Sulfuric Ferric Sulfate | Tensile Properties 0.2% Y.S. MPa | U.T.S. MPa | Elongation in 50.8%MM |
|---|---|---|---|---|---|
| B-2 | 0.1% | Consumed | 396 | 914 | 55 |
| C-276 | 1.0 | 5.7 | 356 | 782 | 61 |

Generally, the tensile and yield strength of alloy B-2 is higher because of its higher Mo content than alloy C-276 as illustrated in Table 2.

Both alloys B-2 and C-276 contain short range order in the annealed condition (otherwise disordered), as indicated by the appearance, in electron diffraction patterns, of extra reflections at all equivalent 1/4 $\{420\}^*_{fcc}$ positions.(7)

This is illustrated in the [001] and [112] electron diffraction patterns of Figure 1. Short range order and/or low stacking fault energy lead to dislocations moving in coplanar groups with little or no tendency to cross slip. This is thought to increase the susceptibility to stress corrosion cracking.(8) A tensile deformation substructure characteristic os alloys B-2 and C-276 after 6 pct. elongtaion at room temprature is shown in Figure 2. Coplanar groupings of dislocations can be seen with evidence of cross slip. The absence of a cellular dislocation structure indicates a "medium" stacking fault energy.(9) As shown later, long range ordering in alloys B-2 and C-276 lowers the stacking fault energy and increases the susceptibility to stress corrosion cracking relative to the disordered state.

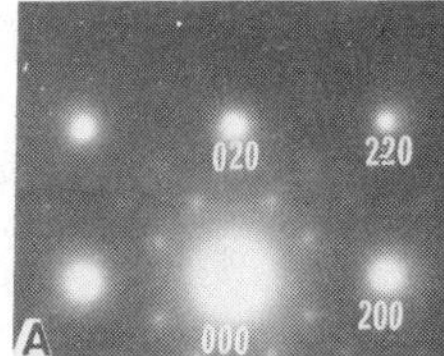


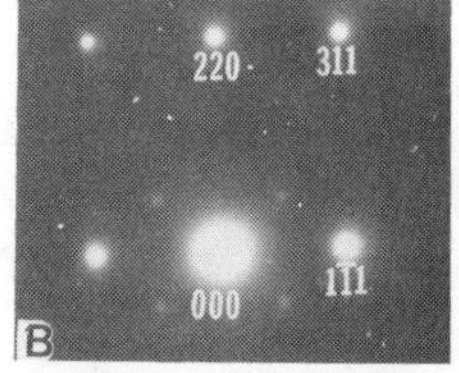


Figure 1: Characteristic electron diffraction patterns of alloys B-2 and C-276 showing short range order reflections at 1/4 {420} positions in the annealed condition. A. [001] B. [$\bar{1}$12]

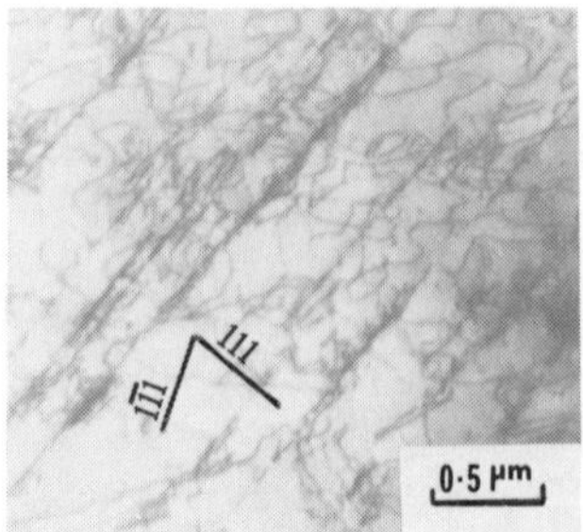


Figure 2: Bright-field TEM micrograph showing tensile deformation substructure characteristic of alloys B-2 and C-276 in the annealed condition (6% strain at room temperature), the micrograph was derived from alloy B-2.

## Ordering Behavior of Alloys B-2 and C-276

Upon exposure to the appropriate temperatures (600°-800°C for alloy B-2 and 400°-600°C for alloy C-276), both alloys B-2 and C-276 undergo long-range ordering reactions which significantly affect the mechanical properties (Table 3). In the case of alloy B-2, the resultant ordered phase is usually $Ni_4Mo$ with $Dl_a$ superlattice (sometimes $Ni_3Mo$ forms in heats with higher Mo contents). The ordered phase in alloy C-276 has a $Pt_2Mo$-type superlattice and can be expected to be of the form $Ni_2$(Cr ,Mo). Figure 3 shows typical microstructures and corresponding [001] electron diffraction patterns of these phases.

Table III
Effect of Long Range Ordering on the Room Temperature Tensile Properties of Alloys B-2 and C-276

| Alloy | Condition | 0.2% Y.S. | U.T.S. | Elongation in 50.8 MM, % |
|---|---|---|---|---|
| B-2 | Disordered | 396 | 914 | 55 |
| | Ordered | 895 | 1195 | 2 |
| C-276 | Disordered | 356 | 792 | 61 |
| | Ordered | 770 | 1240 | 28 |

After given the appropriate exposure time at temperature, $Ni_4Mo$ in alloy B-2 assumes a platelet morphology as shown in Figure 3A. In contrast, $Ni_2$(Cr, Mo) in alloy C-276 maintains near spherical morphology (Figure 3B). This difference in morphology can be attributed to the magnitude of strain energy in relation to surface energy in each case. Both the $Dl_a$ and $Pt_2Mo$-type superlattices can be envisioned as stackings of (420) planes.(10) In the case of the $Dl_a$ superlattice, every fifth plane is occupied by Mo atoms and planes in-between contain only Ni atoms while for the $Pt_2Mo$-type superlattice every third plane is occupied by Mo atoms (and/or Cr). Therefore, the characteristic $Dl_a$ superlattice reflections are located at 1/5 {420} reciprocal lattice vectors and those of the $Pt_2Mo$-type superlattice are located at 1/3 <420> (or 1/3 {220}).

A characteristic feature of the ordered microstructure in both alloys is the presence of randomly distributed (111) stacking faults as indicated in the example of Figure 4. These faults could have formed to accomodate the strain associated with the transformation and indicate that long-range ordering has the effect of lowering the stacking fault energy of these

* All the diffraction data in this paper is interpreted in terms of fcc notations.

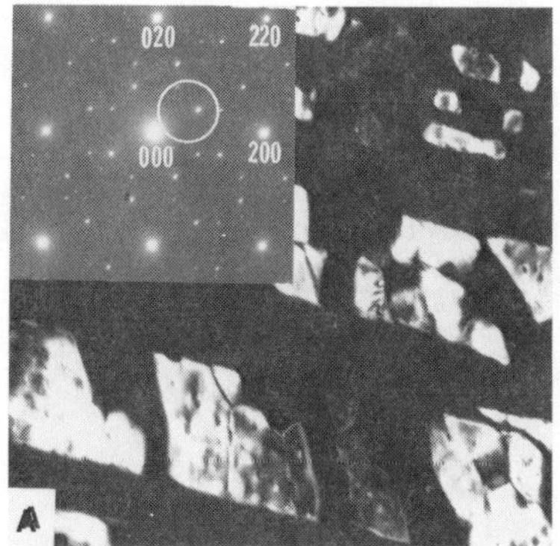


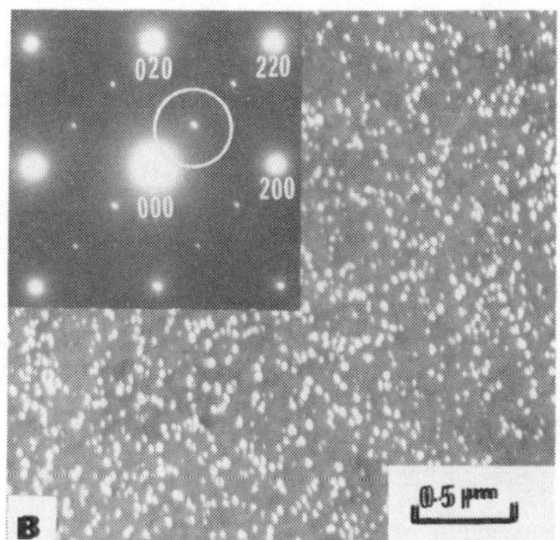


Figure 3: Dark-field images and corresponding $[001]_{fcc}$ electron diffraction patterns of the ordered phases in alloys B-2 and C-276.

A. Alloy B-2, platelets of $Ni_4Mo$ (sample exposed 100 hours at 760°C).

B. Alloy C-276, spheroids of $Ni_2(Cr, Mo)$ (sample exposed 8000 hours at 540°C).

The circles in the electron diffraction patterns mark the superlattice reflections used in forming the images.

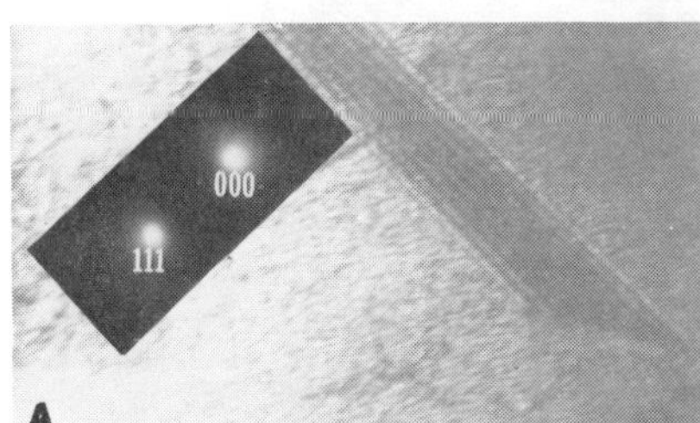


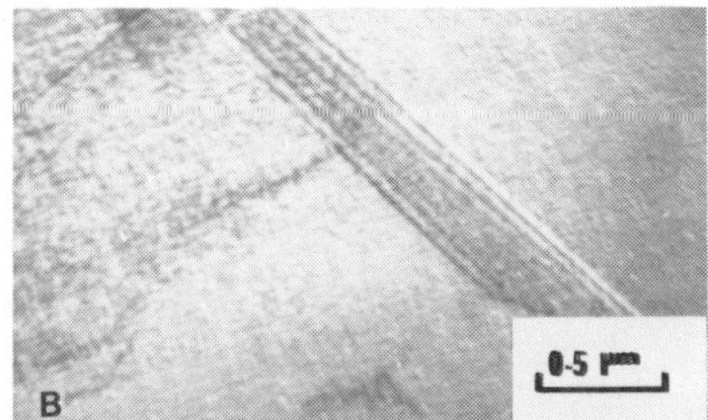


Figure 4: Example illustrating the presence of $(111)_{fcc}$ stacking faults in the ordered microstructures of alloys B-2 and C-276

A. Bright-field image and corresponding electron diffraction pattern obtained under two-beam conditions (near 110).

B. Dark-field image formed with (111) reflection.

alloys. A similar result has been reported in the case of binary Ni-Mo alloys.(11) The effect of long range ordering on the stacking fault energy is reflected by the defomation substructure as shown later.

Effect of Long-Range Ordering on the Corrosion Properties of Alloys B-2 and C-276

The corrosion rate of alloy B-2 in 20% boiling HCl showed a substantial increase after up to 1000 hours of exposure at temperatures in the range 600°

to 800°C. Microstructural examinations in the transmission electron microscope revealed that the only phenomenon which occurs in the above temperature range is long range ordering to $Ni_4Mo$. The effects of up to 1000 hours of exposure at 700°C on the corrosion rate, tensile ductility and mirostructure of alloy B-2 are illustrated in Figure 5. It is to be noted that there is one-to-one correspondence between the increase in corrosion rate and decrease in tensile ductility. The set of dark-field TEM micrographs (all formed with 1/5 <420> $Dl_a$ superlattice relfections) shown in Figure 5 illustrate the evolution of the ordered microstructure of $Ni_4Mo$ with progressive exposure at 700°C. In the early stages of exposure, the microstructure is characterized by a tweed contrast which is followed by the development of aligned arrays of discrete particles which then grow to impingement and ultimately form a mosaic assembly of twin-related platelets of $Ni_4Mo$. These platelets continue to grow with exposure time. It can be seen from Figure 7 that the maximum corrosion rate and minimum tensile ductility are associated with the development of $Ni_4Mo$ platelets. Examinations of corrosion tested samples revealed extensive intergranular corrosion attack as

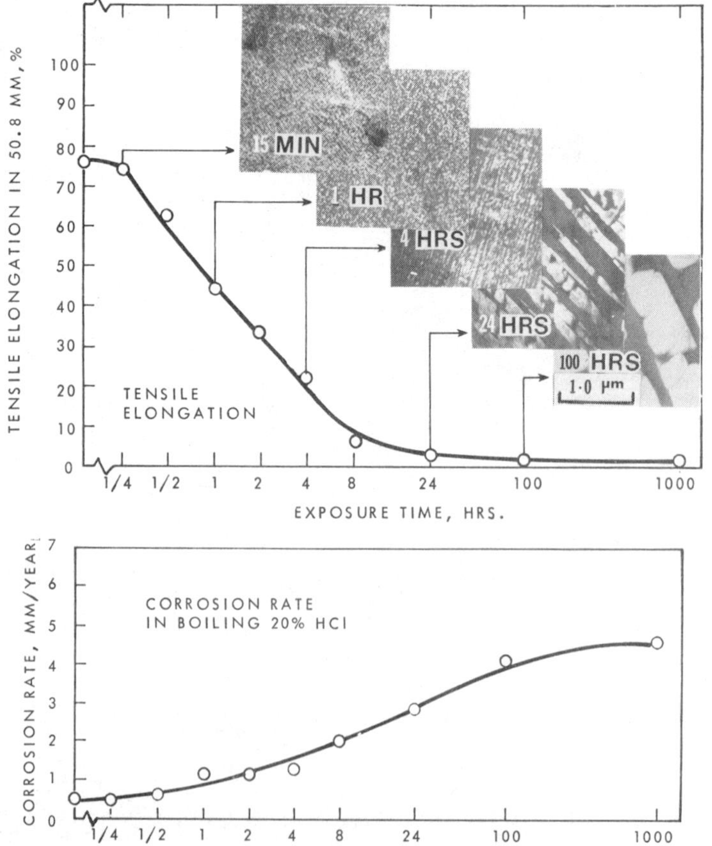


Figure 5: Effect of up to 1000 hours of exposure at 700°C on the corrosion rate in boiling 20% HCl and room temperature tensile elongation of alloy B-2.

illustrated in the optical micrograph of Figure 6. The overall corrosion rate, however, is not only attributed to localized corrosion but also to uniform corrosion due to the presence of Mo-depleted zones in the matrix as described later. Table 4 summarizes the effect of long-range ordering in alloy B-2 on its stress corrosion cracking resistance in 10 pt. HI solution. As can be seen the susceptibility to cracking is increased in the ordered state.

The effects of long-range ordering on the corrosion properties of alloy C-276 in different media are illustrated in Table 5. It can be seen that the corrosion rates in both boiling HCl and ferric sulfate solutions are lower in the ordered state. This indicates that unlike the case of alloy B-2, long range ordering in alloy C-276 has a beneficial effect on the corrosion properties. However, the susceptibility to stress corrosion cracking is increased in the ordered state similar to the case of alloy B-2 as indicated

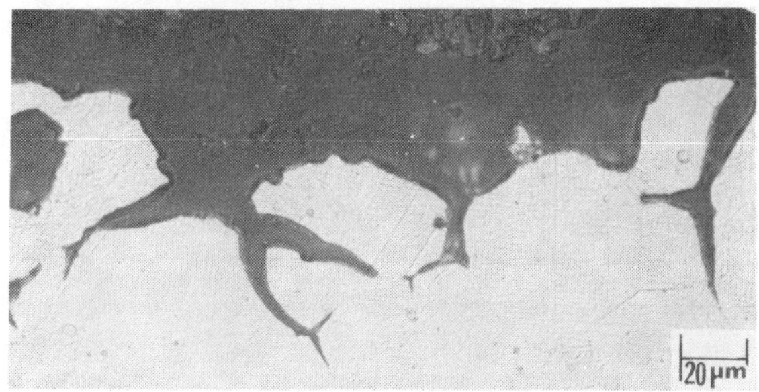


Figure 6: Optical micrograph of a corrosion tested sample of alloy B-2 (20% boiling HCl) in the ordered state (sample exposed 100 hours at 700°C). Note the extensive intergranular attack.

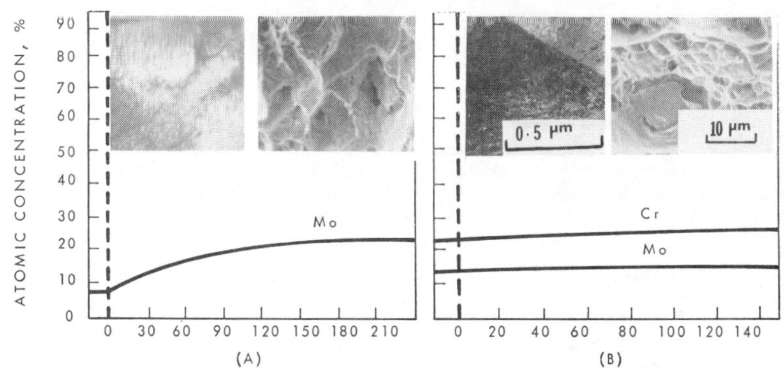


Figure 7: Auger spectrographs showing the concentration of Mo and Mo and Cr as functions of distance from a grain boundary

A. Alloy B-2 (sample exposed 100 hours at 700°C), the insert bright- field TEM micrograph shows heterogeneous ordering reaction at a grain boundary.

B. Alloy C-276 (sample given 60 pct. cold rolling and then exposed for 100 hours at 500°C), the insert bright-field TEM micrograph shows the absence of an ordering reaction at the grain boundary.

Table IV
Effect of Long-Range Ordering on the Stress Corrosion Cracking Properties of Alloy B-2
(100-Hour Exposure; Boiling 10% HI)

| Condition | Number of Specimens Cracked out of Two Specimens Tested |
|---|---|
| Disordered | No Cracking |
| Ordered | 212 |

Table V
Effect of Long-Range Ordering on the Corrosion Rate of Alloy C-276 in Different Media

| Test Medium | Condition | Corrosion Rate MM/Year |
|---|---|---|
| 10% Boiling HCl | Annealed (Disordered) | 6.2 |
| | 60% Cold reduced + 100 hours at 500°C (ordered) | 5.2 |
| Boiling 50% $H_2SO_4$+25% | Annealed (Disordered) | 5.7 |
| $Fe_2(SO_4)_3$ | 60% Cold reduced + 100 hours at 500°C (ordered) | 5.2 |

in Table 6. The difference in behavior between alloys B-2 and C-276 can be explained in terms of the ordering characteristics in each case as described below.

Auger electron spectroscopy was employed to characterize the influence of long-range ordering on the chemical compositions of grain boundaries. Sputtring was utilized to generate Mo and Mo and Cr profiles for alloys B-2 and C-276 respectively. Figure 7 summarizes the results. The inserts show the surfaces of as-fractured samples and TEM micrographs of the grain boundaries. In the case of alloy B-2, the fracture mode is predominantly intergranular. A Mo-depleted zone can be seen to exist alongside the grain boundaries. This can be interpreted in terms of a grain boundary discontinuous ordering reaction which involves boundary migration as illustrated in the TEM micrograph of Figure 7. The fracture mode of alloy C-276 can be seen to be a mixture of intergranular and transgranular. Neither Mo nor Cr depletion seems to exist alongside the grain boundaries. Also, the discontinuous reaction observed in the case of alloy B-2 is absent in alloy C-276. X-ray microanalysis in the scanning transmission electron microscope showed

Table VI
Effect of Long-Range Ordering on the Stress Corrosion Cracking properties of Alloy C-276
(200 Hour Exposure; Boiling $M_gCl_2$ Solution, Temp. = 154°C)

| Condition | Number of Specimens Cracked out of Two Specimens Tested |
|---|---|
| Disordered | No cracking |
| Ordered | 212 |

Mo depletion in the matrix of alloy B-2 in the ordered state and neither Mo nor Cr depletion in the case of alloy C-276 as illustrated in the spectrographs of Figure 8. As can be seen from Figure 8, the residual matrix of alloy B-2 has a lower Mo content than the ordered phase. In contrast, both the matrix and ordered phase in alloy C-276 have similar Mo and Cr contents. The Mo-depleted zones in alloy B-2 can be expected to contribute to the overall corrosion rate. It is possible that the origin of those zones in alloy B-2 is associated with deviation of the overall composition from exact

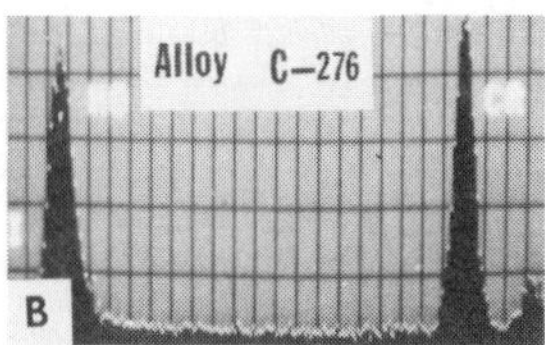


Figure 8: X-ray spectrographs generated in the scanning transmission electron microscope illustrating the compositions of the ordered phases and matrix.

A. Alloy B-2 (sampled exposed 100 hours at 700°C).

B. Alloy C-276 (sample given 60 percent cold rolling and then exposed 100 hours at 500°C).

The dotted peaks represent the composition of the ordered phases.

stoichiometry and/or the slow diffusivity of Mo in Ni. In the case of alloy C-276, the experimental results indicate that the overall composition is at or very near the stoichiometric composition of the ordered phase. The increased susceptibility to stress corrosion cracking in the ordered state of both alloys can be explained in terms of lower stacking fault energy (i.e., high density of stacking faults) and higher strength as a result of ordering (Table 3). Typical tensile deformation substructures corresponding to the ordered microstructures of Figure 3 are shown in Figure 9. In the case of alloy B-2, a high density of {111} stacking faults can be observed. Partial dislocations within the faults can be distinguished (Figure 9A). A high density of {111} deformation twins is observed in the case of alloy C-276 (Figure 9B).

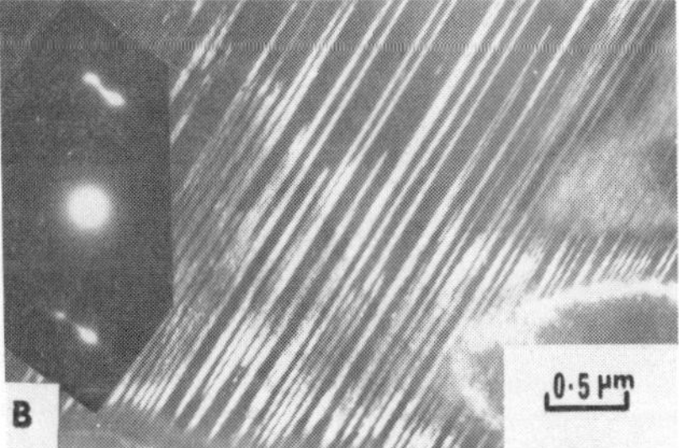


Figure 9: Bright-field TEM micrographs showing typical room temperature tensile deformation substructure in the ordered state.

A. Alloy B-2
B. Alloy C-276

Thin foils were prepared from fractured tensile tested samples and adjacent to the fracture surface 2% elongation in alloy B-2 and 28% elongation in alloy C-276.

SUMMARY

The effects of long range ordering on the corrosion properites of two commercial alloys representing the Ni-Mo (alloy B-2) and Ni-Mo-Cr (alloy C-276) systems have been examined. In the case of alloy B-2, ordering to $Ni_4Mo$ with $D1_a$ superlattice has a detrimental effect on its aqueous corrosion properties. This is attributed to the presence of Mo-depleted zones alongside the grain boundaries and in the matrix as a result of ordering. Ordering in alloy C-276 does not create similar zones and therefore the aqueous corrosion resistance is not adversely affected but rather improved. However, the susceptibility to stress corrosion cracking is increased for both alloys as a result of lower stacking fault energy and higher strength in the ordered state.

## FUTURE PERSPECTIVE

It is plausible to expect that ordered alloys and intermetallic compounds will play significant roles in advanced corrosion applications. As design schemes increase in sophistication, the search continues for alloys that offer high strength and can handle a larger variety of complex agressive media than the conventional alloys. The issue of low ductility remains an important one and therefore manufacturing innovations will play an increasingly important role. However, in a case like alloy C-276, a relatively high ductility level is maintained in the ordered state. Also, extensive work is being done in various laboratories to improve the ductility of ordered alloys and intermetallic compounds. Finally, further laboratory work is needed to better understand the relationship between ordering and corrosion behavior.

## ACKNOWLEDGEMENT

The authors would like to acknowledge the valuable assistance of Drs. J. Horton and C. White of the Metals and Ceramics Division, Oak Ridge National Laboratory in generating some of the STEM (JH) and Auger (CW) data.

## REFERENCES

1. G. Thomas, V. F. Zackay and E. R. Parker, "Structure and High Strength Metals" in Strengthening Mechanisms, Metals and Ceramics, J. J. Burke et al., (eds.), Syracuse University Press, p. 3 (1966).

2. H. L. Logan "The Stress Corrosion of Metals" Wiley, New York, p. 178 (1966).

3. N. S. Stoloff, "Ordered Alloys: Physical Metallurgy and Structural Applications," International Metals Review, Vol. 29, No. 3, p. 123, (1984)

4. N. S. Stoloff and R. G. Davies "Mechanical Properties of Ordered Alloys" Progress in Materials Science, Vol. 13, p. 1 (1966)

5. J. Koits, J.B.C. Wu and A. I. Asphahani, "Highly Alloyed Austenitic Materials for Corrosion Services" Metal Progress, p. 25, September (1983).

6. H. M. Tawancy "Long-Term Aging Characteristics of Some Commercial Ni-Cr-Mo Alloys" J. Materials Science, Vol. 16, p 2883, (1981).

7 C. R. Brooks, J. E. Spruiell and E. E. Stansbury "Physical Metallurgy of Nickel-Molybdenum Alloys," International Metals Review, Vol. 29, No. 3, p. 210, (1984)

8. D. L. Douglass, G. Thomas and W. R. Rosey, "Ordering, Stacking Faults and Stress Corrosion Cracking of Austenitic Alloys", Corrosion, Vol. 20, No. 1, p. 15t (1964).

9. P. S. Kotval, "The Microstructure of Superalloys," Metallography, Vol. 1, p. 251, (1969).

10. P. R. Okamoto and G. Thomas, "Short-Range Order in Ni-Mo Alloys: Structural Considerations," Materials Research Bulletin, Vol. 6, p. 45 (1971).

11. L. A. Nesbit and D. E. Laughlin, "The Deformation Microstructure of the Ni-$Ni_4Mo$ System," Acta Metallurgica, Vol. 28, p. 989 (1980).

# WELDABILITY OF $Ni_3Al$-TYPE ALUMINIDE ALLOYS*

M. L. SANTELLA, S. A. DAVID, AND C. L. WHITE
Metals and Ceramics Division, Oak Ridge National Laboratory, P. O. Box X, Oak Ridge, TN 37831

## ABSTRACT

If the ductile $Ni_3Al$-based aluminides are to achieve their full potential as engineering materials, they must demonstrate a capability for being joined by conventional welding processes. The goal of this study was to identify some of the factors controlling weldability of the ductile aluminides. The alloys selected for this initial study were boron-doped $Ni_3Al$ and $Ni_3Al$ containing 10 at. % Fe. Sheet stock was used to make full penetration autogenous welds by the electron beam and gas tungsten arc processes. The main process variables were travel speed and preheat. The as-welded coupons were examined visually and by normal metallographic methods. Although sound welds were made, initial results indicated weldments of these alloys are susceptible to cracking. Weldability of the ductile $Ni_3Al$-based aluminides was found to be affected by alloy composition and welding speed.

## INTRODUCTION

Recent developments [1,2] have produced $Ni_3Al$-type ordered alloys which are ductile in polycrystalline form, and stimulated interest in the use of these materials for structural applications. Aluminides based on the $Ni_3Al$-$Fe_3Al$ pseudo-binary system are being viewed as candidate materials for a variety of applications in fossil energy conversion systems, since it is thought that their use could substantially improve performance and reliability. Aluminides may also provide the opportunity for conservation of strategically important chromium by replacing austenitic stainless steels in various applications.

The welding behavior of the ductile $Ni_3Al$-type alloys has been studied because good weldability is an essential property of most metallic structural materials. This paper outlines a study which shows that sound welds can be made in some of the $Ni_3Al$-type aluminides, and their weldability is influenced by welding speed and alloy composition. Preliminary testing of defect-free welds indicates that they have good mechanical properties.

## EXPERIMENTAL DETAILS

The nominal compositions for the alloys used in this study are listed in Table I. Ingots (500 g) of the alloys were prepared from pure charge materials and a Ni – 4 wt % B master alloy by vacuum arc melting and drop casting. The 125 × 25 × 9 mm ingots were processed into 0.7-mm-thick sheet by cold rolling.

Weldments were made by the autogenous bead-on-plate technique, using either the electron beam (EB) or the gas tungsten arc (GTA) process. The parameters for each process were adjusted to produce full penetration welds at various travel speeds. Weld beads were typically 2 to 3 mm wide and 35 to 40 mm long. Preheating was accomplished through the use of calrod

---

*Research sponsored by the U.S. Department of Energy, AR&TD Fossil Energy Materials Program and the Office of Energy Utilization Research, Energy Conversion and Utilization Technologies (ECUT) Program under contract DE-AC05-84OR21400 with Martin Marietta Energy Systems, Inc.

 

Table I. Alloy Compositions in Atomic %

| Alloy identification | Composition | | | | |
|---|---|---|---|---|---|
| | Ni | Al | Fe | B | Others |
| IC-15 | 75.8 | 24.0 | | 0.24 (500 wppm) | |
| IC-25 | 69.9 | 18.9 | 10.0 | 0.24 (500 wppm) | 0.5 Ti + 0.5 Mn |
| IC-103 | 70.0 | 18.9 | 10.0 | 0.10 (200 wppm) | 0.5 Ti + 0.5 Mn |

resistance heating elements. The upper and lower as-welded surfaces were examined for the presence of cracks at a magnification of 100×.

Electron beam welds were made in a 6 kW machine with the beam focal spot set for the sheet surface. The welding speed varied from 0.9 to 64 mm/s. For EB welding, beam voltages were 60 to 100 kV and beam currents were 1 to 6 mA. Gas tungsten arc welding parameters were 12 to 14 V and 12 to 35 A, at welding speeds ranging from 0.9–13.5 mm/s.

Specimens for optical metallography were prepared using standard methods outlined elsewhere [3]. Transverse weld fracture surfaces were examined by scanning electron microscopy (SEM). The grain boundary compositions of heat-affected zone (HAZ) microstructures in IC-25 and IC-103 were determined by Auger electron spectroscopy (AES). The HAZ microstructures were simulated by heat-treating specimens with a Gleeble thermomechanical simulator. These specimens were subsequently notched and fractured in-vacuo in the AES chamber.

Specimens with gage sections of 12.7 × 3.6 × 0.7 mm were used for evaluation of EB weldment tensile properties. Tensile tests were conducted at room temperature on an Instron machine using a crosshead speed of $4.2 \times 10^{-2}$ mm/s. Specimens were oriented to have the weld beads either parallel or perpendicular to the tensile axis.

## RESULTS

### Welding and Microstructure of $Ni_3Al$ Alloy

Previous work on B-doped $Ni_3Al$ [3] has shown that a B level of 200 ppm is optimum from a weldability standpoint. It was also shown that, for welding speeds over 13 mm/s, significant weld cracking problems exist for these alloys [3]. Consequently, the present study was limited mainly to low-speed EB and GTA welding.

IC-15 was EB and GTA welded at 0.9, 1.7 and 3.4 mm/s with no preheating, and at 3.4 mm/s with a 400°C preheat. All of these weldments contained visible cracks on their as-welded surfaces. Cracks were generally located in both the fusion zone and the HAZ and were mainly intergranular regardless of location in the weldment. It was not determined whether the fusion zone cracks in IC-15 were solidification cracks, or cracks formed below the solidus temperature which could indicate a weakening of the grain boundaries at elevated temperatures. When HAZ cracks were examined on polished and etched sections, they did not appear to have formed by liquation at grain boundaries or second phases as has often been observed in some iron and nickel-based alloy systems [4,5]. Figure 1 shows cracks which have occurred in the fusion zone, near the fusion line of an EB weld in IC-15.

Electron Beam Welding Ni,Fe Aluminides

IC-25 was EB welded at speeds ranging from 0.9 to 42.3 mm/s; IC-103 was EB welded at speeds ranging from 0.9 to 50.8 mm/s. Cracking frequency has been plotted versus welding speed for those tests, along with some data of David et al. [3], in Fig. 2. These results confirm that IC-25 weldments are very susceptible to cracking at welding speeds exceeding about 13 mm/s. On the other hand, sound crack-free welds were made in IC-103 at speeds up to 50.8 mm/s. Figure 3(a) shows a weldment microstructure for IC-25. Although occasional fusion zone cracks were observed in IC-25, most were confined to the HAZ. All crack paths were intergranular, and HAZ cracks tended to extend in a direction normal to the welding direction. For both alloys, the primary phase in the fusion zone had the same ordered structure as the annealed base metal. The second phase, distributed interdendritically, was martensitic $\beta'$ which forms from the aluminum-rich $\beta$ phase [3]. The volume fraction of $\beta'$ in the fusion zone varied from 0.09 to 0.12 and increased with welding speed. The $\beta'$ phase also formed on the HAZ grain boundaries in both alloys [Fig. 3(b)], particularly at welding speeds below about 13 mm/s. The extent of HAZ $\beta'$ formation increased with decreasing speed. Optical metallography failed to reveal any apparent difference in the details of weldment microstructure between IC-25 and IC-103 other than the absence of HAZ cracks in IC-103.

To examine HAZ microstructures in more detail, coupons of IC-25 welded at 4.2 and 33.9 mm/s were fractured at room temperature by loading parallel to the weld axis. The weld made at 33.9 mm/s broke at preexisting HAZ cracks. These fracture surfaces corresponded to a cross-sectional view of each weldment, and three distinct regions were visible on both: fusion zone, HAZ, and base metal. For both, solidification structure was apparent in the fusion zone, and the base metal showed the ductile transgranular failure mode typical of $Ni_3Al$ fractured at room temperature [6].

Real differences in microstructure existed for these two weldments only in the HAZ. The HAZ of the higher speed weld, which extended the full thickness of the sheet, failed in a brittle intergranular mode [Fig. 4(a)]. Close inspection of the HAZ revealed the fusion line to be very distinct,

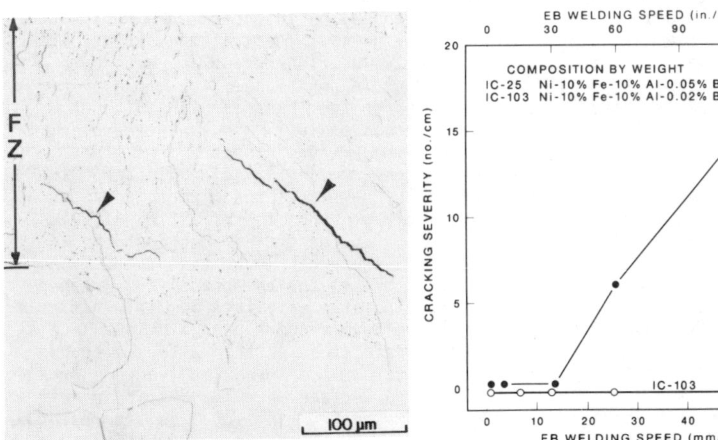


Fig. 1. Microstructure of electron beam weld in IC-15 showing fusion zone cracks.

Fig. 2. Cracking frequency as a function of welding speed for Fe-containing aluminides.

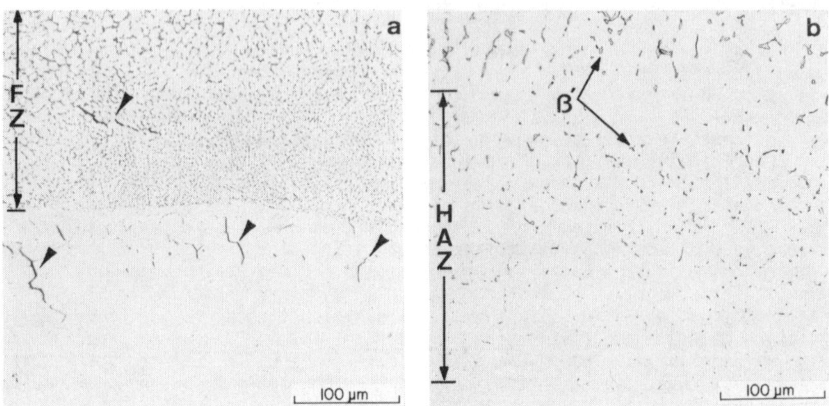


Fig. 3(a). Microstructure of IC-25 welded at 50.8 mm/s.

Fig. 3(b). Microstructure of IC-103 welded at 0.9 mm/s.

and the HAZ grain boundaries to be largely free of any second phase. Lack of evidence of a liquid film on the crack surface was confirmed and this indicated the crack occurred after passage of the molten weld pool, i.e., at some temperature below the solidus. The HAZ region of the lower speed weld, Fig. 4(b), also extended the full thickness of the sheet, but had a much different appearance than the high-speed microstructure. In this case, the HAZ fractured in a relatively ductile manner. While fracture in the low-speed HAZ region had more of an intergranular nature than fracture in the base metal, it also displayed some transgranular cleavage and dimpling. Also, at the lower speed the fusion line was no longer distinct because of significant β´ formation on the HAZ grain boundaries.

Only simulated microstructures corresponding to the HAZ of low-speed (i.e. crack free) welds were analyzed using AES. The fracture surface of the IC-103 specimen [Fig. 5(a)] was nearly 100% intergranular, with much evidence of second phase (β´) on triple grain junctions [Fig. 5(b)]. The IC-25 specimen exhibited mixed intergranular and transgranular failure [Fig. 6(a)], and also had significant second phase (β´) on intergranular portions of the fracture surface [Fig. 6(b)]. Auger spectra from grain faces in both specimens (e.g., the top spectrum in Fig. 7) indicated boron enrichment above bulk concentrations. The degree of boron enrichment can be expressed in terms of the ratio of the 179 eV boron to the 102 eV nickel Auger peak, which was in the range 0.5 to 0.6 for grain faces in both specimens. This B:Ni peak-height ratio is somewhat lower than the 0.8 to 1.0 values reported for IC-15 base metal by Liu and co-workers [6]. Auger electron spectroscopy analysis of triple point precipitate in IC-103 (middle spectrum in Fig. 7) indicated lower boron levels and higher aluminum levels than on precipitate-free grain faces. Grain faces in both alloys also had higher iron levels than either precipitates or transgranular fracture areas.

Isolated areas on the IC-103 fracture surface were found to have very high impurity concentrations. The bottom spectrum in Fig. 7 is from point A in Fig. 5(b), and indicates Auger peaks for S, B, and Ca, in addition to peaks from the elements already mentioned. There is also a peak at ~1000 eV that could be from either zinc (994 eV) or sodium (990 eV). A clear determination could not be made from the available results. Mapping of sulfur and calcium Auger peaks showed that these elements were associated with the

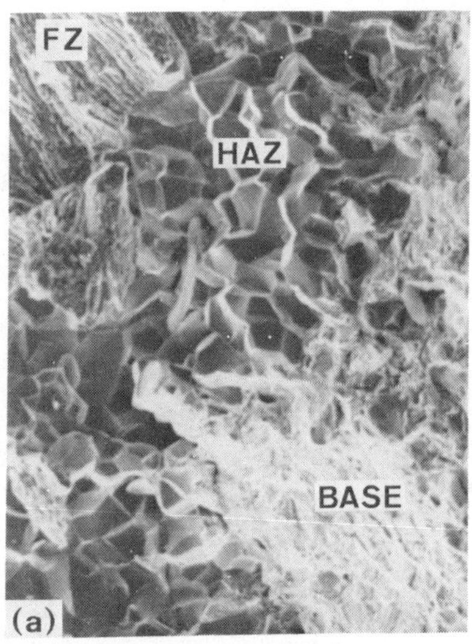


Fig. 4(a). Microstructure of IC-25 welded at 33.9 mm/s.

Fig. 4(b). Microstructure of IC-25 welded at 4.2 mm/s.

(approximately) rectangular feature at point A. Boron and zinc (or sodium) were distributed in the same general area but covering a region several times larger than for sulfur and calcium.

Tensile test results from EB weldments of IC-25 made at a welding speed of 4.2 mm/s are given in Table II. Two specimens were tested in each weld orientation, and they were electropolished and determined to be free of visible defects prior to testing. Both specimens of parallel orientation fractured across the pinhole of one of the grip ends, so the true ultimate tensile strength and elongation values would be higher than those recorded.

Table II. Tensile Test Results of IC-25 EB Welds

| Specimen location | Yield strength (MPa) | Ultimate tensile strength (GPa) | Elongation (%) |
|---|---|---|---|
| Weldment, ⊥ axis | 341.2 | 1.290 | 41.5 |
| Weldment, ∥ axis | 398.5 | >1.205 | >26.5 |
| Base material | 386.1 | 1.214 | 45 |

## GTA Welding of Ni,Fe Aluminides

Results of GTA welding IC-25 are given in Table III. The most frequent type of defect encountered for these welds was crater cracks, although HAZ cracks were also found as welding speed was increased. The crater cracks occurred at the points of initiation or termination of the weld beads. The microstructure of the IC-25 GTA welds was indistinguishable from that of the EB welds.

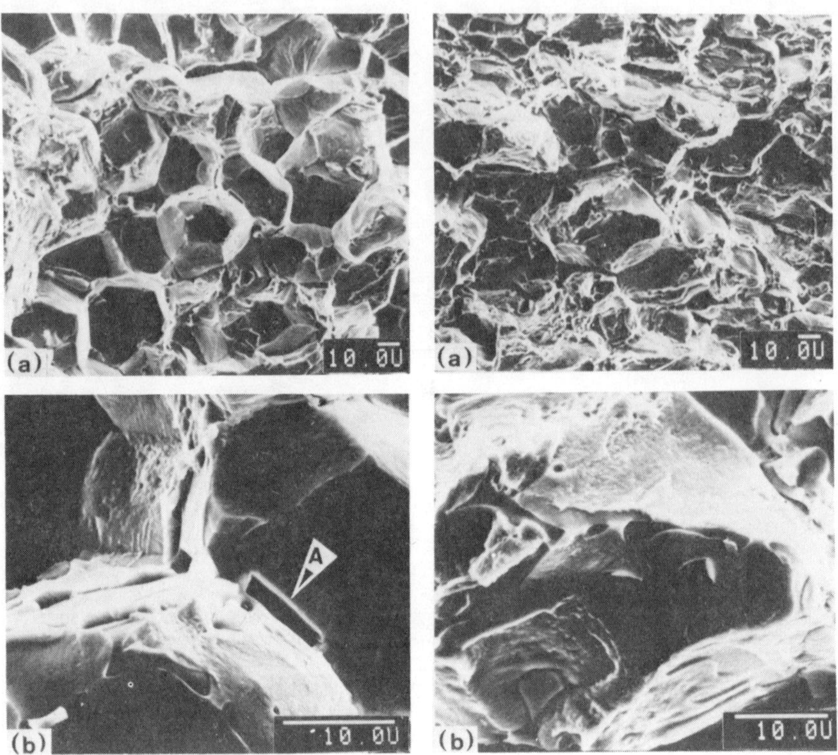


Fig. 5. (a) Microstructure of IC-103 specimen used for AES; (b) grain boundary triple point.

Fig. 6. (a) Microstructure of IC-25 specimen used for AES; (b) intergranular area on fracture surface.

Table III. Results of GTA Welding IC-25

| Welding speed (mm/s) | Cracks observed[a] | | |
|---|---|---|---|
| | No Preheat | 200°C Preheat | 400°C Preheat |
| 0.9 | Crater | Crater | None |
| 1.7 | Crater | Crater + HAZ | |
| 3.4 | Crater | Crater + HAZ | |
| 6.8 | None | None | |
| 13.5 | In HAZ | Crater + HAZ | Crater + HAZ |

[a]Crater cracks at weld start and stop. HAZ indicates cracks in heat-affected zone.

DISCUSSION

Comparison of the results of EB welding IC-15 and IC-25 reveals two major differences. First, sound EB welds were made in IC-25 while none could be made in IC-15 under identical conditions. Second, IC-15 tended to crack in the weldment fusion zone while IC-25 showed a much greater propensity for HAZ cracking. Both of these alloys contain 500 wppm B, with the major difference between them being the 10 at. % Fe added to IC-25. The addition of Fe significantly alters the phase relations in this system [7]. First of all, solidification of $Ni_3Al$ takes place by a eutectic reaction, whereas the Ni,Fe aluminide solidifies by a peritectic reaction in which the β phase is a constituent. The β phase is an equilibrium constituent in the Fe-containing alloys above about 1250°C.

The formation of β phase may be at least partly responsible for the differences between IC-15 and IC-25. The β phase has a higher specific volume than the $L1_2$ γ′ phase or the disordered fcc γ phase [8]. Thus, β formation could help reduce the shrinkage stresses which normally accompany solidification and cooling. Another possibility is that the difference in solidification mode between IC-15 and IC-25 results in different segregation patterns in their respective weld fusion zones. This could result in IC-15 having inherently weaker fusion zone grain boundaries than those of IC-25. These possibilities have not yet been studied in detail.

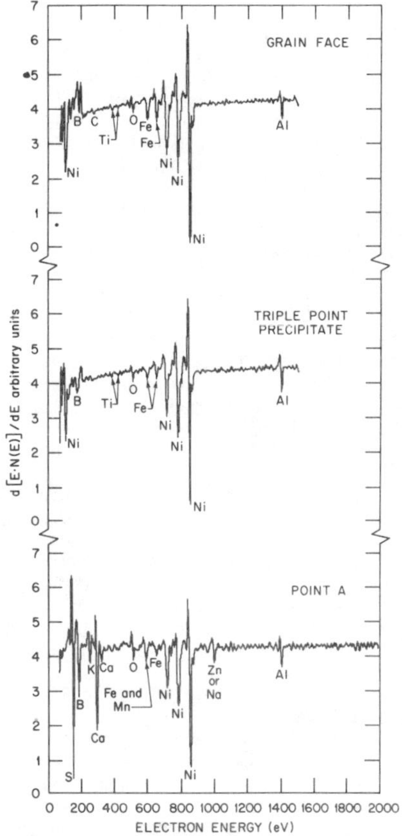


Fig. 7. Auger spectra from Ni,Fe aluminide fracture surfaces.

Comparison of the results for IC-25 and IC-103 shows that sound, crack-free EB welds could be made in both of these alloys using welding speeds up to 13 mm/s. Above this speed IC-25 displayed a marked tendency to develop HAZ cracks, while sound welds were made in IC-103 even at 50.8 mm/s. The only significant difference between these two alloys is B level.

Welding speed affects the thermal cycles experienced by all regions of the weldment [9–11]. High speeds produce very high heating and cooling rates, very steep thermal stress gradients in the fusion zone and HAZ, and reduce the time available for thermally activated processes to occur. The effect of welding speed on thermal cycles explains the variation in amount of HAZ β′ phase. At the higher speeds, there is not enough time above 1250°C for β to form and grow on the HAZ grain boundaries. High welding speeds not only tend to localize stresses but also result in stress distributions which change rapidly with time [10,11]. For a point in the HAZ close to the fusion line, longitudinal compressive stresses are built up as the arc approaches, and then reversed to become tensile in nature after passage of the arc [9,10]. These tensile stresses may be high enough to exceed the local yield point of a material [9,10].

The orientation of HAZ cracking in IC-25 suggest it was produced by the longitudinal stresses developed during EB welding at high speeds. Also, examination of the crack surfaces suggested they occurred after passage of the arc since there was no indication of the presence of liquid films. At that point in time, material in the HAZ would have been in a state of tension. The intergranular nature of the cracking indicates a weakening of the grain boundaries at elevated temperatures [6]. The absence of HAZ cracking in IC-25 at lower welding speeds indicates that in this range the material was able to accommodate welding-generated stresses without fracturing. This may have resulted from delocalization of stresses or reduced rates of stress change. Since IC-103 did not experience HAZ cracking, even at high welding speeds, it may be concluded that it could respond to thermal stresses without failure.

The AES analyses of room temperature fracture surfaces from simulated HAZs in IC-25 and IC-103 failed to indicate any large difference between their grain boundary compositions, although the IC-25 alloy exhibited more transgranular fracture than did IC-103. The boron levels on grain faces in both alloys were below the levels required to inhibit intergranular fracture in $Ni_3Al$ at room temperature, according to the observations of Liu et al. [6]. It should be noted that, although boron segregation has pronounced beneficial effects on grain boundary strength at room temperature, little is known concerning its effects on elevated temperature failure mechanisms such as creep or HAZ cracking. We should also note that the simulated HAZs examined here correspond to welding speeds that yielded sound weldments in both alloys, so perhaps it is not too surprising that no large differences were observed. Comparison between HAZ microstructures corresponding to cracked and uncracked welds is planned in the future. Isolated regions high in S, B, and Ca (and perhaps Zn or Na) were observed in IC-103. Although these regions may be significant with regard to cracking susceptibiity, additional comparisons would be required to establish this.

These important points can also be made regarding the behavior of IC-25 and IC-103 during EB welding. First, the presence of grain boundary $\beta'$ in the HAZ does not appear to be a major factor in HAZ cracking. For instance, IC-103 welds did not crack at low speeds when HAZ $\beta'$ was present or at high speeds when it was not. Also, examination of fracture surfaces of sound EB welds indicated the two-phased HAZ microstructure was not inherently brittle at room temperature. Lastly, tensile testing of sound EB welds showed that weld fusion zones are not inherently brittle, and that weldments of Ni,Fe aluminides can display excellent strength and ductility at room temperature.

The results of GTA welding suggest acceptable welds can also be made in the Ni,Fe aluminides with this process. Preliminary data suggest that, as for EB welding, HAZ crack susceptibility increases with arc travel speed. Better process control should help eliminate the crater cracking problem. Also, some recent work has indicated the aluminides are sensitive to contaminants such as oxygen in the arc shielding gas, as well as to the overall effectiveness of the shielding. Shielding variables may have been a factor in the data presented here.

## SUMMARY

It has been shown that sound defect-free welds can be made in $Ni_3Al$-type aluminides containing 10 at. % Fe. Iron additions appear to improve weldability of the alloy aluminides compared to boron-doped $Ni_3Al$. Boron was also found to affect weldability of the iron-containing aluminides and a boron level of 200 wppm produced much better results than 500 wppm B. Welding procedure also affected weldability with high welding speeds being detrimental in terms of HAZ cracking.

Grain boundary β′ phase in the HAZ of Ni,Fe aluminides did not appear to correlate with the occurrence HAZ cracking. Heat-affected zone cracks in the Ni,Fe aluminides were mainly intergranular, and their nature suggests they are due to a weakening of the grain boundaries at elevated temperatures. Defect-free weldments of a Ni,Fe aluminide were shown to have excellent tensile strength and ductility.

## ACKNOWLEDGMENTS

The authors gratefully acknowledge the encouragement of R. A. Bradley and A. C. Schaffhauser as program managers. We also wish to thank G. M. Goodwin and V. P. Kujanpää for reviewing the manuscript, J. D. Hudson for making the welds, R. A. Padgett for assisting with the Auger spectroscopy, and K. Gardner for typing the manuscript.

## REFERENCES

[1]K. Aoki and O. Izumi, Nippon Kinzaku Gakkaishi **43**, 1190 (1979).
[2]C. T. Liu, C. L. White, C. C. Koch, and E. H. Lee, Proceedings of the Symposium on High Temperature Materials Chemistry — II, ed. by Z. A. Munir and D. Cubicciotti (The Electrochemical Society, Inc., 1983) Vol. 83(7): 32.
[3]S. A. David, W. A. Jemian, C. T. Liu, and J. A. Horton, Weld. J. **64**(1): 22-s—28-s (1985).
[4]W. A. Owczarski, Proceedings of a Symposium on Effects of Minor Elements on the Weldability of High-Nickel Alloys, ed. by B. M. MacPherson (Welding Research Council 1969), p. 6.
[5]J. J. Pepe and W. F. Savage, Weld. J. **46**(9): 422-s—422-s (1967).
[6]C. T. Liu, C. L. White, and J. A. Horton, to be published in Acta. Met.
[7]V. G. Rivlin and G. V. Raynor, Int. Met. Rev., **25**(3): 79 (1980).
[8]E. M. Grala, Mechanical Properties of Intermetallic Compounds, ed. by J. H. Westbrook (Wiley 1960), p. 358.
[9]W. Yeniscavich, Proceedings of a Symposium on Methods of High-Alloy Weldability Evaluation, (Welding Research Council 1970), p. 2.
[10]K. Masubuchi, Analysis of Welded Structures, (Pergamon Press Ltd. 1980).
[11]S. Kou and Y. Lee, Met. Trans. A, **14A**, 2245 (1983).

# PHYSICAL PROPERTIES OF $Ni_3Al$ CONTAINING 24 AND 25 ATOMIC PERCENT ALUMINUM*

R. K. WILLIAMS, R. S. GRAVES, F. J. WEAVER AND D. L. McELROY
Metals and Ceramics Division, Oak Ridge National Laboratory, Oak Ridge, Tennessee 37831

## ABSTRACT

Thermal conductivity, electrical resistivity, Seebeck coefficient and thermal expansion data were obtained on well-annealed $Ni_3Al$ containing 24 and 25 at. % Al. The results span the temperature range 300 to 1000 K. The expansion coefficients did not vary with composition and increased with temperature, reaching values of about 17 x $10^{-6}$ $K^{-1}$ at 1000 K. The thermal conductivity and electrical resistivity changed rapidly with composition, and the thermal conductivity of 24 at. % Al is as much as 30% lower than that for stoichiometric $Ni_3Al$. The electronic Lorenz function of $Ni_3Al$ was obtained by subtracting the estimated phonon conductivity component and found to be within about 5% of the Sommerfeld prediction from 300 to 1000 K. The electrical resistivity results for stoichiometric $Ni_3Al$ are influenced by the loss of ferromagnetic order at lower temperatures and are not adequately described by the Bloch-Grüneisen equation.

## INTRODUCTION

The compound $Ni_3Al$ exists over a significant range of compositions and exhibits ferromagnetism [1] for Al contents less than about 25.7 at. %. This compound may be viewed as a base material which, when alloyed further, may have commercial potential, and an understanding of the binary aluminide is necessary to deal with the more complex multicomponent aluminides. The goal of this study was to determine which factors most influence the following physical properties: thermal conductivity, $\lambda$; electrical resistivity, $\rho$; Seebeck coefficient, S; and thermal expansion coefficient, $\alpha$.

## SAMPLE PREPARATION AND CHARACTERISTICS

The 24 and 25 at. % Al specimens were prepared by repeated arc-melting of high purity Ni and Al stock. Cylindrical ingots, 25 mm diameter and 150 mm long, were drop-cast from the arc-melted buttons. The castings were then homogenized at 1473 K for 20 h in a high-purity argon atmosphere. Annealing at successively lower temperatures (1 h at 1273 K, 2 h at 1173 K, 4 h at 1073 K and 20 h at 973 K) was used to promote ordering. Cylindrical specimens, 10 mm diameter and 75 mm long, were then electrically discharge-machined from the ingots and ground to final size.

The characteristics of these samples (Group 3) have been presented previously [2]. Metallographic examination and X-ray diffraction showed that they were single phase, and impurity analysis indicated about 100 ppm of interstitial and 50 ppm of metallic impurities. Radiographs showed that the samples were free from casting porosity, and electrical

*Research sponsored by the Division of Materials Science, U.S. Department of Energy, under contract DE-AC05-84OR21400 with Martin Marietta Energy Systems, Inc.

resistivity, $\rho$, determinations at different positions along the sample length showed $\rho$ variations of less than 1%. Thus, microcracking was not a serious problem. Residual resistivity determinations showed that the stoichiometric samples were well ordered. These data indicate that the compositional deviations or disordering were at most 0.1 at. %.

## EXPERIMENTAL METHODS

The thermal expansion data were obtained in a quartz push rod dilatometer described by Kollie [3]. In this apparatus, temperature measurement uncertainties are about 0.1 K and length changes as small as $5 \times 10^{-5}$ mm can be detected by an electronic micrometer. Experiments are conducted by determining the sample length at ~ 100 K intervals after allowing the apparatus to reach a thermal steady state (~24 h) at each temperature. The differential expansion of the quartz system was determined by making measurements with an NBS-SRM 739 certified quartz standard. Tests with copper and tungsten standards [3,4] indicate that the expansion coefficient values are uncertain by about ±2%.

The transport properties, $\lambda$, $\rho$, and S, were determined in an absolute longitudinal heat flow apparatus described previously by Moore et al.[5]. The Cu heater and Ni heat sink were electron-beam brazed to the specimens, and a Cu chill block was used to minimize the sample temperature excursion during this operation. Five Pt—Pt 10% Rh thermocouples were welded to the sample and used to determine the temperature, temperature gradients and voltage drops. In this apparatus, the sample is surrounded by a matched-guard cylinder, and the temperature difference at the heater end is nulled to minimize heat losses along the heater and $\rho$ current leads. The annular volume between the sample and guard cylinder is filled with $Al_2O_3$ insulating powder, and corrections for heat exchange across this region are included in the $\lambda$ calculations. Tests with standard specimens [6] have demonstrated that maximum experimental uncertainties in $\lambda$, $\rho$, and S are ±3%, ±0.4%, and ±0.3 μV/K, respectively. The reproducibility of the experimental data were typically much better than these values.

## EXPERIMENTAL RESULTS

The smoothed expansion and Seebeck coefficient results and the resistivity and thermal conductivity data are shown in Table I and Figs. 1—4. Figure 4 also includes lower temperature $\lambda$ data obtained by a comparative method [2]. These comparative $\lambda$ samples were cut from the ingots used in this study, and for one 25 at. % Al sample the $\lambda$ difference (~8%) exceeds the combined experimental uncertainties of the two measurements. The most probable explanation of this discrepancy is that the smaller comparative samples contained some microcracks [1,2].

## DISCUSSION

The thermal expansion coefficient values for Ni-24 and 25 at. % Al are shown in Fig. 1. These results and dilatometric [7] and lattice parameter [8] values from the literature indicate that this property is not sensitive to stoichiometry. This is not surprising since low-temperature studies [9] show that the lattice part of the specific heat is also insensitive to composition. At temperatures above 1300 K, the lattice parameter data for Ni-23 at. % Al [8] indicate an upswing in $\alpha$, possibly due to the onset of sublattice disordering.

Table I. Smoothed Physical Property Values for Ni-24 and 25 at. % Al

| Temperature (K) | $Ni_{0.75}Al_{0.25}$ Electrical Resistivity ($10^{-8}$ Ωm) | $Ni_{0.75}Al_{0.25}$ Thermal Conductivity (W/m•K) | $Ni_{0.75}Al_{0.25}$ Absolute Seebeck Coefficient (μV/K) | $Ni_{0.75}Al_{0.25}$ Thermal Expansion Coefficient[a] ($10^{-6}$ $K^{-1}$) | $Ni_{0.76}Al_{0.24}$ Electrical Resistivity ($10^{-8}$ Ωm) | $Ni_{0.76}Al_{0.24}$ Thermal Conductivity (W/m•K) | $Ni_{0.76}Al_{0.24}$ Absolute Seebeck Coefficient (μV/K) | $Ni_{0.76}Al_{0.24}$ Thermal Expansion Coefficient[a] ($10^{-6}$ $K^{-1}$) |
|---|---|---|---|---|---|---|---|---|
| 4.2 | 1.54 | — | — | — | 9.05 | — | — | — |
| 300 | 32.59 | 28.85 | +5.4 | 12.5 | 51.3 | 21.37 | +9.9 | 12.5 |
| 400 | 39.00 | 33.72 | +4.1 | 13.0 | 58.89 | 24.21 | +10.8 | 13.2 |
| 500 | 44.75 | 35.13 | +1.8 | 13.5 | 64.17 | 26.35 | +7.9 | 13.8 |
| 600 | 50.13 | 36.18 | -0.4 | 14.0 | 68.60 | 28.30 | +5.1 | 14.3 |
| 700 | 55.27 | 37.07 | -2.4 | 14.6 | 72.60 | 30.14 | +2.3 | 14.8 |
| 800 | 60.54 | 37.27 | -4.3 | 15.2 | 76.52 | 31.82 | -0.4 | 15.4 |
| 900 | 66.50 | 36.86 | -5.5 | 16.0 | 80.41 | 33.13 | -2.8 | 16.1 |
| 1000 | 73.02 | 36.12 | -6.3 | 16.8 | 84.24 | 32.70 | -4.8 | 16.8 |

[a] $\left(\frac{1}{L_{300}}\right) \frac{dL}{dT}$

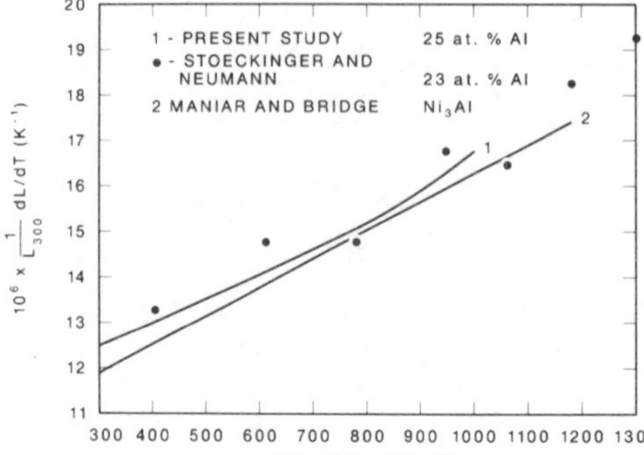


Fig. 1. Instantaneous thermal expansion coefficient values for $Ni_3Al$. Dilatometric data of Maniar and Bridge [7] were apparently reported as mean expansion coefficients. The lattice parameter results of Stoeckinger and Neumann [8] were obtained by computing $(1/a_0)\ \Delta a_0/\Delta T$. Compositions such as "$Ni_3Al$" are as presented by the original authors.

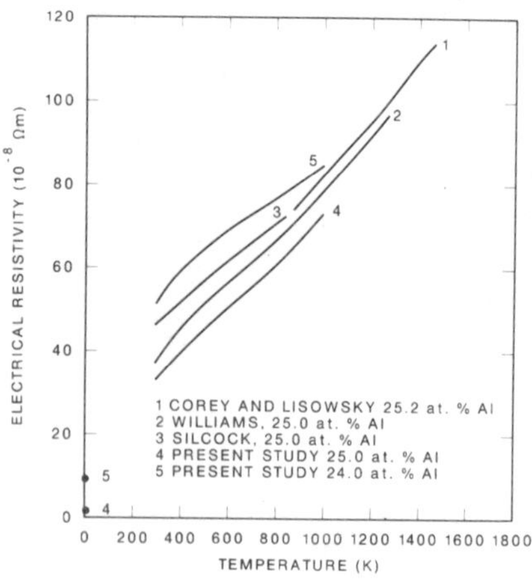


Fig. 2. Electrical resistivity values for Ni-24 and 25 at. % Al. Literature data [10,11,12] are included for comparison.

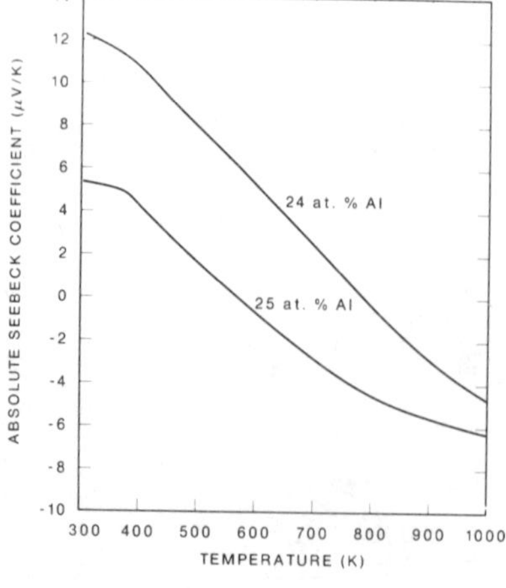


Fig. 3. Temperature dependence of the absolute Seebeck coefficients of Ni-24 and 25 at.% Al.

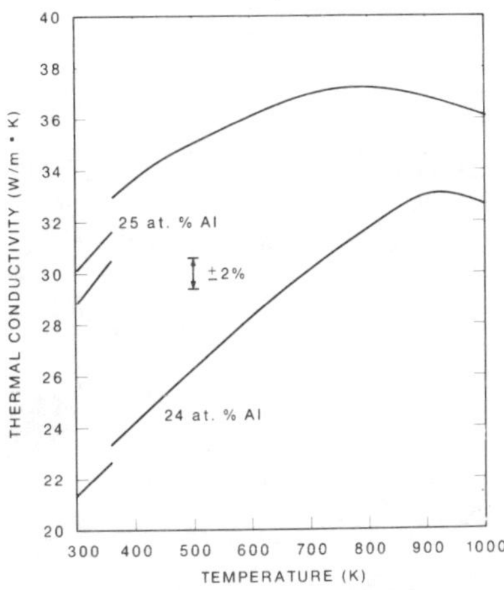


Fig. 4. Thermal conductivity vs temperature for Ni-24 and 25 at.% Al. Lower temperature results were described in a previous study [2].

The resistivity results are shown in Fig. 2, which also includes the available [10,11,12] literature data. This property varies rapidly with composition and most of the differences in the reported $Ni_3Al$ results are probably due to this factor. At temperatures above about 1500 K, it has been reported [12] that the $\rho$ passes through a maximum, and this has been attributed to sublattice disordering. Inspection of Fig. 2 shows that the $\rho$ of $Ni_3Al$ varies rapidly with temperature, much like the behavior of an austenitic stainless steel. Also, the curve does not extrapolate to the origin. This alloy-like behavior is undoubtedly due to the loss of ferromagnetic order [1], since for 25 at. % Al samples at 4.2 K, the measured $\rho$ values are only $1.5 \times 10^{-8}$ $\Omega$m. The disorder scattering has been discussed in the literature [1,13], but the behavior near the Curie temperature does not appear to have been studied.

The absolute Seebeck coefficient values are shown in Fig. 3. These values were calculated from measurements against Pt [14], and also show a strong variation with concentration. The curves must pass through a maximum at lower temperature, and the high-temperature values show that a simple diffusion thermopower [15] explanation probably is not adequate. This difficulty is frequently observed for transition elements, and may indicate unusual behavior of the other transport properties.

The thermal conductivity data are shown in Fig. 4. This property also varies rapidly with composition. At room temperature a change of 1 at. % Al lowers the $\lambda$ of stoichiometric $Ni_3Al$ by 30%. Another interesting feature is that both compositions behave like alloys, i.e., the $\lambda$ curves increase with increasing temperature over most of the range. The source of this behavior is the low-temperature ferromagnetism which, when destroyed at $T_c$, adds a large and approximately constant term to the $\rho$, $\rho_m$. It can easily be seen that a large, temperature-independent resistivity term, $\rho_m$, coupled with a constant electronic Lorenz function ($L_e = \lambda_e \rho/T$) leads to an electronic thermal conductivity, $\lambda_e$, which increases with increasing temperature.

The Wiedemann-Franz ratios ($\lambda\rho/T$) for the two compositions are shown in Fig. 5. The values are normalized to the Sommerfeld value ($L_0$ = $2.443 \times 10^{-8} V^2/K^2$), and the experimental curves exceed $L_0$ by as much as

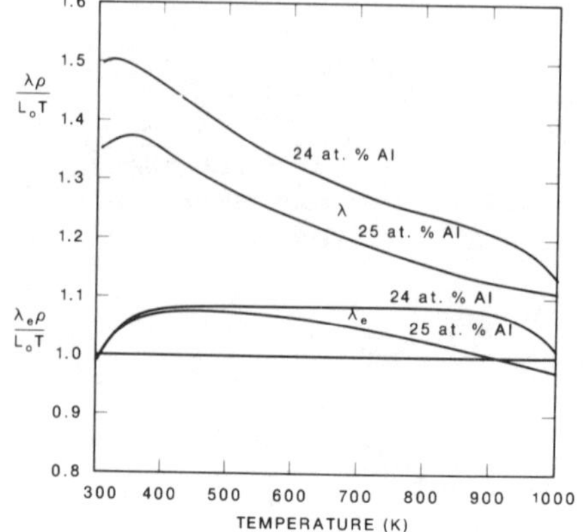


Fig. 5. Normalized ($1/L_o$) ratios for Ni-24 and 25 at. % Al. The lower curves, labeled $\lambda_e$, were obtained by correcting for phonon conduction as described in the text.

50%. Two effects are known to cause this behavior: (1) $L_e$ is larger than $L_o$ [16] and (2) the phonon conductivity, $\lambda_p$, is significant [17]. The composition and temperature variations of the results shown in Fig. 5 are both qualitatively consistent with the phonon conductivity explanation, but this does not exclude the other effect.

The thermal conductivity and electrical resistivity results for the two Al contents can be used to calculate an experimental estimate of $\lambda_p$. The calculations, which have been described elsewhere [18,19], involve assuming that the composition dependence of $\lambda_p$ is small [18] or that it can be estimated from theory [19]. They also require that the compositional change in ρ be related to a corresponding electronic thermal resistance through $L_o$ [18]:

$$\lambda_e^{-1}\ (24\text{ at. \% Al}) = \left( \frac{\rho_{24} - [\rho_{25} - \rho_{25}(0)]}{L_o T} + \frac{\rho_{25} - \rho_{25}(0)}{L_e T} \right)^{-1}$$

In this equation $\rho_{25}(0)$ is the residual resistivity of the stoichiometric aluminide. Previously [2], the 300 K results were shown to yield:

$$L_e = 0.99\ L_o$$

and

$$\lambda_p = 7.5\ \text{W/m·K}\ .$$

The new data at 400 and 500 K yield a similar picture ($1.04L_o$ and 7.3 W/m·K). Note that for the 24 at. % Al sample, $\lambda_p$ apparently contributes about 30% of the measured λ. At higher temperatures, as the differences in λ and ρ decrease, the calculations become increasingly sensitive to experimental errors. As previously noted [2], the $\lambda_p$ values obtained in this way are in remarkably (~10%) good agreement with estimates based on Callaway's formula [20]. This $\lambda_p$ calculation was therefore used to derive $L_e$ values at higher temperatures, and the results are also shown in Fig. 5. There is an indication that $L_e$ is somewhat larger than $L_0$, but the deviations are not nearly as pronounced as those noted for Ni [21]. Also, the deviations are smaller for the stoichiometric sample, and this result is less sensitive to the $\lambda_p$ choice.

The conclusion that $L_e$ is reasonably close to $L_0$ is perhaps surprising in relation to the density of states curves for this material [22,23,24]. These calculations show that the Fermi energy is located in a region in which the density of states changes by about 50% over an energy range equivalent to 1500 K. This structure would be expected to lead to an $L_e$ significantly different from $L_0$, but this does not appear to be the case. This may perhaps be attributed to smearing of the peak due to small deviations from the stoichiometric composition and temperature.

## ACKNOWLEDGMENTS

A. P. Maclin and G. M. Stocks discussed the results of their theoretical calculations prior to publication. The helpful comments of D. W. Yarbrough, R. E. Pawel, and C. T. Liu are much appreciated. This manuscript was prepared by Brenda Hickey.

## REFERENCES

1. J.H.J. Fluitman, R. Boom, P. F. DeChatel, C. J. Schinkel, J.L.L. Tilanus, and B. R. DeVries, J. Phys. F: Met. Phys. **3**, 109 (1973).

2. R. K. Williams, F. J. Weaver, and R. S. Graves, accepted for publication in Proceedings of 18th International Thermal Conductivity Conference, Rapid City, S.D. (1983).

3. T. G. Kollie, Phys. Rev. B **16**, 4872 (1977).

4. T. A. Hahn, J. Appl. Phys. **41**, 5096 (1970). Standard Reference Materials (SRMs) Copper (736) and Tungsten (737) are available from the Office of Standard Reference Materials, National Bureau of Standards, Washington, D.C.

5. J. P. Moore, R. K. Williams, and R. S. Graves, J. Appl. Phys. **48**, 610 (1977).

6. R. S. Graves, R. K. Williams, and J. P. Moore, in Thermal Conductivity 16, edited by D. C. Larsen (Plenum Press, New York, 1983), p. 343.

7. G. Maniar and J. E. Bridge, Jr., Metallog. **5**, 91 (1972).

8. G. R. Stoeckinger and J. P. Neumann, J. Appl. Cryst. **3**, 32 (1970).

9. W. DeDood and P. F. DeChatel, J. Phys. F: Met. Phys. **3**, 1039 (1973).

10. J. M. Silcock, Metal Sci. J. **5**, 182 (1971).

11. R. O. Williams, TAIME **215**, 1026 (1959).

12. C. L. Corey and B. Lisowsky, TAIME **239**, 239 (1967).

13. F. P. DeChatel, F. R. DeBoer, W. DeDood, J.H.J. Fluitman, and C. J. Schinkel, J. De Physiq. Colloq. C1, 999 (1971).

14. J. P. Moore and R. S. Graves, J. Appl. Phys. **44**, 1174 (1973).

15. F. J. Blatt, P. J. Schroeder, C. L. Foiles and D. Greig, in _Thermoelectric Power of Metals_ (Plenum, New York, 1976), p. 202.

16. M. J. Laubitz, High Temp.—High Press. **4**, 379 (1972).

17. R. K. Williams, D. W. Yarbrough, J. W. Masey, T. K. Holder and R. S. Graves, J. Appl. Phys. **52**, 5167 (1981).

18. R. K. Williams and W. Fulkerson, in _Thermal Conductivity 8_, edited by C. Y. Ho and R. E. Taylor (Plenum Press, New York, 1969), p. 389.

19. R. K. Williams, W. H. Butler, R. S. Graves, and J. P. Moore, Phys. Rev. B **28**, 6316 (1983).

20. J. Callaway, Phys. Rev. **113**, 1046 (1959).

21. M. J. Laubitz, T. Matsumura, and P. J. Kelly, Can. J. Phys. **54**, 92 (1976).

22. G. C. Fletcher, Physica **62**, 41 (1972).

23. D. Hackenbracht and J. Kübler, J. Phys. F: Met. Phys. **10**, 427 (1980).

24. J. J. Buiting, J. Kübler, and F. M. Mueller, J. Phys. F: Met. Phys. **13**, L179 (1983).

# MAGNETIC STUDIES OF $Ni_3Al(Fe)$ COMPOUNDS

S. T. Sekula, H. R. Kerchner, J. R. Thompson,[a] and Th. Leventouri[b]
P. O. Box X, Oak Ridge National Laboratory,[c] Oak Ridge, TN 37831

## ABSTRACT

Several samples of the ordered intermetallic compound with the $L1_2$ structure and chemical composition $Ni_{0.75-x}Al_{0.25-x}Fe_{2x}$ (doped with 0.2 at. % B) have been studied over an Fe compositional range $0 \leq 2x \leq 0.20$. Stoichiometric $Ni_3Al$ is known to be a strongly exchange-enhanced material, forming a giant moment system when Fe impurities are introduced in trace amounts. For the greater Fe concentrations employed in this work, more conventional ferromagnetic systems are obtained. Well below the ferromagnetic Curie temperature $T_c$ the samples exhibit a spontaneous moment in small magnetic fields and at higher fields the isothermal magnetization data show a clear tendency toward saturation. Magnetic investigations using vibrating-sample techniques permitted measurements over the temperature interval from 4.2 K to 800 K in fields up to 5.0 T. The Curie temperature was found to increase from ∿50 K for the Fe-free sample in the series to 410 K for the limiting concentration of 12 at. % Fe in the $L1_2$ structure. Samples with higher Fe content up to 20 at. % contain a second, disordered fcc phase with $T_c \simeq 715$ K. Peaks in the temperature dependence of the lattice parameter observed in x-ray studies yield $T_c$ values in reasonable agreement with the magnetic measurements.

## INTRODUCTION

The alloy $Ni_3Al$ is an intermetallic compound with exceptionally interesting magnetic properties. Like $Ni_3Ga$ and Pd metal to a lesser extent, it has a strongly exchange-enhanced magnetic susceptibility, which gives rise to many of its interesting magnetic aspects. In addition to the fundamental interest in this material, it has technical importance as a superalloy possessing many desirable characteristics, including great strength at elevated temperature, oxidation resistance, and so on. These features are not independent, however, for it is well known that the structural stability and solubilities are affected by magnetic contributions to an alloy's energy [1]. Thus, for a series of $Ni_3Al$-based alloys containing Fe, we have investigated magnetic properties including determination of the Curie temperature, the saturation magnetization, and the magnetic moment per Fe atom as a function of iron concentration.

Pure $Ni_3Al$ is highly polarizable and has a critical concentration for ferromagnetism of 74.5 at. % Ni [2]. There have been a number of studies of this material [2,3] which qualitatively and, more recently, quantitatively show it to be a weak itinerant ferromagnet. When a substitutional magnetic species, especially iron, is added, the local magnetic moment polarizes a region around it forming a giant magnetic moment. Previous investigations of $Ni_3Al(Fe)$ have focused on the formation of giant moments in rather dilute alloys containing $\leq 1$ at. % Fe. For example, Mossbauer-effect measurements [4] on samples containing Fe at the parts-per-million level yielded a total moment per Fe atom, $\mu_{Fe}$, of 39 $\mu_B$ (Bohr

[a] Also Department of Physics, University of Tennessee, Knoxville, TN 37996
[b] Permanent address: Solid State Section, Physics Department, Athens University, 104 Solonos, Athens 10680, Greece
[c] Operated by Martin Marietta Energy Systems, Inc. under contract DE-AC05-84OR21400 for the U. S. Department of Energy.

magnetons). Neutron scattering experiments [5] were conducted on materials containing up to 1 at. % Fe. At the higher Fe concentrations used in the work presented here, we are aware of only one other magnetic investigation, a Mössbauer-effect study [6].

In addition, the disordered, binary fcc alloy FeNi is closely related to the ternary alloys investigated here, especially those with higher iron contents. This alloy, γ-FeNi, has also been extensively investigated previously over the whole composition range for which the fcc structure is stable [7,8]. The Curie temperature peaks near 70 at. % Ni, while the low-temperature magnetization increases almost linearly with increasing Fe content over a wide range of compositions until a peak is reached near 40 at. % Ni (almost at the α-γ phase boundary). The strong ferromagnetism of FeNi alloys differs markedly from that of pure $Ni_3Al$. The Curie temperature is hundreds instead of tens of kelvins, and the low-temperature magnetization is of the order of a Bohr magneton per atom roughly ten times that in $Ni_3Al$.

The electronic spins in FeNi are to some extent localized, and neutron-scattering data [9] allow the (composition-dependent) spins associated with each species of atom to be identified. Moreover, analysis of the data for these alloys [10] demonstrate that the interactions between Ni and Fe spins are primarily responsible for the increase in magnetism as iron is added to nickel. The interaction between spins on neighboring Fe atoms encourages antiferromagnetic ordering (neighboring spins having opposite orientation) in the fcc structure.

Our investigation covers the experimentally accessible region of the ternary alloys connecting these two magnetically very different binary alloys. We find a surprisingly smooth transition from the one type of behavior to the other.

## EXPERIMENTAL ASPECTS

As is discussed in a number of articles in this proceedings, the compound $Ni_3Al$ solidifies in the $L1_2$($Cu_3Au$-type) structure for a small range of compositions around stoichiometry. The ordering of Ni and Al atoms in this structure can be preserved with the addition of Fe up to a maximum of 12 at. % Fe so long as one Ni atom and one Al atom are removed for every two Fe atoms added. The materials we investigated were prepared with this stoichiometric substitution of 0, 1, 3, 6, 10, 15, and 20 at. % Fe. These materials were prepared as part of a high-strength-alloy development program at Oak Ridge, and their microstructure and mechanical properties are described elsewhere in this volume by J. A. Horton and coworkers, and by C. T. Liu and C. R. White. (More detail may also be found in Ref. [11].) All except the 1 and 3% alloys and one of the two Fe-free samples we studied contain 0.2 at. % B to improve their mechanical properties. Since we see no evidence - nor do we expect - that this small amount of B affects the magnetic properties of these samples, we disregard its presence in interpreting our data.

Of crucial importance to the magnetic investigation was an anneal after the final fabrication. A few attempts to study unannealed samples (all with relatively low-Fe content) revealed very broad magnetic transitions, making the data generally uninterpretable. The data presented here were obtained from samples annealed for 1 hour at 1000-1050°C.

The magnetic measurements were carried out by using the vibrating-sample technique [12]. Two vibrating-sample magnetometers (VSM) are available in our laboratory: One is used primarily for superconductivity studies and allows samples to be investigated over the temperature range 1.5-100 K in magnetic fields up to 10 T. The other covers the temperature regime 80-800 K and fields up to 7 T. The fundamentals of the techniques are identical in the two facilities. A superconducting solenoid is energized in order to apply a uniform magnetizing field to the sample. Situated symmetrically about the center of symmetry of the field and coaxial with it are two identical pickup coils connected in series opposition. The sample is caused to vibrate in the magnetic field direction about the center of symmetry. When magnetized the vibrating sample induces an ac voltage in the pickup coils proportional to the sample moment.

Figure 1 schematically shows the higher-temperature VSM which is the apparatus primarily employed in this study. The superconducting magnet, a NbTi solenoid located in a liquid-helium dewar, produces a vertical magnetic field. The liquid-helium chamber is isolated on all sides except the very top by a liquid-nitrogen-cooled radiation shield and by a high-vacuum space. For measurements above room temperature, the pickup coils are held near room temperature by convection of helium gas between the experimental region near the bottom of the apparatus and the room-temperature space near the top. In order to reach lower temperatures the pickup-coil chamber is cooled by a liquid-nitrogen bath. The sample is attached to a holder suspended vertically inside the sample chamber from a mechanical vibrator mounted at the top of the cryostat. An electrical heater thermally attached to the outside of the sample chamber is used to raise the sample temperature, which is measured by platinum-film resistance thermometry. Helium exchange gas filling the sample chamber to 100 Torr provides close thermal coupling among the heater, sample, and platinum thermometer. An electronic controller powers the heater so as to maintain a constant sample temperature. Without heat applied both sample and pickup coils come to nearly the same temperature, about 77 K with liquid nitrogen cooling the pickup coils and about 200 K without.

The lower-temperature VSM is cooled by liquid or cold gaseous helium from the helium bath flowing through a capillary. Much lower sample temperatures are of course attainable, and the requirement for thermal isolation between the sample chamber and helium bath is less severe. In addition a composite NbTi-$Nb_3Sn$ superconducting magnet is used in the lower-

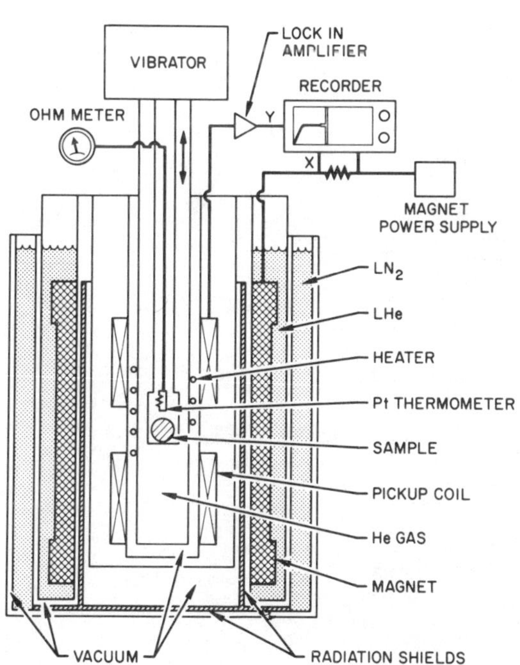


Fig. 1. Schematic diagram of the experimental apparatus used in this work for magnetic measurements in the temperature range 80-800 K.

temperature facility in order to provide the higher fields needed for investigations of high-field superconducting alloys.

The 33 Hz ac-voltage output of the pickup coils is amplified and detected by using a lockin amplifier. We calibrate both systems by measuring the signal due to a nickel magnetic-moment standard. A solenoid calibration by the manufacturer enables us to deduce the applied magnetic field from a measurement of the dc current used to produce it. Finally, standard tables give the temperature from a measurement of the platinum resistance.

## RESULTS

Our investigation shows that magnetism in these $Ni_3Al$-based ternary alloys initially increases strongly with the addition of iron. When the Fe concentration approaches the solubility limit, however, the magnetic properties are relatively insensitive to changes in iron content. Data were acquired in two modes: fixing the temperature, then measuring the alloy magnetization as a function of applied magnetic field (isothermal magnetization); alternately, a fixed field was applied and the temperature varied (isochor).

The isothermal magnetization curves observed are typical of soft ferromagnetic metals. For small applied fields the sample magnetic moment rises linearly in proportion to field with a large, temperature-independent slope. This region is ascribed to the redistribution of microscopic domains, each having the same, spontaneous magnetization but with different directions. After the entire sample is filled with just one domain, further increase in sample moment develops at a progressively slower rate as electron spins are aligned by the field to a greater degree than is associated with the spontaneous magnetization. At fields in the nonlinear (high-field) portion of the magnetization curve, Arrott [13] has expanded the free energy F of a magnetic system as a power series in the magnetization $\sigma$,

$$F = a(T - T_c)\sigma^2 + b\,\sigma^4 + \dots , \qquad (1)$$

where a and b are independent of the temperature T. It follows from the thermodynamic relation $H = dF/d\sigma$ that the (internal) magnetic field H and $\sigma$ are related by

$$H/\sigma = 2a(T - T_c) + 4b\,\sigma^2 + \dots . \qquad (2)$$

Equation (2) is rigorously valid for fields that are not too high and for temperatures near the Curie temperature $T_c$. Thus a plot of experimental values of the ratio $H/\sigma$ versus $\sigma^2$ has a straight region at intermediate H. Extrapolated to the $H/\sigma = 0$ intercept, this straight-line portion of the data yields the square of the spontaneous magnetization when the sample is ferromagnetic. On the other hand when the sample is paramagnetic, such an extrapolation intersects the $H/\sigma$ axis at the reciprocal of the susceptibility $\chi = \sigma/H$. A plot of values of the extrapolated $\sigma^2$ vs T passes linearly through $\sigma^2 = 0$ at the Curie temperature $T_c$.

If Eq. (2) is generalized by replacing $a(T-T_c)$ with an arbitrary function of temperature, the result applies rigorously to weak itinerant ferromagnets such as $Ni_3Al$ at all temperatures, and it reasonably describes at least a portion of the isothermal magnetization curves of all ferromagnets at arbitrary temperature.

The Curie temperature can be determined by a different procedure as well. A small constant field is applied which magnetizes the sample slightly. If the temperature is then varied the induced moment $\sigma$ retains its temperature-independent value, determined by the sample shape, as long as the sample is ferromagnetic. As the temperature is raised through the Curie temperature $T_c$ the moment drops abruptly to the small value $\chi H$ producing a clean, sharp step in a plot of $\sigma$ vs T at the Curie temperature $T_c$.

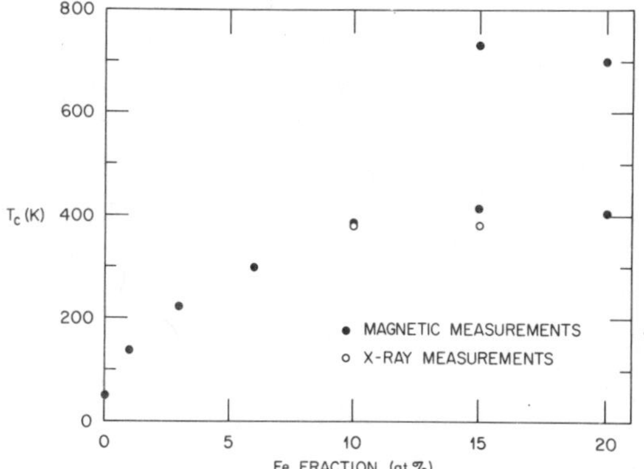


Fig. 2. The measured Curie temperature $T_C$ plotted vs Fe composition. The samples containing 15 and 20 at. % Fe have two Curie temperatures associated with the two phases present in these alloys.

Both methods have been used to determine $T_C$ for several of the samples in order to verify the internal consistency of our interpretation of the data.

The observed Curie temperatures $T_C$ are plotted in Fig. 2 as a function of the iron content for all the samples we investigated. With only small additions of iron, the Curie temperature rises quickly from the low value of 50 K for $Ni_{.76}Al_{.24}$ in this series of alloys. Up to 10 at. % Fe the results are single-valued and fall on a smoothly rising curve. The 15 and 20 at. % alloys both contain two phases according to transmission-electron-microscopy results [11]. Both phases are ferromagnetic, and the Curie temperature of each shows up clearly as steps in plots of $\sigma$ vs T at low field. (The lower $T_C$ could not be determined by the use of Arrott plots.) These two Curie temperatures are almost independent of the overall iron concentration of the samples.

The temperature dependence of the magnetization in a constant, relatively high applied field of 2kOe is shown in Fig. 3 for four of the materials studied. Comparison of these constant-H plots reveals the qualitative difference in the data between the single-phase (6 and 10 at. % Fe) and the two-phase materials (15 and 20 at. % Fe). In these plots, the Curie temperature is located roughly at the inflection point for a single-phase sample. The two-phase samples each have three inflection points - the two Curie temperatures are at the first and third. Extrapolation of the isochor to T = 0 gives a magnetization larger than the spontaneous value by only a relatively small, field-induced magnetization.

The zero-temperature, spontaneous magnetization has been determined for all samples as well. (The lowest-temperature data, measured at 4.2 K, are identical to the zero-temperature values within experimental accuracy.) We have chosen to present these results in Fig. 4 as the total magnetic moment divided by the number of iron atoms present, denoted as $\mu_{Fe}$. This choice is motivated by two considerations: First, the magnetism in these materials can be attributed largely to the presence of iron as evidenced by the rapid increase in $T_C$ with iron addition (see Fig. 2). Second, earlier work on $Ni_3Al$ with small iron additions (less than 1 at. %) demonstrated that the iron atoms take on giant moments, which are consistent with the large values of $\mu_{Fe}$ that we observe for small iron concentrations. The moment per iron atom decreases smoothly as the iron concentration increases. One of the more remarkable results of these data is that no discontinuous change in the smooth downward trend is evident between the single-phase and the two-phase regimes. The average moment per iron atom (and for that matter the moment per average atom if the data were plotted that way) depends only on the iron concentration averaged over the sample.

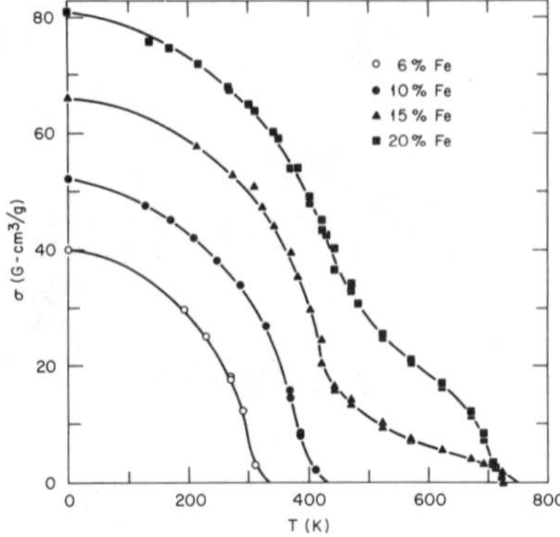


Fig. 3. Isochoric plots of the magnetization per unit mass σ vs temperature T for four samples at an applied field of 2kOe.

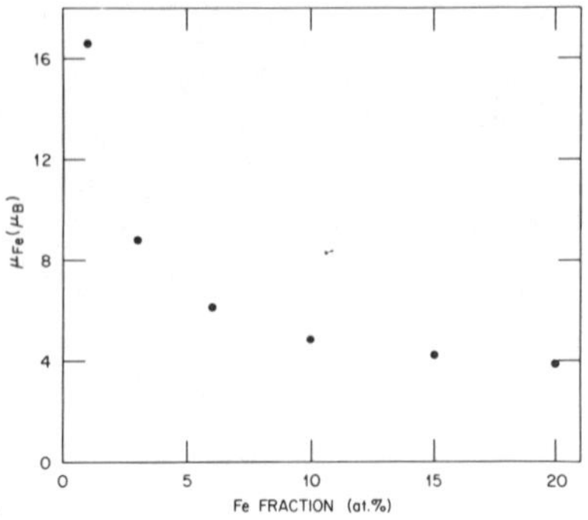

Fig. 4. The average zero-temperature spontaneous moment per Fe atom, $\mu_{Fe}$, plotted vs Fe composition.

DISCUSSION

It is observed in Fig. 2 that the Curie temperature of the ternary alloys increases rapidly with the initial addition of iron. Most previous magnetic studies of Fe-doped $Ni_3Al$ have been concerned with the very low concentration region where $T_c$ is little affected by the Fe content. An exception is Ling and Hicks' investigation [5] of alloys containing up to 1.0 at. % Fe. The variation of $T_c$ observed there is qualitatively similar to the present results, but the actual values of $T_c$ are lower. This is due in part to differing Ni-Al compositions; the Ni content in our 1 at. % Fe alloy is greater, which makes the matrix more polarizable and leads to a higher value for $T_c$. The earlier $T_c$ values [5] may have been affected as well by the necessity to grind the samples, since it is known that this can strongly affect [3] magnetic properties in $Ni_3Al$. The sensitivity to mechanical damage may also have affected the $T_c$ values in a Mossbauer-effect study [6], where fine powder samples with 2.5 and 9.3 at. % Fe were used. In that work, two sets of Curie temperatures bracketing our values are quoted, with substantial differences (∿100 K) between the two sets. Their "approximate" values, estimated from the disappearance of the hyperfine magnetic field [6], lie closer to the $T_c$ values we obtain by more direct measurement.

An x-ray investigation has been performed [14] on two of the alloys containing 10 and 15 at. % Fe, which were prepared from the same ingots as the magnetization samples. The Curie temperature of each was deduced from the ferromagnetic peak in the thermal expansion coefficient and these values are shown in Fig. 2. The agreement between the two measurements is excellent for the 10 at. % sample. The origin of the difference in the 15 at. % Fe alloy, which is somewhat larger than expected, is not understood, but may be associated with the presence of the two phases in this ternary alloy. The zero-temperature spontaneous moment per Fe atom, $\mu_{Fe}$, in stoichiometrically substituted alloys is presented in Fig. 4 as a function of Fe concentration. It is clear that the giant moment persists but its magnitude decreases dramatically with increasing iron. This decrease is qualitatively similar to that observed by Ling and Hicks [5], who found $\mu_{Fe}$ to decrease from 17.7 to 7.7 $\mu_B$ upon increasing the Fe content from 0.2 to 1.0 at. %. As with the $T_c$, we measure a larger value ($\mu_{Fe}$ = 17 $\mu_B$) for 1 at. % Fe because the Ni concentration in our matrix is greater.

Neutron scattering investigations for $Ni_3Al$ [5] and $Ni_3Ga$ [15] have shown that the matrix is polarized by an iron impurity in a volume extending outward for several lattice constants. As the Fe concentration is increased, the polarization clouds overlap very considerably, thereby leading to a reduction in the total magnetic moment associated with each iron atom. For alloys with several at. % Fe, the strong overlap of polarization clouds leads to a comparatively uniform spatial magnetization. This conjecture is consistent with the bulk magnetization of the higher iron-concentration alloys, which is more like conventional than weak itinerant ferromagnets. We note, however, that in the Mössbauer-effect investigation, it was deduced that $\mu_{Fe}$ = 12 $\mu_B$ for both the 2.5 and 9.3 at. % Fe samples [6]. This conclusion is inconsistent with both the present work and previous studies at x < 1 at. % Fe which find that $\mu_{Fe}$ decreases monotonically with increasing Fe content. The origin of this discrepancy is unclear.

Considering now the two-phase regime, we expect that the two phases are nearly unchanged as the average Fe content is increased from 15 to 20 at. %. This expectation is born out by the near equality of both Curie temperatures in these two samples as shown in Fig. 2. We will therefore assume in what follows that there are just two phases in each sample and that they are unchanged. With these assumptions we can learn more about these two phases from the magnetic measurements. (If a third phase is present that somehow escaped recognition by x-ray, transmission-electron-microscopy, and magnetic techniques then our conclusions will have to be altered.)

First, for temperatures above the lower $T_c$, the ratio of magnetizations of the 15 and 20 at.-%-Fe samples is just equal to the ratio of the proportions of high-$T_c$ phase in these two samples. This phase was identified by electron microscopy as disordered $\gamma$ precipitates [11]. Writing $C_{\gamma'}$ for the mass concentration of a species in the $\gamma'$ ($L1_2$, low-$T_c$) phase, $C_\gamma$ for the concentration in $\gamma$, and $C_i$ for the average concentration in each sample, i = 1 (15 at. % Fe) or 2 (20 at. % Fe), two equations for each species relate the compositions and proportions $m_i^p$ of the two phases, p = $\gamma$ or $\gamma'$, in the two samples.

$$C_{\gamma'} m_1^{\gamma'} + C_\gamma m_1^\gamma = C_1 \tag{3a}$$

$$C_{\gamma'} m_2^{\gamma'} + C_\gamma m_2^\gamma = C_2 \tag{3b}$$

Using the ratio $m_2/m_1$ = 2.7 ± 0.3 deduced from the magnetic measurements leads immediately to the composition of $\gamma'$: $C_{\gamma'}$ = 0.13 ± .005 or 12 at. % for Fe, $C_{\gamma'}$ = 0.77 ± .01 or 69 at. % for Ni, and $C_{\gamma'}$ = 0.10 ± .005 or 19 at. % for Al. This result is in good agreement with other determinations [16] of the limiting composition of the $\gamma'$ phase at 1000°C (the temperature of the final anneals).

The measurements of the low-temperature magnetizations $\sigma_i$ of the two-phase samples can be used at least to characterize the magnetization $\sigma_{\gamma'}$ of the $\gamma'$ phase. If $\sigma$ is substituted for C in Eqs. (3), with the experimental values of $\sigma_i$(i = 1,2) we find $\sigma_{\gamma'}$ = 58.4 ± 1.5 G-cm$^3$/g or $\mu_{Fe} = \sigma_{\gamma'}/C_{\gamma'}$ = 4.52 $\mu_B$/Fe atom. Note that this value falls on a smooth interpolation of the data in Fig. 4, again emphasizing the absence of a substantial discontinuity in magnetic properties across the $\gamma'$ phase boundary.

Unique characterization of the $\gamma$ phase requires additional information. The Curie temperature $T_c$ = 715 ± 20 K is just slightly below the value $T_c$ = 790 K [8] for $\gamma$ - $Fe_{.5}Ni_{.5}$ which is the composition consistent with the absence of aluminum in the high-$T_c$ phase. The presence of a small amount of Al requires the Ni (Fe) concentration to be larger (smaller) than 50 at. %. Although we could not find data on the Al-dependence of $T_c$ in $\gamma$-FeNi one might reasonably assume that $T_c$ drops roughly at the rate of 25 K/at. % Al as it does in AlNi [17]. With this figure, the observed Curie temperature requires 3 at. % Al, 53 at. % Ni, and 44 at. % Fe. Since this reduction in iron content tends to increase $T_c$, partially offsetting the effect of the aluminum, this estimate is probably slightly low for Al and high for Fe. With the above composition, Eqs. (3) lead to $m_1$ = .09 and $m_2$ = .25 for the mass fractions of the $\gamma$ phase in the 15 and 20 at. % Fe samples, respectively. Using this information, the $\sigma$ analogue of Eqs. (3) leads finally to $\mu_{Fe}$ = 3.4 $\mu_B$, which is comparable to the local moment on an iron site, 3.0 $\mu_B$, measured by scattering experiments [5,9,15].

Our experimental results provide a consistent picture of the magnetic properties in the ternary fcc AlFeNi system which connects the previously investigated regions of binary FeNi and of ordered $Ni_3Al$ with small Fe concentrations. The giant-moment picture, appropriate to dilute arrays of Fe impurities in $Ni_3Al$, gradually fades with the addition of iron and is finally replaced with a model of strong metallic ferromagnetism appropriate to FeNi. The change from one regime to the other varies smoothly with Fe content - no discontinuity in $\mu_{Fe}$ is observable even at the $\gamma'$-$\gamma$ phase boundary. The absence of any such discontinuity may arise from the fact that the iron, which plays a decisive role in determining the magnetic properties, is disordered in both phases.

## ACKNOWLEDGEMENTS

We are grateful to C. T. Liu, A. Das Gupta, O. B. Cavin, R. K. Williams, and Y. K. Chang for providing the samples studied in this work. We also gratefully acknowledge helpful discussions with J. S. Faulkner, J. W. Cable, C. T. Liu, and R. K. Williams.

## REFERENCES

[1] A. P. Miodownik, Bull. Alloy Phase Diag. 2, 406 (1982).

[2] H. Sasakura, K. Suzuki, and Y. Masuda, J. Phys. Soc. Japan 53, 754 (1983).

[3] F. R. de Boer, C. J. Schinkel, J. Biesterbos, and S. Proost, J. Appl. Phys. 40, 1049 (1969).

[4] P. R. Liddell and R. Street, J. Phys. F: Metal Phys. 3, 1648 (1973).

[5] P. C. Ling and T. J. Hicks, J. Phys. F.: Metal Phys. 3, 697 (1973).

[6] J. R. Nicholls and R. D. Rawlings, Acta. Metall. 25, 187 (1977).

[7] M. Hansen, Constitution of Binary Alloys, 2nd ed., (McGraw-Hill, New York, 1958) pp. 677-684 and references contained therein.

[8] J. Crangle and G. C. Hallen, Proc. Roy. Soc. A272, 119 (1962).

[9] C. G. Shull and M. K. Wilkinson, Phys. Rev. 97, 304 (1955).

[10] J. S. Kouvel, Magnetism and Metallurgy, edited by A. E. Berkowitz and E. Kneller, Vol. 2 (Academic Press, New York, 1969).

[11] C. T. Liu, W. Jemian, H. Inouye, J. V. Cathcart, S. A. David, J. A. Horton, and M. L. Santella, ORNL 6067, Martin Marietta Energy Systems, Oak Ridge National Laboratory, 1984.

[12] S. Foner, Rev. Sci. Instrum. 30, 548 (1959).

[13] A. Arrott, Phys. Rev. 108, 1394 (1957).

[14] Th. Leventouri, O. B. Cavin, and J. S. Faulkner, to be published.

[15] J. W. Cable and H. R. Child, A.I.P. Conf. Proc. No. 10 (American Institute of Physics, New York, 1973) p. 1623.

[16] S. Ochiai, Y. Oya, and T. Suzuki, Acta. Metall. 32, 289 (1984).

[17] M. Hansen, op. cit., p. 119.

SILICON-NICKEL COMPOUNDS BY ION IMPLANTATION

S. G. B. MAYER+, F. F. MILILLO**, AND D. I. POTTER+
+Department of Metallurgy, Institute of Materials Science, University of Connecticut, Storrs, Connecticut 06268
**Department of Mechanical Engineering, Union College, Schenectady, New York 12308

## ABSTRACT

Compounds of silicon and nickel have been formed by implanting silicon ions into nickel at various fluences and temperatures. Temperature during implantation is a major factor controlling implanted ion concentrations and phase developments. The ordered phase $Ni_3Si$, which forms during implantation at 600°C, is replaced by $Ni_5Si_2$ during implantation at 400°C or below. In contrast to aluminum implantation of nickel, crystalline phases form even at high fluences ($>1 \times 10^{18}$ ions/$cm^2$) and at low temperatures (<200°C) in silicon implanted nickel.

## INTRODUCTION

Ion implantation is a technique for changing the chemical composition and properties of materials surfaces without affecting the bulk. By this method a beam of ions that have been accelerated to kinetic energies in the 100-200 keV range penetrates and becomes part of the surface material to a depth of 0.1 micron or less. Phase transformations occur as a result of both the chemical changes and the radiation effects accompanying implantation.

The present research reports the formation of compounds of nickel and silicon by implantation of silicon into nickel. In the nickel-silicon phase diagram [1], a face centered cubic solid solution of silicon in nickel exists up to approximately 10 atomic percent silicon. At higher silicon content there are a series of compounds having very narrowly-defined stoichiometries. $Ni_3Si$, an ordered phase of the $L1_2$ structure, is among the phases produced in the present research by ion implantation. The conditions of fluence, temperature and subsequent aging at elevated temperature leading to this phase or others are reported below.

The problem of phase formation by ion implantation leads to two general questions: What fluence is needed to achieve a given concentration of implanted ions in the near-surface layer? Then, given the concentration of implanted ions, what conditions favor stability of one phase over another?

## EXPERIMENTAL

Pure nickel specimens were prepared in the shape of discs, 0.25 mm thick and 3 mm diameter, chemically polished on one face. These were mounted in an ultrahigh vacuum chamber on a stage equipped for temperature regulation between room temperature and 600°C. The chamber was connected to a Varian/Extrion DF4 type 200 KV accelerator, which provided a beam of $Si^+$ ions at 175 keV and a flux of approximately $1.5 \times 10^{14}$ ions/$cm^2$s. The specimens were implanted on their polished faces. In one experiment, the implantation was conducted at nominal room temperature to fluences between

$7.8 \times 10^{17}$ and $3.1 \times 10^{18}$ ions/cm$^2$. Effects of beam heating, generally about 150°C, were noted using infrared pyrometry. Several specimens at each fluence were subsequently aged at 600°C for one hour, under a vacuum of $10^{-8}$ Torr. Composition profiles were determined by Rutherford backscattering (RBS) and Auger electron spectroscopies (AES). Phases present were identified by transmission electron microscopy (TEM) diffraction patterns, and their crystallographic and morphological relations were observed using TEM. Specimens were prepared for this purpose by chemical thinning from the back to the implanted surface.

In another experiment, nickel specimens were implanted to between $0.31 \times 10^{18}$ and $0.78 \times 10^{18}$ ions/cm$^2$. This fluence range was selected to highlight the relative stability of $Ni_5Si_2$ versus $Ni_3Si$. Beam-heating was eliminated in these specimens, and implantation temperatures were set at intervals between 25°C and 600°C, with heat provided by electron emission heaters on the stage. Specimen temperatures were monitored by infrared pyrometry. These specimens were also examined by RBS and AES for composition profiles and by TEM for characterization of phases and microstructures.

## RESULTS

### 1. Composition Profiles

Figure 1a shows a typical profile of silicon concentration versus depth determined by RBS from a specimen implanted at 25°C. The profile is characterized by a region of aopproximately constant composition extending to about 1500 Å, then decreasing to negligible values at depths less than 3000 Å. The effect of implantation temperature below 400°C was relatively small, as described below.

Figure 1b shows the peak silicon concentration implanted at 150°C, determined by RBS, as a function of fluence. Increasing fluence results in higher maximum concentrations. The relationship between fluence and maximum concentration is not linear. At higher fluences the curve rises more slowly, suggesting that there may be a limiting implanted composition somewhat beyond the maximum achieved in the present experiments.

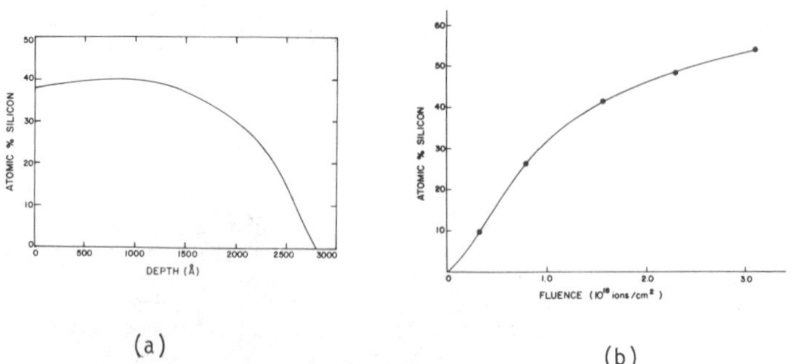


Figure 1: (a) as-implanted concentration profile, $1.5 \times 10^{18}$ $Si^+$/cm$^2$, 25°C, by RBS (b) maximum concentration vs. fluence, specimens implanted at 150°C.

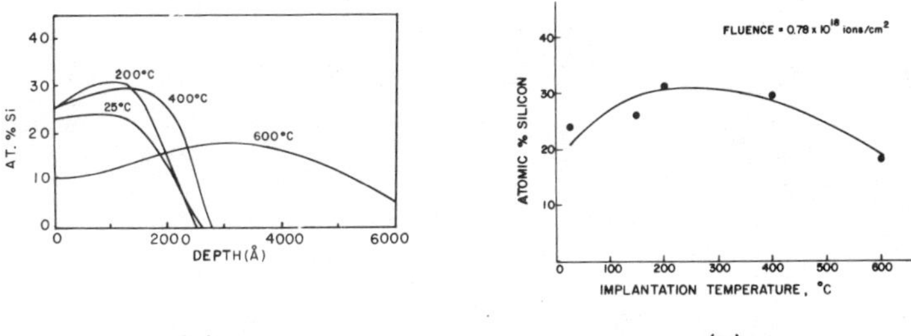


(a) (b)

Figure 2: (a) concentration profiles, $0.78 \times 10^{18}$ ions/cm$^2$, by RBS; (b) maximum concentration vs. temperature.

When the implantations are carried out at elevated temperatures, greater total amounts of silicon are incorporated into the substrate. This is evidenced by the increasing area under the curves in figure 2a. The four concentration profiles of figure 2a were all determined by RBS for the same fluence, $0.78 \times 10^{18}$ ions/cm$^2$. The increasing depth of penetration between 200 and 600°C results in lower maximum silicon concentrations. The maximum silicon concentrations as a function of implantation temperature are shown in figure 2b. Below about 300°C the concentration of silicon implanted increases with temperature. Above 300°C this concentration decreases with increasing temperature.

2. Phases and Microstructure

Beginning at lower fluences ($10^{17}$ ions/cm$^2$), implanting silicon ions into nickel causes a break-up of the original nickel grains into subgrains at the specimen surface. Figure 3 shows results typical of implantation to $0.78 \times 10^{18}$ ions/cm$^2$ at room temperature. The micrograph, figure 3a, displays many subgrains within what was, before implantation, a single grain. Such a microstructure gives rise to several characteristic diffrac-

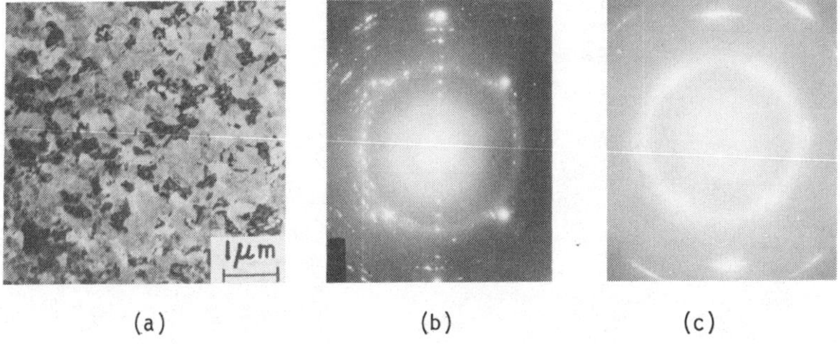


(a) (b) (c)

Figure 3: (a) microstructure, (b) and (c) diffraction patterns after implanting $0.78 \times 10^{18}$ ions/cm$^2$ at 25°C.

tion features, shown in figures 3b and 3c. Figure 3b is consistent with the presence of a second phase. The line of closely spaced spots was indexed and showed that the second phase is $Ni_5Si_2$, an hexagonal material (the spots lie along the [001] direction). The numerous closely-spaced arcs are also $Ni_5Si_2$ reflections, and originate from small, misaligned regions of the compound. Figure 3c shows a strongly textured polycrystalline face centered cubic nickel diffraction pattern. This shows that the nickel grain has been replaced by small subgrains of slightly different orientations. The diffuse ring indicates the presence of an amorphous phase.

Figure 4 shows microstructures and diffraction patterns from a higher fluence specimen, $1.5 \times 10^{18}$ ions/cm$^2$, implanted at 150°C. The silicon concentration is about 41 atomic percent, as measured by RBS. A crystalline phase with parallel arrays of dislocations is observed in figure 4a. Figure 4b shows an amorphous phase. Compositional analysis by energy dispersive x-ray spectroscopy in an analytical electron microscope revealed that the amorphous phase is almost pure silicon. This amorphous phase is generally observed as a film in the thinnest regions of the specimens. It is not observed in compositional profiles by RBS or AES.

The phases which form during ion implantation are not necessarily those predicted on the basis of the equilibrium phase diagram. This is demonstrated in several figures which follow. All three diffraction patterns of figure 5 come from specimens implanted to the same fluence, $0.31 \times 10^{18}$ ions/cm$^2$. The lines of closely spaced reflections in figure 5b, like those in figure 3b, arise from the $Ni_5Si_2$ phase. Under equilibrium conditions this compound would not exist at compositions below ∿25 atom percent silicon; from ∿25 to ∿28 atom percent silicon, $Ni_3Si$ and $Ni_5Si_2$ coexist under equilibrium. During ion implantation at 400°C $Ni_5Si_2$ has formed, mixed with face centered cubic nickel-silicon solid solution,

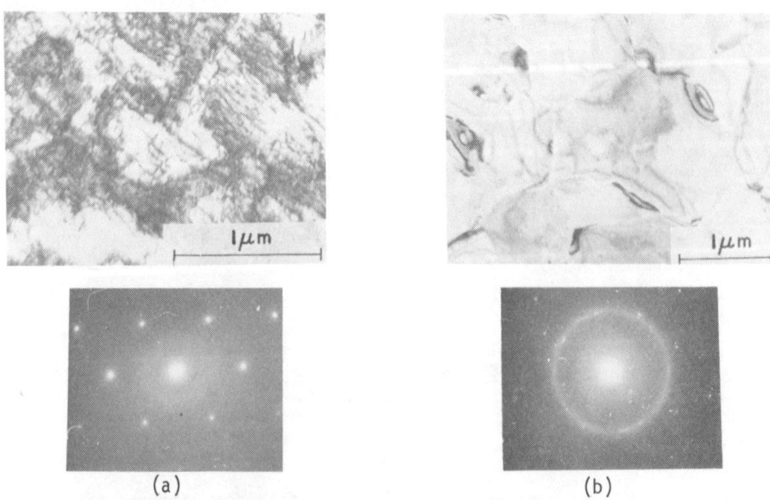


Figure 4: (a) Crystalline phase, and (b) amorphous phase observed after implanting $1.5 \times 10^{18}$ ions/cm$^2$ at 150°C.

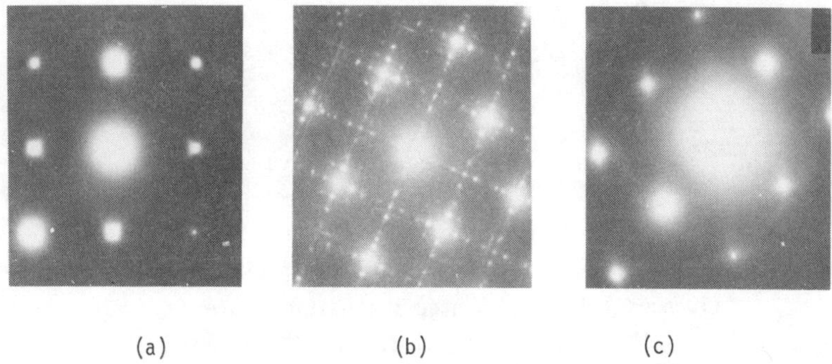

(a) (b) (c)

Figure 5: (a) 200°C, 10 at % Si; (b) 400°C, 14 at % Si; (c) 600°C, 11 at % Si; all at fluence 0.31 x $10^{18}$ ions/$cm^2$.

at only 14 atomic percent silicon overall. For comparison, figures 5a and 5c show diffraction patterns of face centered cubic nickel-silicon solution.

Figure 5 suggests that implantation temperature may be as important as overall composition in favoring formation of $Ni_3Si$ or $Ni_5Si_2$. The role of temperature is demonstrated again in figure 6. The four specimens providing these diffraction patterns were all implanted to the same fluence, 0.78 x $10^{18}$ ions/$cm^2$. The superlattice indicative of $Ni_3Si$ is seen only in figure 6d, the diffraction pattern from the specimen implanted at 600°C.

Specimens that were implanted at 150°C, then aged for one hour at 600°C, also show the presence of ordered $Ni_3Si$. The as-implanted compositions are as noted. The specimen implanted to 2.3 x $10^{18}$ ions/$cm^2$, then aged, shows $Ni_5Si_2$ as well as $Ni_3Si$. The basal plane of $Ni_5Si_2$ is shown in figure 7c, and figure 7d shows a string of spots from the [100] direction of $Ni_3Si$ together with lines of reflections along the [001] direction of $Ni_5Si_2$. Both of these diffraction patterns came from the same specimen.

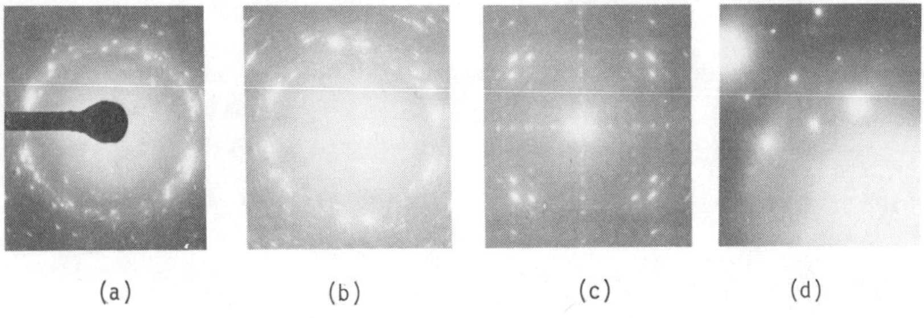

(a) (b) (c) (d)

Figure 6: (a) 25°C, 24 at % Si; (b) 200°C, 31 at % Si; (c) 400°C, 29 at % Si; (d) 600°C, 18 at % Si; all at fluence 0.78 x $10^{18}$ ions/$cm^2$.

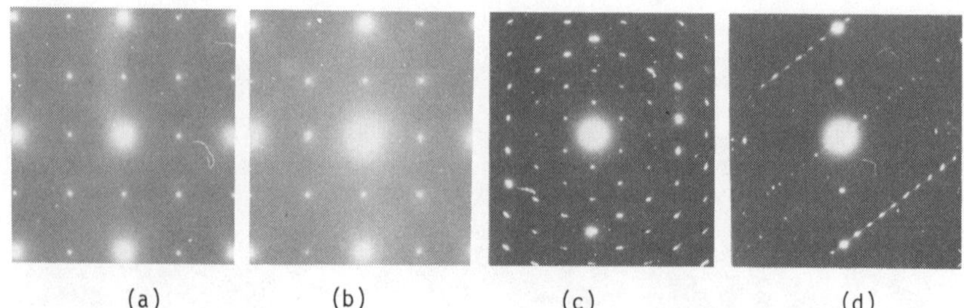

(a) (b) (c) (d)

Figure 7: Effect of aging 1 hour at 600°C on specimens implanted at 150°C; fluences and as-implanted compositions are (a) 0.78 x $10^{18}$ ions/cm$^2$, 25 at % Si; (b) 3.1 x $10^{18}$ ions/cm$^2$, 54 at % Si; (c) and (d) 2.3 x $10^{18}$ ions/cm$^2$, 48 at % Si.

A final feature noted here is the persistence of crystalline phases at low temperatures and relatively high fluences of silicon implanted into nickel. Figure 8 shows diffraction patterns from implantations at 150°C to fluences of 1.5 x $10^{18}$ and 3.1 x $10^{18}$ ions/cm$^2$. They should be viewed in comparison with figures 3b and c. Each of the patterns in figure 8 results from a single crystal contained in the selected area aperture, and the subgrain structure noted in figure 3 is no longer present at these higher fluences. In contrast, nickel implanted with aluminum ions to the same fluence is amorphous [2].

## DISCUSSION

The results show that compounds of silicon and nickel can be made by ion implantation. They also show that implantation temperature and subsequent heat treatment are major factors determining implanted ion concen-

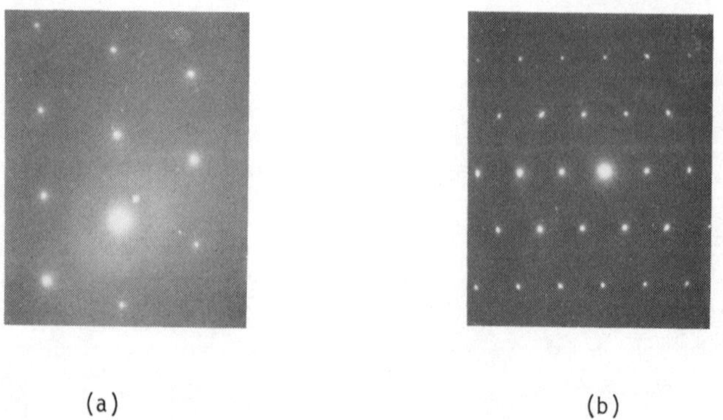

(a) (b)

Figure 8: (a) 1.5 x $10^{18}$ ions/cm$^2$, (b) 3.1 x $10^{18}$ ions/cm$^2$, as-implanted at 150°C.

trations and distributions and the phases which form. These findings may be interpreted in the light of models that have been developed through radiation damage studies of nickel-silicon alloys.

Implanted concentration profiles are considered first. The leveling off of silicon concentration with increasing fluence, figure 1b, indicates that sputtering limits the achievable concentrations. The limiting composition appears to be beyond the highest fluence used in the present study. The role of sputtering is also seen in the temperature dependence of implanted profiles, figure 2a. As temperature increases, the amount of material removed by sputtering appears to decrease. This is evidenced by two points: one, the increased total area under the curves with increasing temperature, and two, the greater maximum concentration and average implanted depth at 200°C compared with 25°C. At 400°C and, notably, 600°C, diffusion of implanted ions into the substrate becomes a major factor, causing deeper penetration of the implanted ions.

A rapid increase in diffusion at temperatures 400°C and above is consistent with phase developments reported above. The main points are as follows: $Ni_3Si$ is absent from specimens implanted at temperatures below 600°C, figures 5 and 6. At these lower temperatures, where the silicon concentration exceeds the solubility limit for the primary solid solution, $Ni_5Si_2$ forms rather than $Ni_3Si$. A comparison between figures 5b and 6d makes this point most clearly. $Ni_5Si_2$ contains about 28 atomic percent silicon, $Ni_3Si$ about 24. The specimen of figure 5b, implanted at 400°C, with a total concentration of 14 atomic percent silicon, contains $Ni_5Si_2$. The silicon concentration in the specimen of figure 6d, implanted at 600°C, is higher - 18 atomic percent. Yet this specimen contains the lower-silicon compound $Ni_3Si$.

This result reflects two competing processes, radiation-induced disordering and thermal re-ordering. Disordered $Ni_3Si$ is essentially a random solution of silicon in fcc nickel. The free energy advantage due to long range order is lost and the solution is greatly supersaturated. Potter [3] has shown that under such circumstances a mixture of fcc solid solution and $Ni_5Si_2$ has the lowest free energy. (Refer to figure 9, p. 536 of that reference for details.) However, if the rate of thermal diffusion is high enough, atoms displaced by radiation damage continuously return to their ordered positions. Indeed it has been shown that there exists a threshold temperature, depending on irradiation conditions, above which $Ni_3Si$ is not disordered by irradiation [4].

The presence of $Ni_3Si$ in aged specimens, figure 7, may be attributed both to reduction in silicon concentrations by increased diffusion into the substrate, and to thermal reordering.

The amorphous phase observed in the present work, figure 4b, appears to be almost pure silicon. This phase is generally observed as a film on the implanted surface in the thinnest regions of the specimens. Radiation damage studies of the nickel-silicon system show that silicon tends to segregate to grain boundaries and free surfaces during irradiation [5]. However, it is not certain that the silicon films observed on the present specimens were formed by radiation induced segregation. The Rutherford Backscattering and Auger electron spectroscopies employed did not detect a silicon film on the surface of any specimen. This suggests that the film is extremely thin (<50 Å). More work is needed to characterize its structure and origin.

## CONCLUSIONS

The results described above lead to the following conclusions:

1. Phases produced by ion implantation are not necessarily equilibrium phases for a given composition. In particular, the ordered phase $Ni_3Si$ does not form at 400°C or below, where thermally activated diffusional processes that promote re-ordering are not sufficient to overcome the disordering due to radiation damage.
2. Temperature plays an important role in determining which phases will form at a given fluence. This role is expressed in two ways. First, temperature affects the concentration profiles of implanted ions and hence their maximum concentrations. Also temperature affects the stability of the ordered phase $Ni_3Si$.
3. Heating the substrate during implantation results in an increase in maximum implanted composition with temperature below 300°C and a decrease with temperature above 300°C. The behavior above 300°C is associated with a rapid increase in diffusion of the implanted ion into the substrate.
4. Amorphous phases were not observed except for a very thin layer of silicon on the specimen surface.

## ACKNOWLEDGMENTS

The authors acknowledge the technical assistance of J. Hampikian, L. McCurdy, L. Witherell and J. Hall. This research was supported by grant number DMR8006084 from the National Science Foundation.

## REFERENCES

[1] American Society for Metals, Metals Handbook, 8th ed, Vol. 8, "Metallography, Structure and Phase Diagrams" (ASM, Metals Park, Ohio, 1973) 325.
[2] D. I. Potter, M. Ahmed and S. Lamond, in "Ion Implantation and Ion Beam Processing of Materials (edited by G. K. Hubler, O. K. Holland and C. R. Clayton), p. 117. North Holland Publishing Co. Inc., New York (1984).
[3] D. I. Potter, in: Phase Stability During Irradiation, eds. R. Holland, L. Mansur and D. Potter, AIME Conference, Pittsburgh, PA (October 5-9, 1980) 521-546.
[4] D. I. Potter and O. G. Hernandez, Acta Metallurgica, 29 (1981) 187.
[5] P. R. Okamoto and L. E. Rehn, J. Nucl. Mater., 83 (1979) 2.

# PSEUDOELASTICITY AND PSEUDOTWINNING IN ORDERED SHAPE-MEMORY ALLOYS

A. ZANGWILL* and R. BRUINSMA**

*Polytechnic Institute of New York, Brooklyn, NY 11201
**University of California, Los Angeles, CA 90024

ABSTRACT

Shape-memory behavior is a complicated phenomenon which intimately relates macroscopic strain recovery with martensitic phase transformation. A vital step in the theoretical understanding of this effect is a clear picture of a superficially rather simpler phenomenon: martensitic pseudo-elasticity. Within the martensitic phase, a number of ordered alloys exhibit deformation strain recovery which is reminiscent of rubber elasticity. The deformation is observed to occur via coherent motion of parallel twin boundaries. In this case, where long range elastic accomodation forces play no role, the origin of the restoring force on the twin boundaries is still unclear. In this work, we present a quantitative theory which generalizes previous suggestions in the literature and unifies this phenomenon with elastic mechanical untwinning. Our description is based on the concept of a "pseudotwin" originally proposed by Laves and later elucidated by Cahn. Here, the motion of a twin boundary generates a new metastable crystallographic structure (the pseudo-twin) of locally higher free energy. This notion not only provides a source for a "volume" restoring force but leads to a natural description of observed stabilization effects. Specific experiments to test our description will be proposed.

## Introduction

The shape-memory effect is a remarkable mechanical phenomenon found in certain metallic alloys wherein a specimen deformed in its low-temperature martensite phase will, when the applied stress is released, regain its original undistorted shape on heating. The alloys which exhibit this behavior typically are ordered and exhibit a crystallographically reversible thermoelastic martensitic phase transformation [1]. Shape recovery occurs during the reverse transformation of the material from the martensite phase to its high temperature equilibrium phase. At the level of the optical microscope, a crystallographic understanding of this effect appears to be in hand [2]. In particular, elastic stress relief by interconversion of martensite variants is an essential feature. Unfortunately, little if anything is known about detailed microscopic mechanisms.

A subclass of shape-memory alloys exhibits an additional unusual effect. Here, the morphology of the martensite is best described as an array of parallel twin bands. External stress causes the twin boundaries to migrate in order to maximize the volume of the twin orientation best able to accomodate the stress. However, stress release results in a return of the boundaries to their original positions. Complete strain recovery of significant magnitude can be achieved since macroscopic excursions (~ 1 mm) of the twin boundaries are typical. Essentially identical effects are found for Au-Cd, Au-Cu-Zn, Cu-Al-Ni and Cu-Zn-Al, all in their low-temperature martensite phase. This behavior is surprising because true crystallographic twins have identical structures so that the system has precisely the same free energy both before and after application of the stress. The integrity of the straight parallel twin boundaries insures that conventional elastic accomodation forces are of negligible importance in this case. The central question is: what is the origin of the restoring force on the twin interfaces?

In the past, a number of explanations have been put forward to acccount for this phenomenon, variously termed "rubber elasticity" [3] and "twinning pseudoelasticity" [4]. For example, Birnbaum and Read [5] suggested that the moving boundaries in Au-Cd generate twinning dislocations that leave behind a trail of untwinned inclusions. The latter exert a restoring force on the released boundary. For the same material, a detailed crystallographic study led Lieberman et. al. [6] to conclude that the conversion of one twin orientation to its mirror image as the boundary moves under stress in fact requires intracell diffusional motion in addition to the twinning shear. The "out-of-position" atoms exert a restoring force on the twin boundary on a cell-by-cell basis due to their higher local free energy. Finally, in their study of rubber elasticity in Cu-Al-Zn, Ahlers et. al. [7] considered the restoring force which would accompany the change in free energy when a twinning shear is applied to a long-range ordered martensite with a particular type of short-range disorder.

The purpose of this work is to demonstrate that all of these mechanisms may be subsumed into a very general picture which focuses on the nature of twinning itself. The main idea, implicit in the work of Cahn [8], is that the conversion of one twin orientation to another under stress is practically impossible for an ordered alloy. Instead, a metastable "pseudo-twin" is formed which is higher in free energy than the equilibrium twin crystal structure. In principle, this is all that is needed to account for a twin boundary restoring force. As a function of time, this metastable state converts to the lower energy "true" twin by atom diffusion. Hence, the existence and time history of pseudoelasticity in any particular case will depend on the detailed nature of the metastable structure and its mobility. In what follows we shall consider a fairly general model which embodies this idea and investigate some of the consequences. In particular, our analysis indicates that the dynamics of twin boundary motion merits greater experimental attention. A shorter version of is work with a somewhat different emphasis has appeared previously [9].

## Pseudotwinning

It was first predicted by Laves [10] that atomic ordering would make deformation twinning of an alloy very difficult if not impossible. For any crystal which must be described as a lattice with a basis a twinning shear may generate the appropriately rotated Bravais lattice without properly arranging the basis atoms within the unit cell [11]. The new structure, while not the true crystallogrphic twin, is a well-ordered structure which must have slightly higher free energy than the equilibrium twin structure. This "pseudotwin" has been explicitly identified in recent diffraction studies of mechanically deformed β Fe-Be [12].

In an elegant paper, Cahn [8] has considered how the symmetry of a crystal is affected by an arbitrary homogeneous strain generated by a twinning shear. Generally, a strain produces a lower symmetry structure as symmetry elements of the original structure are destroyed. The symmetry group of the resultant structure is the largest common subgroup of the original symmetry group and the strain. For the cases of either a perfectly ordered crystal without a basis or a perfectly disordered alloy, particular values of the strain may reconstitute the original symmetry group. This represents creation of the true twin. In certain exceptional cases, a unique strain may create entirely new symmetry elements. This could occur if the strain were applied to a crystal not originally in its equilibrium

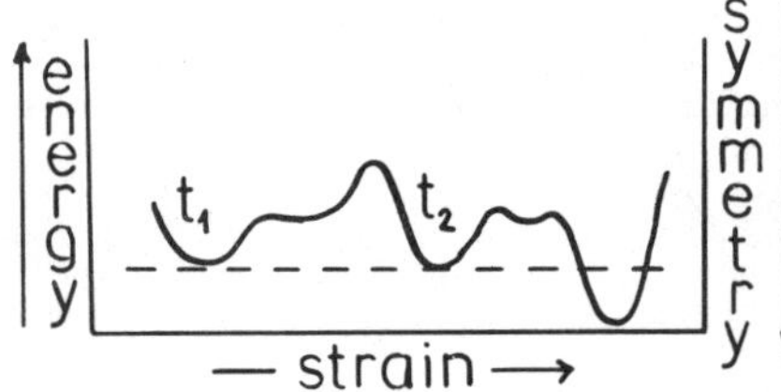


Figure 1

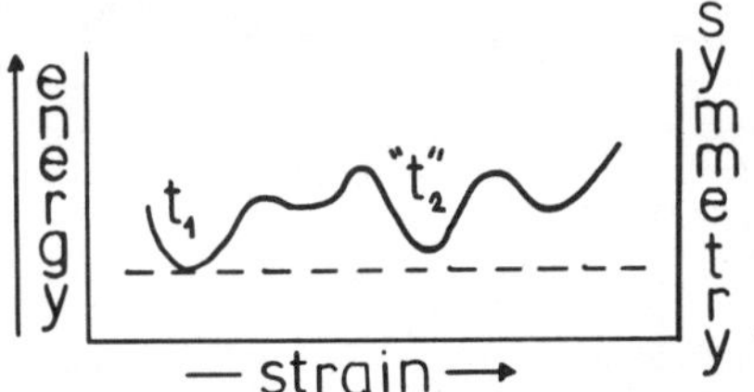


structure. These situations are illustrated in the top panel of Figure 1. Here, $t_1$ and $t_2$ denote two symmetry-related crystallographic twins; a hypothetical lower energy equiblibrium structure is also indicated.

The lower panel of Figure 1 illustrates the situation most often encountered in real alloys: The crystal has multiple atoms/cell and some disorder must be present at non-zero temperature. One begins in the ground state structure, $t_1$, and all other structures which can be reached via a twinning shear have lower symmetry and higher free energy. At best, some structures can be metastable, like the pseudotwin "$t_2$". Again quite generally, one can expect that diffusional atom motion or shuffles will ultimately restore the pseudotwin to the true twin, "$t_2$" $\rightarrow$ $t_2$. If the pseudotwin differs from the true twin only by a non-equilibrium distribution of short range disorder [7,9] a "volume" restoring force will certainly be present. However, additional insight can be gained if we restrict ourselves to the case where each unit cell is sheared to the metastable pseudotwin structure. This is the model for Au-Cd proposed by Lieberman et. al. [6] and also suggested by Tsunekawa [13] to account for pseudoelasticity in the non-metal $NdNbO_4$.

Our first goal is to obtain an expression for the free energy of an alloy which can differentiate the intracell atomic arrangement of the twin from the pseudotwin. In principal, this requires an explicit calculation of both the internal energy and the entropy. In practice, we shall adopt a coarse-grained phenomenological free energy expression which captures the essential physics of the problem [14]:

$$\mathcal{F} = \int \frac{d\vec{r}}{a^3}\left[\frac{1}{2}J(\nabla e)^2 - \frac{1}{2}re^2 + \frac{1}{4}ue^4 - he\right]. \quad (1)$$

The integral is over all of space and $a^3 = \Omega$ is the unit cell volume so the quantity in square brackets is a free energy density. $e(\vec{r})$ is a generalized displacement coordinate which varies continuously as atoms are shuffled within the unit cell. The final three terms, a polynomial in the variable $e(\vec{r})$, is sketched below in Figure 2. $e^+$ represents the atomic arrangement in the equilibrium crystal structure and $e^-$ represents the atomic arrangement in the pseudotwin structure. The diffusion barrier, $E_{ab}$, here represented with the one-dimensional coordinate, $e(\vec{r})$, actually should be imagined in the multidimensional space of all the atomic coordinates. Finally, $\Delta f(T)$ is the temperature dependent increase in free energy/cell as the twin boundary moves one lattice spacing.

To complete the model we need to take account of the elastic coupling between adjacent unit cells. Atomic shuffling in one cell tends to induce shuffling in its neighbors so that $e(\vec{r})$ tries to achieve a uniform value throughout the crystal. The strain gradient term in equation (1) accomplishes this effect. The three phenomenological constants r, u and h are obtained from numerical values of $e^{\pm}$, $E_{ab}$ and $\Delta f$. The former two must be of order 1 Å and 0.5 eV respectively while $\Delta f$ can be read directly from a stress-strain curve since the restoring stress is simply $\sigma_r = \Delta f(T)/\Omega$. Typically this is of order 10 meV. In this model, the dispersion of the optical phonon frequency, $\omega(k)$, is given by,

$$\frac{1}{2} m\omega^2 = r + \frac{1}{2} J k^2, \tag{2}$$

where m is the reduced mass. Hence, typical phonon data for the heavy metals which comprise pseudoelastic alloys yields an estimate of $J \simeq 3$ meV.

Although we have not done so here, one can imagine a first-principles calculation of all of the parameters which enter the free energy functional. In particular, such an approach should shed light on two puzzling questions. First, why do only a specific limited class of alloys exhibit pseudoelastic behavior? Second, why is the phenomenon confined to a very limited concentration range of the specified materials? We regard these as critical questions which deserve further theoretical study. In addition, the temperature dependence of these quantities, particularly $\Delta f(T)$, could be found from a statistical mechanics calculation at varying levels of sophistication [9]. Here, we will take the attitude that the Cahn analysis provides an origin of the restoring force and explore some of the consequences of our model for the dynamics of twin boundary motion.

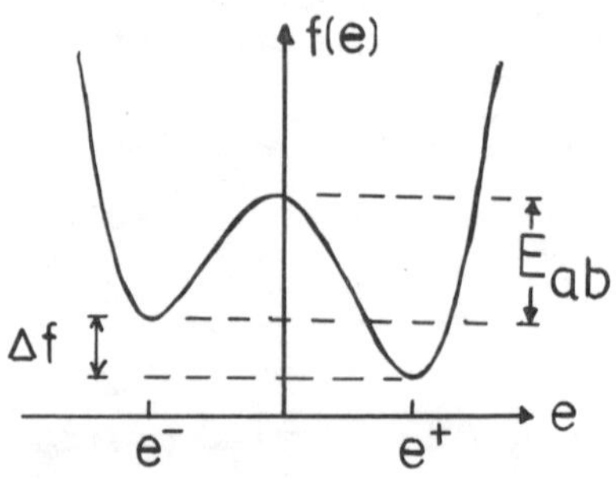


Figure 2

Dynamics of Pseudoelasticity

The experimental fact of rubber elasticity described above is not the entire story of pseudoelasticity in shape-memory alloys. In fact, if the applied stress is held fixed for a sufficiently long time ($t > \tau$) before release the twin boundaries do not return completely, resulting in plastic deformation rather than elastic recovery. The stabilization time is observed [6,13] to depend on temperature as $\exp(\Delta/T)$ with $\Delta \sim 1$ eV. This type of activated behavior is not unexpected in our model. Suppose the diffusion barrier is very low. Then, even though the pseudotwin structure is formed at the moment the twinning shear is applied, the "out-of-place" atoms will rapidly diffuse to their correct twinned positions and arrest any restoring force. As the barrier is raised, some time must elapse before the force on the boundary ceases. Plastic deformation occurs because friction may be expected to rapidly drive the twin boundary velocity to zero. Formally we may write a Langevin-type equation for the time evolution of the generalized atom coordinate:

$$\frac{d}{dt} e(\vec{r}) = -\Gamma \frac{\delta \mathcal{F}(e)}{\delta e(\vec{r})} + \eta(\vec{r},t) . \tag{3}$$

Here, $\Gamma$ is a phenomenological damping constant and $\eta(\vec{r},t)$ is a thermal noise source. This equation is very difficult to solve in general. However, the qualitative behavior to be expected can be illustrated with reference to Figure 3.

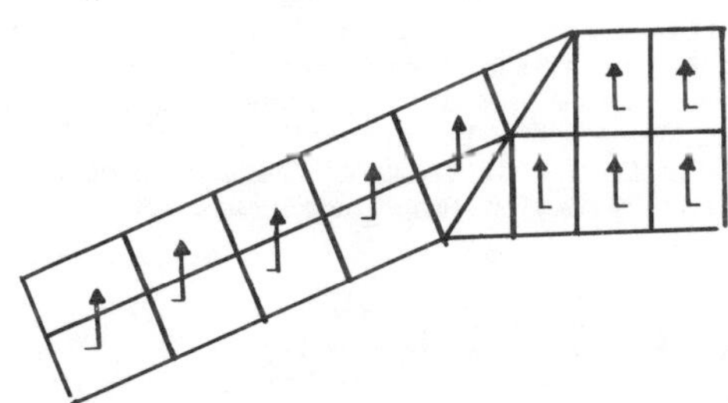

Figure 3

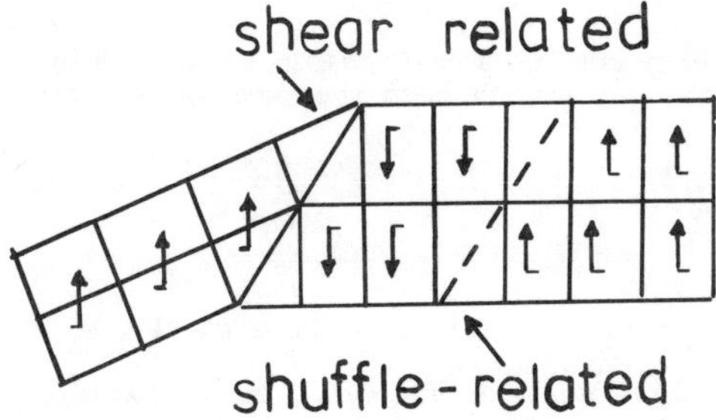

The top panel shows a martensite crystal which consists of only two twins. A modified Ising spin representation is used to indicate that the two twins have exactly the same energy (imagine a uniform magnetic field pointing upward) but differ crystallographically by a mirror reflection. The lower panel shows the distortion of the material under a compressive stress. The dashed line indicates where the twin boundary was prior to stress and the solid line indicates its position immediately after. The reversed spins in the pseudotwin volume reflect the higher energy of this phase. Yet, in the simplest case, the Bravais lattice of the pseudotwin is precisely that of the true twin volume on the right. Hence, there are really two boundaries in the problem. The crystal volumes are shear-related across one boundary and shuffle-related across the other. The conventional twin shear-boundary (solid line) accelerates to the right as long as it is in contact with a cell of the pseudotwin structure. However, this volume is transforming itself via shuffle to the mirror image structure. How does this transformation occur?

There are two ways in which the process "$t_2$" → $t_2$ can proceed. First, the strain gradient term in equation (1) insures that there is a force on the shuffle-boundary which accelerates it to the left, transforming pseudotwin to twin as it moves. Hence, plastic deformation will occur when the two boundaries collide with one another. This process is not obviously activated although presumably Peierls barriers must be surmounted as the motion proceeds. Second, "bubbles" of the true twin can nucleate and grow within the pseudotwin volume. Consider a single unit cell within this volume. The strain gradient term exacts a very high cost for the atoms within this single cell to move from $e^-$ to $e^+$ in Figure 2. However, if thermal activation can induce a group of cells to transform in concert, this "critical droplet" will continue to grow. The analogy to a conventional first-order phase transition is quite precise [14]. (Note: this is not the martensitic transformation alluded to above! The entire volume of the crystal in Figure 3 is always in the martensitic phase)

Let us consider this "bubble" scenario in more detail. The restoring stress on the twin boundary will be proportional to the volume of untransformed pseudotwin: $\sigma_r(T,t) = \Delta f(T)v(t)/\Omega$. The time dependence of the this volume is known from the Avrami-Kolmogorov formula for the kinetics of phase change [15]:

$$v(t) = \exp\left(-\frac{1}{3}\pi G^3 I t^4\right). \tag{4}$$

In this expression, G is the linear growth rate of the bubble and I is the nucleation rate. Equation (4) establishes the stabilization time, τ, to be of order $(G^3 I)^{-1/4}$. The nucleation rate itself is of order $\exp(-\Delta g/T)$, where $\Delta g$ is the free energy cost of a critical droplet. This quantity has been calculated explicitly for our presumed free energy, equation (1). The result is [16]:

$$\Delta g = \frac{\pi}{2^{7/2}} \left(\frac{8^3}{9}\right)^2 E_{ab} \left(\frac{E_{ab}}{\Delta \rho}\right)^2 \left(\frac{J}{r a^2}\right)^{3/2}. \tag{5}$$

which yields a stabilization time activation energy ($\Delta = \Delta g/4$) of 1 eV, comparable to that found in experiment.

A number of experiments suggest themselves as tests of the view of martensitic pseudoelasticity suggested here. Most obviously, time-dependent microbeam x-ray diffraction would establish the existence of the pseudotwin as the origin of the boundary restoring force. For the simplest lattices [9], the pseudotwin has a different superlattice than the true twin and new diffraction spots will be observed. If only small shuffles or short-range order effects dominate one must look to the change in pre-existing superlattice or diffuse intensity. Our model predicts the intensity will vary according to equation (4). Lieberman et.al. [6] initiated such a study for Au-Cd but no diffraction analysis was reported. Given the existence of the pseudotwin, high-contrast dark field electron microscopy could readily distinguish between the two proposed transformation mechanisms: droplet growth vs. shuffle-boundary motion. Indeed, the unique initial conditions represented in Figure 3 may provide a laboratory for the study of the kinetics of first-order phase transitions in general.

## Conclusion

We have presented a general description of pseudoelastic behavior in shape-memory alloys twinned by martensitic transformation which unifies previous discussion in the literature. In our view, a displaced twin boundary sweeps out a volume of metastable Laves-Cahn pseudotwin which provides a restoring force on the boundary. If the pseudotwin can be identified with each unit cell of the system, the time-dependence of boundary motion is controlled by the dynamics of a conventional first-order phase transition and is amenable to direct experimental test.

## References


1. M. Cohen and C.M. Wayman, in Metallurgical Treatises, edited by J.K. Tien and J.F. Elliot (AIME, Warrendale, PA, 1982), p. 445.

2. C.M. Wayman, Metals Forum 4, 135 (1981).

3. K. Otsuka and C.M. Wayman, Reviews on the Deformation Behavior of Materials, edited by P. Feltham (Freund, Tel-Aviv, 1977), p. 81.

4. K. Otsuka and K. Shimizu, Metals Forum 4, 142 (1981).

5. H.K. Birnbaum and T.A. Read, Trans. Met. Soc. AIME, 218, 662 (1960).

6. D.S. Lieberman, M.A. Schmerling and R.S. Karz, in Shape Memory Effects in Alloys, edited by J. Perkins (Plenum, New York, 1975), p. 203.

7. M. Ahlers, G. Barcelo and R. Rapacioli, Scr. Metall. 12, 1075 (1978).

8. J.W. Cahn, Acta Metall. 25, 1021 (1977).

9. A. Zangwill and R. Bruinsma, Phys. Rev. Lett. 53, 1073 (1984).

10. F. Laves, Naturwissenschaften 23, 546 (1952).

11. R.W. Cahn and J.A. Coll, Acta Metall. 9, 138 (1961).

12. M.L. Green and M. Cohen, Acta Metall. 27, 1523 (1979).

13. S. Tsunekawa, Sci. Rep. Res. Inst. Tohoku Univ. 30A, 1 (1981).

14. J.D. Gunton and M. Droz, Introduction to the Theory of Metastable and Unstable States (Springer-Verlag, Berlin, 1983).

15. G.S. Zhdanov, Crystal Physics (Academic, New York, 1965), p. 376.

16. J.S. Langer, Ann. Phys. 41, 108 (1967).

# INFLUENCE OF MICROSTRUCTURE ON THE CORROSION BEHAVIOR OF A NI-SI ALLOY

P. KUMAR
Cabot Corporation, P. O. Box 1296, Reading, PA 19603

## ABSTRACT

Nickel-based alloys with over 7 percent silicon have excellent corrosion resistance to sulfuric acid, oleum and sulfate-ion-containing environments. This alloy is characterized by the presence of intermetallic $Ni_3Si$ (beta) and $Ni_5Si_2$ (gamma) phases in Ni-Si solid solution (alpha) matrix. The corrosion resistance is provided by the coarse intermetallic phases. Size of intermetallic phases is dependent on the processing condition and the heat treatment. The corrosion resistance can be changed by heat treatment.

Results of corrosion tests are explained in terms of microstructural features.

## I. INTRODUCTION

Silicon-bearing alloys are used in sulfuric acid, oleum and sulfate-ion-containing environments. There are several other materials available for sulfuric acid applications, for example, Ta,[1] Zr,[2] and Cr-Mo containing alloys,[3,4,5,6] which are suitable for limited temperature/acid concentrations combinations only. In addition, these are more expensive than silicon-bearing alloys. The silicon-bearing alloys are, by far, the most suitable and economically attractive materials for sulfuric acid applications.

There are two classes of silicon-bearing alloys: iron base and nickel base. The corrosion resistance, of both iron and nickel, increases with an increase in silicon content. Iron with over 14.5 percent silicon shows an excellent corrosion resistance over a wide range of sulfuric acid concentrations/temperature combinations.[4] This would be the most suitable material for industrial applications, excepting for its extreme brittleness. This is so brittle that any thermal stresses induced during the solidification of castings may cause cracking during subsequent handling.[7]

The ductility of another class of silicon-bearing alloys, viz. Ni-Si alloys, allows its fabrication as castings via conventional techniques. The resistance to sulfuric acid and hardness increase with increase in silicon content.[8] Nickel with 9-11 percent silicon is suitable for most of the sulfuric-acid-containing environments.

The marginal ductility of Ni-Si alloys, while adequate for producing castings, makes its hot working impractical.[9] Barker, et al[4] showed that the ductility of these alloys can be improved, while maintaining the good corrosion resistance, by the judicious additions of other alloying elements like Cr, Mo and Ti. But the improvement was not enough. Therefore, these alloys were, hitherto, not produced via wrought processes.

A process has recently been developed,[10] which enables the production of these alloys in wrought form. The process was used to produce the Ni-9Si-3Cu alloy which is commonly known as HASTELLOY® alloy D. The microstructure of these alloys was considerably finer than those of castings. The effect of fine microstructure on the corrosion behavior of HASTELLOY alloy D is not known. Due to the unusual corrosion mechanism in this alloy (reviewed below) it is reasonable to assume that the microstructure would have an effect on its corrosion behavior. That is why this study was undertaken.

## II. CORROSION BEHAVIOR OF NI-SI ALLOYS

The mechanism of corrosion resistance in Ni-Si and Fe-Si alloys is very different from those in many other corrosion alloys. In a typical corrosion-resistant alloy, the chemical homogeneity and the absence of a second phase limit the galvanic cells in aqueous environments. However, the corrosion resistance of Ni-Si and Fe-Si alloys is due to the formation of an impervious inert film. The film has been analyzed by various investigators under different conditions.[11,12,13] It is generally found to be rich in silicon. The exact composition and crystalline structure vary, with the acid concentration, temperature and the alloy composition. For example, Evans and Hart[14,15] found that for Ni-9.5Si-3Ti-3Cu-3Mo alloy, the protective film changed from black (53% $H_2SO_4$) to brown (58% $H_2SO_4$) and then to white (63% $H_2SO_4$) with the increase in $H_2SO_4$ concentrations at its boiling point. The X-ray diffraction analysis was inconclusive at 58 and 63 percent concentrations. The film formed in 53% $H_2SO_4$ was identified to be CuS and S. While the difference in the color and composition of the film was not explained, it is reasonable to assume that it was due to the difference in the oxidizing potential of acid.

To summarize, the corrosion resistance of Ni-Si alloys is due to the formation of passive film. The film was generally found to be enriched in silicon. However, incidences have been reported where the film had sulfur in it.

## III. EXPERIMENTAL PROCEDURE

HASTELLOY alloy D (nominal composition: Ni-9%Si-3%Cu) was produced in shapes suitable for corrosion testing via casting and wrought processing.

The microstructure of wrought alloy was varied by heat treatments as shown in Table 1. Temperatures for the heat treatment were selected on the basis of Ni-Si binary phase diagram, shown in Figure 1.[16] Temperature selections based on Ni-Si-Cu ternary phase diagrams would have been more appropriate. However, these were not available at the time of this work.

® HASTELLOY is a registered trademark of Cabot Corporation.

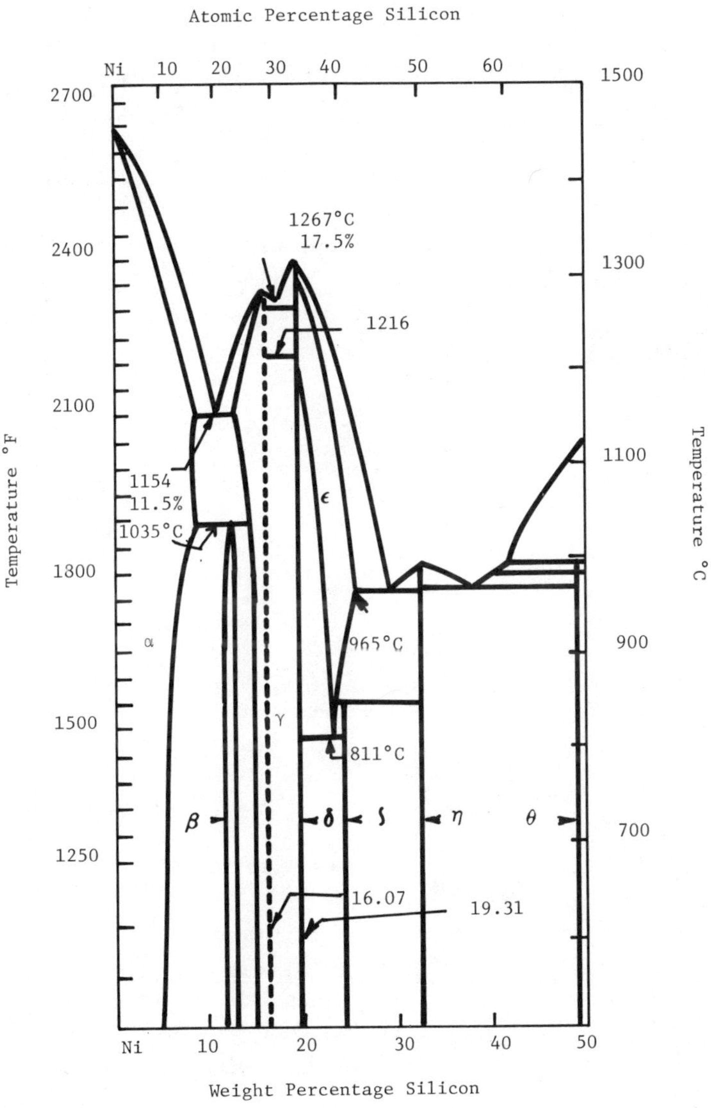

Figure 1. Ni-Si Binary Phase Diagram

TABLE 1: PROCESSING HISTORY OF "HASTELLOY" ALLOY D

| Si No. | Processing Method | Heat Treatment |
|---|---|---|
| 1 | Cast | None |
| 2 | Wrought Processing | None |
| 3 | " | 980°C/8 Hrs./Furnace Cool |
| 4 | " | 980°C/24 Hrs./Furnace Cool |
| 5 | " | 980°C/48 Hrs./Furnace Cool |
| 6 | " | 1120°C/30 Mins./Water Quench |
| 7 | " | 1120°C/30 Mins./Water Quench +980°C/8 Hrs./Furnace Cool |

The corrosion testing was done by immersing the samples in 60 and 77% $H_2SO_4$ at their respective boiling temperatures, 140 and 190°C.[17] These two concentrations were selected so that the corrosion behavior of the alloy could be investigated in both oxidizing (77%) and reducing (60%) acids. The corrosion rate, expressed in millimeters per year, were calculated on the basis of weight losses in consecutive 24-hour periods, density and the original surface area.

Tested samples were evaluated via optical microscopy, electron microscopy and X-ray analyses.

## IV. RESULTS

### Chemical Analysis

Table 2 gives the chemical analyses of cast and wrought samples. It is interesting to note that the agreement between the aim (Ni-9Si-3Cu) and the

TABLE 2: CHEMICAL ANALYSES(1) of "HASTELLOY" ALLOY D

| S. No. | Processing Method | Cu | Fe | Ni | $O_2$ | Si |
|---|---|---|---|---|---|---|
| 1 | Cast | 2.8 | 0.15 | Bal. | 0.1 | 9.5 |
| 2 | Wrought | 3.01 | 0.19 | Bal. | 0.17 | 8.9 |

(1) Concentrations are expressed in weight percent.

actual composition is better for wrought than for the cast samples. Among the metallic elements present in this alloy, silicon has the highest affinity (i.e., highest free energy of oxide formation) for oxygen.[18] Therefore, it is expected that oxygen will be present as silicon oxide and the amount of silicon available for providing resistance to sulfuric acid will be slightly less (~8.75%) than the total silicon content. The published information[8,10] indicates that in cast material this variation (8.9% vs. 9.5%) in silicon will not affect the corroson behavior. Comparison with the cast samples indicates that the difference between oxygen contents of two samples may not be significant.

## Mechanical Properties

Table 3 compares the mechanical properties of cast with wrought material. A significant improvement in ductility and impact strength is apparent. To the best of the author's knowledge, the wrought alloy showed higher strength ductility than ever reported for as-produced Ni-9Si alloy systems.[4,19]

TABLE 3: MECHANICAL PROPERTIES OF "HASTELLOY" ALLOY D

| S. No. | Processing Method | Tensile Properties | | | | Notched Impact Strength |
|---|---|---|---|---|---|---|
| | | Y.S. in MPa | T.S. in MPa | %El. | %Ra | In Joules |
| 1[(a)] | Cast | 793 | 793 | 1 | 1 | < 2.7 |
| (b) | Wrought | 930 | 1260 | 3.6 | 8.8 | 4.8 |

a. Published values[8]

b. Samples were tested without any heat-treatment. These are averages of three tests. Standard deviations for Y.S., T.S., %El and %RA were 1.5, 1.5, 16 and 16.7% of the respective means.

## Corrosion Properties

Table 4 gives results of corrosion tests in 60% and 77% $H_2SO_4$ at their respective boiling points. Corrosion rates of wrought samples, without heat treatment, were considerably higher than those for cast samples in the first test period. The corrosion rates for both cast and wrought alloys decreased precipitously in subsequent periods. While the wrought alloy had a higher corrosion rate, the difference in corrosion rates of cast and wrought alloys generally decreased in the later test periods.

The heat treatment had different effects on the corrosion behavior in 60 and 77% $H_2SO_4$ acids. Aging at 980°C for various times (sample Nos. 3, 4 and 5) did not effect the corrosion rate in 77% $H_2SO_4$ but it had a dramatic-adverse effect in 60% $H_2SO_4$.

Heat-treatment at 1120°C increased the corrosion rate in both 60% and 77% $H_2SO_4$. The subsequent aging at 980°C for eight hours, improved the corrosion resistance to both 60 and 77% $H_2SO_4$. The improvement was different for two acids.

In summary, the corrosion resistance of HASTELLOY alloy D varied with the heat treatment (or microstructure) as follows:

A. In 77% $H_2SO_4$ — As Cast (Best) - As Wrought Processed or As Aged-As Solutionized + Aged - As Solutionized (Worst)

B. In 60% $H_2SO_4$ — As Cast (Best) - As Wrought Processed - As Solutionized + Aged - As Aged or As Solutionized (Worst)

TABLE 4: CORROSION PROPERTIES OF "HASTELLOY" ALLOY D[1]

| S. No. | Condition | Corrosion Rate in Boiling (190°C) 77% $H_2SO_4$ | | | | | Corrosion Rate in Boiling (140°C) 60% $H_2SO_4$ | | | | |
|---|---|---|---|---|---|---|---|---|---|---|---|
| | | I. | II | III | IV | V | I | II | III | IV | V |
| 1 | Cast | 11.9 | 4.98 | 1.73 | 1.4 | 1.47 | 7.77 | 0.64 | 0.51 | 0.46 | 0.48 |
| 2 | Wrought | 18.71 | 4.06 | 3.38 | 2.82 | 2.21 | 10.05 | 0.94 | 0.7 | 0.55 | 0.57 |
| 3 | Same as 2 + 980°C/8 Hr/Fce. Cool[3] | 18.44 | 4.16 | 3.94 | 3.48 | 2.54 | 45.3 | 32.1[2] | | | |
| 4 | Same as 2 + 980°C/24 Hr/Fce. Cool | 18.9 | 4.7 | 3.91 | 3.35 | 2.82 | 43.9 | 21.03[2] | | | |
| 5 | Same as 2 + 980°C/48 Hr/Fce. Cool | 19.1 | 4.45 | 3.94 | 3.3 | 2.44 | 46.3 | 51.97[2] | | | |
| 6 | Same as 2 + 1120°C/30 Min/Water Quench | 18.62 | 11.4 | 14.22 | 16.66[2] | | 70.69 | 54.89[2] | | | |
| 7 | Same as 6 + 980°C/8 Hr/Fce. Cool | 25.07 | 6.17 | 6.43 | 5.03 | 3.60 | 25.6 | 4.29 | 4.32 | 4.32 | 3.94 |

1. Corrosion rates are given for consecutive 24-hour periods in millimeters per year.
2. Test was terminated due to the high corrosion rate.
3. Heat-treatment was done in argon.

## Microstructure

Figures 2-4 give optical micrographs of HASTELLOY alloy D in various conditions. The cast structure, shown in Figure 2, consists of α dendrites and silicon-rich phases ($Ni_3$ Si:β and $Ni_5Si_2$:γ) in the interdendritic regions. The presence of γ indicates that the as-cast structure is not at equilibrium. Precipitates of secondary β in α are visible.

A comparison between cast and wrought microstructures shows that the latter, shown in Figure 3, has a considerably finer structure. For example, the size of silicon-rich (primary β phase is approximately 15 and 70 micrometers for the wrought and cast materials, respectively.

According to the phase diagram (Figure 1), Ni with 9%Si should have only α and secondary (cooling) β. The presence of other phase(s), for example γ ($Ni_5Si_2$) can be explained in terms of nonequilibrium microstructure. Presence of all three phases, α, β, amd γ in the cast and wrought alloys indicates that these are not fully homogenized.

The wrought microstructure is coarsened and homogenized by aging heat-treatment; γ($Ni_5Si_2$; gray area in Figure 3a) is eliminated and the secondary β is coarsened. Primary β did not coarsen in eight hours. However, it shows growth after longer (24- and 48-hour) aging heat-treatments. But even after the long heat-treatments, β is considerably finer than in the cast material.

The solutionizing of wrought material at 1120°C, shown in Figure 4, resulted in grain coarsening (grain size ASTM 8-9) and an equilibrium structure ( α and γ). Subsequent aging at 980°C gave secondary β precipitate in α grains and an incomplete γ to β transformation. We do not know the origin of black regions in these microstructures.

To summarize, the microstructure of wrought HASTELLOY alloy D is considerably finer than that of cast material. Both show the nonequilibrium γ phase. The size of microconstituents in the wrought material can be altered by the heat-treatment.

## Analysis

In order to understand the mechanism of corrosion in HASTELLOY alloy D, the corrosion tested samples were analyzed via scanning electron microscopy and the energy dispersive X-ray analysis (EDX).

Figures 5-7 show various areas which were analyzed via EDX.

Results of analyses are given in Table 5.

It should be noted that these results are only semiquantitative and should only be used for comparisons.

The uncorroded areas of various samples had 10-14 percent silicon. The high silicon phase in the core of the cast sample, Area 4 in Figure 5, had 11.2% silicon. The weight percent of silicon in β($Ni_3Si$) and γ($Ni_5Si_2$) phases is 14 and 16 percent, respectively. While the silicon content of the unattacked area does not agree with that of either

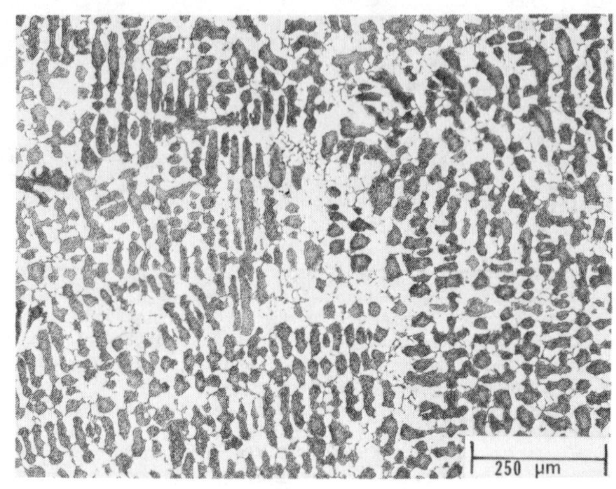


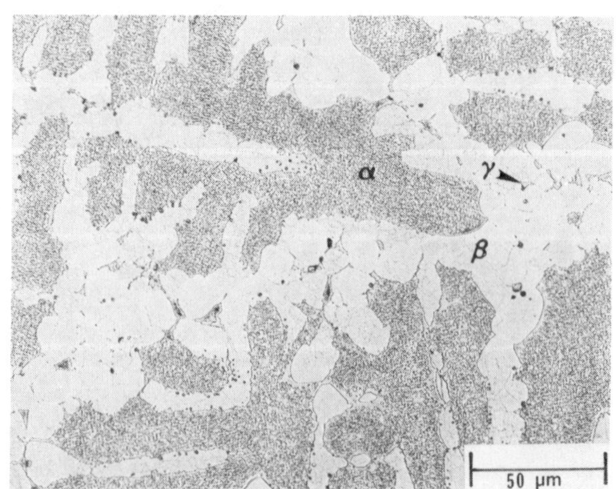


Figure 2. Micrograph of as-cast HASTELLOY alloy D.

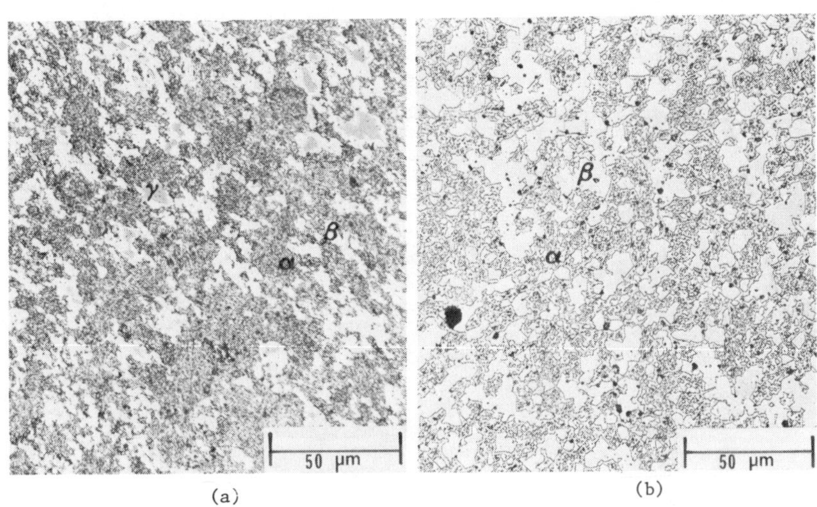


(a) (b)

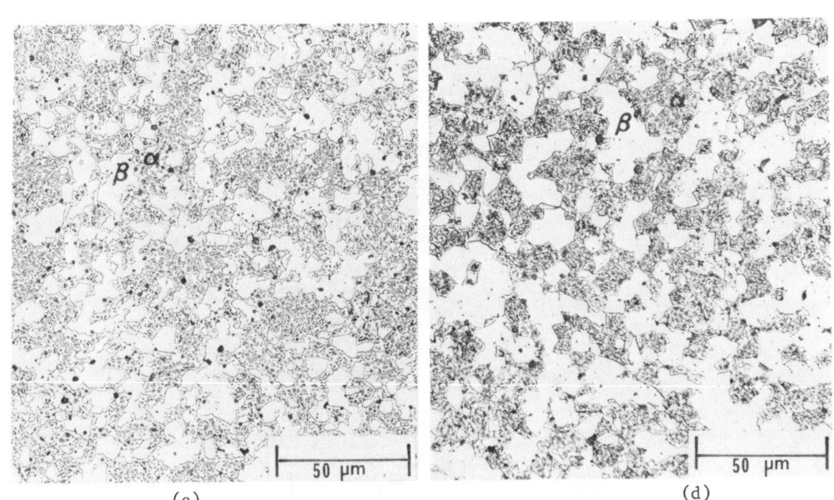


(c) (d)

Figure 3. Micrograph of wrought HASTELLOY alloy D.
a. As produced
b. 980°C/8 hrs/fce. cool
c. 980°C/24 hrs/fce. cool
d. 980°C/48 hrs/fce. cool

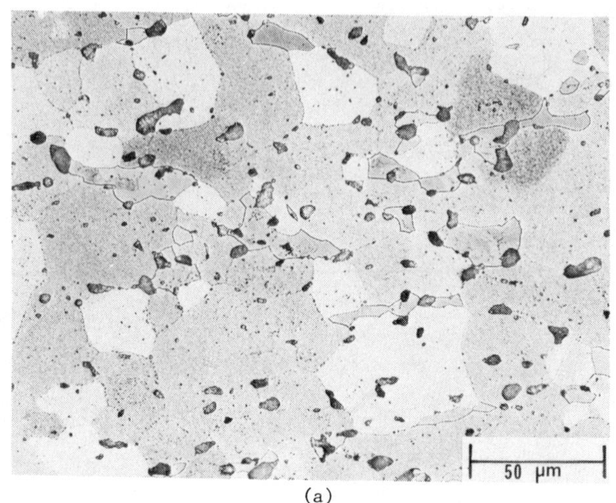


(a)

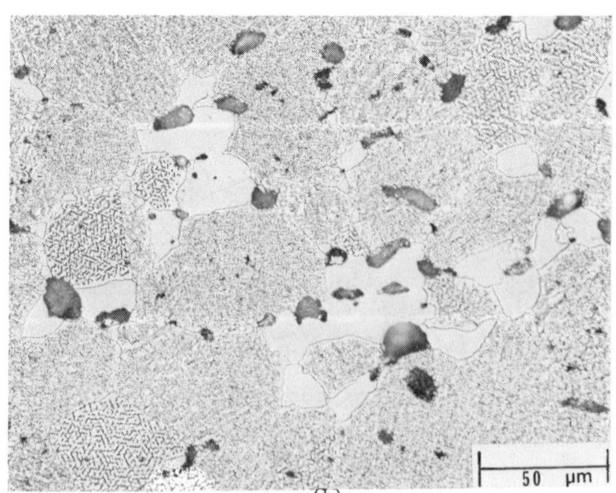


(b)

Figure 4. Micrographs of wrought HASTELLOY alloy D.
a. As solutionized 1120°C/30 min/water quench
b. As aged 1120°C/30 min/water quench + 980°C/8 hrs/fce. cool

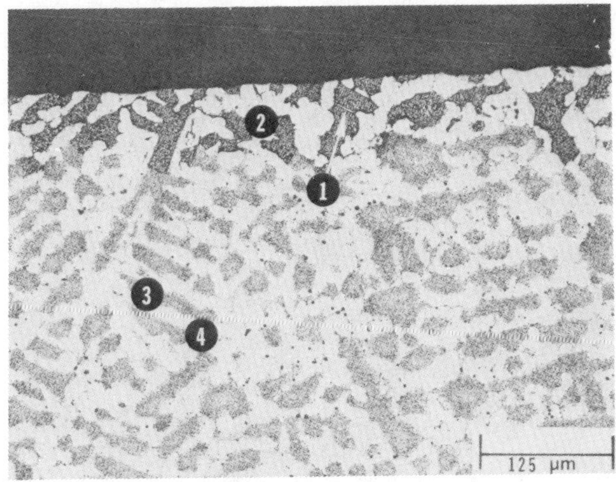


Figure 5. Energy dispersive X-ray analyses of as-cast HASTELLOY alloy D (see Table 5 for results of analyses).

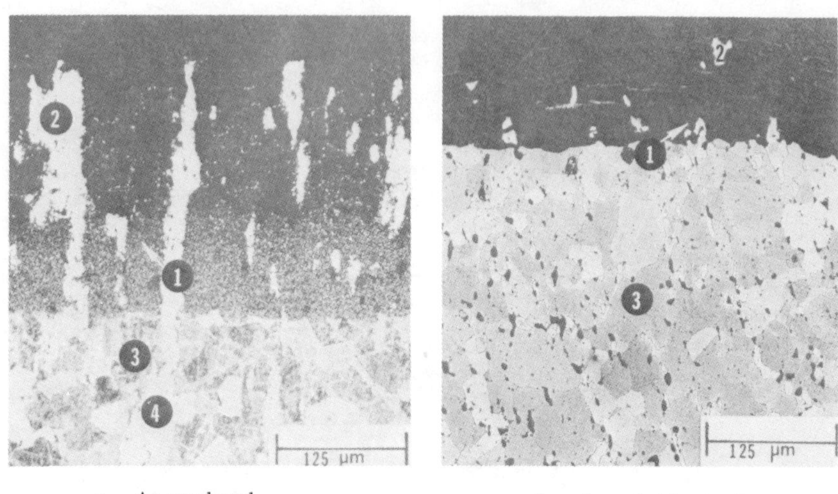


a. As produced

b. As solutionized

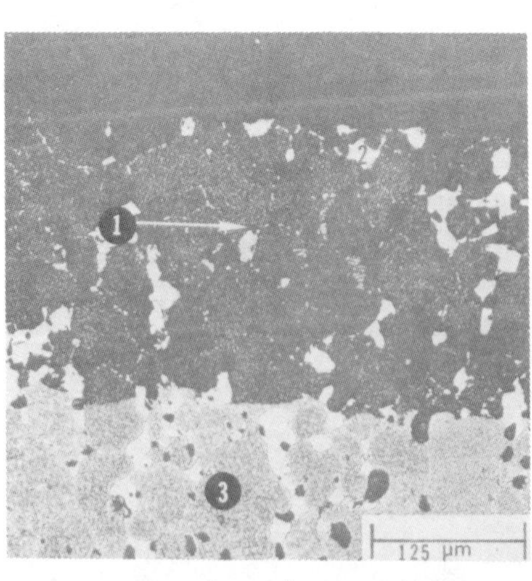


c. As aged

Figure 6. Energy dispersive X-ray analyses of wrought HASTELLOY alloy D

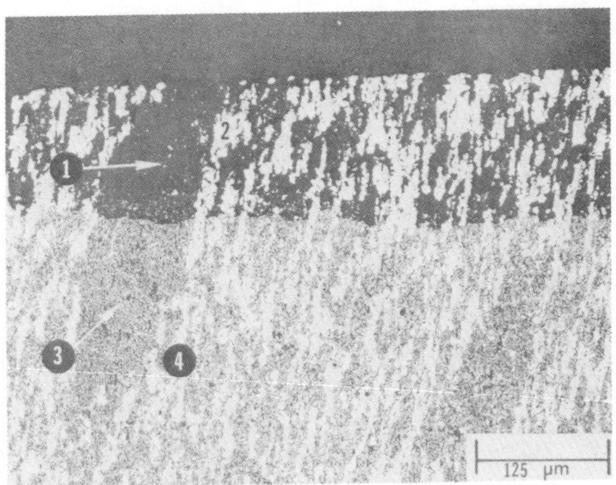


a. 77% $H_2SO_4$

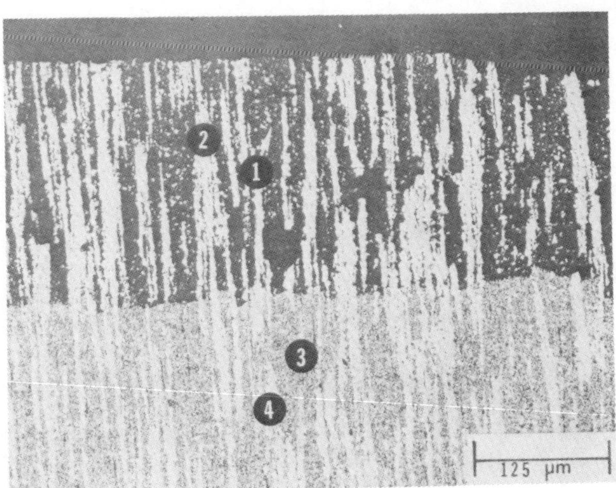


b. 60% $H_2SO_4$

Figure 7. Energy dispersive X-ray analysis of wrought HASTELLOY alloy D after 98°C/24 hrs/fce. cool heat treatment

TABLE 5: RESULTS OF ENERGY DISPERSIVE X-RAY ANALYSIS[1] OF "HASTELLOY" ALLOY D

| S. No.(2) | Condition | Near the Surface | | | | | | | |
|---|---|---|---|---|---|---|---|---|---|
| | | Corroded Area(1) | | | | Unattacked Area(2) | | | |
| | | Si | Cu | S | Ni | Si | Cu | S | Ni |
| 1 | As Cast: Corroded in in 77% $H_2SO_4$ | 20.0 | 1.0 | 0.1 | 79.0 | 10.6 | 1.9 | 0 | 87.5 |
| 2(3) | Wrought Processed: Tested in 77% $H_2SO_4$ | 32.0 | 0 | 0.8 | 67.0 | 12.0 | 0 | 0 | 88.0 |
| 2(3) | Wrought Processed: Tested in 60% $H_2SO_4$ | 30.0 | 0 | 2.0 | 68.0 | 12.0 | 0 | 0 | 88.0 |
| 4 | Wrought Processed +980°C/24 Hr/Fce Cool: Tested in 77% $H_2SO_4$ | 42.0 | 0 | 7.8 | 50.2 | 11.0 | 0 | 0 | 89.0 |
| | Wrought Processed + 980°C/24 Hr/Fce Cool: Tested in 60% $H_2SO_4$ | 62.0 | 6.0 | 5.0 | 27.0 | 11.0 | 0 | 0 | 89.0 |
| 6 | Wrought Processed 1120/30 Min/Water Quench: Tested in 77% $H_2SO_4$ | 91.0 | 0.3 | 2.5 | 6.2 | 14.5 | 2.7 | 0.2 | 82.6 |
| 7(4) | Wrought Processed + 1120/30 Min/Water Quench + 980°C/8 Hr/Fce Cool: Tested in 77% $H_2SO_4$ | 21.0 | | | 79.0 | 11.0 | | | 89.0 |

1. Chemical compositions are semiquantitative. Various areas are identified in Figures 5-7.
2. See Table 4 for corrosion rates.
3. Inconsistent partitioning of copper can not be explained.
4. Sample was not analyzed for Cu or S.

TABLE 5: RESULTS OF ENERGY DISPERSIVE X-RAY ANALYSIS[1] OF "HASTELLOY" ALLOY D CONTINUED

| S. No.[(2)] | Condition | Inside the Sample: Matrix[(3)] Si | Matrix Cu | Matrix S | Matrix Ni | Second Phase Si | Second Phase Cu | Second Phase S | Second Phase Ni |
|---|---|---|---|---|---|---|---|---|---|
| 1 | As Cast: Corroded in in 77% $H_2SO_4$ | 7.7 | 2.1 | 0 | 90.2 | 11.2 | 2.0 | 0 | 86.8 |
| 2[(3)] | Wrought Processed: Tested in 77% $H_2SO_4$ | 5.0 | 2.0 | 0 | 93.0 | 13.0 | 0 | 0 | 87.0 |
| 2[(3)] | Wrought Processed: Tested in 60% $H_2SO_4$ | 6.0 | 0 | 0 | 94.0 | 8.0 | 0 | 0 | 92.0 |
| 4 | Wrought Processed +980°C/24 Hr/Fce Cool: Tested in 77% $H_2SO_4$ | 6.6 | 0 | 0 | 93.4 | 9.0 | 0 | 0 | 91.0 |
|  | Wrought Processed + 980°C/24 Hr/Fce Cool: Tested in 60% $H_2SO_4$ | 6.0 | 2.3 | 0 | 91.7 | 9.0 | 0 | 0 | 91.0 |
| 6 | Wrought Processed 1120/30 Min/Water Quench: Tested in 77% $H_2SO_4$ | 7.6 | 2.3 | 0 | 90.1 | 13.1 | 2.3 | 0 | 84.6 |
| 7[(4)] | Wrought Processed + 1120/30 Min/Water Quench + 980°C/8 Hr/Fce Cool: Tested in 77% $H_2SO_4$ | 7.0 |  |  | 93.0 |  |  |  |  |

1. Chemical compositions are semiquantitative. Various areas are identified in Figures 5-7.
2. See Table 4 for corrosion rates.
3. Inconsistent partitioning of copper can not be explained.
4. Sample was not analyzed for Cu or S.

β or γ, both phases can be seen in Area 2 of Figure 5. The γ phase is darker than β phase. This indicates that the primary β or γ regions in the cast material are not attacked by sulfuric acid.

Figure 6 shows surfaces of wrought, solutionized and solutionized and aged samples. It is interesting to note that corroded areas have a finely divided phase, excepting in the solutionized samples. The corroded area of cast samples also shows a finely divided phase. This phase is expected to be secondary β. These observations imply that (a) fine β does not corrode, and (b) secondary β does not provide corrosion resistance. This means that high corrosion rate of solutionized sample is not because of the lack of secondary β.

On the basis of the binary phase diagram, the amount of silicon-rich phase (γ) in the solutionized sample is expected to be about 25 percent. Its size is finer than that of β (or γ) in as-processed or solutionized + aged samples. Two possible reasons for the high corrosion rates of solutionized sample are amount and size of the silicon-rich phase.

The corroded surfaces of aged samples are shown in Figure 7. Since there was no significant difference in the corrosion rates after 8, 24, and 48 hours of aging, only 24-hour aged samples were analyzed. The silicon content of the corroded area is higher after exposure to 60% $H_2SO_4$ than to 77% $H_2SO_4$. No other significant difference could be observed.

Data in Table 5 indicates that the corroded area in all samples had higher silicon and sulfur and lower copper and nickel than α phase. The concentration of both silicon and sulfur increased with an increase in corrosion rate. It indicates that before the impervious film is formed, nickel and copper are leached out of the surface.

To summarize, EDX indicates that during corrosion in both 60 and 77% $H_2SO_4$, Ni and Cu are dissolved out of α phase, α phase is enriched in Si and S, primary β and γ are not attacked.

## V. DISCUSSION

Superior mechanical properties of wrought HASTELLOY alloy D are attributed to the reduced chemical segregation, and the finer size of α and β (or γ) phases.

The corrosion testing and subsequent analysis of tested samples gave the following information:

1. Corrosion rates in 77% boiling $H_2SO_4$ are lower for the as-cast material than for either as wrought or for wrought + heat-treated material. Solutionizing increased the corrosion rates significantly. Solutioning plus aging, or aging only, did not affect the corrosion rate.

2. Corrosion rates in 60 percent boiling $H_2SO_4$ are lower for the as-cast material than for either as wrought or for wrought plus heat-treated alloy.

3. During corrosion, in both 60 and 77% $H_2SO_4$, at their boiling points, Ni and Cu are dissolved out of α phase; α phase is enriched in S and Si. Secondary β does not provide any corrosion protection, but primary β and γ are not attacked.

It is proposed that the corrosion resistance of HASTELLOY alloy D is controlled by the following interrelated chemical and microstructural features:

1. Size of β (or γ) precipitates: The chemical composition of primary and secondary precipitates will be the same. The only explanation for secondary precipitates not being able to provide the corrosion resistance and the high corrosion rate of wrought material is the size of β particles. The critical size and the electrochemical mechanism for the effect of size on corrosion resistance is yet to be determined.

2. α phase does not have any corrosion resistance irrespective of the amount of Si in it. The highest silicon which can be put into solution in α is 7 percent. This was done by solutionizing the sample. However, it had very high corrosion rates. Certainly the corrosion resistance of α can be expected to increase with an increase in silicon content.[4] But there is no evidence to indicate that it could be as high as that of β. Is the difference in α and β due to silicon content? Probably it is. But another factor which could be contributing to the difference in corrosion behavior is crystal structure of α and β. α is a Ni-Si solid solution, whereas β ($Ni_3$ Si) is an ordered structure. Is the passivity of large, incoherent[20] β related to its low free-energy-ordered state? No information is available on this subject.

The aging heat-treatment has a markedly different effect on corrosion resistance in 60% and 77% $H_2SO_4$. While the effect in 77% $H_2SO_4$ can be explained in terms of size of β, the behavior in 60% $H_2SO_4$ cannot be explained.

## VI. CONCLUSION

The ductility of HASTELLOY alloy D can be improved by producing it via a wrought process. Its corrosion resistance is inferior to that of as-cast material due to finer β precipitates. The corrosion resistance can be changed by heat-treatment.

During corrosion in $H_2SO_4$, Ni and Cu are dissolved out of α; it becomes enriched in S and Si. Secondary β does not provide any corrosion protection.

## VII. ACKNOWLEDGMENTS

The author wishes to acknowledge helpful discussions with Messrs. A. J. Hickl, A. I. Asphahani, R. W. Heckel, and S. M. Soltesz. The assistance of the Corrosion and Processing Laboratory technicians is appreciated. The author is grateful to the High Technology Division of Cabot Corporation for its permission to publish this paper.

## VIII. REFERENCES

1. Badger, F. S., "Industrial Engineering Chemical," 1608, 50, 1958.

2. Rabald, E., "Corrosion Guide," 83, 1968, Publisher: Elseirer, Amsterdam.

3. McClain, G. E., and Muller, W. A., "Corrosion Problem in Acid Flow Control," Chemical Engineering Processing, 48, Feb. 1982.

4. Barker, W., Evans, T. E., and Williams, K. J., "Effects of Alloying Additions on the Microstructure, Corrosion Resistance and Mechanical Properties of Nickel-Silicon Alloys," British Corrosion Journal, 76-86, 5, March 1970.

5. "HASTELLOY alloy G," Publ. No. F-30,267D, Cabot Corporation, Kokomo, Indiana.

6. "HASTELLOY alloy C-276," Publ. No. H-2002, ibid.

7. Collins, H. H., "Silicon-Iron Alloys," 3.98, Corrosion, 1, Editor: L. L. Shreir, Publisher: George Newner Ltd., 1963.

8. "HASTELLOY alloy D," Publ. No. F-30, 127B, Cabot Corporation, Kokomo, Indiana, 1970.

9. Weisert, E. D., Corrosion, 659, 13, 1957.

10. S. M. Soltesz, "Structure-Property Relationships of Ni-Si-Cu Alloys," M. S. Thesis, Michigan Technological University, 1982.

11. Lumiden, J. B., and Stocker, P. J., "Properties of the Protective Film on High Silicon Iron," in Corrosion and Corrosion Protection, Vol. 81-8, Editors: R. P. Frankenthan and F. Manifield, Publisher: The Electrochemical Society, 1981.

12. Enterlein, G., and Feller, H. G., "Investigation of Anodic Dissolution of Nickel-Silicon Alloys," Z. Metallkude, 611, 10, 66, 1975.

13. Blondeau, G., Froelicher, M., Froment, M., Hugot-LeGoff, A., and Vignaud, C., "Growth of Oxide Films During the Anodic Polarization of Ni-Si Alloys in $H_2SO_4$ Solution," C. R. Acad. Sci., 407-20, 9, March 1976.

14. U. S. Patent 3311470, Inco.

15. Evans, T.S., and Hart, A.C., "Corrosion and Passivation of a Nickel-Silicon-Base Alloy in Sulfuric Acid Solution," Electrochemica Acta, 1955-70, 16, 1971.

16. "Metallography, Structures and Phase Diagrams," p. 325, Metals Handbook Vol. 8. Published by American Society for Metals, 1973.

17. "Corrosion Resistance of HASTELLOY Alloys," Publ. No. H-2000, Cabot Corporation, Kokomo, Indiana.

18. "Oxidation of Metals and Alloys," Published by American Society for Metals, 1971.

19. Barker, W., Williams, K. J., and T. E. Evans, "New Ni/Si Base Alloy for Sulfuric Acid," Chemical and Process Engineering, 57-70, September 1970.

20. Rastogi, P. K., and Ardell, A. J., "The Coarsening Behavior of $\gamma$ precipitate in Nickel-Silicon Alloys," Acta Metallurgica, 321-30, 19, April 1971.

# Author Index

# Subject Index